Logic, Methodology and Philosophy of Science

Proceedings of the Thirteenth International Congress

Logic, Methodology and Philosophy of Science

Proceedings of the Thirteenth International Congress

edited by

Clark Glymour,

Wang Wei

and

Dag Westerståhl

© Individual author and College Publications 2009. All rights reserved.

ISBN 978-1-904987-45-1

College Publications
Scientific Director: Dov Gabbay
Managing Director: Jane Spurr
Department of Computer Science
King's College London, Strand, London WC2R 2LS, UK

http://www.collegepublications.co.uk

Original cover design by orchid creative   www.orchidcreative.co.uk
Printed by Lightning Source, Milton Keynes, UK

All rights reserved. No part of this publication may be reproduced, stored in a retrieval system or transmitted in any form, or by any means, electronic, mechanical, photocopying, recording or otherwise without prior permission, in writing, from the publisher.

Preface 1

This volume contains the texts of almost all of the invited lectures and symposia of the 13th International Congress of Logic, Methodology and Philosophy of Science, including the text of the Presidential Address by Adolf Grünbaum. The Congress was held in Beijing, People's Republic of China, in August of 2007. In addition, nearly four hundred contributed papers were presented, representing scholars from nearly every continent (Antarctica had no representative). Enough of the formal stuff, on to the personal.

Beijing was fantastic! The air was clear, the people friendly, charming and interesting, the cuisine incomparable, the lectures fascinating. Morning strolls found men and women exercising, playing Mahjong, working – need I say it – harmoniously. My brilliant former student, Jiji Zhang, guided me to restaurants, apologized to waitresses when I insisted on tipping (it is, he explained, our strange custom in America), and gave me a tour of Beijing University, his Alma Mater. Young people sought to speak English, and to trade tales of my life and theirs. Women of middle years walked with parasols shading summer dresses, reminding me of an America I knew long ago. Go there if you can.

This volume has three editors. The work of sending invitations and assessing contributions was shared, but the production of this volume fell almost entirely to Dr. WANG Wei. Dr. QIAO Ying assisted Dr. Wang in preparing the copy for publication, an enormous job given the length of the volume and the complexity of its contents. I, the contributors to this volume, and its readers, are indebted to them.

Organizing such a meeting is a large task, and the Chinese committees provided a marvelous service. They include:

Advisory Committee
GU Bing-lin, Kim Jae-youl (chair); XIE Wei-he, HU Xian-zhang, SUN Jia-guang (vice-chair); LI Qiang, QIU Ren-zong, WANG Sun-yu.

Organizing Committee
XIE Wei-he (chair), CAI Shu-shan (executive vice-chair), WANG Qi-long, WANG Yu-ping, ZENG Guo-ping, ZOU Chong-li (vice-chair), GUO Gui-chun, XU Fei, JU Shi-er, CHENG Xiao-tang, HUANG Hua-xin, ZHOU Jian-she, LIU Zheng-guang (standing member); Wen Bang-yan, LI Bo-cong, CHEN, Xiao-ping, ZHANG Yi, LV Shi-rong, MENG Qing-wei, HE Xiang-dong, WAN Xiao-long, WANG Qian, LI Cheng-zhi.

Academic Committee
WANG Ning (chair), CAI Shu-shan, CHEN Jia-ying, CHEN Yong-guo, CHENG Ching-ying, LIN Yun-king, LIU Da-chun, WANG Yu-ping, WANG Qi-long, WU Guo-sheng, WU tong, ZHAO Shi-lin, ZOU Chong-li.

Secretariat

WANG Wei (secretary-general), SHENG An-feng, ZHOU Yun-cheng (vice secretary-general), ZHANG Cheng-gang, JIANG Jin-song, YANG Jian, BAO Ou, YU Jing-long, WANG Na, YANG Ren-jie, HE Hua-qing, HU Ming-yan, JIN Ping-yue, LI Dawei, WANG Fang, TIAN Xiao-fei, YUAN Hang, WANG Cheng-wei, LIU Xiaoling.

Clark Glymour

Pittsburgh, Pennsylvania, U.S.A.

May, 2009

Preface 2

We, the other two editors, can only happily concur with the assessment of the Congress and of Beijing by the first Editor. The two of us spent much time in close contact during the years of preparation leading up to the Congress, and we are very pleased that everything went so well. We'd also like to express our thanks to Jane Spurr at King's College Publications for excellent help with the production of the book.

WANG Wei
Dag Westerståhl

Beijing and Gothenburg

May, 2009

Contents

Preface 1   i

Preface 2   iii

President's Address   1

1   Welcome Greeting   3

2   Adolf Grünbaum: Why Is There a Universe AT ALL, Rather Than Just Nothing?   5

3   Farewell Closing Remarks   18

A   LOGIC   19

4   Toshiyasu Arai: Iterating the recursively Mahlo operations   21

5   Rod Downey, et al: Degree Spectra of Unary Relations on $\langle \omega, \leq \rangle$ 36

6   Eric Jaligot: Groups of finite dimension in model theory   56

7   Gerhard Jäger: Operations, sets and classes   74

8   Alexander S. Kechris: Set theory and dynamical systems   97

9   Itay Neeman: Monadic theories of wellorders   108

10   Mark van Atten: On the hypothetical judgement in the history of intuitionistic logic   122

11   Johan van Benthem: Logic, Rational Agency, and Intelligent Interaction   137

12   Giovanna Corsi: "Necessary for"   162

13   Jan Krajíček: A proof complexity generator   185

B   GENERAL PHILOSOPHY OF SCIENCE   191

14   Patrick Maher: Physical Probability   193

15 Theo A.F. Kuipers: Comparative realism as the best response to antirealism 211

16 Gerhard Schurz: Meta-Induction: A Game-Theoretical Approach to the Problem of Induction 241

17 Oliver Schulte: How Particle Physics Cut Nature At Its Joints 267

18 Lawrence E. Blume, et al: Redoing the Foundations of Decision Theory 287

19 Kevin B. Korb, et al: The Philosophy of Computer Simulation 306

20 Thomas Mormann: Updating Classical Mereology 326

21 Soshichi Uchii: An Interpretation of Monadology 344

22 Henk J. M. Bos: Descartes' attempt, in the *Regulae*, to base the certainty of algebra on mental vision – A conjectural reconstruction 354

C PHILOSOPHICAL ISSUES OF PARTICULAR SCIENCES 375

23 Mary Leng: Structuralism, Fictionalism, and Applied Mathematics 377

24 Laura Ruetsche: What's it take to interpret a physical theory? 390

25 Tianjiao Chu: Causal models: A Philosophical Definition Built on Statistical Concepts 408

26 Shushan Cai: Logics in a New Frame of Cognitive Science 427

27 Christopher Cherniak: Minimal Rationality vs Optimized Brain-Wiring 443

28 Michiel van Lambalgen: Logic in the study of autism: reasoning with rules and exceptions 455

29 Michael Eichler: Causal inference from time series 481

30 Kevin D. Hoover: Probability and Structure in Econometric Models 497

31 Peter Spirtes: Probability and Structure in Econometric Models 514

D SCIENCE AND SOCIETY 539

32 Kirsten B. Endres: National Ethics Committees: Diversity in Judgements and the Advice Function 541

33 Norman Daniels: Global Aging and Intergenerational Equity 554

34 Guoping Zeng, et al.: Living Science and Public Scientific Literacy 571

Special Symposium on COSMOLOGY 583

35 Don N. Page: Observational Selection Effects in Quantum Cosmology 585

Special Symposium on FREUD AND PSYCHOANALYSIS 597

36 Edward Erwin: Is Freud Back? 599

Special Symposium on CHINESE TRADITIONAL MEDICINE 613

37 Zhai Xiaomei et al.: The Concept of Disease in Traditional Chinese Medicine: In Comparison with Modern Western Medicine 615

38 Hee-Jin Han: Western Concept of Pulse Diagnosis: Théophile de Bordeu and Organicism 622

39 Ruipeng Lei: Is Traditional Chinese Medicine a Pseudo-Science? – Debates on TCM in China 627

40 Evelyne Micollier: Experimenting on innovative scientific versus traditional treatments: the case of AIDS medical research in China 639

41 Fabrice Gzil: A philosophical analysis of the place of acupuncture in the French health care system 645

42 Anne Fagot-Largeault: Scientific vs. traditional (empirical) medicine - a recurrent debate? 650

Index 655

President's Address

Welcome Greeting
August 9, 2007, 9:30 a.m

Colleagues and friends:

On behalf of the Division of Logic, Methodology and Philosophy of Science of the IUHPS, let me extend a warm welcome to you, who came from far and wide to this quadrennial 13th World Congress.

We are very glad indeed that our Chinese hosts have invited us to hold this gathering here in the great city of Beijing, and that they are providing outstanding hospitality in this formidable Friendship Hotel. This venue makes possible the participation of colleagues from the Far East, whom we have missed in other meeting sites.

We are most grateful to a considerable number of people, whose hard work, admirable competence and dedication have brought this meeting into existence, and who are conducting it.

At the behest of the Executive Committee of our DLMPS, which devised the basic framework of this Congress, Professor Clark Glymour as Chairman and his team of seven international colleagues constituted the Program Committee. We thank them very much for the rich program that resulted from their labors, which included dealing with worrisome financial problems of various sorts.

A vital role was and is being played by the Local Arrangements Committee here in Beijing. It is chaired by Professor Cai Shushan, who is very ably assisted by Dr. Wang Wei, the Secretary General of the local Secretariat and its representative on the Program Committee, as well as by Dr. Li Jingjing.

Please allow me the personal remark that Dr. Wang was most helpful to me and to my wife in our preparations for this event, after we got to know him in 2006 during his year as a Visiting Fellow in our Center for Philosophy of Science at the University of Pittsburgh. And Dr. Li Jingjing helped generously to make our stay in Beijing enjoyable and memorable.

My five fellow members on the current Executive Committee of our Division are the following Professors: The first Vice President Chaim Gaifman of Columbia University in New York; the second V.P. Daniel Andler of the University of Paris; our Secretary General Dag Westerståhl of the University of Göteborg in Sweden; our Treasurer Ulf Schmerl of the University of the German Bundeswehr in Munich; and our Past President Michael Rabin of both Harvard University and the Hebrew University in Jerusalem.

All of them deserve our gratitude for their efforts. But I need to emphasize that we have been especially fortunate to have the highly dedicated, outstanding and versatile services of Dag Westerståhl as the Division's Secretary General. His job is relentlessly, as well as multiply demanding and he is a virtuoso at it, a general factotum and Jack-of-all-trades, including being the EC representative on the Program Committee.

I know that we are indebted to others as well, whom I have omitted inadvertently for which I ask you to forgive me.

It is my hope that our Congress here will provide you with a genuine philosophical feast, and will afford you vigorous intellectual exchanges with your fellow participants.

Adolf Grünbaum

Why Is There a Universe AT ALL, Rather Than Just Nothing?[1]

Adolf Grünbaum

University of Pittsburgh

grunbaum@pitt.edu

ABSTRACT. The titular question of this presentation "Why is There A Universe AT ALL, Rather Than Just Nothing?" is a *fusion* of two successive queries posed by Leibniz in 1697 and 1714. He did so to lay the groundwork for his explanatory theistic answer. My argument here is a sequel to my lengthy (54 page) 2004 article "The Poverty of Theistic Cosmology", which appeared in *The British Journal for the Philosophy of Science* (vol.55, p.561-614). The present paper offers (i) A very unfavorable verdict from my critical scrutiny of the explanatory demand made by Leibniz, and (ii) My argument for the complete failure of his interrogative ontological challenge as a *springboard* for his and Richard Swinburne's creationist theistic answer. I argue under (i) that Leibniz's explanatory demand is an *ill-conceived non-starter* which poses a *pseudo* issue. Thus, his and Swinburne's case for divine creation miscarries altogether. My collateral conclusion: The philosophical enterprise need not be burdened *at all* by Leibniz's ontological query, because it is just a will-o'-the-wisp.

1 Introduction

In his 1697 article "On the Ultimate Origination of Things", Gottfried Wilhelm Leibniz posed a historic question: He demanded "a full reason why there should be any world rather than none" ([13], p.136). In a sequel of 1714, he asked *more generally: "Why is there something rather than nothing?"* [italics in original] ([14], section 7, p.199). And yet he speaks of the answer to this latter question as providing a "sufficient reason" for "the existence of the *universe*", since the something that actually exists is indeed the universe [italics added] ([14], section 8). Thus, presumably, Leibniz's two successive interrogative formulations of 1697 and 1714 can legitimately coalesce into the titular question of my presentation here: "Why is there a *Universe* at all, rather than just nothing?"

In a published *precursor* of 2004 to my impending deliberations, I have used the locution "the Primordial Existential Question" to denote Leibniz's 1714 ontological query "Why is there something rather than nothing?" ([9], p.563). And in my prior essay, I employed the acronym "*PEQ*" to abbreviate the phrase "the

---

[1] Presidential Address, delivered on Thursday, August 9, 2007 at the 13th quadrennial International Congress of Logic, Methodology, and Philosophy of Science (DLMPS) of the International Union of History and Philosophy of Science (IUHPS), held at Tsinghua University, Beijing, China. I dedicated this Address explicitly to the memory of our ex-President Wesley C. Salmon, who was my very dear friend for over 50 years.

Primordial Existential Question". Thus the locution "PEQ" denotes Leibniz's 1714 question "Why is there something rather than nothing?" Yet, I shall extend the designation "PEQ" here to refer *alternatively* to my *more specific* titular question "Why is there a *Universe at all*, rather than just nothing? But that alternative use will not incur the risk of confusion.

Within a Leibnizian framework, his 1714 version of PEQ must be *refined* to preclude its trivialization. Familiarly, Leibniz distinguished between a logically *contingent* entity, on the one hand, and a *necessary* being, on the other: A logically **contingent** object is one whose non-existence is logically possible, and which thus might well not exist. But, for Leibniz, a "necessary being" is one "bearing the reason of its existence within itself", a being whose *non*-existence is thus *logically impossible* ([14], p.199).

But **if** there is a *necessary* being, there can be no question why *it* exists, rather than not, because such a being could not possibly *fail* to exist. Therefore, it would clearly *trivialize* Leibniz's cardinal PEQ, if that question were asked about a "something" that exists necessarily.

Thus, we can formulate Leibniz's *non*-trivial construal of PEQ as follows: "Why is there something *contingent* at all, rather than just nothing *contingent?*" And since Leibniz argued that God exists necessarily, he considered the being of God *compossible* with a putative state in which absolutely nothing *contingent* exists.

Unlike Leibniz, the present-day philosopher Richard Swinburne claims that God exists only contingently. Hence, Swinburne believes that God is *also* absent from a world which is *devoid* of all contingent entities.

Like the philosopher Derek Parfit, I shall speak of the presumed *logical possibility* of there being nothing contingent as "The Null Possibility". And like him, I shall use the label "Null World" to refer to a hypothetical world in which there is nothing contingent at all.

My major concern here will be, in due course, to provide a thorough *critical scrutiny* of Leibniz's time-honored PEQ, and then to develop the important ramifications of that critique. But to lay the groundwork for the complete deflation of PEQ, several preliminary *admonitions* will occupy us beforehand.

2 Is it imperative to explain why the Null Possibility is not instantiated?

First I need to comment on the gloss or twist that Parfit and Swinburne have put upon Leibniz's PEQ. Almost a decade ago, Parfit wrote:

> [W]hy is there a Universe at all? It might have been true that nothing [contingent] ever existed; no living beings, no stars, no atoms, not even space or time. **When we think about this ["Null"] possibility** ([15], p.420), **it can seem astonishing that anything [contingent] exists** [bolding added] ([16], p.24).

Thereupon, Parfit enthrones PEQ on a pedestal, saying: "No question is more sublime than why there is a Universe [i.e., some world or other]: why there is anything rather than nothing" (ibid., column 1). Importantly, Parfit's logical motivation for this cosmic version of PEQ derives largely from the insidious peremptory assumption that the actual existence of a contingent universe *in lieu* of the Null World is *not to be expected*, and that the de facto existence of our world is *therefore inescapably amazing and perplexing*!

Swinburne shares Parfit's astonishment that anything at all exists, declaring: "It remains to me, as to so many who have thought about the matter, a source of *extreme puzzlement* that there should exist anything at all" ([19], p.283). And, more recently, Swinburne opined: "It is extraordinary that there should exist anything at all. Surely *the most natural state of affairs* is simply nothing: no universe, no God, nothing" [italics added] ([20], p.48). Evidently, Swinburne's avowed "*extreme puzzlement*" [my italics] that anything contingent exists at all is driven by the same peremptory *mind-set* as Parfit's astonishment.

The late Paul Edwards, in a 1967 article "Why?" in *The Encyclopedia of Philosophy*, chronicled some of the long history of PEQ and of its cognates ([4], pp.296-301). In 1999, this saga culminated in a tome of over 750 pages by the Swiss philosopher Ludger Lütkehaus, published in German, whose fetching title in English is *Nothing: Farewell to Being, End of Anxiety*. Suffice it just to mention that Lütkehaus deplores the so-called onto-centricity of our culture, the purported paranoid nihilophobia of our supposed ontological greed, and the like.

Turning to Parfit, I challenge his declared astonishment that anything contingent exists at all by asking him: Why should the *mere contemplation* of the Null Possibility reasonably make it "seem astonishing that anything exists?" I claim that it should not do so. Let me point out why it should indeed not.

If some of us were to consider the logical possibility that a person we see might conceivably metamorphose spontaneously into an elephant, for example, I doubt strongly that we would feel even the *slightest* temptation to ask why that *mere logical possibility is not realized*. Why then, I ask Parfit, should anyone reasonably feel astonished at all that the Null Possibility, if genuine, has remained a *mere* logical possibility, and that something does exist *instead? In short, why* **should** *there be just nothing, merely because it is* **logically possible?** This *mere* logical possibility of the Null World, I claim, does *not suffice* to legitimate Parfit's demand for an **explanation** of why the Null World does *not* obtain, an explanation he seeks as a philosophical anodyne for his misguided astonishment that anything at all exists.

3 Christian doctrine as an inspiration of PEQ

It now behooves me to explicate the implicit and explicit *presuppositions* of Leibniz's PEQ. This articulation is vital for a fundamental reason: If one or more of these presuppositions of PEQ is either ill-founded or presumably false, then PEQ is *aborted* as a *non*-starter, because it would be posing a *non*-issue (or a pseudo-problem). And, in that case, the very existence of something contingent, instead of nothing contingent, does *not* require explanation. For example, if a Mr. X never committed a murder, it is ill-conceived to ask him just when he did it, and it is fatuous to blame him for not answering this question.

In earlier writings ([7], p.16; 2000, pp.5,19), I have used the rather pejorative term "pseudo-problem" to reject "a question that rests on an ill-founded or demonstrably false presupposition" ([8], p.19). But, since the German term "*Scheinproblem*" for "pseudo-problem" was given currency by the Vienna Circle,

I now reiterate my caveat that, in my own use of that label, "I definitely do not intend to hark back to early positivist indictments of 'meaninglessness' " (ibid.).

Yet the notion that a question is ill-conceived, because it rests on substantive quicksand, surely ante-dates the logical positivist disparagement of certain traditional philosophical problems as pseudo-questions. Thus, in medieval debates, some issues were dismissed as clearly unproblematic under the Latin rubric of *cadit quaestio*. Despite this ancestry, the challenge from the Vienna Circle was timely, I believe, because sometimes a *seemingly* well-conceived question may not be warranted after all. Thus, a question may be misguided, because it is inappropriately generated by an assumption that was previously unrecognized to be very **misleading** indeed.

One of the main tasks that I have set for myself here is to show precisely how Leibniz's **PEQ is vitiated** by presupposing an altogether dubious corollary of an old Christian doctrine. Elsewhere ([9], pp.561,571), I have formulated that unacceptable corollary as follows: Spontaneously, the world *should* feature nothing contingent at all, and indeed there *would* be nothing contingent in the absence of an overriding external cause (or reason), because that *null* state of affairs is the "most natural" of all!

For brevity, I say that this tribute to the Null World asserts "**the ontological spontaneity of nothingness**" ([8], p.5). And I have introduced the acronym "SoN" to designate the doctrine which avows this ontological spontaneity of the Null World. In this acronym, the "S" stands for "Spontaneity", the "o" for "of", and the "N" for the word "Nothingness". And my reason for having articulated SoN is precisely that its claim will turn out to be a completely unwarranted presupposition of PEQ. Bear in mind that SoN is the thesis that a null state of affairs is "the most natural" of all.

The traditional Christian doctrine which unilaterally entails SoN as a corollary axiomatically makes the following avowal: The *very existence* of any and every *contingent* entity, apart from God himself, is utterly dependent on God at any and all times. Clearly, this tenet of total ontological dependency has two immediate corollaries:

1. The first is SoN, which tells us that, in the absence of a supernatural external cause, the ontologically spontaneous, natural or normal state of affairs is one in which nothing contingent exists at all, and

2. The second corollary is that, without constant divine *creative* support – "so-called" perpetual creation – the world would instantly lapse into nothingness, as claimed by Aquinas, Descartes and many others.

Thus, according to SoN, the actual existence of something contingent or other is a *deviation* from the supposedly spontaneous and natural state of nothingness. And, qua such a deviation, contingently existing objects would clearly require a *creative external cause ex nihilo*, a so-called *ratio essendi*, a reason for existing *at all*.

Yet, such a supposed creative cause must be distinguished, as Aquinas emphasized, from a merely *transformative* cause: Transformative causes produce changes of state in contingent things that *already exist in some form*, or these causes generate new sorts of entities from *previously* existing objects or materials, as when a new house is built from wood and bricks.

Furthermore, in accord with the traditional Christian commitment to SoN, creation *ex nihilo* is required anew at *every* instant at which the world exists, even if it has existed forever. Therefore, traditional Christian theism makes a major claim as follows: If any contingent entity exists, but does so *without* having a *transformative* cause, then its existence must have a creative cause *ex nihilo*, **rather than being externally UNCAUSED**.

However, very importantly, SoN can be strongly challenged by the **counter-question**: "But why should there be nothing contingent, rather than something contingent?" And, indeed, why *would* there be just *nothing contingent*, rather than something contingent? And, moreover, why *would* there be nothing contingent in the absence of an overriding *external* cause?

Unfortunately, in the Christian culture of the Occident, both philosophers and ordinary people have inveterately imbibed SoN with their mother's milk. And it is deeply ingrained even among a good many of those who altogether *reject* its received theological underpinning. But before Christianity molded the philosophical intuitions of our culture, neither Greek philosophy nor most other world cultures featured SoN ([5]). No wonder that Aristotle regarded the material universe as both **uncreated** and **eternal**.

Thus, I am curious to learn from our Chinese colleagues whether their philosophical patrimony has avoided the Occidental SoN dogma. Yet, as we shall see, to Leibniz's credit, when *he* espoused SoN, he tried to *give* a **legitimating** *ontological argument* to support it as part of a *two-fold a priori* justification of his PEQ. I shall soon contend, however, that his effort was entirely unsuccessful.

In 1935, the French philosopher Henri Bergson aptly, though incompletely, sketched SoN, when he rightly deplored its beguiling role in the misguided posing of PEQ. As Bergson put it:

> ...[P]art of metaphysics moves, consciously or not, around the question of knowing why anything exists – why matter, or spirit, or God, rather than nothing at all? But the question presupposes that reality fills a void, that underneath Being lies nothingness, that **de jure there should be nothing, that we must therefore explain why there is de facto something** [bolding added] ([1], pp.239-240).

As shown by my articulation of SoN, Bergson's concise rendition of it needs to be amplified by the further claim that, in the absence of an overriding external cause or reason, the Null World would *spontaneously* prevail ontologically.

How then have the defenders of SoN tried to *justify* it in its own right, rather than just as a logically weaker corollary of the aforestated Christian axiom of the world's total ontological dependence on the Deity?

4 *A priori* justifications of SoN by Leibniz, Swinburne and others

Some philosophers, notably Leibniz and Richard Swinburne, have appealed to the presumed *a priori* simplicity of the Null World to argue that *de jure* there *should* be nothing contingent, so that the *de facto* existence of our world would make an answer to PEQ imperative. However, as I shall contend, the recourse to *simplicity* to defend SoN *a priori* is very unsuccessful, and moreover, significantly, there is no empirical support for **SoN** either. Therefore, this two-fold ill-foundedness of SoN will **undermine PEQ** precisely because PEQ presupposes SoN.

To mount an *a priori* defense of SoN, Leibniz and Swinburne maintained that the Null World is *simpler*, both ontologically and conceptually, than a world containing something contingent or other. This dual assertion of greater simplicity poses two immediate questions: (1) Is the Null World actually *a priori simpler*, and indeed is it the *simplest* world *ontologically* as well as conceptually? And (2) even assuming that the Null World is thus doubly simpler, does its supposed maximum dual **simplicity** *mandate ontologically* that there *should* be just nothing *de jure*, and that, furthermore, there *would* be just nothing in the absence of an overriding cause (reason), as claimed by SoN?

In answer to the first of these two questions, let us assume for the sake of argument, that Leibniz and Swinburne could warrant *a priori* the maximum conceptual and ontological simplicity of the Null World, as Leibniz avowed, when he declared: " 'nothingness' is simpler and easier than 'something' " ([14], sec.7, p.199). Then my emphatically negative answer to the second of these questions is as follows: *Even if the supposed maximum ontological simplicity of the Null World is warranted a priori, that presumed simplicity would not mandate the claim of SoN that de jure the thus simplest world must be spontaneously realized ontologically in the absence of an overriding cause.* After all, having the simplest ontological constitution does not itself make for the actualization or instantiation of the world featuring that constitution! Yet, to my knowledge, neither Leibniz nor Swinburne nor any other author has offered any cogent reason at all to posit such an ontological imperative.

5   Are the philosophical fortunes of Occam's Razor helpful?

Very interestingly, when Leibniz affirmed the ontological simplicity of the Null World, he made no mention at all of the early 17th century enunciations of the injunction of simplicity that came to be known, by mid-19th century, as either "Occam's Razor" or as "The Principle of Parsimony". Yet Leibniz may well have learned of the early 17th century versions of Occam's directive, and may yet have refrained from invoking them, perhaps because he did not think that they would strengthen his case.

Even more noteworthy is that our contemporary Richard Swinburne, who was undoubtedly aware of modern appeals to Occam's Razor, completely passes them over in silence in two pertinent contexts: First, in his 1997 monograph *Simplicity as Evidence of Truth*, and furthermore, in his appeal to the supposed *maximum ontological simplicity* of the Deity as "an argument for God being the cause [ex nihilo] of the existence of the universe" ([23], p.138, n.9).

In a very recent unpublished paper sent to me by the Australian philosopher Daniel King, he cited my aforestated objection to Leibniz's *a priori* defense of SoN, a defense which invoked the ontological simplicity of the Null World, as we saw. And King brought in *Occam's Razor* there as follows: "Grünbaum certainly provides a criticism of *what amounts to an ontological invocation of Occams Razor* [by Leibniz]" [italics added] ([12]).

Although neither Leibniz nor Swinburne mentioned Occam's injunction in their appeal to simplicity, the extant differing versions of the injunction may provide some perspective on their philosophical treatment of simplicity. Thus, let me comment on these versions of *Occam's Razor*.

William of Occam (1285-1349) worked during the first half of the 14th century. But, as reported by J. J. C. Smart ([17], p.118), the formulation of Occam's Razor as "Do not multiply entities beyond necessity" was a 17th century invention in its original Latin enunciation. However, the term "Occam's Razor" *itself* was first introduced in mid-19th century by William Hamilton, who also spoke of it as The Principle of "*Parsimony*".

Smart's ([17]) gloss on Occam's original formulation is that it enjoins theoreticians "against an unnecessary luxuriance of [explanatory] principles or laws or statements of existence". As John Stuart Mill emphasized in this vein, the demand for parsimony is a rule of *methodology* mandating that we have **evidence** for our beliefs (ibid, p.119). But this demand is emphatically not a viable thesis about the workings of nature, as Hamilton had claimed erroneously.

Smart declared only somewhat soberingly, yet with very insufficient caution: "I suspect that it is not possible fully to justify the idea that simple theories are objectively more likely to be true than are complex ones or even that they contain fewer arbitrary elements" (ibid.). However, he spoke very unguardedly and presumably enthymematically, when he referred there *favorably* to "the idea that simple theories are objectively more likely to be true than are complex ones". As he stated it, this notion is surely untenable without at least an articulated proviso: After all, the ancient Greek Thales' monistic hydrochemistry of the chemical universality of water is staggeringly *simpler* than Mendeleyev's 19th century polychemistry, but clearly the polychemistry is overwhelmingly *more likely to be true*.

Smart issued a mere mild *caveat* about not being fully able to justify the expectation that simpler theories are more likely to be true than complex ones. However, he did not know that his cautionary qualification would turn out to be *an understatement with a vengeance* ([10]). Let me explain the context of this very recent development.

6 The failure of Swinburne's simplicity recipe for verisimilitudinous theories

In two books, Swinburne ([21]; [22], chaps.3 and 4) has argued strenuously that simplicity provides probabilistic evidence of truth by being a *tie-breaker* among *conflicting* theories as follows: Greater simplicity is a criterion for "*choosing* among [competing] scientific theories of equal scope [or **content**] fitting equally well with background evidence and yielding the same data ... [italics added] ([22], p.83). This choice of the simpler of two theories purportedly yields the theory that is more likely to be true *in virtue of being simpler*.

However, in a lecture which I delivered in March 2006 at All Souls College in Oxford, and also at the Royal Institute of Philosophy in London, I demonstrated at least two results which fundamentally subvert Swinburne's thesis. To state my results, let me speak of a theory **B** that is more likely to be true than a theory **A** as "having greater *verisimilitude*" than **A**. Then two of my damaging contentions against Swinburne can be stated as follows:

1. His comparative simplicity ratings, which are to yield a verdict of greater verisimilitude, avowedly pertain to rival hypotheses of **equal content** or scope. Yet he, like Karl Popper before him, have left the *implementation* of this crucial *content-parity* requirement glaringly unfulfilled ([6]); and furthermore

2. Just for the sake of argument, let us grant Swinburne's claim that the existence of God, as the creator *ex nihilo*, is the *simplest* of existential hypotheses. Then, I contend, this supposed greater divine simplicity does not sustain his conclusion that theism is inductively more likely to be true as an explanation of the universe than an *atheistic*, scientific account of the facts of the world: Swinburne's inference that theism is more verisimilitudinous than atheism fails, precisely because the requisite content-equality of these two rival hypotheses is entirely hollow. Worse, even if such content-equality could be demonstrated, the supposed greater *a priori* simplicity of the God hypothesis would surely not confer any greater verisimilitude upon it: As we saw in Section 4 above, *a priori* simplicity (and *a priori* probability) are not at all ontologically legislative; thus Swinburne cannot milk any theological capital out of them ([10], p.188) Yet Swinburne's theism is explanatorily omnivorous, avowing very dubiously ([9], section 2) that theism explains "everything we observe" ([20], p.2).

Futhermore, theory **B** might be simpler than theory **A** in *one* respect, while being more complicated in *another*. But inter-theory comparisons of simplicity for assessing relative verisimilitude call for criteria of greater *overall* simplicity. Yet, previously I have used the comparison of Einstein's General Theory of Relativity with Newton's Theory of Gravitation to *impugn* the feasibility of ratings of *comparative* **overall** *simplicity* for rival theories. Nonetheless, differing verdicts from just such ratings are indispensable to Swinburne's prescription for the greater verisimilitude of one of these theories and to Smart's aspiration.

Indeed, if Einstein's theory of gravitation were held to be *more complex overall* than Newton's in virtue of the *non-linearity* of its partial differential field equations, then its presumed greater verisimilitude is the death knell of both Swinburne's prescription and Smart's aspiration.

It has been said that scientists think they do *know* one theory to be simpler than another *overall* as a matter of its greater beauty and elegance. But beauty is in the eye of the beholder, and, as Einstein remarked aptly, elegance had best be left to tailors.

7 The demise of Leibniz's 1714 justification for PEQ

Now, let us come to grips with the specific 1714 *context* in which Leibniz formulated his PEQ, and then tried to *justify* it at once, by relying carefully on *both* of the following two premises: (1) His well-known Principle of Sufficient Reason, to which I shall refer by the acronym "PSR", and (2) his aforestated *a priori* argument from simplicity for the presupposition SoN, which is inherent in PEQ ([14], sections 7 and 8).

Leibniz declared:

> ... *the great principle* of sufficient reason [PSR] ... holds that *nothing takes place without sufficient reason*, that is ... a reason sufficient to determine why it is thus and not otherwise. This principle having been laid down, the first question we are entitled to ask will be: *Why is there something rather than nothing?* For "nothing" [i.e., the Null World] is simpler and easier than "something". Further supposing that things must exist, it must be possible to give a reason *why they must exist just as they do* and not otherwise [italics in original] ([14], sec.7, p.199).

These avowals by Leibniz invite the following set of comments:

1. Right after enunciating his Principle of Sufficient Reason (PSR), he poses PEQ "*Why is there something rather than nothing?*" as "the first question we are entitled to ask". However, *immediately after* raising this question, he relies on the supposed **simplicity** of the Null World to **justify** the presupposition SoN of PEQ, claiming, in effect, that the Null World would be **spontaneously realized ontologically** in the absence of an overriding external cause. As he puts it concisely: "For 'nothing' [i.e., the Null World] is simpler and easier than 'something' " [bolding added]. And clearly, there is *either* something or nothing.
2. Evidently, Leibniz is *not* content to rely on his PSR alone to ask *merely* the *truncated* question "Why is there something contingent?" *without* the accompanying **contrasting** clause "rather than nothing". Instead, he uses SoN as presupposed in this contrasting clause to assert a *dual* thesis: (i) the existence of something contingent **is not to be expected at all**, and (ii) its actual existence therefore **cries out** for explanation. As will be recalled, just this dual thesis was Parfit's rationale for embracing Leibniz's PEQ.

Thus, the soundness of Leibniz's justification of his PEQ evidently turns on the **cogency** of his PSR *as well as of his a priori argument from simplicity for SoN*. But we have already discounted his *a priori* argument for SoN in Section 4. Thus, we can concentrate on appraising his Principle of Sufficient Reason.

Consider within our universe, the grounds for the demise in 20th century quantum theory of the universal causation familiar from Newton's physics, as codified by Laplace's "determinism". This *empirically* well-founded quantum theory features merely probabilistic rather than universal causal laws governing such phenomena as the spontaneous radioactive disintegration of atomic nuclei, yielding emissions of alpha or beta particles, and/or gamma rays.

In this domain of phenomena, there are physically possible particular events that *could* but do *not* actually occur at given times under specified initial conditions. Yet it is *impermissibly legislative* to insist that merely because these unrealized events are thus physically *possible*, there *must* be an explanation entailing their specific *non*-occurrence, and similarly, a deductive explanation of probabilistically governed actually occurring events, as demanded by Leibniz's PSR.

This admonition against PSR was not heeded by Swinburne, who avowed entitlement to *pan*-explainability, declaring: "We expect all things to have explanations" ([19], p.287). In just this vein Leibniz had demanded, for every event, an *explanatory* "reason [cause] sufficient to determine **why it is thus and not otherwise**" [bolding added]. Hence the history of modern quantum physics teaches that PSR, which Leibniz avowedly saw as metaphysical, cannot be warranted *a priori* and indeed is untenable on *empirical* grounds.

Thus, to discover that the universe does not accommodate rigid prescriptions for deterministic explanatory understanding is not tantamount to scientific failure; instead, it is to discover *positive* reasons for identifying certain coveted explanations as phantom.

As we saw, Leibniz had *generated* PEQ by conjoining his PSR with SoN. Yet since his *a priori* defense of SoN via simplicity has failed, it remains to inquire whether his avowed ontological spontaneity of the Null World might possibly be warranted *empirically*. My answer is emphatically *negative* for the following reason: *It turns out, as an induction from various episodes in the history of science, that SoN is altogether ill-founded empirically.*

To *examine* the **empirical** *status* of SoN, it will be useful to *reformulate* it in Richard Swinburne's aforecited words as follows: "Surely *the most natural* state of affairs is simply *nothing*: no universe, no God, nothing" [italics added]. But since our empirical evidence comes, of course, from our own universe **U**, consider the *corollary* of SoN which pertains to our **U**. This corollary asserts that it is *natural* or *spontaneous* for **U** *not* to exist, rather than to exist. As against any *a priori* dictum on what is the "*natural*" ontological behavior of **U**, the verdict on that behavior will now be seen to *depend crucially on* **empirical** *evidence*, and indeed to provide no support for SoN at all.

Two specific cosmological examples spell this empirical moral:

1. The first example is furnished by the *natural evolution* of one of the big bang models of the universe countenanced by general relativistic cosmology, the dust-filled so-called "Friedmann universe". This universe has the following features ([24], pp.100-101):

    (a) It is a spatially closed, 3-dimensional spherical universe (a "3-sphere"), which *expands* from a point-like big bang to a *maximum finite* size, and then *contracts* into a point-like crunch,
    (b) That universe exists *altogether* for *only a finite* span of time, such that *no* instants of time existed *prior* to its finite duration *or* exist *afterward*,
    (c) As a matter of *natural law*, its total rest-mass is *conserved* for the entire time-period of its existence, so that, *during that entire time, there is no need for a supernatural agency to generate it ex nihilo and/or to prevent it from lapsing into nothingness*, contrary to SoN and to Aquinas and Descartes.

    Evidently, the "*natural*" dynamical evolution of the Friedmann big bang universe **as a whole** is specified by Einstein's *empirically supported* cosmology. Thus, the "natural" or spontaneous ontological behavior of big bang worlds is *not* vouchsafed *a priori*.

2. The same epistemic moral concerning the empirical status of *cosmological naturalness* is spelled by the illuminating case of the now largely defunct Bondi and Gold so-called steady-state cosmology of 1948 ([2], 1960).

Their 1948 steady-state theory features a *spatially* and *temporally* **infinite** universe in which the following cosmological principle holds: As a matter of *natural law*, there is large-scale conservation of **matter-density**. Note that this conservation is *not of matter*, but of the **density of matter over time**. The conjunction of this constancy of the density with Hubble's mutual recession of the galaxies from one another then entails a *counter-intuitive* consequence: Throughout space-time, and ***without*** *any matter-generating agency, new matter (in the form of hydrogen) pops into existence completely naturally in violation of* **matter-energy** *conservation*.

Hence the Bondi and Gold world features the **accretion** or formation of *new* matter as its *natural*, normal, spontaneous behavior, yet terrestrially at a very slow rate. And although this accretive formation *is indeed out of nothing*, it is clearly *not* "creation" by an *external* agency. Apparently, if the steady-state world **were** actual, it would discredit the doctrine of the medieval Latin epigram "*Ex nihilo, nihil fit*", which means "from nothing, you *cannot* get *anything*", or more familiarly, "you cannot get blood out of a stone".

The steady-state theory owes its demise to the failure of its predictions and retrodictions to pass *observational* muster in its competition with the Big Bang cosmology. This episode again teaches us that *empirically-based scientific theories are our sole epistemic avenue to the "'natural'" behavior of the universe at large*, though of course only fallibly so.

What then is the *empirical cosmological verdict on the corollary* of SoN which asserts that "It is natural for our universe *not* to exist, rather than to exist"? Apparently, *there is no empirical evidence for this corollary from cosmology, let alone for SoN itself*. Its proponents surely have not even tried to offer any such evidence for SoN, believing mistakenly, as we saw, that it can be vouchsafed *a priori à la* Leibniz.

8 PEQ as a failed springboard for creationist theism: The collapse of Leibniz's and Swinburne's theistic cosmological arguments

Probably every one of us from the Occident has wondered at some time in our lives: "Where did everything come from?" As we know, typically this question is *not* a demand for a statement of the earlier *physical history* of our ***existing*** universe. Instead, the question is driven by the *largely unconscious assumption of SoN*, and is thus simply another version of Leibniz's query PEQ. Yet as I have argued painstakingly, PEQ rests on the *ill-founded* premise SoN, as well as on Leibniz's very questionable PSR. Therefore, PEQ is an *ill-conceived* **non**-starter, which poses a *pseudo*-issue.

But, as we know, both Leibniz and Swinburne raised PEQ as an *imperative* question, and thence they concluded misguidedly that the *answer* to it mandates *divine creation*. Indeed, in the recent May 26, 2007 issue of the German magazine *Der Spiegel*, we are told incoherently: "For most believers, God is neither a person nor a principle, nor yet an existing entity, *but rather an answer to the question why there is something rather than nothing*" [my English translation and italics] ([18], p.66). Thus, PEQ is invoked to generate incoherent theological capital in the culture at large: How, one needs to ask, can God *not* be an "existing entity" and yet be "an answer" to PEQ?

However, **PEQ evidently cannot serve as a viable springboard for** creationist theism, because it is a pseudo-issue based on quicksand! By the same token, *Leibniz's and Swinburne's cosmological arguments for divine creation are fundamentally unsuccessful.*

Hence I say to you: Whatever philosophical problems you have brought to this 13th congress, it is my plaidoyer today that answering Leibniz's PEQ should not engage your curiosity, because PEQ is just a will-o'-the wisp.

9 Coda on the failure to heed this plea

Two widely read atheistic authors, Richard Dawkins ([3], p.155) and Sam Harris ([11], pp.73-74) have succumbed to the guiles of Leibniz's PEQ by countenancing it misguidedly as a searching question that rightly calls for an explanatory answer. Thus, Dawkins allowed a "first cause, the great unknown which is responsible for something existing rather than nothing" (ibid.). And Harris capitulates to PEQ, declaring with very misplaced intellectual humility: "Any intellectually honest person will admit that *he does not know* why the universe exists. Scientists, of course, readily admit their ignorance on this point"[italics in original; [11], p.74]. But surely the failure to answer a pseudo-question does not bespeak ignorance on the part of scientists, philosophers or even the man in the street.

Bibliography

[1] Bergson, H. *The two sources of morality and religion.* Westport, CT: Greenwood Press, 1974.

[2] Bondi, H. *Cosmology.* Cambridge: Cambridge University Press, 2nd edn., 1960.

[3] Dawkins, R. *The God delusion.* New York: Houghton Mifflin, 2006.

[4] Edwards, P. (Ed.). *Why?:The encyclopedia of philosophy,* vol. 8. 1967.

[5] Eliade, M. *Essential sacred writings from around the world.* San Francisco: Harper, 1992.

[6] Grünbaum, A. Can a theory answer more questions than one of its rivals? *The British Journal for the Philosophy of Science,* 27(1):1–23, 1976.

[7] Grünbaum, A. Theological misinterpretations of current physical cosmology. *Philo,* 27(1):15–34, 1998.

[8] Grünbaum, A. A new critique of theological interpretations of physical cosmology. *The British Journal for the Philosophy of Science,* 51(1):1–43, 2000.

[9] Grünbaum, A. The poverty of theistic cosmology. *The British Journal for the Philosophy of Science,* 55(4):561–614, 2004.

[10] Grünbaum, A. Is simplicity evidence of truth? *American Philosophical Quarterly,* 45(2):179–189, 2008.

[11] Harris, S. *Letter to a Christian nation.* New York: A.A. Knopf, 2006.

[12] King, D. Issues arising from grünbaum's critique of the 'primordial existential question'. private communication of May 24, 2007.

[13] Leibniz, G. W. On the ultimate origination of things. In Parkinson, G. H. R. (Ed.), *Leibniz: Philosophical writings,* 136–144. London: J. M. Dent & Sons., 1973.

[14] Leibniz, G. W. Principles of nature and of grace founded on reason. In Parkinson, G. H. R. (Ed.), *Leibniz: Philosophical writings,* 195–204. London: J. M. Dent & Sons., 1973.

[15] Parfit, D. The puzzle of reality: Why does the universe exist? In InWagen, P. v. D. Z. (Ed.), *Metaphysics: The big questions*, 418–427. Malden, MA: Blackwell, 1998.

[16] Parfit, D. Why anything? why this? *London Review of Books*, 20(2):24–27, 1998.

[17] Smart, J. Ockhams razor. In Fetzer, J. (Ed.), *Principles of philosophical reasoning*, 118–128. Totowa, NJ: Rowman & Allanheld, 1984.

[18] Smoltczyk, A. Der Kreuzzug der Gottlosen (the crusade of the godless). *Der Spiegel*, 56–69, May 2007.

[19] Swinburne, R. *The existence of God*. Oxford/New York: Clarendon Press/Oxford University Press, 1991.

[20] Swinburne, R. *Is there a God?* Oxford/New York: Oxford University Press, 1996.

[21] Swinburne, R. *Simplicity as evidence of truth*. Milwaukee, Wisconsin: Marquette University Press, 1997.

[22] Swinburne, R. *Epistemic justification*. Oxford: Clarendon Press, 2001.

[23] Swinburne, R. *The existence of God*. Oxford: Clarendon Press, 2nd edn., 2004.

[24] Wald, R. *General Relativity*. Chicago: University of Chicago Press, 1984.

Farewell Banquet Remarks
August 15, 2007

Colleagues and friends:

Welcome to the Farewell Banquet and Closing Ceremony.

We hope that you feel you had an enlightening and collegial experience during the 7 days of our 13th Congress here in the great city of Beijing. And I hope you will enjoy your dinner and the festivity tonight.

I am happy to announce that our Division has accepted the very welcome invitation from the French delegation to the General Assembly to conduct our 14th Quadrennial Congress in 2011 in the beautiful city of Nancy, France under the leadership of Prof. Gerhard Heinzmann, who is the Director of the renowned Institut Henri Poincaré there.

Let me take this opportunity to thank our very kind Chinese colleagues for their warm and excellent hospitality: Especially Professor Cai Shushan, and Professor Zeng Guo-Ping, and the constantly hard-working Reception staff run by Dr. Wang Wei and Dr. Li Jingjing, both of whom were repeatedly most helpful for us.

Last, but not least I thank our Division's Executive Committee: Our first Vice President Professor Chaim Gaifman of Columbia University in New York, our second Vice President Professor Daniel Andler of the University of Paris-Sorbonne; our Treasurer Prof. Ulf Schmerl of the University of the German Bundeswehr in Munich, and very specially indeed, I thank our outstanding Secretary-General Professor Dag Westerståhl of the University of Göteborg in Sweden, who has given our Division 8 years of very devoted service, and has even volunteered to maintain our divisional Website hereafter. I am also grateful to our Past President Professor Michael Rabin of both Harvard and the Hebrew University in Jerusalem for his service on the Executive Committee.

I call on the Secretary-General now to say his farewell to you.

Adolf Grünbaum

A
LOGIC

Iterating the recursively Mahlo operations[1]

Toshiyasu Arai

Kobe University

arai@kobe-u.ac.jp

ABSTRACT. In this paper we address a problem: How far can we iterate lower recursively Mahlo operations in higher reflecting universes? Or formally: How much can lower recursively Mahlo operations be iterated in set theories for higher reflecting universes?

It turns out that in Π_N-reflecting universes the lowest recursively Mahlo operation can be iterated along towers of Σ_1-exponential orderings of height $N-3$, and that all we can do is such iterations. Namely the set theory for Π_N-reflecting universes is proof-theoretically reducible to iterations of the operation along such a tower.

1 Introduction

For set-theoretic formulas φ,

$$P \models \varphi :\Leftrightarrow (P, \in) \models \varphi.$$

In what follows, let L denote a transitive set, which is a universe in discourse. $P, Q, \ldots$ denotes transitive sets in $L \cup \{L\}$ such that $\omega \in P$.

Let $\mathcal{X}$ be a first-order class of transitive sets. This means that there exists a first-order sentence φ such that $P \in \mathcal{X} \Leftrightarrow P \models \varphi$. Then a set theory T is said to prove $L \in \mathcal{X}$ iff $T \vdash \varphi$.

A Π_i-recursively Mahlo operation for $2 \leq i < \omega$ is then defined through a universal Π_i-formula $\Pi_i(a)$:

$$\begin{aligned} P \in M_i(\mathcal{X}) \quad &:\Leftrightarrow \quad \forall b \in P[P \models \Pi_i(b) \\ &\quad \to \quad \exists Q \in \mathcal{X} \cap P(b \in Q \models \Pi_i(b))] \end{aligned}$$

(read:P is Π_i-reflecting on $\mathcal{X}$.)

Its iteration is defined by transfinite recursion on ordinals β:

$$M_i^\beta := \bigcap \{M_i(M_i^\nu) : \nu < \beta\}.$$

Observe that $M_i(\mathcal{X})$ is Π_{i+1}, i.e., there exists a Π_{i+1}-sentence $m_i(\mathcal{X})$ such that $P \in M_i(\mathcal{X})$ iff $P \models m_i(\mathcal{X})$ for any transitive (and admissible) set P.

A transitive set P is said to be Π_i-*reflecting* if $P \in M_i = M_i^1$.

---

[1] The author is with Graduate School of Engineering, Kobe University, Rokko-dai, Nada-ku, Kobe, 657-8501, Japan.

Let us denote

$$\mathcal{X} \prec_i \mathcal{Y} :\Leftrightarrow \mathcal{Y} \subseteq M_i(\mathcal{X}), \text{ i.e., } \forall P \in \mathcal{Y}(P \in M_i(\mathcal{X})).$$

$P \in M_{i+1}$ is much stronger than $P \in M_i$: Assume $P \in M_{i+1}$ and $P \models \Pi_i(b)$ for $b \in P$. Then $P \in M_i$ and $P \models m_i \wedge \Pi_i(b)$ for the Π_{i+1}-sentence m_i such that $P \in M_i$ iff $P \models m_i$. Hence there exists a $Q \in P$ such that $Q \models m_i \wedge \Pi_i(b)$, i.e., $Q \in M_i \& Q \models \Pi_i(b)$. This means $P \in M_i^2 = M_i(M_i)$, i.e., $M_i \prec_i M_{i+1}$. Moreover $P \in M_i^\triangle$, i.e., $P \in \bigcap\{M_i^\beta : \beta \in ord(P)\}$, $M_i^\triangle \prec_i M_{i+1}$, and so on.

In particular a set theory $\text{KP}\Pi_{i+1}$ for universes in M_{i+1} proves the consistency of a set theory for universes in $M_i^\triangle$.

In this paper we address a problem: How far can we iterate lower recursively Mahlo operations in higher reflecting universes? Or formally: How much can lower recursively Mahlo operations be iterated in set theories for higher reflecting universes? Specifically: What kind of iterations of the lowest operations M_2 do we need to obtain equiconsistent theories for set theories for higher reflecting universes?

2 Iterations of the operation M_i in Π_{i+1}-reflectings

In this section we see that iterations of the operation M_i along Σ_1-relations on ω are too short to resolve Π_{i+1}-reflecting universes provided that the Σ_1-relations are provably wellfounded in $\text{KP}\Pi_{i+1}$.

Definition 1 *1.* $\text{KP}\ell$ denotes a set theory for limits of admissibles. $\text{KP}\Pi_N$ denotes a set theory for universes in M_N.

2. For a definable relation $\prec$ and set-theoretic universe P (admissibility suffices) let

$$P \in M_i(a; \prec) :\Leftrightarrow P \in \bigcap\{M_i(M_i(b; \prec)) : b \prec^P a\},$$

where $b \prec^P a :\Leftrightarrow P \models b \prec a$. Note that $M_i(a; \prec)$ is a Π_{i+1}-class for (set-theoretic) Σ_{i+1} $\prec$.

3. We say that a theory T is *proof-theoretically reducible* to another theory S if T is a Π_1^1 (on ω)-conservative extension of S, and the fact is provable in a weak arithmetic, e.g., the elementary recursive arithmetic EA.

4. For a relation $\prec$ on ω, $TI(a, \prec)$ denotes the transfinite induction schema up to $a \in \omega$:

$$\{\forall x \in \omega[\forall y \prec x \varphi(y) \to \varphi(x)] \to$$
$$\forall x \prec a \varphi(x) : \varphi \text{ is a set-theoretic formula}\}$$

and $TI(a, \prec, \Pi_n)$ its restriction to Π_n-formulas φ. Using a universal Π_n-formula, $TI(a, \prec, \Pi_n)$ is equivalent to a single Π_{n+2}-formula.

5. A relation $\prec$ on ω is said to be *almost wellfounded* in $\text{KP}\ell$ if $\text{KP}\ell$ proves the transfinite induction schema $TI(a, \prec)$ up to *each* $a \in \omega$.

It is easy to see the following lemma using the fact that $M_i(a; \prec)$ is Π_{i+1}.

Lemma 1 Let $\prec$ be a Σ_1 relation on ω. Then $\mathrm{KP\Pi}_{i+1}$ $(i \geq 2)$ proves

$$\forall a \in \omega[TI(a, \prec, \Pi_{i+1}) \to \mathrm{L} \in M_i(a; \prec)].$$

A fortiori $\mathrm{KP\Pi}_{i+1}$ proves

$$\forall a \in \omega[TI(a, \prec, \Pi_{i+1}) \to \mathrm{L} \in M_2(M_i(a; \prec))].$$

In other words, $\mathrm{KP}\ell$ proves

$$P \in M_{i+1} \to \forall a \in \omega[TI(a, \prec^P, \Pi^P_{i+1}) \to P \in M_i(a; \prec)].$$

Therefore $\forall a \in \omega[\mathrm{L} \in M_i(a; \prec)]$ is too weak to reduce $\mathrm{KP\Pi}_{i+1}$ proof-theoretically for any Σ_1 relation $\prec$ on ω, for example

$$\mathrm{KP\Pi}_{i+1} \vdash \mathrm{CON}(\forall a \in \omega[\mathrm{L} \in M_i(a; \prec)])$$

if $\forall a \in \omega[TI(a, \prec)]$ is provable in $\mathrm{KP\Pi}_{i+1}$.

Nonetheless Π_{i+1}-reflecting universes can be approximated by iterations of the operation M_i along well founded Σ_1 relations on ω.

Theorem 1 For each i $(2 \leq i < \omega)$ there exists a Σ_1 almost wellfounded relation $\lhd_i$ in $\mathrm{KP}\ell$ such that $\mathrm{KP\Pi}_{i+1}$ is proof-theoretically reducible to the theory

$$\mathrm{KP}\ell + \{\mathrm{L} \in M_i(a; \lhd_i) : a \in \omega\}.$$

Theorem 1 follows from Lemma 3 and Theorem 4 below. The case $i = 2$ means that the set theory $\mathrm{KP\Pi}_3$ for Π_3-reflecting universes can be resolved by iterations of the recursively Mahlo operations M_2.

Remark. Although $\mathrm{KP}\ell$ is weaker than $\mathrm{KP\Pi}_{i+1}$, $\mathrm{KP\Pi}_{i+1}$ does not prove the soundness of $\mathrm{KP}\ell$: Let Fund denote the axiom schema for Foundation. Then for a $\varphi \in \Sigma_{i+2}$ and a standard provability predicate $\mathrm{Pr}_{\mathrm{Fund}}$ of Fund

$$\mathrm{KP\Pi}_{i+1} \nvdash \forall n \in \omega[\mathrm{Pr}_{\mathrm{Fund}}(\lceil \varphi(\dot{n}) \rceil) \to \varphi(n)]$$

since $\mathrm{KP\Pi}_{i+1} \setminus \mathrm{Fund} \subseteq \Pi_{i+2}$ $(i \geq 2)$.

Hence even if $\mathrm{KP\Pi}_{i+1} \vdash \forall a \in \omega[\mathrm{Pr}_{\mathrm{KP}\ell}(\lceil TI(\dot{a}, \lhd_i, \Pi_{i+1}) \rceil)]$, this does not imply $\mathrm{KP\Pi}_{i+1} \vdash \forall a \in \omega\, TI(a, \lhd_i, \Pi_{i+1})$.

3 Π_3-reflecting on Π_3-reflectings

Our goal is to approximate Π_{i+1}-reflecting universes by iterations of the lowest recursively Mahlo operations M_2. Let us consider first the simplest case: Π_3-reflecting universes on Π_3-reflectings, $M_3^2 = M_3(M_3)$. Universes in M_3^2 are seen to be resolved in terms of iterations of the operation M_2 along a *lexicographic ordering* on pairs.

Definition 2 1. For a Σ_1 relation $\prec$ on ω, $W(\prec)$ denotes the *wellfounded part* of $\prec$:

$$a \in W(\prec) :\Leftrightarrow \forall f \in {}^\omega\omega \exists n \in \omega[f(0) = a \to f(n+1) \not\prec f(n)].$$

$W(\prec)$ is Π_1.

Note that $W(\prec^Q)$ is a *set* in limits of admissibles P for any transitive set $Q \in P$.

2. For two transitive relations $<_1, <_0$ on ω, $<_L :\equiv L(<_1, <_0)$ denotes the lexicographic ordering:

$$\langle n_1, n_0 \rangle <_L \langle m_1, m_0 \rangle :\Leftrightarrow$$
$$n_1 <_1 m_1 \text{ or } (n_1 = m_1 \,\&\, n_0 <_0 m_0).$$

$L(<_1, <_0)$ is Σ_1 if $<_1$ and $<_0$ are Σ_1.

$<_{LW}$ denotes the restriction of $<_L$ to the wellfounded part in the second component:

$$\langle n_1, n_0 \rangle <_{LW} \langle m_1, m_0 \rangle :\Leftrightarrow$$
$$\langle n_1, n_0 \rangle <_L \langle m_1, m_0 \rangle \,\&\, n_0, m_0 \in W(<_0).$$

$<_{LW}$ is Δ_2 if $<_1$ and $<_0$ are Σ_1.

Proposition 1 *Let P be a limit of admissibles and $<$ be a Σ_1 relation on ω. Suppose $P \models a \in W(<)$. Then $a \in W^P(<^Q) = W(<^Q)$ and $Q \models TI(a, <)$ for any $Q \in P$, where*

$$a \in W^P(<^Q) :\Leftrightarrow \forall f \in {}^\omega\omega \cap P \exists n \in \omega[f(0) = a \to f(n+1) \not<^Q f(n)].$$

Proof Since $<$ is Σ_1 and $Q \subseteq P$, we have $<^Q \subseteq <^P$. Hence $a \in W^P(<^P) \subseteq W^R(<^Q)$ for any $R \subseteq P$. Therefore $a \in W^P(<^Q) = W^{Q^+}(<^Q) = W(<^Q)$ for the set $<^Q$ in P, and the next admissible $Q^+ \in P$ above Q. This yields the transfinite induction schema $TI(a, <^Q)$ up to a. ♠

$\mathrm{KP\Pi}_3(\Pi_3)$ denotes a set theory for universes in $M_3(M_3)$.

Lemma 2 *Let $<_1, <_0$ be two Σ_1 transitive relations on ω, and $<_{LW}$ the restriction of the lexicographic ordering defined from these to the wellfounded part in the second component.*
Then $\mathrm{KP\Pi}_3(\Pi_3)$ proves

$$\forall a, \alpha \in \omega[TI(a, <_1, \Pi_3) \to L \in M_2(\langle a, \alpha\rangle; <_{LW})].$$

Proof
Let $L \in M_3(M_3)$. By transfinite induction on a along $<_1$ we show

$$\forall \alpha \in \omega[L \in M_2(\langle a, \alpha\rangle; <_{LW})]$$

where

$$P \in M_2(\langle a, \alpha\rangle; <_{LW}) \Leftrightarrow$$
$$P \in \bigcap\{M_2(M_2(\langle b, \beta\rangle; <_{LW})) : \langle b, \beta\rangle <_{LW}^P \langle a, \alpha\rangle\}$$

and

$$\langle b, \beta\rangle <_{LW}^P \langle a, \alpha\rangle \Leftrightarrow \langle b, \beta\rangle <_L^P \langle a, \alpha\rangle \,\&\, P \models \alpha, \beta \in W(<_0).$$

Suppose that $\forall b <_1 a \forall \beta \in \omega[L \in M_2(\langle b, \beta\rangle; <_{LW})]$, and $\langle b, \beta\rangle <_{LW} \langle a, \alpha\rangle$. We show $L \in M_2(M_2(\langle b, \beta\rangle; <_{LW}))$.

IH yields the case $b <_1 a$. Assume $b = a$ and $\beta <_0 \alpha \in W(<_0)$. Suppose a $\varphi \in \Pi_2$ holds in $L \in M_3(M_3)$. Pick a $Q \in L \cap M_3$ so that $Q \models \varphi$ and $Q \in \bigcap \{M_2(M_2(\langle b,\gamma\rangle;<_{LW})) : Q \models b <_1 a \wedge \gamma \in W(<_0)\}$ by IH.

We claim that $Q \in M_2(\langle a,\beta\rangle;<_{LW})$. By Proposition 1 we have $Q \models TI(\beta, <_0)$. Hence we have $Q \in M_2(\langle a,\beta\rangle;<_{LW})$ by transfinite induction on β. ♠

Theorem 2 *There exist Σ_1 transitive relations $<_1,<_0$ on ω such that $<_1$ is almost wellfounded in $KP\ell$, and $KP\Pi_3(\Pi_3)$ is proof-theoretically reducible to the theory*

$$KP\ell + \{L \in \bigcap\{M_2(M_2(\langle a,\alpha\rangle;<_{LW})) : \alpha \in W(<_0)\} : a \in \omega\}$$

for the restriction $<_{LW}$ of the lexicographic ordering $<_L = L(<_1,<_0)$ defined from these to the wellfounded part in the second components.

For a proof of Theorem 2, see [2].

4 Π_N-reflection

As you expected, an *exponential* structure involves in resolving Π_N-reflecting universes L.

Definition 3 *Let $<_1, <_0$ be two transitive relations on ω.*

1. *The relation $<_E = E(<_1,<_0)$ is on sequences $\langle(n_i^1, n_i^0) : i < \ell\rangle$ of pairs with $<_1$-decreasing first components $(n_{i+1}^1 <_1 n_i^1)$, and is defined by*

$$\langle(n_i^1, n_i^0) : i < \ell_0\rangle <_E \langle(m_i^1, m_i^0) : i < \ell_1\rangle \text{ iff}$$

either
$$\exists k \forall i < k \forall j < 2[n_i^j = m_i^j \,\&\, (n_k^1, n_k^0) <_L (m_k^1, m_k^0)]$$

or
$$\ell_0 < \ell_1 \,\&\, \forall i < \ell_0 \forall j < 2[n_i^j = m_i^j]$$

where $<_L = L(<_1,<_0)$ in Definition 2.2.

Write $\sum_{i<\ell} \pi^{n_i^1} n_i^0$ for $\langle(n_i^1, n_i^0) : i < \ell\rangle$.

2. *Let $dom(<_E)$ denote the domain of the relation $<_E$:*

$$dom(<_E) := \{\sum_{i<\ell} \pi^{n_i^1} n_i^0 : \forall i < \ell \dot{-} 1(n_{i+1}^1 <_1 n_i^1) \,\&\, n_i^1, n_i^0, \ell \in \omega\}.$$

3. *$<_{EW}$ denotes the restriction of $<_E$ to the wellfounded part in the second components:*

$$\alpha = \sum_{i<\ell_0} \pi^{n_i^1} n_i^0 <_{EW} \sum_{i<\ell_1} \pi^{m_i^1} m_i^0 = \beta \text{ iff}$$

$$\alpha <_E \beta \,\&\, \{n_i^0 : i < \ell_0\} \cup \{m_i^0 : i < \ell_1\} \subseteq W(<_0).$$

Lemma 3 Let $<_1, <_0$ be two transitive relations on ω, $<_1$ is Δ_2, $<_0$ is Σ_1, and $<_{EW}$ the restriction of the exponential ordering defined from these to the wellfounded part in the second components. Then $KP\ell$ proves for each $i \geq 2$

$$\forall P \in L \cup \{L\}, \forall a \in \omega,$$
$$\forall \alpha <^P a[P \in M_{i+1}(M_{i+1}(a; <_1)) \to$$
$$P \in M_i(\alpha; <_{EW})]$$

where for $\alpha = \sum_{i<\ell} \pi^{n_i^1} n_i^0 \in dom(<_E^P)$, $\alpha <^P a :\Leftrightarrow n_0^1 <_1^P a$.

Proof We show for any $a \in \omega$ and any $\beta \in dom(<_{EW}^P \uparrow a)$

$$P \in M_{i+1}(M_{i+1}(a; <_1)) \,\&\, P \in M_i(\beta; <_{EW}) \to$$
$$\forall \alpha <^P a\{P \in M_i(\beta + \alpha; <_{EW})\}$$

by main induction on $P \in L \cup \{L\}$ with respect to the relation $\in$, where for $\beta = \sum_{i<\ell_1} \pi^{m_i^1} m_i^0$ and $\alpha = \sum_{i<\ell_0} \pi^{n_i^1} n_i^0$,

$$\beta \in dom(<_{EW}^P \uparrow a) :\Leftrightarrow$$
$$\beta \in dom(<_{EW}^P) \,\&\, (\ell_1 > 0 \to a \leq_1^P m_{\ell_1-1}^1)$$

and $\beta + \alpha = \sum_{i<\ell_1} \pi^{m_i^1} m_i^0 + \sum_{i<\ell_0} \pi^{n_i^1} n_i^0$.

Suppose $\beta \in dom(<_{EW}^P \uparrow a)$, $P \in M_{i+1}(M_{i+1}(a; <_1))$ and $P \in M_i(\beta; <_{EW})$. Pick an $\alpha = \pi^b x + \alpha_0 \in dom(<_{EW}^P)$ so that $\alpha_0 <^P b <_1^P a$ and $x \in W^P(<_0^P)$. We show $P \in M_i(\beta + \alpha; <_{EW})$. It suffices to show $P \in M_i(M_i(\beta + \gamma; <_{EW}))$ for any $\gamma <_{EW}^P \alpha$ by $P \in M_i(\beta; <_{EW})$.

If γ is the empty sequence, then $P \in M_i(M_i(\beta; <_{EW}))$ follows from $P \in M_i(\beta; <_{EW})$, which is Π_{i+1}, and

$$P \in M_{i+1}(M_{i+1}(a; <_1)) \subseteq M_{i+1}.$$

Let $\gamma = \pi^c y + \gamma_0$ with $\gamma_0 <^P c \leq_1^P b$, and $P \models \theta$ for a $\theta \in \Pi_i$. It suffices to find a $Q \in P$ so that $Q \in M_i(\beta + \gamma; <_{EW})$ and $Q \models \theta$.

First consider the case when $c <_1^P b$. By $P \in M_{i+1}(M_{i+1}(a; <_1))$, pick a $Q \in P$ so that $Q \in M_{i+1}(a; <_1)$, $Q \models \theta$, $\beta \in dom(<_{EW}^Q \uparrow a)$, $Q \in M_i(\beta; <_{EW})$ and $dom(<_{EW}^Q) \ni \gamma <^Q b <_1^Q a$.

Then $Q \in M_{i+1}(a; <_1) \subseteq M_{i+1}(M_{i+1}(b; <_1))$, and hence MIH yields $Q \in M_i(\beta + \gamma; <_{EW})$.

Thus we have shown $P \in \bigcap\{M_i(\beta + \delta; <_{EW}) : \delta <^P b\}$, which is Π_{i+1}, and hence

$$P \in M_i(M_{i+1}(a; <_1) \cap \bigcap\{M_i(\beta + \delta; <_{EW}) : \delta < b\}) \tag{1}$$

Second consider the case when $c = b$.

We can find a $Q \in P$ so that $Q \in M_{i+1}(a; <_1)$, $Q \models \theta$, $\beta \in dom(<_{EW}^Q \uparrow a)$, $Q \in \bigcap\{M_i(\beta + \delta; <_{EW}) : \delta <^Q b\}$ by Eq.(1) and $dom(<_{EW}^Q) \ni \gamma \,\&\, b <_1^Q a$. We have $x \in W^P(<_0^P) \subseteq W(<_0^Q)$ by Proposition 1.

Therefore it suffices to show

$$\forall x \in W(<_0^Q), \forall b \in \omega \forall \beta \in dom(<_{EW}^Q \uparrow b)$$
$$[Q \in P \& Q \in M_{i+1}(M_{i+1}(b; <_1)) \&$$
$$Q \in \bigcap \{M_i(\beta + \delta; <_{EW}) : \delta <^Q b\}$$
$$\Longrightarrow \forall \gamma_0 <^Q b \{Q \in M_i(\beta + \pi^b x + \gamma_0; <_{EW})\}]$$

by subsidiary induction on $x \in W(<_0^Q)$.

First assume $\beta + \pi^b y + \delta_0 <_{EW}^Q \beta + \pi^b x + \gamma_0$ with $y <_0^Q x$. SIH yields $Q \in M_i(\beta + \pi^b y + \delta_0; <_{EW})$, and this implies $Q \in M_i(M_i(\beta + \pi^b y + \delta_0; <_{EW}))$ by $Q \in M_{i+1}$.

Therefore we have shown $Q \in M_i(\beta + \pi^b x; <_{EW})$ with $\gamma_0 = 0$. Now let $\gamma_0 = \pi^c y + \gamma_1$ with $c <_1^Q b$. We have $\beta + \pi^b x \in dom(<_{EW}^Q \uparrow c)$, $Q \in M_{i+1}(M_{i+1}(b; <_1)) \& Q \in M_i(\beta + \pi^b x; <_{EW})$ and $Q \in P$. Hence MIH yields $Q \in M_i(\beta + \pi^b x + \gamma_0; <_{EW})$ for $\gamma_0 <^Q b$. ♠

Definition 4 Let $<_i$ ($2 \leq i \leq N-1$) be Σ_1 relations on ω. Define a *tower* relation $<_T$ from these as follows.

Define inductively relations $<_{E_i}$ ($2 \leq i \leq N-1$).

1. $<_{E_{N-1}} :\equiv <_{N-1}$.
2. $<_{E_i} :\equiv E(<_{E_{i+1}}, <_i)$ for $2 \leq i \leq N-2$, cf. Definition 3.

Then let
$$<_T :\equiv <_{E_2}.$$

$<_{TW}$ denotes the restriction of $<_T$ to the wellfounded parts in the second components hereditarily. Namely $<_{TW} = <_{E_2 W}$ and

$$\sum_{n < \ell} \pi^{\alpha_n} x_n \in dom(<_{E_i W}) :\Leftrightarrow$$
$$\forall n < \ell \dot{-} 1 (\alpha_{n+1} <_{E_{i+1} W} \alpha_n) \& \forall n < \ell (x_n \in W(<_i))$$

with $<_{E_{N-1} W} = <_{N-1}$.

For $a \in \omega$ and $\alpha = \sum_{n < \ell} \pi^{\alpha_n} x_n \in dom(<_T)$, define inductively

$$\alpha < a :\Leftrightarrow \forall n < \ell (\alpha_n < a)$$

with $\alpha_n < a :\Leftrightarrow \alpha_n <_{N-1} a$ for $\alpha_n \in \omega$.

Lemmas 3 and 1 yield the following for the set theory $KP\Pi_N$ for universes in M_N.

Theorem 3 *Let $<_i$ ($2 \leq i \leq N-1 < \omega$) be Σ_1 transitive relations on ω. Let $<_{TW}$ denote the restriction of the tower $<_T$ of the exponential orderings $<_{E_i}$ defined from these to the wellfounded parts in the second components hereditarily. Then $KP\Pi_N$ proves that*

$$\forall a \in \omega \forall \alpha < a[TI(a, <_{N-1}, \Pi_N) \to L \in M_2(\alpha; <_{TW})]$$

and hence

$$\forall a \in \omega \forall \alpha < a[TI(a, <_{N-1}, \Pi_N) \to L \in M_2(M_2(\alpha; <_{TW}))].$$

We see an optimality of this resolving of Π_N-reflecting universes in terms of the lowest recursively Mahlo operation M_2.

Theorem 4 *For each N ($2 < N < \omega$) there exist Σ_1 transitive relations $<_i$ ($2 \leq i \leq N-1$) on ω such that $<_{N-1}$ is almost wellfounded in $\mathrm{KP}\ell$, and $\mathrm{KP}\Pi_N$ is proof-theoretically reducible to the theory*

$$\mathrm{KP}\ell + \{\mathrm{L} \in \bigcap \{M_2(M_2(\alpha; <_{TW})) : dom(<_{TW}) \ni \alpha < a\} : a \in \omega\}$$

for the restriction $<_{TW}$ of the tower $<_T$ of the exponential orderings $<_{E_i}$ defined from these to the wellfounded parts in the second components hereditarily.

Theorem 4 is extracted from proof-theoretic analyses of $\mathrm{KP}\Pi_N$ in [1] and [2]. Let me spend some words on *ordinal analyses*, an ordinal informative proof-theoretic investigations in generalities.

5 Background materials from proof theory

Let T be a recursive theory containing ACA_0 [the predicative (and hence conservative) extension of the first order arithmetic PA], and Π_1^1-*sound*, i.e., any T-provable Π_1^1-sentence is true in the standard model.

Then its *proof-theoretic ordinal* $|\mathrm{T}|$ is defined to be the supremum of the order types of the provably recursive well orderings:

$$|\mathrm{T}| := \sup\{\alpha < \omega_1^{CK} : \mathrm{T} \vdash WO[<] \,\&$$
$$\alpha = \text{order type } |<| \text{ of } < \text{ for a recursive ordering } <\}$$

Remark:
The ordinal $|\mathrm{T}|$ is stable if we consider Σ_1^1-orderings and/or add true Σ_1^1-sentences to $\mathrm{T} \supseteq \mathrm{ACA}_0$, an anlogue to the C. Spector's boundedness theorem. For a proof see [7].

It is seen that $|\mathrm{T}|$ is recursive, i.e., $|\mathrm{T}| < \omega_1^{CK}$, and easy to cook up a recursive well ordering $<^T$ whose order type is equal to $|\mathrm{T}|$.

For each $p \in \omega$ let $<_p$ denote a recursive well ordering defined as follows:

1. The case when p is a Gödel number of a proof in T whose endformula is $WO[\prec]$ for a recursive binary relation $\prec$: Then put $<_p := \prec$.
2. Otherwise, let $<_p$ denote an empty ordering, i.e., $dom(<_p) = \emptyset$.

Glue these orderings together to get a recursive ordering $<^T$:

$$\langle n, p \rangle <^T \langle m, q \rangle :\Leftrightarrow [p = q \,\&\, n <_p m] \vee p < q$$

for a bijective pairing function $\langle n, p \rangle$.

Then $<^T$ is a recursive well ordering by the assumptions, and $|<^T| \leq |\mathrm{T}| = \sup\{|<_p| : p \in \omega\} \leq |<^T| < \omega_1^{CK}$ as desired.

Gentzen's celebrated pioneering work yields $|\mathrm{ACA}_0| = \varepsilon_0$. The first achievement for proof theory of impredicative theory was done by G. Takeuti. He designed a recursive notation system of ordinals, which describes the proof theoretic ordinal of, e.g., Π_1^1-Comprehension Axiom. Nowadays Takeuti's proof is understood as for set theories of Π_2-reflecting universes, i.e., for the Kripke-Platek set theory with the Axiom of Infinity, $KP\omega$.

Ordinal analyses for stronger theories are now obtained. Let $\langle O(T), <_T \rangle$ denote a notation system of proof-theoretic ordinal of $T = ACA_0$, $KP\omega$, KPM, $KP\Pi_N$, etc.

Ordinal analyses of theories T show not only the fact $|O(T)| = |T|$ but also more, i.e., some conservative extension results.

Theorem 5 *Let EA denote the elementary recursive arithmetic, a fragment $I\Delta_0 + \forall x \exists y (2^x = y)$ of PA.*

1. *If $\prec$ is an irreflexive, transitive and provably well founded relation in T (not necessarily a total ordering), then there exists an ordinal term $\alpha \in O(T)$ and an elementary recursive function f so that $EA + \forall n, m, k[n \not\prec n \,\&\, (n \prec m \prec k \to n \prec k)]$ proves that*

$$\forall n, k[(n \prec k \to f(n) <_T f(k)) \,\&\, f(n) <_T \alpha]$$

2. *Over EA, $WO[<_T]$ is equivalent to the uniform reflection principle $\mathrm{RFN}_{\Pi_1^1}(T)$ of T for Π_1^1-formulas.*

3. *T is Π_1^1-conservative over the theory $ACA_0 \cup \{WO[<_T \restriction n] : n \in \omega\}$, which is an extension of ACA_0 by augmenting the wellfoundedness of each initial segment $<_T \restriction n$ of the ordering $<_T$.*

4. *Over EA, the 1-consistency $\mathrm{RFN}_{\Pi_2^0}(T)$ of T is equivalent to the fact $ERWO[<_T]$ that there is no elementary recursive descending chain of ordinals in $O(T)$.*

5. *T is Π_2^0-conservative over the theory $EA \cup \{ERWO[<_T \restriction n] : n \in \omega\}$.*

   Therefore provably recursive functions in T are exactly the functions defined by ordinal recursions along initial segments $<_T \restriction n \, (n \in \omega)$.

6. *Over EA, finitely iterated consistency statements $\mathrm{CON}^{(n)}(T)$ of T*

$$\mathrm{CON}^{(0)}(T) :\Leftrightarrow \forall x(0=0);$$
$$\mathrm{CON}^{(n+1)}(T) :\Leftrightarrow \mathrm{CON}(T+\mathrm{CON}^{(n)}(T))$$

   is equivalent to the inference rule

$$\frac{[q(\alpha) <_T \alpha \to A(q(\alpha))] \to A(\alpha)}{A(\alpha)}$$

   where α denotes a variable ranging over $O(T)$, and A [q] is an elementary recursive relation [function], resp.

For a proof of Theorem 5.1, see [7]. Theorem 5.6 is seen from Theorem 5.4 through an Herbrand analysis and a result due to W. Tait [23].

The rest of Theorem 5 is seen from Lemma 4 below, cf. [1–6, 8–10, 13–15]. Also cf. [11, 12, 16–18] for proof theory based on epsilon substitution method.

Lemma 4 *1. T proves that each initial segment $<_T \restriction n$ is wellfounded. The proof is uniform in the sense that*

$$EA \vdash \mathrm{Proof}_T(p(x), WO[<_T \restriction x])$$

for an elementary recursive function $p(x)$ and a canonical proof predicate $\text{Proof}_T(x, y)$ (read: x is a (code of a) T-proof of a (code of a) formula y).

2. We can define a rewrite rule(cut-elimination step) $r(p, n)$ on (finite) T-proofs p of Π_1^1-formulas, and an ordinal assignment $o : p \mapsto o(p) \in O(T)$ so that EA proves

$$\forall n[o(r(p,n)) <_T o(p) \to \text{Tr}_{\Pi_1^1}(end(r(p,n)))] \to \text{Tr}_{\Pi_1^1}(end(p))$$

where $\text{Tr}_{\Pi_1^1}$ denotes a partial truth definition for Π_1^1-sentences, and $end(p)$ the end-formula of a proof p.

For proofs p of Σ_1^0-sentences, the rewrite rule degenerates to be unary, $r(p,n) = r(p,m)$.

NB.
The size of proof-theoretic ordinals is by no means related to consistency strengths of theories. Only when we restrict to initial segments of notation systems $O(T)$, the sizes are relevant. Cf. [19, 20] for some pathological examples on provably well orderings.

Let $\text{CON}(T, n) :\Leftrightarrow \forall x \leq n \neg \text{Proof}_T(x, \lceil 0 = 1 \rceil)$ denote a partial consistency of T up to n.

1. ([22])

   Let $n \prec m$ denote a recursive relation defined as follows:

   $$n \prec m :\Leftrightarrow$$
   $$[\text{CON}(T, \min\{n,m\}) \,\&\, n < m] \vee$$
   $$[\neg \text{CON}(T, \min\{n,m\}) \,\&\, n > m]$$

   Even though $|\prec| = \omega$ since T is assumed to be consistent, $WO[\prec]$ implies $\text{CON}(T)$ finitistically.

2. Modifying the above Kreisel's pathological example, one sees that for any recursive and $Bool(\Pi_1^1)$-sound theory T ($Bool(\Pi_1^1)$ denotes the Boolean combinations of Π_1^1-sentences), there exists a recursive and $Bool(\Pi_1^1)$-sound theory T' such that $|T| < |T'|$ but $T' \not\vdash \text{CON}(T)$: let $<_T$ be any recursive well ordering of type $|T|$, and let

   $$n \prec' m :\Leftrightarrow \text{CON}(T, \max\{n, m\}) \,\&\, n <_T m.$$

   Although $|\prec'| = |<_T|$, $\prec'$ is a finite ordering if T is inconsistent. A fortiori $EA \vdash \neg \text{CON}(T) \to WO[\prec']$. Hence $T \not\vdash WO[\prec'] \to \text{CON}(T)$ by the second incompleteness theorem. Therefore $T' := T \cup \{WO[\prec']\}$ is a desired one.

   Note that if each initial segment of $<_T$ is provably wellfounded in T, then so is for $\prec'$.

6 Collapsing functions iterated

The essential step in cut-elimination for a set theory T is to analyse the axiom expressing an ordinal σ reflects any Π_2-formula φ:

$$\varphi^{L_\sigma}(a) \wedge a \in L_\sigma \to \exists \beta < \sigma[\varphi^{L_\beta} \wedge a \in L_\beta].$$

This means that given a proof figure P of the premise, we have to find an ordinal term $\beta < \sigma$:

$$\begin{array}{c}\vdots P\\ \varphi^{L_\sigma}(a) \land a \in L_\sigma\end{array} \Longrightarrow \begin{array}{c}\vdots\\ \varphi^{L_\beta}(a) \land a \in L_\beta\end{array}$$

This is done by putting $\beta = d_\sigma \alpha < \sigma$ $(o(P) = \alpha \in Od(T))$ for a (Mostowski) collapsing function d.

Let $C(\alpha)$ $(\alpha = o(P))$ denote the set of ordinals which may occur in the reducts of P. Ordinals in $C(\alpha)$ are on the solid lines with gaps here and there in the following figure:

```
0            d_σα              σ          σ+d_σα
|————————————)                 |————————————)    .........
```

By stuffing the gap below σ in the set $C(\alpha)$ up, σ is collapsed down to the least indescribable ordinal $d_\sigma \alpha$. Then ordinals in $C(\alpha)$ cannot discriminate between σ and $d_\sigma \alpha$

$$\gamma < \sigma \Leftrightarrow \gamma < d_\sigma \alpha \, (\gamma \in C(\alpha)),$$

Thus the ordinal $\beta = d_\sigma \alpha$ can be a substitute for σ.

To analyse larger ordinals, e.g., Π_3-reflecting ordinals, the collapsing process has to be iterated.

A Π_3-reflecting ordinal K is understood to be $< \varepsilon_{K+1}$-recursively Mahlo, $L_K \in \bigcap_{\mu < \varepsilon_{K+1}} M_2^\mu$. First K is collapsed to a μ_0-recursively Mahlo ordinal for a $\mu_0 < \varepsilon_{K+1}$: $\kappa_1 = d_K^{\mu_0} \alpha_0 < K$. Then $L_{\kappa_1} \in M_2^{\mu_0}$ is collapsed to a μ_1-recursively Mahlo ordinal: $\kappa_2 = d_{\kappa_1}^{\mu_1} \alpha_1 < \kappa_1$ ($\mu_1 < \mu_0$), etc. In this way a possibly infinite collapsing process is generated: $K = \kappa_0 > d_K^{\mu_0} \alpha_0 = \kappa_1 > d_{\kappa_1}^{\mu_1} \alpha_1 = \kappa_2 > \cdots$ ($\varepsilon_{K+1} > \mu_0 > \mu_1 > \cdots$).

We have designed a recursive notation system $\langle Od(\Pi_N), < \rangle$ of ordinals for proof theoretical analysis of $\text{KP}\Pi_N$, and showed in [1] that $\text{KP}\Pi_N$ is proof-theoretically reducible to the theory $\text{ACA}_0 + \{WO[< |\alpha|] : \Omega > \alpha \in Od(\Pi_N)\}$, where $\Omega \in Od(\Pi_N)$ denotes the least Π_2-reflecting ordinal ω_1^{CK} and $< |\alpha|$ the restriction of the ordering $<$ in $Od(\Pi_N)$ to α. Thus $O(\text{KP}\Pi_N) = Od(\Pi_N)|\Omega$.

On the other side in [2] we have shown that $\text{KP}\Pi_N$ proves $WO[< |\alpha|]$ for *each* $\alpha < \Omega$. Indeed, this wellfoundedness proof is essentially formalizable in a theory $\text{KP}\ell + \{L \in \bigcap \{M_2(M_2(\alpha; <_{TW})) : dom(<_{TW}) \ni \alpha < a\} : a \in \omega\}$ for some Σ_1 relations $<_i$ ($2 \leq i \leq N-1$) on ω such that $<_{N-1}$ is almost wellfounded in $\text{KP}\ell$. This shows Theorem 4.

In the next section we give a sketch of the wellfoundedness proof.

7 Wellfoundedness proof

Our wellfoundedness proof of $Od(\Pi_N)$ is based on the *maximal distinguished class* $\mathcal{W}$ [21], a Σ_1-definable set of integers, and a proper *class* in $\text{KP}\Pi_N$.

To formalize the proof *in* $\text{KP}\Pi_N$, we have to show for each $\eta \in Od(\Pi_N)$ there exists an η-*Mahlo set* on which the maximal distinguished class enjoys the same closure properties as $\mathcal{W}$ up to the given η. The η-Mahlo sets are defined through a ramification process to resolve the reflecting universes in terms of iterations of lower Mahlo operations[2].

7.1 The notation system $Od(\Pi_N)$

The notation system $Od(\Pi_N)$ (an element of $Od(\Pi_N)$ is called an *ordinal diagram*, which is abbreviated o.d.) contains the constants Ω for ω_1^{CK} and π for the least Π_N-reflecting ordinal.

The main constructor is to form an o.d. $d_\sigma^q \alpha < \sigma$ from a symbol d and o.d.'s σ, q, α, where σ denotes a recursively regular ordinal and q a finite sequence of o.d.'s.

$\gamma \prec_2 \sigma$ denotes the transitive closure of $\{(\beta, \sigma) : \exists \alpha, q(\beta = d_\sigma^q \alpha)\}$. The set $\{\tau : \sigma \prec_2 \tau\}$ is finite and linearly ordered by $\prec_2$ for each σ, namely $\{\sigma : \sigma \preceq_2 \pi\}$ is a tree with its root π.

In the diagram $d_\sigma^q \alpha$, q includes some data telling us how the diagram $d_\sigma^q \alpha$ is constructed from $\{\tau : d_\sigma^q \alpha \prec_2 \tau\} = \{\tau : \sigma \preceq_2 \tau\}$.

The main task in wellfoundedness proofs is to show the tree $\{\sigma : \sigma \preceq_2 \pi\}$ to be wellfounded.

Specifically q in $\eta = d_\sigma^q \alpha$ includes some data $st_i(\eta), pd_i(\eta), rg_i(\eta)$ for $2 \leq i < N$. $st_{N-1}(\eta)$ is an o.d. less than $\varepsilon_{\pi+1}$, and $pd_2(\eta) = \sigma$.

A relation $\prec_i$ is defined from $pd_i(\eta)$ as the transitive closure of $\{(\eta, \kappa) : \kappa = pd_i(\eta)\}$. This enjoys $\prec_{i+1} \subseteq \prec_i$. Therefore the diagram $pd_i(\eta)$ is a proper subdiagram of η. $st_i(\eta)$ is an o.d. less than the next admissible κ^+ to a $\kappa = rg_i(\eta) \leq pd_{i+1}(\eta)$. $rg_{N-1}(\eta) = \pi$ for any such $\eta = d_\sigma^q \alpha$.

q determines a sequence $\{\eta_i^m : m < lh_i(\eta)\}$ of subdiagrams of η with its length $lh_i(\eta) = n + 1 > 0$. The sequence enjoys the following property:

$$\eta \preceq_{i+1} \eta_i^0 \prec_{i+1} \eta_i^1 \prec_{i+1} \cdots \prec_{i+1} \eta_i^n < \pi$$

with $st_i(\eta_i^m) < (rg_i(\eta_i^m))^+$.

7.2 Towers derived from ordinal diagrams

Define relations $\ll_i$ for $2 \leq i \leq N - 1$ by

$$\eta \ll_i \rho :\Leftrightarrow \eta \prec_i \rho \,\&\, rg_i(\eta) = rg_i(\rho) \,\&\, st_i(\eta) < st_i(\rho).$$

Extend $\ll_i$ by augmenting the least element 1:

$$1 \ll_i \eta.$$

π^α denotes $\pi^\alpha \cdot 1$.

Let $\lhd_i :\equiv <_{E_i}$ be exponential ordering defined from $\ll_i$ ($2 \leq i \leq N - 1$). Namely $\lhd_{N-1} :\equiv \ll_{N-1}$ and $\lhd_i :\equiv E(\lhd_{i+1}, \ll_i)$, cf. Definition 3.

Extend $\lhd_i$ to $\lhd_i^+$ by adding the successor function $+1$. Namely the domain is expanded to $dom(\lhd_i^+) := dom(\lhd_i) \cup \{a + 1 : a \in dom(\lhd_i)\}$, and define for $a, b \in dom(\lhd_i)$

$$\begin{aligned} a + 1 \lhd_i^+ b + 1 &:\Leftrightarrow a \lhd_i b \\ a + 1 \lhd_i^+ b &:\Leftrightarrow a \lhd_i b \\ a \lhd_i^+ b + 1 &:\Leftrightarrow a \lhd_i b \text{ or } a = b \end{aligned}$$

From the sequence $\{\eta_i^m : 2 \leq i < N - 1, m < lh_i(\eta)\}$ we define a tower $T(\eta) = E_2(\eta)$. The elements of the form $E_i(\eta)(+1)$ are understood to be ordered by $\lhd_i^+$. Let $\lhd_T :\equiv \lhd_2^+$.

$$E_{N-1}(\eta) := \eta$$
$$E_i(\eta) := \sum_{1 \leq m < lh_i(\eta)} \pi^{E_{i+1}(\eta_i^m)} \eta_i^{m-1} + \pi^{E_{i+1}(\eta_i^0)+1} + \pi^{E_{i+1}(\eta)}$$

The sequence $\{\eta_i^m : m < lh_i(\eta)\}$ is defined so that, cf. [2] for a proof,

$$\gamma \prec_i \eta \Rightarrow E_i(\gamma) \triangleleft_i^+ E_i(\eta).$$

In particular

$$\gamma \prec_2 \eta \Rightarrow T(\gamma) \triangleleft_T T(\eta) \tag{2}$$

7.3 Distinguished classes

An elementary fact on the maximal distinguished class $\mathcal{W}$ says that $\mathcal{W}$ is well ordered by $<$ on $Od(\Pi_N)$, and $\mathcal{W}|\Omega$ is included in the wellfounded part of $Od(\Pi_N)$. Therefore it suffices to show $\eta \in \mathcal{W}$ for each $\eta \in Od(\Pi_N)$.

$\mathcal{W}$ is defined to be the union of the distinguished sets,

$$\mathcal{W} = \bigcup \{X \subseteq Od(T) : D[X]\}$$

where $D[X]$(read:X is a distinguished set) is a Δ_1-formula on limits of admissible sets. Hence $\mathcal{W}$ is a Σ_1-definable set of integers, and a proper *class* in $KP\Pi_N$.

Since $D[X]$ is Δ_1 on limits of admissibles, it is absolute: $D[X] \Leftrightarrow P \models D[X]$ for any $X \in P \cap \mathcal{P}(\omega)$. Let $\mathcal{W}^P = \bigcup \{X \in P : P \models D[X]\}$ denote the maximal distinguished class on P.

The following is a key on distinguished sets.

Lemma 5 *There exists a Π_2-formula $g(\eta)$ ($\eta \in Od(\Pi_N)$) for which the following holds for any limits Q of admissibles: Assume $g(\eta)^Q$ and*

$$\forall \gamma \prec_2 \eta \{g(\gamma)^Q \Rightarrow \gamma \in \mathcal{W}^Q\} \tag{3}$$

Then there exists a distinguished class X such that $\eta \in X$ and X is definable in Q.

For some Σ_1 classes U_i on ω, the Σ_1 transitive relations on ω, $<_i$ mentioned in Theorem 4 are now defined to be

$$\eta <_i \rho :\Leftrightarrow \eta \ll_i \rho \,\&\, \eta, \rho \in U_i.$$

By definition $1 \in U_i$ for any i. $<_{N-1}$ is seen to be almost wellfounded in $KP\ell$.

Let $<_{TW}$ denote the restriction of the tower $<_T$ of the exponential orderings $<_{E_i}$ defined from these Σ_1 relations $<_i$ ($2 \leq i \leq N-1$) to the wellfounded parts in the second components hereditarily.

In other words,

$$T(\eta) <_T T(\rho) \Leftrightarrow T(\eta) \triangleleft_T T(\rho) \,\&\, \forall i[\mathcal{K}_i(\eta) \cup \mathcal{K}_i(\rho) \subseteq U_i]$$

and

$$T(\eta) <_{TW} T(\rho) \Leftrightarrow T(\eta) <_T T(\rho) \,\&\, \forall i > 0[\mathcal{K}_i(\eta) \cup \mathcal{K}_i(\rho) \subseteq W(<_i)]$$

where

1. $\mathcal{K}_2(\eta) := \{\eta_2^m : m < lh_2(\eta)\}$.
2. For $2 < i < N-1$, $\mathcal{K}_i(\eta) := \{\rho_i^m : m < lh_i(\rho), \rho \in \mathcal{K}_{i-1}(\eta)\}$.

Lemma 6 *If $P \in M_2(M_2(T(\eta); <_{TW}))$, then $g(\eta)^P \to \eta \in \mathcal{W}^P$.*

Proof By induction on $\in$. Suppose $P \in M_2(M_2(T(\eta); <_{TW}))$ and $g(\eta)^P$. Pick a $Q \in P \cap M_2(T(\eta); <_{TW})$ so that $g(\eta)^Q$.

We show (3). Assume $\gamma \prec_2 \eta$ and $g(\gamma)^Q$. (2) yields $T(\gamma) \triangleleft_T T(\eta)$. On the other side the Π_2 formula $g(\gamma)$ is defined so that

$$g(\gamma)^Q \to \forall i [\mathcal{K}_i(\gamma) \subseteq U_i^Q] \& \forall i > 0 [\mathcal{K}_i(\gamma) \subseteq W^Q(<_i^Q)].$$

Since $\bigcup_i \mathcal{K}_i(\eta)$ is finite, we can assume $\forall i [\mathcal{K}_i(\eta) \subseteq U_i^Q]$, and hence $T(\gamma) <_{TW}^Q T(\eta)$. Therefore $Q \in M_2(M_2(T(\gamma); <_{TW}))$. IH yields $\gamma \in \mathcal{W}^Q$. This shows (3).

By Lemma 5, let X be a distinguished class definable over Q such that $\eta \in X$. Thus $X \in P \& D[X]$, and $\eta \in \mathcal{W}^P$. ♠

Assuming $L \in M_2(M_2(T(\eta); <_{TW}))$ for each η, we have $g(\eta)^L \to \eta \in \mathcal{W}^L = \mathcal{W}$ by Lemma 6. On the other side, it is not hard to show $g(\eta)^L$ for each η in KPℓ.

Therefore the wellfoundedness of $Od(\Pi_N)$ up to each $\eta < \Omega$ follows from $\{L \in M_2(M_2(T(\eta); <_{TW})) : \eta \in Od(\Pi_N)\}$ over KPℓ.

Bibliography

[1] Arai, T. Proof theory for theories of ordinals III:Π_N-reflection. Submitted.

[2] Arai, T. Wellfoundedness proofs by means of non-monotonic inductive definitions II: first order operators. Submitted.

[3] Arai, T. Proof theory for theories of ordinals I: Reflecting ordinals. 1996. Draft.

[4] Arai, T. Systems of ordinal diagrams. 1996. Draft.

[5] Arai, T. Proof theory for theories of ordinals II: Σ_1-stability. 1997. Draft.

[6] Arai, T. Proof theory for theories of ordinals III: Π_1-collection. 1997. Draft.

[7] Arai, T. Some results on cut-elimination, provable well-orderings, induction and reflection. *Ann. Pure Appl. Logic*, 95:93–184, 1998.

[8] Arai, T. Introduction to finitary analyses of proof figures in: Sets and Proofs. In Cooper, S. B., Truss, J. K. (Eds.), *London Mathematical Society Lecture Notes*, vol. 258, 1–25. Logic Colloquium '97-European Meeting of the Association for Symbolic Logic, Leeds, (July 1997), Cambridge University Press, 1999. Invited papers.

[9] Arai, T. Ordinal diagrams for Π_3-reflection. *Jour. Symb. Logic*, 65:1375–1394, 2000.

[10] Arai, T. Ordinal diagrams for recursively Mahlo universes. *Arch. Math. Logic*, 39:353–391, 2000.

[11] Arai, T. Epsilon substitution method for theories of jump hierarchies. *Arch. Math. Logic*, 41:123–153, 2002.

[12] Arai, T. Epsilon substitution method for $ID_1(\Pi_1^0 \vee \Sigma_1^0)$. *Ann. Pure Appl. Logic*, 121:163–208, 2003.

[13] Arai, T. Proof theory for theories of ordinals I: recursively Mahlo ordinals. *Ann. Pure Appl. Logic*, 122:1–85, 2003.

[14] Arai, T. Proof theory for theories of ordinals II: Π_3-Reflection. *Ann. Pure Appl. Logic*, 129:39–92, 2004.

[15] Arai, T. Wellfoundedness proofs by means of non-monotonic inductive definitions I: Π_2^0-operators. *Jour. Symb. Logic*, 69:830–850, 2004.

[16] Arai, T. Epsilon substitution method for $[\Pi_1^0, \Pi_1^0]$-FIX. *Arch. Math. Logic*, 44:1009–1043, 2005.

[17] Arai, T. Ideas in the epsilon substitution method for Π_1^0-FIX. *Ann. Pure Appl. Logic*, 136:3–21, 2005.

[18] Arai, T. Epsilon substitution method for Π_2^0-FIX. *Jour. Symb. Logic*, 71:1155–1188, 2006.

[19] Beckmann, A. A non-well-founded primitive recursive tree provably well-founded for co-r.e. sets. *Arch. Math. Logic*, 41:251–257, 2002.

[20] Beklemishev, L. Another pathological well-ordering. In *Lect. Notes Logic 13, Assoc. Symb. Logic*, 105–108. Logic Colloquium 1998(Prague), 2000.

[21] Buchholz, W. Normalfunktionen und konstruktive Systeme von ordinalzahlen. In Diller, J., Müller, G. H. (Eds.), *Lecture Notes in Mathematics*, vol. 500, 4–25. Proof Theory Symposion, Kiel 1974, Berlin: Springer, 1975.

[22] Kreisel, G. Wie die Beweistheoire zu ihren Ordinalzahlen kam und kommt. *Jber. Deutsch. Math.-Verein*, 78:177–223, 1977.

[23] Tait, W. W. Functionals defined by transfinite recursion. *Jour. Symb. Logic*, 30:155–174, 1965.

Degree Spectra of Unary Relations on $\langle \omega, \leq \rangle$ [1]

Rod Downey
Victoria University
Rod.Downey@mcs.vuw.ac.nz

Bakhadyr Khoussainov
Auckland University
bmk@cs.auckland.ac.nz

Joseph S. Miller
University of Connecticut
joseph.miller@math.uconn.edu

Liang Yu
Nanjing University
yuliang.nju@gmail.com

ABSTRACT. A computable presentation of the linearly ordered set $(\omega, \leq)$, where ω is the set of natural numbers and $\leq$ is the natural order on ω, is any linearly ordered set $\mathbf{L} = (\omega, \leq_L)$ isomorphic to $(\omega, \leq)$ such that $\leq_L$ is a computable relation. Let X be subset of ω and X_L be the image of X in the linear order $\mathbf{L}$ under the isomorphism between $(\omega, \leq)$ and $\mathbf{L}$. The degree spectrum of X is the set of all Turing degrees of X_L as one runs over all computable presentations of $(\omega, \leq)$. In this paper we study the degree spectra of subsets of ω.

1 Introduction

Our interest in this paper falls into part of a long term program in computable model theory where we study the *spectrum* problem for relations on various classes of models. Recall that a **computable structure** is one whose domain and basic relations and predicates are uniformly computable. In particular, if the language (equivalently signature) of the structure is finite then computability of the structure is simply equivalent to saying that its domain and all of its basic relations and operations are computable. If $\mathcal{A}$ is a computable structure isomorphic to $\mathcal{B}$ then we say that $\mathcal{B}$ is **computably presentable** and $\mathcal{A}$ is a **computable presentation (or copy)** of $\mathcal{B}$. Note that computable presentations of a given structure, by definition, are isomorphic structures but can exhibit different computability-theoretic behavior. To capture this behavior the concept of the spectrum of a relation has been introduced that we now explain.

Suppose that R is a relation on a computable structure $\mathcal{A}$. We postulate that R is closed under automorphisms of the structure $\mathcal{A}$, that is, $g(R) = R$

---

[1] Rod Downey, School of Mathematical and Computing Sciences, Victoria University, PO Box 600, Wellington, New Zealand; Bakhadyr Khoussainov, Depart of Computer Science, Auckland University, Private Bag 92019, Auckland, New Zealand; Joseph S. Miller, Department of Mathematics, University of Connecticut, Storrs, CT 06269-3009, USA; Liang Yu, Institute of Mathematical Science, Nanjing University, JiangSu province 210093, China

for all automorphism g of $\mathcal{A}$. If $\mathcal{B}$ is a computable structure isomorphic to $\mathcal{A}$, then we will let $R_\mathcal{B}$ denote the image of R in $\mathcal{B}$. The **spectrum of** R then is defined as the set of all Turing (or perhaps other) degrees of $R_\mathcal{B}$ as we run over all computable $\mathcal{B}$ isomorphic to $\mathcal{A}$. The spectrum of a relation describes the algorithmic properties of the relation in different computable presentations of the structure. The assumption that R is closed under automorphisms is an important one. Indeed, think of R as a unary relation that singles out integers in the linearly ordered set of rational numbers $(Q, \leq)$. This ordering has exactly one computable isomorphism type but the spectrum of R consists of all possible Turing degrees.

Our understanding of possible degree spectra for distinguished relations on various structures has had significant advances in recent years. We know that for many relations the typical spectra we would expect to see would consist of a single element or infinitely many. (See e.g. Harizanov [5–7], Hirschfeldt [8], Goncharov [4], or Moses [12]) We have also made great strides in the construction of models such as graphs, partial orderings, rings, groups, integral domains, etc with many pathological spectra such as 2 element spectra. (See e.g. Hirschfeldt [8], Hirschfeldt, Khoussainov, Shore and Slinko [9].)

In spite of the advances mentioned in the study of degree spectra of relations, our understanding of the spectra is much more limited for *particular* structures and *particular* relations. For instance, whilst we know that there are computable linear orderings L where the adjacency relation, $Adj(L) = \{(x, y) : x$ is adjacent to $y\}$, has either an infinite degree spectrum, or a spectrum consisting of a single element from $\{\mathbf{0}, \mathbf{0'}\}$, (Downey and Moses [3]) or a principal filter computably enumerable Turing degrees (Downey [1]), or every computably enumerable degree *except* $\mathbf{0}$, little else is known. Recall that $\mathbf{0}$ denoted the computable Turing degree. It has recently been shown that the complete degree $\mathbf{0'}$ is always a member of the spectrum of $Adj(L)$ if the spectrum is nontrivial (Downey, Lempp, and Wu [2]).

In this paper we will study the simplest relation on the simplest ordering one could imagine. Namely, we will investigate unary relations X on the standard ordering of the natural numbers $(\omega, \leq)$. Note that the standard natural ordering of the order of natural numbers is characterized by the fact that the adjacency relation is computable. Indeed, any two computable presentations of the order of natural numbers in which the adjacency relations are computable must be computably isomorphic. The ordering $(\omega, \leq)$ is well understood from a computability-theoretic point of view. For example, the ordering is Δ_2^0 stable in the sense that if $(L, \leq_L)$ is a computable ordering isomorphic to $(\omega, \leq)$ then the isomorphism between $(\omega, \leq)$ and is $(L, \leq_L)$ in a Δ_2^0-set.

Let X be a unary relation on $(\omega, \leq)$. If $(L, \leq_L)$ is a computable presentation of $(\omega, \leq)$ then the image of X in $(L, \leq_L)$ can be thought simply as coding of X in $(L, \leq_L)$. We denote this image as X_L. The reader might therefore think very little can really happen. Our results indicate that even in such a simple situation, and even for X that have simple descriptions in computability-theoretic terms, we find interesting and unexpected behaviour. We single out this notation:

If $(L, \leq_L)$ is a computable presentation of $(\omega, \leq)$
then the image of X in $(L, \leq_L)$ is denoted by X_L.

Our paper starts by studying the situation where the unary relation X in $(\omega, \leq)$ is computably enumerable (c.e. for short). One of our results show that the degree spectra of c.e. unary relations in $(\omega, \leq)$ is closed upwards in the Turing degrees. This informally tells us that coding is always possible:

Theorem 1 *For any infinite and coinfinite c.e. sets X and Y if $X \leq_T Y$, then there is a computable presentation $\mathbf{L}$ of $(\omega, \leq)$ so that $Y \equiv_T X_L$. That is, the degree spectra of c.e. unary relations represented in a standard copy of $(\omega, \leq)$ are all closed upwards in the Turing degrees.*

As an immediate corollary we have the following.

Corollary 1 *The degree spectrum of every infinite and co-infinite c.e. set X possesses a Turing complete degree.*

A natural question arises as to whether the spectrum of a c.e. unary relation X contains *all* c.e. Turing degrees. To prove this, it would suffice to show that the image of X in some computable presentation of $(\omega, \leq)$ is computable. It turns out that the situation here is rather surprising as stated in the following two results. For the first result we mention that a degree $\mathbf{a}$ is **low** if $\mathbf{a}' = \mathbf{0}'$ where $\mathbf{0}'$ denotes the Turing degree of the halting problem.

Theorem 2 *Let X be a non-computable c.e. set. The degree spectrum of X possesses a low c.e. degree. In fact, there exists a computable presentation $\mathbf{L}$ such that X_L is a c.e. set, $X_L \leq_T X$, and X_L is low.*

Lowness of a degree does not guarantee that the degree is computable. Hence, one would like to know a sufficient condition at which the degree spectrum of a relation contains the computable degree. This is answered in the next theorem. For the theorem we recall a c.e. degree $\mathbf{a}$ is called *high* if $\mathbf{a}' = \mathbf{0}''$.

Theorem 3 *If X is a c.e. set whose Turing degree is not high, then the degree spectrum of X contains the computable degree. In other words, there exists a computable presentation $(L, \leq_L)$ of $(\omega, \leq)$ in which the copy of X is a computable set.*

We note that Hirschfeldt, R. Miller, and Podzorov independently proved a (weaker) version of this theorem for low$_2$ sets X in place of non-high. Here $\mathbf{a}$ is called low$_2$ iff $\mathbf{a}'' = \mathbf{0}''$. (See [10]).

It turns out that high degrees are source of examples of c.e. sets X whose degree spectra do *not* have the computable degree. The result below is somewhat of a converse to Theorem 3.

Theorem 4 *Every high c.e. degree contains a c.e. set X such that the degree spectrum of X does not contain the computable degree.*

In some sense, Theorem 4 is the best we could hope for in terms of degrees, since we can easily establish the following.

Theorem 5 *Every c.e. m-degree contains a set X such that the degree spectrum of X contains a computable set.*

Degree Spectra of Unary Relations on $\langle \omega, \leq \rangle$

Our remaining results are concerned with spectra of general X. It is not too difficult to construct an X whose degree spectrum contains $\mathbf{0}''$ only. This result can be proved by using the Δ_2^0-stability of $(\omega, \leq)$, and a crude coding of the set $\mathbf{0}''$. However, the situation for Δ_2^0-sets X is much more subtle. Here is one result:

Theorem 6 *There is a Δ_2^0 set X such that the degree spectrum of X is $\{\mathbf{0}'\}$.*

Theorem 6 is just a special case of the following more general result. The result is of interest it says that the Turing degree of any set that computes the halting problem can be realized as the degree spectrum of some unary relation X. Formally:

Theorem 7 *Suppose that for a set $B \subset \omega$ we have $\emptyset' \leq_T B$. Then there is a set X whose degree spectrum is exactly $\{deg(B)\}$.*

The next theorem, in certain respect, refines Theorem 2 for a particular set and particular Turing degrees:

Theorem 8 *There is a Δ_2^0-set X such that images of X in all computable presentations of $(\omega, \leq)$ have low but non-computable degrees.*

Our last result in this paper shows that for every infinite and co-infinite subset X of ω, the degree spectrum of X can not be bounded by the complete Turing degree. An interesting note here is that the unboundedness (by the complete degree) of the degree spectrum is witnessed by two computable linearly ordered sets for *all* infinite and co-infinite subset X. Formally, here is the statement:

Theorem 9 *There exist two computable presentations $\mathbf{L_1}$ and $\mathbf{L_2}$ of the linear order $(\omega, \leq)$ such that for all infinite and coinfinite X, we have $\emptyset' \leq_T X \oplus X_{L_1} \oplus X_{L_2}$.*

2 Notation and Terminology

In this section we introduce a set of notations that will be used throughout the paper. Assume that $\mathbf{L} = (\omega, \leq_L)$ is a computable presentations of the order of natural numbers $(\omega, \leq)$.

Definition 1 *For the linear order $\mathbf{L} = (\omega, \leq_L)$ and for every x in the domain of $\mathbf{L}$ define:*

- $Suc(L)(x) = y$ iff $y >_L x$ and for all y' if $y' >_L x$ then $y' \geq_L y$.
- $Pre(L)(x) = y$ iff $y <_L x$ and for all y' if $y' <_L x$ then $y' \leq_L y$.
- $Ord(L)(x)$ is, by definition, the number of elements y such that $y <_L x$.
- *Given a number $z \in \omega$, we define the isomorphic copy of z in $\mathbf{L}$ to be z_L. Clearly, $Ord(L)(z_L) = z$.*
- *Given a set $X \subseteq \omega$, the isomorphic copy of X in $\mathbf{L}$ is denoted by X_L. It is clear again that $X_L = \{x_L | x \in X\}$.*

Our other notation is standard and follows Soare [13]. For example $\leq_T$ denotes the Turing reduction. For a basic reference on the theory of computable linearly ordered sets we refer the reader to Downey [1]. We now start proving theorems stated in the introduction.

3 The Proof of Theorem 2

3.1 Basic strategies

Given a noncomputable c.e. set X, we will construct a computable presentation **L** of $(\omega, \leq)$ such that X_L is a low c.e. set and $X_L <_T X$. Naturally, we will build **L** in stages, keeping track of X_s. At each stage s we will add one or more elements to X_L that is being built. The overall requirements we must satisfy are the following:

$$Q \ : \ X_L \text{ is c.e.}$$
$$N_e \ : \ X \neq \Phi_e^{X_L}$$
$$M \ : \ X_L \text{ is low}$$
$$P \ : \ X_L \leq_T X.$$

We will break down these tasks into more manageable sub-requirements that are explained below.

Meeting requirement Q. Suppose that there exists an element x such that $X(x)[s+1] \neq X_L(x_L)[s+1]$. Then there are two cases:

1. $X(x)[s+1] = 0$. This case can occur through our efforts to meet other obligations. Since we want X_L to be c.e., in this case we insert a new number between $Pre(L)(x_L)$ and x_L.
2. $X(x)[s+1] = 1$. In this case we either insert a new number between $Pre(L)(x_L)$ and x_L or put x_L into X_L.

We say an element x is moved at stage $s+1$ if $Ord(L)(x)[s+1] \neq Ord(L)(x)[s]$. Now it will suffice to prove that every element $i \in \omega$ has a final position in **L** (i.e. moved only finitely often). Thus, we split the Q-requirement to infinitely many subrequirements:

$$Q_i \ : \ i \in L \text{ will stop moving eventually.}$$

Meeting requirement N_e. To meet N_e we use classical Sacks technique of preserving agreements. To satisfy N_e, we will monitor the length of agreement $\Phi_e^{X_L} = X[s]$. Thus, at stage s, we define:

$$\hat{\ell}_e[s] = \min\{n | X(n)[s] \neq \Phi_e^{X_L}(n)[s]\},$$
$$r_e[s] = \sum_{m \leq \hat{\ell}_e[s]} \phi_e^{X_L}(m)[s],$$
$$R_e[s] = \sum_{e' \leq e} r_e[s].$$

At every stage s, we try to preserve $\hat{\ell}_{e,s}$ by setting up the restraint $R_e[s]$. That is, once we see the length of agreement exceed some value y at some stage s, we will preserve X_L on the use of this computation. In the usual way, we argue that $\ell(e,s) \not\to \infty$, lets X be computable, which it is not.

Of course, the trouble is when a small element say x entered X (whose current corresponding image $x_L[s]$ is below the use of some e-computation to be

Degree Spectra of Unary Relations on $\langle \omega, \leq \rangle$

preserved) at stage $s+1$. Now we have agreed by the restraint, we can not put the element $x_L[s]$ into X_L since $x_L[s]$ less than the restraint. However, x has entered $X[s]$, and hence it needs an image in L. The idea is to shift the isomorphism. We replace $x_L[s]$ with a big number z. That is, we move $x_L[s]$ by defining $x_L[s+1] = z$ and $(x+i)_L[s+1] = (x+i+1)_L[s]$. Were x_L moved infinitely often, then we would fail to satisfy Q since we are destroying a requirement Q_j for some j. A crucial point is if X is noncomputable then it must be infinite and coinfinite. We will argue that this allows us to ensure that every number will stop moving. The reader should note that the introduction of the new element z as the new image for the number entering X could consequentially cause us to need to add a lot of new elements into L to allow us to rearrange the isomorphism whilst preserving the N_e computations, with a sort of "cascade" effect.

Meeting requirement M. Our argument is finite injury and so via the N_e we will be able to argue *post hoc* that X_L is automatically low.

Meeting requirement P. Once we prove that **L** is isomorphic to $(\omega, \leq)$, we can effectively in X decide X_L. To decide x in X_L or not, we just need search a stage s at which $X \restriction (Ord(L)(x)[s]+1) = X[s] \restriction (Ord(L)(x)[s]+1)$. Since **L** is isomorphic to $(\omega, \leq)$, there must exist such a stage. After this stage, x will never be moved. So $X_L \leq_T X$.

There will be conflicts between satisfying requirements Q and N. To give more intuition, we provide an idea of meeting and satisfying one requirement N_e.

3.2 One N requirement

We will concentrate on a single N requirement, say N_0. During the construction, we use the concept of a *crucial points sequence* (c.p.s.) $\{z_i\}_{i \leq m, m \leq \omega}$.

Definition 2 *A crucial points sequence (c.p.s.) is a (finite or infinite) sequence $\{z_i\}_{i \leq m, m \leq \omega}$ which has the following properties:*

1. $\{z_i\}_{i \leq \omega}$ *is a computable set.*
2. $\forall i \leq m(z_i <_L z_{i+1})$.
3. $\forall z \in \omega \exists i \leq m(z \leq_L z_i)$.
4. $\forall i \leq m \forall z \leq z_i(z \leq_L z_i)$.

Item (3) assumes that the sequence is an infinite one.

Clearly, if there is some number z moved infinitely often, then there must be a number z' in the c.p.s. moved infinitely often (just pick up a number $z' > z$ in the c.p.s.). We denote the set $\{z | z_i <_L z \leq_L z_{i+1}\}$ by $(z_i, z_{i+1}]_{\leq_L}$.

Choose a stage s after which N_0 does not receive attention. Then $\forall t \geq s(R(0,t) = R(0,s) = R)$. Suppose $R > 0$.

Suppose there is a number moved infinitely often for the sake of N_0. Then there must be a least number y moved infinitely often. The point is that $y \leq R$ since the only numbers that N_0 moves are below the restraint. For more details, see Lemma 3. So there must a z_i in the c.p.s. so that $y \in (z_i, z_{i+1}]_{\leq_L}$.

Define $Mov = \{z | z$ moved infinitely often$\}$ and $M(y) = Mov \cap (z_i, z_{i+1}]_{\leq_L}$. Then $\forall z \in Mov(z > z_i)$. Select a number $y' \in M(y)$ so that for all $z \in M(y)$, $y' \leq_L z$. Then y' is the least number $\leq R$ moved infinitely often. We will show that this does not happen. Select a stage $t \geq s$ so that for all numbers z with

$z <_L y'$ and $z < R$ are not moved after stage t. Then at any stage $t' > t$, the only action to move y' is inserting a big number between y' and $Pre(L)(y')[t']$. Using the fact the X is infinite, we will argue that we will not move y' infinitely often by Subsection 3.1 Case (1) (that is because of new numbers entering X causing y''s movement.). Without loss of generality, at every stage $t' > t$, for every $z = Ord(L)(y')[t']$, $z \in X$. So X includes the set $\{z | z > Ord(L)(y')[t]\}$ and so is a cofinite set. This contradicts with the assumption that X is noncomputable and so coinfinite. Therefore, we will stop moving y' eventually.

We will see that $X_L \leq_T X$ since we can decide whether any number in the c.p.s. will be moved with the help of X.

3.3 The Construction

We only distribute the priority ordering to N requirements by setting $N_e < N_{e+1}$ for all e. For Q, there is a dynamic priority ordering during the construction. At every stage s, we will construct a linear order $\mathbf{L}[s]$, a finite c.p.s. set $CPS[s] = \{z_i\}$ and restraint function $r_e[s]$. Eventually, $CPS = \cup_s CPS[s]$.

At stage -1, define the c.p.s. set $CPS[-1] = \{0\}$ and $X_L[-1] = \emptyset$. Define $r_e[-1] = -1$ and $\hat{\ell}_e[-1] = -1$, for every $e \geq 0$.

At stage $s+1$, we define a block set by induction on e.

- $BL_0[s+1] = \{z_i | z_i \in CPS[s] \text{ and } z_{i-1} < R_0[s]\}$,
- $BL_{e+1}[s+1] = \{z_i | z_i \notin BL_e[s+1], z_i \in CPS[s] \text{ and } z_{i-1} < R_{e+1}[s]\}$.

Now we have two steps. The first step is devoted to satisfying the requirements, and the second step constructs the c.p.s

Step 1. Suppose $X[s+1] \neq X_L[s]$. Select the least number x so that $X(x)[s+1] \neq X_L(x_L)[s]$. There are two cases.

Case (1). $X(x)[s+1] = 0$. Pick up the least number y which has not yet been ed in the construction. Set $Pre(L)(x_L)[s] <_L y <_L x_L[s]$. Speed up the enumeration of X so that there is a number $z > x$ with $z \in X[s+1]$.

Case (2). $X(x)[s+1] \neq 0$. Suppose $x_L[s] \in (z_i, z_{i+1})_{\leq_L}[s]$ and $z_{i+1} \in BL_e[s+1]$ for some e. If $x_L[s] > R_e[s]$ then we put $x_L[s]$ into $X_L[s+1]$. If $s+1$ has not been used in the construction, then put $s+1$ into the ordering $<_L$ above all current members of the ordering.

Step 2. We construct CPS by induction on substep i. First, let $CPS[s+1][0] = CPS[s]$. Suppose $y = \max\{z | z \in CPS[s][i]\}$. Search the $\leq_L$-least number $z > y$ so that $\forall z' < z(z' <_L z)$. Define $CPS[s+1][i+1] = CPS[s+1][i] \cup \{z\}$. Finally, set $CPS[s+1] = \cup_i CPS[s+1][i]$.

This finishes the construction.

3.4 Verification

Lemma 1 *CPS is a c.p.s.*

Proof

1. CPS is a computable set. To decide $x \in CPS$ or not. Select a stage $s > x$, then every number $\leq x$ has been included L. So $x \in CPS$ iff $x \in CPS[s]$.

   The following facts are easy to see.

Degree Spectra of Unary Relations on $\langle \omega, \leq \rangle$ 43

2. $\forall i (z_i < z_{i+1})$.
3. $\forall z \in \omega \exists i \in \omega (z \leq_L z_i)$. At any stage $s > z$, the $\leq_L [s]$ largest number must be in $CPS[s]$ and so z is L-less than it.
4. $\forall z \leq z_i (z \leq_L z_i)$.

♠

Note we do not claim that CPS is an infinite set although it is. We prove Lemmas 2 and 3 by simultaneous induction on e.

Lemma 2 N_{e+1} is satisfied and $\lim_s R_{e+1}[s] < \infty$.

Proof Select a stage s_0 so that every requirement higher than N_e has been satisfied. Then, by induction (see lemma 3), there exists a stage $s_1 \geq s_0$ so that for all stages $t \geq s$, $BL_e[s] = BL_e[t]$ and $\forall z_{i+1} \in BL_e[s]((z_i, z_{i+1}]_{\leq_L}[s] = (z_i, z_{i+1}]_{\leq_L}[t]))$. So there is a stage $s_2 \geq s_1$ so that we will never put any number in those blocks into X_L. Thus, by the construction, the computation in N_{e+1} will be preserved once set up. That means $\forall s, t(s > t \geq s_2 \Rightarrow R_{e+1,s} \geq R_{e+1,t}$ and $\hat{\ell}_{e+1}[s] \geq \hat{\ell}_{e+1}[t])$. Now, by a standard priority argument, N_{e+1} is satisfied and $\lim_s R_{e+1}[s] < \infty$, lets X be computable. ♠

Lemma 3 There exists a stage s so that for all stage $t \geq s$, $BL_{e+1}[s] = BL_{e+1}[t]$ and $\forall z_{i+1} \in BL_{e+1}[s]((z_i, z_{i+1}]_{\leq_L}[s] = (z_i, z_{i+1}]_{\leq_L}[t]))$.

Proof By lemma 2, there is a stage s_0 so that $\forall t > s_0(R_{e+1}[s] = R_{e+1}[t] = R_{e+1})$. So, we claim that, $\forall t > s_0(BL_{e+1}[s_0 + 1] = BL_{e+1}[t])$.
Suppose that this does not hold. Select

$$z_{i+1} = \min\{z_{i+1} | z_{i+1} \in BL_{e+1} \text{ and } \lim_s |(z_i, z_{i+1}]_{\leq_L}[s]| = \infty\}.$$

Note z_{i+1} is also $\leq_L$-minimal in BL_{e+1} since $z_i \leq z_{i+1}$ iff $z_i \leq_L z_{i+1}$. Define $Mov(e+1) = \{x | Ord(L)(x) = \infty\}$ and $M(z_{i+1}, e+1) = Mov(e+1) \cap (z_i, z_{i+1}]_{\leq_L}$. Choose $y' \in M(z_{i+1}, e+1)$ so that for every $z \in M(z_{i+1}, e+1)$, $y' \leq_L z$. Actually, $y' \leq R_{e+1}$. To prove this, we prove the following claim:

Claim 1

$$\forall x \exists y (x \in Mov(e+1) \Rightarrow$$
$$(y \in Mov(e+1), y \leq R_{e+1} \text{ and } y \leq_L x)).$$

Proof If not, then there is a number $x \in M(z_{i+1}, e+1)$ so that for every y with $y \in Mov(e+1)$ and $y \leq R_{e+1}$, $x <_L y$. Select a stage $s_1 \geq s_0$ so that for every stage $t \geq s_1$ and every number z with $z <_L x$ and $z \leq R$, $Ord(L)(z)[t] = Ord(L)(z)[s_1]$. It follows, that after stage s_1, substep (2.1) in the construction will never apply to any number L-less than x. Define $\hat{M}(x)[s_1] = \{z | z \in (z_i, z_{i+1}]_{\leq_L}[s_1] \cap X_L[t] \text{ and } z \leq_L x\}$. We enumerate $\hat{M}(x)[s_1] = \{x_i\}_{i \leq n}$ with $x_i <_L x_{i+1}$ for every $i < n$. W.l.o.g, suppose $\forall i (x_i > R(e+1))$. Now, for any number x_i, we move it only in the Construction, Case(1). Suppose we move x_0 at some stage $s_2 \geq s_1$, then by the construction, x_0 is the L-least number so that $X(Ord(x_0))[s_2] \neq X_L(x_0)[s_2 - 1]$. Note we will never move any number L-before x_0 by the construction after stage s_2 since we only apply subcase(2.1) to them. Then by speeding up the enumeration of X, there must be a stage $s_3 \geq s_2$ so that $X(Ord(x_0))[s_3] = 1 = X_L(x_0)[s_3 - 1]$. By the construction, we will never move x_0 after this stage. By induction on i, all of the numbers in $\hat{M}(x)[s_1]$ will stop moving. A contradiction. ♠

There are only finitely many numbers less than R_{e+1}, so we just need to select the $\leq_L$-least number y' from the set $\{x | x \leq R_{e+1}\}$. Select a stage $s_1 \geq s_0$ so that for every stage $t \geq s_1$ and every number z with $z <_L y'$ and $z \leq R$, $Ord(L)(z)[t] = Ord(L)(z)[s_1]$. By the same reason as in the claim 1, we also can select a stage $s_2 \geq s_1$ so that we never apply Case (1) at step 1 in the construction to any number L-before y'. So by the construction at step 1, we move y' at a stage $t \geq s_1$ only if $Ord(L)(y')[t] \in X_t$. If there are infinitely many such stages, then $\{z | z \geq Ord(L)(y')[s_1]\} \subseteq X$. But X is noncomputable, so coinfinite. A contradiction. ♠

So $|CPS| = \infty$.

Lemma 4 *The built linear order* **L** *is isomorphic to* $(\omega, \leq)$ *and and* $X_L = \cup_s X_L[s]$.

Proof By lemma 3, every x will stop moving eventually. It means that for every x, $Ord(L)(x) < \infty$. So **L** has order type $\langle \omega, \leq \rangle$ and $X_L = \cup_s X_L[s]$. ♠

Lemma 5 $X_L \leq_T X$.

Proof For any x, there must be a stage s and a number y with $y = Ord(L)(x)[s]$ so that $\{Ord(L)(z)[s] | z \leq_L x \text{ and } z \in X_L[s]\} = X \upharpoonright (y+1) = \{z | z \leq y \text{ and } z \in X\}$. By the construction, $y = Ord(L)(x)[s] = Ord(L)(x)$. So $x \in X_L[s]$ iff $x \in X_L$. ♠

Lemma 6 X_L *is a low set.*

Proof
It is not hard, by induction on e, to prove that we can effectively in $0'$ find a stage s_e so that $\forall t \geq s_e(\hat{\ell}_{g(e)}[s_e] = \hat{\ell}_{g(e)}[t])$.

By $s - m - n$ theorem, there is a computable function g so that

$$\Phi_{g(e)}^{X_L}(x) = \begin{cases} X(0) & : \ \Phi_e^{X_L}(e) \downarrow \text{ and } x = 0, \\ \uparrow & : \ \text{otherwise.} \end{cases}$$

Then $\Phi_e^{X_L}(e) \downarrow$ iff $\lim_s \hat{\ell}_{g(e)}[s] > 0$. Now we just need to find a stage $s_{g(e)}$ so that $\forall t \geq s_e(\hat{\ell}_{g(e)}[s_e] = \hat{\ell}_{g(e)}[t])$. So $\lim_s \hat{\ell}_{g(e)}[s] > 0$ iff $\hat{\ell}_{g(e)}[s_e] > 0$. Thus $X_L' =_T \emptyset'$. ♠

4 The Proof of Theorem 1

4.1 The main idea

The main idea of the proof is somewhat similar to the proof of Theorem 2. The property "infinite and coinfinite " is the crucial point again.

Given an infinite and coinfinite c.e. set X, we try to code set Y into the even numbers column of X_L. That is $x \in Y$ iff $2x \in X_L$.

We start by defining the order $\leq_L$ by declaring the order $\leq_L$ on even numbers x and y is consistent with the natural order. That is if x and y are even numbers then $x \leq y$ implies $x \leq_L y$. Then the construction inserts odd numbers into the order.

We ensure that at any even stages s, if $x \in Y[s]$, we put $2x$ into $X_L[s]$. At odd stages s, suppose there is a number x so that $X(x)[s] \neq X_L(x_L)[s]$ at some stage s, then we handle three cases:

- Case 1: $X(x)[s] = 0$. Select the least (odd) number y has not been placed in the ordering, define the order $Pre(L)(x_L)[s] <_L y <_L x_L[s]$.
- Case 2: $X(x)[s] = 1$ and $x_L[s]$ is odd. Put $x_L[s]$ into $X_L[s]$.
- Case 3: $X(x)[s] = 1$ and $x_L[s]$ is even. Select the least (odd) number y has not been put into the ordering, put it into $X_L[s]$ and define the order $Pre(L)(x_L)[s] <_L y <_L x_L[s]$.

As in the proof of theorem 2, we can stop moving $x_L[s]$ in the first case by speeding up the enumeration of X since X is infinite. In the second case nothing is moved. In the third case, $x_L[s]$ is moved. But X is coinfinite, so this happens at most finitely often.

In addition, $X_L \leq_T X \oplus Y$ will be true since, given an x, we can effectively in $X \oplus Y$ decide whether x in X_L. Let's turn to the formal proof.

4.2 Construction

At stage -1. For all even numbers x,y, declare $x <_L y$ iff $x < y$. Set $X_L[-1] = \emptyset$. We say that every even number has been enumerated into the ordering.

At even stage $2s$, wait (speeding up the enumeration of Y if necessary) for a number $x \in Y[2s] - Y[2s-1]$. Put $2x$ into $X_L[2s]$.

At odd stage $2s+1$, check whether there is a number x so that $X(x)[2s] \neq X_L(x_L)[2s]$. If so, pick up the least one, say x. There are three cases:

1. $X(x)[2s] = 0$. Select the least (odd) number y has not been enumerated into the ordering, define the order $Pre(L)(x_L)[2s] <_L y <_L x_L[2s]$. Thus, y has been enumerated into the ordering. Speed up the enumeration of X so that there is a number $z > x_L[2s]$ and $X_{2s+1}(z) = 1$.
2. $X(x)[2s] = 1$ and $x_L[s]$ is odd. Put $x_L[s]$ into $X_L[2s+1]$.
3. $X(x)[2s] = 1$ and $x_L[s]$ is even. Select the least (odd) number y that has not been enumerated into the ordering, put it into $X_L[2s+1]$ and define the order $Pre(L)(x_L)[2s] <_L y <_L x_L[2s]$. Thus, y has been enumerated into the ordering.

    This finishes the construction.

4.3 Verification

Lemma 7 *Every number is moved finitely often.*

Proof It suffices to prove that every even number is moved finitely often. If not, pick up the least such even number, say x. Select a stage s_0 so that $Y[s_0] \restriction x+1 = Y \restriction x+1$ and no even numbers less than x moved after that. At stage s_0, define $M = \{z \mid \exists y(z = y_L[s_0]$ and $x - 2 <_L y_L[s_0] \leq_L x$ and $X_L(y_l)[s_0] = 1)\}$. Since M is finite, suppose $M = \{z_0 <_L z_1 <_L \ldots <_L z_n\}$. We prove, by induction on i, that every $z_i \in M$ will stop moving.

Indeed, the element z_0 will stop moving since no number $z <_L z_0$ will move after stage s_0 by the construction. But we move z_0 only when case(1) is applicable. However, by the construction, case (1) will be applied to z_0 only finitely often since we speed up the enumeration of X. Once $X(Ord(L)(z_0))[s] = 1 = X_L(z_0)[s]$ at some stage s then z_0 will stop moving forever since case(1) and case(3) can not be applied to any number L-less than z_0. Select the last stage s_1 at which z_0 moved. Note no number can be inserted between z_0 and x before stage s_1 since we always select the L-least number to do at any stage.

Select the least stage s_{i+1} so that no z_j ($j \leq i$) moved after that. And, by the induction, no number can be inserted between z_i and x before stage s_{i+1}. Now we can replace z_0 with z_{i+1} in the proof of above to prove z_{i+1} will stop moving eventually.

So, if $Y(x) = 1$ then $x \in M$. Thus x will stop moving eventually. Otherwise, select the least stage $t \geq s_0$ so that no number $z \in M$ moved after that. Note, by the same reason as above, no number can be inserted between z_n and x before stage t. Actually at stage t, we have the following: $\forall z(z_i <_L z \leq_L x \implies X_L(z)[t] = 0)$. This is because some numbers L-less than z_n always require attention before stage t, and hence we have no time to put anything L-between z_n and x into X_L. So, if $X(Ord(L)(x))[t] = 0$, then x will never be moved. Otherwise, we keep on x moving until a stage $t' \geq t$ so that $X(Ord(L)(x)) = X(Ord(L)(x))[t'] = 0$. But X is coinfinite, there must be such a stage. After stage t', x will stop moving.

♠

Lemma 8 *The linear order* **L** *constructed is isomorphic to* $(\omega, \leq)$. *In addition,* $X_L = \cup_s X_L[s]$.

Proof By Lemma 7, every number will stop moving. So **L** is isomorphic to $(\omega, \leq_L)$ and $X_L = \cup_s X_L[s]$.

♠

Lemma 9 $Y \leq_T X_L$

Proof By Lemma 8, $X_L = \cup_s X_L[s]$. So by the construction, $x \in Y$ iff $2x \in X_L$.

♠

Lemma 10 $X_L \leq_T X \oplus Y$.

Proof Take an x. If x is even then $x \in X_L$ iff $\frac{x}{2} \in Y$. If x is odd then select an even number $y >_L x$ and a stage s so that $Y \restriction (x+1) = Y[s] \restriction (x+1)$. By Lemma 8, there is a stage $t \geq s$ and a number $y' = Ord(L)(y_L)[t]$ so that $\forall z \leq y'(z \in X[t]$ iff $z_L \in X_L[t])$. So after stage t, y will never been moved. So $x \in X_L$ iff $Ord(L)(x_L)[t] \in X$. ♠

5 The Proof of Theorem 3

For the proof we use Martin's characterization of the high degrees. Say that f *dominates* g if $f(n) \geq g(n)$ for all but finitely many n. In [11] Martin proved that X is *high* ($X' \geq_T \emptyset''$) iff there is an $f \leq_T X$ that dominates all computable functions. Therefore, if X is not high, then for any $f \leq_T X$ there is a computable function g such that $(\exists^\infty n)\, g(n) > f(n)$.

If X is computable then the theorem is clearly true. So, we assume that X is not computable. Hence there are infinitely many bit alternations in X.

Construction. Given a nondecreasing computable function g, which we will choose later, we construct a computable linear order $\mathbf{L}$. During the construction we also declare values of X_L Since we want X_L to be computable, we must continuously adjust $\mathbf{L}$ to ensure that values of X_L, once declared, remain correct. These adjustments could easily prevent $\mathbf{L}$ from being isomorphic to $\langle \omega, \leq \rangle$. This is the purpose of g; we will choose it so that, infinitely often, $X[g(s)]$ has stabilized on a sufficiently large initial segment to ensure that all values of X_L that have been declared by the end of stage $s+1$ are permanently correct without the need of further adjustment to $\mathbf{L}$.

Stage 0. Let $L = \emptyset$.

Stage s+1. Assume that the values of X_L have been declared on $L[s]$. Let y be the $\leq_L$-largest element of $L[s]$. For some $m \leq_L y$, it may be the case that $X(Ord(L)(m))[g(s)] \neq X_L(m)[s]$. These values are currently wrong. Check if there are at least $|L[s]| = Ord(L)(y)[s] +1$ bit alternations in $X[g(s)]$. If so, then it is certainly possible to add elements to $\mathbf{L}$ to ensure that $X(Ord(L)(m))[g(s)] = X_L(m)[s+1]$ for all $m \in L[s]$. Do so, as conservatively as possible. Also add an $x \geq_L y$. Declare $X_L(m)[s+1] = X(Ord(L)(m))[g(s)]$ for all new $m \leq_L x$.

Remark. If the initial segment of $X[g(s)]$ with $|L[s]|$ bit alternations has stabilized, then we will never again add elements to L less than y. So all numbers in $L[s]$ have stopped moving by the end of stage $s+1$. If this happens infinitely often, then $\mathbf{L} \cong \langle \omega, \leq \rangle$ and X_L is computable (all declarations that were ever made are correct).

Defining g. It remains to define the computable function g. First, we define an X-computable function f as follows. To define $f(s)$, consider all possibilities for $L[s]$ over all choices of $g(0), \ldots, g(s-1)$. Note that there are only finitely many possibilities because, for any $i < s$, once $g(i)$ is large enough that $X[g(i)]$ has stabilized on an initial segment with $|L[i]|$ alternations, then no larger value of $g(i)$ would change the resulting $L[i+1]$. Furthermore, X can enumerate all such possibilities because it can detect when sufficient initial segments have stabilized. Let n be the maximum value of $|L[s]|$ over all possibilities and define $f(s)$ so that

$X[f(s)]$ has stabilized on an initial segment with at least n bit alternations (and making sure that f is monotone). Then f is an X-computable function.

Because X is not high, there is a computable g, which we used for the construction, such that $(\exists^\infty n)\, g(n) > f(n)$. Hence by the definition of f and the remark above, **L** is isomorphic to $(\omega, \leq)$ and X_L is computable. This is the end of the proof.

6 The Proof of Theorem 4

Let $\{\langle L_e, W_e, V_e\rangle : e \in \omega\}$ be uniformly effective list, where L_e, $e \in \omega$, is a list of computably enumerable subsets of the rational numbers (hence will be a listing of all computable linear orderings), and (W_e, V_e), $e \in \omega$, is a list of all c.e. pairs of disjoint sets of rational numbers. We need to construct $X \subseteq \omega$ such that if L_e is isomorphic to $\langle \omega, \leq \rangle$, then X_{L_e} is not computable. In the construction we must meet requirements of the following form:

$$Q_e : \text{If } L_e \cong \omega, \leq \text{ then } X_{L_e} \text{ is not listed by } W_e.$$

We will have a "master" list of numbers $\{x^i : i \in \omega\}$ that we enumerate into X. As we will see below, other numbers may also be enumerated into X. These numbers will be sufficiently far apart for the following to work. For example, we choose x^i to be $f(i)$ for a reasonably fast growing computable function f. We may add numbers into the intervals $(0, x^0)$ or (x^i, x^{i+1}). This will be done in the reverse order: e.g., we would add $x^{i+1}-1$ first. We allow only Q_0 to enumerate a number into X below x^0, only Q_1 and Q_0 to do this below x_1, etc.

For a fixed e, say $e = 0$, the construction waits till we see a stage s_0 where the pair $(W_e, V_e)[s_0]$ says that the initial segment of $L_e[s_0]$ agrees with the first x^0 many elements of $(\omega, \leq)$. That is, we have $y \in W_e[s_0]$ with $\{z : z <_{L_e} y\} \subset V_e[s_0]$, and all the rational numbers with Gödel numbers $\leq y$ are enumerated into either W_e or V_e at this stage. Thus, it appears that y is the image of x^0. Here we say that x_0 is *realized for* Q_0. (The construction for Q_e will begin with x^e in place of x^0.)

We now enumerate some $p_0 < x^0$ into X_{s_0+1}. The point is that the opponent cannot enumerate any of the current numbers $\{z : z <_{L_e} y\}$ into X_{L_e} since they are already in V_e, and X_{L_e} is supposedly a subset of W_e. Thus, the opponent has to enumerate a new rational number as a potential image of x^0. We conclude that if L_e is isomorphic to $(\omega, \leq)$, and $X_{L_e} \subseteq W_e$, then y is left of the image of x^0 in L_e.

The next action is that $Q_e = Q_0$ would like to move y right of the image of x^1. We wait till we see the initial segment of $(\omega, \leq)[s_1]$ up to some x^1 look the same as the initial segment of L_e of length x^1 at stage $s_1 > s_0$, and this fact is apparently confirmed by W_e and V_e at stage s_1. That is, if there are n elements of X_{s_1} less than or equal to x_1 then there are exactly the same number (and in the same positions) of elements of $L_e[s_1]$ all of which are in W_e and the others are in V_e, and the disjoint union $W_e \sqcup V_e$ computes the appropriate initial segment the natural numbers so that no rational numbers with small Gödel numbers can be enumerated into L_e after stage s_1. Since y has already been declared to be in W_e, we can also ask that y be the apparent image of something in $X[s_1]$. Now one of the following situations will occur for which we describe appropriate actions:

(i) y is already right of the image of x^1 in L_e at stage s_1.

   Action: Do nothing, but move on to Q_e processing x^2.

(ii) y is the apparent image of something in $X[s_1] \upharpoonright x^1$.

   Remark: If Q_e was the only requirement around, that number would be x^1, but other requirements (such as Q_1) might have already put numbers below x_1 into $X[s]$.

   Action: Put into X_{s_1+1} the minimum number of new elements in (x^0, x^1) needed to move y to the right of the image of x^1. Thus if $(x^0, x^1]$ already had k elements in it in X, then we would need to put the greatest k elements not yet in $X[s_1]$ into X_{s_1+1}. (In the actual case of x^1, since only Q_1 could have acted, this number would be either 1 or 2.)

Now the construction proceeds inductively. The goal is for Q_e to drive some $y = y(e)$ to the right of all the potential images of x^i and hence if Q_e acts for all x^i it is not possible for L_e to be isomorphic to $\langle \omega, \leq, X \rangle$ and have witnesses to the computable copy of X in L_e being $W_e \sqcup V_e = \omega$. That is because L_e will have an element $y(e)$ with infinitely many points left of it, so its order type is not ω.

Note that in the interval (x^i, x^{i+1}) since only Q_j for $j \leq i$ will be allowed to enumerate elements into X, it is enough to have the distance between x^i and x^{i+1} be at least $\sum_{j \leq i}(j+1)$.

We now outline the construction above with high permitting. Given a high c.e. set A, we construct a procedure Γ and X meeting the requirements above with $\Gamma^A = X$. We assume by Martin's Theorem that A is enumeration dominant so that for any total computable function g, the principal function of A dominates g. (That is, if $\overline{A}_s = \{a_i^s : i \in \omega\}$ then if $p_A(i) = \mu s[a_i^s = a_i]$, $p_A(i) > g(i)$ for almost all i.)

For a single requirement Q_0, say, the idea is to try to meet Q_0 as above but by using infinitely many potential y_0^n as y's. We then argue that for some n, we meet Q_0 via $y_0^n = y_n$, say. For the sake of Q_0 we will build an auxiliary computable function $g = g_0$.

Initially, we will set $\gamma(i) = a_i^s$ for the procedure below, but this changes in the construction. If we reset $\gamma(i) = a_k^{s+1}[s+1]$ at some stage s, then we would reset $\gamma(j+i) = a_{k+j}[s+1]$ for all j.

We begin as above. For Q_0 we wait till x_0 is realized, and then define $g(0) = s_0$. We would like to put x_0-1 into $X - X_{s_0}$ to move $y = y^0$ right of the current image of x^0. To do this we need a permission from A on $\gamma(0)[s_0]$. Thus, the definition of g challenges the domination of A.

The action at x^0 is *resolved* if at some stage $t_0 > s_0$ A permits $\gamma(0)$. Resolution will allow us to continue the construction. However, whilst we await a resolution at x^0, we start a new strategy based on the belief that no resolution ever occurs at x^0. Thus, we wait for a stage u_1 where till we see the initial segment of $\langle \omega, \leq \rangle [u_1]$ up to some x^1 look the same as the initial segment of L_e of length x^1 at stage $u_1 > s_0$, and this fact is apparently confirmed by W_e and V_e at stage u_1. That is if there are n elements of X_{s_1} less than or equal to x_1 then there are exactly the same number (and in the same positions) elements of $L_e[u_1]$ all of which are in W_e and the others are in V_e, and $W_e \sqcup V_e$ computes the appropriate initial segment the natural numbers so that no rational numbers with small Gödel numbers can be enumerated into L_e after stage u_1. Now we take the apparent image of x^1, y^1, and attempt the construction to satisfy Q_0 using y^1 in the place of y^0.

Thus, we now define $g(1) = u_1$, and await an A permission on $\gamma(1)[u_1]$. We remark that all strategies for Q_0 will be using the *same* computable g. We continue as above inductively, starting strategies at various x^i using y^i.

Say, at some stage s_i we get a permission on some $\gamma(i)$. Then in the interval (x^{i-1}, x^i) we put the necessary number of elements into X to move y^i right of the apparent image of x^i. Next, we will make the use of $\Gamma^A(i+k)[s_i+1]$ bigger. That is, we find a large fresh number d and set $\gamma(i+k, s_i+1) = a_{p+i+k, s_i+1}$ for all $k > 1$. The action is not that all lower priority strategies for Q_0 are initialized at this stage. This entails that a strategy pursuing x^j on the assumption that the i-strategy is stuck is now initialized *forever*, and we would only pursue m-strategies for Q_0 using $m \leq i$ and $m \geq s_i$. The point of "kicking" the uses for $\gamma(i+1, s_i+1)$ is the following. What the i-strategy for Q_0 would like to move y^i right of the image of x^{i+1}. For this it needs a permission from A. We would like to argue that if we ever fail then A is not dominant. But some other strategy might have already defined $g(i+1)$. Now, A might have fulfilled its commitment for dominating $g(i+1)$ and may not actually care to change again, *after* we are ready to move y^i again. Since we have kicked the $i+1$-use to a place that we have not as yet defined g, we can allow that i-strategy to assert control of $g(p+i+1)$. That is, we next wait for L_0 to give us an apparent isomorphism for L_e up to x^{i+1} again, but now at that stage s_2 we will define $g(p+i+1)$ to try to pursue the i-strategy. Of course, the other i'-strategies for $i' < i$ might act before this, initializing (forever) the i-strategy, but there is some i which is immortal. Notice that g is total if we fail to meet Q_0 and would not be dominated by p_A, a contradiction. Also $X \leq_T A$ by permitting on γ.

Notice also that, even for a single Q_0, we would need the intervals (x^i, x^{i+1}) to be larger, since differing Q_0 strategies can enumerate elements into these intervals for their own y^i's.

Now we consider the combination of the requirements. Consider Q_1. The outcomes for the procedure above have order type $\omega + 1$, namely $0, 1, 2, \ldots, f$, where f denotes that Q_0 fails to have infinitely many stages where L_e can be considered as isomorphic to $(\omega, \leq, Q)$ with witness $W_0 \sqcup V_0$, and the others denote the least n where the n-strategy acts infinitely, and hence succeeds.

In a standard way, each outcome has its own version of Q_1. Q_1^f guessing f will simply work on x^i which are untouched by Q_0. Q_1^n guessing n will "know" that n is the correct outcome, and builds its versions of g_1, g_1^n at stages when n looks correct. There are little interactions between the requirements to ensure that $\gamma(i, s)$ has a limit; this also happens in a standard way. The remaining details are a typical tree of strategies argument, and provide little insight but detail.

7 The Proof of Theorem 5

This is a short proof. Given a c.e. set $X \subseteq \omega$ we construct a set Y m-equivalent to X such that Y has a computable copy in a computable copy of $(\omega, \leq)$. The idea is simple. We define a "very spread out and smeared" version of X, so that no point can be pushed to infinity. Thus break the natural numbers into long intervals I_k for $k \in \mathbb{N}$, and interval I_k is has coding locations for $j \leq k$ as follows. The coding location for k is the first element, then there are large (ever increasing) gaps with coding locations for $j < k$ in increasing order, and with large gaps between them. Let $c(j, k)$ denote the coding location for j in interval k.

We will build Y in the primary copy of $(\omega, \leq)$ using only the coding locations. We declare that $j \in X$ iff $c(j, k) \in Y$ for all j. Then clearly $Y \equiv_m X$

The computable copy of Y we build is made by the least effort strategy. Namely, we will have an ordering $\leq_A$ at stage s, and an isomorphism from $\langle \omega, \leq, Y \rangle[s]$ to $A = \langle \omega, \leq_A, Y_A \rangle[s]$. If some element j enters $Y_{s+1} - Y_s$, putting coding markers into all intervals I_k for $k \geq j$, then we add new elements from $\mathbb{N} - \text{dom}(A_s)$. to regain the isomorphism, moving the image of each point as little as possible (at worst to the stage s image of the least point in Y_s above it). Clearly the copy of Y in A must be computable if it exists. By induction, no point in A which is the image of a point in I_k can ever be moved out of the image of $A \upharpoonright (\min\{p : p \in I_{k+2}\} - 1)$, and so the isomorphism exists. This ends the proof.

8 The Proof of Theorem 7

For the proof of the theorem we need a series of lemmas. The first lemma is an easy exercise:

Lemma 11 *Let X_L be a copy of $X \subseteq \omega$ in a computable linear order $\mathbf{L}$ isomorphic to $(\omega, \leq)$. Then $X_L \leq_T X \oplus \emptyset'$ and $X \leq_T X_L \oplus \emptyset'$. Particularly, if both X_L and X compute $\emptyset'$, then $X \equiv_T X_L$.*

Consider the modulus function of the halting problem K: $f(n) = \mu(s)(K_s \upharpoonright n = K \upharpoonright n)$. Define a set $S = \{f(n) | n \in \mathbb{N}\}$. The next lemma is again easy.

Lemma 12 *Given a set $X = \{x_1 < x_2 < ... < x_n < ...\}$ which satisfies the following property:*
$$\exists m \forall n \geq m (a_n \geq f(n)).$$
Then $X \geq_T K$.

Note if X has the property in Lemma 12, then every infinite subset of X has the property too. The proof of the next lemma is more involved.

Lemma 13 *There is a Δ_2^0-set X so that each copy X_L of X in computable presentation $\mathbf{L}$ of $(\omega, \leq)$ has the property in Lemma 12.*

Proof We construct a Δ_2^0-set X which has the property in Lemma 12. Fix an effective enumeration of computable linear orderings $\mathbf{L}_e = (\omega, \leq_{L_e})$. Particularly, define $\mathbf{L}_0 = (\omega, \leq)$.

To decide x_e, we look at $\{\mathbf{L}_i\}_{i \leq e}$. This number needs to be one so that number $x > \sum_{i<e} x_i$ chosen so that $x_{L_i} > f(e)$ for each $i \leq e$. The problem is that we cannot actually decide what a_{L_i} from $\emptyset'$, since $\mathbf{L}_e$ might not even be isomorphic to $\mathbf{L}$. But we can $\emptyset'$-effectively find a number a for which $x_{L_i} > f(e)$ for each $i \leq e$.

This is done using a recursive procedure as follows. For a fixed i and $f(e)$ we can definitely $\emptyset'$ decide if it is i-bad (for $f(e)$). That is, $x_{L_i} \leq f(e)$. Namely for a fixed x we can run the enumeration of $\mathbf{L}_i$ and (using $\emptyset'$) discover that s is a stage where $x_{L_i}[s] = x_{L_i}$ and $x_{L_i} \leq f(e)$. We claim that this is a basis for finding an x

which is i-good. Take some x, run the enumeration of $\mathbf{L}_i$ until we discover that either x is i-bad, in which case we pick a new x and try again with this x, or we get a certification that $x_{L_i}[s] = x_{L_i}$ and $x_{L_i} > f(e)$, or we reach a stage s where for all $y \leq f(e)$,

(i) either $y = b_{L_i}[s] = b_{L_i} \neq x_{L_i}$, or

(ii) $x_{L_i} <_{L_i} y$. (That is y has moved beyond x_{L_i}.)

Then we can see that x is i-good since x_{L_i} cannot be i-bad. Since there are only at most $f(e)$ many i-bad elements for each i, we can run the procedures above for each $i \leq e$ together and hence find an element x which is i-good for $f(e)$ for all $i \leq e$. (Alternatively we could find $ef(e)+1$ many numbers which are 1-good for $f(e)$, of these at least $(e-1)f(e)+1$ must be 2-good for $f(e)$, etc.)

This allows us to define x_e and we put this into A.

If $\mathbf{L}$ is isomorphic to $(\omega, \leq)$ and is the e-th linear ordering, $(x_e)_L \geq f(e)$ for each e by the construction. Rearrange X_L so that $X_L = \{b_1 < b_2 < ...b_n < ...\}$. There is a bijection $g : \omega \to \omega$ so that $b_e = (x_{g(e)})_L$. Take the first $i \leq e$ so that $g(i) \geq e$. Note there must be such an i.

Now we prove, by induction on j, that $b_{i+j} \geq f(i+j)$. For $j = 0$, $b_i = (x_{g(i)})_L \geq f(g(i)) \geq f(e) \geq f(i)$. For $j > 0$, $b_{i+j} = (x_{g(i+j)})_L$. If $g(i+j) \geq i+j$, then $b_{i+j} = (x_{g(i+j)})_L \geq (x_{i+j})_L \geq f(i+j)$. Otherwise, there must be some $k < i+j$ for which $g(k) \geq i+j$. Then $b_{i+j} > b_k = (x_{g(k)})_L \geq f(g(k)) \geq f(i+j)$.
♠

Lemma 14 *For each set $B \geq_T \emptyset'$, there is a set $X \equiv_T B$ so that for each copy X_L of X, $X_L \equiv_T B$.*

Proof Take a Δ_2^0 set $C = \{c_1 < c_2 < ... < c_n...\}$ as in Lemma 13. Define $X = \{c_n | n \in B\}$. Then each copy X_L of X satisfies the property in Lemma 12 and so computes $\emptyset'$. Hence X_L computes C_L for each L. Particularly X computes C and so computes B. Obviously B computes X. So, $B \equiv_T X$. By Lemma 11, each $X_L \equiv_T X \equiv_T B$.
♠

These lemmas prove the theorem.

9 The Proof of Theorem 8

For the proof we need to construct a Δ_2^0-set X so that for each e the set X_e is low. Thus we need to meet the requirements:

$$N_{e,i} : \text{If } \mathbf{L}_e \cong \mathbf{L} \text{ then } \exists^\infty s \Phi_i^{X_{e,s}}(i) \downarrow \to \Phi_i^{X_e}(i) \downarrow,$$

where $\mathbf{L}$ is the structure $(\omega, \leq, X)$ that we are building. Additionally, we must ensure that X has no computable copy. Thus we also need to meet the requirements:

$$P_{e,i} : \text{If } \mathbf{L}_e \cong \mathbf{L} \text{ then } X_e \neq \varphi_i.$$

We meet $P_{e,i}$ in the standard way. In L_e we pick a witness $x = x(e, i)$. We wait for a stage s where $\varphi_i(x) \downarrow$. At such a stage we commit to putting $x \in X_e$, or keep x out of X_e to make sure that $X_e(x) \neq \varphi_i(x)$. Let y_s denote the element of $\mathbf{L}$ corresponding to x at stage s.

For simplicity lets suppose that $\varphi_i(x) = 0$. Then this commitment will ask that we put $y_s \in X_s$. We say that $P_{e,i}$ *asserts control* of y_s. At a later stage t it might be that $y_t \neq y_s$ since a new element has entered $\mathbf{L}_e <_{L_e}$-below x. At this stage t, $P_{e,i}$ releases control of y_s and asserts control of y_t, by putting it into X_t. Note that once $P_{e,i}$ releases control of y_s if will never again assert control of y_s. If $\mathbf{L}_e \cong \mathbf{L}$ then $\lim_s y_s$ exists and we will succeed in meeting $P_{e,i}$.

The only thing that we ensure is that the witnesses are chosen $<_{L_e}$ monotonically. Thus, for $i < j$, if we choose x for $P_{e,i}$ then we can only choose some x' for $x <_{L_e} x'$ as a follower for $P_{e,j}$. This can be done in case $\mathbf{L}_e \cong \mathbf{L}$. If $\mathbf{L}_e \not\cong \mathbf{L}$ it might be possible that only finitely many followers might be chosen for requirements of the form $P_{e,d}$. The reason for this is the following. If $\mathbf{L}_e \not\cong \mathbf{L}$, because some $n \in \mathbf{L}_e$ has infinitely many $<_{L_e}$ predecessors, then should we happen to choose some x where y_s changes infinitely many times, then almost all of the y'_s will also be driven to infinity. This means that there effect on X will be "transitory."

Now we turn to the method of meeting the $N_{e,i}$. At some stage s, we might see $\Phi_i^{X_{e,s}}(i)$. Then we would pursue the usual method of lowness. That is $N_{e,i}$ would assert control of $u(e,i,s) = X_{e,s} \upharpoonright \varphi_i^{X_{e,s}}(i)$. It would ask that this set be preserved forever, and thus we would win.

Again this would entail looking at the pre-image of $u(e,i,s)$ in $\mathbf{L}$ and asking that they not change their status with respect to X. This will mean that there is a set of numbers $z(e,i,s)$ which correspond to the use $u(e,i,s)$ in $\mathbf{L}_s$ whose membership pattern is determined by $\Phi_i^{X_{e,s}}(i)$. At stage s control will be asserted with priority $N_{e,i}$, upon $\Phi_i^{X_{e,s}}(i)$ For a single $N_{e,i}$ if $\mathbf{L}_e \cong \mathbf{L}$, these pre-images will eventually settle down, and we will succeed in meeting $N_{e,i}$, with finite effect. In the case that $\mathbf{L}_e \not\cong \mathbf{L}$, it might be possible that some $\leq_{L_e}$-least member z of $z(e,i,s) \to \infty$. Then a finite number of $\mathbf{L}$ will be preserved at almost all stages, but the overall effect of $N_{e,i}$ will be that those numbers $\leq_{L_e}$ above z below $u(e,i,s)$ will have only transitory effect on X.

Now we need to consider the overall effect of the interactions of the requirements. The first effect is that various higher priority $P_{f,j}$ will be able to injure some $N_{e,i}$. As outlined above some $P_{f,j}$ will be able to assert control over some $y = y_s$ in $\mathbf{L}$, to ask that it enter (or leave) $X_{e,s}$. All is sweet unless y corresponds to some pull back of an element in the use $u(e,i,s)$.

Now the plan is to allow $P_{f,j}$ to injure $N_{e,i}$ if it has higher priority by simply letting it do what it wants. If it injures $N_{e,i}$ then we would need to find a new $\mathbf{L}_e$ configuration for $N_{e,i}$ to preserve, with its own priority. Whilst we are doing this we will remember this desirable $N_{e,i}$ configuration. It may be that this injury by $P_{f,j}$ might be f,j-transitory, in that $y_s \to \infty$, since $\mathbf{L}_f \not\cong \mathbf{L}$.. Once y_s clears the $\mathbf{L}$-zone corresponding to the $u(e,i,s)$ elements, then we would be free to return to that configuration. Note that, thereafter, *unless* $\mathbf{L}_e$ itself changes the region of $\mathbf{L}$ that it needs to control to preserve the $u(e,i,s)$-computation, $P_{f,j}$ will have no further effect on $N_{e,i}$. Thus in this case, *provided that* $\mathbf{L}_e \cong \mathbf{L}$, if the action of $P_{f,j}$ is infinitary, it basically has only finite effect on $N_{e,i}$.

On the other hand, it might well be possible that $P_{f,j}$ might be concerned with some $\mathbf{L}_f \cong \mathbf{L}$ and hence the position of y_s might settle down, and might have a permanent injury to $N_{e,i}$. This case is handled in the obvious way. When $N_{e,i}$ was injured by $P_{f,g}$ as above, then we would look for a new computation involving $\Phi_i^{X_{e,s'}}(i)$ for some $s' > s$ to preserve. In the situation that $P_{f,g}$ settles down, the next configuration would be one that could really be preserved with priority e, i.

Similar comments pertain to $N_{q,r}$ of higher priority than $N_{e,i}$ which again might have their own demands as to the look of X in the region that $N_{e,i}$ wishes to control. Again we note that either this effect is permanent and we would meet $N_{e,i}$ after this injury, or we would be able to resurrect the $u(e, i, s)$ computations once the $N_{q,r}$ demands have permanently cleared the use.

Clearly, arguing by priorities, we would not allow a lower priority $P_{g,d}$ to injure some $N_{e,i}$. If the relevant x had a y_s which would be in some critical region corresponding to $N_{e,i}$ then we would regard this as a (possibly) temporary injury to $P_{g,d}$. and we would be able to meet $P_{g,d}$ with another follower. The same considerations hold for $N_{r,s}$ of lower priority. Really there is guessing going on here, since for any fixed r we could know the outcome of the $\mathbf{L}_e$ for $e < r$, but if $\mathbf{L}_r \cong \mathbf{L}$ then all injuries are finite, and hence we will be able to argue that all the requirements are met. The remaining details are straightforward.

10 The Proof of Theorem 9

For the proof of the theorem we need to exhibit two computable presentations $\mathbf{L}_1$ and $\mathbf{L}_2$ of $(\omega, \leq)$. The definitions of $\mathbf{L}_1$ and $\mathbf{L}_2$ are quite simple. For $\mathbf{L}_1$, first let L_1 be the even numbers under the usual ordering. If n goes into $\emptyset'$, put the least remaining odd number immediately *before* $2n$. For $\mathbf{L}_2$, again let L_2 be the even numbers under the usual ordering. But now if n goes into $\emptyset'$, put the least remaining odd number immediately *after* $2n$.

Assume that X is infinite and coinfinite. To a large extent, the blocks of bits in X are reflected in $X_{L_1} \restriction 2\omega$. The only exception is when X has an isolated bit at position x and x_{L_1} is odd. But if x_{L_1} is odd, then $(x+1)_{L_1} = x_{L_2}$ is even. So $X_{L_1}(x_{L_2}) = X_{L_1}((x+1)_{L_1}) = X(x+1) \neq X(x) = X_{L_2}(x_{L_2})$ and thus we can detect the missing block by looking for a discrepancy between $X_{L_1} \restriction 2\omega$ and $X_{L_2} \restriction 2\omega$.

Once we have located the missing blocks of bits, it is straightforward to compute $\emptyset'$. To determine if $n \in \emptyset'$, find an even $m > 2n$ such that either $X_{L_1}(m) \neq X_{L_1}(m-2)$ or there is a "missing block" before m. Run the construction of $\mathbf{L}_1$ until a stage s such that $X_{L_1}[s] \restriction m+1$ agrees on the even bits with $X_{L_1} \restriction m+1$ *and* every missing block has been accounted for. In other words, if we determined that there is a missing block in $X_{L_1} \restriction 2\omega$ right before $2t \leq m$, then the immediate predecessor of $2t$ in $L_1[s]$ must be odd. Once such a stage s has been reached, no number $\leq n$ can enter $\emptyset'$ because otherwise we would exceed the number of bit alternations—missing and apparent—in $X_{L_1} \restriction 2\omega$ before position m. Therefore, $n \in \emptyset'$ iff $n \in \emptyset'_s$. We have proved the theorem.

Bibliography

[1] Downey, R. G. Computability theory and linear orderings. In Ershov, Y. L., Goncharov, S. S., Nerode, A., Remmel, J. B. (Eds.), *Handbook of Recursive Math-*

ematics (Stud. Logic Found. Math), vol. 138–139, 823–976. Amsterdam: Elsevier Science, 1998.

[2] Downey, R. G., Lempp, S., Wu, G. *On the complexity of the successivity relation in computable linear orderings.* In preparation.

[3] Downey, R. G., Moses, M. F. Recursive linear orderings with incomplete successivities. *Trans. Amer. Math. Soc.*, 320:653–668, 1991.

[4] Goncharov, S. S. On the number of nonautoequivalent constructivizations. *Algebra and Logic*, 16:169–185, 1977.

[5] Harizanov, V. Some effects of ash–nerode and other decidability conditions on degree spectra. *Ann. Pure Appl. Logic*, 54:51–65, 1991.

[6] Harizanov, V. Uncountable degree spectra. *Ann. Pure Appl. Logic*, 54:255–263, 1991.

[7] Harizanov, V. Turing degree of the non-zero member in a two element degree spectrum. *Ann. Pure Appl. Logic*, 60:1–30, 1993.

[8] Hirschfeldt, D. *Degree Spectra of Relations on Computable Structures.* Ph.D. thesis, Cornell University, 1999.

[9] Hirschfeldt, D., Khoussainov, B., Shore, R., Slinko, A. *Degree spectra and computable dimension in algebraic structures.*

[10] Hirschfeldt, D., Miller, R., S., P. Order computable sets. *Notre Dame Journal of Formal Logic*, (3):317–347, 2007.

[11] Martin, D. A. Classes of recursively enumerable sets and degrees of unsolvability. *Z. Math. Logik Grundlag. Math*, 12:295–310, 1966.

[12] Moses, M. F. Relations intrinsically recursive in linear orderings. *Z. Math. Logik Grundlagen. Math.*, 32(5):467–472.

[13] Soare, R. I. Recursively enumerable sets and degrees. In *Perspect. Math. Logic.* Heidelberg: Springer–Verlag, 1987.

Groups of finite dimension in model theory[1]

Eric Jaligot

Université de Lyon - CNRS and Université Lyon 1

jaligot@math.univ-lyon1.fr

ABSTRACT. The Morley rank is the usual notion of dimension in model theory which encapsulates the more classical notion of dimension of algebraic varieties in algebraic geometry. In this paper we give a survey of results concerning the classification of infinite simple groups of finite Morley rank. We emphasize both the developments parallel to the Classification of the Finite Simple Groups and the increasing developments of the subject in infinite combinatorics.

1 Introduction

Since the late sixties, first-order model theory has become more and more involved with "concrete" mathematics. The reason is that first-order properties, despite their weaknesses, mesh well with combinatorics and algebra in general and explain many phenomena at a high level of abstraction.

The first theorem bringing model theory to the most modern considerations in algebra and geometry is certainly Morley's theorem on the categoricity in any uncountable cardinal of first-order theories categorical in one uncountable cardinal [38]. From this point on, the classification theory by Shelah provided a very clear picture of the first-order complexity of mathematical structures, with a substantive study of the case of *stable* theories [50]. This has had important applications to diophantine problems, and the abstract stability theory also has nowadays generalizations to instable contexts, with developments on *simple* theories on the one hand and on *o-minimality* and *dependent* theories on the other.

There are however many open questions left in the stable case, notably questions concerning the existence of certain structures and their exact ranking in the stability hierarchy. Such questions arise in particular for structures with an algebraic flavor, especially for groups, even at the very bottom of the stability hierarchy. Here the so-called Morley rank, the usual model-theoretic notion of dimension which encapsulates the dimension of varieties in algebraic geometry, is finite.

The first theorems describing properties of certain algebraic structures satisfying model-theoretic constraints of that nature were due to Macintyre in the very early seventies. For example, an infinite field of finite Morley rank is algebraically closed [37]. It was also shown by Reineke that a group of Morley

---
[1] The author is with Université de Lyon - CNRS and Université Lyon 1.

rank 1 must be abelian-by-finite [48]. In the late seventies, a conjecture has been suggested by Cherlin and Zilber about simple groups of finite Morley rank.

Algebricity Conjecture *An infinite simple group of finite Morley rank is isomorphic to an algebraic group over an algebraically closed field.*

This can be seen as an infinite version of the Classification of the Finite Simple Groups (CFSG), but contrarily to the finite case this is a conjecture and not a classification considered as true. In fact this is largely open, but undoubtably a major source of inspiration since almost thirty years. A lot of work has been done in this direction, the first one being certainly the result obtained by Cherlin in [11] about groups of small Morley rank. It has been proved there in continuation of the theorem of Reineke that a group of Morley rank 2 must be solvable-by-finite. For connected nonsolvable groups of Morley rank 3, it has been proved that the center is finite and modulo this center one gets a simple group G with the following alternative.

Algebraic case : Either G has a definable subgroup of rank 2, in which case $G \simeq \mathrm{PSL}_2$ over some algebraically closed field, or

Nonalgebraic case : G has a definable connected subgroup B (of rank 1) which is malnormal, i.e. satisfies $B \cap B^g = 1$ for any $g \in G \setminus B$, and whose conjugates cover G.

The existence of a group as in the second alternative remains entirely open, and constitutes the first apparition of the so-called *bad group problem*. Here the term "bad" simply reflects local properties which are very far from those of groups encountered in the algebraic category, and in particular the existence of counterexamples to the main Algebricity Conjecture. Later this problem has been formulated at various levels of generality, the most general one being maybe the following question from [31]. Is there an infinite group G satisfying strong stability properties and having a definable subgroup H such that

(∗) $\qquad\qquad H$ is malnormal in G and $G = H^G$?

A group G with a subgroup H as in condition (∗) is called a *full Frobenius group*. It is easy to see that such a group cannot be finite by counting its elements, and cannot have involutions, i.e. elements of order 2, by using the fact that two involutions in a group always generate a dihedral group. Hence the bad group problem concerns exclusively infinite groups without involutions. A case of particular interest and which we will investigate below is the case H abelian, as in the hypothetical nonalgebraic configuration of a group of Morley rank 3 described above.

At the opposite of the bad group problem, which is a difficult problem at the moment, a lot of progress has been done since the late eighties in the direction of the Algebricity Conjecture. The main impulse in this direction is *Borovik's program*. It aims essentially at continuing the recognition of the Bruhat decomposition of the expected algebraic groups, which was the main point in the algebraic case of the above dichotomy for groups of Morley rank 3, using ideas from CFSG. The latter spreads in its first generation of highly non-inductive proofs over approximately 15000 pages of research papers, and in its "revisionist" second generation of highly inductive proofs over less than (!) 20 expected volumes [24]. Borovik's program also aims at giving a purified version, or rather a version free from finite combinatorics complications and the 26 sporadic finite simple groups, of CFSG.

Most of the research in simple groups of finite Morley rank tend to lead to a kind of dichotomy between bad groups configurations (in a large sense) and algebraic groups. In the present paper we are going to review the little which is known concerning potential constructions of bad groups. Then we will see the main results toward the Algebricity Conjecture in presence of involutions. Finally, we will conclude with some speculations which can be formulated between these two opposite directions.

2 Infinite combinatorics

The first dividing line by Shelah concerning first-order theories was the existence or not of a formula with the *order property* [50]. The existence of such a formula implies the presence in some model of the theory of an infinite order and by compactness of a dense linear order without endpoints. The countable ordered set of rationals has $2^{\aleph_0}$ types, the irrational cuts, and this phenomenon of explosion of the number of types over a certain number of elements propagates to larger cardinals. Realizing or omitting types provides in this case the maximal number of models in uncountable cardinals, and thus a situation at the opposite of Morley's theorem.

The first-order theories for which no formula has the order property are the *stable* ones. Here the number of types is under a certain control. This is why stable theories have been highly investigated, both in pure and applied model theory, with spectacular applications by Hrushovski to number theory such as its solution to the Mordell-Lang conjecture. The "most" stable theories, in which there exists a global finite dimension, are those of finite Morley rank.

A major question in the theory of groups of finite Morley rank is to know whether there exists such a group not occuring "in nature", and in particular satisfying condition (∗) above. We just make a note of a few basic properties of such groups, if they exist.

Fact 1 [31, Propositions 3.3 and 3.4] *Let $H < G$ be a full Frobenius group, with G of finite Morley rank and connected. Then*

(a) *H is definable in the pure group G and connected.*
(b) $\text{rk}(G) \geq 2\text{rk}(H) + 1$.
(c) *There exists a nontrivial definable simple subgroup $\tilde{G}$ of G such that $(\tilde{G} \cap H) < \tilde{G}$ is a full Frobenius group.*

Hence the problem of bad groups reduces to infinite simple groups without involutions. We will concentrate below on the case H abelian, and for that purpose we consider the following class of groups.

A *CSA-group* is a group G in which every maximal abelian subgroup is malnormal. It can be checked easily that free groups are CSA-groups. A finitely generated free group cannot be ω-stable as it contains infinite cyclic groups as definable subgroups (for example the centralizer of a generator), and even not superstable [46, §5.7]. Some natural generalizations of finitely generated free groups, in the sense of finitely presented groups with very few relations between the generators, are the hyperbolic groups defined by Gromov [25]. Torsion-free hyperbolic groups are also CSA-groups [40, Proposition 12] in which maximal abelian subgroups are cyclic [34]. Concerning the first-order theories of these groups, a great deal has been accomplished recently by Sela in continuation of its investigations of definable sets in these groups (see also [36]).

Theorem 1 [49] *Let G be a torsion-free hyperbolic group. Then the first-order theory of G is stable.*

A contemporary question is the generalization of Theorem 1 to free products of stable groups, for example the free product of a free group with, say, a divisible abelian group. We will say a little bit about this at the end of this section.

Theorem 1, together with results announced recently by Bestvina and Feighn on a new purely combinatorial notion of genericity in free groups [5], provides a better understanding of the geometry of algebraic closure in free groups. By a recent work of Pillay [45], this shows similarities with bad groups [43, 44]. All this indicates that free groups and related groups such as hyperbolic groups probably have to do with the bad group problem.

In the context of this problem, the study of CSA-groups is motivated on the one hand by the nonalgebraic configuration arising in Morley rank 3, and on the other hand by the fact that CSA-groups are usually more suitable than arbitrary groups for solving equations, an important step towards understanding their first-order theory.

We can also restrict their torsion as follows. If f is a function from the set of prime integers into $\mathbb{N} \sqcup \{\infty\}$, a CSA_f-*group* is a CSA-group which contains no elementary abelian p-subgroup of rank $f(p)+1$ for any prime p such that $f(p)$ is finite. Once such a function f is fixed, it is easily seen that the first-order class of CSA_f-groups is inductive, and thus contains *existentially closed CSA_f-groups*. If a CSA-group contains an involution, then it must be abelian, a case irrelevant for the bad group problem and of low interest from the model-theoretic point of view, as all abelian groups are known to be stable. Hence we usually work with CSA-groups without involutions, i.e. with CSA_f-groups where $f(2) = 0$.

It is known [35, Sect. 2, 5] in this case that existentially closed CSA_f-groups are simple, in a strong sense that simplicity is provided by a first-order formula, and that their maximal abelian subgroups are conjugate, divisible, and of Prüfer p-rank $f(p)$ for each prime p. One gets thus the elementary algebraic properties of a bad group. But concerning the model theory of existentially closed CSA_f-groups, always assuming $f(2) = 0$, it has also been proved in [35] that such groups are not ω-stable by counting the number of types, more precisely by showing the existence of $2^{\aleph_0}$ types over the empty set. The same method for counting types has later been used in [42] to prove that the first-order theory of existentially closed CSA_f-groups ($f(2) = 0$) is not superstable.

In fact, a somewhat simpler argument has shown that they are far from being stable. If $\phi(\overline{x}, \overline{y})$ is a first-order formula in a given language $\mathcal{L}$ and $\mathcal{C}$ is any (not necessarily elementary) class of $\mathcal{L}$-structures, we say that ϕ has the *independence property relatively to $\mathcal{C}$* if for any positive integer n there exists a structure M_n in $\mathcal{C}$ with sequences of tuples $\overline{x}_1, \ldots \overline{x}_i, \ldots \overline{x}_n$, and $\overline{y}_1, \ldots \overline{y}_\sigma, \ldots \overline{y}_{2^n}$, where the indices σ vary over the set of subsets of $\{1, \ldots, n\}$, such that the following holds

in M_n:
$$M_n \models \phi(\overline{x}_i, \overline{y}_\sigma) \text{ if and only if } i \in \sigma.$$
When the class $\mathcal{C}$ consists of all models of a complete first-order theory T, this definition corresponds to the usual notion of independence property of the formula ϕ relatively to the first-order theory T as defined in [50]. This is a property stronger than the order property as any formula with the independence property has the order property. As we are dealing with groups here, the language is typically that of groups, and the simplest formulas one has to investigate first are merely group equations in two variables, i.e. of the form $w(x,y) = 1$ with $w(x,y)$ a group word in two variables x and y.

Theorem 2 [34] *Let λ be a real in $[0, 1/6]$ and $w(x,y)$ a group word in two variables. Then the probability that the formula $w(x,y) = 1$ has the independence property relatively to the class of torsion-free $C'(\lambda)$-groups tends rapidly to 1 has the length of w tends to the infinity.*

The $C'(\lambda)$ condition is a classical small cancellation condition from combinatorial group theory [41], and the parameter λ essentially measures the length of possible cancellations when forming the product of relators in a finitely presented group. Generally, small cancellation conditions allow one to prove that large traces of the relators remain in all their consequences, and in this sense they provide finitely presented groups with few relations among their generators. For example, $C'(\lambda)$-groups are hyperbolic groups when $\lambda \leq 1/6$ [23, Théorème 33].

To give an idea of the proof of Theorem 2, consider an "arbitrarily chosen" and "sufficiently long" word w, and a (possibly oriented) irreflexive graph Γ on a finite set of elements $a_1, \cdots, a_n$. Then the set of elements $w(a_i, a_j)$ in the free group generated by the a_i's will most probably satisfy the $C'(\lambda)$ condition. Groups presented by these generators and relations will then satisfy the small cancellation condition $C'(\lambda)$, and this implies for $\lambda \leq 1/5$ that they are *aspherical* in the sense that van Kampen diagrams of their presentations cannot be irreducible over a sphere, or equivalently that a relator is not consequence of the others (except the trivial cases) [34, Fact 3.3]. Hence the group G_Γ generated by the a_i's and with relations $w(a_i, a_j) = 1$ whenever a_i and a_j are linked in the graph Γ will be a $C'(\lambda)$-group, and by independence of relations one will get in G_Γ
$$w(a_i, a_j) = 1 \text{ if and only if } (a_i, a_j) \text{ are linked in } \Gamma.$$
For $\lambda \leq 1/6$, G_Γ is hyperbolic, and as the "random" word w is not a power, there is no "obvious" torsion in G_Γ. This together with asphericity implies that G_Γ is torsion-free [39, Lemma 64]. Now it suffices for Theorem 2 to choose a finite graph encoding subsets of a finite set, and as this can be done for finite sets of arbitrarily large size one gets the relative independence property.

The $C'(\lambda)$-groups built in the proof of Theorem 2 are torsion-free hyperbolic groups, and in particular torsion-free CSA-groups. By Theorem 1 one cannot imagine a version of Theorem 2 where the class of groups would consist of groups elementarily equivalent to a given torsion-free hyperbolic group, or more generally to a fixed *finite* set of torsion-free hyperbolic groups. The proof provides however a countable set of hyperbolic groups.

If G is an existentially closed CSA_f-group ($f(2) = 0$), then any CSA-group embeds into a model of the first-order theory of G [35, Corollary 8.2], or more generally into a model of any theory whose universal part coincides with that of the theory of G. As this applies to the preceding $C'(\lambda)$-groups, one gets that for most words w the formula $w(x, y) = 1$ has the independence property relatively to the first order theory of G, or more generally relatively to the first-order theory of any group having the same universal theory as G. Hence in such groups many formulas have the independence property, at least a large majority of the quantifier-free formulas. These groups are in particular very far from being stable.

Other properties of a formula implying the instability of a first-order theory, orthogonal in some sense to the independence property, are those like the *strict order property*. One may wonder whether the first-order theory of an existentially closed CSA-group has such properties also. We provide here some speculations on this question, again for formulas $\phi(x, y)$ of the form $w(x, y) = 1$ for some group word w, and we concentrate on weaker properties such as the *n-Strong Order Property* SOP_n defined in [51, Definition 2.5] for $n \geq 3$. We refer to [51, Sect. 2] for a general discussion about these properties, and we just recall the implications

$$\begin{aligned}
\text{strict order property} &\Rightarrow \cdots \\
&\Rightarrow SOP_{n+1} \\
&\Rightarrow SOP_n \\
&\Rightarrow \cdots \\
&\Rightarrow SOP_3 \\
&\Rightarrow \text{non-simplicity} \Rightarrow \text{instability}
\end{aligned}$$

If w is a "long" and "arbitrarily chosen" cyclically reduced group word in two letters x and y as above, then "most probably" the formula $w(x, y) = 1$ will not exemplify SOP_n. The reason is that the kind of arguments described above for the proof of Theorem 2 provide torsion-free $C'(1/6)$-groups generated by elements $a_1, \ldots, a_n$ in which $w(a_i, a_j) = 1$ if and only if $j = i + 1$ modulo n, and thus the graph defined on an existentially closed CSA_f-group by the formula $w(x, y) = 1$ contains cycles of size n, contrary to one of the two requirements for the condition SOP_n, the other one being the existence in some model of an infinite chain in this graph.

Hence one may be tempted to look at "short" words. As usual in this context of commutative transitive groups, the very short word $[x, y]$ witnessing the commutation of x and y boils down to an equivalence relation, and hence is useless. Incidentally, the formula $\phi(x, y)$ used in [52, Proposition 4.1] to prove that Group Theory "in general" has SOP_3 is

$$(xyx^{-1} = y^2) \wedge (x \neq y).$$

In the more specific context of CSA-groups it implies $y = 1$. Hence the absence of certain triangles (a_1, a_2, a_3) satisfying $\phi(a_1, a_2) \wedge \phi(a_2, a_3) \wedge \phi(a_3, a_1)$ (provided by [53, p. 493] in the context of arbitrary groups) is immediate in the context of CSA-groups, but one cannot hope to find an infinite chain in the graph associated to $\phi(x, y)$. Hence this formula could not exemplify the SOP_3 in our context of CSA-groups.

These discussions mean that a formula examplifying the SOP_n of an existentially closed CSA-group, if it exists, cannot involve only an "arbitrarily chosen long" equation, and it does not seem to involve only "short" equations.

Similarly, the non-simplicity is equivalent to the *tree property* for a formula [56], and the definition involves the inconsistency of a certain formula

$$\phi(x, a_1) \wedge \cdots \wedge \phi(x, a_k)$$

for some parameters a_i's placed on certain levels of a tree. Again the proof of Theorem 2 as such cannot imply the inconsistency of certain sets. As properties similar to the strict order property involve the inconsistency of certain sets over the whole structure, the proof of Theorem 2 cannot yield such properties.

After these considerations on the strict order property and some of its weaker forms, we return to the order property and the independence property. While stable theories are those for which no formula has the order property, the more general *dependent* theories are those for which no formula has the independence property. Both the order and the independence property are local properties, and thus stability and dependence are local hypothesis as it was emphasized already in [27, §5] for stability. From a combinatorial point of view, the order property is essentially the same as the independence property, but instead of coding all subsets of a set one just codes a maximal chain (for inclusion) of such subsets. This is usually read as

"$\phi(x_i, y_j)$ if and only if $i \leq j$"

where both indices i and j vary in $\{1, \cdots, n\}$. This combinatorial difference between stable theories and dependent theories is relatively thin and this is certainly the reason why certain aspects of stability generalize to dependent theories in recent works such as [28] and [29].

Apparently Theorem 2 gives a strongly negative result concerning the existence of new stable non-abelian CSA-groups, but at the same time it gives very strong hints for constructing such groups. If one wants a formula to be without the independence property, it suffices to bound the size of the finite sets for which it will encode all subsets. This can be done in a first-order way, simply with universal quantifiers in front of a boolean combination of that formula. And in fact one can do exactly the same thing for stability directly. The only problem is to have an idea of the bounds, for each formula. But Theorem 1 provides such bounds!

A stable set is a set defined by a formula which do not have the order property. One sees easily with Ramsey's theorem that a boolean combination of stable sets is also a stable set [55, 0.2.10] and that the interpretation of variables by parameters do not affect stability. Hence, if one wants to build a group in which at least quantifier-free (but possibly with parameters) definable sets are stable, one can restrict the class of groups considered by adding universal axioms stipulating the stability of atomic formulas, i.e. group equations without parameters. A remark is in order here. One restricts to quantifier-free definable sets first for two reasons. The first one is that quantifier elimination is in general only partial and a very difficult problem in such contexts. The second reason is to be sure to get existentially closed groups in the class under consideration.

The natural question arising then is to know whether the new restricted universal class is closed when taking certain free products and HNN-extensions. It involves either to rework Theorem 1 to get bounds for atomic formulas depending nicely on those of the original groups in certain free products or HNN-extensions, or merely to get such bounds by applying Theorem 1 and Ramsey theory. Both approaches seem to be very promising, and the existentially closed groups in the restricted universal class would be stable at the level of quantifier-free definable sets and would satisfy condition (∗).

To conclude, the study of existentially closed groups in the full class of CSA-groups provides very strong hints about the restrictions which should be imposed to build stable bad groups, simply by prohibiting all the phenomena which show up in Theorem 2. But, contrarily to the local character of stability, the finiteness of Morley rank, or more generally superstability, demands a global notion of dimension. And it is still very hard to have any idea of what a global dimension should be. This is however needed in Hrushovski's amalgamation method. Constructions produced by this method generally give new superstable structures, or even structures of finite Morley rank. They are based essentially on the study of certain existentially closed models in certain classes, up to certain embeddings [57], and we refer to [26] for a general survey about these constructions together with speculations on further structures they may produce.

Though this cannot be envisioned as clearly as for building new stable groups, a natural possibility is to consider as Ould Houcine the quantifier-free Morley rank, i.e. the Morley rank computed only with quantifier-free formulas (as above, this is mainly because things become too tricky when quantifiers become involved). This provides a robust global notion of dimension at the level of quantifier-free definable sets. In this sharper global context the same questions concerning certain free products and HNN-extensions occur exactly as for the local property of stability.

3 Groups of finite dimension with involutions

At the opposite of bad groups configurations, a large body of work has been done since the early nineties towards the Algebricity Conjecture, notably under the impulse of Nesin and with Borovik's program [7]. The latter runs parallel to CFSG. In the finite case, nonsolvable groups always have involutions by the Feit-Thompson Theorem [18]. A minimal counterexample to that theorem is a *minimal simple* group, i.e. a finite simple group with all proper subgroups solvable, and as it is finite it cannot be a full Frobenius group. Hence there are proper subgroups with nontrivial intersections, which is the starting point for the local analysis for the Feit-Thompson Theorem. This key point is exactly the missing point in the finite Morley rank case, because of the bad group problem [31].

When involutions are present however in a group, many arguments (going back to Brouwer in the finite case) become feasible. The essential trick is that any two such involutions in a group always generate a dihedral group. With this trivial remark, and after the thousands of pages of research papers built on it, CFSG provides the following list of finite simple groups.

1. Cyclic of prime order
2. Alternating type
3. Lie type

4. The 26 sporadics

The Lie type is the most prominent part of the list and groups of this type, which are essentially matrices groups over finite fields, constitutes the main model for the recognition of target groups in all the classification. Infinite groups of alternating type have the independence property with the formula $[x,y] = 1$ ([2, 4]; this was also observed by Zilber in 1972) and are thus far from being stable. This is certainly a strong historical motivation for the Algebricity Conjecture.

CFSG roughly splits into four cases following the double dichotomy

$$\text{characteristic 2 vs. characteristic not 2}$$

and

$$\text{small groups vs. large groups}$$

Boarders between these dichotomies may be fuzzy, and the small/large groups dichotomy, which involves the size of the expected matrices groups, is rather a trichotomy about matrices groups of size 2, 3, and > 3.

In Borovik's program the same architecture is adopted for infinite simple groups of finite Morley rank (with involutions!). The starting point is the existence of a nice Sylow theory for groups of finite Morley rank for the prime $p = 2$. Sylow 2-subgroups of a group of finite Morley rank are conjugate, and if S is one of them then its connected component is a central product, with finite intersection, of a 2-*unipotent* subgroup U, i.e. a definable connected 2-subgroup of bounded exponent, and a 2-torus T, i.e. a divisible abelian 2-subgroup, with both subgroups definably characteristic in S [8]. This two pieces U and T correspond naturally in the category of algebraic groups to the case of a base field of characteristic 2 and different from 2 respectively. Accordingly, a group of finite Morley rank is said to be of the following type depending on the non-triviality of these subgroups.

	$U \neq 1$	$U = 1$
$T \neq 1$	Mixed	Odd
$T = 1$	Even	Degenerate

It is known from [6] that a connected group of finite Morley rank has either trivial or infinite Sylow 2-subgroups, and hence that the degenerate type essentially reduces to the case of groups without involutions. As we saw already in the last section, bad groups are in this secret box.

Contrarily to the algebraic category, the mixed type is a real possibility in the category of groups of finite Morley rank. As the latter is closed under finite direct products, a product of algebraic groups over fields of different characteristics provide a group of finite Morley rank involving different characteristics.

In even type there is a full answer to the Algebricity Conjecture for simple groups.

Theorem 3 [1] *Let G be a simple group of finite Morley rank of even type. Then G is isomorphic to an algebraic group over an algebraically closed field of characteristic 2.*

The full answer to the Algebricity Conjecture in characteristic 2 of Theorem 3 allows one to evacuate entirely simple groups of mixed type, essentially by the argument of [30]. We refer to [1] for the proof of this corollary.

Theorem 3 is at present the most achieved piece of the whole theory of groups of finite Morley rank with involutions, with a complete parallel to CFSG. It goes by induction, assuming thus that infinite definable simple sections of even type of the minimal counterexample are algebraic groups. (But a priori allowing failure of the Feit-Thompson Theorem in such sections, i.e. definable infinite simple sections without involutions and possibly bad groups!) Besides, it takes the strategy small/large groups. Hence a gradual series of recognition theorems are shown, starting from the lengthy and difficult case of the very small group $PSL_2(K)$ with K an algebraically closed field of characteristic 2, then with intermediate groups of Lie rank 2, and then with the generic case of Lie rank ≥ 3.

Most of the study, notably in the very small case of PSL_2, is done via Borel subgroups, i.e. maximal definable connected solvable subgroups. The distinction between small groups and the larger ones is provided by the number of parabolic subgroups which can a priori be present in the group. Here parabolic subgroups are defined by analogy with the algebraic case, where they are the most relevant subgroups of matrices containing a Borel subgroup.

Throughout the analysis, a key ingredient is Zilber's field theorem [7, Theorem 9.1] which gives an interpretable field in any connected solvable nonnilpotent group of finite Morley rank. It is however worth mentioning that great complications are due to the existence "bad" fields of finite Morley rank, i.e. with an infinite proper definable subgroup in the multiplicative group [3]. These pathologies give for example non-algebraic connected solvable groups, possibly missing some expectable torsion, and this is a major complication for the recognition of PSL_2. (Borovik's program started with the simplifying "tameness" assumption which excluded these complications. [30] and [32] were the first papers where the systematic reworking of the project without this assumption was undertaken, at a time when the existence of bad fields just became conceivable.)

It follows from Theorem 3 and its corollary on groups of mixed type that for simple groups of finite Morley rank with involutions only the odd type remains to be analyzed. Here unfortunately serious problems arise. Again, an inductive approach is adopted for groups of odd type. But in this type the recognition of the very small simple group PSL_2, now in characteristic different from 2, has not been fully reached and the potential pathological configurations which occur in this case resist any analysis by "known" methods. We will say a word about this in the final section of this paper.

In finite group theory, Thompson provided after the lengthy local analysis developed for [18] a classification of minimal simple groups with involutions. He later generalized it, with only very few additional groups, to nonsolvable groups in which normalizers of nontrivial solvable subgroups are still solvable [54]. Analogous hypotheses are natural for characterizing "small" groups in the finite Morley rank context and are particularly relevant in the odd type case.

A *minimal connected simple* group of finite Morley rank is an infinite simple group of finite Morley rank in which proper definable connected subgroups are all solvable, and in odd type these groups have been studied in [10, 13–16]. In odd type, an important parameter is the Prüfer 2-rank, i.e. the finite number of copies of the Prüfer 2-group $\mathbb{Z}_{2^\infty}$ involved in the connected component of Sylow

2-subgroups. A bound by 2 is known for minimal connected simple groups of odd type. As in the finite case, this generalizes from minimal connected simple groups to the slightly larger class of "locally solvable" groups.

Theorem 4 [17] *Let G be a connected non-solvable group of finite Morley rank of odd type in which $N°(A)$ is solvable for each nontrivial definable connected abelian subgroup A. Then G has Prüfer 2-rank 1 or 2. Assume furthermore $C°(i)$ solvable for each involution i of G. Then involutions are conjugate, and either*

- *$G \simeq \mathrm{PSL}_2(K)$ for some algebraically closed field K of characteristic different from 2, or*
- *$C°(i)$ is a Borel subgroup for each involution i of G. Moreover, in Prüfer 2-rank 1 the Weyl group has order 1 or 2, and in Prüfer 2-rank 2 it has order 3.*

The important notion of Weyl group shall be explained later, and the three configurations left in the second case of Theorem 4 are the critical ones which cannot be eliminated by standard methods. In [13] they are described at length under the additional assumption of non-appearance of bad fields. They resemble bad groups by many aspects, but are not bad groups in the sense of the preceding section as they contain involutions.

Beyond the tricky case of small groups, the most general result known about groups of odd type is the following. It unfortunately involves an inductive setting for the Algebricity Conjecture.

Theorem 5 [10] *Let G be a minimal counterexample of odd type to the Algebricity Conjecture for simple groups of finite Morley rank. Then G has Prüfer 2-rank 1 or 2.*

Theorem 5 is the result of a collection of papers about groups of odd type, mostly carried out by Burdges in the generic case and which culminates in [10] for the final bound on the Prüfer 2-rank.

The whole analysis of odd type groups has necessitated the creation of a unipotence theory, which a priori does not exist in groups of finite Morley rank as it does for algebraic groups over algebraically closed fields. This was developed mainly in Burdges' thesis [9], and the main complications are due to the existence of bad fields in characteristic 0. The problematic point, in the local analysis of groups of finite Morley rank, is often the action of a torsion-free subgroup onto another one. We recall here the presentation adopted in [22] for this abstract unipotence theory dealing with these phenomena.

We denote by $\mathcal{P}$ the set of all prime numbers. A *decent torus* is a divisible abelian group of finite Morley rank which coincides with the definable hull of its (divisible abelian) torsion subgroup. The latter is known in the finite Morley rank context to be a direct product of finite products of the Prüfer p-group $\mathbb{Z}_{p^\infty}$, with p varying in $\mathcal{P}$, and by divisibility decent tori are connected. If p is a prime, a *p-unipotent* group is a definable connected nilpotent p-group of bounded exponent. A *unipotence parameter* is a couple

$$\tilde{p} = (characteristic\ p,\ unipotence\ degree\ r)$$

Groups of finite dimension in model theory 67

in $(\{\infty\} \cup \mathcal{P}) \times (\mathbb{N} \cup \{\infty\})$ satisfying $p < \infty$ if and only if $r = \infty$. A group of finite Morley rank is a $\tilde{p}$-*group* if it is nilpotent and of the following form, depending on the value of $\tilde{p}$:

- if $\tilde{p} = (\infty, 0)$, a decent torus.
- if $\tilde{p} = (\infty, r)$, with $0 < r < \infty$, generated by its definable indecomposable subgroups A such that $A/\Phi(A)$ is torsion-free and of Morley rank r. (Here an indecomposable group A is an abelian group which is not the sum of two proper definable subgroups, and $\Phi(A)$ is its largest proper definable connected subgroup.)
- if $\tilde{p} = (p, \infty)$, with p prime, a p-unipotent subgroup.

The very technical second type is defined as such in order to control all possible actions between torsion-free subgroups. The main feature of this unipotence theory is that it is entirely gradual, handling everything which can occur in an arbitrary group of finite Morley rank. More precisely, the unipotence degree r controls all possible actions in the sense that the "most" unipotent subgroups cannot act seriously on the "less" unipotent subgroups. Imposing maximality on definable $\tilde{p}$-subgroups also yields some kind of Sylow $\tilde{p}$-subgroups theory, with striking reminiscences of Sylow p-subgroups theory in finite group theory. We refer to [22] for a detailed survey about this unipotence theory and its analogies with Sylow theory.

Incidentally, this theory has had a nice application concerning the existence of particular subgroups which are the natural candidates for generalizing maximal tori in algebraic groups, or more generally the so-called Cartan subgroups. Here there is no inductive assumption. A *Carter* subgroup of a group of finite Morley rank is a definable connected nilpotent subgroup Q such that $N^\circ(Q) = Q$. More generally, if $\tilde{\pi}$ is a set of unipotence parameters, a *Carter $\tilde{\pi}$-subgroup* is a definable connected nilpotent subgroup $Q_{\tilde{\pi}}$ which coincides with the subgroup generated by the definable $\tilde{p}$-subgroups, $\tilde{p} \in \tilde{\pi}$, of its normalizer.

Theorem 6 [22] *Let G be a group of finite Morley rank and $\tilde{\pi}$ a set of unipotence parameters. Let $(p, r) \in \tilde{\pi}$ with r minimal. Then any Sylow (p, r)-subgroup of G is contained in a Carter $\tilde{\pi}$-subgroup of G.*

In particular, if one considers the set of all unipotence parameters, or merely the set of unipotence parameters involved in the ambient group, one gets the existence of Carter subgroups in any group of finite Morley rank.

It is however still a major question to understand whether Carter subgroups allow one to understand the generic element of arbitrary groups of finite Morley rank, as maximal tori do in algebraic groups.

4 Generix' adventures in Groupland

Generix is the french hero of arguments based purely on genericity. It appeared for the first time under this name in [47], though in a context far from the model theory of groups.

In a group of finite Morley rank, a definable subset is *generic* if it has the same dimension as the ambient group, and *generous* if the union of its conjugates is generic in the ambient group. There is a nice theory of generic subsets in groups of finite Morley rank, and more generally in stable groups [55], which connects nicely to a notion of generic type, i.e. of type in which each formula defines a generic subset. By the combination of [5] and [45] these generic subsets are now particularly well understood in free groups.

Despite the known *existence* of such a generic type, the main problem with an arbitrary group of finite Morley rank is that we don't know in general *what* is its generic element. If the Algebricity Conjecture turns out to be true, then this would be known essentially as the veracity of this conjecture implies a good structural description of any group of finite Morley rank in terms of simple algebraic factors, in which the generic element belongs up to conjugacy to a maximal torus. For a bad group $H < G$, the generic element of G is the same as that of H, and hence the question reduces to the generic element of the subgroup H.

In general one expects one of the following successively weaker conjectures to be true.

Conjecture 1 *In any group of finite Morley rank,*

(a) *any Borel subgroup is generous.*
(b) *any Carter subgroup is generous.*
(c) *there is a generous Carter subgroup.*

These are unfortunately widely open conjectures at the moment. The fact that Conjecture 1 (b) is weaker than Conjecture 1 (a) follows from the fact that Carter subgroups are generous in solvable groups and that generosity is transitive [22].

There are other conjectures weaker than Conjecture 1 (c), such that the existence of a definable generic subset all of whose elements are contained in a Carter subgroup, or even weaker the existence of a definable generic subset all of whose elements are contained in a definable connected nilpotent subgroup. The latter is known in the specific case of minimal connected simple groups [6].

Conjecture 1 (a) is by far the strongest one. It would give a very good description of any group of finite Morley rank, even in case of a failure of the Algebricity Conjecture. However it still does not seem to imply easily the conjugacy of Borel subgroups, an important classical result from the algebraic case. Conjecture 1 (a) would also allow one to dispose of the nonalgebraic configurations left in [13] and Theorem 4, as such configurations are known to have non-generous Borel subgroups. Besides, it seems to be the unique way to dispose of such configurations! However, if $C^\circ(i)$ is not assumed solvable in Theorem 4, there are new nonalgebraic configurations appearing, and Conjecture 1 (a) does not seem to be particularly useful for excluding them.

In general one really expects the generic element of a group of finite Morley rank to belong to a (generous) Carter subgroup. Generous Carter subgroups of a group of finite Morley rank can be characterized as follows.

Theorem 7 [33] *Let G be a group of finite Morley rank and Q a Carter subgroup of G. Then the following are equivalent to the generosity of Q in G:*

(a) *There exists a definable generic subset $\tilde{Q}$ of Q such that, for each element of $\tilde{Q}$, Q is the unique maximal definable connected nilpotent subgroup containing that element.*

(b) *Q is generically disjoint from its conjugates.*

This gives a quite clear picture of the generic element in presence of a generous Carter subgroup. It can also be characterized by the following weaker form: there exists a definable generic subset of Q all of whose elements are contained in only finitely many conjugates of Q. These descriptions also yield a conjugacy theorem for generous Carter subgroups.

Theorem 8 [33] *In any group of finite Morley rank, generous Carter subgroups are conjugate.*

Some weak forms of Conjecture 1 are also available in presence of a decent torus. If T is a definable decent torus in a group of finite Morley rank, than $C°(T)$ is generous. This implies a conjugacy theorem for maximal definable decent tori in any group of finite Morley rank [12], and even a finiteness theorem on the number of conjugacy classes in uniformly definable families of decent tori [22, Theorem 6.4].

It has been proved in [19] that in a group of finite Morley rank where no bad groups and no bad fields occur naturally, there always exists a nontrivial decent torus T, and with T maximal $C°(T)$ is a Carter subgroup. (Here only bad groups with H nilpotent are excluded. "Naturally" means as a definable section for bad groups, and by Zilber's field theorem for bad fields). With these two pathologies excluded, the question of the generic element is thus very well understood with Theorems 7 and 8. More recently, the same result has been proved in [21] under a weakening of the preceding strong hypothesis. Indeed, this hypothesis has been reduced to its "core", i.e. the sufficient presence in the ambient group of sections *not* definably isomophic to the additive group of an interpretable field. This is based on the introduction and the analysis of *pseudo-tori*, a notion generalizing that of decent torus and defined in term of non-appearence of additive groups of interpretable fields.

We also note that the theory of Weyl groups is clarified in [21] for these pseudo-tori. It is shown that their centralizers are connected in any connected group of finite Morley rank. For decent tori this easy lemma was kept for the end of the analysis of each configuration of minimal connected simple group left in [13], as it was known from this work that the kind of general protocol developed there for this purpose was working in each case. With [12] it became clear that this protocol was working in general in minimal connected simple groups, and later in arbitrary connected groups. This kind of results on Weyl groups go back to Nesin, originally in the context of bad groups! All these arguments are always based on the consideration of the generic element of certain subsets, generally generous cosets of subgroups.

Finally, we note that a conjugacy theorem has been proved recently for nongenerous Carter subgroups in any minimal counterexample to the Algebricity Conjecture [20]. Despite the caveat of an inductive assumption, it surprisingly gives with Theorem 8 a full conjugacy theorem for Carter subgroups in this inductive context without Conjecture 1.

Now we conclude with some perspectives opened by conjectures in the style of Conjecture 1. The structure of algebraic groups over algebraically closed fields is usually described in terms of (generous) Borel subgroups, maximal tori, root groups, and the Weyl group which is the normalizer of the maximal torus modulo the maximal torus. Root groups are the minimal subgroups normalized by the maximal torus inside the Borel subgroup. The whole structure of the algebraic group is governed by the action of the finite Weyl group on the maximal torus and by the way it permutes a naturally associated finite set of Borel subgroups, together with all symmetries which may be involved between the root groups. (There are apparitions of Euclidian geometry in this classical theory.)

The Algebricity Conjecture postulates at a philosophical level that an abstract notion of dimension forces a geometry, and at a more technical level that for simple groups such a geometry is a classical one. More precisely it postulates that the important Weyl groups are all already known. Now the existence of Carter subgroups in an arbitrary group of finite Morley rank provided by Theorem 6 gives a natural candidate for a Weyl group even if the Algebricity Conjecture fails. This is the finite group $N(Q)/Q$ for some Carter subgroup Q, and it is particularly well defined by Theorem 8 provided Conjecture 1 holds. In a bad group with H nilpotent one has $H = Q$, a selfnormalizing Carter subgroup, and in this case the natural candidate for the Weyl group would be the trivial group.

Hence Conjecture 1 encapsulates the classical case as well as configurations such as bad groups, if they are built one day. In particular it provides strong hints that a classification of simple groups of finite Morley rank in the style of that describing algebraic groups as above might be feasible, even if this class of simple groups of finite dimension contains much more than the classical algebraic groups occuring in nature. But, in any case, one seems to be very far from the end of that story.

Bibliography

[1] Altınel, T., Borovik, A., Cherlin, G. Simple groups of finite Morley rank, 2008. Math. Surveys, Amer. Math. Soc. (expected 2008).

[2] Baldwin, J. T., Saxl, J. Logical stability in group theory. *J. Austral. Math. Soc. Ser. A*, 21(3):267–276, 1976.

[3] Baudisch, A., Hils, M., Martin-Pizarro, A., Wagner, F. Die böse farbe. *J. Inst. Math. Jussieu*, 2007. To appear.

[4] Belegradek, O. V. Unstable group theories. *Izv. Vyssh. Uchebn. Zaved. Mat.*, (8):41–44, 1978. ISSN 0021-3446.

[5] Bestvina, M., Feighn, M. 2006. In preparation.

[6] Borovik, A., Burdges, J., Cherlin, G. Involutions in groups of finite Morley rank of degenerate type. *Selecta Math. (N.S.)*, 13(1):1–22, 2007. ISSN 1022-1824.

[7] Borovik, A., Nesin, A. *Groups of finite Morley rank*. New York: The Clarendon Press Oxford University Press, 1994. ISBN 0-19-853445-0. Oxford Science Publications.

[8] Borovik, A. V., Poizat, B. P. Tores et p-groupes. *J. Symbolic Logic*, 55(2):478–491, 1990. ISSN 0022-4812.

[9] Burdges, J. Odd and degenerate types groups of finite Morley rank, 2004. Doctoral Dissertation, Rutgers University.

[10] Burdges, J., Cherlin, G., Jaligot, E. Minimal connected simple groups of finite Morley rank with strongly embedded subgroups. *J. Algebra*, 314(2):581–612, 2007.

[11] Cherlin, G. Groups of small Morley rank. *Ann. Math. Logic*, 17(1-2):1–28, 1979. ISSN 0003-4843.

[12] Cherlin, G. Good tori in groups of finite Morley rank. *J. Group Theory*, 8(5):613–622, 2005.

[13] Cherlin, G., Jaligot, E. Tame minimal simple groups of finite Morley rank. *J. Algebra*, 276(1):13–79, 2004. ISSN 0021-8693.

[14] Deloro, A. Groupes simples connexes minimaux algébriques de type impair. *J. Algebra*, 2007. To appear.

[15] Deloro, A. Groupes simples connexes minimaux de type impair, 2007. Thèse de doctorat, Université de Paris 7.

[16] Deloro, A. Groupes simples connexes minimaux nonalgébriques de type impair. *J. Algebra*, 2007. To appear.

[17] Deloro, A., Jaligot, E. Groups of finite Morley rank with solvable local subgroups, 2006. In preparation.

[18] Feit, W., Thompson, J. G. Solvability of groups of odd order. *Pacific J. Math.*, 13:775–1029, 1963.

[19] Frécon, O. Carter subgroups in tame groups of finite Morley rank. *J. Group Theory*, 9(3):361–367, 2006. ISSN 1433-5883.

[20] Frécon, O. Conjugacy of Carter subgroups in groups of finite Morley rank, 2007. Submitted.

[21] Frécon, O. Pseudo-tori and subtame groups of finite Morley rank, 2007. Submitted.

[22] Frécon, O., Jaligot, E. Conjugacy in groups of finite Morley rank. In Chatzidakis, Z., Macpherson, H., Pillay, A., Wilkie, A. (Eds.), *Model Theory with applications to algebra and analysis, I and II*. Cambridge: Cambridge University Press, 2008.

[23] Ghys, É., de la Harpe, P. (Eds.). *Sur les groupes hyperboliques d'après Mikhael Gromov*, vol. 83 of *Progress in Mathematics*. Boston, MA: Birkhäuser Boston Inc., 1990. ISBN 0-8176-3508-4. Papers from the Swiss Seminar on Hyperbolic Groups held in Bern, 1988.

[24] Gorenstein, D., Lyons, R., Solomon, R. *The classification of the finite simple groups*. Providence, RI: American Mathematical Society, 1994. ISBN 0-8218-0334-4.

[25] Gromov, M. Hyperbolic groups. In *Essays in group theory*, vol. 8 of *Math. Sci. Res. Inst. Publ.*, 75–263. New York: Springer, 1987.

[26] Hasson, A. Some questions concerning Hrushovski's amalgamation constructions. *Logicum Lugdunensis, J. Inst. Math. Jussieu*, 2007. To appear.

[27] Hrushovski, E. Pseudo-finite fields and related structures. In *Model theory and applications*, vol. 11 of *Quad. Mat.*, 151–212. Aracne, Rome, 2002.

[28] Hrushovski, E., Peterzil, Y., Pillay, A. Groups, measures, and the NIP. *Journal Amer. Math. Soc.*, 2007. To appear.

[29] Hrushovski, E., Pillay, A. On NIP and invariant measures, 2007. Preprint.

[30] Jaligot, E. Groupes de type mixte. *J. Algebra*, 212(2):753–768, 1999. ISSN 0021-8693.

[31] Jaligot, E. Full Frobenius groups of finite Morley rank and the Feit-Thompson theorem. *Bull. Symbolic Logic*, 7(3):315–328, 2001. ISSN 1079-8986.

[32] Jaligot, E. Groupes de rang de Morley fini de type pair avec un sous-groupe faiblement inclus. *J. Algebra*, 240(2):413–444, 2001. ISSN 0021-8693.

[33] Jaligot, E. Generix never gives up. *J. Symbolic Logic*, 71(2):599–610, 2006.

[34] Jaligot, E., Muranov, A., Neman, A. Independence property and hyperbolic groups. *Bull. Symb. Logic*, 2007. To appear.

[35] Jaligot, E., Ould Houcine, A. Existentially closed CSA-groups. *J. Algebra*,

280(2):772–796, 2004. ISSN 0021-8693.

[36] Kharlampovich, O., Myasnikov, A. Elementary theory of free non-abelian groups. *J. Algebra*, 302(2):451–552, 2006. ISSN 0021-8693.

[37] Macintyre, A. On ω_1-categorical theories of fields. *Fund. Math.*, 71(1):1–25. (errata insert), 1971.

[38] Morley, M. Categoricity in power. *Trans. Amer. Math. Soc.*, 114:514–538, 1965.

[39] Muranov, A. Y. Finitely generated infinite simple groups of infinite commutator width. *Internat. J. Algebra Comput.*, 17(3):607–659, 2007.

[40] Myasnikov, A. G., Remeslennikov, V. N. Exponential groups. II. Extensions of centralizers and tensor completion of CSA-groups. *Internat. J. Algebra Comput.*, 6(6):687–711, 1996. ISSN 0218-1967.

[41] Ol'shanskiĭ, A. Y. *Geometry of defining relations in groups*, vol. 70 of *Mathematics and its Applications (Soviet Series)*. Dordrecht: Kluwer Academic Publishers Group, 1991. ISBN 0-7923-1394-1. Translated from the 1989 Russian original by Yu. A. Bakhturin.

[42] Ould Houcine, A. On superstable CSA-groups. *Ann. Pure Appl. Logic*, 2007. To appear.

[43] Pillay, A. The geometry of forking and groups of finite Morley rank. *J. Symbolic Logic*, 60(4):1251–1259, 1995. ISSN 0022-4812.

[44] Pillay, A. A note on CM-triviality and the geometry of forking. *J. Symbolic Logic*, 65(1):474–480, 2000. ISSN 0022-4812.

[45] Pillay, A. Forking in the free group, 2007. ArXiv.org preprint: http://arxiv.org/abs/math/0702134.

[46] Poizat, B. *Groupes stables*. Lyon: Bruno Poizat, 1987. ISBN 2-9500919-1-1. Une tentative de conciliation entre la géométrie algébrique et la logique mathématique. [An attempt at reconciling algebraic geometry and mathematical logic].

[47] Poizat, B. Generix strikes again. *J. Symbolic Logic*, 54(3):847–857, 1989. ISSN 0022-4812.

[48] Reineke, J. Minimale Gruppen. *Z. Math. Logik Grundlagen Math.*, 21(4):357–359, 1975.

[49] Sela, Z. Diophantine geometry over groups VIII: Stability, 2007. Preprint: http://www.ma.huji.ac.il/~zlil/.

[50] Shelah, S. *Classification theory and the number of nonisomorphic models*. Amsterdam: North-Holland Publishing Co., second edn., 1990. ISBN 0-444-70260-1.

[51] Shelah, S. Toward classifying unstable theories. *Ann. Pure Appl. Logic*, 80(3):229–255, 1996. ISSN 0168-0072.

[52] Shelah, S., Usvyatsov, A. Banach spaces and groups—order properties and universal models. *Israel J. Math.*, 152:245–270, 2006. ISSN 0021-2172.

[53] Stallings, J. R. Non-positively curved triangles of groups. In *Group theory from a geometrical viewpoint (Trieste, 1990)*, 491–503. World Sci. Publ., River Edge, NJ, 1991.

[54] Thompson, J. G. Nonsolvable finite groups all of whose local subgroups are solvable. *Bull. Amer. Math. Soc.*, 74:383–437, 1968.

[55] Wagner, F. O. *Stable groups*. Cambridge: Cambridge University Press, 1997. ISBN 0-521-59839-7.

[56] Wagner, F. O. *Simple theories*, vol. 503 of *Mathematics and its Applications*. Dordrecht: Kluwer Academic Publishers, 2000. ISBN 0-7923-6221-7.

[57] Zilber, B. Analytic and pseudo-analytic structures. In *Logic Colloquium 2000*,

vol. 19 of *Lect. Notes Log.*, 392–408. Urbana, IL: Assoc. Symbol. Logic, 2005.

Operations, sets and classes[1]

Gerhard Jäger

Universität Bern

jaeger@iam.unibe.ch

ABSTRACT. Operational set theory, as described below, is an enterprise which consolidates classical set theory with some central concepts of Feferman's *explicit mathematics*. It provides for a careful distinction between operations and set-theoretic functions and as such reconciles set theory with needs arising in constructive environments and even in those enhanced by computer science.

In the following we consider, primarily from a proof-theoretic perspective, the theory OST and some of its most important extensions and determine their consistency strengths by exhibiting equivalent systems in the realm of traditional theories of sets and classes.

1 Introduction

Operational set theory, in the form described below, is a comparatively young enterprise which consolidates classical set theory with some central concepts of Feferman's *explicit mathematics*. It provides for a careful distinction between operations and set-theoretic functions and as such reconciles set theory with needs arising in constructive environments and even in those oriented towards computer science.

The general topic of explicit mathematics originated in Feferman's seminal paper [14], where several formal systems, including the famous theory T_0, where introduced. The original aim of explicit mathematics was to provide an appropriate framework for Bishop-style constructive mathematics, and it can be seen as one specific effort in parallel to rather different work by others; see Feferman [16] for a thorough discussion of this aspect.

However, soon it turned out that explicit mathematics also plays an important role in reductive proof theory and as an axiomatic approach to abstract computability. In Buchholz, Feferman, Pohlers and Sieg [11] important subsystems of T_0 are related to subsystems of second order arithmetic; the question concerning the exact proof-theoretic strength of full T_0 is settled by Jäger [26] and Jäger and Pohlers [33].

Feferman [15] lays the foundations for later work about the connections between explicit mathematics and generalized recursion theory and presents some first important results. In Feferman and Jäger [21] and Jäger and Strahm [35] the proof theory of the non-constructive μ-operator and the Suslin operator in an

---

[1]The author is with Institut für Informatik und angewandte Mathematik, Universität Bern, CH-3012 Bern, Switzerland.

explicit context are studied; Jäger and Strahm [34, 36] deal with various forms of explicit reflections, in particular with Mahloness and analogues of Π_3 reflection.

The proofs of these and many other results about explicit mathematics make heavy use of interesting set-theoretic concepts, at least implicitly. Very often (the intuitive background of) a proof-theoretic argument depends on a subtle interplay between notions in classical set theory and their admissible, constructive or recursive analogues. Operational set theory turns these implicit analogies into explicit generalizations. A central ingredient of this approach is a strict distinction between operations, which may be interpreted as computations or even programs, and functions in the set-theoretic sense, i.e. binary right-unique relations.

Feferman [18] is the starting points of the following considerations and introduces the system OST of operational set theory and a few extensions, motivated by the aim to develop a common language for small large cardinal notions as in classical set theory, admissible set and recursion theory. Feferman [19] presents variants of these systems closer in syntax to original explicit mathematics, and Feferman [20] is a polished up version of parts of [18].

Related work by Cantini and Crosilla [12] is about a constructive set theory with operations COST, which may be considered as a constructive version of OST, and may be regarded as providing a bridge between Aczel's constructive set theory CZF, see Aczel [1–3], and explicit mathematics. As predecessors of present day operational set theory we may consider Beeson [10], presenting an interesting computation system based on set theory and formulated as a theory of sets and rules, and Feferman [17], where some of the central ideas are outlined.

The present article begins with a rough description of the landscape of set theory and then studies OST and its extensions by operational power set and operational unbounded existential quantification. Afterwards we determine their consistency strengths by exhibiting equivalent systems in the realm of traditional set theory and describe an interesting extension of OST which is conservative over ZFC.

2 Aspects of the set-theoretic landscape

No doubt, Zermelo-Fraenkel set theory with the axiom of choice is the generally accepted framework for wide parts of everyday mathematics. From a foundational perspective, however, the general picture is much richer, and – very roughly – we can distinguish three areas:

I. Classical set theories. Clearly, ZF and ZFC are the two most prominent exponents. But there is also von Neumann-Bernays-Gödel set theory NBG, a theory of sets and classes which yields the same results as ZFC. Morse-Kelley set theory MK is a significant strengthening of NBG permitting highly non-elementary class formation. Further natural strengthenings of NBG are obtained by adding reflection principles such as Π_1^1 reflection or strict Π_1^1 reflection.

On the other hand, theories like ZF, ZFC and their extensions have significant drawbacks from a logical perspective. To mention only a few: (i) They heavily violate the principle of parsimony; most mathematical theorems can be proved from much weaker set existence axioms. (ii) They demand that all mathematical objects (also, for example, computer programs) are realized as sets. (iii) They have only very huge models and, for example, no recursive models. (iv) They do not differentiate between levels of existence and between constructively/recursively and classically valid assertions.

II. Constructive set and type theories. A first possible reaction is to replace classical set theories by their constructive variants. Prominent example of those are:

- Myhill's CST (cf. [42]) and the intuitionistic version IZF of ZF.
- Martin-Löf type theories (cf. [39, 40]).
- Aczel's constructive set theory CZF and its extensions à la Aczel and Rathjen (cf. [1–4]).
- Proof development systems and proof assistants such as Coq, HOL and Nuprl (cf. [13, 22, 43]).

Most of these systems use intuitionistic logic and have the disjunction and existence property.

III. Admissible set theories. An alternative is to abide with classical logic, but to weaken the set existence axioms dramatically. Most distinctive along these lines is the system KPω of Kripke-Platek set theory with infinity. In calibrating the proof-theoretic strength of subsystems of second order arithmetic and set theory, theories of iterated admissible sets play an important role; for example:

- KPu and KPi for admissible and recursively inaccessible universes (cf. [23–25, 28]).
- KPm and KPω + (Π_n-Ref) for recursively Mahlo universes and universes satisfying Π_n reflection (cf. [5, 6, 44, 45]).
- admissibility without foundation, i.e. systems like KPi^0, KPm^0 and KPu^0 + (Π_n-Ref) which are obtained from the above theories by dropping $\in$-induction; they represent the wide spectrum of predicative and metapredicative theories (cf. [27, 28, 31, 34, 36, 49, 50]).

Recent proof-theoretic work of Rathjen (cf.,e.g., [46, 47]) about extensions of Kripke-Platek set theory leading to Π_2^1-CA may be subsumed under classical and admissible set theories.

It has already been mentioned in the introduction that there exist interesting connections between notions of classical set theory and their recursive or admissible analogues. A typical example is the notion of regular cardinal which collapses to admissible ordinal if we replace arbitrary set-theoretic functions by recursive functions; see Richter and Aczel [48] for more on this from a recursion-theoretic perspective.

Operational set theory is the approach to isolate the basic common principle underlying each of the areas I –III. The central notion is that of operation which may be interpreted classically, constructively or recursively. In this regard it is very much like Feferman's marriage of convenience for explicit mathematics [15].

3 The theory OST and its relatives

The presentation of the theory OST and its extensions follows Jäger [29, 30]; all unexplained notions and further motivation can be found there.

Let $\mathcal{L}_1$ be a typical language of admissible or classical set theory with a symbol for the element relation as its only relation symbol and countably many set variables $a, b, c, f, g, u, v, w, x, y, z, \ldots$ (possibly with subscripts). The formulas of $\mathcal{L}_1$ are defined as usual.

$\mathcal{L}^\circ$, the language of OST and its extensions, augments $\mathcal{L}_1$ by the binary function symbol $\circ$ for partial term application, the unary relation symbol $\downarrow$ (defined) and the following constants: (i) the combinators k and s; (ii) $\top$, $\bot$, **el**, **non**, **dis**, **e** and **E** for logical operations; (iii) $\$$, $\mathbb{R}$, $\mathbb{C}$ and $\mathbb{P}$ for set-theoretic operations. The meaning of these constants follows from the axioms below.

The *terms* $(r, s, t, r_1, s_1, t_1, \ldots)$ of $\mathcal{L}^\circ$ are inductively generated as follows:

1. The variables and constants of $\mathcal{L}_1$ are terms of $\mathcal{L}^\circ$.
2. If s and t are terms of $\mathcal{L}^\circ$, then so is $\circ(s, t)$.

In the following we often abbreviate $\circ(s, t)$ as $(s \circ t)$, as (st) or (if no confusion arises) simply as st. We also adopt the convention of association to the left so that $s_1 s_2 \ldots s_n$ stands for $(\ldots (s_1 s_2) \ldots s_n)$. In addition, we often write $s(t_1, \ldots, t_n)$ for $s t_1 \ldots t_n$ if this seems more intuitive. Moreover, we frequently make use of the vector notation $\vec{s}$ as shorthand for a finite string $s_1, \ldots, s_n$ of $\mathcal{L}^\circ$ terms whose length is either not important or evident from the context.

Self-application is possible and meaningful, but it is not necessarily total; there may be terms which do not denote an object. We make use of the definedness predicate $\downarrow$ to single out those which do, and $(t\downarrow)$ is read "t is defined" or "t has a value".

The formulas $(A, B, C, D, A_1, B_1, C_1, D_1, \ldots)$ of $\mathcal{L}^\circ$ are inductively generated as follows:

1. All expressions of the form $(s \in t)$ and $(t\downarrow)$ are formulas of $\mathcal{L}^\circ$; the so-called *atomic* formulas.
2. If A and B are formulas of $\mathcal{L}^\circ$, then so are $\neg A$, $(A \vee B)$ and $(A \wedge B)$.
3. If A is a formula and t a term of $\mathcal{L}^\circ$ which does not contain x, then $\exists x A$, $\forall x A$, $(\exists x \in t)A$ and $(\forall x \in t)A$ are formulas of $\mathcal{L}^\circ$.

We will often omit parentheses and brackets whenever there is no danger of confusion. Since we will be working within classical logic, the remaining logical connectives can be defined as usual; equality of sets is introduced by

$$(s = t) := (s\downarrow) \wedge (t\downarrow) \wedge (\forall x \in s)(x \in t) \wedge (\forall x \in t)(x \in s).$$

The free variables of t and A are defined in the conventional way; the closed $\mathcal{L}^\circ$ terms and closed $\mathcal{L}^\circ$ formulas, also called $\mathcal{L}^\circ$ sentences, are those which do not contain free variables.

Given an $\mathcal{L}^\circ$ formula A and a variable u not occurring in A, we write A^u for the result of replacing each unbounded set quantifier $\exists x(\ldots)$ and $\forall x(\ldots)$ in A by $(\exists x \in u)(\ldots)$ and $(\forall x \in u)(\ldots)$, respectively. Suppose now that $\vec{u} = u_1, \ldots, u_n$ and $\vec{s} = s_1, \ldots, s_n$. Then $A[\vec{s}/\vec{u}]$ is the $\mathcal{L}^\circ$ formula which is obtained from A by simultaneously replacing all free occurrences of the variables $\vec{u}$ by the $\mathcal{L}^\circ$ terms

$\vec{s}$; in order to avoid collision of variables, a renaming of bound variables may be necessary. If the $\mathcal{L}^\circ$ formula A is written as $B[\vec{u}]$, then we often simply write $B[\vec{s}]$ instead of $B[\vec{s}/\vec{u}]$. Further variants of this notation will be obvious.

The logic of OST is the classical *logic of partial terms* due to Beeson [8, 9], including the common equality axioms. Partial equality of terms is introduced by

$$(s \simeq t) := (s{\downarrow} \vee t{\downarrow} \to s = t)$$

and says that if either s or t denotes anything, then they both denote the same object.

The non-logical axioms of OST comprise axioms about the applicative structure of the universe, some basic set-theoretic properties, the representation of elementary logical connectives as operations and operational set existence axioms. They divide into four groups.

I. Applicative axioms.

(1) $\mathsf{k} \neq \mathsf{s}$,
(2) $\mathsf{k}xy = x$,
(3) $\mathsf{s}xy{\downarrow} \wedge \mathsf{s}xyz \simeq (xz)(yz)$.

Thus the universe is a partial combinatory algebra. We have λ-abstraction and thus can introduce for each $\mathcal{L}^\circ$ term t a term $(\lambda x.t)$ whose variables are those of t other than x such that

$$(\lambda x.t){\downarrow} \wedge (\lambda x.t)y \simeq t[y/x].$$

As usual we can generalize λ abstraction to several arguments by simply iterating abstraction for one argument. Accordingly, we set for all $\mathcal{L}^\circ$ terms t and all variables $x_1, \ldots, x_n$,

$$(\lambda x_1 \ldots x_n.t) := (\lambda x_1.(\ldots(\lambda x_n.t)\ldots)).$$

Often the term $(\lambda x_1 \ldots x_n.t)$ is also simply written as $\lambda x_1 \ldots x_n.t$. If $\vec{x}$ is the sequence $x_1, \ldots, x_n$, then $\lambda \vec{x}.t$ stands for $\lambda x_1 \ldots x_n.t$ and $t\vec{x}$ for $tx_1 \ldots x_n$.

Furthermore, there exists a closed $\mathcal{L}^\circ$ term fix, a so-called fixed point operator, with

$$\mathsf{fix}(f){\downarrow} \wedge (\mathsf{fix}(f) = g \to gx \simeq f(g,x)).$$

II. Basic set-theoretic axioms. They state that: (i) there is the empty set; (ii) there are unordered pairs and unions; (iii) there exists an infinite ordinal; (iv) $\in$-induction is available for arbitrary formulas $A[x]$ of $\mathcal{L}^\circ$,

$$\forall x((\forall y \in x)A[y] \to A[x]) \to \forall x A[x]. \qquad (\mathcal{L}^\circ\text{-}\mathsf{I}_\in)$$

To increase readability, we will freely use standard set-theoretic terminology; also, if $A[x]$ is an $\mathcal{L}^\circ$ formula, then $\{x : A[x]\}$ denotes the collection of all sets satisfying A; it may be (extensionally equal to) a set, but this is not necessarily the case. In particular, we set

$$\mathbb{B} := \{x : x = \top \vee x = \bot\} \quad \text{and} \quad \mathbb{V} := \{x : x{\downarrow}\}$$

so that $\mathbb{B}$ stands for the unordered pair consisting of the truth values $\top$ and $\bot$, which is a set by the previous axioms. $\mathbb{V}$ is the collection of all sets but not a set itself. The following shorthand notations, for n an arbitrary natural number,

$$(f : a \to b) := (\forall x \in a)(fx \in b),$$
$$(f : a^{n+1} \to b) := (\forall x_1, \ldots, x_{n+1} \in a)(f(x_1, \ldots, x_{n+1}) \in b)$$

express that f, in the operational sense, is a unary and $(n+1)$-ary mapping from a to b, respectively. They do not say, however, that f is a unary or $(n+1)$-ary function in the set-theoretic sense.

In the previous definition the set variables a and/or b may be replaced by $\mathbb{V}$ and/or $\mathbb{B}$. So, for example, $(f : a \to \mathbb{V})$ means that f is an operation which is total on a, and $(f : \mathbb{V} \to b)$ means that f maps all sets into b. If we have $(f : a \to \mathbb{B})$, we may regard f as a *definite predicate* on a; if we have $(f : \mathbb{V} \to \mathbb{B})$, we call f a *total characteristic operation*.

III. Logical operations axioms.

(L1) $\top \neq \bot$.

(L2) $(\mathbf{el} : \mathbb{V}^2 \to \mathbb{B}) \wedge \forall x \forall y (\mathbf{el}(x,y) = \top \leftrightarrow x \in y)$.

(L3) $(\mathbf{non} : \mathbb{B} \to \mathbb{B}) \wedge (\forall x \in \mathbb{B})(\mathbf{non}(x) = \top \leftrightarrow x = \bot)$.

(L4) $(\mathbf{dis} : \mathbb{B}^2 \to \mathbb{B}) \wedge (\forall x, y \in \mathbb{B})(\mathbf{dis}(x,y) = \top \leftrightarrow (x = \top \vee y = \top))$.

(L5) $(f : a \to \mathbb{B}) \to (\mathbf{e}(f,a) \in \mathbb{B} \wedge (\mathbf{e}(f,a) = \top \leftrightarrow (\exists x \in a)(fx = \top)))$.

The Δ_0 formulas of $\mathcal{L}^\circ$ are those $\mathcal{L}^\circ$ formulas which do not contain the function symbol $\circ$, the relation symbols $\downarrow$ or unbounded quantifiers. Hence they are the usual Δ_0 formulas of set theory, possibly containing additional constants. The above logical operations make it possible to represent all Δ_0 formulas by constant $\mathcal{L}^\circ$ terms. For a proof of the following see Feferman [18, 20].

Lemma 1 *Let $\vec{u}$ be the sequence of variables $u_1, \ldots, u_n$. For every Δ_0 formula $A[\vec{u}]$ of $\mathcal{L}^\circ$ with at most the variables $\vec{u}$ free, there exists a closed $\mathcal{L}^\circ$ term t_A such that the axioms introduced so far yield*

$$t_A \downarrow \wedge (t_A : \mathbb{V}^n \to \mathbb{B}) \wedge \forall \vec{x}(A[\vec{x}] \leftrightarrow t_A(\vec{x}) = \top).$$

IV. Operational set-theoretic axioms.

(S1) Separation for definite operations:

$$(f : a \to \mathbb{B}) \to$$
$$(\$(f,a)\downarrow \wedge \forall x(x \in \$(f,a) \leftrightarrow (x \in a \wedge fx = \top))).$$

(S2) Replacement:

$$(f : a \to \mathbb{V}) \to$$
$$(\mathbb{R}(f,a)\downarrow \wedge \forall x(x \in \mathbb{R}(f,a) \leftrightarrow (\exists y \in a)(x = fy))).$$

(S3) Choice:
$$\exists x(fx = \top) \rightarrow (\mathbb{C}f\downarrow \wedge\ f(\mathbb{C}f) = \top).$$

This finishes the description of the non-logical axioms of OST. A significant strengthening OST($\mathbb{P}$) of OST is obtained by adding the operational form of the power set axiom

$$(\mathbb{P} : \mathbb{V} \rightarrow \mathbb{V}) \wedge \forall x \forall y (x \in \mathbb{P}y \leftrightarrow x \subset y). \tag{$\mathbb{P}$}$$

Note that in OST and OST($\mathbb{P}$) we cannot treat unbounded existential quantification operationally. For that we use the constant $\mathbf{E}$ and the additional axiom

$$(f : \mathbb{V} \rightarrow \mathbb{B}) \rightarrow (\mathbf{E}(f) \in \mathbb{B} \wedge (\mathbf{E}(f) = \top \leftrightarrow \exists x(fx = \top))). \tag{$\mathbf{E}$}$$

In the following we write OST($\mathbf{E}, \mathbb{P}$) for OST($\mathbb{P}$) + ($\mathbf{E}$). This is the strongest operational theory we consider in this article.

Call those formulas of $\mathcal{L}^\circ$ which do not contain the function symbol $\circ$ or the relation symbol $\downarrow$ *pure formulas* of $\mathcal{L}^\circ$; they are the same as those of $\mathcal{L}_1$ plus the constants of OST. Then ($\mathbf{E}$) permits the extension of Lemma 1 to pure formulas.

Lemma 2 *Let $\vec{u}$ be the sequence of variables $u_1, \ldots, u_n$. For every pure formula $A[\vec{u}]$ of $\mathcal{L}^\circ$ with at most the variables $\vec{u}$ free, there exists a closed $\mathcal{L}^\circ$ term t_A such that OST($\mathbf{E}, \mathbb{P}$) proves*

$$t_A\downarrow \wedge\ (t_A : \mathbb{V}^n \rightarrow \mathbb{B}) \wedge \forall \vec{x}(A[\vec{x}] \leftrightarrow t_A(\vec{x}) = \top).$$

Its proof is analogous to the proof of Lemma 1; simply use ($\mathbf{E}$) to deal with unbounded quantifiers.

From Feferman [18] and Jäger [29] we know that, provably in the system OST, there exist closed $\mathcal{L}^\circ$ terms $\emptyset$ for the empty set, **uopa** for forming unordered pairs, **un** for forming unions, **p** for forming ordered pairs (Kuratowski pairs) and **prod** for forming Cartesian products. In addition, there are closed $\mathcal{L}^\circ$ terms $\mathbf{p}_L$ and $\mathbf{p}_R$ which act as projections with respect to **p**, i.e.

$$\mathbf{p}_L(\mathbf{p}(a,b)) = a \quad \text{and} \quad \mathbf{p}_R(\mathbf{p}(a,b)) = b.$$

To comply with the set-theoretic conventions, we generally write $\{a, b\}$ instead of $\mathbf{uopa}(a,b)$, $\cup a$ instead of $\mathbf{un}(a)$, $\langle a, b\rangle$ instead of $\mathbf{p}(a,b)$ and $a \times b$ instead of $\mathbf{prod}(a,b)$. Remember that ω is a constant for the first infinite ordinal and belongs to the base language $\mathcal{L}_1$. OST is also fairly strong with respect to definition by cases.

Lemma 3 *There exist closed $\mathcal{L}^\circ$ terms $\mathbf{d}_=$, $\mathbf{d}_\emptyset$ and $\mathbf{d}_\mathbb{B}$ such that OST proves:*

1. $(u = v \wedge \mathbf{d}_=(a,b,u,v) = a) \vee (u \neq v \wedge \mathbf{d}_=(a,b,u,v) = b)$.
2. $(u = \emptyset \wedge \mathbf{d}_\emptyset(a,b,u) = a) \vee (u \neq \emptyset \wedge \mathbf{d}_\emptyset(a,b,u) = b)$.
3. $\mathbf{d}_\mathbb{B}(a,b,\top) = a \wedge \mathbf{d}_\mathbb{B}(a,b,\bot) = b$.

We end this section with an interesting aspect of our axiom about operational choice: it provides for a form of global choice and as such is of some relevance in connection with von Neumann-Bernays-Gödel set theory NBG to be introduced later.

Theorem 1 (Global choice) *There exists a closed $\mathcal{L}^\circ$ term* choice *such that* OST *proves*

$$(\text{choice} : \mathbb{V} \to \mathbb{V}) \wedge \forall x(x \neq \emptyset \to \text{choice}(x) \in x) \wedge \text{choice}(\emptyset) = \top.$$

The above results about definition by cases and this theorem about global choice are proved in Jäger [30]; see also Feferman [20] where it is pointed out that (AC) is provable in OST, but the proof actually demonstrates our statement of global choice.

Recall the system KPω of Kripke-Platek set theory with infinity. It is formulated in $\mathcal{L}_1$ and based on classical first order predicate calculus with equality. Its non-logical axioms are pair, union, infinity, Δ_0 separation, Δ_0 collection and $\in$-induction for arbitrary formulas of $\mathcal{L}_1$. We write (AC) for the axiom of choice and (V=L) for the axiom of constructibility. As well-known from the literature, KPω, KPω + (AC) and KPω + (V=L) are of the same consistency strength, and KPω + (V=L) is conservative over KPω for absolute formulas.

$\mathcal{L}_1(\mathcal{P})$ is first order language obtained from $\mathcal{L}_1$ by adding a new binary relation symbol $\mathcal{P}$. The formulas of $\mathcal{L}_1(\mathcal{P})$ are defined as the formulas of $\mathcal{L}_1$, but with expressions of the form $\mathcal{P}(r, s)$ permitted as additional atomic formulas. The $\Delta_0(\mathcal{P})$ formulas are those formulas of $\mathcal{L}_1(\mathcal{P})$ which do not contain unbounded quantifier; in particular, each $\mathcal{P}(r, s)$ is $\Delta_0(\mathcal{P})$.

The extension KP($\mathcal{P}$) of KPω is formulated in $\mathcal{L}_1(\mathcal{P})$ and characterized by: (i) it encompasses pair, union, infinity, $\Delta_0(\mathcal{P})$ separation and $\Delta_0(\mathcal{P})$ collection; (ii) $\in$-induction is formulated for arbitrary $\mathcal{L}_1(\mathcal{P})$ formulas; (iii) the new axiom ($\mathcal{P}$) provides the meaning of the relation symbol $\mathcal{P}$,

$$\forall x \exists y \mathcal{P}(x, y) \wedge \forall x \forall y (\mathcal{P}(x, y) \leftrightarrow \forall z(z \in y \leftrightarrow z \subset x)). \tag{$\mathcal{P}$}$$

The following theorem, which is proved in Feferman [20] and Jäger [29], is not surprising and establishes a lower bound for the proof-theoretic strength of OST.

Theorem 2 *1. The theory* KPω + (AC) *is contained in* OST.
  2. The theory KP($\mathcal{P}$) + (AC) *is contained in* OST($\mathbb{P}$).

Our next goal is to look for upper proof-theoretic bounds, and the method to find those is to model OST and OST($\mathbb{P}$).

4 Modeling OST and OST($\mathbb{P}$)

There are two principal ways for constructing models of OST: one goes back to Feferman [18, 20] and uses ideas from generalized recursion theory, the other is presented in detail in Jäger [29] and is based on interpreting the application operation of OST via a suitably defined (nonmonotonic) inductive definition. Both will be briefly sketched in the following.

4.1 Feferman's model construction

We follow Feferman [20] and quote from there: The underlying applicative structure of OST is interpreted in the codes for functions that are Σ_1 definable in parameters, obtained by uniformizing the Σ_1 predicates. This proceeds as in Barwise [7], pp. 164-167, which is applicable since under the assumption ($V=L$), the universe is recursively listed in the sense given there. The treatment in Barwise must be modified slightly to account for parameters; this is done as follows. First one constructs a Σ_1 formula $\psi[w,x,y,z]$ such that for each Σ_1 formula $\theta[x,y,z]$ one can effectively find an $e \in \omega$ with $\theta[x,y,z]$ equivalent to $\psi[e,x,y,z]$. Then one uniformizes ψ with respect to y, i.e. produces a Σ_1 formula $\psi^*[w,x,y,z]$ that satisfies

(1) $\psi^*[w,x,y,z] \to \psi[w,x,y,z]$,
(2) $\exists y \psi^*[w,x,y,z] \to \exists! y \psi[w,x,y,z]$.

Given a set parameter p, one takes $\langle e, p \rangle$ to be the code of the partial function

(3) $\langle e, p \rangle(x) = y \leftrightarrow \psi^*[e,x,y,p]$.

One can then define generalized "S-n-m" functions in a straightforward way, and from those give a model of the applicative axioms of OST. The rest of the interpretation proceeds in a straightforward way.

This construction is very elegant and compact. It depends, however, on some features specific of admissible recursion theory, and it seems not so clear whether it can be generalized to a framework for significant strengthenings of OST.

4.2 Inductive model construction

Alternatively, we can provide a direct inductive definition of the application operation. Apart from being more direct, this way of modeling OST has the advantage that it can be directly adapted (see below) to dealing with strong extensions of OST.

We use lower case Greek letters $\alpha, \beta, \gamma, \delta \ldots$ (possibly with subscripts) for ordinals – they are Δ_0 definable in $\mathsf{KP}\omega$ – and write $(\alpha < \beta)$ for $(\alpha \in \beta)$. Furthermore, $(a \in L_\alpha)$ states that the set a is an element of the αth level L_α of the constructible hierarchy, and $(a <_L b)$ means that a is smaller than b according to the well-ordering $<_L$ on L. It is well-known that the assertions $(a \in L_\alpha)$ and $(a <_L b)$ are Δ over $\mathsf{KP}\omega$; see, e.g., Barwise [7] or Kunen [37].

The following approach is similar to those in Feferman and Jäger [21] and Jäger and Strahm [34] and begins with some notational preparations. For any natural number n greater than 0 we select (i) a Δ_0 formula $Tup_n(a)$ formalizing that a is an ordered n-tuple and (ii) a Δ_0 formula $(a)_n = b$ formalizing that b the projection of a on its nth component so that

$$Tup_n(a) \wedge (a)_1 = b_1 \wedge \ldots \wedge (a)_n = b_n \to a = \langle b_1, \ldots, b_n \rangle.$$

Then we fix pairwise different sets $\widehat{\mathsf{k}}, \widehat{\mathsf{s}}, \widehat{\top}, \widehat{\bot}, \widehat{\mathsf{el}}, \widehat{\mathsf{non}}, \widehat{\mathsf{dis}}, \widehat{\mathsf{e}}, \widehat{\mathsf{E}}, \widehat{\mathsf{\$}}, \widehat{\mathbb{R}}, \widehat{\mathbb{C}}$ and $\widehat{\mathbb{P}}$ none of which belongs to the collection of ordered pairs and triples; they will later act as the codes of the corresponding constants of $\mathcal{L}^\circ$. We are going to code the $\mathcal{L}^\circ$ terms $\mathsf{k}x$, $\mathsf{s}x$, $\mathsf{s}xy$, ... by the ordered tuples $\langle \widehat{\mathsf{k}}, x \rangle$, $\langle \widehat{\mathsf{s}}, x \rangle$, $\langle \widehat{\mathsf{s}}, x, y \rangle$, ... of the corresponding form. For example, to satisfy $\mathsf{k}xy = x$ we interpret $\mathsf{k}x$ as $\langle \widehat{\mathsf{k}}, x \rangle$, and "$\langle \widehat{\mathsf{k}}, x \rangle$ applied to y" is taken to be x.

Next let R be a fresh 4-place relation symbol and extend $\mathcal{L}_1$ to the language $\mathcal{L}_1(R)$ with expressions $R(\alpha, a, b, c)$ as additional atomic formulas. We also abbreviate
$$R^{<\alpha}(a,b,c) \; := \; (\exists \beta < \alpha) R(\beta, a, b, c).$$
For finding the required interpretation of the application operation of OST within $\mathsf{KP}\omega + (V{=}L)$ we work with a specific $\mathcal{L}_1(R)$ formula, introduced in the following definition. Afterwards, this formula together with Σ recursion will help to provide what we need.

Definition 1 *We choose $\mathfrak{A}[R, \alpha, a, b, c]$ to be the $\mathcal{L}_1(R)$ formula defined as*
$$\mathfrak{A}[R, \alpha, a, b, c] \; := \; c \in L_\alpha \;\wedge\; \mathfrak{B}[R, \alpha, a, b, c],$$
where $\mathfrak{B}[R, \alpha, a, b, c]$ is an auxiliary $\mathcal{L}_1(R)$ formula given as the disjunction of the following clauses:

(1) $a = \widehat{\mathsf{k}} \,\wedge\, c = \langle \widehat{\mathsf{k}}, b \rangle$,
(2) $\mathrm{Tup}_2(a) \,\wedge\, (a)_1 = \widehat{\mathsf{k}} \,\wedge\, (a)_2 = c$,
(3) $a = \widehat{\mathsf{s}} \,\wedge\, c = \langle \widehat{\mathsf{s}}, b \rangle$,
(4) $\mathrm{Tup}_2(a) \,\wedge\, (a)_1 = \widehat{\mathsf{s}} \,\wedge\, c = \langle \widehat{\mathsf{s}}, (a)_2, b \rangle$,
(5) $\mathrm{Tup}_3(a) \,\wedge\, (a)_1 = \widehat{\mathsf{s}} \,\wedge\,$
    $(\exists x, y \in L_\alpha)(R^{<\alpha}((a)_2, b, x) \,\wedge\, R^{<\alpha}((a)_3, b, y) \,\wedge\, R^{<\alpha}(x, y, c))$,
(6) $a = \widehat{\mathsf{el}} \,\wedge\, c = \langle \widehat{\mathsf{el}}, b \rangle$,
(7) $\mathrm{Tup}_2(a) \,\wedge\, (a)_1 = \widehat{\mathsf{el}} \,\wedge\, (a)_2 \in b \,\wedge\, c = \widehat{\top}$,
(8) $\mathrm{Tup}_2(a) \,\wedge\, (a)_1 = \widehat{\mathsf{el}} \,\wedge\, (a)_2 \notin b \,\wedge\, c = \widehat{\bot}$,
(9) $a = \widehat{\mathsf{non}} \,\wedge\, b = \widehat{\top} \,\wedge\, c = \widehat{\bot}$,
(10) $a = \widehat{\mathsf{non}} \,\wedge\, b = \widehat{\bot} \,\wedge\, c = \widehat{\top}$,
(11) $a = \widehat{\mathsf{dis}} \,\wedge\, c = \langle \widehat{\mathsf{dis}}, b \rangle$,
(12) $\mathrm{Tup}_2(a) \,\wedge\, (a)_1 = \widehat{\mathsf{dis}} \,\wedge\, (a)_2 = \widehat{\top} \,\wedge\, c = \widehat{\top}$,
(13) $\mathrm{Tup}_2(a) \,\wedge\, (a)_1 = \widehat{\mathsf{dis}} \,\wedge\, (a)_2 = \widehat{\bot} \,\wedge\, b = \widehat{\top} \,\wedge\, c = \widehat{\top}$,
(14) $\mathrm{Tup}_2(a) \,\wedge\, (a)_1 = \widehat{\mathsf{dis}} \,\wedge\, (a)_2 = \widehat{\bot} \,\wedge\, b = \widehat{\bot} \,\wedge\, c = \widehat{\bot}$,
(15) $a = \widehat{\mathsf{e}} \,\wedge\, c = \langle \widehat{\mathsf{e}}, b \rangle$,
(16) $\mathrm{Tup}_2(a) \,\wedge\, (a)_1 = \widehat{\mathsf{e}} \,\wedge\, (\exists x \in b) R^{<\alpha}((a)_2, x, \widehat{\top}) \,\wedge\, c = \widehat{\top}$,
(17) $\mathrm{Tup}_2(a) \,\wedge\, (a)_1 = \widehat{\mathsf{e}} \,\wedge\, (\forall x \in b) R^{<\alpha}((a)_2, x, \widehat{\bot}) \,\wedge\, c = \widehat{\bot}$,
(18) $a = \widehat{\mathsf{S}} \,\wedge\, c = \langle \widehat{\mathsf{S}}, b \rangle$,
(19) $\mathrm{Tup}_2(a) \,\wedge\, (a)_1 = \widehat{\mathsf{S}} \,\wedge\, (\forall x \in b)(R^{<\alpha}((a)_2, x, \widehat{\top}) \vee R^{<\alpha}((a)_2, x, \widehat{\bot})) \,\wedge\,$
    $(\forall x \in c)(x \in b \,\wedge\, R^{<\alpha}((a)_2, x, \widehat{\top})) \,\wedge\,$
    $(\forall x \in b)(R^{<\alpha}((a)_2, x, \widehat{\top}) \to x \in c)$,
(20) $a = \widehat{\mathsf{R}} \,\wedge\, c = \langle \widehat{\mathsf{R}}, b \rangle$,
(21) $\mathrm{Tup}_2(a) \,\wedge\, (a)_1 = \widehat{\mathsf{R}} \,\wedge\, (\forall x \in b)(\exists y \in c) R^{<\alpha}((a)_2, x, y) \,\wedge\,$
    $(\forall y \in c)(\exists x \in b) R^{<\alpha}((a)_2, x, y)$,
(22) $a = \widehat{\mathbb{C}} \,\wedge\, R^{<\alpha}(b, c, \widehat{\top}) \,\wedge\, (\forall x \in L_\alpha)(x <_L c \to \neg R^{<\alpha}(b, x, \widehat{\top})) \,\wedge\,$
    $(\forall \beta < \alpha)(\forall x \in L_\beta) \neg R^{<\beta}(b, x, \widehat{\top})$.

We immediately see that $\mathfrak{A}[R,\alpha,a,b,c]$ is Δ over $\mathsf{KP}\omega$ with respect to the language $\mathcal{L}_1(R)$. It is also easy to verify that $\mathfrak{A}[R,\alpha,a,b,c]$ is deterministic in the following sense: from $\mathfrak{A}[R,\alpha,a,b,c]$ we can conclude that exactly one of the clauses (1)–(22) of the previous definition is satisfied for these α, a, b, c.

For any $\mathcal{L}_1$ formula $B[\alpha,a,b,c]$ with at most the indicated free variables we write $\mathfrak{A}[B,\alpha,a,b,c]$ for the $\mathcal{L}_1$ formula resulting by replacing each occurrence of an atomic formula of the form $R(\alpha,r,s,t)$ in $\mathfrak{A}[R,\alpha,a,b,c]$ by $B[\alpha,r,s,t]$. The following theorem is a special case of "Definition by Σ Recursion" as developed in Barwise [7].

Theorem 3 *There exists a Σ formula $B[\alpha,a,b,c]$ of $\mathcal{L}_1$ with at most α, a, b and c free so that $\mathsf{KP}\omega$ proves*

$$B[\alpha,a,b,c] \leftrightarrow \mathfrak{A}[B,\alpha,a,b,c]. \qquad (\Sigma\text{-Rec}/\mathfrak{A})$$

Any such a formula $B[\alpha,a,b,c]$ may be used to describe the αth level of the interpretation of the OST application $(ab \simeq c)$. Accordingly, we proceed as follows.

Definition 2 *Let $B_\mathfrak{A}[\alpha,a,b,c]$ be a Σ formula of $\mathcal{L}_1$ associated to the operator form $\mathfrak{A}[R,\alpha,a,b,c]$ according to (Σ-Rec/$\mathfrak{A}$) of the previous theorem. Then we define*

$$Ap_\mathfrak{A}[a,b,c] := \exists \alpha B_\mathfrak{A}[\alpha,a,b,c].$$

As can be easily checked, $Ap_\mathfrak{A}[a,b,c]$ is functional in its third argument. It is therefore suitable for handling the $\mathcal{L}^\circ$ terms within $\mathsf{KP}\omega + (V{=}L)$. To each term t of $\mathcal{L}^\circ$ we associate a formula $[\![t]\!]_\mathfrak{A}(u)$ of $\mathcal{L}_1$ expressing that u is the value of t under the interpretation of the OST-application via the Σ formula $Ap_\mathfrak{A}[a,b,c]$.

Definition 3 *For each $\mathcal{L}^\circ$ term t we introduce an $\mathcal{L}_1$ formula $[\![t]\!]_\mathfrak{A}(u)$, with u not occurring in t, which is inductively defined as follows:*

1. *If t is a set variable, then $[\![t]\!]_\mathfrak{A}(u)$ is the formula $(t = u)$.*
2. *If t is a constant, then $[\![t]\!]_\mathfrak{A}(u)$ is the formula $(\hat{t} = u)$.*
3. *If t is the term (rs), then we set*

$$[\![t]\!]_\mathfrak{A}(u) := \exists x \exists y ([\![r]\!]_\mathfrak{A}(x) \wedge [\![s]\!]_\mathfrak{A}(y) \wedge Ap_\mathfrak{A}[x,y,u]).$$

Observe that for every term t of $\mathcal{L}^\circ$ its translation $[\![t]\!]_\mathfrak{A}(u)$ is a Σ formula of $\mathcal{L}_1$. By this treatment of the terms of $\mathcal{L}^\circ$, the translation of arbitrary formulas of $\mathcal{L}^\circ$ into formulas of $\mathcal{L}_1$ is predetermined.

Definition 4 *The translation of an $\mathcal{L}^\circ$ formula A into the $\mathcal{L}_1$ formula A^* is inductively defined as follows:*

1. *For the atomic formulas of $\mathcal{L}^\circ$ we stipulate*

$$(t\downarrow)^* := \exists x [\![t]\!]_\mathfrak{A}(x),$$

$$(s \in t)^* := \exists x \exists y ([\![s]\!]_\mathfrak{A}(x) \wedge [\![t]\!]_\mathfrak{A}(y) \wedge x \in y).$$

2. If A is a formula $\neg B$, then A^* is $\neg B^*$.
3. If A is a formula $(B \diamond C)$ for $\diamond$ being the binary junctor $\vee$ or $\wedge$, then A^* is $(B^* \diamond C^*)$.
4. If A is a formula $(\exists x \in t)B[x]$, then

$$A^* := \exists y([\![t]\!]_{\mathfrak{A}}(y) \wedge (\exists x \in y)B^*[x]).$$

5. If A is a formula $(\forall x \in t)B[x]$, then

$$A^* := \forall y([\![t]\!]_{\mathfrak{A}}(y) \to (\forall x \in y)B^*[x]).$$

6. If A is a formula $QxB[x]$ for a quantifier Q, then A^* is $QxB^*[x]$.

This translation of $\mathcal{L}^\circ$ formulas leads directly to an interpretation of OST in $\mathsf{KP}\omega + (V{=}L)$. The corresponding interpretation result is proved in Jäger [29].

Theorem 4 *The theory* OST *is interpretable in* $\mathsf{KP}\omega + (V{=}L)$; *i.e. for all formulas A of $\mathcal{L}^\circ$ we have*

$$\mathsf{OST} \vdash A \quad \Longrightarrow \quad \mathsf{KP}\omega + (V{=}L) \vdash A^*.$$

As remarked earlier, $\mathsf{KP}\omega + (V{=}L)$ is a conservative extension of $\mathsf{KP}\omega$ for absolute formulas. If we combine this with Theorem 2 and Theorem 4, we obtain the following corollary, which settles the question of the consistency strength of OST. This result was first established in Feferman [18].

Corollary 1 *The theory* OST *is conservative over* $\mathsf{KP}\omega$ *for absolute formulas. In particular,* OST *and* $\mathsf{KP}\omega$ *are equiconsistent.*

We establish an upper bound for $\mathsf{OST}(\mathbb{P})$ by an easy modification of the argument in the previous section: only extend the disjunction in Definition 1 by a clause taking care of the constant $\mathbb{P}$.

Definition 5 *We choose $\mathfrak{C}[R, \alpha, a, b, c]$ to be the $\Delta(\mathcal{P})$ formula of $\mathcal{L}_1(\mathcal{P}, R)$ defined as*

$$\mathfrak{C}[R, \alpha, a, b, c] := c \in L_\alpha \wedge (\mathfrak{B}[R, \alpha, a, b, c] \vee (a = \widehat{\mathbb{P}} \wedge \mathcal{P}(b, c))),$$

where $\mathfrak{B}[R, \alpha, a, b, c]$ is the formula introduced in Definition 1.

In $\mathsf{KP}(\mathcal{P})$ we have $\Sigma(\mathcal{P})$ recursion. Completely in the line of the previous section we apply it now, of course, to the operator form $\mathfrak{C}[R, \alpha, a, b, c]$, yielding the following analogue of Theorem 3.

Theorem 5 *There exists a $\Sigma(\mathcal{P})$ formula $B[\alpha, a, b, c]$ of $\mathcal{L}_1(\mathcal{P}, R)$ with at most α, a, b and c free so that $\mathsf{KP}(\mathcal{P})$ proves*

$$B[\alpha, a, b, c] \leftrightarrow \mathfrak{C}[B, \alpha, a, b, c]. \qquad (\Sigma(\mathcal{P})\text{-Rec}/\mathfrak{C})$$

Each $\Sigma(\mathcal{P})$ formula $B[\alpha, a, b, c]$ fulfilling this recursion equation ($\Sigma(\mathcal{P})$-Rec/$\mathfrak{C}$) is now a possible candidate for interpreting the $\mathsf{OST}(\mathbb{P})$ application $(ab \simeq c)$.

Definition 6 Let $B_{\mathfrak{C}}[\alpha, a, b, c]$ be a $\Sigma(\mathcal{P})$ formula of $\mathcal{L}_1(\mathcal{P})$ associated to the operator form $\mathfrak{C}[R, \alpha, a, b, c]$ according to ($\Sigma(\mathcal{P})$-Rec/$\mathfrak{C}$) of the previous theorem. Then we define

$$Ap_{\mathfrak{C}}[a, b, c] := \exists \alpha B_{\mathfrak{C}}[\alpha, a, b, c].$$

It only remains to proceed as in the previous section, but with $Ap_{\mathfrak{A}}[a, b, c]$ replaced by $Ap_{\mathfrak{C}}[a, b, c]$. For each $\mathcal{L}^\circ$ term t, an $\mathcal{L}_1(\mathcal{P})$ formula $[\![t]\!]_{\mathfrak{C}}(u)$ is introduced, saying that u is the value of the term t under the interpretation of the OST($\mathbb{P}$) application via $Ap_{\mathfrak{C}}[a, b, c]$. Finally, following the pattern of Definition 4 and based on these $[\![t]\!]_{\mathfrak{C}}(u)$, each $\mathcal{L}^\circ$ formula A is canonically translated into a formula $A^\sharp$ of $\mathcal{L}_1(\mathcal{P})$.

The interpretation of OST($\mathbb{P}$) in KP($\mathcal{P}$) + ($V{=}L$) is a straightforward extension of Theorem 4. For details see Jäger [29].

Theorem 6 *The theory* OST($\mathbb{P}$) *is interpretable in* KP($\mathcal{P}$) + ($V{=}L$); *i.e. for all formulas A of $\mathcal{L}^\circ$ we have*

$$\text{OST}(\mathbb{P}) \vdash A \quad \Longrightarrow \quad \text{KP}(\mathcal{P}) + (V{=}L) \vdash A^\sharp.$$

Unfortunately, the combination of Theorem 2 and Theorem 6 does not completely settle the question about the consistency strength of OST($\mathbb{P}$) yet. So far we have an interesting lower and an interesting upper bound, but it still has to be determined what the relationship between KP($\mathcal{P}$) and KP($\mathcal{P}$) + ($V{=}L$) is.

5 The theories OST(**E**, $\mathbb{P}$) and OST^r(**E**, $\mathbb{P}$)

This section is dedicated to the extension OST(**E**, $\mathbb{P}$) of OST – baptized OST + (Pow) + (Uni) in Feferman [18, 20] – and its subsystem OST^r(**E**, $\mathbb{P}$). OST(**E**, $\mathbb{P}$) provides for unbounded existential quantification and power set and brings us into the realm of ZFC and beyond.

The subsystem OST^r(**E**, $\mathbb{P}$) of OST(**E**, $\mathbb{P}$) is designed to be a witness of an operational set theory of the same strength as ZFC, sought in Feferman [18]. It is obtained from OST(**E**, $\mathbb{P}$) by simply restricting the schema of $\in$-induction for arbitrary formulas to $\in$-induction for sets.

It should be fairly straightforward to prove in OST(**E**, $\mathbb{P}$) that ZFC is consistent, implying that OST(**E**, $\mathbb{P}$) is stronger than ZFC; actually this also follows from Theorem 8 below. To characterize it in terms of consistency strength it is natural to turn to theories of sets and classes.

5.1 NBG and a bit more

In surveying von Neumann-Bernays-Gödel set theory NBG and its extension NBG$_{<E_0}$ we follow their presentation in Jäger [30]. The formalization of NBG there is based on standard literature, for example Levy [38] and Mendelson [41].

NBG is a theory of sets and classes conservative over the system ZFC of Zermelo-Fraenkel set theory with the axiom of choice. NBG is known to be finitely axiomatizable although the version we are going to present below permits axiom schemas and as such is an infinite axiomatization.

Operations, sets and classes

$\mathcal{L}_2$, the language of NBG, augments $\mathcal{L}_1$ by a second sort of countably many variables $U, V, W, X, Y, Z, \ldots$ (possibly with subscripts) for classes. The set terms of $\mathcal{L}_2$ are the terms of $\mathcal{L}_1$, as class terms we simply have the class variables.

The *formulas* ($A, B, C, D, A_1, B_1, C_1, D_1, \ldots$) of $\mathcal{L}_2$ are inductively generated as follows:

1. If s and t are set terms of $\mathcal{L}_2$ and U is a class variable, then all expressions of the form $(s \in t)$ and $(s \in U)$ are (atomic) formulas of $\mathcal{L}_2$.
2. If A and B are formulas of $\mathcal{L}_2$, then so are are $\neg A$, $(A \vee B)$ and $(A \wedge B)$.
3. If A is a formula and t a set term of $\mathcal{L}_2$ which does not contain x, then $\exists x A, \forall x A, (\exists x \in t)A, (\forall x \in t)A, \exists X A$ and $\forall X A$ are formulas of $\mathcal{L}_2$.

As before, the remaining logical connectives are introduced as abbreviations, and we will often omit parentheses and brackets whenever there is no danger of confusion. Equalities between sets/sets, sets/classes, classes/sets and classes/classes are not atomic formulas of $\mathcal{L}_2$ but defined as

$$(Var_1 = Var_2) := \forall x(x \in Var_1 \leftrightarrow x \in Var_2)$$

where Var_1 and Var_2 denote set or class variables. A formula of $\mathcal{L}_2$ is called *elementary* or a Π_0^1 formula if it does not contain bound class variables; free class variables, however, are permitted. The Σ_1^1 formulas of $\mathcal{L}_2$ are those of the form $\exists X A$ with elementary A.

The logic of NBG is classical two-sorted logic with equality for the first sort. The non-logical axioms of NBG are given in six groups. To increase readability, we freely use standard set-theoretic terminology.

I. Elementary comprehension For any elementary formula $A[u]$ of $\mathcal{L}_2$:

$$\exists X \forall y(y \in X \leftrightarrow A[y]). \tag{ECA}$$

Hence every elementary NBG formula $A[u]$ defines a class, which is typically written as $\{x : A[x]\}$. It may be (extensionally equal to) a set, but this is not necessarily the case. The intersection of a class with a set, however, is always supposed to produce a set by the following principle of Aussonderung.

II. Aussonderung

$$\forall X \forall y \exists z(z = X \cap y). \tag{AUS}$$

From logical reasons, (ECA) and (AUS) we conclude that there is a unique set which has no members; it is denoted by $\emptyset$.

III. Basic set existence

$$\forall x \forall y \exists z(z = \{x, y\}), \tag{Pair}$$

$$\forall x \exists y(y = \cup x), \tag{Union}$$

$$\forall x \exists y \forall z(z \in y \leftrightarrow z \subset x), \tag{Powerset}$$

$$\exists x(\emptyset \in x \wedge (\forall y \in x)(y \cup \{y\} \in x)). \tag{Infinity}$$

As in OST($\mathbf{E}, \mathbb{P}$) we write $\langle a, b \rangle$ for the ordered pair of the sets a and b à la Kuratowski. Class relations are classes which consist of ordered pairs only, and class functions are class relations which assign to every set exactly one set; i.e. for all U we set:

$$Rel[U] := \forall x(x \in U \rightarrow \exists y \exists z(x = \langle y, z \rangle)),$$

$$Fun[U] := Rel[U] \land \forall x \exists! y(\langle x, y \rangle \in U).$$

If U is a function we write $U(x)$ for the uniquely determined y associated to x by U. Replacement states that the range of a set under a function is a set.

IV. Replacement

$$\forall X(Fun[X] \rightarrow \forall y \exists z(z = \{X(u) : u \in y\})). \tag{REP}$$

Global choice is a very uniform principle of choice which claims the existence of a class function which picks an element of any non-empty set.

V. Global choice

$$\exists X(Fun[X] \land \forall y(y \neq \emptyset \rightarrow X(y) \in y)). \tag{GC}$$

Finally, in NBG it is claimed that the element relation is well-founded with respect to classes.

VI. Class foundation

$$\forall X(X \neq \emptyset \rightarrow \exists y(y \in X \land (\forall z \in y)(z \notin X))). \tag{C-I$_\in$}$$

The axioms (Infinity) and (C-I$_\in$) imply that there exists a least infinite ordinal, which we denote by ω, as usual. The elements of ω are identified with the natural numbers in the sense that $0 := \emptyset$, $1 := \{0\}$, $2 := 1 \cup \{1\}$ and so on.

The axioms in groups I - VI supply one possible axiomatization of NBG. According to a well-known result (cf., e.g., Levy [38]) NBG is a conservative extension of ZFC.

Theorem 7 *A sentence of the language $\mathcal{L}_1$ is provable in NBG if and only if it is provable in ZFC.*

A first step in extending NBG is to add the principle of $\in$-induction for arbitrary $\mathcal{L}_2$ formulas $A[u]$,

$$\forall x((\forall y \in x)A[y] \rightarrow A[x]) \rightarrow \forall x A[x]. \tag{$\mathcal{L}_2$-I$_\in$}$$

Besides that, we want to be able to iterate elementary comprehension along all well-orderings which can be constructed from the ordinals and the order type of all ordinals by closing them under addition and ω-exponentiation. For this purpose we introduce a notation system $(E_0, \prec)$ which can be considered as the canonical blowing up of $(\varepsilon_0, <)$ triggered by replacing the natural numbers by the ordinals. In particular:

(i) E_0 is an elementarily definable class, and $\prec$ is an elementarily definable binary class relation on E_0.

Operations, sets and classes

(ii) For any ordinal α there exists a code $\bar{\alpha}$ which belongs to E_0.

(iii) E_0 contains an element Ω such that $(\Omega, \prec)$ is an isomorphic copy of the class of all ordinals.

(iv) There are a binary class function $\oplus$ and a unary class function Exp_ω, both elementary, such that E_0 is closed under $\oplus$ and Exp_ω. These two functions are for the addition and ω-exponentiation of elements of E_0 in the expected sense.

In the following we write $(a + b)$ – or often simply $a + b$ – for $(a \oplus b)$ and ω^a for $Exp_\omega(a)$. For all natural numbers n, the ordinal terms Ω_n are inductively defined by

$$\Omega_0 := \Omega + 1 \quad \text{and} \quad \Omega_{n+1} := \omega^{\Omega_n}.$$

All additional relevant details concerning $(E_0, \prec)$ are worked out in detail in Jäger and Krähenbühl [32]. Amongst other things it is shown there that, for any standard natural number n, the theory $\mathsf{NBG}+(\mathcal{L}_2\text{-}\mathsf{I}_\in)$ proves transfinite induction along $\prec$ up to any Ω_n.

Let $A[U, V, u, v]$ be an elementary $\mathcal{L}_2$ formula with at most the indicated variables free. Then we write $Hier_A[a, U, V]$ for the elementary $\mathcal{L}_2$ formula

$$(\forall b \prec a)((V)_b = \{x : A[U, \Sigma(V, b), b, x]\})$$

where $\Sigma(V, b)$ stands for the class $\{\langle x, c \rangle \in V : c \prec b\}$ representing the disjoint union of the projections of V up to b. This formula states that V codes the hierarchy generated by iterating comprehension via A with class parameter U along $\prec$ up to a.

$\mathsf{NBG}_{<E_0}$ is defined to be the theory of sets and classes which consists of NBG, full $\in$-induction ($\mathcal{L}_2\text{-}\mathsf{I}_\in$) plus the additional axioms

$$\forall X \exists Y\, Hier_A[\Omega_n, X, Y]$$

for all standard natural numbers n and all elementary formulas $A[U, V, u, v]$ of $\mathcal{L}_2$ with at most the variables U, V, u, v free. So it permits iteration of elementary class comprehension along each (standard) initial segment of E_0.

In operational set theory we call an operation f a *total characteristic operation* if is totally defined and takes values in $\mathbb{B}$ only;

$$TCO(f) := \forall x(fx \in \mathbb{B}).$$

By regarding the class variables of $\mathcal{L}_2$ as variables of $\mathcal{L}^\circ$ ranging over total characteristic operations and by translating an atomic $\mathcal{L}_2$ formula $(a \in U)$ into $(fa = \top)$, where f is the $\mathcal{L}^\circ$ variable associated to the $\mathcal{L}_2$ variable U, every $\mathcal{L}_2$ formula A canonically translates into an $\mathcal{L}^\circ$ formula A°. This translation is such that

$$\exists X A[X]^\circ = \exists x(TCO(x) \wedge A^\circ[x]),$$
$$\forall X A[X]^\circ = \forall x(TCO(x) \to A^\circ[x]),$$

always modulo a renaming of the variables if necessary. This translation leads to the following interpretation theorem.

Theorem 8 1. *The theory* NBG *is interpretable in* $\mathsf{OST}^r(\mathbf{E}, \mathbb{P})$.
2. *The theory* $\mathsf{NBG}_{<E_0}$ *is interpretable in* $\mathsf{OST}(\mathbf{E}, \mathbb{P})$.

The second part of this theorem is proved in Jäger [30]; the first part follows directly from an inspection of this proof. A result equivalent to the first assertion can be found in Jäger [29] where it is shown that ZFC can be embedded into $\mathsf{OST}^r(\mathbf{E}, \mathbb{P})$.

5.2 Inductive extensions of ZF

By the previous theorem we have lower proof theoretic bounds for $\mathsf{OST}^r(\mathbf{E}, \mathbb{P})$ and $\mathsf{OST}(\mathbf{E}, \mathbb{P})$, and what remains is to show that these bounds are sharp. This will be achieved by utilizing inductive model constructions again. They can be carried out in the auxiliary theories $\mathsf{E}^r(\mathsf{ZFW})$ and $\mathsf{E}^r(\mathsf{ZFW}) + (\mathcal{L}_S\text{-}\mathsf{I}_\in)$, depending on how much induction we have to model, which are reducible to NBG and $\mathsf{NBG}_{<E_0}$, respectively.

When building up the inductive model of $\mathsf{OST}(\mathbf{E}, \mathbb{P})$, we have to handle the choice axiom (S3). For this end it is convenient to have a global well-ordering of the set-theoretic universe at our disposal. Therefore, let $\mathcal{L}_1(\mathcal{W})$ be the extension of $\mathcal{L}_1$ by the fresh binary relation symbol $\mathcal{W}$, and let ZFW be the extension of ZF which comprises all axioms of ZF – formulated, of course, with respect to the new language $\mathcal{L}_1(\mathcal{W})$ – plus the following well-ordering axiom

$$\forall x \exists! \alpha \mathcal{W}(x, \alpha). \qquad (\mathcal{W})$$

From axiom ($\mathcal{W}$) the desired well-ordering of the universe of sets is canonically obtained if we set

$$a <_\mathcal{W} b := \exists \alpha \exists \beta (\mathcal{W}(a, \alpha) \wedge \mathcal{W}(b, \beta) \wedge \alpha < \beta). \qquad (<_\mathcal{W})$$

Analogously to Section 4.2 we pick an n-ary relation symbol R which does not belong to the language $\mathcal{L}_1(\mathcal{W})$ and write $\mathcal{L}_1(\mathcal{W}, R)$ for the extension of $\mathcal{L}_1(\mathcal{W})$ by R. An $\mathcal{L}_1(\mathcal{W}, R)$ formula which contains at most $a_1, \ldots, a_n$ free is called an n-ary operator form, and we let $\mathfrak{F}[R, a_1, \ldots, a_n]$ range over such forms.

Based on a model $\mathcal{M}$ of ZFW with universe $|\mathcal{M}|$, any n-ary operator form $\mathfrak{F}[R, \vec{a}]$ gives rise to subsets $I_\mathfrak{F}^\zeta$ of $|\mathcal{M}|^n$ generated inductively for all ordinals ζ (not only those belonging to $|\mathcal{M}|$) by

$$I_\mathfrak{F}^{<\zeta} := \bigcup_{\eta < \zeta} I_\mathfrak{F}^\eta \quad \text{and} \quad I_\mathfrak{F}^\zeta := \{\langle \vec{x} \rangle \in |\mathcal{M}|^n : \mathcal{M} \models \mathfrak{F}[I_\mathfrak{F}^{<\zeta}, \vec{x}]\}.$$

These sets $I_\mathfrak{F}^\zeta$ are the *stages* of the inductive definition induced by $\mathfrak{F}[R, \vec{a}]$, relative to $\mathcal{M}$; for many models $\mathcal{M}$, operator forms $\mathfrak{F}[R, \vec{a}]$ and ordinals ζ the $I_\mathfrak{F}^\zeta$ are not elements of $|\mathcal{M}|$. We now enrich ZFW so that we can speak about such stages.

The theory $\mathsf{E}^r(\mathsf{ZFW})$ is formulated in the language $\mathcal{L}_S$ which extends $\mathcal{L}_1(\mathcal{W})$ by adding a new sort of so called stage variables $\rho, \sigma, \tau, \ldots$ (possibly with subscripts) as well as new binary relation symbols $\prec$ and $\equiv$ for the less and equality relation for stage variables, respectively. Moreover, $\mathcal{L}_S$ includes an $(n+1)$-ary relation symbol $Q_\mathfrak{F}$ for each operator form $\mathfrak{F}[R, a_1, \ldots, a_n]$.

Operations, sets and classes

The set terms of $\mathcal{L}_S$ are the set terms of $\mathcal{L}_1$, and the atomic formulas of $\mathcal{L}_S$ are the atomic formulas of $\mathcal{L}_1(\mathcal{W})$ plus all expressions $(\sigma \prec \tau)$, $(\sigma \equiv \tau)$ and $Q_{\mathfrak{F}}(\sigma, \vec{s})$ for each n-ary operator form $\mathfrak{F}[R, \vec{a}]$. Usually we write $Q_{\mathfrak{F}}^\sigma(\vec{s})$ instead of $Q_{\mathfrak{F}}(\sigma, \vec{s})$.

The formulas $(A, B, C, A_1, B_1, C_1, \ldots)$ of $\mathcal{L}_S$ are generated from these atomic formulas by closure under negation, conjunction and disjunction, bounded and unbounded quantification over sets, bounded stage quantification $(\exists \sigma \prec \tau)$ and $(\forall \sigma \prec \tau)$ as well as unbounded stage quantification $\exists \sigma$ and $\forall \sigma$. The Δ_0^S formulas are those $\mathcal{L}_S$ formulas that do not contain unbounded stage quantifiers. An $\mathcal{L}_S$ formula A is is called Σ^S if all positive occurrences of unbounded stage quantifiers in A are existential and all negative occurrences of unbounded stage quantifiers in A are universal; it is called Π^Ω if all positive occurrences of unbounded stage quantifiers in A are universal and all negative occurrences of unbounded stage quantifiers in A are existential.

Further, we write A^σ to denote the $\mathcal{L}_S$ formula which is obtained from A by replacing all unbounded stage quantifiers $Q\tau$ in A by bounded stage quantifiers $(Q\tau \prec \sigma)$. Additional abbreviations are

$$Q_{\mathfrak{F}}^{\prec \sigma}(\vec{s}) := (\exists \tau \prec \sigma) Q_{\mathfrak{F}}^\tau(\vec{s}) \quad \text{and} \quad Q_{\mathfrak{F}}(\vec{s}) := \exists \sigma Q_{\mathfrak{F}}^\sigma(\vec{s}).$$

Clearly, any formula of $\mathcal{L}_1(\mathcal{W})$ is a (trivial) Δ_0^S formula, and A^σ is Δ_0^S for any $\mathcal{L}_S$ formula A.

The theory $\mathsf{E}^r(\mathsf{ZFW})$ is formulated in classical two sorted predicate logic with equality in both sorts; in addition it contains as non-logical axioms all ZFW-axioms of the language $\mathcal{L}_1(\mathcal{W})$, some axioms about stage variables and operator forms, reflection for Σ^S formulas, separation and replacement for Δ_0^S formulas plus induction along $\in$ and along $\prec$ for Δ_0^S formulas.

I. ZFW-axioms. All axioms of the theory ZFW formulated in the language $\mathcal{L}_1(\mathcal{W})$; they do not refer to stage variables or relation symbols associated to operator forms.

II. Linearity axioms. For all stage variables ρ, σ and τ:

$$\sigma \not\prec \sigma \wedge (\rho \prec \sigma \wedge \sigma \prec \tau \to \rho \prec \tau) \wedge (\sigma \prec \tau \vee \sigma \equiv \tau \vee \tau \prec \sigma).$$

III. Operator axioms. For all operator forms $\mathfrak{F}[R, \vec{u}]$ and all set terms $\vec{s}$:

$$Q_{\mathfrak{F}}^\sigma(\vec{s}) \leftrightarrow \mathfrak{F}[Q_{\mathfrak{F}}^{\prec \sigma}, \vec{s}].$$

IV. Σ^S reflection. For all Σ^S formulas A:

$$A \to \exists \sigma A^\sigma. \qquad (\Sigma^S\text{-Ref})$$

V. Δ_0^S Separation. For all Δ_0^S formulas $A[u]$ and all set terms s:

$$\exists x (x = \{y \in s : A[y]\}). \qquad (\Delta_0^S\text{-Sep})$$

VI. Δ_0^S Replacement. For all Δ_0^S formulas $A[u, v]$ and all set terms s:

$$(\forall x \in s) \exists! y A[x, y] \to \exists z \forall y (y \in z \leftrightarrow (\exists x \in s) A[x, y]). \qquad (\Delta_0^S\text{-Rep})$$

VII. Δ_0^S induction along $\in$ and $\prec$. For all Δ_0^S formulas $A[u]$:

$$\forall x((\forall y \in x)A[y] \to A[x]) \to \forall x A[x], \tag{$\Delta_0^S\text{-}\mathsf{I}_\in$}$$

$$\forall \sigma((\forall \tau \prec \sigma)A[\tau] \to A[\sigma]) \to \forall \sigma A[\sigma]. \tag{$\Delta_0^S\text{-}\mathsf{I}_\prec$}$$

The theory $\mathsf{E}^r(\mathsf{ZFW})$ is a restricted system (hence the superscript "r") in the sense that the axioms in groups V, VI and VII are restricted to Δ_0^S formulas. By $\mathsf{E}^r(\mathsf{ZFW}) + (\mathcal{L}_S\text{-}\mathsf{I}_\in)$ is meant $\mathsf{E}^r(\mathsf{ZFW})$ extended by the schema of $\in$-induction for arbitrary $\mathcal{L}_S$ formulas.

It is important to observe that the stage variables do not belong to the collection of sets; they constitute a different entity which is used to "enumerate" the stages of the inductive definition associated to each operator form. However, in the form of Δ_0^S separation and Δ_0^S replacement they can nevertheless help to constitute new sets in a carefully restricted way.

Following the pattern of Section 4.2 we are now going to introduce a specific inductive definition which will lead to a suitable treatment of the application operation of $\mathsf{OST}^r(\mathbf{E}, \mathbb{P})$ and $\mathsf{OST}(\mathbf{E}, \mathbb{P})$. The decisive new aspect, see clauses (24) and (25), is the treatment of operational unbounded existential quantification. Also, the well-ordering of the set-theoretic universe generated by the axiom $(V = L)$, as it is used in Definition 1, is replaced by $<_W$.

Definition 7 *The operator form $\mathfrak{F}[R, a, b, c]$ is defined to be the disjunction of the following clauses:*

(1) $a = \widehat{\mathsf{k}} \wedge c = \langle \widehat{\mathsf{k}}, b \rangle$,

(2) $Tup_2(a) \wedge (a)_1 = \widehat{\mathsf{k}} \wedge (a)_2 = c$,

(3) $a = \widehat{\mathsf{s}} \wedge c = \langle \widehat{\mathsf{s}}, b \rangle$,

(4) $Tup_2(a) \wedge (a)_1 = \widehat{\mathsf{s}} \wedge c = \langle \widehat{\mathsf{s}}, (a)_2, b \rangle$,

(5) $Tup_3(a) \wedge (a)_1 = \widehat{\mathsf{s}} \wedge \exists x \exists y (R((a)_2, b, x) \wedge R((a)_3, b, y) \wedge R(x, y, c))$,

(6) $a = \widehat{\mathsf{el}} \wedge c = \langle \widehat{\mathsf{el}}, b \rangle$,

(7) $Tup_2(a) \wedge (a)_1 = \widehat{\mathsf{el}} \wedge (a)_2 \in b \wedge c = \widehat{\top}$,

(8) $Tup_2(a) \wedge (a)_1 = \widehat{\mathsf{el}} \wedge (a)_2 \notin b \wedge c = \widehat{\bot}$,

(9) $a = \widehat{\mathsf{non}} \wedge b = \widehat{\top} \wedge c = \widehat{\bot}$,

(10) $a = \widehat{\mathsf{non}} \wedge b = \widehat{\bot} \wedge c = \widehat{\top}$,

(11) $a = \widehat{\mathsf{dis}} \wedge c = \langle \widehat{\mathsf{dis}}, b \rangle$,

(12) $Tup_2(a) \wedge (a)_1 = \widehat{\mathsf{dis}} \wedge (a)_2 = \widehat{\top} \wedge c = \widehat{\top}$,

(13) $Tup_2(a) \wedge (a)_1 = \widehat{\mathsf{dis}} \wedge (a)_2 = \widehat{\bot} \wedge b = \widehat{\top} \wedge c = \widehat{\top}$,

(14) $Tup_2(a) \wedge (a)_1 = \widehat{\mathsf{dis}} \wedge (a)_2 = \widehat{\bot} \wedge b = \widehat{\bot} \wedge c = \widehat{\bot}$,

(15) $a = \widehat{\mathsf{e}} \wedge c = \langle \widehat{\mathsf{e}}, b \rangle$,

(16) $Tup_2(a) \wedge (a)_1 = \widehat{\mathsf{e}} \wedge (\exists x \in b) R((a)_2, x, \widehat{\top}) \wedge c = \widehat{\top}$,

(17) $Tup_2(a) \wedge (a)_1 = \widehat{\mathsf{e}} \wedge (\forall x \in b) R((a)_2, x, \widehat{\bot}) \wedge c = \widehat{\bot}$,

(18) $a = \widehat{\$} \wedge c = \langle \widehat{\$}, b \rangle$,

(19) $Tup_2(a) \wedge (a)_1 = \widehat{\$} \wedge (\forall x \in b)(R((a)_2, x, \widehat{\top}) \vee R((a)_2, x, \widehat{\bot})) \wedge$
    $\forall x(x \in c \leftrightarrow x \in b \wedge R((a)_2, x, \widehat{\top}))$,

(20) $a = \widehat{\mathbb{R}} \wedge c = \langle \widehat{\mathbb{R}}, b \rangle$,

(21) $Tup_2(a) \wedge (a)_1 = \widehat{\mathbb{R}} \wedge (\forall x \in b)(\exists y \in c)R((a)_2, x, y) \wedge$
$(\forall y \in c)(\exists x \in b)R((a)_2, x, y)$,
(22) $a = \widehat{\mathbb{C}} \wedge R(b, c, \widehat{\top}) \wedge \forall x(x <_W c \rightarrow \neg R(b, x, \widehat{\top})) \wedge \forall x \neg R(\widehat{\mathbb{C}}, b, x)$,
(23) $a = \widehat{\mathbb{P}} \wedge \forall x(x \in c \leftrightarrow x \subset b)$,
(24) $a = \widehat{\mathbf{E}} \wedge \exists x R(b, x, \widehat{\top}) \wedge c = \widehat{\top}$,
(25) $a = \widehat{\mathbf{E}} \wedge \forall x R(b, x, \widehat{\bot}) \wedge c = \widehat{\bot}$.

Clearly, $Q_{\mathfrak{F}}(a, b, c)$ is functional in its third argument. All we have to do now is to follow Section 4.2 again, this time with $Ap_{\mathfrak{A}}[a, b, c]$ replaced by $Q_{\mathfrak{F}}(a, b, c)$. In parallel to Definition 3 an $\mathcal{L}_S$ formula $[\![t]\!]_{\mathfrak{F}}(u)$ is assigned to any $\mathcal{L}^\circ$ term t, saying that u is the value of the term t under the interpretation of the $\mathsf{OST}^r(\mathbf{E}, \mathbb{P})$ and $\mathsf{OST}(\mathbf{E}, \mathbb{P})$ application via $Q_{\mathfrak{F}}$.

Employing these $[\![t]\!]_{\mathfrak{F}}(u)$, each $\mathcal{L}^\circ$ formula A is translated into a formula $A^\Diamond$ of $\mathcal{L}_S$ in the obvious way, simply by following Definition 4. Please keep in mind that A and $A^\Diamond$ are identical in the case that A is an $\mathcal{L}_1$ formula. With the exception of the treatment of operational $\mathbf{E}$, the following interpretation result is as Theorem 6 and proved in Jäger [30].

Theorem 9 *The two theories $\mathsf{OST}^r(\mathbf{E}, \mathbb{P})$ and $\mathsf{OST}(\mathbf{E}, \mathbb{P})$ are interpretable in $\mathsf{E}^r(\mathsf{ZFW})$ and $\mathsf{E}^r(\mathsf{ZFW}) + (\mathcal{L}_S\text{-}\mathsf{I}_\in)$, respectively; i.e. for all formulas A of $\mathcal{L}^\circ$ we have:*

1. $\mathsf{OST}^r(\mathbf{E}, \mathbb{P}) \vdash A \quad \Longrightarrow \quad \mathsf{E}^r(\mathsf{ZFW}) \vdash A^\Diamond$.
2. $\mathsf{OST}(\mathbf{E}, \mathbb{P}) \vdash A \quad \Longrightarrow \quad \mathsf{E}^r(\mathsf{ZFW}) + (\mathcal{L}_S\text{-}\mathsf{I}_\in) \vdash A^\Diamond$.

If we are able to reduce $\mathsf{E}^r(\mathsf{ZFW})$ and $\mathsf{E}^r(\mathsf{ZFW}) + (\mathcal{L}_S\text{-}\mathsf{I}_\in)$ to NBG and $\mathsf{NBG}_{<E_0}$, respectively, the consistency strength of $\mathsf{OST}^r(\mathbf{E}, \mathbb{P})$ and that of $\mathsf{OST}(\mathbf{E}, \mathbb{P})$ are determined. This task is dealt with in Jäger [29, 30] by means of the following conservation result.

Theorem 10 *Let A be a formula of the language $\mathcal{L}_1$. Then we have:*

1. $\mathsf{E}^r(\mathsf{ZFW}) \vdash A \quad \Longrightarrow \quad \mathsf{NBG} \vdash A$.
2. $\mathsf{E}^r(\mathsf{ZFW}) + (\mathcal{L}_S\text{-}\mathsf{I}_\in) \vdash A \quad \Longrightarrow \quad \mathsf{NBG}_{<E_0} \vdash A$.

All together, Theorem 8, Theorem 9 and Theorem 10 complete the analysis of the theories $\mathsf{OST}^r(\mathbf{E}, \mathbb{P})$ and $\mathsf{OST}(\mathbf{E}, \mathbb{P})$.

Corollary 2 *The theory $\mathsf{OST}^r(\mathbf{E}, \mathbb{P})$ is equiconsistent with NBG and ZFC, the theory $\mathsf{OST}(\mathbf{E}, \mathbb{P})$ equiconsistent with $\mathsf{NBG}_{<E_0}$. For all formulas A of the language $\mathcal{L}_1$ we have:*

1. $\mathsf{OST}^r(\mathbf{E}, \mathbb{P}) \vdash A \quad \Longleftrightarrow \quad \mathsf{NBG} \vdash A \quad \Longleftrightarrow \quad \mathsf{ZFC} \vdash A$.
2. $\mathsf{OST}(\mathbf{E}, \mathbb{P}) \vdash A \quad \Longleftrightarrow \quad \mathsf{NBG}_{<E_0} \vdash A$.

Actually, Jäger [29] introduces a theory ZFL_Ω^r which is closely related to $\mathsf{E}^r(\mathsf{ZFW})$. It is shown there that $\mathsf{OST}^r(\mathbf{E}, \mathbb{P})$ can be embedded into ZFL_Ω^r and that ZFL_Ω^r can be reduced to $\mathsf{ZF} + (V = L)$, thus providing a proof of the first assertion of the previous theorem.

Bibliography

[1] Aczel, P. The type theoretic interpretation of constructive set theory. In MacIntyre, A., Pacholski, L., Paris, J. (Eds.), *Logic Colloquium '77*, Studies in Logic and the Foundations of Mathematics, 55–66. North-Holland, 1978.

[2] Aczel, P. The type theoretic interpretation of constructive set theory: choice principles. In Troelstra, A. S., van Dalen, D. (Eds.), *The L.E.J. Brouwer Centenary Symposium*, Studies in Logic and the Foundations of Mathematics, 1–40. North-Holland, 1982.

[3] Aczel, P. The type theoretic interpretation of constructive set theory: inductive definitions. In Marcus, R. B., Dorn, G. J. W., Weingartner, P. (Eds.), *Logic, Methodology, and Philosophy of Science VII*, Studies in Logic and the Foundations of Mathematics, 17–49. North-Holland, 1986.

[4] Aczel, P., Rathjen, M. Notes on constructive set theory. Tech. Rep. 40, Institut Mittag-Leffler, 2001.

[5] Arai, T. Proof theory for theories of ordinals I: recursively Mahlo ordinals. *Annals of Pure and Applied Logic*, 122:1–85, 2003.

[6] Arai, T. Proof theory for theories of ordinals II: Π_3-reflection. *Annals of Pure and Applied Logic*, 129:39–92, 2004.

[7] Barwise, K. J. *Admissible Sets and Structures*. Perspectives in Mathematical Logic. Springer, 1975.

[8] Beeson, M. J. *Foundations of Constructive Mathematics: Metamathematical Studies*. Ergebnisse der Mathematik und ihrer Grenzgebiete. Springer, 1985.

[9] Beeson, M. J. Proving programs and programming proofs. In Marcus, R. B., Dorn, G. J. W., Weingartner, P. (Eds.), *Logic, Methodology, and Philosophy of Science VII*, Studies in Logic and the Foundations of Mathematics, 51–82. North-Holland, 1986.

[10] Beeson, M. J. Towards a computation system based on set theory. *Theoretical Computer Science*, 60:297–340, 1988.

[11] Buchholz, W., Feferman, S., Pohlers, W., Sieg, W. *Iterated Inductive Definitions and Subsystems of Analysis: Recent Proof-Theoretical Studies*, vol. 897 of *Lecture Notes in Mathematics*. Springer, 1981.

[12] Cantini, A., Crosilla, L. Constructive set theory with operations. In Andretta, A., Kearnes, K., Zambella, D. (Eds.), *Logic Colloquium 2004*, vol. 29 of *Lecture Notes in Logic*, 47–83. Cambridge University Press, 2007.

[13] COQ. http://coq.inria.fr/.

[14] Feferman, S. A language and axioms for explicit mathematics. In Crossley, J. N. (Ed.), *Algebra and Logic*, vol. 450 of *Lecture Notes in Mathematics*, 87–139. Springer, 1975.

[15] Feferman, S. Recursion theory and set theory: a marriage of convenience. In Fenstad, J. E., Gandy, R. O., Sacks, G. E. (Eds.), *Generalized Recursion Theory II, Oslo 1977*, Studies in Logic and the Foundations of Mathematics, 55–98. North-Holland, 1978.

[16] Feferman, S. Constructive theories of functions and classes. In Boffa, M., van Dalen, D., McAloon, K. (Eds.), *Logic Colloquium '78*, Studies in Logic and the Foundations of Mathematics, 159–224. North-Holland, 1979.

[17] Feferman, S. Gödel's program for new axioms: Why, where, how and what? In Hájek, P. (Ed.), *Gödel '96*, vol. 6 of *Lecture Notes in Logic*, 3–22. Springer, 1996.

[18] Feferman, S. Notes on operational set theory, I. Generalization of "small" large cardinals in classical and admissible set theory. http://math.stanford.edu/

~feferman/papers/OperationalST-I.pdf, 2001.
[19] Feferman, S. Operational theories of sets and classes, 2005. Draft.
[20] Feferman, S. Operational set theory and small large cardinals. http://math.stanford.edu/~feferman/papers/ostcards.pdf, 2006.
[21] Feferman, S., Jäger, G. Systems of explicit mathematics with non-constructive μ-operator. Part I. *Annals of Pure and Applied Logic*, 65:243–263, 1993.
[22] HOL. http://www.cl.cam.ac.uk/research/hvg/HOL/.
[23] Jäger, G. *Die konstruktible Hierarchie als Hilfsmittel zur beweistheoretischen Untersuchung von Teilsystemen der Mengenlehre und Analysis*. Ph.D. thesis, Mathematisches Institut, Universität München, 1979.
[24] Jäger, G. Iterating admissibility in proof theory. In Stern, J. (Ed.), *Logic Colloquium '81. Proceedings of the Herbrand Symposion*, 137–146. North-Holland, 1982.
[25] Jäger, G. Zur Beweistheorie der Kripke-Platek-Mengenlehre über den natürlichen Zahlen. *Archiv für mathematische Logik und Grundlagenforschung*, 22:121–139, 1982.
[26] Jäger, G. A well-ordering proof for Feferman's theory T_0. *Archiv für mathematische Logik und Grundlagenforschung*, 23:65–77, 1983.
[27] Jäger, G. The strength of admissibility without foundation. *Journal of Symbolic Logic*, 49:867–879, 1984.
[28] Jäger, G. *Theories for Admissible Sets: A Unifying Approach to Proof Theory*, vol. 2 of *Studies in Proof Theory*. Bibliopolis, 1986.
[29] Jäger, G. On Feferman's operational set theory OST. *Annals of Pure and Applied Logic*, 150:19–39, 2007.
[30] Jäger, G. Full operational set theory with unbounded existential quantification and power set. *Annals of Pure and Applied Logic*, to appear.
[31] Jäger, G., Kahle, R., Setzer, A., Strahm, T. The proof-theoretic analysis of transfinitely iterated fixed point theories. *Journal of Symbolic Logic*, 64:53–67, 1999.
[32] Jäger, G., Krähenbühl, J. Σ_1^1 choice in a theory of sets and classes, 2009. Draft.
[33] Jäger, G., Pohlers, W. Eine beweistheoretische Untersuchung von $(\Delta_2^1\text{-CA}) + (\text{BI})$ und verwandter Systeme. *Sitzungsberichte der Bayerischen Akademie der Wissenschaften, Mathematisch-naturwissenschaftliche Klasse*, 1–28, 1982.
[34] Jäger, G., Strahm, T. Upper bounds for metapredicative Mahlo in explicit mathematics and admissible set theory. *Journal of Symbolic Logic*, 66:935–958, 2001.
[35] Jäger, G., Strahm, T. The proof-theoretic strength of the Suslin operator in applicative theories. In Sieg, W., Sommer, R., Talcott, C. (Eds.), *Reflections on the Foundations of Mathematics: Essays in Honor of Solomon Feferman*, vol. 15 of *Lecture Notes in Logic*, 270–292. Association for Symbolic Logic, 2002.
[36] Jäger, G., Strahm, T. Reflections on reflections in explicit mathematics. *Annals of Pure and Applied Logic*, 136:116–133, 2005.
[37] Kunen, K. *Set Theory. An Introduction to Independence Proofs*. Studies in Logic and the Foundations of Mathematics. North-Holland, 1980.
[38] Levy, A. The role of classes in set theory. In Müller, G.-H. (Ed.), *Sets and Classes. On the Work by Paul Bernays*, Studies in Logic and the Foundations of Mathematics, 277–323. North-Holland, 1976.
[39] Martin-Löf, P. An intuitionistic theory of types: predicative part. In Rose, H. E., Shepherdson, J. (Eds.), *Logic Colloquium '73*, Studies in Logic and the Foundations of Mathematics, 73–118. North-Holland, 1975.
[40] Martin-Löf, P. *Intuionistic Type Theory*, vol. 1 of *Studies in Proof Theory*. Bib-

liopolis, 1984.

[41] Mendelson, E. *Introduction to Mathematical Logic*. Chapmann & Hall, 1997 (fourth edition).

[42] Myhill, J. Constructive set theory. *Journal of Symbolic Logic*, 40:347–382, 1975.

[43] Nuprl. http://www.cs.cornell.edu/info/projects/nuprl/.

[44] Rathjen, M. Proof-theoretic analysis of KPM. *Archive for Mathematical Logic*, 30:377–403, 1991.

[45] Rathjen, M. Proof theory of reflection. *Annals of Pure and Applied Logic*, 68:181–224, 1994.

[46] Rathjen, M. An ordinal analysis of stability. *Archive for Mathematical Logic*, 44:1–62, 2005.

[47] Rathjen, M. An ordinal analysis of parameter free Π^1_2-comprehension. *Archive for Mathematical Logic*, 44:263–362, 2005.

[48] Richter, W., Aczel, P. Inductive denitions and reflecting properties of admissible ordinals. In Fenstad, J., P.Hinman (Eds.), *Generalized Recursion Theory*, 301–381. North-Holland, 1974.

[49] Strahm, T. First steps into metapredicativity in explicit mathematics. In Cooper, S. B., Truss, J. (Eds.), *Sets and Proofs*, 383–402. Cambridge University Press, 1999.

[50] Strahm, T. Wellordering proofs for metapredicative mahlo. *Journal of Symbolic Logic*, 67:260–278, 2002.

Set theory and dynamical systems[1]

Alexander S. Kechris

California Institute of Technology

kechris@caltech.edu

ABSTRACT. We give an introduction to a recent direction of research in set theory, developed primarily over the last 15–20 years, and discuss its connections with aspects of dynamical systems and in particular rigidity phenomena in the context of ergodic theory.

1 Introduction

The general context of this work is the development of a theory of complexity of classification problems in mathematics. From another point of view it can be thought of as the study of "definable" or Borel cardinality theory of quotient spaces (vs. the "classical" or Cantor cardinality theory).

Classification Problems. A classification problem is given by:

- A collection of objects X.
- An equivalence relation E on X.

A *complete classification* of X up to E consists of:

- A set of invariants I.
- A map $c : X \to I$ such that $xEy \Leftrightarrow c(x) = c(y)$.

For this to be of any interest both I, c must be as explicit and concrete as possible. Here are some examples of classification problems and their complete invariants:

Example 1 *Classification of finitely generated (f.g.) abelian groups up to isomorphism.*
Invariants: *(Essentially) finite sequences of integers.*

Example 2 *Classification of Bernoulli automorphisms up to conjugacy by entropy* (Ornstein).
Invariants: *Reals.*

Example 3 *Classification of increasing homeomorphisms of $[0,1]$ up to conjugacy.*
Invariants: *(Essentially) countable linear orderings up to isomorphism.*

---

[1] The author is with the Department of Mathematics, California Institute of Technology, Pasadena, CA 91125. Research partially supported by NSF Grant DMS-0455285. I would like to thank Ben Miller for his help in making the figures.

Most often the collection of objects we try to classify can be viewed as forming a "nice" space, namely a standard Borel space, i.e., a Polish (complete separable metric) space with its associated Borel structure and the equivalence relation E turns out to be *Borel* or *analytic* (as a subset of X^2). We will concentrate primarily below on Borel equivalence relations.

The theory of Borel equivalence relations studies the set-theoretic nature of possible (complete) invariants and develops a mathematical framework for measuring the complexity of classification problems.

The following simple concept is basic in organizing this study.

Definition 1 *Let $(X, E), (Y, F)$ be Borel equivalence relations. Then E is (Borel) reducible to F, in symbols*
$$E \leq_B F,$$
if there is Borel map $f : X \to Y$ such that
$$x \mathrel{E} y \Leftrightarrow f(x) \mathrel{F} f(y).$$

The intuitive meaning of this concept can be expressed in two ways:

- The classification problem represented by E is at most as complicated as that of F.
- F-classes are complete invariants for E.

Definition 2 *E is bi-reducible to F if E is reducible to F and vice versa. Let*
$$E \sim_B F \Leftrightarrow E \leq_B F \text{ and } F \leq_B E.$$

We also put:

Definition 3
$$E <_B F \Leftrightarrow E \leq_B F \text{ and } F \not\leq_B E.$$

Let us now discuss the previous and some further examples in the light of this concept.

Example 4 *(Isomorphism of f.g. abelian groups) $\sim_B (=_\mathbb{N})$.*

Example 5 *(Conjugacy of Bernoulli automorphisms) $\sim_B (=_\mathbb{R})$.*

Example 6 *(Isomorphism of torsion-free abelian groups of rank 1) $\sim_B E_0$ (Baer), where E_0 is the equivalence relation on $2^\mathbb{N}$ given by*
$$x \mathrel{E_0} y \Leftrightarrow \exists n \forall m \geq n (x_m = y_m).$$

Example 7 *(Conjugacy of discrete spectrum measure preserving transformations) $\sim_B E_c$ (Halmos-von Neumann), where E_c is the equivalence relation on $\mathbb{T}^\mathbb{N}$ given by*
$$(x_n) \mathrel{E_c} (y_n) \Leftrightarrow \{x_n : n \in \mathbb{N}\} = \{y_n : n \in \mathbb{N}\}.$$

Example 8 *(Conjugacy of increasing homeomorphisms of $[0,1]$) $\sim_B$ (Isomorphism of countable linear orderings).*

Set theory and dynamical systems

Borel cardinality theory. The preceding concepts can be also interpreted as the basis of a "definable" or Borel cardinality theory for quotient spaces.

- $E \leq_B F$ means that there is a Borel injection of X/E into Y/F, i.e., an injection that has a Borel lifting to X, Y. This can be understood as saying that X/E has Borel cardinality less than or equal to that of Y/F, in symbols
$$|X/E|_B \leq |Y/F|_B.$$

- $E \sim_B F$ means that X/E and Y/F have the same Borel cardinality, in symbols
$$|X/E|_B = |Y/F|_B.$$

- $E <_B F$ means that X/E has strictly smaller Borel cardinality than Y/F, in symbols
$$|X/E|_B < |Y/F|_B.$$

2 The Borel reducibility hierarchy

Below X stands for the equality relation on X, $=_X$. We clearly have:

$$1 <_B 2 <_B 3 \cdots <_B \mathbb{N} <_B E$$

and this is an initial segment of the Borel reducibility hierarchy. The first non-trivial result is now the following:

Theorem 1 (Silver [18]) *For every Borel E, either $E \leq_B \mathbb{N}$ or $\mathbb{R} \leq_B E$.*

Thus we have the following continuation of the hierarchy:

$$1 <_B 2 <_B 3 \cdots <_B \mathbb{N} <_B \mathbb{R} <_B E$$

Note that $E \leq_B \mathbb{R}$ means that there is a standard Borel space Y and a Borel map $f : X \to Y$ such that $x \mathrel{E} y \Leftrightarrow f(x) = f(y)$. Such E are called *concretely classifiable* or *smooth*. A canonical example of a non-smooth E is the equivalence relation E_0 defined above. So $\mathbb{R} <_B E_0$.

The next step in the hierarchy is given by the following result:

Theorem 2 (Harrington-Kechris-Louveau [7]) *For any Borel E, either $E \leq_B \mathbb{R}$ or $E_0 \leq_B E$.*

This is called the *General Glimm-Effros Dichotomy* because its first special instances where discovered by Glimm [6] and Effros [3] in connection with work in operator algebras.

Thus we have:

$$1 <_B 2 <_B 3 \cdots <_B \mathbb{N} <_B \mathbb{R} <_B E_0 <_B E$$

and this is an initial segment of the reducibility hierarchy.

The proofs of these two dichotomies, which are about very simple classical concepts of descriptive set theory, i.e., Borel sets and functions, use methods of *effective descriptive set theory*, which are based on computability theory, i.e., the theory of algorithms, Turing machines, etc. No "classical" type proofs are known.

This hierarchy of Borel cardinalities looks so far like the wellordered hierarchy of Cantor cardinalities. However the linearity of $\leq_B$ breaks down after E_0. Various examples have been discovered rather early in this theory. Here are some relatively more recent ones:

Example 9 (Kechris-Louveau [15]) *The following equivalence relations on $\mathbb{R}^{\mathbb{N}}$ are incomparable:*

$$(x_n)\ E_1\ (y_n) \Leftrightarrow \exists n \forall m \geq n (x_m = y_m)$$

$$(x_n)\ E_2\ (y_n) \Leftrightarrow \lim_{n \to \infty} (x_n - y_n) = 0$$

So the picture is as follows:

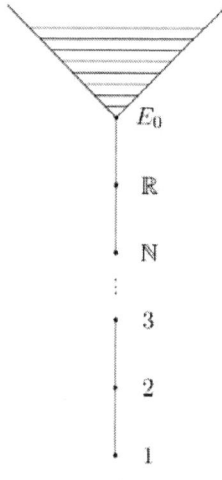

Figure 1:

The Borel equivalence relations above E_0 that have been analyzed so far fall into exactly 4 types and it may be that they all do. This is partially supported by a series of results of Hjorth, Kechris, Louveau and others, see, e.g., Hjorth-Kechris [11] for more detailed explanations. Below we use the following definitions.

Definition 4 *For a Polish group G, Polish space X, and a continuous or Borel action of G on X, we denote by E_G^X the induced (orbit) equivalence relation. (Equivalence relations of the form E_G^X are analytic but not necessarily Borel.)*

Definition 5 *S_∞ is the infinite symmetric group.*

Definition 6 *Γ denotes an arbitrary countable (discrete) group.*

Definition 7

$$(x_n)\ E_0\ (y_n) \Leftrightarrow \exists n \forall m \geq n (x_m = y_m),\ on\ 2^{\mathbb{N}}$$

$$(x_n) \ E_1 \ (y_n) \Leftrightarrow \exists n \forall m \geq n (x_m = y_m), \ on \ \mathbb{R}^\mathbb{N}$$

$$(x_n) \ E_2 \ (y_n) \Leftrightarrow \lim_{n \to \infty} (x_n - y_n) = 0, \ on \ \mathbb{R}^\mathbb{N}$$

$$E_3 = (E_0)^\mathbb{N}, \ on \ (2^\mathbb{N})^\mathbb{N}$$

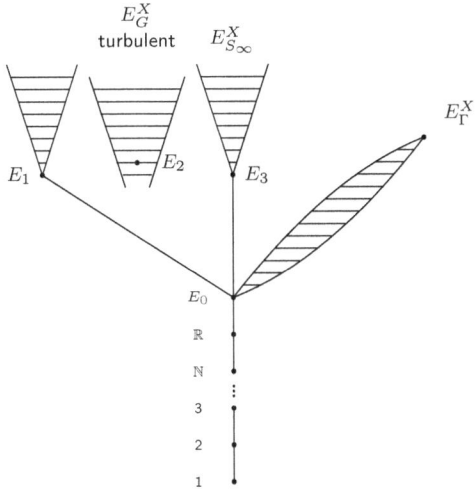

Figure 2:

This figure shows the division of Borel equivalence relations (above E_0) into four types which we label (1)–(4), starting from the left. Class (1) consists of all $E \geq E_1$. All the other classes (2)–(4) consist of E that satisfy $E \leq E_G^X$, for some G, X as in Definition 4. Class (2) consists of all such E that are above, in the sense of $\leq_B$, some particularly complex E_G^X, i.e., those induced by the so-called *turbulent actions*, see Hjorth [8]. Class (3) consists of the equivalence relations E that are $\leq_B E_{S_\infty}^X$ for some action of S_∞ and are also $\geq_B E_3$. Finally, class (4) consists of all E that are $\leq_B E_\Gamma^X$ for some countable group Γ.

We emphasize that this is only a conjectural picture. Although counterexamples may be found that do not belong to any of these classes, the picture helps organize the study of Borel equivalence relations and it may very well be the case that most natural examples fall into one of these four classes.

3 Countable Borel equivalence relations

We will now concentrate on the Borel equivalence relations in the fourth class of Figure 2, i.e., those that are Borel reducible to equivalence relations of the form E_Γ^X, for some countable group Γ. The latter are also called *countable* in view of what follows.

Definition 8 *E is* countable *if every E-class is countable.*

Example 10 *Any equivalence relation, E_Γ^X, induced by a Borel action of a countable group Γ on X.*

We actually have:

Theorem 3 (Feldman-Moore [4]) *Every countable Borel equivalence relation E is of the form E_Γ^X.*

A totally different example of a countable Borel equivalence relation familiar to logicians is the following:

Example 11 *Turing equivalence.*

There are also many Borel equivalence relations, which although not literally countable, fall in this domain, since they are Borel bireducible to countable ones. Here are some representative examples.

Example 12 (Kechris [14]) E_G^X *for G a second countable locally compact group (e.g., a Lie group).*

Example 13 (Hjorth-Kechris [9]) *Isomorphism of countable structures that are of "finite type", e.g., finitely generated groups, locally finite trees, finite rank torsion-free abelian groups, finite transcendence degree fields, etc.*

Example 14 (Hjorth-Kechris [10]) *Conformal equivalence of Riemann surfaces*

We will now consider the structure of $\leq_B$ on the countable Borel equivalence relations. We refer the reader to Dougherty–Jackson–Kechris [2] and Jackson–Kechris–Louveau [13] for more information.

The simplest countable equivalence relations are the smooth ones, which have a trivial structure. The next more complicated ones are the so-called hyperfinite.

Definition 9 *E is hyperfinite if $E = \bigcup_n E_n$, with E_n Borel, increasing and finite (i.e., having equivalence classes that are finite).*

Theorem 4 (Slaman-Steel [19], Weiss [22]) *E is hyperfinite iff it is of the form $E_\mathbb{Z}^X$.*

Which groups always give hyperfinite equivalence relations? A necessary condition is that they have to be amenable (i.e., admit a left-invariant finitely additive probability measure). The following asks whether the converse is true.

Problem 1 (Weiss [22]) *If Γ is amenable, is E_Γ^X hyperfinite?*

For finitely generated groups, the next result is essentially all that is known so far.

Theorem 5 (Jackson–Kechris–Louveau [13]) *If Γ is finitely generated of polynomial growth, then E_Γ^X is hyperfinite.*

Very recently, Gao-Jackson [5] proved that this is also true for any abelian group Γ.

The hyperfinite equivalence relations have been classified both under bireducibilty and isomorphism.

Theorem 6 (Dougherty–Jackson–Kechris [2]) *i) Up to Borel bireducibility, there is only one non-smooth, hyperfinite equivalence relation, namely E_0.*

ii) Up to Borel isomorphism, there are exactly countably many non-smooth, aperiodic (i.e., having infinite classes), hyperfinite equivalence relations, namely

$$E_t, E_0, 2E_0, 3E_0, \ldots, nE_0, \aleph_0 E_0, E_s.$$

Here nE_0 is the direct sum of n copies of E_0 for $1 \leq n \leq \aleph_0$, E_t is the *tail equivalence relation* on $2^\mathbb{N}$, i.e., $x E_t y \Leftrightarrow \exists n \exists m \forall k (x_{n+k} = y_{m+k})$ and E_s is the aperiodic part of the *shift equivalence relation* on $2^\mathbb{Z}$.

The hyperfinite equivalence relations are the simplest non-trivial countable equivalence relations. At the other end are the most complex ones, the so-called *universal* ones.

Theorem 7 (see [13]) *There is a universal countable Borel equivalence relation, E_∞, i.e., one that satisfies $E \leq_B E_\infty$, for all countable E.*

Example 15 $E_\infty \sim_B$ *(the shift equivalence relation on 2^{F_2}).*

Here F_n is the free group with n generators. The following shows that E_∞ is not hyperfinite.

Theorem 8 (see [13])
$$E_0 <_B E_\infty.$$

There are countable equivalence relations that are neither hyperfinite nor universal.

Theorem 9 (see [13]) *There exist intermediate countable Borel equivalence relations E, i.e.,*
$$E_0 <_B E <_B E_\infty.$$

Example 16 $E = $ *(the free part of the shift equivalence relation on 2^{F_2}).*

This is a typical example of a *treeable* equivalence relation. These were first studied by S. Adams in ergodic theory.

Since the early 1990's only a small finite number of intermediate equivalence relations were known and they were linearly ordered under $\leq_B$. This lead to the following basic problems:

- Are there infinitely many?
- Does non-linearity occur here?

These were answered by the following result.

Theorem 10 (Adams-Kechris [1]) *Every Borel partial order embeds into $\leq_B$ on the countable equivalence relations.*

In Figure 3 we give a schematic picture of the structure of countable Borel equivalence relations. On the left side of the figure we have listed some representative examples of classification problems whose complexity is measured by equivalence relations in this domain. For example, the isomorphism problem for finitely generated groups is bireducible to E_∞ (Thomas–Velickovic [21]).

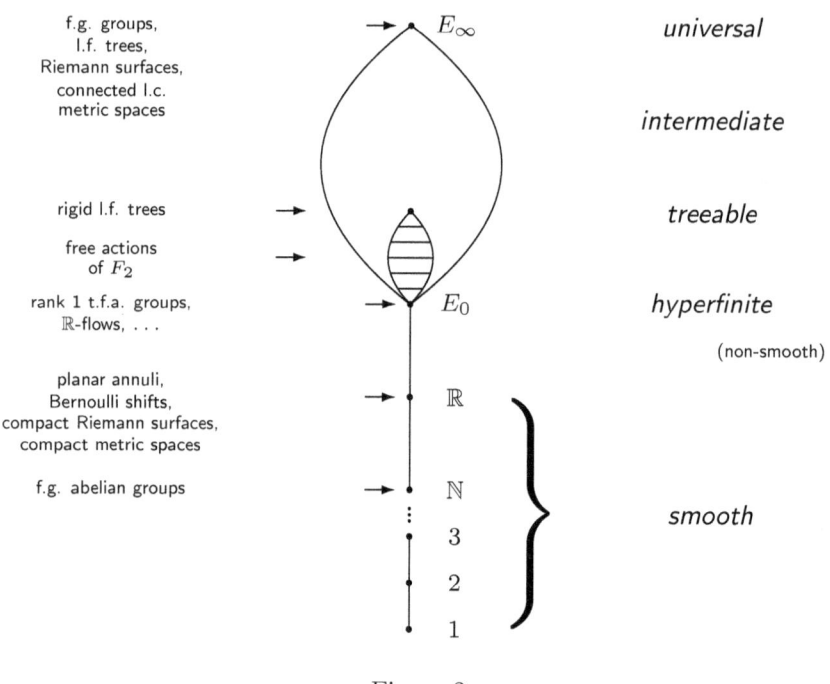

Figure 3:

4 Set theoretic rigidity

The proof of the preceding theorem of Adams-Kechris used *Zimmer's cocycle superrigidity theory* for ergodic actions of linear algebraic groups and their lattices, see Zimmer [23].

The key point is that there is a phenomenon of *set theoretic rigidity* analogous to the *measure theoretic rigidity* phenomenon discovered by Zimmer.

- **(Measure theoretic rigidity)** Under certain circumstances, when a countable group acts preserving a probability measure, the equivalence relation associated with the action together with the measure "encode" or "remember" a lot about the group (and the action).
- **(Set theoretic rigidity)** Such information is simply encoded in the Borel cardinality of the (quotient) orbit space.

To illustrate this, let us mention some set theoretic rigidity results.

Set theory and dynamical systems

Theorem 11 (Adams-Kechris [1])

$$|\mathbb{T}^m/\mathrm{GL}_m(\mathbb{Z})|_B = |\mathbb{T}^n/\mathrm{GL}_n(\mathbb{Z})|_B \Leftrightarrow m = n.$$

Here $\mathrm{GL}_m(\mathbb{Z})$ is the group of $n \times n$ matrices in $\mathbb{Z}$ with determinant ± 1. It acts in the obvious way on $\mathbb{T}^n$. This result shows that the dimension is coded in the Borel cardinality of the quotient space. It also implies the existence of infinitely many distinct up to $\sim_B$ countable Borel equivalence relations.

Below $\Gamma_p = \mathrm{SO}_7(\mathbb{Z}[1/p])$, p prime. Also E_p is the free part of the shift equivalence relation on 2^{Γ_p}.

Theorem 12 (Adams-Kechris [1])

$$E_p \leq_B E_q \Leftrightarrow p = q.$$

In particular this shows that there are infinitely many incomparable under $\leq_B$ countable Borel equivalence relations.

Below let $\cong_n$ be isomorphism of torsion-free abelian groups of rank at most n, i.e., subgroups of $(\mathbb{Q}^n, +)$. This can be seen to be (up to $\sim_B$) a countable Borel equivalence relation.

Theorem 13 (S. Thomas, [20])

$$(\cong_m) \sim_B (\cong_n) \Leftrightarrow m = n.$$

Thus the rank is encoded in the Borel cardinality of the isomorphism types. This result has important implications for the classical classification problem for finite rank torsion-free abelian groups.

Recently Hjorth-Kechris [12] developed a set theoretic rigidity theory for product groups that has several applications in the study of countable Borel equivalence relations – but also in ergodic theory. They also use ergodic theoretic methods, like cocycle reduction techniques, actions on boundaries, etc. (Also, independently, Monod-Shalom [16] and Popa [17] have recently proved important rigidity results for product groups in the context of ergodic theory – it is yet unclear what is the relationship between these theories.)

Here are a few results from the work of Hjorth-Kechris. Below, for any group Γ, we let E_Γ be the free part of the shift equivalence relation on 2^Γ.

Theorem 14 (Hjorth–Kechris [12])

$$E_{(\mathbb{Z}_p \star \mathbb{Z}_p) \times \mathbb{Z}} \leq_B E_{(\mathbb{Z}_q \star \mathbb{Z}_q) \times \mathbb{Z}} \Leftrightarrow p = q.$$

Note however that:

$$E_{(\mathbb{Z}_p \star \mathbb{Z}_p)} \sim_B E_{(\mathbb{Z}_q \star \mathbb{Z}_q)}.$$

The next result concerns the distinction between the equivalence relation $E_{F_2^n}$ induced by the shift action of the product of n copies of F_2 (*shift of the product*) and the product equivalence relation of n copies of the shift action of F_2, i.e., $(E_{F_2})^n$ (*product of the shift*). It can be best summarized in a picture (see Fig.4).

Finally an application to ergodic theory.

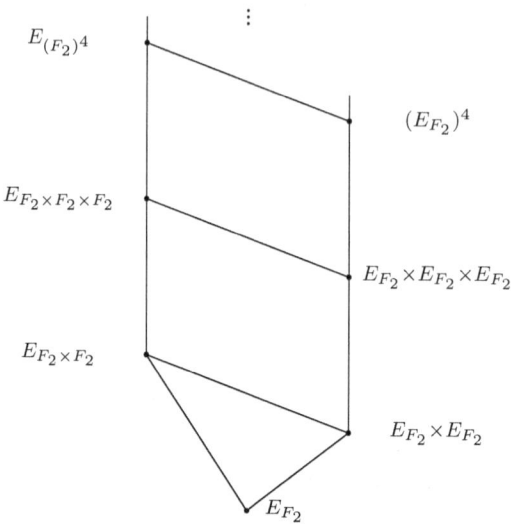

Figure 4:

Theorem 15 (Hjorth–Kechris [12]) *Suppose H_0, H_1 are non-amenable, torsion-free, hyperbolic groups and Δ_0, Δ_1 are infinite amenable groups. Let each $H_i \times \Delta_i$ act freely on X_i with invariant, probability measure, so that the action is ergodic on Δ_i, $i = 1, 2$. If the action of $H_0 \times \Delta_0$ is (stably) orbit equivalent to the action of $H_1 \times \Delta_1$, then $H_0 \cong H_1$.*

We conclude with the following remark about methodology. The theory of countable Borel equivalence relations points to an interesting phenomenon. Although one is dealing here with very simple set theoretic notions (countable Borel equivalence relations and Borel reducibility) most basic questions about them (like existence of intermediate or incomparable ones) have been answered by using rather sophisticated ergodic theory methods, and this certainly represents an interesting application of ergodic theory to set theory. At this time no other methods to study these problems are known.

Bibliography

[1] Adams, S., Kechris, A. Linear algebraic groups and countable Borel equivalence relations. *J. Amer. Math. Soc.*, 13(4):909–943, 2000.

[2] Dougherty, R., Jackson, S., Kechris, A. The structure of hyperfinite Borel equivalence relations. *Trans. Amer. Math. Soc.*, 341:193–225, 1994.

[3] Effros, E. Transformation groups and c^*-algebras. *Ann of Math*, 81:38–55, 1965.

[4] Feldman, J., Moore, C. Ergodic equivalence relations and von Neumann algebras, I. *Trans. Amer. Math. Soc.*, 234:289–324, 1977.

[5] Gao, S., Jackson, S. *Countable abelian group actions and hyperfinite equivalence relations*. 2007, preprint.

[6] Glimm, J. Locally compact transformation groups. *Trans. Amer. Math. Soc.*, 101:124–138, 1961.

[7] Harrington, L., Kechris, A., Louveau, A. A Glimm-Effros dichotomy for Borel equivalence relations. *Journal of the Amer. Math. Soc.*, 3(4):903–928, 1990.

[8] Hjorth, G. Classification and Orbit Equivalence Relations. *Math. Surveys and Monographs,* (Amer. Math. Soc.), 75, 2000.

[9] Hjorth, G., Kechris, A. Borel equivalence relations and classifications of countable models. *Ann. Pure Appl. Logic*, 82:221—272, 1996.

[10] Hjorth, G., Kechris, A. The complexity of the classification of Riemann surfaces and complex manifolds. *Illinois J. Math*, 44:104–137, 2000.

[11] Hjorth, G., Kechris, A. Recent developments in the theory of Borel reducibility. *Fund. Math*, 170:21–52, 2001.

[12] Hjorth, G., Kechris, A. Rigidity theorems for actions of product groups and countable Borel equivalence relations. *Men. Amer. Math. Soc.*, 177(833), 2005.

[13] Jackson, S., Kechris, A., Louveau, A. Countable Borel equivalence relations. *J. Math. Logic*, 2(1):1–80, 2002.

[14] Kechris, A. Countable sections for locally compact group actions. *Ergodic Theory Dynam. Systems*, 13:283–295, 1992.

[15] Kechris, A., Louveau, A. The classification of hypersmooth Borel equivalence relations. *J. Amer. Math. Soc.*, 10:215–242, 1997.

[16] Monod, N., Shalom, Y. Orbit equivalence rigidity and bounded cohomology. *Ann. of Math*, 164(3):825–878, 2006.

[17] Popa, S. Deformation and rigidity for group actions and von Neumann algebras. In *International Congress of Mathematicians*, vol. I, 445–477. Eur. Math. Soc., 2007.

[18] Silver, J. Counting the numbers of equivalence classes of Borel and coanalytic equivalence relations. *Ann. Math. Logic*, 18:1–28, 1980.

[19] Slaman, J., Steel, J. Definable functions on degrees. In *Cabal Seminar 81–85, Lecture Notes in Math*, vol. 1333, 37–55. Springer-Verlag, 1988.

[20] Thomas. The classification problem for torsion-free abelian groups of finite rank. *J. Amer. Math. Soc.*, 16(1):233–258, 2003.

[21] Thomas, S., Velickovic, B. On the complexity of the isomorphism relation for finitely generated groups. *J. Algebra*, 217(1):352–373, 1999.

[22] Weiss, B. Measurable dynamics. *Contemp. Math*, 26:395–421, 1984.

[23] Zimmer, R. *Ergodic Theory and Semisimple Groups*. Birkhäuser, 1984.

Monadic theories of wellorders[1]

Itay Neeman
University of California Los Angeles
ineeman@math.ucla.edu

ABSTRACT. The article is a partial survey of connections between the monadic second order theory of wellorders and finite state automata. Results presented include equivalences of automata and monadic formulas on finite ordinals, on ω, on countable ordinals, on ordinals below ω_2, and finally on all ordinals uniformly. Other results presented include applications of the equivalences to monadic decidability and definability over wellorders.

This article is a partial survey of work on the monadic second order theory of wellorders, concentrating on connections with finite state automata. We present a progression of results, starting with the case of finite wellorders and ending with a general connection between monadic truth and automata on all ordinals. We give proofs and proof sketches at the initial levels, to illustrate some of the ideas in the work connecting automata and monadic truth. At higher levels the proofs are substantially more complicated and beyond the scope of this article. Our exposition follows the most direct mathematical route, and should not be taken as either a complete or an historical account. We refer the reader to Gurevich [7] for a survey on monadic theories, to Khoussainov–Nerode [9] for a comprehensive account of fundamental results on finite state automata, and to the papers by Vardi and Thomas in this volume for specific applications of automata theory in computer science.

Recall that the monadic second order language, monadic language for short below, has two kinds of variables: first order variables which range over elements of the structure, and second order variables which range over subsets of the structure. The atomic formulas in the monadic language are the usual first order atomic formulas (in first order variables), and formulas of the form $v \in U$ where v is a first order variable and U is a second order variable. General formulas are built from atomic formulas using negations, conjunctions, and existential quantifications over both first and second order variables. In many cases the monadic language provides a nice balance of expressivity and feasibility. Feasibility here is a vague term, and can mean many different things, for example that the corresponding theory is decidable, that the theory can be described in terms of a theory in a more limited language, or that definability can be described. It is in proving these kinds of feasibility that we make use of finite state automata.

We are concerned specifically with theories of wellorders, namely of structures of the form $(\alpha; <)$, where $<$ is a wellorder of the set α. Without loss of generality

---

[1] The author is with Department of Mathematics, University of California Los Angeles, Los Angeles, CA 90095-1555. This material is based upon work supported by the National Science Foundation under Grant No. DMS-0556223.

Monadic theories of wellorders

we may assume that α is an ordinal, and $<$ is the membership relation restricted to α.

1 Finite ordinals

Consider to begin with the case of finite α. This case serves as a simple illustration of the connection between the monadic theory and finite state automata.

The basic core of a finite state automaton with (finite) alphabet Σ is a finite *set of states* S, a smaller set $I \subseteq S$ of *initial states*, and a *transition table* $T \subseteq S \times \Sigma \times S$. The automaton takes as input a string $X \colon \alpha \to \Sigma$. A *run* of the automaton on X is a string of states $s \colon \alpha + 1 \to S$ which satisfies the rules

$$s(0) \in I, \text{ and} \qquad \text{(Initial)}$$
$$\langle s(\xi), X(\xi), s(\xi+1) \rangle \in T \qquad \text{(Succ)}$$

for all $\xi < \alpha$.

The automaton is *deterministic* if I is a singleton and the transition table is the graph of a function from $S \times \Sigma$ into S, meaning that for each $b \in S$ and $\sigma \in \Sigma$ there is a unique $b^* \in S$ so that $\langle b, \sigma, b^* \rangle \in T$. Abusing notion we then refer to I as a state and to T as a function. In the case of a deterministic automaton, for finite α at least, conditions (Initial) and (Succ) determine a unique run of the automaton on X. The run is produced by setting $s(0) = I$ and then successively setting $s(n+1) = T(s(n), X(n))$. Non-deterministic automata in contrast may have many runs on an input X, and may also have none.

In addition to the basic core, the automaton has a set F of *accepting final states*. A run $s \colon \alpha + 1 \to S$ is *accepting* if $s(\alpha) \in F$. The automaton *accepts* input $X \colon \alpha \to \Sigma$ just in case that there is an accepting run of the automaton on X. Note the existential quantifier that is built into the definition. We shall make good use of it with a non-deterministic automaton soon. But first let us quickly describe a coding of elements and subsets of α by strings which may be taken as inputs for automata.

For a set $A \subseteq \alpha$ define $\chi_A \colon \alpha \to 4$ by $\chi_A(\xi) = 1$ if $\xi \in A$ and $\chi_A(\xi) = 0$ otherwise. For an ordinal $a \in \alpha$ define $\chi_a \colon \alpha \to 4$ by $\chi_a(\xi) = 3$ if $\xi = a$ and $\chi_a(\xi) = 2$ otherwise. For a tuple $\langle e_1, \ldots, e_k \rangle$ with each e_i either an element of α or a subset of α, define $\chi_{\langle e_1, \ldots, e_k \rangle} \colon \alpha \to {}^k 4$ by $\chi_{\langle e_1, \ldots, e_k \rangle}(\xi) = \langle \chi_{e_1}(\xi), \ldots, \chi_{e_k}(\xi) \rangle$. $\chi_{\langle e_1, \ldots, e_k \rangle}$ is then a string of length α in the alphabet $\Sigma = {}^k 4$, and codes the tuple $\langle e_1, \ldots, e_k \rangle$. The domain α is suppressed in the notation, and is typically understood from the context.

The coding above lets us view tuples of elements and subsets of α as possible inputs for automata. We say that an automaton with alphabet $\Sigma = {}^k 4$ *accepts* the tuple $\langle e_1, \ldots, e_k \rangle$ iff it accepts $\chi_{\langle e_1, \ldots, e_k \rangle}$.

An automaton $\mathcal{A}$ is *equivalent* to a monadic formula $\varphi(v_1, \ldots, v_k)$ on structure $(\alpha; <)$ just in case that for every tuple $\langle e_1, \ldots, e_k \rangle$ of elements and subsets of α which match the orders of variables of φ, $\mathcal{A}$ accepts $\langle e_1, \ldots, e_k \rangle$ iff $(\alpha; <) \models \varphi[e_1, \ldots, e_k]$.

Theorem 1 *For every monadic formula φ, there is a deterministic automaton $\mathcal{A}$ which is equivalent to φ on all structures $(\alpha; <)$ with α finite.*

The theorem is part of a large body of work analyzing finite state automata and regular languages. Its proof given below is very direct. For a more complete account which includes the related work we refer the reader to Khoussainov–Nerode [9, Chapter 2].

Proof The proof is by induction on the complexity of φ.

If φ is atomic then it is easy to explicitly define an automaton witnessing the theorem. Let us only go over one example, the formula $v_1 \in v_2$, with v_1 a first order variable and v_2 a second order variable. The following automaton is equivalent to this formula on finite structures: The automaton has three states, true, false, and unknown. The initial state is unknown, and the only accepting state is true. $T(\text{unknown}, \langle 3, 2 \rangle) = \text{false}$, so that if a ξ is reached so that $\xi = e_1$ and $\xi \notin e_2$ the automaton falls into the state false. $T(\text{unknown}, \langle 3, 1 \rangle) = \text{true}$, so that if a ξ is reached so that $\xi = e_1$ and $\xi \in e_2$ the automaton falls into the state true. $T(\text{unknown}, \sigma) = \text{unknown}$ for all other σ, and $T(\text{false}, \sigma) = \text{false}$ and $T(\text{true}, \sigma) = \text{true}$ for all σ.

If φ is a negation $\neg \psi$, take an automaton $\bar{\mathcal{A}}$ witnessing the theorem for ψ, and define the automaton $\mathcal{A}$ to have the same set of states, the same transition table, the same initial state, and the inverse set of final states, namely $F = S - \bar{F}$. Then $\mathcal{A}$ witnesses the theorem for φ. Notice that it is important here that we are dealing with deterministic automata, so that every input string leads to a final state uniquely determined by the string. $\bar{\mathcal{A}}$ accepts if this final state belongs to $\bar{F}$, and $\mathcal{A}$ accepts if it does not.

If φ is a conjunction $\psi_1 \wedge \psi_2$, take automata $\mathcal{A}_1$ and $\mathcal{A}_2$ witnessing the theorem for ψ_1 and ψ_2, and define an automaton $\mathcal{A}$ which simulates a simultaneous run of $\mathcal{A}_1$ and $\mathcal{A}_2$. The set of states S of $\mathcal{A}$ is $S_1 \times S_2$, the transition function T is defined by $T(\langle b_1, b_2 \rangle, \sigma) = \langle T(b_1, \sigma), T(b_2, \sigma) \rangle$, the initial state I is $\langle I_1, I_2 \rangle$, and the set of final states F is $F_1 \times F_2$. It is clear that $\mathcal{A}$ accepts X iff both $\mathcal{A}_1$ and $\mathcal{A}_2$ accept X.

Suppose finally that φ is an existential formula $(\exists v_k) \psi(v_1, \ldots, v_k)$. Let $\bar{\mathcal{A}}$ witness the theorem for ψ. It is easy, modifying $\bar{\mathcal{A}}$, to define a *non-deterministic* automaton $\mathcal{A}^{nd}$ which is equivalent to φ. The automaton $\mathcal{A}^{nd}$ uses non-determinism to guess the characteristic function of v_k. Suppose for definiteness that v_k is second order. Define S^{nd} to be $\bar{S} \times \{0, 1\}$ where $\bar{S}$ is the set of states of $\bar{\mathcal{A}}$. Set $\langle \langle b, i \rangle, \sigma, \langle b^*, i^* \rangle \rangle \in T^{nd}$ just in case that $T(b, \sigma^\frown \langle i \rangle) = b^*$. The definition is such that if s^{nd} is a run of $\mathcal{A}^{nd}$ on $\langle e_1, \ldots, e_{k-1} \rangle$, then $s^{nd}(\xi)$ has the form $\langle \bar{s}(\xi), i_\xi \rangle$ where, setting $e_k = \{\xi < \alpha \mid i_\xi = 1\}$, $\bar{s}$ is a run of the original automaton $\bar{\mathcal{A}}$ on $\langle e_1, \ldots, e_{k-1}, e_k \rangle$. In that sense the part $\langle i_\xi \mid \xi < \alpha \rangle$ of a run of $\mathcal{A}^{nd}$ is a guess by the non-deterministic automaton for a characteristic function of a set that can be substituted for v_k.

Continuing to define $\mathcal{A}^{nd}$, set the initial states to be $\langle \bar{I}, 0 \rangle$ and $\langle \bar{I}, 1 \rangle$, and let the set of final states F^{nd} be $\bar{F} \times \{0, 1\}$. Recall that existential quantification over runs was built into the definition of acceptance for automata. Inspecting that definition and the definition of $\mathcal{A}^{nd}$ it is easy to check that the non-deterministic $\mathcal{A}^{nd}$ accepts $\langle e_1, \ldots, e_{k-1} \rangle$ iff there exists $e_k \subseteq \alpha$ so that $\bar{\mathcal{A}}$ accepts $\langle e_1, \ldots, e_{k-1}, e_k \rangle$, namely iff there exists $e_k \subseteq \alpha$ so that $(\alpha; <) \models \psi[e_1, \ldots, e_{k-1}, e_k]$. It follows that $\mathcal{A}^{nd}$ is equivalent to $\varphi = (\exists v_k) \psi$.

Of course $\mathcal{A}^{nd}$ is not deterministic. To complete the proof of the theorem we have to convert it to a deterministic automaton, and this can be done using:

Lemma 1 (Rabin–Scott [17]) *Every non-deterministic automaton is equivalent to a deterministic automaton on finite domains. Precisely, for every non-deterministic automaton $\mathcal{A}^{nd}$ there is a deterministic automaton $\mathcal{A}$, so that for every input string X of finite length, $\mathcal{A}$ accepts X iff $\mathcal{A}^{nd}$ accepts X.*

Proof Runs of $\mathcal{A}$ keep track of all possible states which may be reached by runs of $\mathcal{A}^{nd}$, from each initial state. More precisely, states of $\mathcal{A}$ are subsets W of $I^{nd} \times S^{nd}$, the initial state I is the set $\{\langle b_0, b_0 \rangle \mid b_0 \in I^{nd}\}$, and the transition function T is defined by $T(W, \sigma) = \{\langle b_0, b^* \rangle \in I^{nd} \times S^{nd} \mid (\exists b \in S^{nd})(\langle b_0, b \rangle \in W \wedge \langle b, \sigma, b^* \rangle \in T^{nd})\}$. With this definition it follows by induction on α that, if s is a run of $\mathcal{A}$ on an input string X of finite length α, then $\langle b_0, b \rangle \in s(\alpha)$ iff there is a run s^{nd} of $\mathcal{A}^{nd}$ on X with $s^{nd}(0) = b_0$ and $s^{nd}(\alpha) = b$. Setting $F = \{W \subseteq I^{nd} \times S^{nd} \mid W \cap (I^{nd} \times F^{nd}) \neq \emptyset\}$ it then easy to check that $\mathcal{A}$ accepts X iff $\mathcal{A}^{nd}$ accepts X. This completes the proofs of Lemma 1 and Theorem 1. ♠

Theorem 1 is constructive, and gives rise to a recursive map $\varphi \mapsto \mathcal{A}_\varphi$ which assigns to each monadic formula φ an equivalent automaton $\mathcal{A}_\varphi$. Already at the level of finite domains this association can be used to prove decidability results, for example:

Corollary 1 *The set of monadic sentences φ so that $(\exists \alpha < \omega)(\alpha; <) \models \varphi$ is decidable.*

For a stronger result, on the decidability of the fragment of the monadic theory of ω involving only finite sets, see Büchi [1] and Elgot [6].

Proof Fix a sentence φ. We describe how to decide whether or not $(\exists \alpha < \omega)(\alpha; <) \models \varphi$. Consider the automaton $\mathcal{A}_\varphi$. Since φ has no free variables, the alphabet of this automaton is $^04 = \{\emptyset\}$. Its transition function T_φ may therefore be viewed simply as a directed graph. The vertices are states, and the graph has an edge from b to b^* iff the automaton transitions from b to b^*, namely if $T_\varphi(b, \emptyset) = b^*$. The automaton accepts the (unique) input string of length α iff the graph has a path of length α from I_φ to a vertex in F_φ. So $(\exists \alpha < \omega)(\alpha; <) \models \varphi$ iff there is a vertex in F_φ which is reachable from I_φ. The graph is finite, and the question of reachability in finite graphs is decidable. ♠

2 Countable ordinals

Büchi [2] discovered that there is a parallel of Theorem 1 to $\alpha = \omega$. Let $\mathcal{A}$ be a finite state automaton with alphabet Σ. Let $X \colon \omega \to \Sigma$ be an input string of length ω. Conditions (Initial) and (Succ) give rise to a notion of runs of the automaton on X, but of length ω rather than $\omega + 1$. A run is a sequence $s \colon \omega \to S$ which satisfies condition (Initial), and satisfies condition (Succ) for each $\xi < \omega$. Since the run does not provide a final state $s(\omega)$, the notion of acceptance requires an additional definition. Büchi equipped each of his automata with a set G of states, which we call *good states*, and defined a run s to be accepting iff $\{\xi < \omega \mid s(\xi) \in G\}$ is infinite. He then proved:

Theorem 2 (Büchi) *For every monadic formula φ, there is a (non deterministic) Büchi automaton which is equivalent to φ on domain $\alpha = \omega$.*

Proof Again the proof is by induction on the complexity of φ. The cases of atomic φ and of conjunctions are similar to the corresponding cases in the proof of Theorem 1. Since the automata in Theorem 2 are non-deterministic, the case of existential quantification is easy, similar to the corresponding case in the proof of Theorem 1 but without the need to prove the equivalence in Lemma 1. (This equivalence fails for Büchi automata.) It is the case of negations which is difficult. The proof in this case makes a clever use of Ramsey's theorem.

Say $\varphi = \neg\psi$. Let $\bar{\mathcal{A}}$ be a Büchi automaton witnessing the theorem for ψ, consisting of a set of states $\bar{S}$, a set of initial states $\bar{I}$, a transition table $\bar{T}$, and a set of good states $\bar{G}$. Suppose $X\colon \omega \to \Sigma$ is an input string for $\bar{\mathcal{A}}$. We have to define $\mathcal{A}$ (independently of X) so that $\mathcal{A}$ accepts X iff $\bar{\mathcal{A}}$ does not.

For $n < m < \omega$ let $C_X(n,m)$ be the set of pairs $\langle b, b^*\rangle \in \bar{S} \times \bar{S}$ so that $\bar{\mathcal{A}}$ can get from state b at n to state b^* at m. Precisely, $\langle b, b^*\rangle \in C_X(n,m)$ if there is a sequence $s\colon [n,m] \to \bar{S}$ which satisfies condition (Succ) for $\xi \in [n,m)$, with $s(n) = b$ and $s(m) = b^*$. Let $C_X^g(n,m)$ be the set of pairs $\langle b, b^*\rangle$ so that $\bar{\mathcal{A}}$ can get from state b at n to state b^* at m, with the additional requirement of passing through the set of good states $\bar{G}$. Precisely, $\langle b, b^*\rangle \in C_X^g(n,m)$ if there is s as above with the added requirement that $s(k) \in \bar{G}$ for some $k \in [n,m)$.

C_X and C_X^g are functions from ω^2 into the finite set $\mathcal{P}(\bar{S} \times \bar{S})$. By applications of Ramsey's theorem there is an infinite set $H \subseteq \omega$, and *fixed* D, E, and E^g, so that $C_X(0,n) = D$, $C_X(n,m) = E$, and $C_X^g(n,m) = E^g$ for all $n,m \in H$.

Note that knowledge of D and E^g suffices to determine whether $\bar{\mathcal{A}}$ has an accepting run on X. Such a run exists iff there are states $b_0 \in \bar{I}$ and $b \in \bar{S}$ so that $\langle b_0, b\rangle \in D$, and $\langle b, b\rangle \in E^g$. (Given such b_0 and b one can construct an accepting run $\bar{s}$ of $\bar{\mathcal{A}}$ on X with $\bar{s}(0) = b_0$ and $\bar{s}(n) = b$ for $n \in H$. Conversely, given an accepting run $\bar{s}$, set $b_0 = \bar{s}(0)$ and set b equal to any state which $\bar{s}$ repeats infinitely many time on the infinite set H.)

Let J be the set of all triples $\langle D, E, E^g\rangle \in \mathcal{P}(\bar{S} \times \bar{S})^3$ so that $\langle b_0, b\rangle$ as above do *not* exist. Define a non-deterministic automaton $\mathcal{A}$ so that a run of $\mathcal{A}$ on X does the following: (a) guess, in the very first state, a triple $\langle D, E, E^g\rangle \in J$; (b) guess, during the entire infinite run, a characteristic function χ_H of a set $H \subseteq \omega$; and (c) verify that $C_X(0,n) = D$, $C_X(n,m) = E$, and $C_X^g(n,m) = E^g$ for all $n < m$ both in H. If all three condition can be achieved for input X with the set H *infinite*, then X is not accepted by the original automaton $\bar{\mathcal{A}}$, and vice versa. This follows from the conclusion of the previous paragraph. Let G, the set of good states for the new automaton $\mathcal{A}$, be the set of states at which $\mathcal{A}$ guesses value 1 for χ_H, so that a run of $\mathcal{A}$ is accepting iff the set H it guesses is infinite. Then $\mathcal{A}$ accepts X iff $\bar{\mathcal{A}}$ does not, completing the proof.

As for the actual construction of the automaton $\mathcal{A}$, conditions (a) and (b) are simple, and an automaton whose runs verify the part $C_X(0,n) = D$ in condition (c) can be defined using ideas similar to those in the proof of Lemma 1. The verification that $C_X(n,m) = E$ and $C_X^g(n,m) = E^g$ for all $n < m$ both in H must be done indirectly, since a finite state automaton cannot at stage m keep track of $C_X(n,m)$ and $C_X^g(n,m)$ for unboundedly many $n < m$. One defines the automaton to only verify the simpler requirement that $C_X(n,m) = E$ and $C_X^g(n,m) = E^g$ for n equal to the immediate predecessor of m in H, and adds the initial demand that the guess of E and E^g for condition (a) must satisfy compositional properties which give the full requirement from the simpler one, and conversely follow from the full requirement. For example, the initial guess must satisfy $(\exists b^*)(\langle b, b^*\rangle \in E \wedge \langle b^*, b^{**}\rangle \in E)$ iff $\langle b, b^{**}\rangle \in E$. It is the use of the compositional properties that forces us to involve E and C_X in the definition of the automaton, as the properties for E^g rely on E. ♠

With Theorem 2 at hand, Büchi obtained the following Corollary. Its proof is similar to that of Corollary 1, relying on Theorem 2 instead of Theorem 1.

Corollary 2 (Büchi) *The monadic theory of $(\omega; <)$ is decidable.*

The result can be extended to all countable ordinals. But first let us pass to a class of automata which is more flexible already in the case of inputs of length ω. With the more flexible class we will be able to recover the equivalence between deterministic and non-deterministic automata. The equivalence fails in the case of Büchi automata, and in fact Theorem 2 would fail if "Büchi automaton" were replaced by "deterministic Büchi automaton."

Given $s\colon \gamma \to S$, with γ a limit ordinal, define $\mathrm{cf}(s)$ to be the set of states which occur cofinally along s. Precisely, $b \in \mathrm{cf}(s)$ iff $(\forall \xi < \gamma)(\exists \delta)(\xi < \delta < \gamma \wedge s(\delta) = b)$. $\mathrm{cf}(s)$ is an element of $\mathcal{P}(S)$.

A finite state automaton with a *countable-limit condition* consists of the usual core, S, I, and T as before, and a function $\Psi_{ctbl}\colon \mathcal{P}(S) \to S$. A *run* of the automaton on input $X\colon \alpha \to \Sigma$ of countable length α is a sequence $s\colon \alpha + 1 \to S$ which satisfies conditions (Initial), (Succ) for all $\xi < \alpha$, and

$$s(\gamma) = \Psi_{ctbl}(\mathrm{cf}(s\restriction \gamma)) \qquad \text{(LimCtbl)}$$

for all limit ordinals $\gamma \leq \alpha$.

As usual the automaton is *deterministic* if I is a singleton and the transition table is the graph of a function. In the case of a deterministic automaton, for countable input length α, conditions (Initial), (Succ), and (LimCtbl) determine a unique run of the automaton on X. The run is produced by setting $s(0) = I$, using condition (Succ) to uniquely determine $s(\xi + 1)$ for successor ordinals $\xi + 1 \leq \alpha$, and using condition (LimCtbl) to uniquely determine $s(\gamma)$ for limit ordinals $\gamma \leq \alpha$.

Reverting to the initial approach to acceptance, we equip the automaton with a set $F \subseteq S$ of accepting final states. A run s on input $X\colon \alpha \to \Sigma$ is *accepting* if $s(\alpha) \in F$, and the automaton *accepts* X if it has an accepting run on X.

Acceptance in the case of $\alpha = \omega$ is determined via a table on the basis of $\mathrm{cf}(s\restriction \omega)$; accepting runs are those with $\mathrm{cf}(s\restriction \omega) \in \Psi^{-1}(F)$. This method is due to Muller. A Muller automaton consists of the usual core, S, I, and T, and an *acceptance table* $B \subseteq \mathcal{P}(S)$. A run of the automaton on input $X\colon \omega \to \Sigma$ is a sequence $s\colon \omega \to \Sigma$ satisfying conditions (Initial) and (Succ) for $n < \omega$. The run is accepting iff $\mathrm{cf}(s) \in B$.

It is clear that every Büchi automaton is equivalent to a Muller automaton. It is not hard to see that the converse is also true, for the non-deterministic case. But deterministic Muller automata are more expressive than deterministic Büchi automata. Indeed, the more flexible acceptance condition in Muller automata allows a determinising construction, due to McNaughton [13], producing a deterministic automaton equivalent to a given non-deterministic automaton.

Theorem 3 (McNaughton) *Every Muller automaton is equivalent to a deterministic Muller automaton.*

The proof is very intricate, substantially more intricate than the proofs given above. We refer the reader to Khoussainov–Nerode [9, §3.8]. The determinism construction has an element of uniformity that is not present in Büchi's original proof of decidability of the monadic theory of $(\omega; <)$. That uniformity allowed Büchi [3] to generalize the construction to all countable domains:

Theorem 4 (Büchi) *Every automaton with countable-limit condition is equivalent to a deterministic automaton with countable-limit condition on countable domains. (The equivalent automaton is independent of the domain.)*

Using Theorem 4 one can very directly imitate the proof of Theorem 1 and obtain:

Theorem 5 (Büchi) *For every monadic formula φ, there is a deterministic automaton with a countable-limit condition, $\mathcal{A}$, which is equivalent to φ on countable domains.*

Corollary 3 (Büchi) *The set of monadic sentences φ so that $(\exists \alpha < \omega_1)$ $(\alpha; <) \models \varphi$ is decidable.*

Proof Similar to the proof of Corollary 3, but this time searching for *generalized paths* in the directed graph of the automaton. A generalized path is a sequence of vertices $\{b_i\}$ so that for each i either (1) the graph has an edge from b_i to b_{i+1}; or (2) $b_i = b_k$ for $k \leq i$ (this gives rise to a *loop*) and $b_{i+1} = \Psi_{ctbl}(\{b_k, \ldots, b_i\})$. The second condition corresponds to a use of condition (LimCtbl), which allows the automaton to reach state b_{i+1} if it had generated a limit-length sequence of states that repeats $\{b_k, \ldots, b_i\}$. ♠

With a more careful analysis one can prove further that for each (non-zero) countable ordinal α, the monadic theory of $(\alpha; <)$ is decidable, and depends only on the remainder obtained when dividing α by ω^ω.

3 The first and second uncountable cardinals

When dealing with monadic theories of uncountable ordinals one has to take into account notions from set theory including clubs, stationarity, and to some extent cofinality.

Recall that an ordinal β is a *limit point* of a set of ordinals A if $A \cap \beta$ is unbounded in β, meaning that $(\forall \xi < \beta)(\exists \delta)(\xi < \delta < \beta \wedge \delta \in A)$. A set $C \subseteq \kappa$ is *closed unbounded* in an ordinal κ, club in κ or simply club for short, if: (a) C is unbounded in κ, and (b) C is closed in κ, meaning that every limit point $\beta < \kappa$ of C is an element of C.

A function on ordinals $f \colon \tau \to \kappa$ is *cofinal* in κ if its range $\{f(\xi) \mid \xi < \tau\}$ is unbounded in κ. The *cofinality* of κ, denoted $\mathrm{cof}(\kappa)$, is the smallest ordinal τ so that there is a function $f \colon \tau \to \kappa$ cofinal in κ. An ordinal τ is *regular* if $\mathrm{cof}(\tau) = \tau$. Regular ordinals are in fact cardinals, and if $\tau = \mathrm{cof}(\kappa)$ then τ is regular. Thus the cofinality of an ordinal κ is always a cardinal.

For κ of uncountable cofinality, any two club subsets of κ have non empty, and in fact club, intersection. For such κ we say that a set $A \subseteq \kappa$ is *stationary* in κ if it meets—meaning it has non-empty intersection with—every club subset of κ. The notion is non trivial as every club in κ is stationary. Using the axiom of choice every stationary set can be split into two disjoint sets which are both stationary, so there are stationary sets which are not club.

For $\tau \geq \omega_1$, if κ has cofinality $\geq \tau$ then every club in κ has points of all cofinalities $< \tau$, and vice versa. In the next paragraph we write $(*)$ to refer to this equivalence.

The notions limit point, club, and stationary are clearly expressible in the monadic language. Cofinality need not be expressible in the monadic language, since its definition uses functions. But using the equivalence $(*)$, and the fact that $\text{cof}(\kappa) \geq \omega_0$ iff κ is a limit ordinal, it is easy to define by recursion, for finite n, monadic formulas $\varphi_{\text{cof} \geq \omega_n}$ so that $(\text{On}; <) \models \varphi_{\text{cof} \geq \omega_n}[\kappa]$ iff $\text{cof}(\kappa) \geq \omega_n$.

Work with a cardinal κ of cofinality $\geq \omega_1$. For sets $A, B \subseteq \kappa$ define $A \sim B$ iff A and B are equal on a club, meaning that there is a club $C \subseteq \kappa$ so that $A \cap C = B \cap C$. Since the intersection of two club subsets of κ is itself club, $\sim$ is an equivalence relation. We write $[A]$ to denote the equivalence class of A. Let $\mathcal{B}_\kappa$ be the set of equivalence classes of $\sim$. The basic operations on sets, union, intersection, and difference, extend naturally to operations on the equivalence classes. Abusing notation slightly we use the same symbols to denote these operations on classes, writing for example $[A] \cap [B] = [C]$ if $A \cap B = C$. The structure $(\mathcal{B}_\kappa; \cup, \cap, -, \emptyset, \kappa)$ is a Boolean algebra, and since the notions used to define it are all expressible in the monadic language, the first order theory of $(\mathcal{B}_\kappa; \cup, \cap, -, \emptyset, \kappa)$ is computable from the monadic theory of $(\kappa; <)$.

In fact much more may be computed from the monadic theory. For $A \subseteq \kappa$ define $R(A)$ to be the set of $\alpha < \kappa$ so that $A \cap \alpha$ is stationary in α. Such α are *reflection points* of A. R is trivial in the case of $\kappa = \omega_1$, but its behavior is highly non-trivial, and indeed independent of ZFC, already at $\kappa = \omega_2$. R extends to act on $\sim$ equivalence classes, and it is clear that the theory of $(\mathcal{B}_\kappa; R, \cup, \cap, -, \emptyset, \kappa)$ is also computable from the monadic theory of $(\kappa; <)$.

Shelah [18] used intricate model theoretic arguments to provide converses to these observations. He showed that the monadic theory of $(\omega_1; <)$ can be reduced to the first order theory of $(\mathcal{B}_\kappa; \cup, \cap, -, \emptyset, \kappa)$, and the monadic theory of $(\omega_2; <)$ can be reduced to the first order theory of $(\mathcal{B}_\kappa; R, \cup, \cap, -, \emptyset, \kappa)$. His results are more general, and reduce the monadic theories of higher cardinals κ to first order theories of structures that extend $(\mathcal{B}_\kappa; \cup, \cap, -, \emptyset, \kappa)$ with operations additional to R, collecting more operations as κ increases.

$(\mathcal{B}_{\omega_1}; \cup, \cap, -, \emptyset, \omega_1)$ is an atomless Boolean algebra and its theory is decidable. Thus it follows from Shelah's reduction that the monadic theory of $(\omega_1; <)$ is decidable. This had been proved previously by Büchi [4], using automata that he defined acting on sequences of length ω_1. In later work Büchi–Zaiontz [5] proved the following theorem, and in fact characterized completely the monadic theories of ordinals below ω_2.

Theorem 6 (Büchi–Zaiontz) *For every $\alpha < \omega_2$, the monadic theory of $(\alpha; <)$ is decidable.*

At ω_2 matters change drastically. Shelah's reduction shows that the monadic theory of $(\omega_2; <)$ and the first order theory of $(\mathcal{B}_{\omega_2}; R, \cup, \cap, -, \emptyset, \omega_2)$ are each computable from the other. But the complexity of the first order theory of $(\mathcal{B}_{\omega_2}; R, \cup, \cap, -, \emptyset, \omega_2)$ is independent from ZFC.

It is helpful to divide ω_2 into two parts, C_0 and C_1, with C_i consisting of the ordinals $\alpha < \omega_2$ of cofinality ω_i. Both are stationary. Subsets of C_1 do not reflect, so in analyzing the operation R on $\mathcal{B}_{\omega_2}$ we are concerned only with its behavior on subsets of C_0.

Assuming mild large cardinals Magidor [12] showed it is consistent to have $R(A) = C_1$ (modulo the equivalence relation $\sim$) for every $A \subseteq C_0$. In this case R is trivial on $\mathcal{B}_{\omega_2}$, and the first order theory of $(\mathcal{B}_{\omega_2}; R, \cup, \cap, -, \emptyset, \omega_2)$ is decidable.

There are other behaviors of R that result in a decidable theory. Shelah [19] shows, assuming just the consistency of ZFC, that it is consistent that for every $A \subseteq C_0$ and every stationary $\hat{B}, \hat{C}$ with $\hat{B} \cup \hat{C} = R(A)$, there are stationary $B, C \subseteq A$ so that $R(B) = \hat{B}$ and $R(C) = \hat{C}$. (Again equality is modulo $\sim$.) From this principle too it follows that the theory of $(\mathcal{B}_{\omega_2}; R, \cup, \cap, -, \emptyset, \omega_2)$ is decidable, though R is not trivial.

On the other hand Gurevich–Magidor–Shelah [8] construct models where the theory of $(\mathcal{B}_{\omega_2}; R, \cup, \cap, -, \emptyset, \omega_2)$ has arbitrary Turing degree, assuming mild large cardinals. Lifsches–Shelah [10] construct models, assuming just the consistency of ZFC, where the theory is arbitrarily complicated.

In short, the theory of $(\mathcal{B}_{\omega_2}; R, \cup, \cap, -, \emptyset, \omega_2)$, and equivalently the monadic theory of $(\omega_2; <)$, cannot be determined from ZFC, and:

Theorem 7 *(Shelah, Gurevich–Magidor–Shelah, Lifsches–Shelah) The decidability of the monadic theory of $(\omega_2; <)$ is independent of* ZFC.

Reflection principles may affect not just decidability, but also definability. Using the monadic formulas $\varphi_{\text{cof} \geq \omega_n}$ defined above it is easy to see that each of the cardinals ω_n, for finite n, is definable by a monadic formula over $(\text{On}; <)$: ω_n is the least ordinal of cofinality $\geq \omega_n$. Magidor [12] construct from large cardinals a model in which ω_{n+1} is also definable. In Magidor's model every stationary subset of $\omega_{\omega+1}$ reflects. This universal reflection fails on the cardinals ω_n (the set of points of cofinality ω_{n-1} does not reflect), and since the property is expressible in the monadic language, one can write a monadic formula which in Magidor's model defines $\omega_{\omega+1}$. Then using the equivalence $(*)$ above and a recursive definition similar to the definition of the formulas $\varphi_{\text{cof} \geq \omega_n}$, it follows that $\omega_{\omega+1+n}$ is definable for each finite n.

One can build on this argument to construct models where other regular cardinals are definable. Note that the argument skips the singular cardinal ω_ω. We shall see in the next section that singular cardinals can never be defined by a monadic formula over $(\text{On}; <)$.

4 Automata capturing monadic truth on all domains

The *almost-all* language, which we shall use below, is obtained from the monadic language by removing the first order quantifiers, and adding instead the quantifiers $(\forall^* \xi)$ and $(\forall^* \xi < \delta)$. The semantics of the first quantifier is given by the following condition: $(\alpha; <) \models (\forall^* \xi) \varphi(\xi)$ iff

1. $\mathrm{cof}(\alpha) \geq \omega_1$; and
2. There is a club $C \subseteq \alpha$ so that $(\alpha; <) \models \varphi[\xi]$ for all $\xi \in C$.

(In writing the condition we suppressed the instantiated variables of $(\forall^*\xi)\varphi(\xi)$, for notational simplicity.) The semantics of the second quantifier is given by a similar condition requiring that $\mathrm{cof}(\delta) \geq \omega_1$, and that $(\alpha; <) \models \varphi[\xi]$ on a club $C \subseteq \delta$.

We saw already that the properties of having uncountable cofinality and of being club can be expressed in the monadic langauge. It follows from this that the almost-all language is a fragment of the monadic language. The almost-all language is strictly less expressive than the monadic language. For example, the truth value of an almost-all formula about sets $A_1, \ldots, A_k \subseteq \alpha$ depends only on the restriction of these sets to a club in α. In other words the language cannot distinguish between tuples $\langle A_1, \ldots, A_k \rangle$ and $\langle A'_1, \ldots, A'_k \rangle$ provided $A_i = A'_i$ on a club. In particular the almost-all language cannot express equality on sets. (In the full monadic language equality can be expressed: $A = B$ iff $(\forall \xi)(\xi \in A \leftrightarrow \xi \in B)$.) Nonetheless it is expressive enough to capture an important essence of the monadic truth. For example, the results in the previous section may be viewed as reducing full monadic truth on ω_1 and ω_2 to truth in the almost-all language, and then reasoning about the decidability or undecidability of this almost-all truth.

More uniformly, almost-all truth can be used as a foundation for decision making at limit stages, in a class of automata that capture monadic truth on all domains. In this section we define the class, and see how it is connected to monadic truth on ordinals.

We begin with some preliminary definitions. The notation $f \colon \alpha \rightharpoonup S$ indicates that f is a *partial* function from α into S. Assuming S is finite the function can be coded by a tuple of subsets of α as follows. Let $b_0, \ldots, b_{n-1}$ be the elements of S, and let $A_i = \{\xi < \alpha \mid f(\xi) = b_i\}$ for $i < n$. Then $\langle A_0, \ldots, A_{n-1} \rangle$ codes f. When we write $\varphi(\ldots, f, \ldots)$ below, with φ an almost-all formula and $f \colon \alpha \rightharpoonup S$, we mean $\varphi(\ldots, A_0, \ldots, A_{n-1}, \ldots)$.

Define $\mathrm{cd}(f) \colon \alpha \rightharpoonup \mathcal{P}(S)$, the *cofinal-state derivative* of f, by $\mathrm{cd}(f)(\gamma) = \mathrm{cf}(f \restriction \gamma)$ for each limit ordinal $\gamma < \alpha$, and $\mathrm{cd}(f)(\gamma)$ undefined otherwise. $\mathrm{cd}(f)$ too can be coded by a tuple $\langle D_0, \ldots, D_{n-1} \rangle$ of n subsets of α, setting $\gamma \in D_i$ iff $b_i \in \mathrm{cd}(f)(\gamma)$ iff $b_i \in \mathrm{cf}(f \restriction \gamma)$. When we write $\varphi(\ldots, \mathrm{cd}(f), \ldots)$, with φ an almost-all formula and $f \colon \alpha \rightharpoonup S$, we mean $\varphi(\ldots, D_0, \ldots, D_{i-1}, \ldots)$.

Given a finite sequence of formulas $\vec{\psi} = \langle \psi_0, \ldots, \psi_{l-1} \rangle$ in the almost-all language, and functions $s \colon \alpha \to S$ and $r \colon \alpha \rightharpoonup S$, define $\mathrm{Tf}_{\vec{\psi}}(s, r)$, the $\vec{\psi}$ *fragment of the almost-all truth table of s, $\mathrm{cd}(s)$, and r*, to be the set of $i < l$ so that $(\alpha; <) \models \psi_i[s, \mathrm{cd}(s), r]$. This is the restriction of the set of almost-all formulas which are true of $\langle s, \mathrm{cd}(s), r \rangle$, to the finite fragment specified by $\vec{\psi}$. Given further a function $\Psi \colon \mathcal{P}(l) \to S$, define $(\Psi \oplus \vec{\psi})(s, r)$ to be $\Psi(\mathrm{Tf}_{\vec{\psi}}(s, r))$.

We are ready now to define our final class of finite state automata. We work as usual with a an alphabet Σ. A finite state automaton with *full-limit condition*, $\mathcal{A}$, consists of the objects in conditions (A1)–(A4) below. The objects in conditions (A1) and (A2) are similar to objects we have seen before, for the basic core of an automaton, and for countable limits which in our context will generalize to all limits of countable cofinality. The objects in condition (A3) will be used to determine the state $s(\lambda)$ for limit λ of uncountable cofinality. The objects in condition (A4) will be used to determine the extra component r of a run of $\mathcal{A}$.

A1. A finite set of states S, a set of initial states $I \subseteq S$, and a transition table $T \subseteq S \times \Sigma \times S$.

A2. A lower-limit function $\Psi_{lo} \colon \mathcal{P}(S) \to S$.

A3. A finite sequence of formulas $\vec{\psi} = \langle \psi_0, \ldots, \psi_{l-1} \rangle$ in the almost-all language, and a higher-limit function $\Psi_{hi} \colon \mathcal{P}(l) \to S$.

A4. A finite set P of *pebbles*, and two functions which will be used in placing and removing pebbles, $u \colon S \to \{U \mid U \subsetneq P\}$, and $h \colon S \to P$ with $h(b) \in P - u(b)$ for all $b \in S$.

A *run* of the automaton on input $X \colon \alpha \to \Sigma$, is a pair of functions $s \colon \alpha+1 \to S$ and $r \colon \alpha \to S$ satisfying the following conditions. The first three we have seen before, and they will now apply to the initial state, successor states, and all limit states of countable cofinality. The fourth condition will apply to limits of uncountable cofinality, and the fifth will determine the additional component r of the run. We shall say more about these rules below.

$$s(0) \in I \qquad \text{(Initial)}$$
$$\langle s(\xi), X(\xi), s(\xi+1) \rangle \in T \qquad \text{(Succ)}$$
$$s(\lambda) = \Psi_{lo}(\mathrm{cf}(s \upharpoonright \lambda)) \qquad \text{(LimCtbl)}$$
$$s(\lambda) = (\Psi_{hi} \oplus \vec{\psi})(s \upharpoonright \lambda, r \upharpoonright \lambda) \qquad \text{(LimHi)}$$

If there exists some $\gamma > \xi$ so that $h(s(\xi)) \notin u(s(\gamma))$ then $r(\xi) = s(\gamma)$ for the least such γ, and otherwise $r(\xi)$ is undefined. (Peb)

As usual the automaton is deterministic if I is a singleton and T is the graph of a function. It is conceptually easier to explain how the conditions above govern the behavior of a deterministic automaton, so we do this first, and comment on the natural extension to non-deterministic automata later.

A deterministic automaton $\mathcal{A}$ should be viewed as running over input $X \colon \alpha \to \Sigma$ and producing a run $\langle s, r \rangle$ through a transfinite sequence of stages. In the initial stage the automaton sets $s(0)$ equal to the unique element of I. In each subsequent stage β the automaton determines $s(\beta)$ through one of the conditions (Succ), (LimCtbl), and (LimHi), depending on whether β is a successor, a limit of countable cofinality, or a limit of uncountable cofinality. If β is a successor, say $\xi+1$, then the automaton sets $s(\xi+1) = T(s(\xi), X(\xi))$, determining the state $s(\xi+1)$ on the basis of the state $s(\xi)$ and input $X(\xi)$, as usual. If β is a limit ordinal of countable cofinality, then the automaton sets $s(\beta) = \Psi_{lo}(\mathrm{cf}(s \upharpoonright \beta))$, determining $s(\beta)$ on the basis of the cofinal set of the run $s \upharpoonright \beta$ produced so far. Finally, if β is a limit of uncountable cofinality then the automaton sets $s(\beta) = (\Psi_{hi} \oplus \vec{\psi})(s \upharpoonright \beta, r \upharpoonright \beta)$, determining $s(\beta)$ on the basis of a fragment of the almost-all theory of the run $s \upharpoonright \beta$ produced so far, its cofinal-state derivative $\mathrm{cd}(s \upharpoonright \beta)$, and the auxiliary sequence $r \upharpoonright \beta$. The sequence $\vec{\psi}$ defines the window of formulas to be consulted, and Ψ_{hi} converts the resulting fragment of the theory into a state.

The auxiliary sequence r is determined using the objects P, u, and h in condition (A4), subject to condition (Peb). We think of P as a finite set of pebbles. The functions h and u are used to place and remove pebbles as follows. Having determined the state $s(\beta)$, the automaton places a pebble $p = h(s(\beta))$ on the ordinal β. The pebble p remains on β until a later stage β^* is reached with $p \notin u(s(\beta^*))$. At the first such stage β^* the automaton removes the pebble from β, and sets $r(\beta) = s(\beta^*)$. This is expressed precisely in condition (Peb). $r(\beta)$ remains undefined until the pebble placed on β is removed, and may indeed remain undefined throughout, if the pebble is not removed at all during the run. The use of pebbles therefore introduces a delay into part of the construction of a run. This delay is essential in the proofs of Theorems 8 and 9 below.

The value of $r \restriction \beta$ known by stage β, call it $(r \restriction \beta)^{\text{local}}$, is not the same as the final value $r \restriction \beta$ known by the end of the run, after stage α, as there may be ordinals $\xi < \beta$ so that the pebble $h(s(\xi))$ placed on ξ is removed at a stage $\gamma \geq \beta$. But there may only be finitely many such ordinals, since the number of pebbles is finite and since no pebble is ever located on two ordinals at the same stage (to see this use the restriction $h(b) \notin u(b)$ in condition (A4)). Thus $(r \restriction \beta)^{\text{local}}$ and $r \restriction \beta$ may only differ on a finite set.

When reaching a limit stage β the automaton looks at the value of $r \restriction \beta$ known by stage β, setting $s(\beta)$ equal to $(\Psi_{hi} \oplus \vec{\psi})(s \restriction \beta, (r \restriction \beta)^{\text{local}})$. This assignment satisfies condition (LimHi) since $(r \restriction \beta)^{\text{local}}$ and $r \restriction \beta$ differ only on a finite set, and the almost-all theory cannot distinguish such a difference.

Runs of non-deterministic automata are governed by conditions (LimCtbl), (LimHi), and (Peb) in exactly the way described above. Deterministic and non-deterministic automata differ only in the initial and successor stages, where conditions (Initial) and (Succ) require a non-deterministic automaton to make a choice.

As usual we equip the automaton with a set F of *accepting final states*. A run $\langle s, r \rangle$ on input $X \colon \alpha \to \Sigma$ is *accepting* iff its last state $s(\alpha)$ belongs to F. As usual the automaton *accepts* X iff there is an accepting run of the automaton on X.

The main result connecting this class of automata to monadic truth, and the purpose behind the definition of the class, is the following theorem:

Theorem 8 (Neeman) *For every monadic formula φ, there is a deterministic automaton with a full-limit condition, $\mathcal{A}$, which is equivalent to φ on all ordinal domains.*

The theorem is a corollary to the following result on determinism, in much the same way that Theorem 1 is a corollary to Lemma 1.

Theorem 9 (Neeman) *Every automaton with full-limit condition is equivalent to a deterministic automaton with full-limit-condition. The equivalence holds on all ordinal domains, and the equivalent automaton is independent of the domain.*

Theorem 9 extends the work of Büchi [3] and Büchi–Zaiontz [5] to automata acting on inputs of lengths ω_2 and greater. The specific details of the definition of automata with full-limit condition above are of course important to the proof of the theorem. The proof, and the uses of the various aspects of the definition, are beyond the scope of this paper. We refer the reader to Neeman [16].

Since automata may refer to the almost-all theory of the run constructed to set limit states, Theorem 8 may be viewed as reducing questions about the monadic theory to questions about the almost-all theory. This kind of reduction can also be made using model theoretic techniques on individual cardinals, see

Shelah [18], with increasing complexity as one moves to higher cardinals. What makes Theorem 8 particularly useful is its uniformity. To each monadic formula φ it assigns a single deterministic automaton which performs the reduction to the almost-all theory on *all* domains. This uniformity, and properties of the almost-all theories of ordinals, for example the fact that the almost-all theory of $(\alpha; <)$ depends only on the cofinality of α, allow deriving the following result from Theorem 8.

Theorem 10 (Neeman) *No singular cardinal is definable over* $(\text{On}; <)$ *by a monadic formula.*

The result was extended in Neeman [14], again using Theorem 8, to show further:

Theorem 11 (Neeman) *An ordinal is definable over* $(\text{On}; <)$ *by a monadic formula iff it can be obtained, using ordinal addition and multiplication, from regular cardinals which are definable over* $(\text{On}; <)$ *by monadic formulas.*

The uniform reduction from monadic theory to almost-all theory given by Theorem 8 may also help obtain results on the monadic theory of ordinals in future forcing extensions which manipulate the almost-all theory of all regular cardinals. At the moment though we only know how to force useful almost-all theories (useful for the purpose of connections with monadic theories) at low cardinals. For example Neeman [15] forces an almost-all theory making the monadic theory of $(\omega_3; <)$ decidable. The model involved satisfies $2^{\omega_2} = \omega_4$, and the construction does not generalize to higher cardinals even if one were willing to let 2^{ω_2} rise further.

Much more remains to be discovered on the monadic theory of ordinals. Is it consistent that for every ordinal α the monadic theory of α is decidable? Is it consistent that $\aleph_{\omega+1}$ is not definable by a monadic formula over $(\text{On}; <)$? By Theorem 8, both questions are in fact questions about the almost-all theories of the mentioned ordinals.

Theorem 8 uses the axiom of choice. In fact already at the level of ω_1 known proofs of decidability of the monadic theory use some fragment of the axiom of choice, see Litman [11]. What happens when this fragment of the axiom of choice fails? In particular, is the monadic theory of $(\omega_1; <)$ decidable under the axiom of determinacy? Very little is known that may help with this question.

Second order theories are typically very complicated. But within second order theories the monadic theories should be relatively manageable, and it is not unreasonable to hope that further research should shed light on the questions above.

Bibliography

[1] Büchi, J. R. Weak second-order arithmetic and finite automata. *Z. Math. Logik Grundlagen Math.*, 6:66–92, 1960.

[2] Büchi, J. R. On a decision method in restricted second order arithmetic. In *Logic, Methodology and Philosophy of Science (Proc. 1960 Internat. Congr.)*, 1–11. Stanford, Calif.: Stanford Univ. Press, 1962.

[3] Büchi, J. R. Decision methods in the theory of ordinals. *Bull. Amer. Math. Soc.*, 71:767–770, 1965.

[4] Büchi, J. R. The monadic second order theory of ω_1. In *The monadic second order theory of all countable ordinals (Decidable theories, II)*, 1–127. Lecture Notes in Math., Vol. 328. Berlin: Springer, 1973.

[5] Büchi, J. R., Zaiontz, C. Deterministic automata and the monadic theory of ordinals $< \omega_2$. *Z. Math. Logik Grundlag. Math.*, 29(4):313–336, 1983. ISSN 0044-3050.

[6] Elgot, C. C. Decision problems of finite automata design and related arithmetics. *Trans. Amer. Math. Soc.*, 98:21–51, 1961. ISSN 0002-9947.

[7] Gurevich, Y. Monadic second-order theories. In *Model-theoretic logics*, Perspect. Math. Logic, 479–506. New York: Springer, 1985.

[8] Gurevich, Y., Magidor, M., Shelah, S. The monadic theory of ω_2. *J. Symbolic Logic*, 48(2):387–398, 1983. ISSN 0022-4812.

[9] Khoussainov, B., Nerode, A. *Automata theory and its applications*, vol. 21 of *Progress in Computer Science and Applied Logic*. Boston, MA: Birkhäuser Boston Inc., 2001. ISBN 0-8176-4207-2.

[10] Lifsches, S., Shelah, S. The monadic theory of $(\omega_2, <)$ may be complicated. *Arch. Math. Logic*, 31(3):207–213, 1992. ISSN 0933-5846.

[11] Litman, A. On the monadic theory of ω_1 without A.C. *Israel J. Math.*, 23(3-4):251–266, 1976. ISSN 0021-2172.

[12] Magidor, M. Reflecting stationary sets. *J. Symbolic Logic*, 47(4):755–771 (1983), 1982. ISSN 0022-4812.

[13] McNaughton, R. Testing and generating infinite sequences by a finite automaton. *Information and Control*, 9:521–530, 1966. ISSN 0890-5401.

[14] Neeman, I. Monadic definability of ordinals. To appear, Computational Prospects of Infinity, Part II: Presented Talks, Vol. 15, Lecture Notes Series, Institute of Mathematical Sciences, National University of Singapore.

[15] Neeman, I. Monadic theory of ω_3. To appear.

[16] Neeman, I. Finite state automata and monadic definability of singular cardinals. *Journal of Symbolic Logic*, 73:412–438, 2008.

[17] Rabin, M. O., Scott, D. Finite automata and their decision problems. *IBM J. Res. Develop.*, 3:114–125, 1959. ISSN 0018-8646.

[18] Shelah, S. The monadic theory of order. *Ann. of Math. (2)*, 102(3):379–419, 1975. ISSN 0003-486X.

[19] Shelah, S. A weak generalization of MA to higher cardinals. *Israel J. Math.*, 30(4):297–306, 1978. ISSN 0021-2172.

On the hypothetical judgement in the history of intuitionistic logic [1]

Mark van Atten

IHPST(CNRS/Paris 1/ENS)

Mark.vanAtten@Univ-Paris1.fr

ABSTRACT. The claim will be defended here that, unlike the name BHK-interpretation suggests, there never was one single understanding of logical connectives that Brouwer, Heyting and Kolmogorov all agreed on. The argument is centered around the hypothetical judgement in general and Ex Falso in particular. Brouwer's logic turns out to be a relevance logic. Some brief comments are made on positions of C.I. Lewis, Freudenthal, Griss, Becker, Johansson, Troelstra/Van Dalen, Martin-Löf, and Dummett.

1 Introduction

The BHK-interpretation[2] of the propositional fragment of intuitionistic logic, as presented by Troelstra and Van Dalen [34, p.9], is given by the following clauses:

(H1) A proof of $A \wedge B$ is given by presenting a proof of A and a proof of B.

(H2) A proof of $A \vee B$ is given by presenting either a proof of A or a proof of B (plus the stipulation that we want to regard the proof presented as evidence for $A \vee B$).

(H3) A proof of $A \rightarrow B$ is a construction which permits us to transform any proof of A into a proof of B.

(H4) Absurdity $\bot$ (contradiction) has no proof; a proof of $\neg A$ is a construction which transforms any hypothetical proof of A into a proof of a contradiction.

Depending on how one understands the notions employed here such as those of construction and of the presentation of a proof, these clauses can be used to interpret classical logic just as well [34, p.9,pp.32–33]. Of course, intuitionists understand these notions constructively. Troelstra and Van Dalen comment that

---

[1] IHPST (CNRS/Paris 1/ENS), 13 rue du Four, 75006 Paris. This is a somewhat reorganised version of the talk presented at the LMPS. Earlier versions were presented in 2005 at meetings in Paris, Lille, Jerusalem and Vienna. I thank the organisers of these meetings and of the LMPS in Beijing for inviting me, and the audiences for their patience, questions, and comments. In addition, I thank Dirk van Dalen and Göran Sundholm for our many discussions of the topic; van Dalen's paper is a reply to his paper on Brouwer and Kolmogorov [36].

[2] BHK for 'Brouwer, Heyting, Kolmogorov'; formerly also 'Brouwer, Heyting, Kreisel', e.g. [32, p.977].

the BHK-interpretation may be regarded as implicit in Brouwer's writings [...] Heyting made the interpretation explicit [...] Heyting came to regard Kolmogorov's interpretation as essentially the same as his own. [34, p.31]

The claim that will be defended here is that, unlike the name BHK-interpretation suggests, there never was one single understanding of logical connectives that Brouwer, Heyting and Kolmogorov all agreed on. The argument is centered around the hypothetical judgement in general and Ex Falso in particular.[3] It proceeds by making three points:

1. In his dissertation from 1907, Brouwer gave an account of the hypothetical judgement that served him all his life. On that account, hypothetical judgements may in certain cases have false antecedents, but there is no justification of the general principle Ex Falso Sequitur Quodlibet. Neither is the familiar derivation of Ex Falso using the disjunctive syllogism acceptable on Brouwer's view of logic. A systematic conclusion, then, is that Brouwer's logic is a relevance logic.
2. Although Heyting and Kolmogorov both accept Ex Falso as an axiom, Heyting's justification of it would not be acceptable to Kolmogorov; Heyting's claim that his interpretation and Kolmogorov's are essentially the same is not correct; and Kolmogorov's own justification is the more reasonable one of the two.
3. Neither Heyting's nor Kolmogorov's justification of Ex Falso fits into Brouwer's conception of logic.

The discussion of the views of Brouwer, Heyting and Kolmogorov will provide natural occasions briefly to comment also on arguments of C.I. Lewis, Freudenthal, Griss, Becker, Johansson, Troelstra/Van Dalen, Martin-Löf, and Dummett.[4]

2 Brouwer

Already in Brouwer's dissertation, one finds the general view on logic that he would hold on to for the rest of his career: logic seeks and describes the patterns in the linguistic descriptions of how one mathematical system (i.e., objects and relations) fits into (is constructed starting from) another. As Brouwer later put it, 'Formal language accompanies mathematics as the weather-map accompanies the atmospheric processes' [9, p.451].

The following three consequences of this view should be noted.

First, mathematics is not dependent on logic, but the other way around. Logic, to the extent that it is a systematic language, is an application of mathematics. By the use of logic, one will not obtain mathematical truths that are not obtainable by a direct mathematical construction.

Second, logic as Brouwer sees it is a relevance logic: for an inference to be valid, it has to be guaranteed that it accompanies a mathematical procedure that

---

[3]Throughout this paper, the concern is with meaning explanation; I will therefore not discuss here, for example, the formal translation of intuitionistic logic into minimal logic [29], or the question whether Ex Falso is actually useful or needed [31].

[4]Although it does not affect the three main points to be made, an omission of this paper is that it does not discuss some (too) short notes of Destouches-Février, e.g. [10].

constructs the system described by the conclusion out of the system described by the premises.

Third, a familiar argument in favour of Ex Falso, a special case of the hypothetical judgement, is not valid:[5]

1. $\bot$ (assumption)
2. $\bot \vee A$ (1, weakening)
3. $\neg \bot$ (tautology)
4. A (2,3 disjunctive syllogism)
5. $\bot \to A$ (1,4 $\to$-introduction)

Brouwer does not discuss this argument, but we can look at it from his point of view. The problem with the argument is then seen to be neither the application of weakening, nor that of the disjunctive syllogism, but the supposed transitivity of the deduction relation.

The linguistic transition from $\bot$ to $\bot \vee A$ which constitutes the application of weakening here, is one of the many that accompany the mathematical operation of leaving unchanged a mathematical construction to which the description $\bot$ applies; the second description is just less informative.

The application of the disjunctive syllogism is not problematic either. For if $A \vee B$ is a description that applies to a mathematical construction, this means that we have a mathematical method that, when carried out, will show that the description A applies, or that the description B applies; a proof of $\neg B$ then simply tells us that the outcome of that method will be a proof of A. But then we also know that we would have obtained A as a description of the mathematical construction in question if no independent proof of $\neg B$ had been available to us. The disjunctive syllogism, then, accompanies the mathematical operation of leaving the construction described by A as is.

The problem is rather with the composition of these two inferences. The first inference requires that the mathematical construction being described is one for $\bot$; the second that it is one for A. As in general A and $\bot$ will not be equivalent descriptions, there is no general guarantee that when $\bot$ describes a mathematical construction, A describes it as well. This means that there is no guarantee that the linguistic figures in Lewis' argument accompany a mathematical procedure. To use Brouwer's term, Lewis' deduction is therefore 'unreliable', or, as one would say today, it does not respect the notion of relevance defined by the priority of mathematics over its linguistic description, which logic in turn depends on.

Brouwer's dissertation is also the one place in his published work where he discusses the interpretation of the hypothetical judgement in general [8, pp.72–73]. What is peculiar to the hypothetical judgement, Brouwer says, is that there the priority of mathematics over logic seems to be reversed. Among the examples Brouwer refers to are the proofs found in elementary geometry to the problems of Apollonius. Here is one of them. Given three circles, defined by their centres and their radii, construct a fourth circle that is tangent to each of the given three. The way this is usually solved is to assume that such a fourth circle exists, then to set up equations that express how it is related to the three given circles, and then, via algebraic manipulations, arrive at explicit definitions of its centre and radius, and, from there, corresponding constructions. So it seems that here one

---

[5]The argument is well-known from its presentation by C.I. Lewis, but goes back to the Middle Ages.

first assumes the existence of the required circle, then uses logic to make various judgements about it, and only thereby arrives at a construction for it.

Brouwer's general interpretation of such cases is as follows. Having first remarked that logical reasoning accompanies or mirrors mathematical activity which is at least conceptually prior to that reasoning, Brouwer then says:

> There is a special case, [...] which really seems to presuppose the hypothetical judgement from logic. This occurs where a structure in a structure is defined by some relation, without it being immediately clear how to effect its construction. Here one seems to *assume* to have effected the required construction, and to deduce from this hypothesis a chain of hypothetical judgements. But this is no more than apparent; what one is really doing in this case is the following: one starts by constructing a system that fulfills part of the required relations, and tries to deduce from these relations, by means of tautologies, other relations, in such a way that in the end the deduced relations, combined with those that have not yet been used, yield a system of conditions, suitable as a starting-point for the construction of the required system. Only by this construction will it then have been proved that the original conditions can indeed be satisfied.
>
> 'But,' the logician will retort, 'it might have happened that in the course of these reasonings a contradiction turned up between the newly deduced relations and those that had been kept in store. This contradiction, to be sure, will be observed as a logical figure, and this observation will be based upon the principium contradictionis.' To this we can reply: 'The words of your mathematical demonstration merely accompany a mathematical *construction* that is effected without words. At the point where you announce the contradiction, I simply preceive that the construction no longer *goes*, that the required structure cannot be imbedded in the given basic structure. And when I make this observation, I do not think of a principium contradictionis.' [8, pp.72–73, trl. modified]

Brouwer's qualm with 'the hypothetical judgement from logic' is that it seems to be a case where purely logical judgements are necessary to obtain a certain mathematical construction, starting from a merely supposed one; however, Brouwer argues, this is not what really happens.

According to Dirk van Dalen, Brouwer's passage bears on the implication $A \to B$ in the following way:

> Brouwer points out in the above lines that if the conditions and specifications for A are given, then we try to add more information in such a way that, after a certain amount of constructional activity, we can really carry out a construction of A which respects the specifications. Once this is accomplished, we can turn to the 'implication' construction for B, which yields the construction for B and to the required embedding of the structure for A into the structure for B. [36, pp.250–251]

And, in a more succinct form, Van Dalen then proposes the following interpretation α:[6]

> (α) In order to establish $A \to B$, one has to carry out two tasks, namely, (i) find a construction for (the structure specified by) A, (ii) find a construction for (the structure specified by) B that departs from the first construction. [36, p.251]

According to interpretation α, $A \to B$ just means $A \wedge B$ with the extra information that the construction for B was obtained from that for A. On this reading $A \to B$ can be asserted only after a construction for A has been found. The idea is clear: namely, to avoid hypothetical constructions by insisting that a construction be supplied that proves the antecedent. But, as Van Dalen also notices, it is also in effect a rejection of the hypothetical judgement in the general case where one does not know whether there is a construction for A. Moreover, it forbids accepting Ex Falso.

A point that anyone who ascribes this strategy to Brouwer (at the time of his dissertation) should address is that Brouwer, only a few pages after his general discussion of the hypothetical judgement, and in a different context, explicitly accepts 'If a function is not differentiable, then it is not differentiable' [8, p.75]. This is of course a general hypothetical judgement, but each of its instances would be a hypothetical judgement that according to interpretation α should not be accepted, unless a proof has been supplied that the function in question is indeed not differentiable. It can be ruled out that the hypothetical judgement Brouwer makes here is a slip of the pen, for he uses it to exemplify his interpretation of the Principle of the Excluded Middle (PEM) as valid but utterly trivial. At the time, Brouwer still held that 'A function is either differentiable or not differentiable' is equivalent to the hypothetical judgement 'If a function is not differentiable, then it is not differentiable', therefore uninformative, and certainly reliable. (PEM would earn the predicate 'unreliable' only the next year, in 1908 [3].)

One might suggest that this discloses a particular incoherence in Brouwer's dissertation, but instead I will now argue that the incoherence is rather in interpretation α. Assume we set out to establish $A \to B$, for some specified A and B. The account instructs us first to find a construction for A. Assume that we do not already have a prior construction for A at our disposal. It may of course turn out to be impossible, but that is something we can only find out along the way. Now it seems that in our attempt to find a construction for A we will have to make at least some hypothetical judgements of the form 'If A, then C', where C refers to some particular property of the construction intended by A. (Such a property may well be relational, so as to embed A in a larger mathematical context.)

For example, suppose that we are as yet ignorant of the fact that the concept of a square circle is contradictory, and consider a hypothetical judgement of the form 'If P is a square circle, then P has property so-and-so'. Certainly our attempt to construct a square circle as mentioned in the antecedent would have to begin by separately deducing the two requirements that the sought construction should be a square and that it should be a circle. But to do so involves making the hypothetical judgements 'If P is a square circle, then P is a square', and 'If P is a square circle, then P is a circle'.

---

[6]Kuiper's interpretation [26, p.235] is essentially the same.

Before we can actually make and use these hypothetical judgements, we would, according to interpretation α, first of all have to find a construction of their antecedent. But their antecedent is the same as that of the hypothetical judgement to be accounted for. So if we proceed this way, we never get started. On the other hand, if we do not proceed this way, and avoid making new judgements of this type, then how can we direct our attempt at finding a construction for the antecedent? Without an answer to that question, interpretation α cannot be made to work.

But even if it could, a problem remains. In our attempt to find a construction for A, we may find that no construction for A is possible. On interpretation α, it then also is impossible to establish $A \to B$.[7] In turn, on Brouwer's conception of negation, one would thereby also establish $\neg(A \to B)$. As a consequence, any hypothetical judgement of which the antecedent asks for an impossible construction would be false. For example, 'If P is a square circle, then P is a square' would be false; and even 'If P is a square circle, then P is a square circle' would. Although this is a position that has been held by some, it may well be regarded as a reductio ad absurdum of this interpretation of the hypothetical judgement, and in any case does not go together well with Brouwer's reading of $A \vee \neg A$ as 'If not A, then not A'; for one should like to affirm $A \vee \neg A$ also in case A is true.

However, I think an alternative reading can be suggested which does not give rise to these difficulties. As we saw in the quotation from his dissertation, Brouwer argues that arguments that seem to involve supposed constructions in reality do not. But then what is it that they *do* involve? Interpretation α suggested that they involve only *actual* constructions. However, this suggestion does not seem to agree with what Brouwer says here. There is a notion in Brouwer's passage that is left out in interpretation α, the notion of a condition. Brouwer says that instead of supposing that a certain construction has been effected, that is, instead of working with a hypothetical construction, one builds a 'system of conditions'. What to me seems essential here is that such a system of conditions is itself actual and therefore ontologically unproblematic; on the reading I have in mind, Brouwer avoids commitment to hypothetical constructions by considering, not constructions, but conditions on constructions. Instead of a 'chain of hypothetical judgements' that one seems to make, according to Brouwer one is really making a chain of transformations in which from required relations further relations are derived. That Brouwer calls the end result, in the second half of Brouwer's passage, 'a system of conditions', shows that the 'required relations' are themselves to be taken as conditions, for they are what that system is made up from. The same is also clear from the fact that 'the original conditions' that Brouwer mentions at the end must refer to the 'required relations' in the sentence preceding it.

Brouwer says that transformations of conditions into further conditions are made 'by means of tautologies', which in his use of that term means 'by fixing one's attention on different substructures of the mathematical system' [8, p.72n2]; so tautologies are, in the simplest case, of the form 'Substructure A is present, therefore so is substructure B'. Of course, a tautology is only intuitionistically valid if the way it proceeds from one system to another corresponds to the construction methods available to the intuitionistic mathematician; it is here that

---
[7]Van Dalen says that if no construction for A is possible, then 'the implication fails to be realized' [p.252].

the expected constraint is put on hypothetical thinking in abstraction from the truth of the antecedent. Hence, tautologies preserve constructibility.[8] It is of course a further step to come up with an actual construction that satisfies the sytem of conditions finally arrived at; in fact, such a construction may turn out to be impossible. But to engage in hypothetical judgement precisely relieves us from the obligation to consider this question of satisfiability; that is what makes it hypothetical. Note also that the fact that tautologies preserve constructibility means that, if the system of conditions obtained at the end turns out not to specify a possible construction, the fault is not in the transformation steps but in the original conditions. This is implicit in Brouwer's claim that if the construction 'no longer goes', this means 'that the required structure cannot be embedded in the given basic structure'.

So I suggest that interpretation α be replaced by interpretation β:

(β) In order to establish $A \to B$, one has to conceive of A and B as conditions on constructions, and to show that from the conditions specified by A one obtains the conditions specified by B, according to transformations whose composition preserves mathematical constructibility.

The transformations have to be composable in order be 'tautologies' in Brouwer's sense and thereby meet the relevance requirement explained at the beginning of this section.

As for how such transformations are obtained (that is, in Brouwer's term, how tautologies are obtained), I propose that they come from carrying out successful constructions of a mathematical system out of another and then observing on what properties of the original system the success of this construction did not depend (as in Kant's schematism).

As an example, consider Heyting's claim from 1936 that it is obviously correct to say that if π contains the sequence 0123456789, then it contains the sequence 012345678 [19]. At the time it was not known yet that the antecedent is true.[9] The reason why Heyting's claim is so convincing, even when one doesn't know that the antecedent is true, seems to be that we are able actually to construct the sequence 0123456789, and then from that the sequence 012345678, while observing that the operation of dropping the final 9 does not depend on how the original sequence was constructed.

Heyting devised this example as a reply to Freudenthal [12], who had denied that it is possible to assert any hypothetical judgement with an unprovable or even merely unproved antecedent. Freudenthal held that the meaning of an assertion is only determined by an actual proof of it. Then it obviously makes no sense to ask for a consideration of the content of a proposition in abstraction from a proof of it, as the hypothetical judgement requires. For a detailed discussion of the debate between Heyting and Freudenthal, and their correspondence on the matter, see [33].

The referee report on the two papers by Heyting and Freudenthal for Compositio Mathematica was written by Brouwer, and reads:

---

[8]The theme of transformations that preserve mathematical constructibility figures prominently in [3].

[9]Borwein has computed that at the 17.387.594.880th digit after the decimal point in π, a sequence 0123456789 begins [2]. One easily changes Heyting's example appropriately.

> Report on the discussion between Freudenthal and Heyting: Interesting discussion on the sense of one sentence being implied by another when nothing is known about the correctness of the latter. L.E.J. Brouwer (signed)'.[10]

Perhaps the reason why Brouwer has so little to say about the matter is given by a remark he published later in 'Points and Spaces' (1954):

> [T]he wording of a mathematical theorem has no sense unless it indicates the construction either of an actual mathematical entity or of an incompatibility (e.g., the identity of the empty two-ity with an empty unity) out of some constructional condition imposed on a hypothetical mathematical system. [7, p.3]

Indeed, a hypothetical judgement can only be directly informative about what can be mathematically constructed in a negative sense, that is, if it is of the form $A \to \bot$. But a sense, even if not a purely mathematical one, needs to be given to judgements $A \to B$ where nothing is known about the constructibility of A: for such judgements will occur in proofs that lead up to $A \to \bot$, where they are instrumental in arriving at the subsequent contradiction, but are therefore themselves made at a moment when it is not yet known that A cannot be constructed (i.e., is incorrect).

Griss [13] of course was right that a commitment to hypothetical constructions would not go well with the fundamental tenets of intuitionism. On that ground, he rejected the hypothetical judgement, and in particular negation ($\neg A := A \to \bot$), altogether. But that seems to go to far: according to interpretation β, Brouwer had already provided an account of the hypothetical judgement (including negation) without ontological commitment to hypothetical constructions.[11]

We have seen that Brouwer, in his justification of PEM in 1907, makes the (general) hypothetical judgement 'If a function is not differentiable, then it is not differentiable' [8, p.75]. On interpretation α, Brouwer would not have been justified in making it. On interpretation β, reasoning by transforming conditions into other conditions is all that is required to be able to make hypothetical judgements, whether or not a construction for the antecedent has been given. The machinery that Brouwer presents to deal with the case where a construction for the antecedent has been given, is equally capable of dealing with the case where such a construction has not been given. Any instance of Brouwer's judgement about differentiable functions is of course a simple hypothetical judgement. The transformation that validates each of these instances (and thereby justifies the general judgement) is the identity transformation; hence Brouwer's comment that this judgement 'says nothing'. To be uninformative, however, is not to be illegitimate.

The account also allows for specific judgements such as 'If the decimal expansion of π contains the sequence 0123456789, then it contains the sequence

---

[10] Brouwer Archive, Utrecht. The director, Dirk van Dalen, kindly granted permission to publish this document. The translation from the German is mine.

[11] Brouwer supported Griss' career by communicating, between 1946 and 1951, three papers in which Griss developed his negationless intuitionistic mathematics to the Royal Dutch Academy of Sciences; but in the middle of that period, in his own paper 'Essentieel negatieve eigenschappen' of 1948 [6], Brouwer showed an intuitionistic construction of a real number with a property that, he claimed, could not be stated without using negation.

012345678' (the tautology used here being the transitivity of containment of sequences of digits), or even the counterfactual 'If the decimal expansion of π has an initial segment 3.123 then it has an initial segment 3.12' (for the same reason). In these two examples, we are able to indicate how to transform the conditions specified by the antecedent into those specified by the consequent, in a way that preserves constructibility. It is solely in virtue of the ability to do this that these judgements are intuitionistically legitimate. In particular, it is for this reason that the second example, where the antecedent is clearly false, is a valid judgement. But we cannot expect to be able to come up with the required transformation for just any given pair of a false antecedent and a consequent, let alone that we have a uniform method to do this. As a consequence, there is on this account of the hypothetical judgement no ground to accept the Ex Falso principle in its full generality.

In 1923, 16 years after the dissertation, Brouwer gave a proof of the hypothetical judgement $\neg\neg\neg A \leftrightarrow \neg A$ [4]. On the account suggested here, this does not indicate a change in Brouwer's attitude towards hypothetical judgements (as it would if one accepts interpretation α); the short argument Brouwer gives proceeds by tautologies. Note, in particular, that this bi-implication also holds if we know that in fact $\neg A$ is false. And Brouwer in 1927 proved the bar theorem [5], which is indeed, as Troelstra says, 'a striking instance of hypothetical reasoning' [33, p.198].[12]

3 Kolmogorov

In 1925, Kolmogorov presented a formal system for a fragment of intuitionistic logic. Ex Falso was excluded from this fragment: Kolmogorov said that, just like the principle of excluded middle, the principle of Ex Falso 'has no intuitive foundation' [37, p.419]. In particular, he says (p.421) that Ex Falso is unacceptable for the reason that it asserts something about the consequences of something impossible.

In 1932, Kolmogorov again published on the formalization of intuitionistic logic. This time, he did see a way to make Ex Falso acceptable, by interpreting propositions as 'problems' and their proofs as 'solutions'. He said that 'As soon as $\neg a$ is solved, then the solution of a is impossible and the problem $a \to b$ is without content' [25, p.331], and proposed that 'The proof that a problem is without content [owing to an impossible assumption] will always be considered as its solution' [25, p.329]. It seems not altogether unreasonable to extend the meaning of the term 'solution' this way, for, just like a genuine solution, an impossibility proof also provides what might be called 'epistemic closure': like a genuine solution, it provides a completely convincing reason to stop working on a certain problem. (This kind of 'higher-order' solution is also familiar from Hilbert's Program, e.g. [23, p.297].) Note that this justification of Ex Falso makes no attempt to describe a counterfactual mathematical construction process; thus, Kolmogorov's solution from 1932 is not incompatible with his view from 1925 that one cannot constructively assert consequences of something impossible. Rather,

---

[12]In Brouwer's proof of the bar theorem, the necessary condition on proofs of the antecedent that Brouwer formulates depends on their ontological status of mental objects; the proof is a transcendental argument. For an elaborate discussion of this issue, see [30].

the solution from 1932 introduces a stipulation to achieve completion of the logical theory for its own sake.

Although Kolmogorov's stipulation is neither unreasonable nor unmotivated, on Brouwer's descriptive conception of logic there is of course no place for stipulation. Another difference between Kolmogorov and Brouwer as I interpret him, is that for Brouwer it is not necessarily the case that as soon as we have a proof of $\neg a$, the problem $a \to b$ is without content. We know that π does not begin with 3.123, but still we can indicate a mathematical construction that would transform a counterfactual proof that π begins with 3.123 into one that it begins with 3.12. In that sense, this construction can be said to give content to the implication.

4 Heyting

In 1927, without knowing Kolmogorov's work, Heyting devised a formalization of intuitionistically valid principles of logic (in Dutch; published (revised), in German, in 1930 [14–16]). In particular, it included the axiom $\neg p \to (p \to q)$. Heyting published separately an interpretation to justify his formal axioms as axioms of intuitionistic logic. In 1931, he described the guiding principle of his interpretation in the following terms:

> We here distinguish between propositions and assertions. An assertion is the affirmation of a proposition. A mathematical proposition expresses a certain expectation. For example, the proposition, 'Euler's constant C is rational', expresses the expectation that we could find two integers a and b such that $C = a/b$. Perhaps the word 'intention', coined by the phenomenologists, expresses even better what is meant here ... The affirmation of a proposition means the fulfilment of an intention. [17, pp.58–59]

In particular, the affirmation of an implication $p \to q$ means the fulfilment of an intention directed at a construction that transforms any proof of p into one of q.

But Heyting did not include a discussion of Ex Falso in these terms. He then came in contact with Kolmogorov, and in his book from 1934, *Mathematische Grundlagenforschung*, elucidated the axiom in the terminology of Kolmogorov. Heyting states the axiom, and comments:

> In fact, an adequate interpretation of the notion of 'reducing' is this: to demonstrate the impossibility of solving a comes down to reducing the solution of any problem whatsoever to that of a. [20, p.18, trl. mine]

Clearly there is a difference between Kolmogorov's own explanation and Heyting's explanation in Kolmogorov's terms. Where Heyting says that a proof of $\neg a$ establishes a reduction of the solution of any problem to that of a, Kolmogorov had said that it established that the problem of reducing the solution of any problem to that of a has become without content. Be that as it may, by then Heyting had come to believe, after an initial hesitation that has nothing to do with Ex Falso,[13] that Kolmogorov's interpretation in terms of problems and

---

[13]Heyting at the time distinguished between 'p' and 'p is provable'.

his own in terms of intentions were 'substantially equivalent' [22, p.107]. Kolmogorov seems to have believed the same, as in a review in Zentralblatt of the exchange between Freudenthal and Heyting, he consistently speaks of 'Intention oder Aufgabe' [24]. Oskar Becker, in a letter to Heyting of September 1934, had remarked that Heyting's interpretation is a generalisation of Kolmogorov's, as a 'problem' and its 'solution' are special cases of an intention and its fulfilment. 'Die int.[uitionistische] Logik ist also eine "Intentionsrechnung".'[14]

Shortly after, in correspondence with Ingebrigt Johansson in 1935, Heyting again justifies the axiom in Kolmogorov's terms. Johansson had objected to the inclusion of Ex Falso (and by excluding it arrived at his minimal logic):

> $\neg a \supset \cdot a \supset b$ means, that when $\neg a$ is shown, b at once becomes derivable from a, even when this had not been demonstrated before. And that goes against my intuition in the most powerful way.[15]

In a note for his reply (and presumably also in the letter actually sent) Heyting said:

> I interpret $a \supset b$ as 'to reduce the solution of b to that of a, or to show the impossibility of solving a'. This interpretation is adequate [doelmatig].[16]

This of course is Kolmogorov's interpretation of 1932; we have already noted two reasons why Brouwer would not have accepted it. In 1955, in the French translation [20] of his book *Grundlagenforschung* from 1934 [18], Heyting adds a note saying that Johansson's interpretation of the implication is different from his own. But Heyting must have developed second thoughts not unrelated to Johansson's insistence that Ex Falso can only be accepted if a construction be indicated that, on the assumption $\neg a$, transforms proofs of a into proofs of b. For in 1956, in his book *Intuitionism. An introduction*, Heyting gave the following justification of Ex Falso:

> [The axiom $\neg p \rightarrow (p \rightarrow q)$] may not seem intuitively clear. As a matter of fact, it adds to the precision of the definition of implication. You remember that $p \rightarrow q$ can be asserted if and only if we possess a construction which, joined to the construction p, would prove q. Now suppose that $\vdash \neg p$, that is, we have deduced a contradiction from the supposition that p were carried out. Then, in a sense, this can be considered as a construction, which, joined to a proof of p (which cannot exist) leads to a proof of q. [21, p.102]

It is hard to understand how this account is supposed to work. In its attempt to provide, 'in a sense', a construction, it is clearly not of the Kolmogorov type; the latter's interpretations from 1925 and 1932 steered clear of such an attempt.

---

[14] In his letter, Becker does not give the exact date. The Becker-Heyting correspondence, to the extent to which it seems to have been preserved, is published in [35].

[15] Johansson to Heyting, September 23, 1935, p.1. And later that year (November 15, 1935, p.3): 'My being unwilling to accept 4.1 [$\neg a \supset \cdot a \supset b$] is in the end simply based on the circumstance that I prefer to work with a more *strict* implication.' Both trl. from the German mine. The letters are in the Heyting papers kept at the Rijksarchief Noord-Holland, Haarlem, The Netherlands.

[16] Note by Heyting to Johansson's letter of September 23, 1935. Trl. from the Dutch mine.

But it does not fit into Heyting's original interpretation of logic in terms of intentions directed at constructions and the fulfilment of such intentions either. For to fulfill an intention directed toward a particular construction we will have to exhibit that construction; we will have to exhibit a construction that transforms any proof of p into one of q. But how can a construction that from the assumption p arrives at a contradiction, and therefore generally speaking not at q, lead to q? It will not do to say that such a construction exists 'in a sense'. A construction that is a construction 'in a sense', as Heyting helps himself to here, is no construction.

This also shows that, in spite of Heyting and Becker's claims, the interpretation in terms of intentions is not equivalent to that of Kolmogorov in terms of problems. Kolmogorov's approach seems the more reasonable one. Note also that neither Heyting nor Kolmogorov justifies Ex Falso by giving Lewis' argument.

5 Troelstra/Van Dalen, Martin-Löf, and Dummett

Troelstra and Van Dalen in volume I of *Constructivism in mathematics* give a justification of $\bot \to A$ that is closely related to Heyting's but more coherent. As in Heyting's explanation from 1956, the observation that there are no proofs of $\bot$ plays an important role, but they drop the attempt to arrive at a construction (in whatever sense) of the specific consequent A:

> Since there is no proof of $\bot$, $\lambda a.a$ (or any other mapping) may count as a proof of $\bot \to A$, since it has to be applied to the empty domain. [34, p.10]

In other words, their argument is that in the BHK-interpretation the proposition Ex Falso is vacuously true. Martin-Löf argues similarly for the validity of Ex Falso as a rule of inference:

> When you infer by this rule, you undertake to verify the proposition C when you are provided with a proof that $\bot$ is true, that is, by the definition of truth, with a method of verifying $\bot$. But this is something that you can safely undertake, because, by the definition of falsehood, there is nothing that counts as a verification of $\bot$. [...] The undertaking that you make when you infer by the rule of falsehood elimination is therefore like saying, 'I shall eat up my hat if you do such and such', where such and such is something of which you know, that is, are certain, that it cannot be done. [28, p.52]

Yet this type of justification raises a question as well. It surely is a reasonable condition on a general account of the hypothetical judgement that it explain why, if we assume that we know that the antecedent is true, the consequent follows; also if we should in fact know the antecedent to be false. It was likely a wish to meet this condition that led Heyting to his curious account of Ex Falso of 1956. That account seems to be incoherent; but while the accounts of Troelstra/Van Dalen and Martin-Löf are not, the price to pay is that they do not meet the stated condition. Here is an example. Consider the proposition $\bot \to 1 = 2$, and apply the account of Troelstra/Van Dalen. There is a choice as to whether to take $\bot$

as a primitive, or to define it, for example as $0 = 1$. In any case, we take, as the account suggests, the identity transformation as the one that will transform any proof of the antecedent into one of the consequent. Of course the observation that one never will actually have to apply this transformation is correct, as there will never be a proof of the antecedent. But assume, counterfactually, that we do have a proof of $\bot$ or of $0 = 1$, respectively. (Some object to the consideration of counterfactuals in logic and mathematics; but then Ex Falso would be objectionable for that very reason.) Clearly, the identity transformation is not going to transform that proof into a proof of $1 = 2$, if only because the conclusions of the two proofs are different. One easily modifies this argument if a different transformation than the identity transformation is proposed, or a different definition of $\bot$.

There are specific contexts, such as those of formalised primitive recursive arithmetic PRA and of Heyting Arithmetic HA where defining $\bot$ as $0 = 1$ enables one formally to derive $\bot \to A$ for any A in the language of the system. Note that the demonstration of this fact [34, pp.121, 126-7] does not depend on knowing that there is no proof of $0 = 1$, and that the demonstration makes it easy to see how in these systems from a formal proof of $0 = 1$ one would obtain a formal proof of any given A.

Dummett has proposed that, in contexts where it is not obvious whether, as is the case in arithmetic, from $0 = 1$ any statement can be proved, we 'shall count any proof of $0 = 1$ as being simultaneously a proof of any other statement' [11, p.9]. This would seem to amount to the same as Troelstra and Van Dalen's explanation if one there insists on the identity transformation, but Dummett is clear that his proposal is not an explanation but a stipulation. From a Brouwerian point of view, one would respond to this in the same way as to Kolmogorov's proposal of 1932: such a stipulation may not be unreasonable, but it does not fit into a descriptive conception of logic.

Bibliography

[1] Benacerraf, P., Putnam, H. (Eds.). *Philosophy of mathematics: selected readings.* Cambridge: Cambridge University Press, 2nd edn., 1983.

[2] Borwein, J. M. Brouwer-Heyting sequences converge. *Mathematical Intelligencer*, 20:14–15, 1998.

[3] Brouwer, L. E. J. De onbetrouwbaarheid der logische principes. *Tijdschrift voor Wijsbegeerte*, 2:152–158, 1908. English translation in [8], 107–111.

[4] Brouwer, L. E. J. Intuïtionistische splitsing van mathematische grondbegrippen. *KNAW verslagen*, 2:877–880, 1923. English translation in [27], 286–289.

[5] Brouwer, L. E. J. Über Definitionsbereiche von Funktionen. *Mathematische Annalen*, 97:60–75, 1927. English translation in [37], 457–463.

[6] Brouwer, L. E. J. Essentieel negatieve eigenschappen. *Indagationes Mathematicae*, 10:322–323, 1948. English translation in [8], 478–479.

[7] Brouwer, L. E. J. Points and spaces. *Canadian Journal of Mathematics*, 6:1–17, 1954.

[8] Brouwer, L. E. J. *Collected works I. Philosophy and foundations of mathematics.* Amsterdam: North-Holland Publisher Co., 1975.

[9] Brouwer, L. E. J., van Eeden, F., van Ginneken, J., Mannoury, G. Signifische

dialogen. *Synthese*, 2:168–174,261–268,316–324, 1937. English translation in [8], 447–452.

[10] Destouches-Février, P. Connexions entre les calculs des constructions, des problèmes, des propositions. *Comptes rendus de l'Académie des Sciences (Paris)*, 228:31–33, 1949.

[11] Dummett, M. *Elements of intuitionism*. Oxford: Clarendon Press, 2nd edn., 2000.

[12] Freudenthal, H. Zur intuitionistischen Deutung logischer Formeln. *Compositio Mathematica*, 4:112–116, 1936.

[13] Griss, G. F. C. Negationless intuitionistic mathematics I. *Indagationes Mathematicae*, 8:675–681, 1946.

[14] Heyting, A. Die formalen Regeln der intuitionistischen Logik I. *Sitzungsberichte der Preussischen Akademie der Wissenschaften*, 42–56, 1930. English translation in [27], 311–327.

[15] Heyting, A. Die formalen Regeln der intuitionistischen Logik II. *Sitzungsberichte der Preussischen Akademie der Wissenschaften*, 57–71, 1930.

[16] Heyting, A. Die formalen Regeln der intuitionistischen Logik III. *Sitzungsberichte der Preussischen Akademie der Wissenschaften*, 158–169, 1930.

[17] Heyting, A. Die intuitionistische Grundlegung der Mathematik. *Erkenntnis*, 2:106–115, 1931. English translation in [1], 52–61.

[18] Heyting, A. *Mathematische Grundlagenforschung, Intuitionismus, Beweistheorie*. Berlin: Springer, 1934.

[19] Heyting, A. Bemerkungen zu dem Aufsatz von Herrn Freudenthal "Zur intuitionistischen Deutung logischer Formeln". *Compositio Mathematica*, 4:117–118, 1936.

[20] Heyting, A. *Les fondements des mathématiques. Intuitionnisme. Théorie de la démonstration*. Paris: Gauthier-Villars, 1955.

[21] Heyting, A. *Intuitionism, an introduction*. Amsterdam: North–Holland, 1956.

[22] Heyting, A. Intuitionism in mathematics. In Klibansky, R. (Ed.), *La philosophie au milieu du vingtième siècle*, vol. 1, 101–115. Firenze: La nuova Italia, 1958.

[23] Hilbert, D. Mathematische Probleme. Vortrag, gehalten auf dem internationalen Mathematiker-Kongreß zu Paris 1900. In *Gesammelte Abhandlungen (zweite Auflage)*, vol. III, 290–329. Berlin: Springer, 1935.

[24] Kolmogorov, A. Zbl 0015.24201 [review of [12] and [19]]. *Zentralblatt der Mathematik*, 15:242, 1937.

[25] Kolmogorov, A. N. Zur Deutung der intuitionistischen Logik. *Mathematische Zeitschrift*, 35:58–65, 1932. English translation in [27], 328–334.

[26] Kuiper, J. *Ideas and explorations. Brouwer's road to intuitionism*. Ph.D. thesis, Utrecht University, 2004. Quaestiones Infinitae vol. XLVI.

[27] Mancosu, P. *From Brouwer to Hilbert. The debate on the foundations of mathematics in the 1920s*. Oxford: Oxford University Press, 1998.

[28] Martin-Löf, P. On the meanings of the logical constants and the justifications of the logical laws. *Nordic Journal of Philosophical Logic*, 1(1):11–60, 1996.

[29] Prawitz, D., Malmnås, P.-E. A survey of some connections between classical, intuitionistic and minimal logic. In Schmidt, H. A., Schütte, K., Thiele, H. J. (Eds.), *Contributions to mathematical logic*, 215–229. Proceedings of the Logic Colloquium, Hannover 1966, Amsterdam: North-Holland, 1968.

[30] Sundholm, G., van Atten, M. The proper explanation of intuitionstic logic: on Brouwer's demonstration of the Bar Theorem. In van Atten, Boldin, Bourdean, Heinzmann (Eds.), *One hundred years of intuitionism*, 60–77. Basel: Birkhäuser,

2008.

[31] Tennant, N. Intuitionistic mathematics does not need ex falso quodlibet. *Topoi*, 13(2):127–133, 1994.

[32] Troelstra, A. S. Aspects of constructive mathematics. In Barwise, J. (Ed.), *Handbook of mathematical logic*, 973–1052. Amsterdam: North-Holland, 1977.

[33] Troelstra, A. S. Logic in the writings of Brouwer and Heyting. In Abrusci, V. M., Casari, E., Mugnai, M. (Eds.), *Atti del Convegne Internazionaledi Storia della Logica*, 193–210. San Gimignano, 4–8 dicembre 1982, Bologna: CLUEB, 1983.

[34] Troelstra, A. S., van Dalen, D. *Constructivism in mathematics, I, II*. Amsterdam: North–Holland Publ. Co., 1988.

[35] van Atten, M. The Becker-Heyting correspondence. In Peckhaus, V. (Ed.), *Oskar Becker und die Philosophie der Mathematik*, 119–142. München: Wilhelm Fink Verlag, 2005.

[36] van Dalen, D. Kolmogorov and Brouwer on constructive implication and the contradiction rule. *Russian Mathematical Surveys*, 59(2):247–257, 2003.

[37] van Heijenoort, J. (Ed.). *From Frege to Gödel: A sourcebook in mathematical logic, 1879–1931*. Cambridge, MA: Harvard University Press, 1967.

Logic, Rational Agency, and Intelligent Interaction [1]

Johan van Benthem

University of Amsterdam & Stanford University

johan@science.uva.nl

ABSTRACT. This paper records a lecture given in successive versions for many audiences, including the *Workshop on Knowledge and Rationality*, 18th European Summer School in Logic, Language and Information, Malaga, August 2005, the *Dynamic Logic Workshop* at UQAM Montréal, June 2007, and the Philosophical Logic Section, *Thirteenth DLMPS Congress* Beijing, August 2007. We make a plea for recasting logic as a theory of interactive agency, and show how this perspective fits both old achievements and new broader ambitions for the field.

1 Logic as information flow: a new perspective, but also an old one

The restaurant: entangled informational processes When asked at public occasions to explain what logic is, I often use the following evergreen scenario, showing our discipline at work every day in our city. You are in a café with two friends, and the three of you have ordered a beer, a wine, and a water. Now some new person comes back with three glasses. What will happen? Everyone agrees that three things occur in sequence:

> First the waiter asks "Who has the wine?", say, and then puts that glass. Then, he asks who has the beer, and puts that glass in its place. And then, he does not ask any more, but just puts the remaining glass. Two questions, and then one inference!

When he puts that third glass without asking, you observe a logical inference in action: the information in the two answers received allows the waiter to deduce where the third one must go.[2] One can spell out this final stage in terms of a valid propositional inference schema

$$A \vee B \vee C, \neg A, \neg B \Rightarrow C$$

---

[1] The author is currently Weilun Professor of Philosophy, Tsinghua University, Beijing.

[2] When hearing this example, the President of Amsterdam University gave me a warning: "Johan, you should be more careful and avoid low-class cafés. When I order something, I am not paying all that money to just have my glass put in front of me, while the others also get a question." Indeed, good waiters put the last glass with a smile, or they will even say: "So this must be you". The calculus of politeness has its own laws on top of logic.

but this is not my main point here. To me, this old example cries out for a new twist. There is a natural unity to this scenario. The waiter first obtains the relevant information by communication and perhaps observation, and then, once enough data have accumulated, he infers an explicit solution. Now on the traditional line, only the latter deductive step is the proper domain of logic, while the former are at best 'pragmatics'. But in my view, both informational processes are on a par, and both should be within the compass of logic, which is about information flow in general, not just deductive elucidation. In my book, asking a question and processing an answer is just as 'logical' an activity as drawing an inference! And accordingly, logical systems should be able to account for both, as observation, communication, and inference occur entangled in most meaningful activities. But what is involved in this ambitious task?

The logical challenge: 'social dynamics' First of all, there is the information flow itself in the café scenario. The initial part is a sequence of *update actions* on information states, viewed as sets of live options at the current stage. Initially, there are 6 ways in which three glasses can be distributed over three people. The first answer reduces this uncertainty from 6 to 2, and the second answer reduces it to 1, i.e., the actual situation:

$$\text{⑥ } answer1 > \text{② } answer2 > \text{①}$$

That is why no third question is needed: one just spells out the situation. This *dynamics of informational actions* is an obvious target for logical theory. And to make this work, we must also give an account of the underlying statics: the information states that the actions work over. I will argue that this can be done, though much remains to be understood. [3] But there is another striking feature to the informational activity going on in the restaurant. Questions and answers typically involve more than one agent, and hence the dynamics is social, having to do with what people know about each other. In particular, the waiter asks us, because he knows that we know what we ordered.

A historical pedigree after all Is all this merely new-fangled tinkering with the good old core values of logic? I do not think so. The ideas put forward here are themselves ancient and obvious. For instance, traditional Indian logic distinguished three principled ways of getting information. The easiest route is to observe, when that is possible. The next method is inference, in case observation is impossible or dangerous, as with a coiled object in a room where we cannot see whether it is a piece of rope, or a cobra. And if these two methods fail, we can still resort to communication, and ask some expert. Similar ideas occur in medieval Western logic, and the Restaurant scenario shows that the same natural combination occur today. [4] Moreover, the social interactive aspect of

---

[3] For instance, the waiter's final act of *inference* also produces 'new information' – but surely, not the same kind as that produced by the initial observational updates. An explicit model for this inferential information would employ syntactic fine-structure (cf. the survey of information paradigms inside logic in [44]), but we will only touch on this issue lightly in Section 8 below.

[4] At a 2005 Winter School at IIT Bombay, Mumbai (cf. [13]), I once presented this point to students with the question whether there was beer on campus. I had tried observation, inspecting most buildings for alcohol outlets the night before. I had tried deduction, reading all the conference material through and through. And now I was reduced to asking experts, viz. the students. No answer was forthcoming right then, but in the evening two students arrived

information flow is just as ancient, going back to the very roots of logic. While many people see Euclid's *Elements* as the paradigm for logic, with its crystalline structure of mathematical proofs and eternal insights, the true origin of the discipline may be closer to Plato's *Dialogues*, an argumentative practice with clear patterns of confirmation and refutation between participants. It has been claimed that logic arose originally out of political and legal debate in all its three main traditions: Chinese, Indian, and Western. [5]. And this multi-agent interactive view has emerged anew in modern times. A beautiful case are the so-called *dialogue games* from the mid 1950s [20], which explained logical validity in terms of winning strategies for a proponent arguing the conclusion against an opponent granting the premises. So, it seems that logical activity is deeply interactive, and that its theory should reflect this. Once again, some colleagues find this alarming, as social multi-agent aspects are dangerously reminiscent of gossip, status, and Sartre's famous warning that "Hell is the Others". Maybe the best way of dispelling such fears is taking a look at what all this entails:

2 Information flow in dynamic epistemic logic

Questions and answers For a start, consider the striking multi-agent phenomenon of communication. A lot of interesting logical structure already shows in very simple question/answer examples, the ubiquitous building blocks of interaction. Consider the following dialogue:

Me: "Is this the Forbidden City? "
You: "No."
You: "It is the Friendship Hotel."

What this certainly conveys are facts about the current location. But there is much more going on. By asking the question, at least in a normal scenario (not, say, a competitive game), I indicate that I do not know the answer. And by asking you, I also indicate that I think that you may know the answer, again under normal circumstances. [6] Moreover, your answer and the follow-up statement do not just transfer the bare facts to me. They also make sure that you know that I know, that I know that you know that I know, and in the limit of epistemic iterations like this, they achieve so-called *common knowledge* of the relevant facts in the group consisting of you and me. This common knowledge is not a by-product of the fact transfer. It rather forms the basis of our mutual expectations about future behaviour. Thus, keeping track of 'higher-order' information about others is crucial in many disciplines, from philosophy (interactive epistemology) and linguistics (communicative paradigms of meaning) to computer science (multi-agent systems) and cognitive psychology ('theory of mind'). Indeed, the ability

---

carrying a plastic bag. What they said was this: "Sir, the answer to your question is 'No'. However, there is a liquor store right outside the campus gate, and since we thought you were asking the question because you needed beer, we bought you three bottles."

[5] E.g., the Mohist literature in early China discusses the Law of Non-Contradiction as an interactive principle of rational conversation: 'resolve contradictions with others', 'avoid contradicting yourself'. Cf. [19], which incidentally, also cites a Mohist view similar to the pluralism advocated here, namely, that all knowledge comes from three sources: questions, proofs, and consulting experience.

[6] All such presuppositions are off in a classroom with a teacher quizzing students.

to move through an informational space keeping track of what other participants do and do not know, including the crucial ability to switch and view things from other people's perspective, seems characteristic of human intelligence.

While this social dynamics sounds forbidding, very simple logical systems exist which shed some light on it. We present one here, just to show how the ambitions in Section 1 can be realized without abandoning logical systems as we know them today. This will also quickly give us some further material for more concrete discussion of relevant issues.

Epistemic logic dynamified Let us quickly review the static base of our dynamic logic-to-be. To model the basics of the preceding question-answer scenario (and much more), we can use models for *epistemic logic*, proposed by Hintikka in the 1960s [15] as a way of analyzing the philosopher's conception of knowledge. In what follows, however, we will read the knowledge found in this system in a modern spirit as what is true 'to the best of agents' information', with that information viewed as ranges of worlds that are still candidates for the actual situation. (Cf. [44] for much further background on this conceptual move.) The language has a classical propositional base with modal operators $K_i\phi$ ('i *knows that* ϕ') and $C_G\phi$ ('ϕ' is *common knowledge* in group G'):

$$p \mid \neg\phi \mid \phi \vee \psi \mid K_i\phi \mid C_G\phi$$

We write $<i>\phi$ for the dual modality $\neg K_i\neg\phi$: 'agent i considers ϕ possible'. The dual of $C_G\phi$ is $<C_G>\phi$. Models M are triples $(W, \{\sim_i \mid i \in G\}, V)$, with W a set of worlds, the $\sim_i$ binary accessibility relations between worlds, and V a propositional valuation. [7] The epistemic truth conditions are:

$$M, s \models K_i\phi \quad \text{iff} \quad \text{for all } t \text{ with } s \sim_i t : M, t \models \phi$$
$$M, s \models C_G\phi \quad \text{iff} \quad \text{for all } t \text{ that are reachable from } s \text{ by some}$$
$$\text{finite sequence of } \sim_i \text{ steps } (i \in G) : M, t \models \phi$$

Now comes what is arguably the common sense view of information flow: incoming new information eliminates possibilities from the current epistemic range. In particular, public announcements of true propositions P provide 'hard information', which changes the current model stepwise as follows:

> For any model M, world s, and formula P true at s, $(M|P, s)$ (in words: '(M, s) *relativized* to P') is the sub-model of M whose domain is the set of worlds $\{t \in M | M, t \models P\}$.

In a picture, one goes:

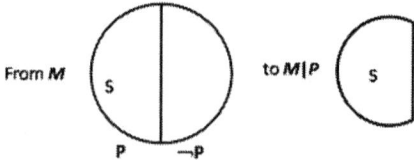

---

[7] One often takes the relations to be equivalence relations, though this is optional.

Logic, Rational Agency, and Intelligent Interaction

Crucially, truth values of formulas may change in such an update step: in particular, agents who did not know that P do after the announcement.

Public announcement logic Now we bring the dynamics of successive update steps into static epistemic logic. Here is how (cf. [34] for a survey of the following system and its technical properties). The language of *public announcement logic PAL* is the epistemic language with added action expressions:

Formulas $\qquad P: \quad p \mid \neg \phi \mid \phi \vee \psi \mid K_i \phi \mid C_G \phi \mid [A]\phi$
Action expressions $\quad A: \quad !P$

The semantic clause for the dynamic action modality is as follows:

$$M, s \models [!P]\phi \quad \text{iff} \quad \text{if } M, s \models P, \text{ then } M|P, s \models \phi$$

In particular, this language allows us to make typical assertions like

$$[!P]K_i\phi \qquad\qquad (\#)$$

which state what agent i knows after having received the hard information that P. Such formulas neatly high-light the combination of ideas from diverse fields involved in logical dynamics. Speech acts $!P$ come from linguistics and philosophy, knowledge assertions $K_i\phi$ from philosophical logic, computer science, and economics. And the dynamic effect modality $[!P]$ combining these actions and assertions comes from program logic in computer science.[8]

Reasoning about information flow in update steps like this revolves around formulas of the formula (#).

In particular, we need to analyze what such an assertion means in terms of a dynamic *recursion equation* telling us what the new knowledge is in terms of the old knowledge which the agent had before the update took place. Here is the relevant valid principle of update, which can be verified using the above truth clauses, and the above picture for concreteness[9]:

$$[!P]K_i\phi \quad \leftrightarrow \quad P \rightarrow K_i(P \rightarrow [!P]\phi)$$

We will discuss this key principle of information flow or conversation later. Here is how it functions in a complete calculus of public announcement:

---

[8] The notorious 'culture gap' between humanities and exact sciences is bridged in logic.

[9] Here is a formal analysis for this recursion equation. Compare the two models (M, s) and $(M|P, s)$ before and after the update. [Drawing a picture juxtaposing both helps.] The formula $[!P]K_i\phi$ says that, in $M|P$, all worlds $\sim_i$-accessible from s satisfy ϕ. The corresponding worlds in M are those $\sim_i$-accessible from s *that satisfy* P. Given that truth values of formulas may change in an update step, the right description of these worlds in M is not that that they satisfy ϕ (which they do in $M|P$), but rather $[!P]\phi$: they become ϕ after the update. Finally, $!P$ is a partial function: P must be true for its public announcement. Thus, we need to make our assertion on the right conditional on $!P$ being executable, i.e., P being true. Putting this together, $[P!]K_i\phi$ says the same as $P \rightarrow K_i(P \rightarrow [!P]\phi)$. The latter can be simplified to the equivalent formula $P \rightarrow K_i[!P]\phi$ usually found in the literature.

Theorem 1 *PAL is axiomatized completely by the usual laws of epistemic logic over our chosen static model class* [10] *plus the following recursion axioms:*

$$[!P]q \leftrightarrow P \to q, \text{ for atomic facts } q$$
$$[!P]\neg\phi \leftrightarrow P \to \neg[!P]\phi$$
$$[!P](\phi \land \psi) \leftrightarrow [!P]\phi \land [!P]\psi$$
$$[!P]K_i\phi \leftrightarrow P \to K_i(P \to [!P]\phi)$$
$$[!P]C_G^\phi\psi \leftrightarrow (P \to C_G^{P \land [!P]\phi}[!P]\psi)$$

Taken together, these axioms analyze the effects of incoming new hard information compositionally. As a result, they reduce all assertions containing dynamic action modalities to basic epistemic statements about the initial model, which 'pre-encode' future effects of the dynamics. Both this recursive type of analysis and the pre-encoding power needed to make it work return in many areas of cognitive action, including belief revision, as we shall see below.

The theorem should suffice to show that logics dealing with information flow look just like systems that we know, and can be developed maintaining the same technical standards. [11] Often, the dynamic super-structure can be fitted onto an already existing logical system describing properties of the static 'snapshots' of the relevant informational process. Indeed, there is a growing literature on the model theory, proof theory, and computational complexity of public announcement logic and its more sophisticated variants.

3 Agents: the dynamic fine-structure of inference and observation

The logic of public announcement is fully capable of dealing with the information flow in our original Restaurant example. Moreover, it performs the two basic tasks involved there in tandem, describing information flow through both observations and inferences by agents. [12] Now there is a more general ambition at play here. The dynamic logics of this paper develop the notion of a *rational agent*, as a much richer counterpart to the austere and solitary paradigmatic proof systems or computational devices in traditional logic. One important theme is then what logical notion of agency emerges from our considerations.

Idealized agents What sort of agents populate *PAL*? For a start, epistemic logic makes some sweeping idealizations. Agents are *omniscient*: their knowledge

---

[10] Here the binary epistemic operator $C_G(\phi, \psi)$ of 'conditional common knowledge' has a technical function of 'pre-encoding' explained in detail in [46].

[11] Note that relating this system to specific applications involves the choice of a model and a set of proposition letters. Thus, what are the relevant 'possible worlds' is itself a process of representation, and it might even change in the course of a conversation. For instance, with my question about the Forbidden City, originally just one proposition letter plays a role, but your coda about the Friendship Hotel now makes a second one relevant, transforming the model. This additional 'dynamics of representation' is less-understood.

[12] I use the term 'observation' for the basic ability now, since the communication in our simple examples may be viewed as the special case of observation of what the others say. Dynamic-epistemic logics are usually presented with an emphasis on communication, but I myself feel they are best understood as logics of observation and experiment in multi-agent settings.

is closed under all inference rules of the system, and on the usual semantics also *introspective*: they know when they know (and they also know when they do not know). Our dynamic analysis has nothing to say per se about these two idealizations. [13] Instead, we have added one more!

The recursive equation in the *PAL* axiom for $[!P]K_i\phi$ embodies a further idealized ability of agents in the dynamics of sequential actions, viz. *perfect memory*. World elimination encodes what has taken place in the current state, in a wholly transparent manner clear to all. Technically, we can see this as follows. Disregarding some syntax, the *PAL* axiom essentially performs an operator switch between $[A]K$ and $K[A]$. And such switches encode strong assumptions on memory and observation. Consider the putative logical principle

$$K[a]\phi \to [a]K\phi,$$

'if I know that doing a will produce ϕ, then after doing a, I know that ϕ holds'. While this is correct for transparent publicly observable actions by agents with Perfect Recall of their previous states, it fails for actions which impair epistemic abilities ([22]). In particular, what is at stake here is memory. I know now that after drinking, I get boring. But the tragedy of drinking is that, after I have drunk, I do not know that I am boring. [14] Likewise, the converse of the preceding axiom, also present in the *PAL* recursion equation, expresses a learning principle called 'No Miracles': if I am uncertain now between two worlds, then seeing the same action in both is not going to remove my uncertainty.

Diversity and parametrized powers The upshot of our discussion so far is this. Information flow essentially happens to *agents* who use it in various ways. But current dynamic epistemic logics do not just give a neutral description of arbitrary participants in this process, they state normatively what idealized agents should be able to observe and infer. And this raises a question. The reality of life is *diversity of agents*, with different bounds on their inferential, introspective, and observational powers, as well as different bounds on memory. Indeed, one hallmark of rational behaviour seems to be our ability to function successfully in environments with agents of very different skills and inclinations. And correspondingly, we may want our logical systems 'parametrized', so that the interplay of different agents can be accommodated smoothly.

While some attempts exist in this direction, including dynamic logics making different assumptions about memory capacity (cf. [18]), there is no standard way so far of doing this all across the board. In particular, while there is some highly suggestive proof-theoretic literature on bounded agents manipulating syntactic 'evidence' (cf. [2]), we lack a canonical way of representing inferential abilities of agents in a parametrized fashion. More generally, unlike the case of computation and Turing Machines, and the Automata Hierarchy providing the fine-structure underneath them, we still lack a universal model of how an agent works, let alone one that can be parametrized for varying abilities. In what follows, we merely explore some bits and pieces of this area. Of course, these are tantalizing bits and pieces, otherwise we might just as well stop here.

---

[13] If we make the inferential process explicitly dynamic, however, with suitable syntactic information states modified by inference steps, then omniscience can be blocked. Section 8 has some references on how to do this – but there is no consensus in the literature.

[14] Thus, our Restaurant scenario was tricky. At least, logical waiters should not drink.

Full dynamic epistemic logic One encouraging fact is that dynamic epistemic logics do have a full-fledged account of diversity in *powers of observation* (cf. Footnote 12). Often, agents cannot fully observe a situation – witness the earlier Indian 'coiled rope, or cobra'. Moreover, different agents can have different observational access to a situation. Think of a card game, where you draw a card from the stack, but the other players do not see which one. The total effect of such mixtures can be hard to describe, witness the complications arising from using emails with lots of tags *cc* and especially, the more hidden *bcc*. In truthful public announcement, there is just one event, publicly visible to all, and the precondition for it to happen is that the announced proposition be true. This line of thinking can be generalized to scenarios involving many possible events, like drawing different cards from a stack, where agents may not be able to distinguish them in the same way (cf. [3]).

This form of update requires dynamic structures much like epistemic models. *Event models* are structures

$$A = (E, \{\sim_i | i \in G\}, \{PRE_e | e \in E\})$$

consisting of a set of relevant events, and relations $\sim_i$ over these encoding what agents cannot distinguish. Events e also have *preconditions* PRE_e for their successful execution: the red card lying on top for my drawing it, your knowing the answer to my question, etc. These provide the core information when we observe the event. Now here is the general *Update Rule*:

> For any epistemic model (M, s) and any event model (A, e), the product model $(MxA, (s, e))$ has
>
> **Domain** $\{(s, e) | s$ a world in M, e an event in A, $(M, s) \models PRE_e\}$
> **Accessibility** $(s, e) \sim_i (t, f)$ iff both $s \sim_i t$ and $e \sim_i f$
>
> The valuation for atomic proposition letters p at (s, e) is just that at s in M. [15]

Product update models a great many scenarios. It deals with misleading actions as well as truthful ones, and with *belief* as well as knowledge. In particular, the smooth course of world elimination is now much more tortuous. Epistemic models can easily get *larger* as product update proceeds, as happens in realistic parlour games, or in email scenarios with *bcc*'s.

The corresponding language is again an epistemic one plus this time new action modalities containing descriptions of event models, interpreted as follows:

$$M, s \models [A, e]\phi \text{ iff if } M, s \models PRE_e, \text{then } MxE, (s, e) \models \phi$$

The logic for this system is effectively axiomatizable and decidable. And again, the key is a recursion equation for knowledge after a complex epistemic event:

$$[A, e]K_i\phi \quad \leftrightarrow \quad PRE_e \rightarrow \wedge\{K_i[A, f]\phi))| f \sim_i e \in A\}$$

There is much literature on this system ([5, 46, 48]), including topics like addition of common knowledge and other forms of group knowledge, extensions to rich modal fixed-point languages, interdisciplinary connections to security and process algebra, epistemic temporal logic, game theory, and so on.

---

[15] This stipulation can be generalized to deal with genuine changes in the world.

4 Warm-blooded agents: acts of revision and processes of learning

Agents who correctly record all information from their observations, and industriously draw the right conclusions from their evidence, may be rational in some Olympian sense. At the same time, they are just cold-blooded recording devices. But rationality does not reside in always being cautious, and always being right. It can be argued, with Popper, that its peak moments only occur with 'warm-blooded agents', who are opinionated, make mistakes, and then: *correct themselves*. [16] Thus, rationality is also about the dynamics of revision and learning. In a concrete setting, revision comes to the fore in conversation, our original example. People contradict each other, and then something more spectacular has to happen than mere update. Maybe one of them was wrong, maybe they all were, and they have to adjust. Modeling this involves a distinction between information coming from some source, and agents' various attitudes and responses to it. Standard references for 'belief revision theory' are [9, 27], while in what follows here, we mainly take the dynamic logic-based line of [36].

Belief and plausibility order Agents can have other attitudes toward propositions than our earlier 'knowledge', in particular, *beliefs* that can turn out incorrect. Standard logics of belief analyze assertions $B_i \varphi$ for '*agent i believes that φ*'. Their semantics adds a new idea to the flat information ranges in our epistemic modeling so far. We now assume further gradations, in the form of a *plausibility ordering* of worlds as seen from some vantage point:

$\leq_{i,s} xy$ in world s, agent i considers y at least as plausible as x.

Thus, while the earlier ranges of epistemic alternatives corresponded to the hard information that we have, the same ranges ordered by plausibility give finer gradations of *soft information*. In particular, we now define the attitude of belief semantically as '*truth in the most plausible options*':

$M, s \models B_i \phi$ iff $M, t \models \phi$ for all worlds t that are maximal in the plausibility ordering $\lambda xy. \leq_{i,s} xy$. [17]

An elementary example is a model with two worlds that are epistemically accessible, but the one with $\neg P$ considered more plausible than the other:

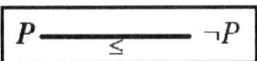

At the actual world with P, the agent does not know whether P, but she does (mistakenly!) believe that $\neg P$. It is crucial that our beliefs can be false. As with epistemic logic, there are complete doxastic logics for these models, and a whole model theory around them (cf. [7]). [18]

---

[16] Compare a lecture with a mathematician writing a proof on a blackboard to a research colloquium with people guessing, spotting problems, and making brilliant recoveries...

[17] There are some complications making this work in infinite models, but this is the idea.

[18] Most logics also analyze the interplay between knowledge and belief in information models with two relations $\sim_i, \leq_j$ entangled in various ways, reflecting a stand on whether knowledge implies belief, or whether one knows one's beliefs. While relations between attitudes toward information are an important topic, we just focus on belief in what follows.

Next, in doxastic logic, one soon finds that mere beliefs are not sufficient for explaining agents' behaviour. We want to know what they would believe were they to receive new information. This *pre-encoding*, in our earlier sense, requires a stronger notion of *conditional belief*:

$M, s \models B_i^\psi \phi$ iff $M, t \models \phi$ for all worlds t which are *maximal* for $\lambda xy \leq_{i,s} xy$ in the set $\{u|\ M, u \models \psi\}$.

Conditional beliefs $B_i^\psi \phi$ are like logical conditionals in the general sense of Ramsey, Lewis, and others (cf. [17]), in that they express what might happen under different circumstances from where we are now. [19]

Changing beliefs under hard and soft information Next, combining logics of knowledge and belief suggest a richer picture than what we had so far. Our *hard information* lies encoded in the current range of epistemically accessible worlds. But this range also carries fine-structure via plausibility orders. The latter encode *soft information* about propositions P, making them more plausible perhaps, but without ruling out $\neg P$-worlds.

The dynamic perspective on information change of *PAL* and *DEL* also applies to our beliefs, and how to revise them triggered by new information. This process involves changes, not in the range of available worlds or epistemic accessibility, but rather in the *plausibility orderings* $\leq_{i,s} xy$ among worlds. First, when we receive hard information $!P$, update proceeds by world elimination as before. We now get new beliefs related to our earlier conditional beliefs, and indeed, the crucial recursion equation driving the logic of belief revision will even say just how new conditional beliefs are acquired:

Theorem 2 *The logic of conditional belief under public announcements is axiomatized completely by*

1. *any complete base logic of $B_i^\psi \phi$ for one's chosen models,*
2. *PAL reduction axioms, plus*
3. *a reduction axiom for conditional beliefs:*

$$[!P]B_i^\psi \phi \leftrightarrow P \to B_i^{P \wedge [!P]\psi}[!P]\phi$$

Hard information already involves non-trivial phenomena. For instance, true information can be misleading to rational agents! Consider the following three-world model with actual world 1, where all worlds are epistemically accessible, with a plausibility ordering $1 \leq 2 \leq 3$. Here the agent believes that p in 1, but for the wrong reason, as she considers the non-actual world 3 most plausible:

$$\boxed{1p,q \ \leq \ 2r \ \leq \ 3p,s}$$

Now suppose that a true public announcement $!\neg s$ is made, which eliminates world 3. Now, in the actual world of the remaining model with domain $\{1, 2\}$, the agent has now come to believe, incorrectly, that $\neg p$!:

---

[19] The analogy is so close that conditional belief on reflexive transitive plausibility models satisfies exactly the laws of the so-called 'minimal conditional logic' (cf. [51]).

$$\boxed{1p, q \ \leq \ 2r}$$

Following earlier ideas of Stalnaker, Baltag & Smets 2006 have emphasized that there is another natural attitude here, in between knowledge and belief, viz. of *safe beliefs* which cannot be changed by new true information. This provides an additional robustness, moving them closer to knowledge. Technically, safe beliefs are ordinary modalities applying to those formulas ϕ which hold *in all worlds that are least as plausible as the current one*. Thus, a dynamic perspective on information change can also suggest new static epistemic-doxastic operators.

Genuine belief revision Next, consider a 'seasoned learner' attuned to finer distinctions in input, who receives *soft information* concerning a proposition P. This just increases her 'preference' for P-worlds, without totally ruling out the others. Thus, incoming soft information leads to plausibility change, not world elimination. This can come in various sorts, reflecting another source of diversity for agents. A quite typical 'belief revision policy' is *lexicographic upgrade* $\Uparrow P$ [29] which replaces the current ordering relation $\leq$ between worlds by the following:

all P-worlds become more plausible than all the $\neg P$-worlds, while within those two zones, the old ordering remains.

Belief changes under such policies can be axiomatized completely. The logic for $\Uparrow P$ revision is in [36] with the following key recursion equation: [20]

$$[\Uparrow P]B^\psi \phi \ \leftrightarrow \ ((\Diamond(P \wedge [\Uparrow P]\psi) \wedge B^{P \wedge [\Uparrow P]\psi}[\Uparrow P]\phi)$$
$$\vee (\neg(\Diamond(P \wedge [\Uparrow P]\psi) \wedge B^{[\Uparrow P]\psi}[\Uparrow P]\phi)))$$

This formula looks complex, and we will not explain it. But after all, we are describing a more subtle informational process now than mere epistemic update.

Richer dynamic doxastic logics handle many further policies, such as softer variants placing just the *most plausible* P-worlds on top, leaving all others in their old position. Again, a general move can be made here similar to the one from *PAL* to *DEL* in Section 2. Instead of charting different policies one by one, we can also 'enrich the trigger' for revision, making information come in the form of 'plausibility event models of signals' to which agents assign different plausibilities. Cf. [4, 5] for the resulting uniform logical view of belief revision via one 'Priority Rule' applied to a variety of inputs.

Toward real learning Our main point in this section has been that belief revision for self-correcting agents is moving within the scope of dynamic logic. While more satisfying than ad-hoc accounts, even this story is far from complete. Genuine *learning* is not just single revision steps triggered by a current signal. It is about longer-term methods producing responses to input streams over time, and these are studied in formal Learning Theory [16]. A merge of this research area with our dynamic logic still needs to be made (see [14] for first attempts).

---

[20] Here $\Diamond$ is the existential epistemic modality 'in some world of the current range'.

5 Interaction over time and games

Temporal perspective also comes in with the next crucial feature of agents, their engaging in purposeful behaviour over time, responding to, and influencing others. Even the simplest form of conversation, our original example, involves saying more things than one, and choosing assertions depending on what others say. This interactive aspect extends the scope of our dynamic logics in many ways.

Program structures Conversation has crucial timing aspects. We say things in a certain order, what we say may depend on circumstances, and we may have to keep repeating assertions until some intended effect obtains, as happens in flattery or threats. Now conversational plans or programs use three well-known operations of computer programs:

1. *sequential composition*   ;
2. *guarded choice* $IF \ldots THEN \ldots ELSE \ldots$
3. *guarded iterations* $WHILE \ldots DO \ldots$

And if we allow participants to speak simultaneously, not just in turn, then conversational scenarios will even involve forms of *parallel program composition*. Now PAL does have some relevant validities here. For instance, its equivalence

$$[!P][!Q]\phi \leftrightarrow [!(P \wedge [!P]Q)]\phi$$

appears to say that one need not say more things than one, as a single assertion will do. But this will no longer be true when we consider arbitrary iterations.

This richer dynamic logic of conversation over time resembles propositional dynamic logic *PDL*–though it is also still like *PAL* in crucial ways. [21] But nevertheless, there is a surprise when putting these two decidable systems together. The longer-term logic of potentially unbounded conversation crosses a computational threshold in terms of the complexity of validity [21]:

Theorem 3 *PAL with common knowledge and all PDL program operations added to the action part of the language is undecidable, and even non-axiomatizable.*

The reason is that models for this logic have two dimensions, one 'forward' in time (where Kleene iteration * acts as an unbounded future modality), and one 'sideways' in information models (where common knowledge provised unlimited access – and together these encode high-complexity Tiling Problems. [22] [23]

But even with this complexity, programs are still single-agent affairs, like a set of notes for a speech, where the audience is a largely passive recipient of your eloquence. In reality, both conversation and more general processes of learning involve a back and forth type of interaction, which is typically found in *games*.

True interaction and games To sample the spirit ot true interactions, consider a not unrealistic game played between a Student and a Teacher. The Student is located at position S in the following diagram, but wants to reach the position of escape E below, whereas the Teacher wants to prevent him from

---

[21] For instance, all formulas are still invariant for epistemic bisimulation.

[22] Technically, validity for *PAL* with Kleene star added is even $\prod_1^1$ complete.

[23] Dynamic epistemic logic seemed a smash summer hit in Beijing 2007, given the many posters for a pop singer 'Update* Jane', though the star was worrisome.

getting there. Each line segment is a path that can be traveled. At each round of the game, the Teacher cuts one connection anywhere in the diagram, while the Student can, and must travel one link still open to him at his current position:

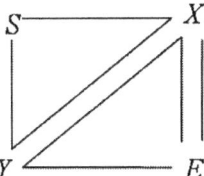

If Teacher is greedy, and starts by cutting a link $S - X$ or $S - Y$ right in front of the Student, then it is easy to see that Student can reach E. However, teacher does have a *winning strategy* for preventing the Student from reaching E, by first cutting one line between X and E, and then letting his cutting be guided by where Student goes subsequently. Here *strategies* for players are rules telling them what to do in every eventuality. Solving games like this can be complex, emphasizing the non-trivial nature of interaction. [24] [25]

Scenarios like this are close to the origins of modern game theory [23]. *Zermelo's Theorem* says that extensive two-player games of finite depth with perfect information and zero-sum outcomes are *determined*: one of the two players has a winning strategy. This result is still close to logic. Let the universal quantifier $\forall$ range over all moves in the game, as chosen by the beginning player A, while the existential quantifier $\exists$ stands for moves by the other player E. With game depth k, the logical *law of excluded middle* then states the following disjunction

$$\forall \exists \ldots \forall \exists \text{ (k times) '} E \text{ wins'} \lor \exists \forall \ldots \exists \forall \text{ (k times) } \neg \text{ '} E \text{ wins'}$$

Given that E's not winning means A's winning, this is determinacy. [26]

Of logical importance here is not just the existence of winning strategies, but the explicit plans and strategies that interacting agents have for dealing with each other. Strategies have strong links with logical notions [37], and they are an important ingredient of rational agency, also in the philosophy of action.

Infinity and temporal logic   While the above games are finite, as is true of many rational activities, populations of rational agents also engage in potentially infinite processes, such as 'language use' or 'reproduction'. Describing long-term behaviour of agents over time then requires statements about histories that may go on forever. Here the dynamic epistemic logics discussed above meet with *temporal* logics that describe properties of histories, which also have versions with knowledge and belief (a discussion with extensive references is found in [41, 45]). Such temporal systems have existed for a long time in philosophical logic and in computer science. This infinite arena is also the natural habitat of Learning

---

[24] [26] shows that solving 'sabotage graph games' like this is Pspace-complete.

[25] The reader will get an even better feel for the difficulty of interaction by considering the following variant. This time, the Teacher wants to force the Student to end up in E without any possibility of escape. Who of the two has the winning strategy this time?

[26] Zermelo proved his result with a view to Chess. His result was rediscovered later by Max Euwe, the only world champion in Chess ever produced by The Netherlands. For much more on the 'logic games' connection behind all this early Game Theory, cf. [31].

Theory, and even more complex structures in Game Theory such as 'type spaces' for extensive games [23]. And finally, this temporal setting is also the realm of Dynamical Systems Theory, the mathematics for biology, evolutionary game theory, as well as neural nets in brain research. In other words, logic in its dynamic guise lives in a larger mathematical landscape where many exciting confluences are still to be expected.

6 Preferences, goals, and social choice

Having gone all the way to infinity, let us now return to the basics of interaction. The Indian students of Footnote 4 did not just answer my question whether there was beer at IIT Bombay. They also asked themselves *why* I had asked my question, and then responded to that. Indeed, behind every communicative interaction, there is a *Why question* concerning agents' goals. And 'making sense' of the interaction does not just involve meaning and information, but also mutually getting clear on those goals. This brings in another level of structure familiar from decision theory and game theory.

Preference logic   To account for goal-directed behaviour, our dynamic logics for informational events need to bring out further structure beyond knowledge and belief, namely, agents' preferences between situations, and the ways they evaluate courses of action. Right now, there is a vigorous development of logics with preference operators that describe preference structure [42] as well as intentions [28]. These logics come in both static and dynamic variants, with the latter describing episodes of preference change, triggered by commands or other events with value-bestowing force [12, 43]. Preference is also at the heart of social choice theory and game theory – and interesting contacts can be observed between all these areas today, with logic providing 'fine-structure'.

Logic and game theory   Indeed, active interfaces between logic and game theory [6, 31, 47] abound, and have to do with epistemic analysis of equilibrium solutions in terms of rational action, or logical analysis of games as a paradigm for interactive computation [1, 24]. [27] Here we merely mention two examples showing how analysis of games naturally connects up with dynamic update logics of the sort we have discussed here. In particular, rational behaviour of players in games combines reasoning about action, belief, and preference.

Games and dynamic logics   Consider the solution procedure of *Backward Induction* for extensive games, a generalization of the algorithm behind Zermelo's Theorem. Starting from outcome preferences on leaves, nodes get evaluated through the tree, representing players' intermediate beliefs as to expected outcomes and values, given that both are acting 'rationally'. As is well-known, Backward Induction often produces 'bad equilibria' representing some socially undesirable outcome. An example occurs to the left in the following picture, where the bad equilibrium $(1,0)$ predicted by reasoning about players' 'rationality' makes both hugely worse off than the cooperative outcome $(99, 99)$.

One way of doing something about this [38] is by making *promises* which change the current game through public announcements of intentions. ***E*** might

---

[27] Two Nobel Prize winners in Economics, Nash and Aumann, have strong interests in logic.

Logic, Rational Agency, and Intelligent Interaction

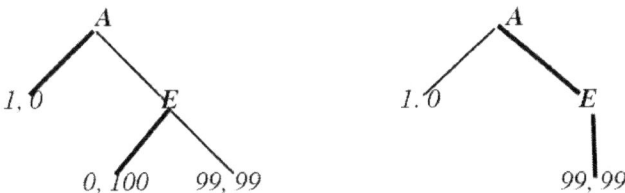

promise that she will not go left, changing the game to the new one depicted on the right – and the new equilibrium (99, 99) results, making both players better off. [49] has a dynamic logic of players strategic powers and preferences, where games can change by announcing intentions. Complete logics then intertwine *PAL* with modal logics of actions and preferences in a straightforward manner.

Excursion: strategies Logic of rational agency is also about strategies themselves. If we use propositional dynamic logic *PDL* to define strategies in games, as is done in [32], adding game change leads to a joint logic *PDL+PAL* adding public announcements [!A]. It is easy to show that *PDL* is closed under the latter, both in its propositional and its program parts, but the crucial recursion equation now also uses an operation $\pi|A$ for relativized programs π using tests ?A. *PDL+PAL* is then axiomatized completely by merging the separate laws of these systems, while adding the equivalence

$$[!A]\{\sigma\}\phi \leftrightarrow (A \to \{\sigma|A\}[!A]\phi)$$

There is more structure to games than just moves and strategies, and *PAL*-style scenarios also make sense with richer epistemic preference languages.

'Rational Dynamics' Here is one final example at the current logic games interface. [33] uses iterated public announcement (Section 5) to analyze another major arena of game theory. *Strategic games* induce epistemic models M of strategy profiles with preferences and uncertainty relations for players who know their own strategy, but not that of the others – and these come with their own solution procedures. Here a combined modal preference language can formulate statements of Weak Rationality ("no player chooses a move which she knows to be worse than some other available one") and Strong Rationality ("every player chooses a move which she thinks may be the best possible one"). When announced, these propositions eliminate worlds where they fail, and iterating these announcements to the limit, there is a smallest sub-model where *WR* or *SR* are now common knowledge. Iterated announcement of *WR* is the well-known solution concept of Iterated Removal of Strictly Dominated Strategies; and its sub-model is defined in M by a formula of a modal μ-calculus with inflationary fixed-points. [28] The same holds for iterated announcement of *SR* and game-theoretic 'Rationalizability'.

In this scenario of internal deliberation about a game, players keep recalling their 'rationality'. A similar analysis applies to extensive games. Backward Induction is obtained through repeatedly announcing that "no player chooses a move all of whose further histories end worse than all histories after some other available move". The procedure ends in largest sub-games where players

---

[28] If A has 'existential-positive' syntax (*SR* does), the definition is in standard μ-calculus.

have common belief of rationality. But one can also announce other types of joint agency. [33] considers history-oriented versions, where players remind themselves of the *legitimate rights of others*, because of 'past favours received'.

This string of examples may suffice to illustrate the lively interface between logic and game theory as congenial accounts of rational agency. There is no canonical theory yet, [29] but one can see contours of an interactive logic.

7 Dynamic logic and intelligent interaction

The claim of this paper has been that logic can move beyond the standard paradigm of abstract consequence relations, or at best, a lonely theorem prover or computer, to become an account of rational agents who observe, infer, communicate, learn, and interact. In a series of dynamic logics and pointers to the literature in which these lie embedded, we indicated how this might be done while sticking to the canons of logic as we know it.

This may be seen as part of a general movement. There is an emerging view today in many disciplines that theories of information, intelligence and rationality need to be *dynamic*, high-lighting human actions of communication, observation and decision, and *social*, high-lighting the interplay of several actors, often even congregating in groups. Thus, the paradigm of rational behaviour shifts from lonely thinkers writing down proofs and contemplating the ultimate truth, to the noisy realities of dialogue, debate, and colliding opinions and preferences as the locus where "la vérité s'éclate". Even within a single discipline, such as philosophy, it is striking how this interactive turn is affecting many areas in parallel, from pragmatics in the philosophy of language to interactive epistemology, and from second-person ethics to philosophy of action and social philosophy. Accordingly, we are now led to think of 'rational agents', not as solitary utility maximizers in the sense of Dickens' Scrooge – but rather as people who thrive in interactive environments, and contribute to their successful functioning.

We conclude this paper by inquiring into the status of all this, and the challenges which it poses. The logical developments presented so far look promising, with a growing body of real work. There are conferences and workshops on 'Knowledge and Rationality', journals on 'Knowledge, Rationality, and Action', and rational agency is in the air everywhere. Many of us like this 'Interactive Turn' – but what can it really achieve? I will discuss some perspectives on this issue – mostly taking the earlier-mentioned developments in logic as my frame of reference. This will also serve as a way of comparing interactive logic with the traditional face of the field.

8 The actors: what is an intelligent agent?

The foundational program The preceding goals may be all to the good, but what *is* a rational agent really, and what is the task we have set ourselves as theorists of intelligent interaction? One cannot consult some standard text or manifesto as our guide, because there are none.

---

[29] For instance, dynamic logics still have to link up with 'type spaces' for games.

Logic, Rational Agency, and Intelligent Interaction

To see the issue, think of an earlier turn which changed the face of logic, viz. the mathematization by Frege, Russell and others, and the matching hopes for the foundations of mathematics. *Hilbert's Program* provided an appealing set of goals here. The formalization would describe mathematical theories, establish their consistency, and where possible completeness, while the logic driving all this would be simple, perhaps decidable. Thus, the goal was to clarify the methodology of exact mathematical reasoning once and for all, and scientific practice would be the better for it ever after. This was a technical enterprise with something important at stake. The great foundational discoveries of the 1930s demonstrated the infeasibility of this program, in the form of Gödel's Incompleteness Theorems, and to a lesser extent, Turing and Church's undecidability results for simple natural computational and logical problems. This course of events may be seen as a Popperian virtue by itself. At least, the foundational research program of the early 20th century had not immunized itself to criticism in a facile manner, and it had made a refutable claim. But far beyond this, the refutation had positive spin-off. Like Vergilius' Romans after the fall of Troy, logicians spread all over the academic world, and founded an empire based on positive follow-up. Just think of the Turing Machine as a universal model for computation, and recursion theory, or of insights into mathematical proofs turning into proof theory, and likewise, limits on expressive power for theories turning into model theory as the study of the variety of models allowed by logical formalisms. [30]

Toward a common model Intelligent interaction is clearly a more ambitious goal than standard computation. So, what are those actors that we have placed at centre stage in our logical enquiry? The Universal Turing Machine quickly became the general computing model, as a lucid analysis of the key features of a human computer doing sums with pencil and paper. And even though nothing similar has been found yet for the more intensional notion of an 'algorithm', this first simple rallying point focused the whole study of computation. So, suppose that our counterpart to this computational device is the Generic Rational Agent, what would be its defining skills and properties, beyond pencil and paper sums? Clearly, that analysis must be more complex than Turing's, since agents engage in such a wide array of activities, far beyond computation. [31]

I have asked many colleagues for the features they consider constitutive of rationality. Answers were lively, but not at all conclusive. Thus, I will list some ingredients and issues which I myself find appealing and worthy of study.

Idealized or bounded processing powers? In line with the tradition, our dynamic logics so far idealize rational agents endowing them with unlimited inferential and observational powers, and with ample memory to store all the fruits of all these talents. But I am also attracted by the opposite tendency in the literature, stressing the huge limitations on all these powers that human cognition operates under. In that case, the heart of rationality would be optimal

---

[30] One might even say that *computer science* was born out of the debris of the foundational era, leading to a much richer agenda. Over the years, this has come to include distributed societies of computing agents that communicate and pursue goals, whose study combines ideas from mathematical, philosophical, and computational logic. One new name for this enterprise, which borders on cognitive science, is *informatics*. Restated in this setting, our paper asks: what is the notion of agency here, and what can logic contribute?

[31] On the other hand, admittedly, computation in the modern sense has turned out to cover such diverse activities as solving puzzles, parsing natural language, or image processing.

performance given heart-breaking constraints. Beautiful examples of surprising optimal behaviour in that setting are found in Gigerenzer's highly original book *Simple Heuristics that Make Us Smart* [11]. This provides a 'tension': we need to explain how our logical systems can function in such a setting.

Which core tasks? Nexts, powers to what end? I said in the above that there is not one single core task which rational agents are called upon to perform, whereas the Turing Machine was supposed to just compute. But I also said that, under suitable encoding of data, 'computing' turned out to cover many more activities than one might think. So here is a major question. Do rational agents have a 'core business' that they must be good at? Is it perhaps *reasoning* – as a 'normal form' reducing all other intelligent activities? I do not think so. Reasoning is indeed one important category, and we should take it in a broad sense. For instance, decision-theoretic views look 'forward' at how agents predict the future, and plan their actions. But some colleagues responding to my request for a 'core list' emphasized a 'backward-looking' talent of rational agents, viz. explaining and *rationalizing* what has already happened. Either way, next to reasoning-related tasks, other crucial abilities of rational agents such as acumen in perception and observation, and talents for successful interaction, do not reduce to 'reasoning' in any illuminating way. [32]

Revision and learning In particular, as in Section 4, I do not think that informational 'soundness': being right all the time, is a hall-mark of rational agents. To me, the peak performances of rational agents are in spotting problems, and then trying to solve them. Rationality is constant *self-correction*. This reflects my general take on the foundational collapse in the 1930s. The most interesting issue in science and mathematics is not guarantees for consistency and safe foundations, but the dynamic ability of repairing theories, and coming up with creative responses to challenges. Thus belief revision and general *learning* are the true tests of rationality in my view, rather than flawless update.

Communication and interaction But the preceding criteria of reasoning and learning are still too restricted. Both apply to a single agent. But the core phenomenon that we are after is intelligent interaction. A truly intelligent agent can perform tasks directed toward others: ask the right questions, explain things, convince, persuade, understand strategic behaviour, synchronize beliefs and preferences with other agents, and so on. Almost paradoxically, I would state the following desideratum, which has no counterpart for Turing Machines (unless one wanted to do a Turing-style analysis for distributed computation):

A rational agent is someone who interacts rationally with other agents!

Diversity Here is another aspect of rational interaction, related to the preceding points. Agents are not all the same, and they form groups whose members have diverse abilities, strategies, and so on. Understanding this diversity is a nontrivial tasks for logic (cf. our discussion of 'parametrization' in logics for agents

---

[32] One might say that even my dynamic logics do just that: reducing a wide range of rational activities to a standard formalism where validity reigns supreme. But I doubt this is the best way of thinking about what these logics achieve. And anyway, it would be a meta-level reduction of rational behaviour, rather than an object-level account of what it does.

in Section 3). Moreover, successful behaviour has to do with functioning well in a wide-range environment of agents with different capacities and habits. [33]

Switching Next to diversity, there is also a rational ability which glues us together. It is *the ability to put yourself in somebody else's place*. In its bleakest form, this is the logician's 'role switch' in a game, i.e., the interactive reading of 'negation'. But in a more concrete form, it is the ability to see social scenarios through other people's eyes, as in Kant's Categorical Imperative: "Treat others as you would wish to be treated by them."

Intelligent groups My criteria are not all exclusive, but here I list a final one. Humans typically tend to form new entities, viz. *groups*, which take on lives of their own. Indeed, our identity is made up of many layers of 'belonging' to various groups. In game theory, this leads to the study of coalitions, and on the logical side, this has led to work on common knowledge and other forms of knowledge typical for groups. [34] The earlier logical systems can help describe the fine-structure of such processes, but we are now after more global levels of rationality between individuals, groups, and even larger institutions. Thus the agenda will go all the way to analyzing, perhaps even designing, procedures for procedural justice and deliberative democracy. [35] The formation of 'rational "we"'s' and intelligent organizations generally needs to be acknowledged.

All this does not add up to a universal model of rational agency. Intelligent interaction is still waiting for its Turing. But I do think that these defining questions should be asked, whatever the outcome – and also, that the above list stakes out a territory which needs to be covered by any eventual model.

But what is the agenda? We have staked out a set of topics concerning rational agency. We have banners like 'intelligent interaction'. We have insights and techniques from philosophy, logic, computer science, game theory, social sciences, and so on. But what is the new agenda, providing focus and unity? It is not as if we are landing on virgin shores. Like most Promised Lands in history, this one is already densely populated by many inhabitants, viz. the disciplines mentioned – to which one could add even more. Could there be an analogue to Hilbert's Program in store for us, setting a worthy goal for interactive logicians to march toward? I will not attempt to answer this intriguing question here, but rather end with some general unifying themes that run across the area.

9 Logical systems for intelligent agency: some integrating trends

Instead of having all its core definitions and long-term goals agreed on, a field might also form around a shared modus operandi. In particular, the logic of rational agency requires a process of combination. As we have shown in this

---

[33] It is said that Marx disliked Bakunin, because his own narrow social range was getting on with German intellectuals like himself. By contrast, Bakunin who came from the Russian nobility, got on well with everyone, from intellectuals to simple workers.

[34] Maybe the most intriguing agendas are in the philosophy of action, with 'shared agency' and common intentions, and in social choice theory and judgment aggregation, which describe intelligent aggregation mechanisms for preferences and opinions.

[35] Parikh has popularized this view under the heading of 'Social Software' [25].

paper, we have component logics for many separate tasks for agents – but in cognitive reality, these all work together. So, can we now form one grand logic of intelligent interaction? This is no simple matter.

Putting the pieces of together? There is not even a standard model for logical agents which describes just their inferential and observational powers in tandem. This requires integrating the different notions of information that occur in logic, semantic and deductive. While there are some proposals, with joint semantic-syntactic structures that can be updated (cf. [44], and the new dynamic-epistemic take in [40]), no consensus has emerged so far. But clearly, this is a substantial problem that must be solved. One current attempt is the new monograph [30] that deals with all aspects of information dynamics and intelligent interaction discussed here in one dynamic-epistemic framework.

In addition to conceptual issues, there are computational pitfalls. The *complexity of combined logics* can be much higher than that of the components. Decidable components may create undecidable logics when the mode of combination is complex. We have seen this in the temporal epistemic logic of agents with *perfect memory*. As for plausible modes of combination, we are far from having charted all the *conceptual entanglements* that rational agents may exhibit. [36] But again, we have to combine, and we do.

Framework integration Another source of coherence are integrative technical paradigms. Now, the area of intelligent interaction is replete with competing systems, schools, and sects. But there are some encouraging technical trends toward convergence. Gradually, contours are emerging of a common framework of 'epistemic-temporal logics' in a broad sense (cf. [45]). A common methodology seems as powerful a source of intellectual identity as a shared language.

An interesting analogy is again with the foundational era. The 1930s saw many competing paradigms for defining computation. But eventually, it became clear that, at a well-chosen level of input-output behaviour, these all described the same computable functions. *Church's Thesis* then proclaimed the unity of the field, saying all approaches described the same notion of computability – despite 'intensional differences' making one or the other more suitable for particular applications. This led to a common field of Recursion Theory, everyone got a place in the joint history, and internal sniping was replaced by external vigour. Something similar might happen in the study of intelligent interaction. If we do not have a Hilbert or Turing, we might at least have a Church.

Shared transformations A third unifying trend are systematic transformations of classical computational problems into broader problems of intelligent interaction. [35] discusses three of these. *Epistemization* turns algorithmic tasks into versions involving information: When does a robot 'know how to stop' once in its goal region? When does an agent have the 'know-how' to achieve her goals? *Dynamization* turns static descriptions of agents into processes: normative actions that induce changes in preference are a good example of a new issue brought into our scope then. Finally, *gamification* turns algorithmic tasks into multi-player games: the above Teacher-Student game was a transformation on what would normally be a single-agent reachability problem. These transformations of classical problems might be birth pangs of a theory of rational agency.

---

[36] Here is one: Preference is what is best for me in the worlds which I believe possible. Now an action of belief revision induces a preference change – and the reverse also happens ...

One final unifying force is the undeniable *empirical reality* of intelligent interaction, an independent sanity check for whatever theory we come up with.

10 Repercussions all around

To conclude, let us go back to our logics with a richer picture of an information-processing agent than just deductive prowess. Even at the preliminary stage sketched here, this stance has a great many repercussions.

Teaching The new logics are concrete, easily taught, and in my experience with speaking for a wide spectrum of audiences, easily understood and appreciated. If things 'fit' in teaching, there must be something to them.

Logic and information Our systems also raise new foundational questions about logic itself. For instance, is there a unifying notion of 'information' underlying all the agent abilities we have discussed? [44] is a sustained discussion, bringing together logical views of information as range, as correlation, and as code. The question is open whether these are complementary perspectives, or signs of a grand unification.

Epistemology and methodology While traditional logics seek a definition of knowledge as true belief with some further static ingredient (evidence, 'counterfactual stiffening'), our dynamic perspective provides a new approach. Knowledge is the true belief that survives interactive processes of communication and debate. This fits with views in the philosophy of science where the essence of rational enquiry is interactive, sometimes modeled by games. Thus, the opposition between formal methodology and 'sociological' views might dissolve, as there is so much formal structure to social intercourse.

Language and communication Our perspective also fits with newer views of linguistic meaning. Standard truth conditions involve no agency at all. Then 'dynamic semantics' in the 1980s introduced the single-agent idea that the meaning of a text resides in the changes it brings about in a hearer or reader. But modern game-theoretic approaches [10, 50] describe meanings as Nash equilibria in two-person communication games. While this lies below the surface of our dynamic logics, where meanings of formulas are already fixed, there are bridges to the dynamic logics of this paper. [37]

Social choice and groups Group formation also fits naturally with logic. There is an incipient literature on logical analysis of social choice theory. Indeed, [39] analyzes belief revision itself as a process of preference aggregation, between competing signals, and characterizes the update rules of our earlier systems in terms of well-known postulates from social choice theory. Thus, what used to be single rational agents themselves turn into communities of past, present, and future signals and intentions.

Our final illustration returns to the classical heartland of logic.

Foundations of mathematics Well-understood, interactive multi-agent aspects have always existed right in the classical phase of logic. For instance, intuitionistic logic is about enquiring agents over time, and its analysis in Lorenzen

---

[37] The communication games of [8] are about the contextual information that linguistic expressions convey over and above their literal meanings.

dialogue games even made it a theory of interaction – even though this did not become part of the common understanding of constructivism. The same is true even more for linear logic as a theory of abstract interaction, particularly in its contemporary manifestation as compositional game semantics [1]. [38]

Now recall Hilbert's Program, mentioned several times already. The foundational era had pathological fears of inconsistency. Frege says that, if a single contradiction were to be discovered in mathematics, "the whole building would collapse like a House of Cards". But this claim is an artefact of the wrong metaphor. Mathematics is not a house with foundations bearing the whole weight. It is rather a *planetary system* of theories with many relationships, happily spinning together in logical space. And there, contradictions are never the end of a story. To the contrary, one of the most striking ability of scientists is not to create infallible theories, but rather, having creative ways of coping with problems once they arise. [39] The history of science is replete with inventive strategies for revision. And that, of course, was the motivation for the logics for belief revision discussed as an essential part of rational agency in the above. We see human intelligence at its finest when we correct ourselves, learn from mistakes, and create something new out of broken dreams and refuted expectations. And there is more to intelligent interaction in this arena. Mathematics is also about invention of new language, precisation when clarity is requested, but also conversely, of inspired informal paraphrasing when communicating the essence of a proof to an audience. Dynamic logic should go for all of these.

Bibliography

[1] Abramsky, S. Information, processes and games. In Adriaans, P., van Benthem, J. (Eds.), *Handbook of the Philosophy of Information*. Computing Lab, Oxford University, Amsterdam: Elsevier, 2007, to appear.

[2] Artemov, S. Logic of proofs. *Annals of Pure and Applied Logic*, 69:29–59, 1994.

[3] Baltag, A., Moss, L., Solecki, S. The logic of public announcements, common knowledge and private suspicions. In *Proceedings TARK 1998*, 43–56. Los Altos: Morgan Kaufmann Publishers, 1998.

[4] Baltag, A., Smets, S. Dynamic belief revision over multi-agent plausibility models. In *Proceedings LOFT 2006*. Department of Computing, University of Liverpool, 2006.

[5] Baltag, A., van Ditmarsch, H., Moss, L. Epistemic logic and information update, Departemtns of computer science, Universities of Indiana, Otago, and Oxford. In van Benthem, J., Adriaans, P. (Eds.), *Handbook of the Philosophy of Information*. Amsterdam: Elsevier Science Publishers, 2007.

---

[38] Wilfrid Sieg pointed me to a little-known passage in Turing where he, too, stresses the social aspects of learning and human intelligence as a challenge for mathematical theory.

[39] In a wonderful study from the mid 1960s [52], the Czech philosopher Ota Weinberger charted persistent strategies removing inconsistencies in common sense reasoning (disagreements in conversation) and in science. Most go back to medieval logic and beyond. One can give up assumptions, the way ZF set theory traded Cantor's Full Comprehension for the Separation Axiom. Other ploys make distinctions between kinds of objects identified before, like 'sets' vs. 'classes' in NBG set theory. Another powerful strategy are 'hidden variables', such as contextual arguments: 'I am tall for a human, but not tall for an animal.'

[6] de Bruin, B. *Explaining Games.* Dissertation. ILLC, University of Amsterdam, 2004.

[7] Fagin, R., Halpern, J., Moses, Y., Vardi, M. *Reasoning about Knowledge.* Cambridge (Mass.): The MIT Press, 1995.

[8] Feinberg, Y. Meaningful talk. In van Benthem, J., Ju, S., Veltman, F. (Eds.), *A Meeting of the Minds*, 41–54. Proceedings LORI Beijing 2007, London: College Publications, 2007.

[9] Gärdenfors, P., Rott, H. Belief revision. In Gabbay, D. M., Hogger, C. J. (Eds.), *Handbook of Logic in Artificial Intelligence and Logic Programming 4.* Oxford: Oxford University Press, 1995.

[10] Gärdenfors, P., Warglien, M. Semantics, conceptual spaces, and the meeting of minds. LUCS Cognitive Science Centre, University of Lund, 2007.

[11] Gigerenzer, G., Todd, P., Group, A. (Eds.). *Simple Heuristics that Make Us Smart.* New York: Oxford University Press, 1999.

[12] Grüne-Yanoff, T., Hansson, S.-O. *Preference Change: Approaches from Philosophy, Economics and Psychology.* Heidelberg: Springer, to appear.

[13] Gupta, A., Parikh, R., van Benthem, J. (Eds.). *Logic at the Cross-Roads.* Proceedings First Indian Winter School in Logic and its Interdisciplinary Environment, 2005, IIT Mumbai: Allied Publishers, 2007.

[14] Hendrick, V. Active agents. *Journal of Logic, Language and Information*, 2:469–495, 2003.

[15] Hintikka, J. *Knowledge and Belief.* Ithaca: Cornel University Press, 1962.

[16] Kelly, K. *The Logic of Reliable Inquiry.* New York: Oxford University Press, 1996.

[17] Lewis, D. *Counterfactuals.* Oxford: Blackwell, 1973.

[18] Liu, F. Diversity of agents, Research Report, Institute for Logic, Language and Computation, University of Amsterdam. *Journal of Logic, Language and Information*, 2006.

[19] Liu, F., Zhang, J. Some thoughts on mohist logic. In van Benthem, J., Ju, S., Veltman, F. (Eds.), *A Meeting of the Minds*, 79–96. Proceedings LORI Beijing 2007, London: College Publications, 2007.

[20] Lorenzen, P. *Einführung in die Operative Logik und Mathematik.* Berlin: Springer, 1955.

[21] Miller, J., Moss, L. The undecidability of iterated modal relativization. *Studia Logica*, 79(3):373–407, 2005.

[22] Moore, R. A formal theory of knowledge and action. In Hobbs, J., Moore, R. (Eds.), *Formal Theories of the Commonsense World*, 319–358. Ablex Publishing Corp, 1985.

[23] Osborne, M., Rubinstein, A. *A Course in Game Theory.* Cambridge (Mass.): The MIT Press, 1994.

[24] Parikh, R. The logic of games and its applications. *Annals of Discrete Mathematics*, 24:111–140, 1985.

[25] Parikh, R. Social software. *Synthese*, 132:187–211, 2002.

[26] Rohde, P. *On Games and Logics over Dynamically Changing Structures.* Dissertation, Institute of Informatics, RWTH Aachen, 2005.

[27] Rott, H. Information structures in belief revision. In van Benthem, J., Adriaans, P. (Eds.), *Handbook of the Philosophy of Information.* Amsterdam: Elsevier Science Publishers, 2007.

[28] Roy, O. *Logic, Intentionality, and Decision.* Institute for Logic, Language and

Computation, University of Amsterdam, 2007.

[29] Segerberg, K. Belief revision from the point of view of dynamic doxastic logic. In *Bulletin of the IGPL 3*, 534–553. 1995.

[30] van Benthem, J. *Logical Dynamics of Information and Interaction.* Cambridge: Cambridge University Press. To appear.

[31] van Benthem, J. *Logic in Games, Lecture Notes.* Institute for Logic, Language and Computation, University of Amsterdam, 1999 and subsequent years. A book version will appear in the new series *Texts in Logic and Games*.

[32] van Benthem, J. Extensive games as process models. *Journal of Logic, Language and Information*, 11:289–313, 2002.

[33] van Benthem, J. Rational dynamics and epistemic logic in games. In Vannucci, S. (Ed.), *Logic, Game Theory and Social Choice III*, 13–45. Singapore: World Scientific, 2003. Also in International Game Theory Review 9:1.

[34] van Benthem, J. One is a lonely number: on the logic of communication. In Chatzidakis, Z., Koepke, P., Pohlers, W. (Eds.), *Logic Colloquium '02*, 96–129. Wellesley MA: ASL & A.K. Peters, 2006.

[35] van Benthem, J. Computation as conversation. In Cooper, B., Löwe, B., Sorbi, A. (Eds.), *New Computational Paradigms: Changing Conceptions of What is Computable.* Heidelberg: Springer, 2006, to appear.

[36] van Benthem, J. Dynamic logic of belief revision. *Journal of Applied Non-Classical Logics*, 17:129–155, 2007.

[37] van Benthem, J. In praise of strategies. In van Eijck, J., Verbrugge, R. (Eds.), *Games, Logic, and Social Software.* Report on a NIAS Project, London: College Publications, 2007.

[38] van Benthem, J. Rationalizations and promises in games. In *Philosophical Trends, special issue on logic.* Chinese Academy of Social Sciences, Beijing, 2007.

[39] van Benthem, J. The social choice behind belief revision, Working Paper presented at Dynamic Logic Montreal 2007. Institute for Logic, Language and Computation, University of Amsterdam, 2007.

[40] van Benthem, J. Merging observation and access in dynamic epistemic logic. *Studies in Logic*, 1(1):1–6, 2008.

[41] van Benthem, J., Gerbrandy, J., Pacuit, E. Merging frameworks for interaction: Del and etl. In *Proceedings TARK 2007.* University of Namur: ILLC Amsterdam & Informatics Torino, 2007.

[42] van Benthem, J., Girard, P., Roy, O. *Everything Else Being Equal. A Modal Logic Approach to Ceteris Paribus Preferences.* Institute for Logic, Language and Computation,University of Amsterdam, 2007.

[43] van Benthem, J., Liu, F. Dynamic logics of preference upgrade. *Journal of Applied Non-Classical Logics*, 17:157–182, 2007.

[44] van Benthem, J., Martinez, M.-C. The stories of logic and information, Research Report, Institute for Logic, Language and Computation, University of Amsterdam. In Adriaans, P., van Benthem, J. (Eds.), *Handbook of the Philosophy of Information.* Amsterdam: Elsevier Science Publishers, 2007.

[45] van Benthem, J., Pacuit, E. The tree of knowledge in action. In *Proceedings Advances in Modal Logic.* ANU Melbourne, 2006.

[46] van Benthem, J., van Eijck, J., Kooi, B. Logics of communication and change. *Information and Computation*, 204(11):1620–1662, 2006.

[47] van der Hoek, W., Pauly, M. Modal logic for games and information. In Blackburn, P., van Benthem, J., Wolter, F. (Eds.), *Handbook of Modal Logic*, 1077–1148.

Amsterdam: Elsevier, 2006.

[48] van Ditmarsch, H., van der Hoek, W.and Kooi, B. *Dynamic Epistemic Logic*. Dordrecht: Springer, 2007.

[49] van Otterloo, S. *A Strategic Analysis of Multi-Agent Protocols*. Dissertation DS-2005-05, ILLC, University of Amsterdam & University of Liverpool, 2005.

[50] van Rooij, R. Signalling games select horn strategies. *Linguistics and Philosophy*, 27:493 – 527, 2004.

[51] Veltman, F. *Logics for Conditionals*. Dissertation, Philosophical Insitute, University of Amsterdam, 1985.

[52] Weinberger, O. *Der Relativisierungsgrundsatz und der Reduktionsgrundsatz – zwei Prinzipien des dialektischen Denkens*. Prague: Nakladatelství Ceskoslovenské akademie Ved, 1965.

"Necessary for"[1]

Giovanna Corsi
Università di Bologna
giovanna.corsi@unibo.it

ABSTRACT. A new language for quantified modal logic is presented in which the modal operators are indexed by terms : "it is necessary for $t_1, \ldots, t_n$". Systems of quantified modal logic are defined in that language and shown to be complete with respect to transition semantics. Formulas such as the Barcan formula, the Ghilardi formula, the necessity of identity can be expressed in a natural way in the new language and are shown to correspond to particular properties of the transition relation.

1 Introduction

Quantified modal logic, as usually understood, is the study of theories in a first-order language plus the box-operator. Typically, $\Box P(x)$ is read

'it is necessary that $P(x)$'.

A considerable variety of such theories have been studied from the pioneering work of Rudolf Carnap and Ruth Barcan to more recent publications such as [6] and [1]. Some dissatisfaction is still felt in particular when one tries to analyse natural language or to deal with semantical structures more general than Kripke frames. Attempts to build richer modal languages by modifying the underlying first-order language have been made in two directions:

- by adding the λ-abstraction operator so as to distinguish, e.g., between *de re* vs *de dicto* sentences, $\lambda x \Box P(x).i$ vs $\Box P(i)$, 'The first pilot was necessarily a pilot' vs 'Necessarily, the first pilot was a pilot'.

- by the introduction of a language with types. A wff $\Box A : n$ is of type n when the free variables occurring in it, either implicitly or explicitly, are $x_1, \ldots, x_n$. Moreover $\Box A : n$ is going to be satisfied or not satisfied by n-tuples of elements of the domain. See [2], [4] and [1].

We introduce a new language which combines features of languages with λ-abstraction operator and languages with types. $\Box P(x)$ is not a well-formed formula anymore since x is free in $P(x)$ and it has to be replaced by

$$| x | P(x)$$

to be read as

---
[1]The author is with Dipartimento di Filosofia, Università di Bologna, via Zamboni, 38–I–40126 Bologna.

"Necessary for"

'it is necessary for x to be $P(x)$'.

$|x|$ is a box-operator indexed by x. A more complex form of the box-operator is the following one

$$|\overset{i}{x}|P(x)$$

The notation has two roles:

- it binds the variable x
- it says that it is necessary for the individual i to have the property $\lambda x.P(x)$.

Dually,

$$\langle \overset{i}{x} \rangle P(x)$$

says that it is possible for i to have the property $\lambda x.P(x)$. Again,

$$|\overset{i}{x}\overset{j}{y}|R(x,y)$$

says that 'it is necessary for i and j to stand in the relation $\lambda x \lambda y.R(x,y)$'. This reading emphasizes that the modal operator depends on i and j and it is alternative to 'i and j stand in the relation $\lambda x \lambda y.\Box R(x,y)$'.

Some examples:

$|x\,y|G(x)$: it is necessary for x and y that x gets a job.

$|x\,y\,z|G(x)$: it is necessary for x, y and z that x gets a job.

$|\overset{m}{x}\overset{j}{y}|G(x)$: it is necessary for Mary and John that she gets a job.

$|x,y|\langle y \rangle \exists w F(y,w)$: it is necessary for x and y that it is possible for y to have a friend.

$|x,y|\exists w \langle y,w \rangle F(y,w)$: it is necessary for x and y that there is someone of whom y is possibly a friend.

2 A language with indexed modalities

A language $\mathcal{L}$ with indexed modalities is a standard first-order language with identity whose logical symbols are $\bot$, $\rightarrow$, $\forall$, $|\overset{t_1}{x_1} \ldots \overset{t_n}{x_n}|$, $n \geq 0$, where $x_1,\ldots,x_n$ are pairwise distinct variables and $t_1,\ldots,t_n$ are terms. When $n = 0$ we write $|\star|$.

Definition 1 Well-formed formulas and free variables occurring in a wff A, $fv(A)$.

- $\bot$ $fv(\bot) = \emptyset$

- $P^n(t_1,\ldots,t_n)$ $fv(P^n(t_1,\ldots,t_n)) = fv(t_1) \cup \cdots \cup fv(t_n)$

- $A \rightarrow B$ $fv(A \rightarrow B) = fv(A) \cup fv(B)$

- $|\overset{t_1}{x_1}\ldots\overset{t_n}{x_n}|A$, where $fv(|\overset{t_1}{x_1}\ldots\overset{t_n}{x_n}|A) = fv(t_1) \cup \cdots \cup fv(t_n)$

  $fv(A) \subseteq \{x_1,\ldots,x_n\}$

- $\forall x A$ $fv(\forall x A) = fv(A) - \{x\}$

$\neg A$, $A \vee B$, $A \wedge B$, $A \leftrightarrow B$, $\exists x A$, $\langle {t_1 \atop x_1} \ldots {t_n \atop x_n} \rangle A$ are defined as usual, $|x_1 \ldots x_n| A$ and $\langle x_1 \ldots x_n \rangle A$ stand for $| {x_1 \atop x_1} \ldots {x_n \atop x_n} | A$ and $\langle {x_1 \atop x_1} \ldots {x_n \atop x_n} \rangle A$, respectively.

Advantages:

- *de re* / *de dicto* distinction

  $| {i \atop x} | P(x)$ is a *de re* sentence, 'it is necessary for i to be $P(x)$', whereas $| \star | P(i)$ is a *de dicto* sentence, 'it is necessary that $P(i)$'.

- substitution

  As we shall see in a moment, $| {t \atop x} | A$ is nothing but $(| x | A)[t/x]$; substitution is indicated inside the modality, it is not carried out in A. Substitution does not commute in general with modalities; actually, the modal operators prevent substitution from being performed in the formula that follows them.

- a richer language

  In a language with λ operator, $\lambda y(\lambda x \Box P(x).m).j$ is equivalent to $\lambda x \Box P(x).m$ by λ-conversion, whereas their corresponding wffs $| {j \atop y}, {m \atop x} | P(x)$ and $| {m \atop x} | P(x)$ are not equivalent.

3 Transition semantics, t-semantics.

Given a *frame* $\mathcal{F} = \langle W, R \rangle$, where $W \neq \emptyset$ and $R \subseteq W^2$, a *system of domains* over $\mathcal{F}$[2] is a triple $\langle W, R, D \rangle$, where D is a function such that $D_w \neq \emptyset$, for each $w \in W$. D_w is said to be the *domain* of w. Domains are interrelated by the *transition relation*

$$\text{if} \quad w R v \quad \text{then} \quad \mathcal{T}_{\langle w,v \rangle} \subseteq D_w \times D_v$$

If $a \, \mathcal{T}_{\langle w,v \rangle} \, b$, then b is said to be an *inheritor* of a in v, or a *counterpart* of a in v.

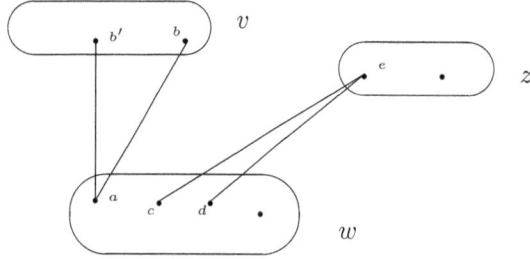

Figure 1:

Definition 2 *A transition frame or a t-frame*, $\mathcal{F}^t$, *is a quadruple* $\langle W, R, D, \mathcal{T} \rangle$ *where* $\langle W, R, D \rangle$ *is a system of domains and* $\mathcal{T} = \biguplus_{w,v \in W} \{\mathcal{T}_{\langle w,v \rangle}\}$, *where* $\mathcal{T}_{\langle w,v \rangle}$ *is defined as above.*

---

[2] This terminology is taken from [6].

Particular cases of $\mathcal{T}$:

$\mathcal{T}$ is a	total	relation	Kripke bundles
	surjective	relation	
	partial	function	
	total	function	Kripke sheaves
	1-1	function	
	inclusion		Kripke frames with increasing domains

Definition 3 *A t-model $\mathcal{M}$ for $\mathcal{L}$ based on a t-frame $\mathcal{F}^t = \langle W, R, D, \mathcal{T}\rangle$ is a pair $\langle \mathcal{F}^t, I\rangle$, where I is a function such that for all $w \in W$, I_w is an* interpretation function relative to w *such that:*

- *for all relations P^n, $I_w(P^n) \subseteq (D_w)^n$*
- *$I_w(=) = \{\langle a, a\rangle : a \in D_w\}$*
- *for all constants i, $I_w(i) \in D_w$*
- *for all functions f^n, $I_w(f^n) : (D_w)^n \to D_w$.*

Rigid designators

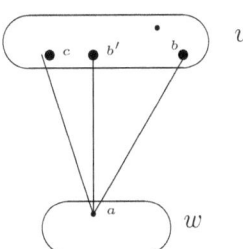

In the context of a t-model, an individual constant i is a *rigid designator* iff if $I_w(i) = a$ and wRv, then $I_v(i)$ is one of the inheritors of a in v. In the above example, $I_v(i)$ is one among $\{c, b', b\}$.[3] So terms are rigid designators iff:

- if wRv then $I_w(i)\, \mathcal{T}_{\langle w,v\rangle}\, I_v(i)$
  and

- if $a_i\, \mathcal{T}_{\langle w,v\rangle}\, b_i$, $1 \leq i \leq n$, then $(I_w(f^n))(a_1, \ldots, a_n)\, \mathcal{T}_{\langle w,v\rangle}\, (I_v(f^n))(b_1, \ldots, b_n)$

Definition 4 *Assignments are world-relative functions $\sigma : VAR \to D_w$. Where σ is a w-assignment, by $\sigma^{x \triangleright d}$ we denote the w-assignment that behaves exactly like σ except that the variable x is mapped to $d \in D_w$.*

Definition 5 *Interpretation of terms. Given a w-assignment σ, the interpretation of t in w under σ, $I_w^\sigma(t)$, is so defined*

- *$I_w^\sigma(x) = \sigma(x)$*
- *$I_w^\sigma(i) = I_w(i)$*
- *$I_w^\sigma(f(t_1, \ldots, t_n)) = I_w(f)(I_w^\sigma(t_1), \ldots, I_w^\sigma(t_n))$.*

---

[3] Recall that in Kripke semantics for all constants i, if wRv then $I_w(i) = I_v(i)$.

When w and I are clear from the context, we write $\sigma(t)$ instead of $I_w^\sigma(t)$.

Definition 6 Simultaneous substitution for terms. *Given a term t containing the free variables $x_1, \ldots, x_k$, we define the term $t[s_1/x_1 \ldots s_k/x_k]$ where s_i is substituted for x_i, $1 \leq i \leq k$. Let $[\mathbf{s}/\mathbf{x}] =_{df} [s_1/x_1 \ldots s_k/x_k]$.*

- $t = y$

$$y[\mathbf{s}/\mathbf{x}] = \begin{cases} y & \text{if } y \neq x_i, \text{ for all } i, 1 \leq i \leq k \\ s_i & \text{if } y = x_i \text{ for some } i, 1 \leq i \leq k \end{cases}$$

- $t = i$

$$i[\mathbf{s}/\mathbf{x}] = i$$

- $t = f(t_1, \ldots, t_n)$

$$f(t_1, \ldots, t_n)[\mathbf{s}/\mathbf{x}] = f(t_1[\mathbf{s}/\mathbf{x}], \ldots, t_n[\mathbf{s}/\mathbf{x}])$$

Lemma 1 Interpretation and substitution for terms. *Let t and s be terms and σ be a w-assignment. Then*

$$\sigma(t[s/x]) \quad = \quad \sigma^{x \triangleright \sigma(s)}(t)$$

If z doesn't not occur in t,

$$\sigma^{z \triangleright a}(t[z/x]) \quad = \quad \sigma^{x \triangleright a}(t)$$

Proof By induction on t.

$t = x$ \qquad $\sigma^{x \triangleright \sigma(s)}(x) = \sigma(s) = \sigma(x[s/x])$

$t = z \neq x$ \qquad $\sigma^{x \triangleright \sigma(s)}(z) = \sigma(z) = \sigma(z[s/x])$

$t = i$ \qquad $\sigma^{x \triangleright \sigma(s)}(i) = \sigma(i) = \sigma(i[s/x])$

$t = f(t_1, \ldots, t_n)$ \qquad $\sigma^{x \triangleright \sigma(s)}(f(t_1, \ldots, t_n))$ $=$
$(I_w(f))(\sigma^{x \triangleright \sigma(s)}(t_1), \ldots, \sigma^{x \triangleright \sigma(s)}(t_n)))$ $=$
$(I_w(f))(\sigma(t_1[s/x]), \ldots, \sigma(t_n[s/x]))$ $=$
$\sigma(f(t_1[s/x], \ldots, t_n[s/x])) = \sigma(f(t_1, \ldots, t_n)[s/x])$

Let z not occur in t. $\sigma^{z \triangleright a}(t[z/x]) = \sigma^{z \triangleright a, x \triangleright \sigma^{z \triangleright a}(z)}(t) = \sigma^{z \triangleright a, x \triangleright a}(t) = \sigma^{x \triangleright a}(t)$, since z doesn't occur in t. ♠

Definition 7 Satisfaction for formulas. *We define when a wff A is satisfied at w under σ in a t-model $\mathcal{M}$, $\sigma \models_w^{\mathcal{M}} A$.*

$\sigma \not\models^{\mathcal{M}}_w \bot$

$\sigma \models^{\mathcal{M}}_w P^k(t_1 \ldots t_k)$ iff $\langle \sigma(t_i), \ldots, \sigma(t_k) \rangle \in I_w(P^k)$

$\sigma \models^{\mathcal{M}}_w B \to G$ iff $\sigma \not\models^{\mathcal{M}}_w B$ or $\sigma \models^{\mathcal{M}}_w G$

$\sigma \models^{\mathcal{M}}_w \forall x G$ iff   for all $d \in D_w$, $\sigma^{x \triangleright d} \models^{\mathcal{M}}_w G$

$\sigma \models^{\mathcal{M}}_w |{}^{t_1}_{x_1} \ldots {}^{t_n}_{x_n}|G$ iff   for all v, wRv and all v-assignments τ, such that $\sigma(t_i) \, \mathcal{T}_{\langle w,v \rangle} \, \tau(x_i)$, $1 \leq i \leq n$, $\tau \models^{\mathcal{M}}_v G$

Consequently,

$\sigma \models^{\mathcal{M}}_w \langle {}^{t_1}_{x_1} \ldots {}^{t_n}_{x_n} \rangle G$ iff   for some v, wRv and for some v-assignment τ such that $\sigma(t_i) \, \mathcal{T}_{\langle w,v \rangle} \, \tau(x_i)$, $1 \leq i \leq n$, $\tau \models^{\mathcal{M}}_v G$

When no ambiguity can arise, we write $\sigma \models_w A$ instead of $\sigma \models^{\mathcal{M}}_w A$.
A is *true at w in $\mathcal{M}$*, $\models^{\mathcal{M}}_w A$, iff for all w-assignments σ, $\sigma \models^{\mathcal{M}}_w A$.
A is *true in $\mathcal{M}$*, $\models^{\mathcal{M}} A$, iff $\models^{\mathcal{M}}_w A$ for all $w \in W$.
A is *valid on a t-frame $\mathcal{F}^t$*, $\mathcal{F}^t \models A$, iff $\models^{\mathcal{M}} A$ for all models $\mathcal{M}$ based on $\mathcal{F}^t$.
A is *t-valid*, $t \models A$, iff $\mathcal{F}^t \models A$, for all t-frames $\mathcal{F}^t$.

An idea which is at the basis of the above definition of satisfaction is that only the worlds where an individual exist or its inheritors exist do matter in order to establish its modal properties, for

$$\sigma \models^{\mathcal{M}}_w |{}^i_x| P(x)$$

iff all the inheritors of $\sigma(i)$ in all related worlds satisfy $P(x)$. Worlds where there are no inheritors of $\sigma(i)$ are not taken into consideration. It turns out that inferences such as

$$\frac{\sigma \models_w |x\,y|Q(x,y) \qquad \sigma \models_w |x\,y|(Q(x,y) \to A(y))}{\sigma \models_w |y|A(y)}$$

are not valid. Let $\sigma(x) = a$ and $\sigma(y) = b$. Suppose it is true in w that "a always quarrels with b" and that "every time that a quarrels with b, then b gets angry", but from this it doesn't follows that "b is always angry", for b may not be angry in those worlds where a is absent.[4]

de re vs *de dicto* modalities

There is an intuitive sense according to which the truth conditions for $|{}^i_x|P(x)$ are different from those for $|\star|P(i)$: in one case it is said that "it is necessary for i to have the property $P(x)$", in the other, the necessity of a sentence is asserted. In transition semantics we do justice to this difference in the following obvious way: in the first case, first we interpret i in the actual world (or the world we are in) and then we see if all its inheritors in all accessible worlds (where they exist) do satisfy the property $P(x)$, in the second case, we first consider all worlds accessible from the actual one and then check if the interpretation of i in those worlds satisfies

---

[4]See [2, p.12]

$P(x)$. This semantical analysis parallels that of Fitting, [5], p.114: "In short, there are two basic actions: letting i designate, and moving to an alternative world. These two actions commute only if i is a rigid designator. Ordinary first-order modal syntax has no machinery to distinguish the two alternative readings of $\Box P(i)$. Consequently when non-rigid designators have been treated at all, one of the readings has been disallowed, thus curtailing expressive power.". According to Fitting, if i is a rigid designator then $|_x^i|P(x) \leftrightarrow \Box P(i)$ holds, or, in his notation, the equivalence $\lambda x(\Box P(x))(i) \leftrightarrow \Box[\lambda x.P(x)(i)]$ holds. We are going to disagree on this point, for we shall show that the failure of the equivalence $|_x^i|P(x) \leftrightarrow |\star|P(i)$ does not depend on i being a non-rigid designator: in transition semantics this equivalence does not hold for rigid designators either.

Rigid designators

If i is a rigid designator, the implication $|_x^i|P(x) \to |\star|P(i)$ is t-valid, whereas $|\star|P(i) \to |_x^i|P(x)$ admits of countermodels. The t-validity of $|_x^i|P(x) \to |\star|P(i)$ is shown as follows. $\sigma \models_w |_x^i|P(x)$ iff for all v such that wRv and for all v-assignment τ such that $\tau(x)$ is an inheritor of $I_w(i)$ in D_v, $\tau \models_v P(x)$. Since i is a rigid designator, $I_v(i)$ is one of the inheritors of $I_w(i)$, therefore if τ is such that $\tau(x) = I_v(i)$, then $\tau(x) \in I_v(P)$, hence $I_v(i) \in I_v(P)$, so for all v, wRv and all v-assignment τ, $\tau \models_v P(i)$, consequently $\sigma \models_w |\star|P(i)$.

A countermodel for $|\star|P(i) \to |_x^i|P(x)$ can be readily constructed: assume that v is the only world related to w and that $I_v(i) \in I_v(P)$, so for all $v.wRv$. and all v-assignment τ, $\tau \models_v P(i)$, therefore $\sigma \models_w |\star|P(i)$. Assume moreover that $I_w(i)$ has two distinct inheritors in v, namely $I_v(i)$ and c and that $c \notin I_v(P)$, consequently there is a v-assignment τ such that $\tau(x) = c$ hence $\tau \not\models_v P(x)$, and so $\sigma \not\models_w |_x^i|P(x)$.

Variables are rigid designators so, in particular

$$|_{x_1}^y \ldots _{x_n}^y|P(x_1, \ldots, x_n) \to |y|(P(y, \ldots, y))$$

is t-valid.

Let wRv and τ be a v-assignment. If $\sigma \models_w |_{x_1}^y \ldots _{x_n}^y|P(x_1, \ldots, x_n)$, then $P(x_1, \ldots, x_n)$ is satisfied in v by any n-tuple of inheritors of $\sigma(y)$, therefore it is satisfied in v by the n-tuple $\langle \tau(y), \ldots, \tau(y) \rangle$, for some particular inheritor $\tau(y)$.

Stable designators

The validity of $|\star|P(i) \to |_x^i|P(x)$ requires the assumption that $I_w(i)$ has at most one inheritor in any related world v and that the inheritor (if any) in v of $I_w(i)$ coincides with $I_v(i)$.

An individual constant i is *stable* iff

- if wRv and $I_w(i) \, T_{\langle w,v \rangle} \, c$ then $I_v(i) = c$.

If an individual constant i is stable, then in particular

$$|x_1 \ldots x_n|A(x_1, \ldots, x_n, i) \to |x_1 \ldots x_n \, _x^i|A(x_1, \ldots, x_n, x)$$

is t-valid.

Definition 8 Simultaneous substitution for formulas. *Given a wff A containing the free variables $x_1, \ldots, x_k$, we define the wff $A[s_1/x_1 \ldots s_k/x_k]$ where s_i is substituted for x_i, $1 \leq i \leq k$. Let $[\mathbf{s}/\mathbf{x}] =_{df} [s_1/x_1 \ldots s_k/x_k]$.*

- $\bot \, [\mathbf{s}/\mathbf{x}] = \bot$
- $(P^n t_1, \ldots, t_n)[\mathbf{s}/\mathbf{x}] = P^n t_1[\mathbf{s}/\mathbf{x}], \ldots, t_n[\mathbf{s}/\mathbf{x}]$
- $(A \to B)[\mathbf{s}/\mathbf{x}] = (A[\mathbf{s}/\mathbf{x}] \to B[\mathbf{s}/\mathbf{x}])$
- $(\forall y A)[\mathbf{s}/\mathbf{x}] =$

$$= \begin{cases} \forall y A & \text{if } y \in \mathbf{x} \\ \forall z((A[z/y])[\mathbf{s}/x]) & \text{if } y \notin \mathbf{x} \text{ and } y \in \mathbf{s} \\ & \text{where } z \text{ doesn't occur in } \forall y A \\ \forall y (A[\mathbf{s}/x]) & \text{if } y \notin \mathbf{x} \text{ and } y \notin \mathbf{s} \end{cases}$$

- $(\mid \begin{smallmatrix} t_1 \\ y_1 \end{smallmatrix} \ldots \begin{smallmatrix} t_n \\ y_n \end{smallmatrix} \mid A)[\mathbf{s}/\mathbf{x}] = \mid \begin{smallmatrix} t_1[\mathbf{s}/\mathbf{x}] \\ y_1 \end{smallmatrix} \ldots \begin{smallmatrix} t_n[\mathbf{s}/\mathbf{x}] \\ y_n \end{smallmatrix} \mid A$, *in particular*
- $(\mid x_1 \ldots x_k \mid A)[s_1/x_1 \ldots s_k/x_k] = \mid \begin{smallmatrix} s_1 \\ x_1 \end{smallmatrix} \ldots \begin{smallmatrix} s_k \\ x_k \end{smallmatrix} \mid A$

Lemma 2 *Let A be a wff and z a variable that doesn't occur in A. For all t-models $\mathcal{M}$ and w-assignments σ,*

$$\sigma^{x \triangleright a} \models_w A \quad iff \quad \sigma^{z \triangleright a} \models_w A[z/x].$$

Proof By induction on A.
$\sigma^{z \triangleright a} \models_w P(t_1, \ldots, t_n)[z/x]$ iff $\sigma^{z \triangleright a} \models_w P(t_1[z/x], \ldots, t_n[z/x])$ iff $\langle \sigma^{z \triangleright a}(t_1[z/x]), \ldots, \sigma^{z \triangleright a}(t_n[z/x]) \rangle \in I_w(P)$ iff by lemma 1, $\langle \sigma^{x \triangleright a}(t_1), \ldots, \sigma^{x \triangleright a}(t_n) \rangle \in I_w(P)$ iff $\sigma^{x \triangleright a} \models_w P(t_1, \ldots, t_n)$.

$\sigma^{z \triangleright a} \models_w (\mid \begin{smallmatrix} t_1 \\ x_1 \end{smallmatrix} \ldots \begin{smallmatrix} t_n \\ x_n \end{smallmatrix} \mid A)[z/x]$ iff $\sigma^{z \triangleright a} \models_w \mid \begin{smallmatrix} t_1[z/x] \\ x_1 \end{smallmatrix} \ldots \begin{smallmatrix} t_n[z/x] \\ x_n \end{smallmatrix} \mid A$ iff $\tau \models_v A$, where $\sigma^{z \triangleright a}(t_1[z/x]) \mathcal{T} \tau(x_1) \ldots \sigma^{z \triangleright a}(t_n[z/x]) \mathcal{T} \tau(x_n)$, therefore, by lemma 1, $\sigma^{x \triangleright a}(t_1) \mathcal{T} \tau(x_1) \ldots \sigma^{x \triangleright a}(t_n) \mathcal{T} \tau(x_n)$, so $\sigma^{x \triangleright a} \models_w \mid \begin{smallmatrix} t_1 \\ x_1 \end{smallmatrix} \ldots \begin{smallmatrix} t_n \\ x_n \end{smallmatrix} \mid A$ ♠

Lemma 3 *(Alphabetic change of bound variables) Let A be a wff and z be a variable not occurring in A.*

$$\sigma \models_w \forall x A \quad iff \quad \sigma \models_w \forall z (A[z/x])$$

Proof $\sigma \models_w \forall x A$ iff $\sigma^{x \triangleright a} \models_w A$ for all $a \in D_w$ iff (by lemma 2) $\sigma^{z \triangleright a} \models_w A[z/x]$ for all $a \in D_w$ iff $\sigma \models_w \forall z (A[z/x])$. ♠

Lemma 4 *Substitution and satisfaction for formulas. Let σ be a w-assignment.*

$$\sigma \models_w A[s/x] \quad iff \quad \sigma^{x \triangleright \sigma(s)} \models_w A$$

Proof By induction on A.

- $A = P^n(t_1, \ldots, t_n)$

$\sigma^{x \triangleright \sigma(s)} \models_w P^n(t_1, \ldots, t_n)$ iff $\langle \sigma^{x \triangleright \sigma(s)}(t_1), \ldots, \sigma^{x \triangleright \sigma(s)}(t_n) \rangle \in I_w(P^n)$ iff $\langle \sigma(t_1[s/x]), \ldots, \sigma(t_n[s/x]) \rangle \in I_w(P^n)$ iff $\sigma \models_w P^n(t_1[s/x], \ldots, t_n[s/x])$ iff $\sigma \models_w P^n(t_1, \ldots, t_n)[s/x]$.

- $A = \forall y B$

    $\sigma^{x \triangleright \sigma(s)} \models_w \forall y B$ iff for all $d \in D_w$, $\sigma^{x \triangleright \sigma(s), y \triangleright d} \models_w B$ iff for all $d \in D_w$, $\sigma^{y \triangleright d, x \triangleright \sigma(s)} \models_w B$ iff, by induction hypothesis, for all $d \in D_w$, $\sigma^{y \triangleright d} \models_w B[s/x]$ iff $\sigma \models_w \forall y(B[s/x])$ iff by def. of substitution $\sigma \models_w (\forall y B)[s/x]$.

- $A = |{}^{t_1}_{y_1} \ldots {}^{t_n}_{y_n}|B$

    $\sigma^{x \triangleright \sigma(s)} \models_w |{}^{t_1}_{y_1} \ldots {}^{t_n}_{y_n}|B$ iff for all v-assignment τ such that $\sigma^{x \triangleright \sigma(s)}(t_i) \, \mathcal{T}_{\langle w, v \rangle} \, \tau(y_i)$, $1 \leq i \leq n$, $\tau \models_v B$ iff for all v-assignment τ such that $\sigma(t_i[s/x]) \, \mathcal{T}_{\langle w, v \rangle} \, \tau(y_i)$, $1 \leq i \leq n$, $\sigma \models_v |{}^{t_1[s/x]}_{y_1} \ldots {}^{t_n[s/x]}_{y_n}|B$ iff $\sigma \models_v (|{}^{t_1}_{y_1} \ldots {}^{t_n}_{y_n}|B)[s/x]$.

♠

3.1 Relevant formulas

PRM (Permutation)

$$|x_1 \ldots x_n|A \leftrightarrow |x_{i_1} \ldots x_{i_n}|A$$

for any permutation $x_{i_1} \ldots x_{i_n}$ of $x_1 \ldots x_n$.

RG (Rigidity of terms)

$$|{}^{t_1}_{x_1} \ldots {}^{t_n}_{x_n}|A \to |v_1 \ldots v_k|(A[t_1/x_1 \ldots t_n/x_n])$$

where $v_1 \ldots v_k$ are all the variables occurring in $t_1 \ldots t_n$.

RG^v (Rigidity of variables)

$$|{}^{y_1}_{x_1} \ldots {}^{y_n}_{x_n}|A \to |y_1 \ldots y_k|(A[y_1/x_1 \ldots y_n/x_n])$$

where $y_1 \ldots y_k$ are the variables $y_1 \ldots y_n$ without repetitions.

RNM (Renaming)

$$|x_1 \ldots x_n|A(x_1 \ldots x_n) \to |{}^{x_1}_{y_1} \ldots {}^{x_n}_{y_n}|(A[y_1/x_1 \ldots y_n/x_n])$$

where $y_1 \ldots y_n$ are pairwise distinct variables.

BF (Barcan Formula)

$$\forall z |x_1 \ldots x_n, z|A \to |x_1 \ldots x_n|\forall z A$$

CBF (Converse of Barcan Formula)

$$|x_1 \ldots x_n|\forall z A \to \forall z |x_1 \ldots x_n, z|A$$

GF (Ghilardi Formula)

$$\exists z |x_1 \ldots x_n, z|A \to |x_1 \ldots x_n|\exists z A$$

CGF (Converse of Ghilardi Formula)
$$|x_1 \ldots x_n| \exists z A \to \exists z |x_1 \ldots x_n, z| A$$

SHRT (Shortening)
$$|x_1 \ldots x_n, z| A \to |x_1 \ldots x_n| A$$

LNGT (Lenghtening)
$$|x_1 \ldots x_n| A \to |x_1 \ldots x_n, z| A$$

CRG (Converse of RG)
$$|v_1 \ldots v_k| (A[t_1/x_1 \ldots t_n/x_n]) \to |\begin{smallmatrix}t_1\\x_1\end{smallmatrix} \ldots \begin{smallmatrix}t_n\\x_n\end{smallmatrix}| A$$

where $v_1, \ldots, v_k$ are all the variables occurring in $t_1, \ldots, t_n$.

CRG^v (Converse of RG^v)
$$|y_1 \ldots y_k| (A[y_1/x_1 \ldots y_n/x_n]) \to |\begin{smallmatrix}y_1\\x_1\end{smallmatrix} \ldots \begin{smallmatrix}y_n\\x_n\end{smallmatrix}| A$$

where $y_1 \ldots y_k$ are the variables $y_1 \ldots y_n$ without repetitions.

SIV (Substitution that Identifies Variables)
$$|v_1 \ldots v_k| (A[y/x_1, y/x_2, t_3/x_3 \ldots t_n/x_n]) \to |\begin{smallmatrix}y\\x_1\end{smallmatrix} \begin{smallmatrix}y\\x_2\end{smallmatrix} \begin{smallmatrix}t_3\\x_3\end{smallmatrix} \ldots \begin{smallmatrix}t_n\\x_n\end{smallmatrix}| A$$

where $v_1, \ldots, v_k$ are all the variables occurring in $y, t_3, \ldots, t_n$.

FCS (Full Commutativity of Substitution)
$$|v_1 \ldots v_k| (A[t_1/x_1 \ldots t_n/x_n]) \leftrightarrow |\begin{smallmatrix}t_1\\x_1\end{smallmatrix} \ldots \begin{smallmatrix}t_n\\x_n\end{smallmatrix}| A$$

where $v_1, \ldots, v_k$ *include* all the variables occurring in $t_1, \ldots, t_n$.

NI (Necessity of Identity)
$$x = y \to |x, y| (x = y)$$

ND (Necessity of Distinction)
$$x \neq y \to |x, y| (x \neq y)$$

LBZ (Leibniz's law)
$$t = s \to (A[t/x] \to A[s/x])$$

4 A quantified modal logic with indexed modalities: $Q.K_{im}$.

Axioms

Tautologies

PRM $\quad |x_1\ldots x_n|A \leftrightarrow |x_{i_1}\ldots x_{i_n}|A$
for any permutation $x_{i_1}\ldots x_{i_n}$ of $x_1\ldots x_n$

K $\quad |x_1\ldots x_n|(A \to B) \to (|x_1\ldots x_n|A \to |x_1\ldots x_n|B)$

UI $\quad \forall x A \to A$

LNGT $\quad |x_1\ldots x_n|A \to |x_1\ldots x_n, z|A$

RG^v $\quad |{}^{y_1}_{x_1}\ldots {}^{y_n}_{x_n}|A \to |y_1\ldots y_k|(A[y_1/x_1\ldots y_n/x_n])$
where $y_1\ldots y_k$ are the variables $y_1\ldots y_n$ without repetitions.[5]

ID $\quad x = x$

LBZ $\quad t = s \to (A[t/x] \to A[s/x])$

Inference rules

$$\frac{A \quad A \to B}{B} \quad \textit{Modus Ponens}\,(\text{MP})$$

$$\frac{A}{|x_1\ldots x_n|A} \quad \textit{Necessitation}\,(\text{N}), \text{provided } \{x_1,\ldots,x_n\} \supseteq fv(A).$$

$$\frac{A \to B}{A \to \forall x B} \quad \textit{Universal Generalization}\,(\text{UG}), \text{provided } x \notin fv(A).$$

$$\frac{A}{A[s/x]} \quad \textit{Substitution for Free Variables}\,(\text{SFV})$$

Theorem 1 *(Soundness.) Every theorem of $Q.K_{im}$ is t-valid. Every theorem of $R.K_{im} = Q.K_{im} + RG$ is true in all t-models with rigid designators based on any t-frame.*

Some derivations

---

[5]Axiom $\mathbf{RG^v}$ could be formulated in a more general form so as to imply axiom **LNGT**: $|{}^{y_1}_{x_1}\ldots {}^{y_n}_{x_n}|A \to |v_1\ldots v_k|(A[y_1/x_1\ldots y_n/x_n])$, where $v_1\ldots v_k$ *include* all the different variables among $y_1\ldots y_n$.

"Necessary for"

$Q.K_{im} \vdash RNM$

$|_{x_1}^{y_1} \ldots _{x_n}^{y_n}|A \to |y_1 \ldots y_n|(A[y_1/x_1 \ldots y_n/x_n])$ RG^v

$(|_{x_1}^{y_1} \ldots _{x_n}^{y_n}|A)[x_1 \ldots x_n/y_1 \ldots y_n] \to (|y_1 \ldots y_n|(A[y_1/x_1 \ldots y_n/x_n]))[x_1 \ldots x_n/y_1 \ldots y_n]$

$|x_1 \ldots x_n|A \to |_{y_1}^{x_1} \ldots _{y_n}^{x_n}|(A[y_1/x_1 \ldots y_n/x_n])$

where $y_1 \ldots y_n$ are pairwise distinct variables not occurring in A.

$Q.K_{im} + LNGT \vdash CBF$

$\forall x A(\vec{x}, x) \to A(\vec{x}, x)$ UI
$|\vec{x}, x|\forall x A(\vec{x}, x) \to |\vec{x}, x|A(\vec{x}, x)$ N
$|\vec{x}|\forall x A(\vec{x}, x) \to |\vec{x}, x|A(\vec{x}, x)$ LNGT
$|\vec{x}|\forall x A(\vec{x}, x) \to \forall x|\vec{x}, x|A(\vec{x}, x)$ UG

$Q.K_{im} + CBF \vdash LNGT$

$A(\vec{x}) \to A(\vec{x})$ ID
$A(\vec{x}) \to \forall x A(\vec{x})$ $x \notin A$ UG
$|\vec{x}|A(\vec{x}) \to |\vec{x}|\forall x A(\vec{x})$ N
$|\vec{x}|A(\vec{x}) \to \forall x|\vec{x}, x|A(\vec{x})$ CBF
$|\vec{x}|A(\vec{x}) \to |\vec{x}, x|A(\vec{x})$ UI

$Q.K_{im} + SHRT \vdash GF$

$A(\vec{x}, x) \to \exists x A(\vec{x}, x)$
$|\vec{x}, x|A(\vec{x}, x) \to |\vec{x}, x|\exists x A(\vec{x}, x)$ N
$|\vec{x}, x|A(\vec{x}, x) \to |\vec{x}|\exists x A(\vec{x}, x)$ SHRT
$\exists x|\vec{x}, x|A(\vec{x}, x) \to |\vec{x}|\exists x A(\vec{x}, x)$

$Q.K_{im} + GF \vdash SHRT$

$\neg A \to \neg A$
$\neg A \to \forall x \neg A$ $x \notin A$
$\exists x A(\vec{x}) \to A(\vec{x})$
$|\vec{x}|\exists x A(\vec{x}) \to |\vec{x}|A(\vec{x})$ N
$\exists x|\vec{x}, x|A(\vec{x}) \to |\vec{x}|A(\vec{x})$ GF
$|\vec{x}, x|A(\vec{x}) \to \exists x|\vec{x}, x|A(\vec{x})$ from UI
$|\vec{x}, x|A(\vec{x}) \to |\vec{x}|A(\vec{x})$ trans.

$Q.K_{im} + SIV \vdash NI$

$x = x$	ID
$\lvert x \rvert (x = x)$	N
$\lvert x \rvert ((x = y)[x/x, x/y]) \to \lvert {}^{x,\,x}_{x,\,y} \rvert (x = y)$	SIV
$\lvert x \rvert (x = x) \to \lvert {}^{x,\,x}_{x,\,y} \rvert (x = y)$	
$\lvert {}^{x,\,x}_{x,\,y} \rvert (x = y)$	MP
$\lvert {}^{x,\,x}_{x,\,y} \rvert (x = y) \to (x = y \to \lvert {}^{x,\,y}_{x,\,y} \rvert (x = y))$	LBZ
$x = y \to \lvert x, y \rvert (x = y)$	MP

$Q.K_{im} + NI \vdash SIV$

Let $B(x, y)$ be given, and, for simplicity's sake, let us assume it to be atomic.

$x = y \to (B(x, x) \to B(x, y))$	LBZ
$\lvert x\,y \rvert (x = y) \to (\lvert x\,y \rvert B(x, x) \to \lvert x\,y \rvert B(x, y))$	N
$(x = y) \to \lvert x\,y \rvert (x = y)$	NI
$(x = y) \to (\lvert x\,y \rvert B(x, x) \to \lvert x\,y \rvert B(x, y))$	
$(x = y)[x/x, x/y] \to (\lvert x\,y \rvert (B(x,x))[x/x, x/y] \to$	SFV
$\qquad (\lvert x\,y \rvert B(x,y))[x/x, x/y])$	
$(x = x) \to (\lvert {}^{x\,x}_{x\,y} \rvert B(x, x) \to \lvert {}^{x\,\,x}_{x\,y} \rvert B(x, y))$	
$(x = x)$	ID
$\lvert {}^{x\,x}_{x\,y} \rvert B(x, x) \to \lvert {}^{x\,x}_{x\,y} \rvert B(x, y)$	MP
$\lvert x \rvert B(x, x) \to \lvert x\,y \rvert B(x, x)$	LNGT
$\lvert {}^{x}_{x} \rvert B(x, x) \to \lvert {}^{x\,x}_{x\,y} \rvert B(x, x)$	SFV
$\lvert {}^{x}_{x} \rvert B(x, x) \to \lvert {}^{x\,x}_{x\,y} \rvert B(x, y)$	trans.

$Q.K_{im} + CRG \vdash SIV$

$\lvert v_1 \ldots v_k \rvert (A[y/x_1, y/x_2, t_3/x_3 \ldots t_n/x_n]) \to \lvert {}^{y}_{x_1}\,{}^{y}_{x_2}\,{}^{t_3}_{x_3} \ldots {}^{t_n}_{x_n} \rvert A$	CRG

where $v_1, \ldots, v_k$ are all the variables occurring in $y, t_3, \ldots, t_n$

SIV is a particular case of CRG, exactly when $t_1 = t_2 = y$ so the same variable y is substituted for x_1 and x_2.

$Q.K_{im} + FCS \vdash LNGT$

$\lvert {}^{x_1}_{x_1} \ldots {}^{x_n}_{x_n} \rvert A \to \lvert x_1 \ldots x_n, z \rvert (A[x_1/x_1 \ldots x_n/x_n])$	FCS

where $x_1, \ldots, x_n, z$ *include* all the variables among $x_1, \ldots, x_n$

$\lvert x_1 \ldots x_n \rvert A \to \lvert x_1 \ldots x_n, z \rvert A$

"Necessary for"

$Q.K_{im} + FCS \vdash SHRT$

$|x_1 \ldots x_n, z|(A[x_1/x_1 \ldots x_n/x_n]) \to |^{x_1}_{x_1} \ldots ^{x_n}_{x_n}|A$ FCS

where $x_1, \ldots, x_n, z$ *include* all the variables among $x_1, \ldots, x_n$

$|x_1 \ldots x_n, z|A \to |x_1 \ldots x_n|A$

Trivially, $Q.K_{im} + FCS \vdash RG, CRG, NI$. In the presence of the principle of full commutativity of substitution, indexed modalities are unnecessary, in fact every box-operator can be thought of as implicitly indexed by the variables of the formula that follows it. This yelds that the standard modal language will do, but, as we shall see, we are confined to t-frames where the transition relation is a totally defined function. See [3].

A Quinean sentence: 'Necessarily the number of planets is greater than 7.'

Let i denote 'the number of planets'. Then, according to Quine the following derivation:

1. $\Box(7 < 9)$
2. $i = 9$
3. $\Box(7 < i)$

transforms the truth $\Box(7 < 9)$ into the falsehood $\Box(7 < i)$. We want to point out that the conclusion is not obtained merely by an application of the substitution of identical terms, but rather it relies on the acceptance of strong principles about substitution. The above inference can be analyzed in a language with indexed modalities as follows:

$$\cfrac{i = 9 \qquad \cfrac{\cfrac{|\star|(7<9)}{|^7_x,^9_y|(x<y)} CRG \qquad \cfrac{|^7_x,^9_y|(x<y) \to (i=9 \to |^7_x,^i_y|(x<y))}{i=9 \to |^7_x,^i_y|(x<y)} LBZ}{|^7_x,^i_y|(x<y)} MP}{\cfrac{|^7_x,^i_y|(x<y)}{|\star|(7<i)} RG}$$

Even if we can accept that 9 and 7 are stable designators and so CRG holds for them, i can hardly be called a rigid designator.

5 Correspondence

$\boxed{BF}$

$\mathcal{F}^t \models \forall x|x_1 \ldots x_n x|A \to |x_1 \ldots x_n|\forall x A$ iff $\mathcal{T}$ is surjective.

We show that if $\mathcal{T}$ is not surjective then $\mathcal{F}^t \not\models \forall x|x_1 \ldots x_n x|A \to |x_1 \ldots x_n|\forall x A$, where $\mathcal{T}$ is *surjective* iff for all w, v, if $b \in D_v$ then there is an $a \in D_w$ such that $a\mathcal{T}_{\langle w,v\rangle}b$.

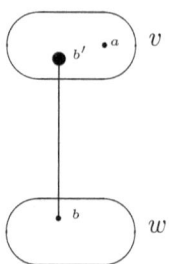

Let $a \notin I_v(P)$, $b' \in I_v(P)$ and $\sigma(x) = a$. Then $\sigma \models_w \forall x |x| P(x)$, $\tau \not\models_v \forall x P(x)$, $\sigma \not\models_w \forall x |x| P(x) \to | \star |\forall x P(x)$.

$\boxed{GF}$

$$\mathcal{F}^t \models \exists x |x_1 \ldots x_n\, x| A \to |x_1 \ldots x_n| \exists x A \qquad \text{iff} \qquad \mathcal{T} \text{ is totally defined.}$$

We show that if $\mathcal{T}$ is not totally defined then $\mathcal{F}^t \not\models \exists x |x_1 \ldots x_n\, x| A \to |x_1 \ldots x_n| \exists x A$, where $\mathcal{T}$ is *totally defined* iff for all w, v, if $a \in D_w$ then there is an $b \in D_v$ such that $a \mathcal{T}_{\langle w,v \rangle} b$.

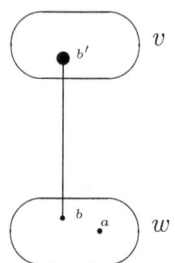

Let $b' \notin I_v(P)$ and $\sigma(x) = a$. Then $\sigma \models_w |x| P(x)$, so $\sigma \models_w \exists x |x| P(x)$, therefore $\sigma \not\models_w \exists x |x| P(x) \to | \star |\exists x P(x)$.

$\boxed{NI}$

$$\mathcal{F}^t \models x = y \to |x\, y|(x = y) \qquad \text{iff} \qquad \mathcal{T} \text{ is a partial function.}$$

We show that if $\mathcal{T}$ is not a partial function then $\mathcal{F}^t \not\models x = y \to |x\, y|(x = y)$, where $\mathcal{T}$ is a *partial function* if for all w, v, if $a \mathcal{T}_{\langle w,v \rangle} b$ and $a \mathcal{T}_{\langle w,v \rangle} c$ then $b = c$.

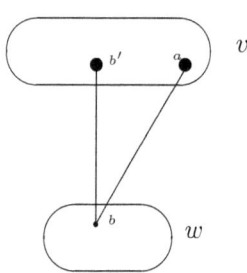

Let $\sigma(x) = \sigma(y) = b$. Then $\sigma \models_w x = y$, but $\sigma \not\models_w |x, y|(x = y)$, so $\sigma \not\models_w x = y \to |x\, y|(x = y)$.

"Necessary for"

$\boxed{ND}$

$$\mathcal{F}^t \models x \neq y \to |xy|(x \neq y) \quad \text{iff} \quad \mathcal{T} \text{ is not convergent.}$$

We show that if $\mathcal{T}$ is convergent then $\mathcal{F}^t \not\models x \neq y \to |xy|(x \neq y)$, where $\mathcal{T}$ is *not convergent* iff for all w, v, if $a\mathcal{T}_{\langle w,v\rangle}c$ and $b\mathcal{T}_{\langle w,v\rangle}c$ then $a = b$.

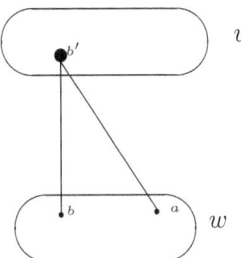

$\boxed{FCS}$

$$\mathcal{F}^t \models |v_1 \ldots v_k|(A[t_1/x_1 \ldots t_n/x_n]) \leftrightarrow |{}^{t_1}_{x_1} \ldots {}^{t_n}_{x_n}|A$$
iff
$\mathcal{T}$ is a totally defined function,

where $v_1, \ldots, v_k$ *include* all the variables occurring in $t_1, \ldots, t_n$.

6 Completeness theorem for $R.K_{im}$

We start by considering the modal logic $R.K_{im} = Q.K_{im} + RG$, where RG is the axiom of rigidity of terms.

6.1 Preliminaries

First we define a classical first-order language $\mathcal{L}^c$ that mimics the modal language $\mathcal{L}$.[6]

- $\mathcal{L}^c$ contains all the predicate and function symbols of $\mathcal{L}$,
- for each wff of $\mathcal{L}$,

$$|x_1 \ldots x_n|A$$

$\mathcal{L}^c$ contains the n-**ary predicate symbol**

$$P_{|x_1 \ldots x_n|A}$$

To every modal formula A of $\mathcal{L}$ we assign a classical formula $A^c \in \mathcal{L}^c$

$$\begin{aligned}
(P^n t_1, \ldots, t_n)^c &= P^n t_1, \ldots, t_n \\
(A \sharp B)^c &= A^c \sharp B^c \\
(\forall x A)^c &= \forall x (A^c) \\
(|{}^{t_1}_{x_1} \ldots {}^{t_n}_{x_n}|A)^c &= P_{|x_1 \ldots x_n|A}(t_1 \ldots t_n)
\end{aligned}$$

---

[6] The proof we present here is based on Ghilardi's completeness proof in [1].

We can easily see that if A contains no modal operators, then A^c is just A and that every formula B of $\mathcal{L}^c$ is equal to A^c for some $A \in \mathcal{L}$.

Second, we define the classical theory $C_{R.K_{im}}$ whose axioms are

$$\{A^c : R.K_{im} \vdash A\}$$

and whose inference rules are MP, UG and SFV.

Lemma 5 $X \vdash_{R.K_{im}} A$ iff $X^c \vdash_{C_{R.K_{im}}} A^c$.

Proof It is easy to see that $\vdash_{R.K_{im}} B_1 \wedge \cdots \wedge B_n \to A$ iff $\vdash_{C_{R.K_{im}}} B_1^c \wedge \cdots \wedge B_n^c \to A^c$, where $B_1, \ldots, B_n \in X$.
$\Rightarrow$ holds by definition of $C_{R.K_{im}}$.
$\Leftarrow$ holds because the axioms of $C_{R.K_{im}}$ are the c-translation of the theorems of $R.K_{im}$ and the inference rules of $C_{R.K_{im}}$ are also inference rules of $R.K_{im}$. ♠

$C_{R.K_{im}}$ is a first order theory, so models of $C_{R.K_{im}}$ are pairs $w = \langle D_w, I_w \rangle$ composed on a non-empty domain D_w and an interpretation function I_w such that the universal closure of all the theorems of $C_{R.K_{im}}$ is true in them.

We use the letters $w, v, \ldots$ to denote $C_{R.K_{im}}$ models. By $\langle \sigma, w \rangle \models A^c$ we denote that A^c is satisfied in the model $w = \langle D_w, I_w \rangle$ under the w-assignment σ.

An *admissible relation* $T_{\langle w, v \rangle}$ among $C_{R.K_{im}}$-models w and v is a relation $T_{\langle w, v \rangle} \subseteq D_w \times D_v$ satisfying the following two requirements

(A) for every term t, for every w-assignment π and for every v-assignment μ,

if $\pi(y_1) T_{\langle w, v \rangle} \mu(y_1), \ldots, \pi(y_k) T_{\langle w, v \rangle} \mu(y_k)$ then $\pi(t) T_{\langle w, v \rangle} \mu(t)$

where t contains at most the variables $y_1, \ldots, y_k$.

(B) for every formula A of $\mathcal{L}$, for every w-assignment π and for every v-assignment μ,

if $\pi(y_1) T_{\langle w, v \rangle} \mu(y_1), \ldots, \pi(y_k) T_{\langle w, v \rangle} \mu(y_k)$,

then

$$\langle \pi, w \rangle \models P_{|y_1, \ldots, y_k|A}(y_1, \ldots, y_k) \quad \text{only if} \quad \langle \mu, v \rangle \models A^c,$$

where A contains at most the variables $y_1, \ldots, y_k$.

Lemma 6 Let w, v be $C_{R.K_{im}}$-models and σ and τ be assignments in w and v, respectively. If for every formula A of $\mathcal{L}$ containing at most the variables $x_1, \ldots, x_n$,

$$\langle \sigma, w \rangle \models P_{|x_1, \ldots, x_n|A}(x_1, \ldots, x_n) \quad \text{only if} \quad \langle \tau, v \rangle \models A^c,$$

then there is an admissible relation $T_{\langle w, v \rangle} \subseteq D_w \times D_v$ such that

$$\sigma(x_1) T_{\langle w, v \rangle} \tau(x_1), \ldots, \sigma(x_n) T_{\langle w, v \rangle} \tau(x_n).$$

Proof Define $\mathcal{T}_{\langle w,v\rangle}$ as follows:

$e\,\mathcal{T}_{\langle w,v\rangle}\,e'$ iff there is a term s containing at most the variables $x_1,\ldots,x_n$, such that $\sigma(s) = e$ and $\tau(s) = e'$.

Trivially $\sigma(x_1)\mathcal{T}_{\langle w,v\rangle}\tau(x_1),\ldots,\sigma(x_n)\mathcal{T}_{\langle w,v\rangle}\tau(x_n)$. We show that condition (A) holds. Let t be a term containing the variables $y_1,\ldots,y_k$, and let π and μ be w and v assignments, respectively, such that

$$\pi(y_1)\,\mathcal{T}_{\langle w,v\rangle}\,\mu(y_1),\ldots,\pi(y_k)\,\mathcal{T}_{\langle w,v\rangle}\,\mu(y_k)$$

then we have to show that

$$\pi(t)\,\mathcal{T}_{\langle w,v\rangle}\,\mu(t).$$

This amounts to show that there is a term s contaning at most the variables $x_1,\ldots,x_n$ such that

$$\sigma(s) = \pi(t) \quad \text{and} \quad \tau(s) = \mu(t).$$

By the definition of $\mathcal{T}_{\langle w,v\rangle}$ above, we know that for each i, $1 \leq i \leq k$, there is a term s_i containing the variables $x_1,\ldots,x_n$, such that

$$\sigma(s_i) = \pi(y_i) \quad \text{and} \quad \tau(s_i) = \mu(y_i)$$

Let $s = t[s_1/y_1,\ldots,s_k/y_k]$. Then $\sigma(s) = \sigma(t[s_1/y_1,\ldots,s_k/y_k]) =$ (by lemma 1) $= \sigma^{y_1 \triangleright \sigma(s_1),\ldots,y_k \triangleright \sigma(s_k)}(t) = \pi(t)$, since $\sigma(s_i) = \pi(y_i)$, $1 \leq i \leq k$.

As to condition (B), let A be a formula of $\mathcal{L}$ and let us assume that its free variables are among $y_1,\ldots,y_k$. Let π and μ be assignments in w and v, respectively, such that

- $\pi(y_1)\,\mathcal{T}_{\langle w,v\rangle}\,\mu(y_1),\ldots,\pi(y_k)\,\mathcal{T}_{\langle w,v\rangle},\mu(y_k)$, $1 \leq i \leq k$
- $\langle \pi, w \rangle \models P_{|y_1,\ldots,y_k|A}(y_1,\ldots,y_k)$

We have to show that $\langle \mu, v \rangle \models A^c$. By the definition of $\mathcal{T}_{\langle w,v \rangle}$, there are terms s_i containing at most the variables $x_1,\ldots,x_n$, such that $\sigma(s_i) = \pi(y_i)$ and $\tau(s_i) = \mu(y_i)$, $1 \leq i \leq k$. So

$$\langle \pi^{y_1 \triangleright \sigma(s_1),\ldots,y_k \triangleright \sigma(s_k)}, w \rangle \models P_{|y_1\ldots y_k|A}(y_1,\ldots,y_k),$$

and consequently

$$\langle \sigma^{y_1 \triangleright \sigma(s_1),\ldots,y_k \triangleright \sigma(s_k)}, w \rangle \models P_{|y_1\ldots y_k|A}(y_1,\ldots,y_k),$$

since all the free variables are among $y_1,\ldots,y_k$. Then by lemma 4,

$$\langle \sigma, w \rangle \models P_{|y_1\ldots y_k|A}(s_1,\ldots,s_k).$$

Since

$$R.K_{im} \vdash |^{s_1\ldots s_k}_{y_1\ldots y_k}|A \rightarrow |x_1\ldots x_n|(A[s_1/y_1,\ldots,s_k/y_k]), \text{(axiom } RG\text{)},$$

then

$$C_{R.K_{im}} \vdash P_{|y_1 \ldots y_k|A}(s_1, \ldots, s_k) \to P_{|x_1 \ldots x_n|(A[s_1/y_1, \ldots, s_k/y_k])}(x_1, \ldots, x_n),$$

so

$$\langle \sigma, w \rangle \models P_{|x_1 \ldots x_n|(A[s_1/y_1, \ldots, s_k/y_k])}(x_1, \ldots, x_n).$$

By the hypothesis of the lemma

$$\langle \tau, v \rangle \models (A[s_1/y_1, \ldots, s_k/y_k])^c,$$

i.e.

$$\langle \tau, v \rangle \models A^c[s_1/y_1, \ldots, s_k/y_k],$$

therefore by lemma 4

$$\langle \tau^{y_1 \triangleright \tau(s_1), \ldots, y_k \triangleright \tau(s_k)}, v \rangle \models A^c,$$

then

$$\langle \mu, v \rangle \models A^c,$$

since

$$\tau(s_1) = \mu(y_1), \ldots, \tau(s_k) = \mu(y_k).$$

♠

Lemma 7 *Let w be a $C_{R.K_{im}}$-model and $|x_1 \ldots x_m|A$ be a formula of $\mathcal{L}$ such that $\langle \sigma, w \rangle \not\models P_{|x_1 \ldots x_m|A}(x_1, \ldots, x_m)$. Then*

1. *the set of classical formulas*

$$\Gamma = \{B^c : \langle \sigma, w \rangle \models P_{|x_1 \ldots x_m|B}(x_1 \ldots x_m)\} \cup \{\neg A^c\}$$

   is $C_{R.K_{im}}$-consistent, where B contains at most the variables $x_1 \ldots x_m$,
2. *there is a classical model v of Γ and a v-assignment τ such that*

$$\langle \tau, v \rangle \models \Gamma,$$

3. *there is an admissible relation $T_{\langle w, v \rangle}$ such that*

$$\sigma(x_1) \, T_{\langle w, v \rangle} \, \tau(x_1), \ldots, \sigma(x_m) T_{\langle w, v \rangle} \, \tau(x_m).$$

Proof

1. Assume by *reductio* that

$$C_{R.K_{im}} \vdash B_1^c \wedge \cdots \wedge B_r^c \to A^c$$

   Then

$$R.K_{im} \vdash B_1 \wedge \cdots \wedge B_r \to A$$

$$R.K_{im} \vdash |x_1 \ldots x_m|B_1 \wedge \cdots \wedge |x_1 \ldots x_m|B_r \to |x_1 \ldots x_m|A \quad \text{by N}$$

$$C_{R.K_{im}} \vdash P_{|x_1\ldots x_m|\tilde{B}_1}(x_1\ldots x_m) \wedge \cdots \wedge P_{|x_1\ldots x_m|B_r}(x_1\ldots x_m) \rightarrow$$
$$P_{|x_1\ldots x_m|A}(x_1,\ldots,x_m).$$

Therefore
$$\langle \sigma, w \rangle \models P_{|x_1\ldots x_m|A}(x_1,\ldots,x_m)$$
contrary to the fact that
$$\langle \sigma, w \rangle \not\models P_{|x_1\ldots x_m|A}(x_1,\ldots,x_m).$$

2. By classical model theory.
3. By lemma 6.

♠

Subordination model. A *subordination model* is a tree $\langle S, \Sigma \rangle$ each node of which is (associated to) a classical model $w = \langle D_w, I_w \rangle$ together with an assignment $\sigma : VAR \to D_w$, so any element of S (any node of the tree) is a triple $\langle \sigma, D_w, I_w \rangle$. Given the node $\langle \sigma, D_w, I_w \rangle$ an *immediate subordinate* node $\langle \tau, D_v, I_v \rangle$, i.e. one for which the relation $\langle \sigma, D_w, I_w \rangle \Sigma \langle \tau, D_v, I_v \rangle$ holds, is defined according to the following procedure.

1. For each formula $\exists x A \in \mathcal{L}$ such that $\langle \sigma, D_w, I_w \rangle \models \exists x A^c$, consider a triple $\langle \sigma^{x \triangleright a}, D_w, I_w \rangle$ such that $\langle \sigma^{x \triangleright a}, D_w, I_w \rangle \models A^c$, for some $a \in D_w$.
   We say that $\langle \sigma, D_w, I_w \rangle \Sigma \langle \sigma^{x \triangleright a}, D_w, I_w \rangle$.
2. For each formula $\exists x A \in \mathcal{L}$ such that $\langle \sigma, D_w, I_w \rangle \not\models \exists x A^c$, consider all the triples $\langle \sigma^{x \triangleright a}, D_w, I_w \rangle$ such that $\langle \sigma^{x \triangleright a}, D_w, I_w \rangle \not\models A^c$, for any $a \in D_w$.
   We say that $\langle \sigma, D_w, I_w \rangle \Sigma \langle \sigma^{x \triangleright a}, D_w, I_w \rangle$, for all $a \in D_w$.
3. For each formula $|x_1 \ldots x_m| A \in \mathcal{L}$ such that $\langle \sigma, D_w, I_w \rangle \not\models P_{|x_1\ldots x_m|A}$, consider a triple $\langle \tau, D_v, I_v \rangle$ such that
   $\langle \tau, D_v, I_v \rangle \models \{B^c : \langle \sigma, D_w, I_w \rangle \models P_{|x_1\ldots x_n|B}(x_1,\ldots,x_n)\} \cup \{\neg A^c\}$.
   We say that $\langle \sigma, D_w, I_w \rangle \Sigma \langle \tau, D_v, I_v \rangle$ and that $\sigma(x_1) \mathcal{T}_{\langle w,v \rangle} \tau(x_1), \ldots, \sigma(x_m) \mathcal{T}_{\langle w,v \rangle} \tau(x_m)$.

Steps 1 and 2 are feasible thanks to classical model theory, step 3 thanks to lemma 6.

Lemma 8 *Let $R.K_{im} \not\vdash A$. Then there is a t-model $\mathcal{M} = \langle W, R, D, \mathcal{T}, I \rangle$ with rigid terms such that $\mathcal{M} \not\models A$.*

Proof Let us first build a subordination model $\langle S, \Sigma \rangle$ having at its root a node $\langle \sigma, D_w, I_w \rangle$ such that $\langle \sigma, D_w, I_w \rangle \models \neg A^c$. Then we define a transition model $\mathcal{M} = \langle W, D, R, \mathcal{T}, I \rangle$ as follows:

- $W = \{\langle D_w, I_w \rangle : \text{for some } \sigma, \langle \sigma, D_w, I_w \rangle \in S\}$
- D is such that $D(\langle D_w, I_w \rangle) = D_w$
- $R \subseteq W^2$ is such that $\langle D_w, I_w \rangle R \langle D_v, I_v \rangle$ iff $\langle \sigma, D_w, I_w \rangle \Sigma \langle \tau, D_v, I_v \rangle$ for some σ and τ

- $\mathcal{T} = \{\langle a, b \rangle :$ for some $\langle \sigma, D_w, I_w \rangle$ and $\langle \tau, D_v, I_v \rangle$, $a \in D_w$, $b \in D_v$, $\langle \sigma, D_w, I_w \rangle \Sigma \langle \tau, D_v, I_v \rangle$, $a = \sigma(x), b = \tau(x)$, and $\sigma(x) \mathcal{T}_{\langle w,v \rangle} \tau(x) \}$
- I is such that $I(\langle D_w, I_w \rangle) = I_w$

In the following, we write w instead of $\langle D_w, I_w \rangle$ and $\langle \sigma, w \rangle \models D^c$ instead of $\langle \sigma, D_w, I_w \rangle \models D^c$. It remains to show that

$$\sigma \models_w^{\mathcal{M}} D \quad iff \quad \langle \sigma, w \rangle \models D^c$$

for all $w \in W$ and all formulas $D \in \mathcal{L}$.

By induction on D. We examine just one case.

$$D = |_{y_1}^{t_1} \ldots _{y_n}^{t_n}| A$$

where $(fv(t_1) \cup \cdots \cup fv(t_n)) = \{x_1, \ldots, x_m\}$.

If

$$\sigma \not\models_w^{\mathcal{M}} |_{y_1}^{t_1} \ldots \ldots _{y_n}^{t_n}| A$$

then by lemma 4

$$\pi \not\models_w^{\mathcal{M}} |y_1 \ldots y_n| A$$

where $\pi = \sigma^{y_1 \triangleright \sigma(t_1), \ldots, y_n \triangleright \sigma(t_n)}$. Then by definition of satisfaction there is a v and a v-assignment τ, such that $\tau \not\models_v^{\mathcal{M}} A$, and $\sigma(t_i) \mathcal{T} \tau(y_i)$, $1 \leq i \leq n$. By induction hypothesis $\langle \tau, v \rangle \not\models A^c$, whence $\langle \pi, w \rangle \not\models P_{|y_1 \ldots y_n| A}(y_1, \ldots, y_n)$, because of condition (B). Consequently $\langle \sigma, w \rangle \not\models P_{|y_1 \ldots y_n| A}(t_1, \ldots, t_n)$.

If

$$\langle \sigma, w \rangle \not\models P_{|y_1 \ldots y_n| A}(t_1, \ldots, t_n),$$

then by lemma 4

$$\langle \pi, w \rangle \not\models P_{|y_1 \ldots y_n| A}(y_1, \ldots, y_n)$$

where $\pi = \sigma^{y_1 \triangleright \sigma(t_1) \ldots \ldots y_n \triangleright \sigma(t_n)}$. Then by lemma 7 there is a model v of $\Gamma = \{B^c : \langle \pi, w \rangle \models P_{|y_1 \ldots y_n| B}(y_1, \ldots, y_n)\} \cup \{\neg A^c\}$ and a v-assignment τ such that $\langle \tau, v \rangle \models \Gamma$ and $\sigma(t_i) \mathcal{T}_{\langle w,v \rangle} \tau(y_i)$, $1 \leq i \leq n$. Hence

$$\langle \tau, v \rangle \not\models A^c,$$

therefore by induction hypothesis $\tau \not\models_v^{\mathcal{M}} A$, so

$$\pi \not\models_w^{\mathcal{M}} |y_1 \ldots y_n| A.$$

Consequently

$$\sigma \not\models_w^{\mathcal{M}} |_{y_1}^{t_1} \ldots \ldots _{y_n}^{t_n}| A.$$

7 Completeness theorem for $Q.K_{im}$

The completenes theorem for $Q.K_{im}$ is easily obtained from the corresponding theorem for $R.K_{im}$. A relation $\mathcal{T}_{\langle w,v\rangle}$ among $C_{Q.K_{im}}$-models w, v is an *admissible relation* iff condition (B) is satisfied. In the proof of Lemma 6, define $\mathcal{T}_{\langle w,v\rangle}$ as follows:

$$e\, \mathcal{T}_{\langle w,v\rangle}\, e'$$

iff there is a variable $x_i \in \{x_1, \ldots, x_n\}$, such that $\sigma(x_i) = e$ and $\tau(x_i) = e'$.
Trivially $\sigma(x_i)\, \mathcal{T}_{\langle w,v\rangle}\, \tau(x_i)$, $1 \leq i \leq n$.
As to condition (B), take any modal formula A of $\mathcal{L}$ containing at most the free variables $y_1, \ldots, y_k$, and a pair of assignments π and μ in w and v, respectively, such that

- $\pi(y_1)\mathcal{T}_{\langle w,v\rangle}\mu(y_1), \ldots, \pi(y_k)\mathcal{T}_{\langle w,v\rangle}\mu(y_k)$
- $\langle \pi, w\rangle \models P_{|y_1,\ldots,y_k|A}(y_1, \ldots, y_k)$

We have to show that $\langle \mu, v\rangle \models A^c$. By the definition of $\mathcal{T}_{\langle w,v\rangle}$, there are variables $x_1^\star, \ldots x_k^\star$ among $x_1, \ldots, x_n$ such that $\sigma(x_i^\star) = \pi(y_i)$ and $\tau(x_i^\star) = (\mu(y_i)$, $1 \leq i \leq k$. So

$$\langle \pi^{y_1 \triangleright \sigma(x_1^\star),\ldots,y_k \triangleright \sigma(x_k^\star)}, w\rangle \models P_{|y_1\ldots y_k|A}(y_1, \ldots, y_k)$$

and consequently

$$\langle \sigma^{y_1 \triangleright \sigma(x_1^\star),\ldots,y_k \triangleright \sigma(x_k^\star)}, w\rangle \models P_{|y_1\ldots y_k|A}(y_1, \ldots, y_k)$$

since all the free variables are among $y_1, \ldots, y_k$. Then by lemma 4

$$\langle \sigma, w\rangle \models P_{|y_1\ldots y_k|A}(x_1^\star, \ldots, x_k^\star).$$

Since

$$Q.K_{im} \vdash |_{y_1\ldots y_k}^{x_1^\star \ldots x_k^\star}|A \to |x_1 \ldots x_n|(A[x_1^\star/y_1, \ldots, x_k^\star/y_k]) \text{ (axiom } RG^v),$$

then

$$C_{Q.K_{im}} \vdash P_{|y_1\ldots y_k|A}(x_1^\star, \ldots, x_k^\star) \to P_{|x_1\ldots x_n|(A[x_1^\star/y_1,\ldots,x_k^\star/y_k])}(x_1, \ldots, x_n),$$

so

$$\langle \sigma, w\rangle \models P_{|x_1\ldots x_n|(A[x_1^\star/y_1,\ldots,x_k^\star/y_k])}(x_1, \ldots, x_n).$$

By the hypothesis of the lemma

$$\langle \tau, v\rangle \models (A[x_1^\star/y_1, \ldots, x_k^\star/y_k])^c,$$

i.e.

$$\langle \tau, v\rangle \models A^c[x_1^\star/y_1, \ldots, x_k^\star/y_k],$$

therefore by lemma 4

$$\langle \tau^{y_1 \triangleright \tau(x_1^\star),\ldots,y_k \triangleright \tau(x_k^\star)}, v\rangle \models A^c,$$

therefore since $\tau(x_1^\star) = d_1, \ldots, \tau(x_k^\star) = d_k$,

$$\langle \tau^{y_1 \triangleright d_1, \ldots, y_k \triangleright d_k}, v \rangle \models A^c,$$

whence
$$\langle \mu, v \rangle \models A^c(y_1, \ldots, y_k)$$
since
$$\mu(y_1) = d_1, \ldots, \mu(y_k) = d_k.$$

Acknowledgements

My deep gratitude to Professor Dag Prawitz for having discussed with me at length a preliminary version of this paper.

Bibliography

[1] Braüner, T., Ghilardi, S. First-order Modal Logic. In *Handbook of Modal Logic*, 549–620. Elsevier, 2006.

[2] Corsi, G. Counterparts and possible worlds. A study on quantified modal logic. *Preprint, Università di Bologna, Dipartimento di Filosofia*, 21:1–61, 2001.

[3] Corsi, G. A unified completeness theorem for quantified modal logics. *The Journal of Symbolic Logic*, 67:1483–1510, 2002.

[4] Corsi, G. BF, CBF and Lewis semantics. *Logique & Analyse*, 181:103–122, 2003.

[5] Fitting, M., Mendelsohn, R. L. *First-Order Modal Logic*. Kluwer AP, 1999.

[6] Gabbay, D., Shehtman, V., Skvortsov, D. *Quantification in Nonclassical Logic*. Elsevier, 2009.

A proof complexity generator[1]

Jan Krajíček

Academy of Sciences and Charles University

krajicek@math.cas.cz

ABSTRACT. We define a map $g : \{0,1\}^n \to \{0,1\}^{n+1}$ such that all output bits are defined by 2DNF formulas in the input bits, and such that g has the following hardness property. For any $b \in \{0,1\}^{n+1} \setminus Rng(g)$, formula $\tau(g)_b$ naturally expressing that $b \notin Rng(g)$ requires exponential size proofs in any proof system for which the pigeonhole principle is exponentially hard. We define a class of generators generalizing g and show that there is a universal one in this class.

1 Introduction

Consider a map $g : x \in \{0,1\}^n \to y \in \{0,1\}^m$ defined by conditions

$$y_i \equiv \varphi_i(x)$$

where $\varphi_i(x)$ are propositional formulas in $x = (x_1, \ldots, x_n)$ and $m > n$. As the domain of g is smaller than $\{0,1\}^m$ there are $b \in \{0,1\}^m \setminus Rng(g)$. For any such b the formula $\tau(g)_b(x)$:

$$\bigvee_{i \in [m]} b_i \not\equiv \varphi_i(x)$$

expresses that $b \notin Rng(g)$ in the sense that $\tau(g)_b$ is a tautology iff $b \notin Rng(g)$.

Our aim is to define g for which the τ-formulas are hard to prove. When all $\tau(g)_b$ require super-polynomial (resp. exponential) size proofs in a proof system P we say (following [22]) that g **is hard (resp. exponentially hard) proof complexity generator for P**. The τ-formulas have been defined in [8] and independently in [2], and their theory is being developed (see [9–11, 14, 15, 21, 22]); the introductions to [11] or [22] offer a more comprehensive exposition. The property "$b \notin Rng(g)$" can be expressed by a tautology even for maps g with output bits defined by non-uniform $\mathcal{NP} \cap co\mathcal{NP}$ conditions on the input bits. Such a generality allowed Razborov [22] to formulate an intriguing conjecture about Extended Frege system EF (see also [10]). We do not need such a generality here.

The map g we define is exponentially hard for proof systems for which the pigeonhole principle is exponentially hard. This includes, for example, constant

---

[1] The author is with Academy of Sciences and Charles University, Prague. The paper is partially supported in part by grants A1019401, AV0Z10190503, MSM0021620839, 201/05/0124, and LC505. Mailing address: Mathematical Institute, Academy of Sciences, Žitná 25, Prague 1, CZ - 115 67, The Czech Republic.

depth Frege systems F_d, polynomial calculus PC or the system from [6] combining the two.

Finally we show that in a class of generators generalizing g, we call them **gadget generators**, there is a universal one.

Exponentially hard generators were previously constructed for resolution R (see [11, Thm.4.2] and [22, Thms.2.10,2.20]). Maps yielding hard τ-formulas for polynomial calculus and a system combing PC with R were constructed in [2] but under the assumption of a particular encoding used in the definition of the τ-formulas (linear encoding, see [2]).

Hard generators are also known to exists (assuming the hardness of factoring) for proof systems admitting feasible interpolation, and our construction applies to systems for which the pigeonhole principle is hard. Note that these two categories of proof systems (not mutually exclusive) cover virtually all[2] proof systems for which a super-polynomial lower bound is known.

The paper is organized as follows. We define the generator in Section 2 and in Section 3 we prove that it is (exponentially) hard for proof systems for which the pigeonhole principle is (exponentially) hard. The class of gadget generators is defined in Section 4 where we construct a universal one in the class.

We assume that the reader has a basic knowledge of proof complexity. In particular, we do not repeat definitions of the proof systems we write about. This background can be found[3] in [3, 4, 18] or in the papers cited at the respective places. However, we do not presume a prior knowledge of proof complexity generators.

Notation: $[n]$ is $\{1, \ldots, n\}$.

2 The generator

Definition 1 *Let $k \geq 1$ and put $t := k^2 + k + 1$ and*

$$n := k(k+1) + kt = k^3 + 2k^2 + 2k .$$

For an n of this form define map $g_n : \{0,1\}^n \to \{0,1\}^{n+1}$ as follows. Input string x of length n is interpreted as

$$x = (v, u^1, \ldots, u^t)$$

where

$$v = (v_{ij})_{i \in [k+1], j \in [k]} \quad and \quad u^s = (u^s_j)_{j \in [k]}$$

for $s = 1, \ldots, t$. We call v the **gadget**[4] *variables.*

The output string y of length $n + 1$ is defined as $y := (y^1, \ldots, y^t)$ where

$$y^s_i := \bigvee_{j \in [k]} (v_{ij} \wedge u^s_j)$$

---

[2] An example of an exception is a constant depth Frege system augmented by PHP as an additional axiom scheme.

[3] This paper accompanies my lecture "Proof complexity and proof search" at the 13th International Congress of Logic, Methodology and Philosophy of Science, Beijing (August 2007). It contains one of the new results mentioned in the talk but not the expository part of the talk. Some of that material can be found in [5, 7, 12, 13].

[4] I borrow this term from Razborov's comment.

for $s = 1, \ldots, t$ and $i \in [k+1]$.

Remarks: (1) We could have defined the generator by conditions

$$y_i^s := \bigwedge_{j \in [k]} (v_{ij} \rightarrow u_j^s) \ .$$

These conditions are equivalent to the original ones assuming that v_{ij}'s satisfy formula $Fn(v)$:

$$\bigwedge_{i \in [k+1]} \bigvee_{j \in [k]} v_{ij} \ \wedge \ \bigwedge_{i \in [k+1]} \bigwedge_{j_1 \neq j_2 \in [k]} \neg v_{ij_1} \vee \neg v_{ij_2}$$

expressing that $\{(i,j) \mid v_{ij} = 1\}$ is a graph of a function from $[k+1]$ to $[k]$. If we postulate that the output of the map is the zero vector whenever $Fn(v)$ fails, the two definitions would be literally equivalent as, assuming $Fn(v)$, $\neg v_{ij}$ is equivalent to $\bigvee_{j' \neq j} v_{ij'}$.

(2) We could have introduced $\ell \cdot k$ gadget variables v_{ij} with intended meaning to represent maps from $[\ell]$ into $[k]$, for $\ell \gg k$ and not just for $\ell = k+1$. The resulting generator would have the output/input ratio about ℓ/k, and its hardness would follow for proof systems where the weak pigeonhole principle PHP_k^ℓ is hard in the same way as Theorem 1. However, for proof systems where so weak PHP is hard to prove generators are known already, cf.[19, 22].

(3) In order to treat algebraic proof systems like polynomial calculus PC we can define the generator by degree 2 polynomials. In particular, put

$$y_i^s := \sum_{j \in [k]} v_{ij} \cdot u_j^s \ .$$

Note that these equations define the same map as the original condition assuming $Fn(v)$.

3 The hardness of generator g

Theorem 1 Let $d \geq 2$. Then for $k = 1, 2, \ldots$ and $n := k^3 + 2k^2 + 2k$ map $g_n : \{0,1\}^n \rightarrow \{0,1\}^{n+1}$ is an exponentially hard proof complexity generator for constant depth Frege systems F_d.

Let $b := (b^1, \ldots, b^t) \in \{0,1\}^{n+1}$ be an arbitrary string, b^s blocks of length $k+1$. Substitute in an alleged F_d-proof π of $\tau(g_k)_b$ everywhere

$$u_j^s := \bigvee_{i \in [k+1]} v_{ij} \wedge b_i^s \ .$$

The substitution depends on v so we shall denote it $u_j^s(v)$. Denote the substituted proof π'.

Let $\neg PHP_k^{k+1}(v)$ be the formula expressing that v defines a graph of a function violating the pigeonhole principle from $[k+1]$ into $[k]$ (not necessarily bijective):

$$Fn(v) \wedge \bigwedge_{i_1 \neq i_2 \in [k+1]} \bigwedge_{j \in [k]} (\neg p_{i_1 j} \vee \neg p_{i_2 j}) \ .$$

Then it is easy to see that there is a size $n^{O(1)} = k^{O(1)}$ F_d-proof σ of

$$\neg PHP_k^{k+1}(v) \to g_k(v, u^1(v), \ldots, u^t(v)) = b \; .$$

Combining σ and π' gives a proof of $PHP_k(v)$. However, by [1, 16, 17] any such proof must have size exponential in k. Hence π' (and so π too) must have exponential size too.

Remarks:

(1) We concentrate on F_d as these are the most important proof systems for which no hard generators were previously known. However, the argument utilizes just the hardness of PHP and so it applies to any proof system where PHP is hard and which supports the simple proof manipulations involved (or one of a variety of alternative formalizations of the argument).

(2) In particular, the argument can be modified for polynomial calculus PC. The $\tau(g_n)_b$ formula is represented by the following set of polynomial equations to be refuted:

$$b_i^s = \sum_{j \in [k]} v_{ij} \cdot u_j^s \; . \qquad (1)$$

The measure of complexity of proofs in PC is its degree. Using the dense notation for polynomials, degree d polynomials in n variables are encoded by strings of $O(n^d)$ of field elements. Hence an exponential lower bound on the size corresponds to an $n^{\Omega(1)}$ lower bound on the degree.

The hardness of g_k in PC is derived using the $k/2$ degree lower bound for PC refutations of $\neg PHP_k^{k+1}$ (even with any number $\ell > k$ of pigeons) from [20].

4 Gadget generators

One can consider maps of a general form similar to that of generator g: gadget v is simply a string of $\ell = \ell(k) = k^{O(1)}$ bits and each output block $y^s \in \{0,1\}^{k+1}$ is computed from $u^s \in \{0,1\}^k$ by a fixed polynomial time function f:

$$y^s := f(v, u^s) \; .$$

In fact, one can take for f the circuit-value function

$$CV_{\ell,k}(v, u)$$

that takes ℓ bits v describing a circuit C with k input bits and $k+1$ output bits and $u \in \{0,1\}^k$, and outputs the string $C(u)$.

For the generator to output more bits than its input has, e.g. $t := \ell + 1$ copies of blocks u^s suffice to swallow the gadget. Thus the only non-canonical part of this construction is the size of the gadget, the parameter ℓ. The observation we want to make is that, in fact, it is possible without a loss of generality to assume that $\ell \leq k^{1+\epsilon}$, any fixed $\epsilon > 0$. This is seen as follows.

Each string $CV_{\ell,k}(v, u^s)$ is computed by a circuit of size $O(\ell + k)$ and the whole generator in size $O(t \cdot (\ell + k))$. Thus for any $\epsilon > 0$ we can take $t > \ell$ large enough but still $t = k^{O(1)}$ such that the generator is computed in time $n^{1+\epsilon}$. Let us call such a generator G.

For the following argument we need a property of a generator stronger than the hardness: we need that not only is each $\tau(g)_b$ hard to prove but that also any disjunction $\bigvee_i \tau(g)_{b^i}$ needs long proofs. This is implied by the **iterability** of g, a concept defined in [11] (we refer the reader for details there).

It is easy to see that the (exponential) hardness of G in a proof system P follows if:

(*) There is any (exponentially) iterable map from $\{0,1\}^k$ to $\{0,1\}^{k+1}$ computed by a circuit of size $\leq \ell(k)^{1/2}$ (in particular, described by $\leq \ell(k)$ bits).

In fact, G is then also (exponentially) iterable. Hence we can now repeat the same construction again, taking blocks u^s of size n and gadgets v of size $n^{1+\epsilon}$. Call this map H.

Hypothesis (*) then implies that there is a size $n^{1+\epsilon}$ gadget describing a circuit computing G for which the corresponding instance of H, and hence H itself, is (exponentially) iterable too.

A general construction like this is unlikely to be useful for lower bounds for specific proof systems. However, a similar universal construction where the gadgets describe the data (a 0-1 matrix and a Boolean function) needed to define the Nisan-Wigderson generator (considered first in this context in [2]) can be helpful.

Acknowledgments

This paper owns its existence to the encouragement from A. A. Razborov (Princeton) to publish the simple construction.

Bibliography

[1] Ajtai, M. The complexity of the pigeonhole principle. In *Proc. IEEE 29th Annual Symp. on Foundation of Computer Science*, 346–355. 1988.

[2] Alekhnovich, M., Ben-Sasson, E., Razborov, A. A., Wigderson, A. Pseudorandom generators in propositional proof complexity. *Electronic Colloquium on Computational Complexity*, (23), 2000. Ext. abstract in: *Proc. of the 41st Annual Symp. on Foundation of Computer Science*, 43–53, 2000.

[3] Krajíček, J. Propositional proof complexity I., lecture notes available at. URL http://www.math.cas.cz/~krajicek/ds1.ps.

[4] Krajíček, J. *Bounded arithmetic, propositional logic, and complexity theory*, vol. 60. Cambridge University Press, 1995.

[5] Krajíček, J. A fundamental problem of mathematical logic. In *Annals of the Kurt Gödel Society*, vol. 2, 55–64. Collegium Logicum, Springer-Verlag, 1996.

[6] Krajíček, J. Lower bounds for a proof system with an exponential speed-up over constant-depth frege systems and over polynomial calculus. In Prívara, I., Růžička, P. (Eds.), *Lecture Notes in Computer Science 1295*, 85–90. The 22nd Inter. Symp. Mathematical Foundations of Computer Science(Bratislava, Aug.1997), Springer-Verlag, 1997.

[7] Krajíček, J. On methods for proving lower bounds in propositional logic. In Dalla, M. L., Chiara, et al. (Eds.), *Logic and Scientific Methods*, vol. 1, 69–83.

Proc. of the Tenth International Congress of Logic, Methodology and Philosophy of Science, Florence (August 19-25, 1995), Synthese Library, Vol.259, Dordrecht: Kluwer Academic Publisher, 1997.

[8] Krajíček, J. On the weak pigeonhole principle. *Fundamenta Mathematicae*, 170(1–3):123–140, 2001.

[9] Krajíček, J. Tautologies from pseudo-random generators. *Bulletin of Symbolic Logic*, 7(2):197–212, 2001.

[10] Krajíček, J. Diagonalization in proof complexity. *Fundamenta Mathematicae*, 182:181–192, 2004.

[11] Krajíček, J. Dual weak pigeonhole principle, pseudo-surjective functions, and provability of circuit lower bounds. *Journal of Symbolic Logic*, 69(1):265–286, 2004.

[12] Krajíček, J. Hardness assumptions in the foundations of theoretical computer science. *Archive for Mathematical Logic*, 44(6):667–675, 2005.

[13] Krajíček, J. Proof complexity. In Laptev, A. (Ed.), *European congress of mathematics (ECM)*, 221–231. Stockholm, Sweden (Jun.27–July 2, 2004), Zurich: European Mathematical Society, 2005.

[14] Krajíček, J. Structured pigeonhole principle, search problems and hard tautologies. *Journal of Symbolic Logic*, 70(2):619–630, 2005.

[15] Krajíček, J. Substitutions into propositional tautologies. *Information Processing Letters*, 101(4):163–167, 2007.

[16] Krajíček, J., Pudlák, P.and Woods, A. An exponential lower bound to the size of bounded depth frege proofs of the pigeonhole principle. *Random Structures and Algorithms*, 7(1):15–39, 1995.

[17] Pitassi, T., Beame, P., Impagliazzo, R. Exponential lower bounds for the pigeonhole principle. *Computational complexity*, 3:97–308, 1993.

[18] Pudlák, P. The lengths of proofs. In Buss, S. R. (Ed.), *Handbook of Proof Theory*, 547–637. Elsevier, 1998.

[19] Razborov, A. Unprovability of lower bounds on the circuit size in certain fragments of bounded arithmetic. *Izvestiya of the R.A.N.*, 59(1):201–224, 1995.

[20] Razborov, A. A. Lower bounds for the polynomial calculus. *Computational complexity*, 7(4):291–324, 1998.

[21] Razborov, A. A. Resolution lower bounds for perfect matching principle. In *Proc. of the 17th IEEE Conf. on Computational Complexity*, 29–38. 2002.

[22] Razborov, A. A. *Pseudorandom generators hard for k-DNF resolution and polynomial calculus resolution*. 2003. Preprint.

B
GENERAL PHILOSOPHY OF SCIENCE

Physical Probability [1]

Patrick Maher

University of Illinois at Urbana-Champaign

patrick@maher1.net

ABSTRACT. By "physical probability" I mean the empirical concept of probability in ordinary language. It can be represented as a function of an experiment type and an outcome type, which explains how non-extreme physical probabilities are compatible with determinism. Two principles, called specification and independence, put restrictions on the existence of physical probabilities, while a principle of direct inference connects physical probabilities with inductive probability. This account avoids a variety of weaknesses in the theories of Levi and Lewis.

1 My account

I will present my account of physical probability in this section and then I will compare it with the theories of Levi and Lewis in Sections 2 and 3.

1.1 Identification of the concept

Suppose a coin is about to be tossed and you are told that it either has heads on both sides or else has tails on both sides; if I ask you to state the probability that the coin will land heads, there are two natural answers: (i) 1/2; (ii) either 0 or 1 but I don't know which. Although these answers are incompatible, there is a sense in which each is right, so "probability" is ambiguous in ordinary language. I call the sense of "probability" in which (i) is right *inductive probability* and I call the sense in which (ii) is right *physical probability*.

I say that a probability concept is *empirical* if some elementary statements for it are synthetic. Physical probability is empirical; for example, the physical probability of a coin landing heads depends on contingent facts about the coin. On the other hand, inductive probability isn't empirical, as I have argued elsewhere [13]. Therefore, physical probability can be defined as the empirical concept of probability in ordinary language.

1.2 Form of statements

By an "experiment" I mean an action or event such as tossing a coin, weighing an object, or two particles colliding. I distinguish between experiment *tokens* and experiment *types*; experiment tokens have a space-time location whereas experiment types are abstract objects and so lack such a location. For example, a

---

[1] The author is with Department of Philosophy, University of Illinois at Urbana-Champaign.

particular toss of a coin at a particular place and time is a token of the experiment type "tossing a coin"; the token has a space-time location but the type does not.

Experiments have *outcomes* and here again there is a distinction between tokens and types. For example, a particular event of a coin landing heads that occurs at a particular place and time is a token of the outcome type "landing heads"; only the token has a space-time location.

Now consider a typical statement of physical probability such as:

> The physical probability of heads on a toss of this coin is 1/2.

Here the physical probability appears to relate three things: tossing this coin (an experiment type), the coin landing heads (an outcome type), and 1/2 (a number). This suggests that elementary statements of physical probability can be represented as having the form "The physical probability of X resulting in O is r," where X is an experiment type, O is an outcome type, and r is a number. I claim that this suggestion is correct.

I will use the notation "$pp_X(O) = r$" as an abbreviation for "the physical probability of experiment type X having outcome type O is r."

1.3 Unrepeatable experiments

The types that I have mentioned so far can all have more than one token; for example, there can be many tokens of the type "tossing this coin." However, there are also types that cannot have more than one token; for example, there can be at most one token of the type "tossing this coin at noon today." What distinguishes types from tokens is not repeatability but rather abstractness, evidenced by the lack of a space-time location. Although a token of "tossing this coin at noon today" must have a space-time location, the type does not have such a location, as we can see from the fact that the type exists even if there is no token of it. It is also worth noting that in this example the type does not specify a spatial location.

This observation allows me to accommodate ordinary language statements that appear to attribute physical probability to token events. For example, if we know that a certain coin will be tossed at noon today, we might ordinarily say that the physical probability of getting heads on that toss is 1/2, and this may seem to attribute a physical probability to a token event; however, the statement can be represented in the form $pp_X(O) = r$ by taking X to be the unrepeatable experiment type "tossing this coin at noon today." Similarly in other cases.

1.4 Compatibility with determinism

From the way the concept of physical probability is used, it is evident that physical probabilities can take non-extreme values even when the events in question are governed by deterministic laws. For example, people attribute non-extreme physical probabilities in games of chance, while believing that the outcome of such games is causally determined by the initial conditions. Also, scientific theories in statistical mechanics, genetics, and the social sciences postulate non-extreme physical probabilities in situations that are believed to be governed by underlying deterministic laws. Some of the most important statistical scientific theories were

Physical Probability

developed in the nineteenth century by scientists who believed that *all* events are governed by deterministic laws.

The recognition that physical probabilities relate experiment and outcome *types* enables us to see how physical probabilities can have non-extreme values in deterministic contexts. Determinism implies that, if X is sufficiently specific, then $pp_X(O) = 0$ or 1; but X need not be this specific, in which case $pp_X(O)$ can have a non-extreme value even if the outcome of X is governed by deterministic laws. For example, a token coin toss belongs to both the following types:

X: Toss of this coin.

X': Toss of this coin from such and such a position, with such and such force applied at a such and such a point, etc.

Assuming that the outcome of tossing a coin is governed by deterministic laws, $pp_{X'}(\text{head}) = 0$ or 1; however, this is compatible with $pp_X(\text{head}) = 1/2$.

1.5 Specification

I claim that physical probabilities satisfy the following:

Specification Principle (SP) If it is possible to perform X in a way that ensures it is also a performance of the more specific experiment type X', then $pp_X(O)$ exists only if $pp_{X'}(O)$ exists and is equal to $pp_X(O)$.

For example, let X be tossing a normal coin, let X' be tossing a normal coin on a Monday, and let O be that the coin lands heads. It is possible to perform X in a way that ensures it is a performance of X' (just toss the coin on a Monday), and $pp_X(O)$ exists, so SP implies that $pp_{X'}(O)$ exists and equals $pp_X(O)$, which is correct.

It is easy to see that SP implies the following; nevertheless, all theorems are proved in Section 5.

Theorem 1 *If it is possible to perform X in a way that ensures it is also a performance of the more specific experiment type X_i, for $i = 1, 2$, and if $pp_{X_1}(O) \neq pp_{X_2}(O)$, then $pp_X(O)$ does not exist.*

For example, let B be an urn that contains only black balls, W an urn that contains only white balls, and let:

$$\begin{aligned} X &= \text{selecting a ball from either } B \text{ or } W \\ X_B &= \text{selecting a ball from } B \\ X_W &= \text{selecting a ball from } W \\ O &= \text{the ball selected is white.} \end{aligned}$$

It is possible to perform X in a way that ensures it is also a performance of the more specific experiment type X_B, likewise for X_W, and $pp_{X_B}(O) = 0$ while $pp_{X_W}(O) = 1$, so Theorem 1 implies that $pp_X(O)$ does not exist, which is correct.

Let us now return to the case where X is tossing a normal coin and O is that the coin lands heads. If this description of X was a complete specification of the experiment type, then X could be performed with apparatus that would precisely fix the initial position of the coin and the force applied to it, thus determining the outcome. It would then follow from SP that $pp_X(O)$ does not exist. I think this consequence of SP is clearly correct; if we allow this kind of apparatus, there is not a physical probability of a toss landing heads. So when we say—as I have said—that $pp_X(O)$ does exist, we are tacitly assuming that the toss is made by a normal human without special apparatus that could precisely fix the initial conditions of the toss; a fully explicit specification of X would include this requirement. The existence of $pp_X(O)$ thus depends on an empirical fact about humans, namely, the limited precision of their perception and motor control.

1.6 Independence

Let X^n be the experiment of performing X n times and let $O_i^{(k)}$ be the outcome of X^n which consists in getting O_i on the kth performance of X. I claim that physical probabilities satisfy the following:

Independence Principle (IN) If $pp_X(O_i)$ exists for $i = 1, \cdots, n$ then
$$pp_{X^n}(O_1^{(1)} \ldots O_n^{(n)}) \text{ exists and equals } pp_X(O_1) \ldots pp_X(O_n).$$

For example, let X be shuffling a normal deck of 52 cards and then drawing two cards without replacement; let O be the outcome of getting two aces. Here $pp_X(O) = (4/52)(3/51) = 1/221$. Applying IN with $n = 2$ and $O_1 = O_2 = O$, it follows that:
$$pp_{X^2}(O^{(1)}O^{(2)}) = [pp_X(O)]^2 = 1/221^2.$$

This implication is correct because X specifies that it starts with shuffling a normal deck of 52 cards, so to perform X a second time one must replace the cards drawn on the first performance and reshuffle the deck, and hence the outcome of the first performance of X has no effect on the outcome of the second performance.

For a different example, suppose X is defined merely as drawing a card from a deck of cards, leaving it open what cards are in the deck, and let O be drawing an ace. By fixing the composition of the deck in different ways, it is possible to perform X in ways that ensure it is also a performance of more specific experiment types that have different physical probabilities; therefore, by Theorem 1, $pp_X(O)$ does not exist. Here the antecedent of IN is not satisfied and hence IN is not violated.

The following theorem elucidates IN by decomposing its consequent into two parts.

Theorem 2 *IN is logically equivalent to: if $pp_X(O_i)$ exists for $i = 1, \cdots, n$ then both the following hold.*

(a) $pp_{X^n}(O_1^{(1)} \ldots O_n^{(n)})$ exists and equals $pp_{X^n}(O_1^{(1)}) \ldots pp_{X^n}(O_n^{(n)})$.

(b) $pp_{X^n}(O_i^{(i)})$ exists and equals $pp_X(O_i)$, for $i = 1, \cdots, n$.

Here (a) says outcomes are probabilistically independent in pp_{X^n} and (b) asserts a relation between pp_{X^n} and pp_X.

Physical Probability

1.7 Direct inference

I will now discuss how physical probability is related to inductive probability. The arguments of inductive probability are two propositions or sentences and I will write "$ip(A|B)$" for the inductive probability of proposition A given proposition B.

Let an *R-proposition* be a consistent conjunction of propositions, each of which is either of the form "$pp_X(O) = r$" or else of the form "it is possible to perform X in a way that ensures it is also a performance of X'." Let "Xa" and "Oa" mean that a is a token of experiment type X and outcome type O, respectively. In what follows, "R" always denotes an R-proposition while "a" denotes a token event. Inductive probabilities satisfy the following:

Direct Inference Principle (DI) If R implies that $pp_X(O) = r$ then $ip(Oa|Xa.R) = r$.

For example, let X be tossing this coin, let X' be tossing it from such and such a position, with such and such a force, etc., let O be that the coin lands heads, and let R be "$pp_X(O) = 1/2$ and $pp_{X'}(O) = 1$." Then DI implies $ip(Oa|Xa.R) = 1/2$ and $ip(Oa|X'a.R) = 1$. Since $Xa.X'a$ is logically equivalent to $X'a$, it follows that $ip(Oa|Xa.X'a.R) = 1$.

As it stands, DI has no practical applications because we always have more evidence than just Xa and an R-proposition. However, in many cases our extra evidence does not affect the application of DI; I will call evidence of this sort "admissible." More formally:

Definition 1 *If R implies that $pp_X(O) = r$ then E is admissible with respect to (X, O, R, a) iff $ip(Oa|Xa.R.E) = r$.*

The principles I have stated imply that certain kinds of evidence are admissible. One such implication is:

Theorem 3 *E is admissible with respect to (X, O, R, a) if both the following are true:*

(a) *R implies it is possible to perform X in a way that ensures it is also a performance of X', where $X'a$ is logically equivalent to $Xa.E$.*

(b) *There exists an r such that R implies $pp_X(O) = r$.*

For example, let X be tossing this coin and O that the coin lands heads. Let E be that a was performed by a person wearing a blue shirt. If R states a value for $pp_X(O)$ and that it is possible to perform X in a way that ensures the tosser is wearing a blue shirt, then E is admissible with respect to (X, O, R, a). In this example, the X' in Theorem 3 is tossing the coin while wearing a blue shirt.

We also have:

Theorem 4 *E is admissible with respect to (X, O, R, a) if both the following are true:*

(a) *$E = Xb_1 \ldots Xb_n.O_1b_1 \ldots O_mb_m$, where $b_1, \cdots, b_n$ are distinct from each other and from a, and $m \leq n$.*

(b) *For some r, and some $r_i > 0$, R implies that $pp_X(O) = r$ and $pp_X(O_i) = r_i$, $i = 1, \cdots, m$.*

For example, let X be tossing a coin and O that the coin lands heads. Let a be a particular toss of the coin and let E state some other occasions on which the coin has been (or will be) tossed and the outcome of some or all of those tosses. If R states a non-extreme value for $pp_X(O)$, then E is admissible with respect to (X, O, R, a). In this example, the O_i in Theorem 4 are all either O or $\sim O$.

Theorems 3 and 4 could be combined to give a stronger result but I will not pursue that here.

2 Comparison with Levi

I will now compare the account of physical probabilities that I have just given with the theory of chance presented by Levi [6, 8].

2.1 Identification of the concept

Levi does not give an explicit account of what he means by "chance" but there are some reasons to think he means physical probability. For example, he says:

> The nineteenth century witnessed the increased use of notions of objective statistical probability or chance in explanation and prediction in statistical mechanics, genetics, medicine, and the social sciences.[8, p.120]

This shows that Levi regards "chance" as another word for "objective statistical probability," which suggests its meaning is a sense of the word "probability." Also, the nineteenth century scientific work that Levi here refers to used the word "probability" in a pre-existing empirical sense and thus was using the concept of physical probability.

However, there are also reasons to think that what Levi means by "chance" is *not* physical probability. For example:

- Levi [8, pp.117,120] speaks of plural "conceptions" or "notions" of chance, whereas there is only one concept of physical probability.

- Levi [8, p.142] criticizes theories that say chance is incompatible with determinism by saying "the cost is substantial and the benefit at best negligible." This criticism, in terms of costs and benefits, would be appropriate if "chance" meant a newly proposed concept but it is irrelevant if "chance" means the pre-existing ordinary language concept of physical probability. If "chance" means physical probability then the appropriate criticism is simply that linguistic usage shows that physical probability is compatible with determinism—as I argued in Section 1.4.

So, it is not clear that what Levi means by "chance" is physical probability. Nevertheless, I think it worthwhile to compare my account of physical probability with the account that is obtained by interpreting Levi's "chance" as if it meant physical probability. I will do that in the remainder of this section.

Physical Probability

2.2 Form of statements

Levi [8, p.120] says:

> Authors like Venn [16] and Cournot [2] insisted that their construals of chance were indeed consistent with respect to underlying determinism ... The key idea lurking behind Venn's approach is that the chance of an event occurring to some object or system—a "chance set up," according to Hacking [3], and an "object," according to Venn [16, chap.3]—is relative to the kind of trial or experiment (or "agency," according to Venn) conducted on the system.

Levi endorses this "key idea." The position I defended in Section 1.2 is similar in making physical probability relative to a type of experiment, but there is a difference. I represented statements of physical probability as relating three things: An experiment type (e.g., a human tossing a certain coin), an outcome type (e.g., the coin landing heads), and a number (e.g., 1/2). On Levi's account, chance relates four things: A chance set up (e.g., a particular coin), a type of trial or experiment (e.g., tossing by a human), an outcome type, and a number. Thus what I call an "experiment" combines Levi's "chance set up" and his "trial or experiment."

An experiment (in my sense) can often be decomposed into a trial on a chance set up in more than one way. For example, if the experiment is weighing a particular object on a particular scale, we may say:

- The set up is the scale and the trial is putting the object on it.
- The set up is the object and the trial is putting it on the scale.
- The set up is the object and scale together and the trial is putting the former on the latter.

These different analyses make no difference to the physical probability. Therefore, Levi's representation of physical probability statements, while perhaps adequate for representing all such statements, is more complex than it needs to be.

2.3 Specification

Since SP is a new principle, Levi was not aware of it. I will now point out two ways in which his theory suffers from this.

2.3.1 A mistaken example

To illustrate how chance is relative to the type of experiment, Levi [8, p.120] made the following assertion:

> The chance of coin a landing heads on a toss may be 0.5, but the chance of the coin landing heads on a toss by Morgenbesser may, at the same time, be 0.9.

But let X be tossing a (by a human), let X' be tossing a by Morgenbesser, and let O be that a lands heads. It is possible to perform X in a way that ensures it is also a performance of X' (just have Morgenbesser toss the coin), so SP implies that if $pp_X(O) = 0.5$ then $pp_{X'}(O)$ must have the same value. Levi, on the other hand, asserts that it could be that $pp_X(O) = 0.5$ and $pp_{X'}(O) = 0.9$.

Intuition supports SP here. If the physical probability of heads on a toss of a coin were different depending on who tosses the coin (as Levi supposes) then, intuitively, there would not be a physical probability for getting heads on a toss by an unspecified human, just as there is not a physical probability for getting a black ball on drawing a ball from an urn of unspecified composition. Thus, Levi's example is mistaken.

2.3.2 An inadequate explanation

Levi [6, p.264] wrote:

> Suppose box a has two compartments. The left compartment contains 40 black balls and 60 white balls and the right compartment contains 40 red balls and 60 blue balls. A trial of kind S is selecting a ball at random from the left compartment and a trial of kind S' is selecting a ball at random from the right compartment ... Chances are defined for both kinds of trials over their respective sample spaces [i.e., outcome types].
>
> Consider trials of kind $S \vee S'$. There is indeed a sample space consisting of drawing a red ball, a blue ball, a black ball, and a white ball. However, there is no chance distribution over the sample space.
>
> To see why no chance distribution is defined, consider that the sample space for trials of kind $S \vee S'$ is such that a result consisting of obtaining a [black] or a [white] ball is equivalent to obtaining a result of conducting a trial of kind S ... Thus, conducting a trial of kind $S \vee S'$ would be conducting a trial of kind S with some definite chance or statistical probability.
>
> There is no a priori consideration precluding such chances; but there is no guarantee that such chances are defined either. In the example under consideration, we would normally deny that they are.

Let O be that the drawn ball is either black or white. I agree with Levi that $pp_{S \vee S'}(O)$ doesn't exist. However, Levi's explanation of this is very shallow; it rests on the assertion that $pp_{S \vee S'}(S)$ doesn't exist, for which Levi has no explanation. It also depends on there not being balls of the same color in both compartments, though the phenomenon is not restricted to that special case; if we replaced the red balls by black ones, Levi's explanation would fail but $pp_{S \vee S'}(O)$ would still not exist.

SP provides the deeper explanation that Levi lacks. The explanation is that it is possible to perform $S \vee S'$ in a way that ensures S is performed, likewise for S', and $pp_S(O) \neq pp_{S'}(O)$, so by Theorem 1, $pp_{S \vee S'}(O)$ does not exist. In Levi's example, $pp_S(O) = 1$ and $pp_{S'}(O) = 0$; if the example is varied by replacing the red balls with black ones then $pp_{S'}(O) = 0.4$; the explanation of the non-existence of $pp_{S \vee S'}(O)$ is the same in both cases.

2.4 Independence

Levi considers a postulate equivalent to IN and argues that it doesn't hold in general. Here is his argument:

> [A person] might believe that coin a is not very durable so that each toss alters the chance of heads on the next toss and that how it alters the chance is a function of the result of the previous tosses. [The person] might believe that coin a, which has never been tossed, has a .5 chance of landing heads on a toss as long as it remains untossed. Yet, he might not believe that the chance of r heads on n tosses is $\binom{n}{r}(.5)^n$. [6, p.272]

The latter formula follows from IN and $pp_X(\text{heads}) = 0.5$.

Levi here seems to be saying that the chance of experiment type X giving outcome type O can be different for different tokens of X. He explicitly asserts that elsewhere:

> Sometimes kinds of trials are not repeatable on the same object or system ... And even when a trial of some kind can be repeated, the chances of response may change from trial to trial.[8, p.128]

But that is inconsistent with Levi's own view, according to which chance is a function of the experiment and outcome types.

In fact, IN is not violated by Levi's example of the non-durable coin, as the following analysis shows.

- We may take X to be starting with the coin symmetric and tossing it n times. Here repetition of X requires starting with the coin again symmetric, so different performances of X are independent, as IN requires. This is similar to the example of drawing cards without replacement that I gave in Section 1.6.
- We may take X to be tossing the coin once when it is in such-and-such a state. Here repetition of X requires first restoring the coin to the specified state, so again different performances of X are independent.
- Levi seems to be taking X to be tossing the coin once, without specifying the state that the coin is in. In that case, $pp_X(\text{heads})$ does not exist, so again there is no violation of IN.

I conclude that Levi's objection to IN is fallacious.

2.5 Direct inference

Levi endorses a version of the direct inference principle; the following is an example of its application:

> If Jones knows that coin a is fair (i.e., has a chance of 0.5 of landing heads and also of landing tails) and that a is tossed at time t, what degree of belief or credal probability ought he to assign to the hypothesis that the coin lands heads at that time? Everything else being equal, the answer seems to be 0.5. [8, p.118]

As this indicates, Levi's direct inference principle concerns the degree of belief that a person ought to have. By contrast, the principle DI in Section 1.7 concerns inductive probability.

To understand Levi's version of the principle we need to know what it means to say that a person "ought" to have a certain degree of belief. Levi doesn't give any adequate account of this, so I am forced to make conjectures about what it means.

One might think that a person "ought" to have a particular degree of belief iff the person would be well advised to adopt that degree of belief. But if that is what it means, then Levi's direct inference principle is false. For example, Jones might know that coin a is to be tossed 100 times, and that the tosses are independent, in which case Levi's direct inference principle says that for each r from 0 to 100, Jones's degree of belief that the coin will land heads exactly r times ought to be $\binom{100}{r}(0.5)^{100}$. However, it would be difficult (if not impossible) to get one's degrees of belief in these 101 propositions to have precisely these values and, unless something very important depends on it, there are better things to do with one's time. Therefore, it is not always advisable to have the degrees of belief that, according to Levi's direct inference principle, one "ought" to have.

Alternatively, one might suggest that a person "ought" to have a particular degree of belief iff it is the only one that is justified by the person's evidence. But what does it mean for a person's degree of belief to be justified by the person's evidence? According to the deontological conception of justification, which Alston [1, p.60] said is used by most epistemologists, it means that the person is not blameworthy in having this degree of belief. On that account, the suggestion would be that a person "ought" to have a particular degree of belief iff the person would deserve blame for not having it. However, there need not be anything blameworthy about failing to have all the precise degrees of beliefs in the example in the preceding paragraph; so on this interpretation, Levi's direct inference principle is again false.

For a third alternative, we might say that a person "ought" to have a particular degree of belief in a particular proposition iff this degree of belief equals the inductive probability of the proposition given the person's evidence. On this interpretation, Levi's direct inference principle really states a relation between inductive probability and physical probability, just as DI does; the reference to a person's degree of belief is a misleading distraction that does no work and would be better eliminated.

So, my criticism of Levi's version of the direct inference principle is that it is stated in terms of the unclear concept of what a person's degree of belief "ought" to be, that on some natural interpretations the principle is false, and the interpretation that makes it true is one in which the reference to degree of belief is unnecessary and misleading. These defects are all avoided by DI.

2.6 Admissible evidence

As I noted in Section 1.7, DI by itself has no practical applications because we always have more evidence than just the experiment type and an R-proposition. For example, Jones, who is concerned with the outcome of a particular toss of coin a, would know not only that coin a is fair but also a great variety of other facts. It is therefore important to have an account of when additional evidence is admissible.

Levi's [6, p.252] response is that evidence is admissible if it is known to be "stochastically irrelevant," i.e., it is known that the truth or falsity of the evidence does not alter the physical probability. That is right, but to provide any substantive information it needs to be supplemented by some principles about what sorts of evidence are stochastically irrelevant; Levi provides no such principles.

By contrast, Theorems 3 and 4 provide substantive information about when evidence is admissible. Those theorems were derived from SP and IN, neither of which is accepted by Levi, so it is not surprising that he has nothing substantive to say about when evidence is admissible.

3 Comparison with Lewis

I will now discuss the theory of chance proposed by Lewis [10, 11]. A related theory was proposed earlier by Mellor [14], and other writers have subsequently expressed essentially the same views (Loewer [12], Schaffer [15]), but I will focus on Lewis's version. The interested reader will be able to apply what I say here to those other theories.

3.1 Lewis's theory

According to Lewis [11, pp.96–97], chance is a function of three arguments: a proposition, a time, and a (possible) world. He writes $P_{tw}(A)$ for the chance at time t and world w of A being true.

Lewis [11, pp.95–97] says that the *complete theory of chance* for world w is the set of all conditionals that hold at w and are such that (1) the antecedent is a proposition about history up to a certain time, (2) the consequent is a proposition about chance at that time, and (3) the conditional is a "strong conditional" of some sort, such as the counterfactual conditional of [9]. He uses the notation T_w for the complete theory of chance for w. He also uses H_{tw} for the complete history of w up to time t. Lewis [11, p.97] argues that the conjunction $H_{tw}T_w$ implies all truths about chances at t and w.

Lewis's version of the direct inference principle, which he calls the *Principal Principle*, is:

> Let C be any reasonable initial credence function. Then for any time t, world w, and proposition A in the domain of P_{tw}, $P_{tw}(A) = C(A|H_{tw}T_w)$. [11, p.97]

Lewis [11, p.127] argues that if H_{tw} and the laws of w together imply A, then $H_{tw}T_w$ implies $P_{tw}(A) = 1$. It follows that if w is deterministic then P_{tw} cannot have any values other than 0 or 1. For example, in a deterministic world, the chance of any particular coin toss landing heads must be 0 or 1. Lewis accepts this consequence.

> If a determinist says that a tossed coin is fair, and has an equal chance of falling heads or tails, he does not mean what I mean when he speaks of chance. [11, p.120]

Nevertheless, prodded by Levi [7], Lewis proposed an account of what a determinist does mean when he says this; he called it "counterfeit" chance. I will now explain this concept.

For any time t, the propositions $H_{tw}T_w$, for all worlds w, form a partition that Lewis [11, p.99] calls the *history-theory partition* for time t. Another way of expressing the Principal Principle is to say that the chance distribution at any time t and world w is obtained by conditioning any reasonable initial credence function on the element of the history-theory partition for t that holds at w. Lewis [11, pp.120–121] claimed that the history-theory partition has the following qualities:

> (1) It seems to be a natural partition, not gerrymandered. It is what we get by dividing possibilities as finely as possible in certain straightforward respects.
>
> (2) It is to some extent feasible to investigate (before the time in question) which cell of this partition is the true cell; but
>
> (3) it is unfeasible (before the time in question, and without peculiarities of time whereby we could get news from the future) to investigate the truth of propositions that divide the cells.

With this background, Lewis states his account of counterfeit chance:

> Any coarser partition, if it satisfies conditions (1)–(3) according to some appropriate standards of feasible investigation and of natural partitioning, gives us a kind of counterfeit chance suitable for use by determinists: namely, reasonable credence conditional on the true cell of that partition. Counterfeit chances will be relative to partitions; and relative, therefore, to standards of feasibility and naturalness; and therefore indeterminate unless the standards are somehow settled, or at least settled well enough that all the remaining candidates for the partition will yield the same answers. [11, p.121]

So we can say that for Lewis, physical probability (the empirical concept of probability in ordinary language) is reasonable initial credence conditioned on the appropriate element of a suitable partition. It may be chance or counterfeit chance, depending on whether the partition is the history-theory partition or something coarser. I will now criticize this theory of physical probability.

3.2 Form of statements

Lewis says that chance is a function of three arguments: a proposition, a time, and a world. He does not explicitly say what the arguments of counterfeit chance are but, since he thinks this differs from chance only in the partition used, he must think that counterfeit chance is a function of the same three arguments, and hence (to put it in my terms) that physical probability is a function of these three arguments.

Let us test this on an example. Consider again the following typical statement of physical probability:

H: The physical probability of heads on a toss of this coin is $1/2$.

Lewis [11, p.84] himself uses an example like this. However, H doesn't attribute physical probability to a proposition or refer to either a time or a possible world. So, this typical statement of physical probability does not mention any of the things that Lewis says are the arguments of physical probability.

Of course, it may nevertheless be that the statement could be analyzed in Lewis's terms. Lewis did not indicate how to do that, although he did say that when a time is not mentioned, the intended time is likely to be the time when the event in question begins [11, p.91]. So we might try representing H as:

H': For all s and t, if s is a token toss of this coin and t is a time just prior to s then the physical probability at t in the actual world of the proposition that s lands heads is $1/2$.

But there are many things wrong with this. First, "s lands heads" is not a proposition, since s is here a variable. Second, H' is trivially true if the coin is never tossed, though H would still be false if the coin is biased, so they are not equivalent. Third, the physical probability of a coin landing heads is different depending on whether we are talking about tossing by a human, with no further specification (in which case H is probably true), or about tossing with such and such a force from such and such a position, etc. (in which case H is false), but H' doesn't take account of this. And even if these and other problems could be fixed somehow (which has not been done), the resulting analysis must be complex and its correctness doubtful. By contrast, my account is simple and follows closely the grammar of the original statement; I represent H as saying that the physical probably of the experiment type "tossing this coin" having the outcome type "heads" is $1/2$.

I will add that, regardless of what we take the other arguments of physical probably to be, there is no good reason to add a possible world as a further argument. Of course, the value of a physical probably depends on empirical facts that are different in different possible worlds, but this does not imply that physical probably has a possible world as an argument. The simpler and more natural interpretation is that physical probably is an empirical concept, not a logical one; that is, even when all the arguments of physical probably have been specified, the value is in general a contingent matter.

Lewis himself sometimes talks of physical probably in the way I am here advocating. For instance, he said that counterfeit chance is "reasonable credence conditional on the *true* cell of [a] partition" (emphasis added); to be consistent with his official view, he should have said that counterfeit chance *at w* is reasonable credence conditional on the cell of the partition *that holds at w*. My point is that the former is the simpler and more natural way to represent physical probably.

So, Lewis made a poor start when he took the arguments of physical probably to be a proposition, a time, and a world. That representation has not been shown to be adequate for paradigmatic examples, including Lewis's own, and even if it could be made to handle those examples it would still be needlessly complex and unnatural. The completely different representation that I proposed in Section 1.2 avoids these defects.

3.3 Reasonable credence

In Lewis's presentation of his theory, the concept of a "reasonable initial credence function" plays a central role. Lewis says this is "a non-negative, normalized, finitely additive measure defined on all propositions" that is

> reasonable in the sense that if you started out with it as your initial credence function, and if you always learned from experience by conditionalizing on your total evidence, then no matter what course of

> experience you might undergo your beliefs would be reasonable for
> one who had undergone that course of experience. I do not say what
> distinguishes a reasonable from an unreasonable credence function to
> arrive at after a given course of experience. We do make the distinction, even if we cannot analyze it; and therefore I may appeal to
> it in saying what it means to require that C be a reasonable initial
> credence function. [11, p.88]

However, there are different senses in which beliefs are said to be reasonable and Lewis has not identified the one he means. A reasonable degree of belief could be understood as one that a person would be well advised to adopt, or that a person would be not be blameworthy for adopting, but on those interpretations Lewis's theory would give the wrong results, for the reasons indicated in Section 2.5. Alternatively, we might say that a reasonable degree of belief is one that agrees with inductive probably given the person's evidence, but then reasonable degrees of belief would often lack precise numeric values [13] whereas Lewis requires a reasonable initial credence function to always have precise numeric values.

I think the best interpretation of Lewis here is that his "reasonable initial credence function" is a probability function that is a precisification of inductive probability given no evidence. This is compatible with the sort of criteria that Lewis [11, p.110] states and also with his view [11, p.113] that there are multiple reasonable initial credence functions.

Although Lewis allows for multiple reasonable initial credence functions, his Principal Principle requires them to all agree when conditioned on an element of the history-theory partition. So, if a reasonable initial credence function is a precisification of inductive probability, Lewis's theory of chance can be stated more simply and clearly using the concept of inductive probability, rather than the concept of a reasonable initial credence function, as follows:

> The chance of a proposition is its inductive probability conditioned on the
> appropriate element of the history-theory partition.

This shows that the concept of credence does no essential work in Lewis's theory of chance; hence Lewis's theory isn't subjectivist and [10] is mistitled.

What goes for chance also goes for counterfeit chance, and hence for physical probability in general. Thus Lewis's theory of physical probability may be stated as:

> The physical probability of a proposition is its inductive probability conditioned on the appropriate element of a suitable partition.

Again, the concept of credence is doing no essential work in Lewis's theory and clarity is served by eliminating it.

3.4 Partitions

We have seen that according to Lewis, physical probability is inductive probability conditioned on the appropriate element of a suitable partition. Also, suitable partitions are natural partitions such that it is "to some extent feasible to investigate (before the time in question) which cell of this partition is the true cell" but "unfeasible" to investigate the truth of propositions that divide the cells. Lewis says the history-theory partition is such a partition and using it gives genuine chance. Coarser partitions, using different standards of naturalness and feasibility, give what Lewis regards as counterfeit chance. I will now argue that Lewis is wrong about what counts as a suitable partition, both for chance and counterfeit chance.

I begin with chance. Let t be the time at which the first tritium atom formed and let A be the proposition that this atom still existed 24 hours after t. The elements of the history-theory partition specify the chance at t of A. But let us suppose, as might well be the case, that the only way to investigate this chance is to observe many tritium atoms and determine the proportion that decay in a 24 hour period. Then, even if sentient creatures could exist prior to t (which is not the case), it would not be feasible for them to investigate the chance at t of A, since there were no tritium atoms prior to t. Therefore, the history-theory partition does not fit Lewis's characterization of a suitable partition.

Now consider a case of what Lewis calls counterfeit chance. Suppose that at time t I bend a coin slightly by hammering it and then immediately toss it; let A be that the coin lands heads on this toss. If I assert that coin tossing is deterministic but the physical probability of this coin landing heads is not 0 or 1 then, according to Lewis, the physical probability I am talking about is inductive probability conditioned on the true element of a suitable partition that is coarser than the history-theory partition. Lewis has not indicated what that partition might be but this part of his theory is adapted from Jeffrey, who indicates [5, p.206] that the partition is one whose elements specify the limiting relative frequency of heads in an infinite sequence of tosses of the coin. However, there cannot be such an infinite sequence of tosses and, even if it existed, it is not feasible to investigate its limiting relative frequency prior to t. On the other hand, it is perfectly feasible to investigate many things that divide the cells of this partition, such as what I had for breakfast. Lewis says different partitions are associated with different standards of feasibility, but there is no standard of feasibility according to which it is feasible prior to t to investigate the limiting relative frequency of heads in an infinite sequence of non-existent future tosses, yet unfeasible to investigate what I had for breakfast. Hence this partition is utterly unlike Lewis's characterization of a suitable partition.

So, Lewis's characterization of chance and counterfeit chance in terms of partitions is wrong. This doesn't undermine his theory of chance, which is based on the Principal Principle rather than the characterization in terms of partitions, but it does undermine his theory of counterfeit chance. I will now diagnose the source of Lewis's error.

Lewis's original idea, expressed in his Principal Principle, was that inductive probability conditioned on the relevant chance equals that chance. That idea is basically correct, reflecting as it does the principle of direct inference. Thus what makes the history-theory partition a suitable one is not the characteristics that Lewis cited, concerning naturalness and feasibility of investigation; it is rather that each element of the history-theory partition specifies the value of the relevant chance. We could not expect the Principal Principle to hold if the conditioning proposition specified only the history of the world to date and not also the relevant chance values for a world with that history. Yet, that is essentially what Lewis tries to do in his theory of counterfeit chance. No wonder it doesn't work.

So if counterfeit chance is to be inductive probability conditioned on the appropriate element of a suitable partition, the elements of that partition must specify the (true!) value of the counterfeit chance. But then it would be circular to explain what counterfeit chance is by saying that it is inductive probability conditioned on the appropriate element of a suitable partition. Therefore, counterfeit chance cannot be explained in this way—just as chance cannot be explained by saying it is inductive probability conditioned on the appropriate element of the history-theory partition. Thus the account of counterfeit chance, which Lewis adopted from Jeffrey, is misguided.

The right approach is to treat what Lewis regards as genuine and counterfeit chance in a parallel fashion. My account of physical probability does that. On my account, Lewis's chances are physical probabilities in which the experiment type specifies the whole history of the world up to the relevant moment, and his counterfeit chances are physical probabilities in which the experiment type is less specific than that. Both are theoretical entities, the same principle of direct inference applies to both, and we learn about both in the same ways.

4 Conclusion

In Section 1 I identified what I mean by physical probability and gave an account of some of its fundamental properties, namely:

- It can be represented as having an experiment type and an outcome type as its arguments.

- This explains how non-extreme values are compatible with determinism.

- The existence of physical properties is governed by principles of specification and independence.

- Physical probability is related to inductive probability by a principle of direct inference.

- Generalizations about admissible evidence follow from the preceding principles.

This is not a complete theory but it is enough to avoid a variety of weaknesses in the theories of Levi and Lewis, as I showed in Sections 2 and 3. I do not know of any other account of physical probability that is successful in these ways.

5 Proofs

5.1 Proof of Theorem 1

Suppose it is possible to perform X in a way that ensures it is also a performance of the more specific experiment type X_i, for $i = 1, 2$. If $pp_X(O)$ exists then, by SP, both $pp_{X_1}(O)$ and $pp_{X_2}(O)$ exist and are equal to $pp_X(O)$; hence $pp_{X_1}(O) = pp_{X_2}(O)$. So, by transposition, if $pp_{X_1}(O) \neq pp_{X_2}(O)$, then $pp_X(O)$ does not exist.

5.2 Proof of Theorem 2

Assume IN holds and $pp_X(O_i)$ exists for $i = 1, \ldots, n$. By letting O_j be a logically necessary outcome, for $j \neq i$, it follows from IN that $pp_{X^n}(O_i^{(i)})$ exists and equals $pp_X(O_i)$; thus (b) holds. Substituting (b) in IN gives (a).

Now assume that $pp_X(O_i)$ exists for $i = 1, \ldots, n$ and that (a) and (b) hold. Substituting (b) in (a) gives the consequent of IN, so IN holds.

5.3 Proof of Theorem 3

Suppose (a) and (b) are true. Since SP is a conceptual truth about physical probability, it is analytic, so R implies:

$$pp_{X'}(O) = pp_X(O) = r.$$

Therefore,

$$ip(Oa|Xa.R.E) = ip(Oa|X'a.R), \text{ by (a)}$$
$$= r, \text{ by DI.}$$

Thus E is admissible with respect to (X, O, R, a).

5.4 Proof of Theorem 4

Assume conditions (a) and (b) of the theorem hold. I will also assume that $m = n$; the result for $m < n$ follows by letting $O_{m+1}, \ldots, O_n$ be logically necessary outcomes.

Since IN is analytic, it follows from (b) that R implies:

$$pp_{X^{n+1}}(O_1^{(1)} \ldots O_n^{(n)}.O^{(n+1)}) = pp_X(O_1) \ldots pp_X(O_n) pp_X(O)$$
$$= r_1 \ldots r_n r. \quad (1)$$

Using obvious notation, $ip(O_1 b_1 \ldots O_n b_n.Oa|Xb_1 \ldots Xb_n.Xa.R)$ can be rewritten as:

$$ip(O_1^{(1)} \ldots O_n^{(n)} O^{(n+1)}(b_1 \ldots b_n a)|X^{n+1}(b_1 \ldots b_n a).R).$$

Since R implies (1), it follows by DI that the above equals $r_1 \ldots r_n r$. Changing the notation back then gives:

$$ip(O_1 b_1 \ldots O_n b_n.Oa|Xb_1 \ldots Xb_n.Xa.R) = r_1 \ldots r_n r. \quad (2)$$

Replacing O in (2) with a logically necessary outcome, we obtain:

$$ip(O_1 b_1 \ldots O_n b_n|Xb_1 \ldots Xb_n.Xa.R) = r_1 \ldots r_n. \quad (3)$$

Since $r_1 \ldots r_n > 0$ we have:

$$ip(Oa|Xa.R.E) = ip(Oa|Xa.R.Xb_1 \ldots Xb_n.O_1 b_1 \ldots O_n b_n)$$
$$= \frac{ip(O_1 b_1 \ldots O_n b_n.Oa|Xb_1 \ldots Xb_n.Xa.R)}{ip(O_1 b_1 \ldots O_n b_n|Xb_1 \ldots Xb_n.Xa.R)}$$
$$= r, \text{ by (2) and (3).}$$

Thus E is admissible with respect to (X, O, R, a).

Bibliography

[1] Alston, W. P. Concepts of epistemic justification. *The Monist*, 68:57–89, 1985.

[2] Cournot, A. A. *Essai sur les fondements de nos connaissances et sur les charactères de la critique philosophique*, 1851. Translated by Moore, M. H. *as Essay on the Foundations of our Knowledge*. New York: Macmillan, 1956.

[3] Hacking, I. *The Logic of Statistical Inference*. Cambridge: Cambridge University Press, 1965.

[4] Jeffrey, R. C. (Ed.). *Studies in Inductive Logic and Probability*. Berkeley: University of California Press, 1980.

[5] Jeffrey, R. C. *The Logic of Decision*. University of Chicago Press, 2nd edn., 1983.

[6] Levi, I. *The Enterprise of Knowledge*. Cambridge: MA: MIT Press, 1980. Paperback edition with corrections 1983.

[7] Levi, I. Review of [4]. *Philosophical Review*, 92:116–121, 1983.

[8] Levi, I. Chance. *Philosophical Topics*, 18:117–149, 1990.

[9] Lewis, D. *Counterfactuals*. Cambridge: MA: Harvard University Press, 1973.

[10] Lewis, D. A subjectivist's guide to objective chance. In Jeffrey, R. C. (Ed.), [4], 263–293. 1980. Reprinted with postscripts in [11].

[11] Lewis, D. *Philosophical Papers*, vol. 2. New York: Oxford University Press, 1986.

[12] Loewer, B. David Lewis's Humean theory of objective chance. *Philosophy of Science*, 71:1115–1125, 2004.

[13] Maher, P. The concept of inductive probability. *Erkenntnis*, 65:185–206, 2006.

[14] Mellor, D. H. *The Matter of Chance*. Cambridge: Cambridge University Press, 1971.

[15] Schaffer, J. Deterministic chance? *British Journal for the Philosophy of Science*, 58:113–140, 2007.

[16] Venn, J. *The Logic of Chance*. 4th edn., 1866.

Comparative realism as the best response to antirealism [1]

Theo A.F. Kuipers
University of Groningen
T.A.F.Kuipers@rug.nl

ABSTRACT. Arguments for and against scientific realism usually presuppose as the main epistemic claim about theories that we may have good reasons to conclude that they are true, or at least approximately true. The antirealist charges against this claim are not easy to counter. In this paper it is argued that the defense of realism is much easier if we relativize its epistemic claim in the light of (theories of) truth approximation. From this *comparative realist* perspective the main epistemic claim becomes that we may have good reasons to conclude that successor theories are closer to the truth than their predecessors. For example, although Einstein's theory of general relativity may still be false, and not even approximately true, we have good reasons to assume that it is closer to the truth than Newton's theory of gravitation. A similar relativization of claims that theoretical terms refer to things in the world is argued for in terms of 'being closer to the referential truth'. For both purposes it is also plausible to relativize 'being empirically successful' to the comparative notion of 'being (persistently) empirically more successful'.

Comparative realism hence is realism guided by the comparative perspective on success and on truth approximation, that is, the notions of 'more successful' and 'closer to the (observational, referential, and theoretical) truth', and their mutual relations. This approach is defended against the antirealist charges and compared with the main other realist responses. The main positive claim of comparative realism is that (theoretical) truth approximation provides the (stratified) default explanation and prediction of empirical progress between non-empirically equivalent theories and of 'aesthetic progress' between empirically equivalent theories. Here 'aesthetic progress' is understood in terms of the prevailing 'aesthetic canon' in the relevant field and period, that is, the prevailing non-empirical virtues of theories.

1 Introduction

Arguments for and against scientific realism usually presuppose as the main epistemic claim about theories that we may have good reasons to conclude that they are true, or at least approximately true. The antirealist charges against this claim are not easy to counter. In this paper it will be argued that the defense

---

[1] The author is with the Faculty of Philosophy, University of Groningen, Oude Boteringestraat 52, 9712 GL Groningen, Netherlands. A Dutch version with more emphasis on 'Inference to the Best Explanation', entitled "Comparatief realisme: het beste antwoord op anti-realisme", appeared as target paper in *Algemeen Nederlands Tijdschrift voor Wijsbegeerte (ANTW)*, 100.3, 2008, 173-200. T.A.F.Kuipers@rug.nl, http://www.rug.nl/filosofie/Kuipers

of realism is much easier if we relativize its epistemic claim in the light of (theories of) truth approximation. From this comparative realist perspective the main epistemic claim becomes that we may have good reasons to conclude that successor theories are closer to the truth than their predecessors. For example, although Einstein's theory of general relativity may still be false, and not even approximately true, we have good reasons to assume that it is closer to the truth than Newton's theory of gravitation. A similar relativization of claims that theoretical terms refer to things in the world will be argued for in terms of 'being closer to the referential truth'. For both purposes it is also plausible to relativize 'being empirically successful' to the comparative notion of 'being (persistently) empirically more successful'.

Comparative realism hence is realism guided by the comparative perspective on success and on truth approximation, that is, the notions of 'more successful' and 'closer to the (observational, referential, and theoretical) truth', and their mutual relations. This approach will be defended against the antirealist charges and compared with the main other realist responses.

In Section 3 I will draw heavily upon the survey of (non-comparative) arguments pro and contra realism presented by James Ladyman in his textbook of [12] and his handbook exposition [13] of ontological, epistemological, and methodological positions. In particular, the *pessimistic meta-induction*, the *no-miracles argument* and *inference to the best explanation* and their counter arguments will be analyzed in detail and will lead to a number of conditions of adequacy for explicating the crucial expressions.

There are at least two ways to specify the truth approximation perspective, the quantitative approach of Niiniluoto [15, 16] and my own qualitative, comparative approach [6]. For the present purposes the latter, in a sense more strict, and hence more cautious, approach is sufficient. More specifically, in Section 2 I will sketch some of its main lines and in Section 4 I will show to what extent this approach fulfils the conditions of adequacy that have been gathered in Section 3. I will conclude with a final comparison of comparative realism with the main kinds of non-comparative realism that have been suggested to meet the antirealist charges. It concerns in particular the restriction of realism to entities, structures, 'mature' theories, theories with 'novel' predictive success, and 'essential' parts/aspects of theories, or stretching realism by stretching the underlying (causal) theory of reference.

The core of *comparative* realism becomes that the (comparative) phenomenon that one theory persistently is empirically more successful than another provides a good reason for the claim that the first theory is closer to the (theoretical and referential) truth than the second; the good, analytical, reason being that this very claim amounts to the generic *default* explanation for that comparative empirical phenomenon. Among empirically equivalent theories, it may occur that one theory persistently is more successful than another according to some non-empirical, notably aesthetic criteria. In this case there may be empirically justified reasons, however weak, for the truth approximation claim and the corresponding default explanation.

I conclude this introduction with a puzzling observation about the realism-antirealism debate. One important success of the first decades of (constructive) analytical philosophy was the discovery, notably by Russell, Carnap, Hempel, and Beth, of relations as a means to solve age-old problems by refined concept explication. This concerned not least asymmetric relations, constitutive of comparative concepts, such as 'longer than', 'caused by', etc. Of course, conceptions of 'better than', hence of improvement or progress, are comparative as well.

Now it is very surprising that in the supposedly analytic realism – antirealism debate there is almost no sign of awareness of the possible relevance of this insight. One continues to talk about true and false theories and reference claims versus empirically adequate and inadequate theories, that is, observationally true and false theories, respectively. The retreat to 'approximately true' is of no help, for that remains basically non-comparative and hence it cannot capture progress. Moreover, it requires a necessarily arbitrary threshold. On the other hand, from the point of view of relations, it is highly plausible to think in terms of comparative notions, such as 'closer to the truth' and 'more successful than'. To be sure, in the comparative approach theories will frequently not be straightforwardly comparable; but for 'mixed' cases the 'principle of dialectics' (that is, try to improve both) and the quantitative approach are plausible concretizations.

2 Truth approximation, some basics for comparative realism

In this short introduction to my favorite, qualitative theory of truth approximation, I will mainly restrict myself to its basic form, that is, without a (O-/T-) distinction between observational and theoretical terms and without other refinements. It is best represented within the structuralist theory of theory representation. Starting from a fixed vocabulary and a suitable similarity type of structures, let M_p indicate the set of structures of that type, also called the *conceptual possibilities* or *potential models* of the theory. Let the subset X of M_p indicate the set of models of theory X. Finally, assuming that our target is a fixed domain of physically or, more broadly, nomically possible constellations and events, let T indicate this domain 'as seen through M_p', hence a subset of M_p, and be called the set of (intended) nomic possibilities or the domain of intended applications. According to the Nomic Postulate, we assume that such a unique fixed set exists, given M_p and an intended domain. Note that we do not yet suppose that we dispose of a general characterization of T as a subset of M_p. What we know is only that each intended application can be represented as a potential model. A general characterization of T is 'the great unknown' which theories, as represented by their models, are looking for. More formally, a theory is a triple of the form $< M_p, X, T >$, together with the weak claim that T is a subset of X ($T \subseteq X$) and the strong claim that $T = X$. A theory is said to be true (false) in the weak sense when the weak claim is true (false). It is easy to check that (a general characterization of) T represents the strongest true weak claim, and hence may be called 'the truth' in this context, that is, the truth about the given domain in the given vocabulary.

Now it is plausible to define what it means, for fixed $< M_p, T >$, that one theory Y is closer (or more similar) to the truth than another X, and hence what it means to say that Y amounts to truth approximation relative to X. Intuitively, when Y is moving from X in the direction of T. Formally, when $Y - T \subseteq X - T$ (Ø2-area empty in Fig.1) and $T - Y \subseteq T - X$ (Ø1-area empty) and at least once it should be a proper subset relation (#1-area and/or #2-area non-empty). In terms of symmetric differences ($A \Delta B =_{df} (A \Delta B) \cup (B - A)$), $Y \Delta T$ should be a proper subset of $X \Delta T$. In [6] I have argued that, among other formulations, the two clauses of this definition amount to: (relative to T) Y has more true consequences than X and Y has more correct models than X, respectively. Of course, we get 'at least as close to the truth', when we omit the proper-subset conditions. We get an asymmetric form of 'closer to' by only requiring in addition that the first relation is a proper subset relation (#1- area empty). Knowing what

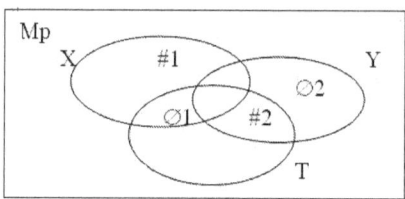

Figure 1: Theory Y is closer to the truth T than theory X, where T represents the domain as seen through M_p

it means that one theory is closer to the truth than another is one thing, in view of the fact that we don't know the truth, judging that this is in fact the case, in the light of our evidence is quite another. Although there is, of course, no theorem guaranteeing that (empirical) more-successfulness entails closer-to-the-truth, there is a theorem, the *Success Theorem*, guaranteeing almost the reverse entailment, viz., at-least-as-close-to-the-truth entails at-least-as-successfulness.

Let R represent the set of realized (described) applications at a certain time, that is, the experimentally or otherwise realized nomic possibilities at that time. Let S indicate the strongest law induced on the basis of R. Of course, R will partly be the result of testing hypothetical laws. If no descriptive mistakes have been made in representing the realized possibilities, R is not only a subset of M_p, but even of T, for nomic impossibilities can't be realized, by definition. Moreover, in the structuralist representation S is also a subset of M_p, such that it has been concluded, inductively, and hence at most provisionally, that conceptual possibilities outside S, that is, in $M_p - S$, are nomically impossible. Finally, if our inductive jump from R to S is correct, S has to be a superset of T. To sum up: $R \subseteq T \subseteq S \subseteq M_p$, provided no descriptive and inductive mistakes have been made.

Let us call R/S the data set. We define what it means that Y is, relative to the data set R/S, (empirically) more successful than X as follows: $X \cap R \subseteq Y \cap R$ and $S \cup Y \subseteq S \cup X$. The first condition amounts to 'no loss of established examples', to be called 'instantially at least as successful', and the second to 'no loss of explained established laws', to be called 'explanatorily at least as successful'.

Now the already announced *Success Theorem* states, assuming that R/S is correct, that is, $R \subseteq T \subseteq S$, that Y is (empirically) at least as successful relative to R/S as X if Y is at least as close to the truth as X. This theorem is such that *persistently* being-more-successful is functional or instrumental for truth approximation, though not guaranteeing it, in the sense that it is very difficult, and without the O-/T-distinction even impossible, for X to remain at least as successful as Y when Y is in fact closer to the truth than X. More in detail we may argue as follows. Let Y be more successful than X relative to R/S at a certain time. This suggests the *Comparative Success Hypothesis*, according to which it is hypothesized that this will remain the case, whatever experiments we design and perform. Testing of this hypothesis may result in the conclusion, at least for the time being, that this is in fact the case: Y persistently remains more successful than X. Note that this conclusion is a comparative inductive generalization, providing the paradigm situation for speaking of empirical progress, consisting of 'instantial' and 'explanatory' progress. That is, the paradigm case for applying the, essentially instrumentalist, 'rule of success', viz. replace, for the time being, X in favor of Y. Here Y may well be known to be false in view of R/S, for which reason the rule is called instrumentalist rather than falsificationist. Assuming empirical progress in this sense, it is very hard to imagine that Y is, despite appearances, not closer to the truth than X. More specifically, it holds that 1) the 'closer to the truth hypothesis', the *TA-hypothesis*, predicts and explains the empirical progress, 2) it cannot be the reverse, that is, it is impossible that X is closer to the truth than Y, and 3) if Y is in fact not closer to the truth than X, there is a specific burden to explain the supposed empirical progress so far. In the last case, without theoretical terms, hence without O-/T-distinction, the apparent empirical progress must have resulted from an unhappy test history of the comparative success hypothesis because the empirical progress conclusion can be broken by appropriate experiments. With O-/T-distinction, see below, the breaking of this conclusion is only excluded when Y, though presumably not closer to 'the theoretical truth' than X, happens to be nevertheless at least as close to 'the observational truth' as X. In sum, the TA-hypothesis provides the *'default explanation and prediction'* of empirical progress.[2]

So far I have presented the naive or basic structuralist theory of truth approximation. It is basic in the sense that several idealizations have been made, requiring refinement or concretization. In [11] I have extensively illustrated the (philosophical) method of concept explication by idealization and concretization by the example of truth approximation. Let us review the main concretizations as they have been elaborated or indicated in [6].

One idealization was that we implicitly assumed that any established counterexample of a theory, that is, a realized possibility not being a model of the theory, is as bad for one theory as for any other. However, in chap.10 I have introduced the ternary relation of '(more) structurelikeness', that is, the idea that one structure may be more similar to another than a third one. In this way, the possibility arises that a counterexample is less dramatic for one theory than for another, because the former has a model that is more similar to the 'counter model' (i.e., the potential model representing the counterexample), than any model of the latter theory. Adapted definitions of more-successfulness, empirical progress, and closer-to-the-truth lead again to the conclusion that the TA-hypothesis provides the default explanation and prediction of empirical progress. As a matter of fact we only know of real life scientific examples of (potential) truth approximation, e.g. the Law of Van der Waals as successor of the ideal gas law, when this concretization is introduced. Without this, we only know of toy examples.

---

[2]Surprisingly enough, the expression 'default explanation', let alone 'default prediction', does not seem to be in use in the philosophy of science. Here it is specifically used as an alternative to the expression 'functional for truth approximation'. As Gerhard Schurz has remarked, the default explanation 'closer to' is also a kind of abductive explanation, with the interesting restriction that the genuine abductive step in this explanation is only from 'at least as successful' to 'at least as close to', for 'not equally close' is a deductive consequence of 'more successful'.

Another idealization was the assumption that there is just one domain. In [6, chap.13] I have indicated that it is possible to introduce a domain vocabulary, as a subvocabulary of the (observational) vocabulary, such that it is possible to define the domain explicitly in that vocabulary, leaving the further behavior of the applications as the unknown to be specified. The advantage of this is that domains become comparable sets, that is, it enables us to say that an intended application of one theory is or is not an intended application of another. This opens the possibility of dealing with truth approximation by domain change besides by theory change, as indicated above, and hence by the combination of domain and theory change [10].

Returning to theory change, it is evident that theories will seldom be comparable in terms of 'more successful' and 'closer to the truth'. These notions apply only in ideal cases. In practice it is important that in the case of 'divided success' between two theories, the crucial challenge is to improve both theories, that is, to design a third theory, a synthesis, that is more successful than both and, if so, plausibly closer to the truth than both[3]. This is called the 'principle of dialectics' in [6, chap.6]. In the theoretical context of the realism – antirealism debate, the move to the quantitative approach is more useful to deal with divided success. The basic form of being quantitatively closer to the truth is the condition, assuming finite M_p, $|Y \Delta T| < |X \Delta T|$. For a refined quantitative approach the best proposal so far is of Ilkka Niiniluoto [15, 16], who has defined the so-called min-sum real-valued distance of a theory to the truth, which always enables comparison. Since the only point of the coming exposition is the principled possibility of straightforward empirical progress and truth approximation I will not go here into details of this approach.

Finally, we did not make a distinction between (relatively) observational and theoretical terms, but such a O-/T-distinction is of course crucial for the realism-antirealism debate,. In chap.9 of [6] I have spelled out how this distinction works out, with the main conclusion that empirical progress, of course, in observational terms, remains functional for truth approximation on the theoretical level, though with some greater risk of being wrong in this tentative conclusion.

For present purposes I will introduce the main features of this type of 'stratified (basic) truth approximation'. Let M_p now indicate the set of theoretical (cum observational) conceptual possibilities of the vocabulary. Let subset T_t of M_p indicate the unique subset of nomic theoretical possibilities, the theoretical truth. Let M_{pp} indicate the set of partial or observational conceptual possibilities and T_o its unique subset of nomic observational possibilities, the observational truth. We assume that T_t and T_o can be characterized in some way. Let π project (subsets of) M_p onto (subsets of) M_{pp}, stripping the (clauses involving) theoretical components. There are good reasons to assume that $T_o = \pi(T_t)$. Of course, theories X and Y are empirically equivalent when $\pi(X) = \pi(Y)$. The relations of 'empirically more successful', and hence of 'empirical progress', and of 'closer to the (observational) truth' between theories X and Y now refer to these notions between $\pi(X)$ and $\pi(Y)$ as defined above, but now on the level of M_{pp}. On the other hand, 'Y is closer to the (theoretical) truth than X' remains defined on the

---

[3] At first sight, a more basic problem of comparability seems to arise from so-called incommensurable vocabularies. However, as long as there is the suspicion that two terms may have a different meaning it is plausible and possible to include both terms, with indices, in a fusion of vocabularies.

(theoretical cum observational) level of M_p, by the otherwise unchanged basic, that is, set theoretic definition. Finally, it is possible to define 'the referential truth' as, roughly, the conjunction of (un)negated reference claims of theoretical terms that are correct according to T_t, that is, those terms that do play or do not play a role in the shaping of T_t relative to T_o. To each theory, a (total) reference claim is associated by a similar condition. In sum, the above assumptions enable definitions of 'closer to the observational, referential, and theoretical truth', and hence of 'observational, referential, and theoretical truth approximation'.

Besides the previously mentioned Success Theorem, now restricted to the observational level, some more TA-theorems and -conjectures become relevant, notably the (conditional) Projection Theorem and the Stratified Success Theorem, which is a kind of combination. In Section 4, I will give some more information. Together they support the main claim of comparative realism: theoretical truth approximation provides the (stratified) default explanation and prediction of empirical progress between non-empirically equivalent theories and of 'aesthetic progress' between empirically equivalent theories. Here 'aesthetic progress' will be defined in terms of the prevailing 'aesthetic canon' in the relevant field and period, that is, the prevailing non-empirical virtues of theories.

3 Antirealist arguments to be met

In this section I will heavily draw upon James Ladyman's [12, 13] in looking for the main arguments that have been put forward against realism. I'll start with a brief preview of the arguments and the responses that naturally follow from the comparative realist point of view. Recall, that this point of view conceives realism guided by the comparative perspective on success and on truth approximation, that is, the notions of 'more successful' and 'closer to the truth', and their mutual relations.

Laudan's so-called pessimistic meta-induction is rebutted by the claim that it is not at all a problem that previously the best theories turned out to be false (and even did not refer). For the main question is whether they remained, at least as a rule, more successful than their predecessor theories and hence can still be argued to be closer to the truth than (and refer at least as well as) their predecessors. Hence, similarly for our currently the best theories, they will presumably turn out to be false, but the interesting question remains whether they will continue to be more successful than their predecessors, so that we can still hold them as closer to the truth.

Regarding the arguments against Putnam's so-called no-miracles argument in favor of realism, the comparative response is that the former will miss their point when the latter is replaced by the observation that the success of science would be miraculous on anything but a scientific realist view in the sense that specific occurrences of (persistent) empirical progress would be miraculous if not, as a rule, due to truth approximation. This, notwithstanding Van Fraassen's Darwinian analogy that, as a rule, the best theories survive because they are selected for that reason.

Finally, the arguments (notably of Van Fraassen) against inference to the best explanation (IBE) are rebutted by the claim that IBE has so far not been adequately explicated. The proposed comparative explication reads: inference to the best explanation as the closest to the truth among the available explanations or theories, even among empirically equivalent theories.

In the course of the presentation, there will be derived a number of conditions of adequacy an acceptable explication of comparative realism, including IBE, will have to satisfy. In Section 4, it will be investigated to what extent they can be satisfied.

3.1 Arguments from theory change

3.1.1 The pessimistic meta-induction

One of the main arguments against realism is known as 'the pessimistic meta-induction' and is due to Larry Laudan. Ladyman [12, pp.236–237], [13, p.345] presents it as follows (my italics):

> (i) There have been *many empirically successful* theories in the history of science which have subsequently been rejected and whose central theoretical terms do not refer according to our best current theories.
> (ii) Our best current theories are no different in kind from those discarded theories and so we have no reason to think they will not ultimately be replaced as well.
>
> So, by induction we have positive reason to expect that our best current theories will be replaced by new theories according to which some of the central theoretical terms of our best current theories do not refer, *and hence, we should not believe in the approximate truth or the successful reference of the theoretical terms of our best current theories*.

According to Ladyman the most common response is that the realist claims should be restricted to mature theories and/or theories having novel predictive success, for which (i) would not apply, and hence the inductive argument would be blocked. This retreat has its own problems, see below. Another way to block this argument is to liberalize the realist elements in the argument according to the comparative perspective:

> (iCR) There have been (some, several, many[4]) theories in the history of science *that were and still are empirically more successful than their predecessor theories*, which have *nevertheless* subsequently been rejected and whose central theoretical terms do not *all* refer according to our best current theories.
> (iiCR) (= (ii)) Our best current theories are no different in kind from those discarded theories and so we have no reason to think they will not ultimately be replaced as well.
>
> So, by induction we have positive reason to expect that our best current theories will be replaced by new theories according to which some of the central theoretical terms of our best current theories do not refer, *however we may still* (instead of: and hence, we should not)

---
[4]Depending on whether we will use qualitative or quantitative versions of 'more successful' and 'closer to the truth'.

> *believe that our best current theories are closer to the truth than their predecessor theories or that their theoretical terms more successfully refer than those of their predecessor theories.*

Hence, it is not plausible anymore to think that the second part of the suggested, partially bracketed conclusion of the CR-version of the inductive argument, starting with "(and hence, we should not) believe that ..." is still 'valid'. By consequence, it seems plausible to think that, starting from the basic explications in Section 2, when making a distinction between (relatively) observational and theoretical terms, refined explications can be given of the crucial phrases, viz. 'empirically more successful', 'closer to the truth' and 'more successfully refer', such that the following conditions of adequacy are satisfied, in which we add 'novel predictive success' in the second one for later purposes.

CA1 The explications of the crucial phrases should leave room for the possibility that false theories, even with some non-referring theoretical terms, are not only (persistently) empirically more successful but even closer to the (theoretical) truth than and refer at least as well as other theories.

CA2 The explications should be such that 'being (persistently) empirically more successful' is, *as a rule*, due to 'being closer to the theoretical truth', which on its turn entails, as a rule, 'referring at least as well' and 'novel predictive success'.

Or, in a plausible 'progress version':

CA2 The explications should be such that empirical progress is, *as a rule*, due to theoretical truth approximation, which on its turn entails, as a rule, referential truth preservation and novel predictive success.

In Section 2 we have already indicated some basic explications that allow 'false theories closer to the truth than other ones' and therefore 'being as least as successful', however, not yet with the distinction between observational and theoretical terms. In Section 4 we will try to meet these conditions while taking this distinction into account.

3.2 The antirealist attack on the no-miracles argument

As mentioned, according to Ladyman, the usual realist response to the pessimistic meta-induction is the retreat to mature theories and/or theories having novel predictive success. The weakness of both, *prima facie* rather ad hoc, retreats is, of course, that they require precise non-ad hoc definitions of 'mature theories' and 'novel predictive success', both of which tasks have so far not generated a proposal that is agreed upon. Note that comparative realism remains to take all theories seriously.

Besides such technical problems with these rejoinders there is a more general problem with such retreats. The antirealist counter argument to such a retreat is that one does not need the (unrestricted) inductive argument to undermine the so-called no-miracles argument in favor of realism, that is, "the idea that the success of science would be miraculous on anything but a scientific realist view" [13, p.342]. Illustrated by the ether theory of light and the caloric theory of heat one may claim, according to [12, p.244], [13, p.346] (as quoted, with my bracket insertions [] and deletion ()):

> (a) Successful reference of its central theoretical terms is a necessary condition for the approximate truth of a theory.
> (b) There are [at least two] examples of theories that (were mature and) had novel predictive success but which are not approximately true.
> (c) [Hence,] Approximate truth and successful reference of central theoretical terms is not a necessary condition for the novel-predictive success of scientific theories
>
> So, the no-miracles argument is undermined since, if approximate truth and successful reference are not available to be part of the explanation of some theories' novel predictive success, there is no reason to think that the novel predictive success of other theories has to be explained by realism.

I have some reservation against this argument as such. Although (a), (b) and (c) may be granted[5], the conclusion drawn from them seems too strong. Why would maturity and novel predictive success not be possible without approximate truth and successful reference? One could imagine that there are logical gaps leaving room for such cases. Of course, in such cases the novel predictive success would have to be explained in another, specific way. However, the suggestion that such cases prevent the use of approximate truth and successful reference in other specific cases for the explanation of novel predictive success seems unjustified or, at least no such justification is reported by Ladyman. In other words, why is it excluded that *prima facie* similar cases have different explanations?

However this may be, the current realist responses to the antirealist rejoinder are twofold according to [12, pp.245–248], [13, pp.346-348]. (I) Stretch realism by stretching the (causal) theory of reference such that the relevant abandoned theoretical terms refer after all or (II) Restrict realism to those theoretical claims about unobservables that feature in an *essential* way in the derivation of novel predictions. For example, Stathis Psillos [19] uses (I) to save the ether theory as referring after all, viz. to the electromagnetic field, and (II) to explain the success of the caloric theory without 'caloric' as a referring term.

The comparative perspective does not need such interventions, although even within this perspective there may be *other* good reasons for such moves. However, in the present context the point is whether the abandoned theories were and are closer to the truth than their respective predecessor theories. And this may well be the case in several historical examples, notably not only for the ether theory but also for the caloric theory.

Hence, let us turn to the CR-versions of the antirealist premises (a) – (c) enabled by CA1&2. This will lead to a defensible CR-version of the no-miracles argument.

(aCR) Being closer to the truth entails, as a rule, referring at least as well, and novel predictive success ('novel', now in whatever favorite sense).

---

[5] However, even with respect to (a), there may be room for doubt, for it very much depends on the definition of 'approximate truth' whether there is the logical possibility of an approximately true theory without reference of (all of) its theoretical terms.

(bCR) Exceptionally there (may) have been examples of theories that (were mature and) had novel predictive success relative to their predecessor theories but which were not closer to the truth and even not referring at least as well.

(cCR) Being (persistently) more successful is, as a rule, due to being closer to the truth, even more so when novel predictive success is involved.

Note that (aCR) and (cCR) are taken care of by CA2. Note also that the caloric and the ether theory may well fit in (aCR) and (cCR) relative to their predecessors. But, by the insertion 'as a rule', (aCR) and (cCR) leave room for other exceptions (bCR) and suggest a cautious CR-version of the no-miracles argument.

CR-no-miracles argument:
As a rule, (persistently) empirically more successful theories are closer to the truth than, and refer at least as well as their predecessors.

If this were not the case, the regular occurrence of persistently empirically more successful theories, including some novel predictive success, would be miraculous.

Occasionally, other, case-specific, explanations of persistent empirically more successfulness, novel or not, may be appropriate.

Hence, the CR-no-miracles argument suggests that truth approximation provides a kind of *default explanation and prediction* of (persistent) empirical more successfulness, including some novel predictive success. It is attractive to reformulate the CR-version in terms of progress.

CR-no-miracles argument, progress version:
As a rule, (persistent) empirical progress is due to theoretical truth approximation and referential truth preservation, if not referential truth approximation.

If this were not the case, the regular occurrence of (novel) empirical progress would be miraculous.

Occasionally, other, case-specific, explanations of (novel) empirical progress may be appropriate.

And, hence, the argument suggests that (theoretical) truth approximation provides a kind of *default explanation and prediction* of (novel) empirical progress.

For standard realism the ether theory of light and the caloric theory of heat give rise to problems because in both cases the crucial theoretical terms do not refer in a straightforward sense according to our present lights, whereas the theories were mature and had quite some novel predictive success. It surely is an intriguing question how these theories can combine both features. However, they may not be so very relevant for the realism debate since these problems do not arise for comparative realism. What matters in this perspective are only comparative questions, that is, questions in comparison with previous and successor theories. For example, in the case of the ether theory the focus in the debate is on (Young and) Fresnel's revised version of Huygens' wave theory of light in ether, by replacing his longitudinal waves by transversal waves. Instead of trying to answer, as a defense of realism - however interesting it may be - the question as to how Fresnel's theory could have impressive novel predictive success, by our present lights, without reference of the crucial theoretical term 'ether', some of the plausible CR-questions are:

> Was Fresnel's theory empirically more successful than its predecessor theory, viz. that of Huygens, and its competitor theory, the particle emission theory of Newton?

As is well known, at least with respect to Foucault's quasi-crucial velocity of light experiment, the result was in favor of Fresnel's theory relative to both Huygens' version of the wave theory and Newton's particle alternative. Hence, the next comparative realist questions are:

> If so, could this be explained by the hypothesis that Fresnel's theory is closer to the truth than, and hence, presumably, refers at least as well as, those of Huygens and Newton?

Not knowing the truth, such questions can only be answered in the abstract. However, assuming that the relativistic version of Maxwell's electromagnetic theory, that is, Einstein's special theory of relativity encompassing Maxwell's theory, is the truth (or at least a close approximation of the truth), we can form potential answers to these questions.

> Can the greater empirical success of Fresnel's theory be explained by the fact that Fresnel's theory is closer to the theory of Maxwell (-Einstein) than those of Huygens and Newton, and hence, presumably, refers at least as well?

Assuming precise enough explications, these questions can be answered. It is at least plausible to think that Fresnel's theory is closer to Maxwell's theory than Huygens' theory. Moreover, it may well be that the sequence of theories of Huygens, Fresnel, Maxwell, Maxwell-Einstein is a matter of truth approximation. However, it does not seem plausible that the sequence of theories of Newton, (Huygens), Fresnel, Maxwell (-Einstein) can be reconstrued as such. Newton's theory is indeed incommensurable with versions of the wave theory, at least without the leap to the first versions of quantum physics, suggesting that the photon theory of light is closer to the truth than both Newton's theory and (non-quantum) versions of the wave theory.[6] I leave it as a challenge to raise similar questions about the caloric theory of heat.

3.3 Inference to the best explanation

So-called inference to the best explanation (IBE) plays an important role in the defense of standard realism. Ladyman [12, p.209], [13, p.341] describes this (putative) rule of inference as follows:

---

[6] Several more encompassing truth approximation questions are suggested by the set of success comparisons of electrodynamic theories that is provided by a table in [18, p.282] (reprinted and commented upon in [6, p.118] and [7, p.236]). It represents the records in the face of 13 general experimental facts of the special theory of relativity (STR) and six alternative electrodynamic theories that were largely developed and defended in the 19th Century, viz., three versions of the ether theory and three emission theories.

"... where we have a range of competing hypotheses all of which are empirically adequate to the phenomena in some domain we should infer the truth of the hypothesis which gives us the best explanation of those phenomena."

Here it is understood that one hypothesis provides a better explanation than another when either the first explains more of these phenomena than the second or when they explain the same phenomena but the first satisfies some additional non-empirical evidential criteria, or epistemic virtues, whereas the second does not satisfy all of them or to a lesser degree. Ladyman (p.340) lists as such non-empirical criteria simplicity, non-ad hocness, novel predictive power, elegance, and explanatory power, [3] mentions simplicity, elegance, inner beauty, fertility, coherence with background theories and/or metaphysical suppositions.[7] Following [14], I will call such non-empirical criteria *aesthetic criteria* when they are put forward by scientists for evaluative purposes. McAllister deals in particular with: symmetry, simplicity, and visualizability. [23] deals in his chap.6, entitled "Beautiful theories", not only with simplicity and symmetry, but also with inevitability or rigidity.

According to the (standard) reading of IBE suggested by Ladyman, the competing hypotheses are not only assumed to be empirically adequate to the phenomena in the relevant domain in the sense that they are compatible with them, that is, they have not been falsified by one of them, but all of these hypotheses are even assumed to explain *all* phenomena in the domain. In this case, the best explanation is that one among those hypotheses that satisfies the non-empirical criteria the best.

IBE is of course not deductively valid, but it is used as a kind of inductive or abductive argument at two levels. At the local level it is used to escape in specific cases of empirically equivalent theories from the so-called '(strong) underdetermination by empirical data'. It is supposed to enable specific theoretical truth and reference claims. At the global level it is used to defend realism as the best explanation of the overall success of scientific theorizing. (According to the strong version of the no-miracles argument it is even the only explanation.) The global use can be defended against Van Fraassen's Darwinian analogy [21, p.40] that, as a rule, the best theories survive because they are selected for that reason, by noting that realism predicts and explains the survival so far but also the continuation of success. Against the charge of circularity the global use has been defended by David Papineau and Stathis Psillos in a similar way as Braithwaite and Carnap defended the inductive defense of induction against the circularity charge, viz. by claiming that it is only 'rule circular', not 'premise circular' [12, p.218], [13, p.343].

However, IBE is debated in general as a rule of inference, where the following explication[8] is more or less presupposed, with a weak and a strong version.

---

[7] See [17] for a discussion of ten empirical and non-empirical virtues of theories.

[8] In earlier publications (e.g. [6, 9]) I have introduced a distinction between the 'the best theory' and the stronger notion of 'the best explanation', viz., being the best theory which has not yet been (empirically) falsified. Although it may sound a bit strange to talk about the best explanation even when it is known to be false, it makes the presentation much easier. Moreover, since I will now take non-empirical criteria into account for empirically equivalent theories it now meets the main criticism of [3, note 9, p.291] of my previous version of 'inference to the best theory'.

IBE-E1 If an explanation has so far proven to be the best one among the available empirically adequate theories, then conclude, for the time being, that

- it is true, that is, true as a hypothesis about the domain, or even
- it is the truth, that is, the strongest true hypothesis about the domain.

Elsewhere (e.g. [6]) I have put forward three general objections to this 'implicit explication':

(1) It is restricted to unfalsified theories, for empirically adequate theories have by definition only true observational consequences. Hence there is no possibility of conclusions to progress by false, but less false, theories.
(2) It is asymmetric in that it deals in the premises with the best of the available theories and in the conclusion with being 'true/the truth' simpliciter, that is, without reference to the available theories. In other words, the premises are comparative, whereas the conclusion is not.
(3) It lacks justification of the conclusion

(3.1) when there is just one empirically the best theory available, let alone
(3.2) when there are empirically equivalent (EE-) theories among empirically the best ones (i.e. in the case of strong underdetermination).

Van Fraassen has raised three specific objections that may be seen as specifications of (3.2), hence dealing with EE-theories. They are known as:

(i) The Argument from Indifference
(ii) The Argument from the Best of a Bad Lot
(iii) The Argument from Bayesianism

and will soon be paraphrased.

These objections suggest the following extra condition of adequacy for the comparative realist explication tasks extended with IBE:

CA-IBE An explication of IBE should also deal with the case that all available theories are false or even have been falsified (1), it should be symmetric between the formal nature of the premise(s) and the conclusion (2), and the (truth-related) conclusion should have some analytical justification and/or allow some empirical justification (3), when there is only one best theory available (3.1), and when there are EE-ones (3.2)

It is easy to check that the following revised explication of IBE satisfies at least CA-IBE1 and CA-IBE2.

IBE-E2 If an explanation has so far proven to be empirically the best one among the available theories, then conclude, for the time being, that it is the closest to the truth among the available theories.

As announced before, Section 4 will indicate to what extent CA1 and CA2 can be satisfied and hence CA-IBE3.1, for that is essentially implied by them. However, IBE-E2 does not yet take EE-best theories into account, hence, it does not yet meet CA-IBE3.2.

There are some famous examples of EE-theories discussed by scientists themselves. Van Fraassen [21] has also discussed some rather artificial examples. The best-known and by no means artificial case is of course provided by the various so-called interpretations of quantum mechanics. The first one being the Copenhagen interpretation (put together principally by Niels Bohr) followed by the hidden variable interpretation of David Bohm. Several other interpretations are discussed today. Another case is provided by the existence of at least four EE-versions of classical mechanics, which will be discussed in Section 4.1.3. Hence, there is a serious problem of how to choose between EE-theories. IBE has been put forward not least for this problem of choice. Hence, let us look at Van Fraassen's objections, as formulated by Ladymain [13, pp.344-345][9], and evaluate them.

(i) The Argument from Indifference: "... since there are many ontologically incompatible yet empirically equivalent theories, we have no reason to choose among them and identify one of them as true. ..."

This argument is to be rejected, for although there may be no compelling reasons, it disqualifies any role of non-empirical criteria as truth indicators.

(ii) The Argument from the Best of a Bad Lot: "... we are to think that the collection of hypotheses that we have under consideration will include the true theory. The best explanatory hypothesis we have may just be the best of a bad lot, all of which are false. ..."

This is evidently a variant of our general asymmetry objection (2), but now specifically intended, though this remains here implicit, for EE-theories. However, also for EE-theories it is rather plausible. It is taken care of by CA-IBE2.

(iii) The argument from Bayesianism: "... any rule for the updating of belief that goes beyond the rules of Bayesian conditionalisation [...] will lead to probabilistic incoherence."

This is also plausible and hence should be respected.

So let us reformulate CA-IBE3.2 in detail and in accordance with these evaluations, in particular, in contrast to (i).

CA3 As far as EE-explanations are concerned, an explication of IBE should only take those prevailing aesthetic criteria into account, *if any*, for which there are empirical or analytical reasons, however weak, to assume that they are truth-conducive in the field.

Note for the moment already that the prevailing aesthetic criteria in a certain field at a certain time constitute what [14] has aptly called the aesthetic canon, arrived at by, and equally aptly called, *aesthetic induction*.

In accordance with (iii) we will require:

CA4 In a probabilistic version of an adequate explication of IBE 'EE-updating' of the relevant belief should remain within the confines of Bayesian conditionalisation.

---

[9] It is to be noted that the formulation of the Argument from Indifference in [12, p.219] is a bit confusing, for it seems to coincide with the Argument from the Best of a Bad Lot (p.220). The Argument from Bayesianism is not dealt with in [12].

A merely probabilistic version will not do in general, for probabilistic updating makes only differential sense for unfalsified theories. However, probabilistic estimation of the distance from the truth [15, 16], using Bayesian updating, is generally applicable.

Of course, IBE-E2 trivially satisfies CA3 and CA4 when there are no EE-best theories. For the case that there are we extend it with an extra clause

IBE-E3

(3.1) (=IBE-E2). If an explanation has so far proven to be empirically the best one among the available theories, then conclude, for the time being, that it is the closest to the truth among the available theories.

(3.2) If some empirically equivalent explanations have so far proven to be empirically the best ones among the available theories, and if among them there is one that is according to the relevant aesthetic canon the best one, then conclude, for the time being, that it is the closest to the truth among the available theories.

It is easy to check that (3.2) satisfies, like (3.1), CA-IBE1 and CA-IBE2. Section 3 will indicate to what extent it also satisfies CA-IBE3.2, that is, CA3 & CA4.

This concludes our survey of the main antirealist arguments and which conditions of adequacy comparative realist explications of the crucial notions have to satisfy in order to meet them. The resulting conditions of adequacy to be satisfied are on the one hand CA1 and CA2 and on the other hand CA3 and CA4, for CA-IBE was argued to be taken care of by IBE-E3 (meeting CA-IBE1 and CA-IBE2) and the resulting four conditions (meeting CA-IBE3).

4 Evaluation of comparative realism

In the previous section we have derived four desiderata for an adequate explication of a comparative realist stance that is claimed to be better defensible than other responses to the discussed antirealist charges. In this section the first task is the separate evaluation of comparative realism, that is, to show to what extent realism guided by the comparative perspective on success and on truth approximation, fulfils these desiderata. The second task is its comparative evaluation, that is, to what extent is it superior to the other responses to antirealism.

4.1 In defense of the main claim of comparative realism

Recall that we have summarized the way in which comparative realism is claimed to fulfill the desiderata a couple of times by its 'main claim': truth approximation provides the default explanation and prediction of empirical and aesthetic progress. This claim is supposed to be realized by the stratified (basic) explication of the notions of 'more successful' and 'closer to the truth', and the resulting mutual relations. Let us start with noting a number of terminological equivalencies that will be presupposed in the defense of the claim.

Empirical progress combines instantial progress and explanatory progress, and a new theory amounts to empirical progress relative to its predecessor when it is concluded on the basis of 'sufficient' comparative testing that the new theory is persistently empirically more successful than the old one. Empirical success preservation provides weak versions of these kinds of progress.

Aesthetic progress of a new theory relative to an old one, according to the prevailing aesthetic canon and in particular among empirically equivalent (EE-)theories amounts to an increase of having desirable-aesthetic properties and a decrease of having undesirable ones. Below I will indicate that empirical progress and aesthetic progress have in some formal sense the same structure. Aesthetic success preservation is again a weak version.

In the expression 'closer to the truth', 'the truth' always refers to the strongest true theory about a given domain in a given vocabulary and 'truth approximation' (TA) is short for 'getting closer to the truth'. In the stratified context we get the equivalences: getting closer to the theoretical/ observational/ referential truth = theoretical/ observational/ referential truth approximation. In this context, 'truth approximation' and TA without qualification are supposed to refer to 'theoretical truth approximation'. The weak version of 'theoretical truth approximation' is called 'theoretical truth preservation'. Similarly, the weak version of 'referential truth approximation' is 'referential truth preservation'.[10]

We will first deal with CA1 and CA2, in which EE-theories and aesthetic progress are not yet at stake, and later with CA3 and CA4. In both cases the claim is of course that the crucial notions can be explicated such that (it is reasonable to assume that) these conditions of adequacy are satisfied. In the first case the justification is primarily analytical, allowing empirical illustrations. In the second case the justification will be a combination of empirical and analytical arguments. The analytical justification will be based on a number of 'TA-theorems and -conjectures'. The empirical justification will be twofold. Moreover, I will first restrict the attention to the cautious, very demanding, explications of the crucial notions, that is, the stratified (basic) explications.

4.1.1 Empirical progress

Let us first focus on matters of empirical progress in relation to theoretical truth approximation, and address matters of referential truth preservation later. For convenience, I repeat the first two conditions of adequacy restricted in this way:

CA1-EP The explications of the crucial phrases should leave room for the possibility that false theories are not only (persistently) empirically more successful but even closer to the (theoretical) truth.

CA2-EP The explications should be such that empirical progress is, *as a rule*, due to theoretical truth approximation, which on its turn entails, as a rule, novel predictive success.

In Section 2 I have already specified the stratified basic definitions of 'empirically more successful', hence of 'empirical progress', and of 'closer to the theoretical truth'. From these definitions it is easy to see that CA1-EP is satisfied, for the definitions do not make a crucial difference between theories that are true or false, in the (weak) sense, that is, theories that don't exclude nomic theoretical possibilities and theories that do, respectively.

---

[10] Note that we do not call 'theoretical, referential, observational truth approximation' progress of that kind, for we like to reserve the term 'progress' for kinds of progress that do not presuppose to know the relevant truth, such as empirical and aesthetic progress.

Regarding CA2-EP, recall, the (empirical) Success Theorem, which can be formulated for the stratified case as follows:

Stratified Empirical Success Theorem
Theoretical truth *preservation* of one theory relative to another entails explanatory success preservation and, as a rule, instantial success preservation. Exceptions to instantial success preservation concern instantial pseudo successes[11], to be explained, of the old theory, dropped by the new theory.

Theoretical truth *approximation* entails in addition, assuming empirically non-equivalent theories, that at least some extra (empirical) success can be obtained, and hence the prospect of empirical progress including novel facts.

The proof of the explanatory part in the first clause is straightforward, as in the unstratified case. However, the instantial part is now not unconditional, for in the stratified situation there may be 'instantial pseudo successes' of X relative to Y. Such a success amounts to an observational nomic possibility that is allowed by X and not by Y (hence breaking observational truth approximation of Y relative to X), which, however, on the theoretical level must be due to the allowance by X of a (theoretical) nomic *imposibilility*, which is not allowed by Y. Of course, any instantial success of X or Y may be a pseudo success, which is something which we cannot observationally distinguish from a genuine instantial success. However, this is only obscuring comparative theory evaluation as far as *extra* instantial successes are concerned. By the way, the epistemic asymmetry between explanatory and instantial successes may well explain why explanatory successes of proper theories (theories with theoretical terms) seem more impressive than counterexamples.

From this theorem we may conclude that theoretical truth approximation provides the realist default explanation and prediction of empirical progress. For, assuming at least some initial extra explanatory success, the hypothesis of theoretical truth approximation, the TA-hypothesis, enables the explanation and prediction of explanatory progress, including some (differential) novel predictive success, as we will see. Moreover, the hypothesis enables the conditional explanation and prediction of instantial progress, including some (differential) novel predictive successes, as we also will see. The condition being that the old theory does not have or get pseudo successes. If the TA-hypothesis is in fact false, new experiments will break the empirical progress conclusion in the long run. However, as far as breaking the instantial progress conclusion is concerned this may be due to pseudo successes.

In some more detail, the situation is as follows, assuming empirical progress, that is, a well confirmed comparative success hypothesis (CSH) of some theory Y, relative to X, on the basis of evidence R/S. If Y is in fact not closer to the theoretical truth than X, then there is some theoretical nomic possibility, say x, excluded by Y but included by X and/or some theoretical nomic impossibility, say y, allowed by Y, but not by X. In the first case there is room for an experimental counterexample of Y which is not a counterexample of X, viz. when the projection of x is also excluded by Y, and hence is no pseudo success of Y. In

---

[11] Called 'extra lucky observational hits' in previous work.

the second case an observational law, according to which at least y is impossible, can be experimentally established that is an explanatory success of X, but not of Y, provided the projection of y is also excluded by (the projection of) X, and hence no pseudo success of X.

Regarding the claim of some (differential) novel predictive success the situation is as follows. It is rather difficult to imagine that, at the start of testing a CSH, see Section 2, where the success superiority of Y relative to X is based on previous findings (R/S), these findings exhaust every kind of thing there is to know of the observable differences between X and Y. In other words, except in finite cases, and perhaps some very simple infinite cases, at least some room for differential 'novel predictive success' is plausible, in whatever sense 'novel' precisely is meant.

The foregoing not only shows that the TA-hypothesis provides the default explanation and prediction of empirical progress when the crucial notions are explicated in the stratified basic way, but also that CA2-EP then is satisfied.

4.1.2 Referential truth preservation

The first two conditions of adequacy also deal with reference claims, specifically:

CA1-Ref The explications of the crucial phrases should leave room for the possibility that false theories, even with some non-referring theoretical terms, are closer to the (theoretical) truth and refer at least as well as other theories.

CA2-Ref The explications should be such that theoretical truth approximation entails, as a rule, referential truth preservation.

In Section 2 I have already indicated the definition of 'reference' and of 'closer to the referential truth'. Here I will elaborate these indications a bit further, partly because of some weak spots in earlier formulations, but the main ideas remain the same.

In the definition of a subset X of M_p, constituting theory X, with strong claim "$X = T_t$", a theoretical term may or may not play a substantial role in the sense that if we project X on the conceptual possibilities generated by the M_p-vocabulary minus that term, and then recombine these possibilities again with that term in all possible ways, the resulting subset of M_p may or may not be a proper superset of X. If it is, X is said to claim that the term does refer, for it makes a difference. If it is not, X is said not to refer, for it does not make a difference.

The definition still needs some qualification. It seems adequate primarily for theoretical terms of which X claims that they refer to attributes (properties, relations, and functions); attribute terms for short. For theoretical terms claimed by X to refer to classes of entities, used as domain-sets for attribute terms - entity terms for short - this definition does not seem to work. However, it is plausible to say that a (theoretical) entity term refers according to X if and only if there is at least one (theoretical) attribute term that is claimed to refer and that uses the entity term as (one of) its domain-set(s). The consequence is that an entity term does not refer according to X if there are no attribute terms using it as domain-set. This is plausible, for in this case it is difficult to see how that theoretical domain could play a substantial role in the definition of X, which is precisely the reason for the detour via attribute terms. If this detour is not possible, the theoretical entities hang in the air according to X as unconstrained entities, not distinguishable from genuine fictitious entities, at least not with the means provided by the vocabulary. For example, the Higgs-particles, postulated for theoretical reasons by the so-called Standard Model in elementary particle physics, are supposed to have theoretical properties with observational consequences. One expects to observe the latter in the upcoming experiments in 2008 with the Large Hadron Collider (LHC) at CERN, near Geneva.

Of course, assuming that T_t can be defined with the means of M_p, T_t also claims, according to this definition of reference claims, of certain theoretical terms that they refer and of others that they do not, but now always rightly of course. Note that this definition of (non-)reference amounts to a direct reference criterion for attribute terms and an indirect one for entity terms. Note in passing that the very possibility to define reference of entity terms on the basis of the definition of reference of attribute terms is a good reason to extend the idea of 'entity realism' to 'referential realism'. Note finally that the given definition of reference is explicitly T_t-relative.

According to the *Nomic Postulate*, applied to M_{pp} as well as M_p, we have assumed that unique fixed sets T_o and T_t exist, given an intended domain. Moreover, we have implicitly assumed that both can be characterized in some way by the available vocabularies. Regarding T_t, however, the Nomic Postulate is somewhat dubious, for it seems to presuppose that the reference of the theoretical terms is given beforehand. More cautiously than we have done before, we have to assume that, if we would have some kind of superhuman observational access to M_{pp}-relevant theoretical entities and attributes it is possible to couple none, some or all theoretical terms constituting M_p to them such that T_t arises. Of course, it is not at all evident that this can be done only in one way. But we may assume that the number of terms that can be meaningfully coupled at once has a maximum and that this maximal coupling is unique and gives rise to T_t. In other words, we assume that there is unique maximal referential use of the theoretical vocabulary and that this is which we are looking for. The conjunction of the (non-)reference claims of T_t is called 'the referential truth' and the conjunction of all (non-)reference claims of theory X is called its (total) referential claim. Finally, theory Y is closer to the referential truth than theory X if Y's referential claim agrees on more terms with the referential truth than that of X. Referential truth preservation is defined as: theory Y is at least as close to the referential truth as theory X if Y's referential claim agrees on at least the same terms with the referential truth as that of X.

Let us call two theories referentially equivalent when they have the same (total) referential claim, and let us call them referentially semi-equivalent when they have the same number of positive reference claims. The uniqueness assumption about T_t now amounts to the assumption that a referentially semi-equivalent theory to T_t is referentially equivalent to T_t. Of course, referential (semi-) equivalence does not imply nor exclude empirical equivalence.

From the above it is clear that being (in the weak sense) true or false, even with some non-referring theoretical terms, does not play some crucial role in the definitions of theoretical truth approximation and referential truth preservation. Hence, CA1-Ref is satisfied. Regarding CA2-Ref, we submit the following (diagnostic) conjecture:

> **Referential Truth Conjecture**:
> Theoretical truth approximation entails, as a rule, referential truth preservation. If not, the old theory has *extra lucky* reference claims.

Note first that we do not add a second clause to the conjecture, similar to that of the Stratified Empirical Success Theorem, which would now amount to: theoretical truth *approximation* entails in addition (to theoretical truth preservation), as a rule, referential truth approximation. In the present case, it is difficult to argue for it, and plausibly so, because this would (almost) exclude the possibility of improving theories without improving its reference.

Let me explain what lucky reference claims, mentioned in the conjecture, are. One might think that the assumption that Y is closer to T_t than X does entail the claim that Y is referentially at least as close to the truth as X. A specific reason to expect this is the fact that the theoretical claim of a theory, e.g. "$X = T_t$", entails its referential claim "$V_r(X) = V_r(T_t) = V_r$", where $V_r(X)$ indicates the set of theoretical terms that refer according to X, and V_r is the one corresponding to T_t. However, a proof for the suggested unconditional conjecture is not possible, for interesting reasons. Suppose that Y is closer to T_t than X. What we would like to prove amounts to the following: all τ in $V_r(Y) - V_r(X)$ belong to V_r and all τ in $V_r(X) - V_r(Y)$ do not belong to V_r. Let us suppose that τ in $V_r(Y) - V_r(X)$ does not belong to V_r. Hence, Y wrongly claims that it does, whereas X rightly claims that it does not. However, X's claim may be based on a wrong aspect of its theoretical claim "$X = T_t$", whereas Y may not yet be so good that its theoretical claim implies the reference claim for the right reasons. It is important to note that the suggested proof already fails in the case of just one theoretical term. Similar arguments can be given for the possibility that τ in $V_r(X) - V_r(Y)$ belongs to V_r. In both cases we might say that X has lucky reference claims relative to Y, that is, X has true reference claims, whether or not precisely for the right reasons, where Y has false reference claims.

In sum, we have to admit the possibility that X has such lucky (non-) reference claims relative to Y. This does not mean, of course, that 'to be as close to the theoretical truth' does not provide good arguments for believing 'to be at least as close to the referential truth'. As can be learned from the suggested proof attempt, the suggested entailment is only violated when a theory bases (part of) its referential claim on wrong reasons in precisely the right direction, which will be rather exceptional because it is rather artificial.

So far, for the analytical justification of the claim that the stratified (basic) definition of truth approximation satisfies to a considerable extent the conditions of adequacy regarding empirical progress and referential truth preservation, viz. CA1 and CA2. After dealing with aesthetic progress, I will come back on the possibility of empirical support of this claim.

4.1.3 Aesthetic progress

For the moment we will focus on CA3, for CA4, dealing with probabilistic versions of IBE, evidently exceeds the boundaries of the qualitative approach.

CA3 As far as EE-explanations are concerned, an explication of IBE should only take those prevailing aesthetic criteria into account, *if any*, for which there are empirical or analytical reasons, however weak, to assume that they are truth-conducive in the field.

As we already noted, the prevailing aesthetic criteria in a certain field at a certain time constitute what James McAllister [14] has called the aesthetic canon, arrived at by aesthetic induction. In contrast to observational criteria, where we may assume, and have argued in some detail in the first subsection, that they are, as a rule, truth-conducive in the comparative sense, empirical information will have to support that certain aesthetic criteria are truth-conducive. However, before we enter the empirical justification, we will first show that there may well be truth-conducive non-empirical criteria at all. We will show this by a theorem that is a general version of (the explanatory part of) the success theorem. Of course, 'the truth' has not only observational consequences, but also theoretical (and mixed) consequences. A (purely) theoretical consequence of T amounts to a superset C of T, of which the projection equals M_{pp}, it does not exclude something observational. Note that every consequence of T can also be interpreted as a property or feature of all nomic possibilities, to be called a nomic property[12].

Nomic Theorem
Theoretical truth preservation (whether or not among EE-theories) entails being at least as successful regarding nomic properties.

A generally interesting question is which non-empirical properties are nomic, hence truth-conducive, properties. However, since observational differences will have priority in truth approximation evaluations, the use of this theorem naturally focuses on EE-theories. Hence, the main question becomes which non-empirical properties may be or have been supposed to be truth-conducive in the indicated sense.

We have already suggested, following McAllister and, for example, Weinberg, calling the non-empirical properties of interest aesthetic criteria, for such properties are as a rule also aesthetically appreciated. The empirical justification of claims to the effect that certain aesthetic criteria are truth-conducive in the field may be given along two different lines, and their combination. In the first line, due to Igor Douven, the claim is tested with hindsight, here called 'aesthetic testing, in the second line, the claim is based on 'aesthetic induction' as introduced by McAllister.

Douven [1-3] has described a way to (bootstrap) test applications of IBE on EE-theories with hindsight, which amounts briefly to the following. Although at the time of the application of IBE the theories were empirically equivalent, this may change in the course of time. New developments, e.g. the development of a sophisticated type of microscopes or the building of higher energy particle colliders, may lead to "a shift in the boundary between what can and cannot be observed" [3, p.297], which he aptly called an OUD-shift in his [1], where OUD is short for 'Observable-Unobservable Distinction'. In our approach this amounts to enriching M_{pp} to some M_{pp}'. Assuming that the new microscope or collider is veridical, which is the bootstrap condition of the testing, the interesting question

---

[12]Similarly, we could introduce 'strongly non-nomic' properties, properties corresponding to consequences of the negation of T, i.e. supersets of the complement of T, enabling a kind of mirror theorem to the nomic theorem. See [7, chap.10] for a general exposition of truth approximation in terms of two kinds of features.

becomes whether the previously best EE-theory, based on certain non-empirical properties, turns out to be empirically the most successful theory according to the new observational evidence acquired with the new means. If so, this confirms the claim that the (still) non-empirical properties were in fact truth-conducive.

The second line of empirical justification is provided by the 'naturalistic-cum-formal' analysis of the relation between aesthetic criteria, empirical success and truth [8]. It starts with the notion of 'aesthetic induction'. As McAllister has shown in his very inspiring book *Beauty and Revolution in Science* [14], our aesthetic judgments are subject to change. We are not only inclined to find the heliocentric worldview of Copernicus more beautiful, because it is simpler, than the geocentric view of Ptolemy, but we are also inclined to find Kepler's elliptic planetary orbits at least as beautiful as Copernicus' circular orbits. However, ellipses are undoubtedly more complicated than circles, and this is precisely the reason why they were found less beautiful, if not problematically ugly, at the time. Moreover, I would like to add, aesthetic criteria not only change with time within a discipline, but may also differ greatly between disciplines. For example, after expressing in an interview series [5] his agreement with Steven Weinberg [23] about the importance of beauty considerations, Stephen Gould hastens to stress that his criteria for beauty totally differ from those of Weinberg. Whereas Weinberg mentions 'inevitability' of desired consequences as his dominant aesthetic criterion – as exemplified by Einstein's theory, which, in contrast to Newton's theory, made the inverse square in the law of gravitation inevitable – Gould stresses that, besides diversity, unrepeatable contingencies and irregularities are the sources of his ultimate aesthetic satisfaction.[13]

Having observed the variation of aesthetic criteria, McAllister's main claims are as follows. First, scientists normally use aesthetic criteria in addition to empirical criteria for theory evaluation. Second, and most importantly, the aesthetic criteria of the time, the 'aesthetic canon', is based on 'aesthetic induction' regarding non-empirical features of paradigms of empirically successful theories which scientists have come to appreciate as beautiful. Third, aesthetic criteria can play a crucial, schismatic role in scientific revolutions. Since they may well be wrong, they may, in the hands of aesthetic conservatives, retard empirical progress, but this does not occur in the hands of aesthetically flexible, 'revolutionary' scientists.

My earlier analysis [8] of the relation between the empirical success and the beauty of scientific theories elaborates and supports McAllister's claims. Like McAllister, I focus on non-empirical aesthetic features, that is, features with aesthetic value but without empirical content, although empirical features may also be aesthetically valued. In the spirit of naturalized epistemology, one may first argue that the phenomenon of aesthetic induction may be a variant of the so-called 'mere-exposure effect' [24], viz. non-conscious affective priming [25]. In this perspective one may decompose the notion of aesthetic induction into aesthetic induction proper, or affective induction, and a related cognitive (meta-) induction. Together they lead to correlations between non-empirical features that are found beautiful on the one hand and empirically successful theories on the other. Such correlations may be called 'beauty-success correlations'. The corresponding received or 'canonical' aesthetic features are non-empirical features that have ac-

---

[13]This is not to suggest that standard examples of aesthetic features mentioned by physicists do not play a role in biology. For example, Gould mentions order a number of times and Sober [20] points out that simplicity in the form of parsimony plays a considerable role in taxonomy.

quired (positive) aesthetic value and (empirical success related) meta-inductive support. This makes it plausible to explicate the notion of an 'aesthetic feature' more precisely as an aesthetically (positively) valued non-empirical feature. Now it is possible to argue that the co-production of the two types of induction is functional for empirical progress as far as the cognitive meta-induction is reliable. That is, aesthetic features may rightly have become indicative of empirical success.

In view of the nomic theorem and our formal analysis of the relation between (more) empirical success and truth approximation, indicated in the first subsection, such features become indirectly indicative of truth approximation, that is, truth-conducive, provided they are of a certain formal nature. Since aesthetic features correlate with empirical success and empirical success with truth approximation, it has some plausibility to submit that such features of EE-theories are nomic properties, in which case they strongly correlate with truth approximation. However, this formal account only holds for aesthetic features of a certain formal kind, to be called distributed features. In order to possibly be a nomic property, such a feature should apply to all models of the relevant theory, in which case they can also be represented as a consequence of the theory. To be sure, the link between distributed aesthetic features and truth approximation is weak, but it makes some sense. It essentially consists in the combination of the stratified empirical success theorem and the nomic theorem. Below we will briefly discuss which aesthetic features are or may be distributed.

Of course, aesthetic induction will cover (positive) results of previous aesthetic testing, whether or not such testing has been undertaken explicitly for this purpose. Hence, like aesthetic induction, aesthetic testing contributes to the relevant 'aesthetic canon'. This strengthens the conclusion that there are good, though not strong, reasons to assume that the aesthetic canon, as far as consisting of distributed features, may rightly hypothesize what are truth-conducive theoretical nomic properties, and hence for the following conjecture:

> *Aesthetic Success Conjecture*
> Theoretical truth *preservation* among EE-theories entails preservation of theoretical nomic properties, and hence, in view of the nature of aesthetic testing and aesthetic induction, as a *weak* rule, aesthetic success preservation according to the aesthetic canon as far distributed. If not, which may appear by new aesthetic testing or a major breaking of aesthetic induction, previous aesthetic testing and induction prove to have their limits.
>
> Theoretical truth *approximation* among EE-theories entails increase of theoretical nomic properties, and hence, as an at least as weak rule, aesthetic progress according to the aesthetic canon as far as distributed.

The important remaining question is which aesthetic features are, or can have a distributed form. Let us review the features that have been mentioned earlier, now grouped in five categories. Each of them may have other examples.

- inner beauty, elegance,
- simplicity, symmetry, visualizability, inevitability or rigidity, coherence with metaphysical suppositions,

Comparative realism as the best response ...

- explanatory power, novel predictive power, fertility,
- coherence with background theories,
- non-ad hocness, parsimony.

I will deal with these categories in the listed ordering. The *first* category lists features that ask for some kind of objective specification, such as those of the other categories. The *second* category typically represents features that are non-empirical and admit an objective specification. Moreover, although they can apply to theories as wholes they can also apply, or not, to separate conceptual possibilities, and hence to all models of a theory, or not, that is, they can distributively apply to a theory, or not. Finally, their precise nature and their aesthetic value may typically vary historically and between disciplines. In sum, they typically form the kind of features to which the aesthetic success conjecture is oriented. The *third* category typically contains features that first of all promise to be accounted for in empirical success: 'explanatory power' in explanatory success, 'novel predictive power' in novel predictive success, instantial or explanatory, and finally 'fertility' frequently is used to express the two previous kinds of power. In all three cases, it may well be that the aesthetic value they have, derives from these promises of empirical virtues. Although such promises may turn out to be realizable, it is important to note that such empirical pay-offs can only arise when there are such promises. All this holds, whether or not the features may apply in a distributed form, which seems to be possible for 'novel predictive power', and hence for a certain kind of fertility. The *fourth* category is a kind of mirror category of the previous one. Incoherence with background theories is a 'promise' of empirical problems, the stronger the more established the background theories are. Hence, this type of incoherence may well have become negatively valued aesthetically as a predictor of empirical problems. However, it may also be a (very weak) predictor of a scientific revolution. Finally, 'non-adhocness' in the *fifth* category, is generally valued positively because adhocness is disvalued, probably because of similar reasons as the previous type of incoherence. The saving of a theory by an ad hoc repair frequently turns out to have its empirical price. However, in this case it is important to note that it may also turn out to be a smooth kind of truth approximation. This applies *mutatis mutandis* to parsimony, for again non-parsimoniousness may show to have its empirical, notably explanatory, price, but it may also be a step in the good direction, the truth.

Let us finally confront the proposed way of dealing with EE-theories in accordance with CA3 with the provoking paper of Roger Jones [4], entitled 'Realism about what?' (see also, [12, pp.252-257], from which I will borrow crucial formulations). In this paper, Jones reminds us of four alternative formulations of classical mechanics, as for example applied to planetary motion. Though empirically equivalent, they suggest nevertheless a quite different ontology and metaphysics. Newton's own approach, based on the three laws of motion and force laws, suggests an ontology of point particles and forces acting at a distance. The gravitational field approach assumes an ontology of action by contact in conformity with a principle of local causality, but postulates a new type of entity, the field. In the curved space approach, the gravitational field is absorbed into the structure of space (analogous to the way in which this field is absorbed into the structure of spacetime in general relativity). In this approach the field is represented by the local degree of curvature and space itself is made part of the fundamental ontology having a causal efficacy of its own. Finally, the analytical mechanics approach, in which the force laws and the laws of motion can be derived from so-called minimum principles, seems to be independent of causal thinking, but requires instead a kind of teleology for particle behavior according to which it is the properties of complete paths between points of space that determine the motion of a body.

The question is of course which ontology (cum metaphysics) is suggested by a realist attitude to these EE-versions of classical mechanics. The suggestion of our analysis is that this depends on which formulations score the best according to the prevailing relevant aesthetic canon, dealing with ontological, epistemological, methodological and pragmatic aspects. However, apart from the difficult question as to which canon is prevailing and relevant, given so long a period of time in developing these formulations, the preliminary question is whether the suggested ontologies are really as different as they seem at first sight. The first approach (with forces acting at a distance) and the field approach may well be seen as compatible ontologies, provided the forces are not assumed to act at a distance instantaneously. With this (non-Newtonian) assumption they are related as the manifest global or macro-ontology to the theoretical local or micro-ontology as seems, for example, the standard view in response to Eddington's non-existence claim of his famous table. The relation between the field approach and the (non-relativistic) curved space approach does not seem to be a matter of compatible ontologies. However, looking back from general relativity theory, and its empirical success, the aesthetic appreciation of a curved spacetime and hence presumably of a curved space (both of which can be distributed properties of a theory), seem to have become (much) higher than that of the (relevant) field approach. Hence, the curved space version of the classical theory of gravitation seems to deliver the most plausible realist ontology, and hence would seem so far to be the best theory in accordance with CA3, and hence for inferring, by IBE-E3.2, that it is the closest to the theoretical as well as to the referential truth. This judgment will not be changed by the last, minimum principles formulation of mechanics, for that delivers from a realist point of view at most an as if metaphysics. However, as such it does no theoretical work, let alone observational work, for in the present deterministic setting this kind of 'as if metaphysics' is purely a matter of logico-mathematical equivalences.

This concludes our articulation and defense of the claim that CA3 can be satisfied, at least to some extent.

As we stated already, CA1 and CA2 and our main claim have primarily analytical justification. Of course, this does not exclude empirical support. On the contrary, their justification asks for real life illustrations. In particular for CA2 one may think of the kind of (implicit or explicit) testing proposed by Douven on the basis of OUD-shifts: do previous 'empirical progress' and 'referential truth preservation' conclusions survive after the shift or even, when *all* theoretical terms have become observable, can we now conclude to 'referential truth approximation'?[14] Unfortunately, the demands for stratified basic truth approximation are very high, so that we will have to extend the analysis to 'refined qualitative'

---

[14] Of course, OUD-shifts will also lead to reconsideration of previous aesthetic inductions: do they survive or has the canon (to be) changed. According to McAllister, a change of the canon is a characteristic for scientific revolutions.

or 'quantitative versions'. This brings us to CA4.

Regarding CA4, the situation is not difficult. Recall:

CA4 In a probabilistic version of an adequate explication of IBE 'EE-updating' of the relevant belief should remain within the confines of Bayesian conditionalisation.

The strictures of the 'argument from Bayesianism', that is, resulting from non-Bayesian conditionalization, can be avoided in at least one way, viz. by taking aesthetic features only into account in the (expert) prior distribution. As far as I know there is no other way[15]. As I already remarked in Section 3.3, a merely probabilistic version will not do in general, for probabilistic updating makes only differential sense for unfalsified theories. However, if we leave behind the strict conditions of stratified basic truth approximation, probabilistic estimation of the distance from the truth [15, 16], using Bayesian updating, is generally applicable.

4.2 Review and extension of the comparison with other realist responses to antirealism

Some preliminary remarks are in order before we can take up the comparison of the comparative approach with other realist responses. First, there are, of course, no *compelling* reasons neither for comparative realism nor for antirealism, whether the latter concerns theoretical, inductive or even experiential skepticism. However, Van Fraassen's adagium "what is rational to believe includes anything that one is not rationally compelled to disbelieve" [22, pp.171-172] is a license for all kinds of wishful thinking, blocking a 'good reasons' debate. Although there is not much hope to convert Van Fraassen, a paradigm believer in antirealism, more specifically, constructive empiricism, into a believer in comparative realism, I have some hope that analytical philosophers with realist inclinations but doubts based on the antirealist charges, may become comparative realists instead of remaining adherent of one of the other retreats from full blown realism.

Let me first review the realist responses already dealt with so far. The retreat to claims of 'approximately true theories' instead of 'true theories' was put aside at the start as of no help, for that remains basically non-comparative. Although we certainly prefer giant leaps to the truth, any step in the direction of the truth, from whatever to whatever distance to the truth should be welcomed by the realist as a kind of progress. Moreover, the retreat to 'approximate truth' requires a necessarily arbitrary threshold. Further we came across restricting realism to 'mature' theories, or to theories with 'novel' predictive success or to the 'essential' parts and aspects of theories. The general tenet of our criticism of these moves was a kind of redundancy. From the comparative perspective all theories may remain in the game, no ban of certain kinds or aspects of theories are required. This is not to say that some of the suggested distinctions may not be useful for other purposes, related or not. The same holds for the one opposite kind of response we have met, viz. stretching realism by stretching the (causal) theory of reference underlying many conceptions of realism: it is not necessary

---

[15]However, Niiniluoto reminded me of the fact that Hintikka's α-λ-system of inductive probabilities converges to the simplest generalization that is still compatible with the evidence. Hence, it need not be necessary to have to take explicitly care of aesthetic considerations in order to be respected.

for defending realism. As we have seen, such a defense is possible by using a non-standard criterion of reference, which may well be compatible with the causal theory of reference, or some stretched refined version.

Let me now turn to the extension of the comparison, in particular, with other realist responses by restricting realism. To begin with, 'entity realism', that is, realism with respect to the referential claim attached to theoretical entity terms, has been motivated, at least partially, by the idealizational, hence truth-abusing, nature of scientific practice, in particular model construction. From the comparative perspective this motive for restricting realism in order to save the important idealization practice is not relevant, for the question is whether of two successive theories, the second, being a concretization of the first, idealized theory, is closer to the truth than the first. In [6, chap.10], I have shown and illustrated that the refined version of the truth approximation theory leaves perfect room for a positive answer to this question, roughly due to the fact that the second theory has models that are more similar to correct models than the models of the first. Of course, the same argument can be applied to the less restrictive version of entity realism, called 'referential realism', that is, realism with respect to the referential claim of all theoretical terms, entity as well as attribute terms. Recall that I have remarked in passing that the transition from entity to referential realism is particularly plausible in view of the non-standard criterion of reference.

Finally, another retreat, which has become much debated in the last decade, is so-called structural realism, according to which typical realist features of successful theories consist of their structural features as expressed by the formalism specifying the formal relations. This retreat is mainly motivated by frequent referential failures, as highlighted by the pessimistic meta-induction. Again, it is an unnecessary retreat from the comparative perspective, now in view of the possibility of referential truth approximation.

5 Concluding remarks

I would like to conclude this paper with a general remark and a survey of epistemological positions.

All realist responses to the antirealist charges are retreats of realism of a non-comparative nature. As already suggested by my puzzling question in the introduction, it seems that all responses are non-comparative because of the lack of awareness that in the present debate relations, in particular, comparative ones, may be crucial and should be taken into account. Of course, comparative realism is also a kind of retreat. However, since it is a kind of piecemeal realism, it fits perfectly well into what may be seen as the core idea of realism, namely that our theories succeed more and more in giving 'a literally true story of what the world is like', to use Van Fraassen's well-known characterization of realism [21, p.8].

In my *From Instrumentalism to Constructive Realism* [6], I have presented a hierarchical survey of epistemological positions. In the course of time I have refined and revised it at some places, but the core is still the same. I would like to present here the latest version, for it captures the main lines of thought in this paper rather well. It is structured by a number of epistemological questions, however, starting with an ontological one (as Tab.I). A non-standard kind of summary of the main lines in this paper, some of it implicit, is as follows,

Q0:	independent natural world?	No ⇒	ontological idealism
	⇓ * Yes: ontological realism		
Q1:	true claims about it possible?	No ⇒	epistemological skepticism
	⇓ * Yes: epistemological realism		- experiential skepticism
			- inductive skepticism
Q2:	beyond the observable?	No ⇒	theoretical skepticism
	⇓ * Yes: scientific realism		instrumentalism
			- constructive empiricism
Q3.1	beyond reference?	No ⇒	referential realism
			⇒ entity realism
Q3.2	beyond structures?	No ⇒	structural realism
	⇓ * 2 × Yes: theory realism		
Q4:	ideal conceptualization?	No ⇒ *	constructive realism
	⇓ Yes: essentialistic realism		

Table I: Hierarchy of epistemological positions

in the scheme indicated by the starred arrows. There is a human-independent natural world, about which true claims may be (non-compellingly) justified, not merely restricted to what is observable, but also with respect to theoretical terms and statements, provided we take the comparative piecemeal perspective, being a core feature of comparative 'theory realism'. However, as to the question whether there is some ideal vocabulary fitting the natural world, and hence leading to a kind of realism, my answer is negative, for which reason I speak of 'constructive' realism, e.g. in the title of my book. Finally, I would like to stress that constructive realism is a kind of nomic realism. As became clear in the present paper, according to my view, theory formation and revision is not directed at truth approximation with respect to the actual world but at the nomic world, that is, the realm of what is physically possible.

Acknowledgements

I like to thank a number of people for various useful comments received at the try-out in Groningen and in Beijing: David Atkinson, Luis-Alberto Cordero-Lecca, Richard Dawid, Lars-Göran Johansson, James Joyce, Barteld Kooi, Ulianov Montano, Ilkka Niiniluoto, Robert Nola, Jeanne Peijnenburg, Isabelle Peschard, Gerhard Schurz, Jan-Willem Romeijn, Allard Tamminga, Sjoerd Zwart. Finally, I like to thank the Netherlands Institute for Advanced Study (NIAS) in Wassenaar for the opportunity to spend two weeks to complete this and another paper.

Bibliography

[1] Douven, I. *In Defence of Scientific Realism*. Ph.D. thesis, University of Leuven, Leuven, 1996.

[2] Douven, I. Testing inference to the best explanation. *Synthese*, 120:355–377, 2002.

[3] Douven, I. Empirical equivalence, explanatory force, and the inference to the best theory. In Festa, R. Aliseda, A., Peijnenburg, J. (Eds.), *Confirmation, Empirical Progress, and Truth Approximation (Poznan Studies in the Philosophy of the Sciences and the Humanities)*, 281–309. Amsterdam/New York, NY: Rodopi, 2005.

[4] Jones, R. Realism about what? *Philosophy of Science*, 58:185–202, 1991.

[5] Kayzer, W. *Het boek over de schoonheid en de troost*. Amsterdam: Contact, 2000.
[6] Kuipers, T. *From Instrumentalism to Constructive Realism. On some relations between confirmation, empirical progress, and truth approximation (Synthese Library)*, vol. 287. Dordrecht: Kluwer Academic Publishers, 2000.
[7] Kuipers, T. *Structures in Science (Synthese Library)*, vol. 301. Dordrecht: Kluwer Academic Publishers, 2001.
[8] Kuipers, T. Beauty, a road to the truth. *Synthese*, 131:291–328, 2002.
[9] Kuipers, T. Inference to the best theory, rather than inference to the best explanation. kinds of abduction and induction. In Stadler, F. (Ed.), *Induction and Deduction in the Sciences*, 25–51. Dordrecht: Dordrecht: Kluwer Academic Publishers, 2004.
[10] Kuipers, T. Theories looking for domains. fact or fiction? In Magnani, L. (Ed.), *Model-based reasoning in science and engineering*, Studies in Logic, vol. 2, 33–55. London: College Publications, King's College, 2004.
[11] Kuipers, T. Introduction: Explication in philosophy of science. In Kuipers, T. (Ed.), *General Philosophy of Science: Focal Issues (Handbook of the Philosophy of Science)*, vol. 1, vii–xxiii. Amsterdam: Elsevier, 2007.
[12] Ladyman, J. *Understanding Philosophy of Science*. London: Routledge, 2002.
[13] Ladyman, J. Ontological, epistemological, and methodological positions. In Kuipers, T. (Ed.), *General Philosophy of Science: Focal Issues(Handbook of the Philosophy of Science)*, vol. 1, 303–376. Amsterdam: Elsevier, 2007.
[14] McAllister, J. *Beauty and Revolution in Science*. Ithaca: Cornell UP, 1996.
[15] Niiniluoto, I. *Truthlikeness (Synthese Library)*, vol. 185. Dordrecht: Reidel, 1987.
[16] Niiniluoto, I. *Critical Scientific Realism*. Oxford: Oxford University Press, 1999.
[17] Niiniluoto, I. Evaluation of theories. In Kuipers, T. (Ed.), *General Philosophy of Science: Focal Issues (Handbook of the Philosophy of Science)*, vol. 1, 175–217. Amsterdam: Elsevier, 2007.
[18] Panofsky, W., Phillips, M. *Classical electricity and magnetism*. London: Addison-Wesley, 2nd edn., 1962.
[19] Psillos, S. *Scientific Realism: How Science Tracks Truth*. London: Routledge, 1999.
[20] Sober, E. *Philosophy of Biology*. Oxford: Oxford University Press, 2000.
[21] van Fraassen, B. *The Scientific Image*. Oxford: Clarendon Press, 1980.
[22] van Fraassen, B. *Laws and Symmetry*. Oxford: Clarendon Press, 1989.
[23] Weinberg, S. *Dreams of a Final Theory*. London: Vintage, 1993.
[24] Zajonc, R. Attitudinal effects of mere exposure. Monograph supplement 9 of *The Journal of Personality and Social Psychology*, 1968.
[25] Zajonc, R. Evidence for nonconscious emotions. In Ekman, P., Davidson, R. (Eds.), *The Nature of Emotion*, 293–297. Oxford: Oxford University Press, 1994.

Meta-Induction: A Game-Theoretical Approach to the Problem of Induction[1]

Gerhard Schurz

University of Düsseldorf

schurz@phil-fak.uni-duesseldorf.de

ABSTRACT. In this paper I suggest a 'best alternative'-justification of induction which is based on *meta-induction*. The meta-inductivist applies the principle of induction at the level of prediction methods: she tries to derive an optimal prediction from the predictions and the observed success rates of all prediction methods which are *accessible* to her. I investigate the question whether there exist meta-inductive prediction strategies whose predictive success is *optimal*. While *one-favorite* meta-inductive strategies, though being quite efficient, turn out to be not universally optimal, one can construct *weighted average* meta-inductive strategies whose success is demonstrably optimal in arbitrary prediction games (or possible worlds). The justification of meta-induction which follows from my theorems is mathematical-analytical or *a priori*. It implies, however, an *a posteriori* justification of object-induction based on the experiences in our world.

1 Introduction

I understand the notion of an *inductive inference* in the Humean sense: an inference in which a property, regularity, or frequency is transferred from the observed to the unobserved, or from the past to the future. Other forms of non-deductive inferences such as inferences 'to the best explanation' are not considered. The *problem of induction* (or Hume's problem) is the following: how can we *rationally justify* inductive inferences?

David Hume has shown that all standard methods of justification seem to fail when applied to the task of justifying induction. (1) Obviously, inductive inferences cannot be justified by deductive logic, since it is logically possible that the future is completely different from the past. (2) More importantly, induction cannot be justified by induction from observation, by arguing that induction has been successful in the past, whence – by induction – it will be successful in the future. For this argument is *circular*, and circular arguments are without any justificatory value. Salmon [11, p.46] has shown that also anti-induction may be pseudo-justified in such a circular manner. (3) It is also impossible to demonstrate that the conclusion of an inductive inferences is at least highly probable – for in order to show this, one must presuppose that the relative event frequencies observed so far can be transferred to the unobserved future, which

---

[1] Prof. Dr. Gerhard Schurz is with Department of Philosophy, University of Düsseldorf. Universitaetsstrasse 1, Geb. 23.21, D-40225 Duesseldorf, Germany.

is nothing but an inductive inference. These were the reasons which led Hume to the skeptical conclusion that induction cannot be rationally justified, but is merely the result of psychological habit. There have been several attempts to solve or dissolve Hume's problem, which cannot be discussed here. It seems that so far, none of these attempts has been successful in giving a *positive* solution to the problem of induction, which establishes in a *non-circular* manner that the inductive method is a superior prediction method in terms of its success frequencies. Let me emphasize that the superiority of scientific induction as a prediction method is by no means obvious. Millions of people do in fact believe in superior non-inductive methods, be it God-guided inner intuition, clairvoyance, or other super-natural abilities. Therefore a satisfying justification of induction would not only be of epistemological importance; it would also have cultural significance in the enterprise of enhancing scientific rationality.

Given that it is impossible to demonstrate that induction *must* be successful (Hume's lesson), and that there are various *alternative* prediction methods, then it seems to follow that the only approach to Hume's problem for which one can at least uphold the *hope* that it *could succeed* if it were adequately developed is Reichenbach's *best alternative* approach (Reichenbach [9, Section 91], Salmon [12]). This approach does not try to show that induction must be successful, but it attempts to establish that induction is an *optimal* prediction method – its success will be maximal among all competing methods in arbitrary possible worlds. Or in simplified words: if any method of prediction will work, then the inductive method will work (Rescher 1980, 207ff). It must be emphasized that in demonstrating optimality one must allow *all* possible worlds, in particular all kinds of *para-normal* worlds in which perfectly successful clairvoyants or anti-inductivistic demons do indeed exist. Restricting the set of worlds to 'normal' or uniform worlds would completely *destroy* the enterprise of justifying induction. For then we would have to justify inductively that our real world is one of these 'normal' worlds, and we would end up in that kind of circle or infinite regress in which according to the *Humean skeptic* all attempts of justifying induction must end up.

In this paper I study the success rates of competing prediction methods in the form of prediction games which consist of infinite sequences of (binary) events and finite sets of players predicting these events by certain methods (or strategies). By *object-induction* (abbreviated as OI) I understand methods of induction applied at the level of *events* (the 'object level'). The major critique to Reichenbach's approach lies in the fact that it fails to establish an optimality argument for induction with respect to the goal of *single event predictions*. Reichenbach demonstrated an optimality argument with respect to the goal of *approximating* the frequency limit of an infinite event sequence in the long run. With respect to that goal, his argument was *almost trivial*: if the event sequence has a frequency limit, then the inductive straight rule (see below) *must* approximate this limit in the long run, while other non-inductive methods may or may not approximate the limit; but if the sequence of events does not have a frequency limit, then *no* method can find the limit (Reichenbach [9, p.474f]). However, our ability to approximate correct frequency limits in the long run is practically not significant. What *is* of practical significance is our success in true *predictions*. In this respect, Reichenbach's approach fails. Nothing in Reichenbach's argument excludes that a clairvoyant may be predictively much more successful than a scientific object-inductivist (cf. Kading [6]). A perfect future-teller may have 100% success in predicting random tossings of a coin, while the object-inductivist can only have a predictive success of 0.5 in this case. Reichenbach remarked that if successful future-teller would indeed exist, then the inductivist could recognize this by applying induction to the success of prediction methods ([8, p.358f], [9, p.476f]). But Reichenbach did neither show nor even attempt to show that by this meta-inductivistic observation the inductivist could have equally high predictive success as the future-teller (cf. Skyrms [13, chap.III.4]).

Generally speaking, the problem of Reichenbach's account lies in the fact that it is impossible to demonstrate that object-induction is an optimal or at least nearly optimal prediction method (which is also a lesson of formal learning theory, see below). In contrast to Reichenbach's approach, my approach is based in the idea of *meta-induction*. The meta-inductivist (abbreviated as MI) applies the inductive method at the level of competing prediction methods. More precisely, the meta-inductivist bases his predictions on the predictions and the observed success rates of the other (non-MI) players and tries to derive therefrom an 'optimal' prediction. The simplest type of MI simply predicts what the presently best prediction method predicts, but one can construct much more refined kinds of meta-inductivistic prediction strategies.

One should expect that for meta-induction the chances of demonstrating optimality are much better than for object-induction. The crucial question of this paper will be: is it possible to design a version of meta-induction which can be proved to be an optimal prediction method? The significance of this question for the problem of induction is this: if the answer is positive, then at least meta-induction would have a rational and non-circular justification based on mathematical-analytic argument. But this analytic justification of meta-induction would at the same time yield an *a posteriori justification of object-induction* in our real word: for we know *by experience* that in our real world, non-inductive prediction strategies have *not* been successful *so far*, whence it would be meta-inductively justified to favor object-inductivistic strategies. In other words: the common-sense argument in favor of object-induction which is based on its *past* success record would no longer be circular, *given* that we had a non-circular justification of meta-induction.

If (a version of) meta-induction would turn out to be optimal, then this alone would be compatible with the existence of other equally optimal prediction methods. But insofar one can reasonably assume that meta-induction is the *only* method for which one can *demonstrate* its optimality (cf. Reichenbach ([9, p.475f]), the justification of meta-induction by its optimality would already yield a sufficiently strong kind of justification. Of course, it would be desirable to prove that meta-induction is even a (weakly) *dominant* prediction method (in the sense of [14, p.10]), i.e., no other prediction method is optimal. We will see in Section 6 that dominance of meta-induction can be demonstrated only w.r.to 'purely' non-inductive strategies, but not w.r.t. 'mixed' strategies.

Let me finally discuss two possible objections to my approach:

A first objection might complain that I presuppose that the predictions of the other players for the next time are *accessible* to the meta-inductivist. There might be possible worlds in which alternative players do not *give away* their predictions but keep them secret. Indeed, this is possible, and so I have to restrict my claim to *accessible methods*. What I intend to show is that among all prediction methods (or strategies) who's *output* is *accessible* to a given person, the meta-inductivistic strategy is a optimal. But I argue that this restriction is not a drawback. For methods whose output is not accessible to a person are not among her *possible actions* and, hence, are without relevance for the optimality argument.

A second objection could point out that my proposed justification of (meta-) induction (if it were successful) is not an epistemic but a practical justification, and hence is not a true solution to Hume's problem. I think this diagnosis is incorrect. For the goal underlying the optimality argument is maximization of true predictions, and this is clearly an *epistemic* and not a practical goal. Although an optimality justification is weaker than a reliability justification (i.e., an argument showing that the predictive success is guaranteed to be sufficiently high), it is nevertheless an epistemic justification. Moreover, Hume did not only argue that that induction cannot be demonstrated to be reliable, but much stronger that no epistemically rational justification of induction is possible. Therefore I regard optimality arguments as a way towards a genuine (though weak) solution to Hume's problem.

2 Prediction Games

A prediction game $((e), \Pi)$ consists of the following components:

(1) An infinite sequence (or 'stream') $(e) := (e_n : n \in \mathbb{N}^+)$ of binary events $e_n \in \{1, 0\}$ of a given type $\pm E$; hence each e_i is either the event '1' (for 'E obtains') or '0' (for 'E does not obtain'). Each discrete time unit $n = 1, 2, \ldots$ corresponds to one *round* of the game. For example, (e) may be a sequence of coin tossings, or of daily weather conditions under a binary description, etc. In Section 5 we consider also real-valued events.

(2) A set of players $\Pi = \{P_1, P_2, \ldots, \text{xMI}\}$, whose task is to predict, for any time n, whether or not E will occur at the next time $n+1$. The success of these players is evaluated by their predictive success rates $suc_n(P_i)$ (the number of P_i's correct predictions until time n divided through n). We will loosely identify players with their prediction methods (unless stated otherwise). The players include:

  (2.1) the object-inductivist OI := P_1 (OI has index 1), whose prediction method is explained below. OI has informational access to the past events; his first prediction (at $n = 1$) is a guess.

  (2.2) a subset of alternative players – for example, persons who rely on their instinct, God-guided future-tellers, etc. These alternative players may have any success and any information you want, including informations about future events (clairvoyants) or informations about the meta-inductivists favorites (deceivers).
  Players of type (2..1) or (2..2) are called *non-MI-players*.

  (2.3) one or several meta-inductivists of a certain *type*, abbreviated as 'xMI', where 'x' is a variable (possibly empty) expression specifying the *type* of the meta-inductivist. At each time n, the meta-inductivist has access to the past events and the past and present predictions of the other players. Hence, each round n of the game consists of the following steps: (i) the world event e_n is determined; next (ii) the non-

MI-players deliver their predictions for time $n+1$; and then (iii) xMI makes her prediction for time $n+1$.

The simplest type of meta-inductivist from which I start my inquiry is abbreviated as MI. At each time, MI predicts what the non-MI-player with the presently highest predictive success rate predicts. If P is this presently best player, then we say that P is MI's present *favorite*, or that MI *favors P*. If there are several best players, MI chooses the first best player in an assumed ordering, say, from left to right. MI changes his favorite player only if another player becomes *strictly* better; otherwise he sticks to his present favorite. MI's first favorite is OI. I assume that MI has always access to OI: even if no person different from MI plays OI, MI constantly simulates the predictions of OI, and uses them if their virtual success supersedes that of the alternative players.

The simplest object-inductive prediction method, abbreviated as OI, consists of a combination of the *straight rule* and the *maximum rule*. The straight rule transfers the relative frequency of the given event type E from the observed events to the conjectured frequency-limit in the indefinite future (cf. Salmon [12, pp.89–95], Rescher [10, chap. VI.3]). In what follows, $f_n(E)$ stands for the relative frequency of E until time n, and $p(E) := \lim_{n\to\infty} f_n(E)$ denotes the frequency-limit, that is, the statistical probability of E in the reference sequence. The definition implies that the straight rule is a method of approximating the frequency limit $p(E)$ in the long run, *provided* that the event sequence converges to a limit. The straight rule does no tell what *predictions* one should make. The maximum rule says that one should always predict an event (out of a given partition of events) whose probability conjectured by straight rule is maximal. For binary events these two rules generate the following prediction strategy of OI: if E's so-far observed frequency is $\geq \frac{1}{2}$ OI predicts E, and otherwise OI predicts $\neg E$. Under the assumption that the reference sequence is *random* one can prove that this rule *maximizes* the predictive success among all prediction rules of the form "predict E in $r\%$ and $\neg E$ in $1 - r\%$ of cases" (cf. Greeno [4, p.95], Reichenbach [8, p.310f]). In this case, OI's success rate will converge against the maximum of $p(E)$ and $p(\neg E)$. For non-random sequences *refined* object-inductivistic prediction strategies exist, whose success dominates OI's success. They can be reduced to the strategy OI with the help of Reichenbach's principle of the narrowest reference class (see Section 6).

I identify prediction games with *possible worlds*. Apart from the definition of a prediction game, I make no assumptions about these possible worlds. My approach does not depend on some (problematic) distinction between 'uniform' and 'non-uniform' worlds (such a distinction is notoriously difficult; cf. Skyrms [13, p.34f]). The stream of events (e) can be arbitrary. Should (e) be non-random, then more refined object-inductivistic strategies may exist (as explained), but their players are assumed to be among the alternative players – nothing which concerns the behaviour of xMI hangs on that question. I also do not assume a fixed list of players – the list of players may vary from world to world, except that it always contains xMI and the (virtual) OI. I make the realistic assumption that xMI has *finitely bounded* computational means. On this reason I restrict my investigation to prediction games with *finitely* many players.

Prediction games are not interactive games in the *narrow* sense because the predictive success rate of each player P is solely determined by 'nature' (the stream of events) and by P's own actions, i.e. predictions. But prediction games are interactive in a *wider* sense, because the actions of the meta-inductivist depend on the actions of the non-MI-players and their successes, and the actions of certain alternative non-MI-players (e.g., systematic deceivers) depend on MI's choice of favorites. Note also that the collective meta-inductivists introduced in Section 5 provide a basis for game-theoretic effects on the narrow sense.

According to my knowledge, prediction games have so far not been studied in the philosophical literature. There are, however, three related approaches in related fields:

1. In *formal learning theory* (cf. Kelly [7]) only *one* 'player', the inductive scientist, plays against a stream of events, and it is investiagted under which conditions on the stream of events which cognitive tasks can be achieved in a reliable way. Concerning inductive prediction tasks the general result is negative, because of the possibility of 'demonic' streams of events which at every time n produce the opposite of what was predicted. This is nothing but a variant of Hume's lesson; and in general, the results of formal learning theory support the fact that for object-inductive methods, the realm of reliably achievable tasks is rather limited.[2] In contrast, my prediction games consist of several prediction methods playing against each other, and my investigation does not focus on the question of the reliability but of the optimality of methods. Even if for every meta-inductive prediction method there exist suitably chosen 'demonic' streams of events for which its predictive success is zero, such a method may still be optimal, provided one can prove that in all 'demonic' cases also all other methods which are accessible to the meta-inductive method must have zero success. Indeed, this will be a consequence of my central theorems.

2. A second field which comes very close to our approach, although it has not been related to the problem of induction, is the theory of *universal prediction* which has been developed in the fields of mathematical decision theory and machine learning (for an overview cf. Cesa-Bianchi and Lugosi [1]). In universal prediction theory, one considers *online predictions based on expert advice*: a forecaster (who corresponds to our meta-inductivist) predicts an arbitrary event sequence based on the predictions of a set of experts (who correspond to our 'non-MI-players'). One speaks of 'universal' prediction theory because, as in our apoach, the event-sequences is not assumed to be generated by a stochastic process but may be arbitrary; and the setting is called *online learning* because at every time n the players have simultaneously to learn from past events to to make new predictions. In Section 5 we will make use of a central theorem achieved in this field.

3. A third related field is the comparative investigation of the efficiency of prediction methods by Gigerenzer and the ABC-research group ('<u>A</u>daptive <u>B</u>ehavior and <u>C</u>ognition). Although this approach focuses on object-inductive prediction methods (which are based on 'cues' of the environment) and does not assume the setting of online learning, I will show that my results entail

---

[2] Kelly's major result about prediction tasks is the following: an infinite stream of events is correctly predictable by a scientific method (a method whose predictions are a function of past events) after some finite time if and only if this infinite data stream is among a recursively enumerable set of possible data streams ([7, p.260ff]). In formal learning theory one considers especially hypotheses evaluation tasks which are not considered here; but also in this field, the results are in conformity with Hume's lesson.

important consequences for some of the central claims of the ABC research group.

I finally introduce the formal definitions of those notions which I need permanently. (Other notions such as 'prediction method' could also be defined formally; but I skip this because they are sufficiently clear.)

(1) With $pred_P(n)$ I denote P's prediction for time n which P makes at time $n-1$, and which (in binary games) is either '1' (for 'E will obtain') or '0' (for 'E will not obtain').
(2) P's score for time n is defined in the obvious way: $score_n(P) = 1$ if $pred_n(P) = e_n$; else $= 0$.
(3) $a_n(P) := \sum_{1 \leqslant i \leqslant n} score_n(P)$ is P's *absolute* success at time n.
(4) $suc_n(P) := \frac{a_n(P)}{n}$ is P's (relative) *success rate* at time n; $limsuc(P) := \lim_{n\to\infty} suc_n(P)$ is P's limit success (rate), provided P's success rate converges to a limit, and $maxsuc_n$ is the maximal success rate of the non-MI-players at time n.
(5) Finally, $fav_n(\text{MI})$ is MI's favorite *for* time n, i.e. the player with the first-best success-rate *at* time $n-1$ among the non-MI-players. So $pred_n(\text{MI}) = pred_n(fav_n(\text{MI}))$.

3 The Simple Meta-Inductivist, Take-the-Best, and Its Limitations

In this paper I present my major mathematical and philosophical results about prediction games. I have also performed computer simulations, but for reasons of space I must omit them here. The performance of a type of meta-inductivist has always two sides: (i) its *long-run* performance, which is of central significance, and (ii) its *short-run* performance, which is also important: although one should be willing to buy *some* short-run losses of a prediction method for sake of its long-term optimality, these short-run losses should not be too large, and in particular, they should be under rational control.

The performance of the simple meta-inductivist MI (defined in the previous section) is described by the Theorem 1.

Theorem 1 *For each prediction game $((e), \{P_1, \ldots, P_m, \text{MI}\})$ in which (a) the success rate of each non-MI-player converges (to a limit success) and (b) there exists a unique non-MI-player P^+ with maximal limit success, it holds:*

(1.1) *(Long-run:)* MI's success rate approximates the maximal limit success (from below): $limsuc(P^+) - limsuc(\text{MI}) = 0$.
(1.2) *(Short-run:) For all times $n > n^*$, $suc_n(\text{MI}) \geqslant maxsuc_n - \frac{n^*}{n}$, where n^* is the earliest time point after which each non-MI-player's success rate differs from its limit success by less than the half of the difference between the best and the second best limit success of the non-MI-players.*

Proof Let δ be the difference between $limsuc(P^+)$ and the limit success of the second-best non-MI-player in Π. By the convergence assumption there exists time point $n^* \in \mathbb{N}$ such that for all times $m \geqslant n^*$, the success rate $suc_m(P)$ of each non-MI-player P deviates from $limsuc(P)$ by less than $\frac{\delta}{2}$. Therefore after time point n^*, P^+'s success rate will be greater than the success of all other non-MI-players. Until time point n^*, MI may have earned a loss compared to P^+, but after time point n^*, MI favors P^+ forever, and so this loss vanishes in the limit. Since $suc_{n^*}(\text{MI})$ is zero in the worst case, we obtain that $\forall n > n^*, suc_n(\text{MI}) \geqslant (suc_n(P^+) - \frac{n^*}{n})$, which yields Theorem (1.2), and and hence limsuc(MI) = limsuc(P^+), which gives us Theorem (1.1). ♠

Theorem (1.1) establishes MI's long-run optimality under the restricted conditions (a) and (b). If one of these conditions is not satisfied, counterexamples to MI's optimality can easily be constructed. In this section we concentrate on counterexamples due to violations of condition (b), the existence of a unique best non-MI-player, while counterexamples due to violations of condition (a), the convergence of the non-MI-player's success rates, are investigated in the next section. Theorem (1.2) informs us about the maximal short-run loss of MI compared to the unique best player. Since the time point n^* may come arbitrarily late, MI's short run loss may be arbitrarily high. Hence the worst-case short run behavior of MI is not very good. Nevertheless, the result of Theorem (1.2) is at least *something*, because it shows that the short-run loss of MI is caused by a low speed of convergence of the non-MI-players: if their success rates converge fast, then MI's short run loss will be small. We *summarize* the limitations of the strategy MI as follows: MI's optimality is restricted to possible worlds (prediction games) whose non-MI-strategies converge sufficiently fast and which contain a unique best strategy. In this sense, MI is a *locally* but not *universally* optimal meta-induction method.

The losses of meta-inductive prediction methods compared to the best non-MI-method result from the fact that in order to predict for time $n + 1$, the meta-inductivist can only take into account the non-MI-players success rates until time n. More specifically, MI looses exactly one (absolute success) point when MI switches between favorite methods. For whenever MI recognizes that his present favorite P_1 has lost one point compared to some player P_2, then MI has also lost this one point compared to P_2, before MI decides to switch to P_2. These losses because of switching favorites may accumulate. Assumptions (a) and (b) of Theorem 1 exclude that MI can have more than finitely many losses due to switching favorites; so these losses must vanish in the limit.

In prediction games where condition (b) of Theorem 1 does not hold, MI comes into trouble when there are two (or more) best players who converge in their success against the same limit success by way of *oscillating* around each other with constant period and with diminishing amplitude. I call this situation a *convergent oscillation*. Here MI looses one success point in every half oscillation period. In the *worst* case, two alternative players A and B oscillate around each other with the smallest possible period of 4 time units as follows, where MI's favorite is *underlined*:

A-scores iterated: ... |0̲ 0̲ 1 1| ...
B-scores iterated: ... |1 1 0̲ 0̲| ...
MI-scores iterated: ... |0 0 0 0| ...

Convergent oscillations are a case of *deception* of the meta- inductivist by alternative players. The alternative players predict incorrectly exactly when they are in the position of being MI's favorite. In the result, the success rates of the two alternative players converges against $\frac{1}{2}$, while the meta-inductivist's success remains zero for all time. The addition of an object-inductivist to this scenario cannot avoid MI's breakdown. For example, if we add an OI with limit success $\frac{1}{2}$ (in a stream of events with $p(E) = \frac{1}{2}$) and assume that the limit success of the two oscillating players is slightly greater than $\frac{1}{2}$, then MI's limit success will be just slightly greater than zero. Moreover, $limsuc(\text{OI}) = lim(E)$ does only hold in 'normal' event-sequences whose events are not correlated with OI's predictions. As we have remarked in Section 2, there exist 'demonic' streams of events with $p(E) = \frac{1}{2}$ which permanently deceive OI, defined by $e_n = 1$ iff $pred_n(\text{OI}) = 0$; in such a case MI will always favor one of the two osciallting alternative players and MI's success will be turned down to zero in spite of the presence of an OI.

Theorem 1 can be generalized to the rule TTB ('Take the Best') of Gigerenzer and the ABC-research group ([2], chap.4–8). The predictions of the non-MI-players correspond to the *cues* in Gigerenzer's setting, and the TTB strategy works like MI except that it is assumed that a cue (or non-MI-player) need not make a prediction at every time. Thus, TTB chooses that non-MI-player as her favorite for time n who delivers a prediction for time n and has the first-best success rate among all non-MI-players who deliver a prediction for time n. Gigerenzer assumes that the success probabilities (the 'ecological validities'; cf. p.130) of the cues are estimated by independent random sampling procedures, instead of being learned online. Given the situation of online-learning, Theorem 1 can be generalized to TTB as follows: under the assumptions (a) and (b) of Theorem 1, TTB's limit success will converge to a weighted average of the limit successes of the non-MI-players, weighted by their probabilities of making a prediction. More importantly, the TTB-meta-inductivist can be deceived in the same way as MI by deceiving non-MI-players with endlessly oscillating success rates.

The ABC-group is especially interested in comparing the success of TTB with the success of certain *refined* (meta-)inductive strategies such as linear regression or Bayes rule. In this paper I consider such refined rules only in the margin (see Section 6) because they do not affect my general results on meta-induction. Gigerenzer has argued repeatedly that in spite of its simplicity, TTB is almost always as good as these refined prediction strategies. Hogarth and Karelaia ([5]) have shown that there exist scenarios (involving highly 'compensatory' cures) in which TTB is inferior, but they point out these scenarios are rare. My results reveal another restriction: TTB can only be optimal in scenarios in which the event fequencies and success rates of the cues converge suffienly fast towards a limit. This is assumption is implicitly made by the setting of the ABC group, because estimations of the cues' success rates are based on repeated random samplings from finite domains, which are 'inductively safe' modulo random errors. In scenarios of online learning with oscillating event frequencies and success rates, as for example in predictions of the stock market, 'inductive safety' cannot be assumed. In such a case it would be a bad recommendation to put one's money only on the presently most successful stock, instead of distributing it according to a weighted average of stocks in form of a stock portfolio.

4 The ϵ-Meta-Inductivist and Systematic Deceivers

There exists a robust defense strategy to deceptions by convergent oscillations: don't switch favorites if their success difference is *practically insignificant*. I call this new type of meta-inductivist the ϵ-*meta-inductivist* eMI: eMI switches her favorite only if the success difference between her present favorite and a new better favorite exceeds a small threshold ϵ which is considered as practically insignificant. This small change enables the proof of a stronger theorem, namely Theorem (2.1), which assures that eMI ϵ-approximates the best non-MI-player in the long run, *provided only* that the success frequencies of all non-MI-players converge. Theorem (2.2) informs us about eMI's short-run performance.

Theorem 2 *For each prediction game* $((e), \{P_1, \ldots, P_m, \text{eMI}\})$ *in which the success rates of all non-MI-players converge, the following holds:*

(2.1) (Long-run): eMI*'s limit success ϵ-approximates the maximal limit success of the other players,* $limsuc^+$, *from below:* $(limsuc^+ - limsuc(\text{eMI})) \leqslant \epsilon$.

(2.2) (Short-run): For all times $n > n^*$, $suc_n(\text{eMI}) \geqslant maxsuc_n - \frac{n^*+1}{n} - 2 \cdot \epsilon$, *where* n^* *be the earliest time point after which each non-MI-player's success differs from his limit success by at most* $\frac{\epsilon}{4}$.

Proof Let n^* be the earliest time point such that for all $m \geqslant n^+$ and all non-MI-players P, $suc_n(P)$ deviates from $limsuc(P)$ by at most $\frac{\epsilon}{4}$. It follows that (*) the success rates of a non-MI-player P at two time points later than time n^* can deviate from each other by at most $\frac{\epsilon}{2}$ (the sum of two $\leqslant \frac{\epsilon}{4}$-deviations is $\leqslant \frac{\epsilon}{2}$). I prove that after time n^* at most *one* switch of eMI's favorite can occur. Assume the first switch of eMI's favorite after time n^* occurs at time $s \geqslant n^*$, and $fav_{s+1}(\text{eMI})$ (eMI's new favorite) $= A$. Hence for every other non-eMI-player B it must hold that $suc_s(B) \leqslant suc_s(A)$. By (*) above, B's success rate after time s can increase by at most $\frac{\epsilon}{2}$ above $suc_s(B)$, and A's success rate can decrease by at most $\frac{\epsilon}{2}$ below $suc_s(A)$. So for all $n \geqslant s$, $suc_n(B) \leqslant suc_n(A) + \epsilon$ must hold, whence eMI will never switch from A to any other non-MI-player. So eMI's success converges against the limit success of A from below (the initial losses vanish in the limit) and, hence, ϵ-approximates the maximal limit success of the non-MI-players from below, which gives us Theorem (2.1).

For Theorem (2.2) we calculate a lower bound of eMI's success rate as follows. The absolute success of eMI until time n^* is zero in the worst case, because eMI may be deceived by alternative players whose success rates oscillate around each other with amplitudes $> \epsilon$ (see the examples below). Let s be the time point of eMI's last switch of favorites (which exists by the argument for Theorem (2.1)), and let A be eMI's last favorite. There are two cases to consider:

Case 1, $s \leqslant n^*$: Then $\forall n > n^*$, $suc_n(\text{MI}) \geqslant suc_n(A) - \frac{n^*}{n}$, and because $\forall n > n^*$, $suc_n(A) \geqslant maxsuc_n - \epsilon$, it follows that $\forall n > n^*$: $suc_n(\text{eMI}) \geqslant maxsuc_n - \frac{n^*}{n} - \epsilon$.

Case 2, $s > n^*$: Between times n^* and s eMI has only one favorite (recall the argument for Theorem (2.1)); let us call him B. By reasoning as in case 1 we obtain:

$\forall n$ with $n^* < n \leqslant s$:

$$suc_n(\text{eMI}) \geqslant suc_n(B) - n^*/n, \qquad (1)$$

and hence
$\forall n$ with $n^* < n \leqslant s$:

$$suc_n(\text{eMI}) \geqslant maxsuc_n - n^*/n - \epsilon. \qquad (2)$$

At time s, A becomes eMI's new favorite, which means for the absolute successes that $a_{s-1}(A) \leqslant a_{s-1}(B) + \epsilon \cdot (s-1)$, $a_s(A) = a_{s-1}(A) + 1$, $a_s(B) = a_{s-1}(B)$ (A has earned score 1 and B score 0 at time s), and therefore

$$a_s(A) \leqslant a_s(B) + \epsilon \cdot (s-1) + 1. \qquad (3)$$

It follows from Eq.1+3 that

$$suc_s(\text{eMI}) \geqslant suc_s(A) - (n^*+1)/s - \epsilon \cdot (s-1)/s. \qquad (4)$$

Beginning with time $s+1$ eMI earns the same scores as A. Therefore (and by Eq.4) we obtain

$$\forall n > s,\ suc_n(\text{eMI}) \geqslant suc_n(A) - \frac{n^*+1}{n} - \epsilon \cdot \left(\frac{s-1}{n}\right). \qquad (5)$$

Since $\forall n > s, suc_n(A) \geqslant maxsuc_n - \epsilon$, result Eq.5 gives us

$$\forall n > s,\ suc_n(\text{eMI}) \geqslant maxsuc_n - \frac{n^*+1}{n} - \epsilon \cdot \left(1 + \frac{s-1}{n}\right). \qquad (6)$$

Because $\frac{s-1}{n}$ is close to 1 in the worst case, it follows from Eq.6 that

$$\forall n > s,\ suc_n(eMI) \geqslant maxsuc_n - (n^*+1)/n - 2 \cdot \epsilon. \qquad (7)$$

From Eq.2+7 of Case 2 and the result of Case 1 we obtain as a lower bound which is *independent from s* the claim of Theorem (2.2). ♠

The short-run behavior of eMI is not especially good. Theorem (2.2) tells us that the possible loss of eMI at time n is bounded by $\frac{n^*+1}{n} + 2 \cdot \epsilon$. Since n^* is the time of the non-MI-players' $\frac{\epsilon}{4}$-convergence, this means that eMI's short run loss increases with a low speed of convergence of the non-MI-players. The reason why Theorem (2.2) assigns to eMI an additional loss of at most $2 \cdot \epsilon$ is that we want its bound to be independent from eMI's last switch time s which does not depend on the non-MI-players' speed of convergence and may occur arbitrarily late. The lower bound which is stated in Eq.6 of the proof of Theorem (2.2) is better for times $> s$ than that of Theorem (2.2), but it involves the dependence on s.

eMI is long-run optimal in a much broader class of possible worlds than MI, on the cost that its optimality is not a *strict* but an *ε-approximative* optimality. Is approximative optimality good enough to count as a justification? I think *yes*, because in every world with converging success rates, the loss of eMI's limit success compared to the maximal limit success in this world is smaller than ϵ. Moreover, for all practical purposes there exists a choice of ϵ which is small enough to count as *practically insignificant*, and Theorem 2 holds for *all* choices of ϵ. Therefore, approximative optimality is 'almost as good' as its strict counterpart.

We summarize the results of Theorem 2 as follows: eMI predicts sufficiently optimal in all worlds whose non-MI-players' success rates converge sufficiently fast. It is easy to see that eMI's optimality is stable against the scenario of two non-MI-players with convergently oscillating success rates, which has caused MI's breakdown in the previous section: for as soon as the amplitude difference between the success rates of the two alternative players has become smaller than ϵ, eMI stops to switch favorites but sticks to one of the two player forever, with the result that MIe's relative success recovers and ϵ-approximates the maximal success of the two alternative player.

We now turn to the study of prediction games in which the condition of converging success rates does not hold. It is here where we can find the *worst cases* for meta-induction. If the success rates of two or more alternative players oscillate around each other in a *non-convergent* manner with a nondiminishing amplitude of $\alpha > \epsilon$, then also eMI can be deceived. It is easy to see that the minimal periods of such non-convergent oscillations must grow exponentially in time. The worst case are (what we call) *systematic deceivers*, who are assumed to know whether the meta-inductivist has chosen them as favorite and use this information to deceive him. The definition is very simple: P is a *systematic deceiver* (of a meta-inductivist xMI) iff for all times $n \in \mathbb{N}^+$ the following holds: if P is xMI's favorite for time n, then P predicts the wrong result for time n; otherwise P predicts the right result for time n.

Two systematically deceiving players must oscillate around each other with an amplitude approximating ϵ from above. Of course, a systematic deception strategy is partially at the cost of the deceiver's own success. But if eMI is playing against k deceivers, then at each time there will be $k-1$ deceivers which predict correctly because they are not eMI's favorite. As long as a one deceiver, say D_1, is eMI's favorite, D_1 predicts the wrong result until D_1's success is more than ϵ below the success of some deceiver D_2. At this time eMI switches his favorite from D_1 to D_2, D_1 starts to predict correctly and D_2 starts to predict wrong results, until the next switch of eMI occurs, etc. In this way, eMI's success rate is turned down to zero, while the limit success of the k deceivers is $\frac{k-1}{k}$, because in the long run each deceiver is eMI's favorite (and hence predicts incorrectly) in 1 out of k times.

We generalize this negative result by the following definition. A meta-inductive prediction strategy xMI is called a *one favorite* strategy iff at any time point n, xMI chooses one of the non-MI-players as favorite for time $n+1$ (according to some computational rule) and predicts what this favorite predicts. MI and eMI are examples of one-favorite meta-inductive strategies, and many more such strategies are conceivable. Then our negative result is this: all one-favorite meta-inductive strategies will *fail to be optimal* when playing against $k \geq 2$ systematic deceivers, because in that case they have zero-success while the deceivers will have a limit success of $\frac{k-1}{k}$. Note that the 'demonic' streams of events mentioned in Section 1 do not produce additional complexities: a prediction game whose stream of events permanently deceives a one-favorite meta-inductivist is indistinguishable from one in which all favorites of the meta-inductivist are sys-

Meta-Induction: A Game-Theoretical Approach ...

tematic deceivers.

The natural question arises: does eMI play at least optimal against non-deceivers? To answer this question, we define the notion of an *ultimate (non-) deceiver*. The success rate of a non-MI-player P *conditional* on times *for* which P was the favorite of eMI, abbreviated as $suc_n(P|\text{eMI})$, is defined as

$$suc_n(P|\text{eMI}) := \frac{\sum\{score_j(P) : 1 \leqslant j \leqslant n,\ fav_j(\text{eMI}) = P\}}{|\{j : 1 \leqslant j \leqslant n,\ fav_j(\text{eMI}) = P\}|},$$

provided the denumerator is greater zero; otherwise we set by convention $suc_n(P|\text{eMI}) := 1$. We say that a non-MI-player P deceives at time n iff $suc_n(P) - suc_n(P|\text{eMI}) > \epsilon_d$, i.e. if the difference between P's success and P's favorite-conditional success is practically significant. The value ϵ_d is the so-called *deception-threshold*. For the sake of the proof of the next theorem we have to require that $\epsilon_d < \epsilon$, i.e. the deception threshold is strictly smaller than the switching threshold. Since ϵ_d can be as close to ϵ as one wants, this requirement is practically insignificant. Note that by our convention a player counts as a non-deceiver as long as he has never been a favorite. In the initial stage of a prediction game when success fluctuations are usually very high it may occur that a player P is deceiving by accident. But when the game goes on, a true non-deceiver should stop to deceive after some time. Thus, we call a non-MI-player an *ultimate non-deceiver* (in a given prediction game) iff there exists a time point u_P from which on P never is a deceiver any more, i.e. $\forall n \geqslant u_P : suc_n(P) - suc_n(P|\text{eMI}) \leqslant \epsilon_d$. We call u_P the *ultimate time* of P. Otherwise P is called an *ultimate deceiver*; hence P is an ultimate deceiver iff P deceives at *infinitely many* times in a prediction game.

Theorem (3.2) establishes that eMI is indeed approximately long-run optimal in all prediction games whose non-MI-players are ultimate non-deceivers. This is a much stronger result than Theorem (2.1), because ultimate non-deceivers need not have converging success rates, while non-MI-player with converging success rate cannot be ultimate deceivers of eMI. Note that since success rates do not converge, we define long-run optimality in Theorem (3.2) by convergence of *success differences*. For expressing Theorem 3, we need some additional terminology: (i) $f_n(P)$ denotes the relative frequency of times $k \leqslant n$ when P was eMI's favorite. (ii) A *switch-point* s (of eMI) is defined as a time point s such that eMI changes his favorite between time $s - 1$ and s; thus $fav_s(\text{eMI}) \neq fav_{s+1}(\text{eMI})$. (iii) $s_1 < s_1 < \cdots < s_n < \ldots$ denote the consecutive switch points of eMI. Finally, (iv) $X_n := \frac{s_1 + \ldots + s_n}{s_n}$ is the sum of all switch points divided through the last switch point. Because the switch intervals may grow even super-exponentially, X_n need not grow with n, but for $n > 1$ it must be > 1. Theorem (3.1) informs us about a *general* long-run result for eMI, not restricted to ultimate non-deceivers. Attention: don't confuse 's_{n+1}' (the next switch point after s_n) with '$s_n + 1$' (the next time point after s_n). Based on Theorem (3.1), Theorem (3.2) establishes the announced result for eMI.

Theorem 3 *For every prediction game* $((e), \{P_1, \ldots, P_m, \text{eMI}\})$:

(3.1) *For every switch-point s_n of eMI, the following holds:*

$$suc_{s_n}(fav_{s_n+1}(\text{eMI})) > \left(\sum_{1\leqslant i\leqslant m} f_{s_n}(P_i) \cdot suc_{s_n}(P_i|\text{eMI})\right) + X_n \cdot \epsilon.$$

(3.2) *If every P_i is an ultimate non-deceiver, then eMI ϵ-approximates the maximal success rate (from below): $\lim_{n\to\infty}(maxsuc_n - suc_n(\text{eMI})) \leqslant \epsilon$.*

The proof of Theorem 3 is involved and therefore removed into the *appendix*. In the proof of Theorem (3.2) it is shown that the success rates of ultimate non-deceivers cannot produce endless switchings of eMI's favorites, for that would eventually force them to become deceivers. The class of worlds for which Theorem (3.2) establishes eMI's long-run optimality is the broadest class of worlds for which we were able to establish an optimality result so far. Unfortunately, eMI's short-run performance is not good when playing against arbitrary ultimate non-deceivers, because the ultimate time $u := max(\{u_{P_i} : 1 \leqslant i \leqslant m\})$ after which all non-MI-players stop to deceive may occur arbitrarily late, and until this time eMI's success rate may be zero in the worst case.

Let us finally consider the question whether we can construct a one-favorite meta-inductivist who is able to *detect* ultimate deceivers and to avoid favorizing them. We call such a strategy an *avoidance* meta-inductivist. Of course, an avoidance meta-inductivist cannot not predict optimal in regard to ultimate deceivers, because after some time the ultimate deceivers will never be chosen as favorites and hence are free to predict better than the avoidance meta-inductivist. Nevertheless one might argue that a 'perfect' avoidance meta-inductivist would constitute already a sufficiently general justification of meta-induction, because prediction strategies who are deceiving whenever one favors them are not 'truly' accessible, insofar their predictive success is not 'truly' accessible.

Unfortunately it is impossible to construct such a 'perfect' avoidance meta-inductivist. For the time point u_P after which an ultimate non-deceiver P stops to deceive is unknown to the avoidance meta-inductivist. What the avoidance meta-inductivist can do is *not to favor* any non-MI-player P for time $n + 1$ who is a deceiver at time n. But this may not be enough, for one may invent refined versions of deceivers of increasing complexity – for example deceivers who start to lower their unconditional success rate as soon as they are recorded as deceivers, until they are no longer recorded as deceivers and can start to deceive again, etc. A meta-inductivist furnished with a simple avoidance technique could maybe deceived by these more refined deceivers.

What can be done in defense against refined deceivers is to introduce an avoidance rule which avoids also all *putative* (ultimate) deceivers. A non-MI-player P is called a *putative deceiver at time n* if either P deceives at time n or P has switched his deception-status from a non-deceiver to a deceiver for more than p times (in my programmings I have chosen $p = 5$). We abbreviate the avoidance meta-inductivist who avoids putative deceivers as aMI. In informal words, aMI does not only avoid players who are deceivers at the given time n, but also players who are non-deceivers at time n but have switched their deception status in the past so often that the meta-inductivist does not trust them any

Meta-Induction: A Game-Theoretical Approach ...

longer. Of course, 'trust' is an inductive assumption which may fail, and hence, a putative deceiver P could nevertheless be an ultimate non-deceiver in the sense that P *would* stop to deceive aMI if aMI *would* continue to favor P after P has switched his deception status for more than p times. Thus by defending him against putative deceivers aMI may loose access to some ultimate non-deceivers. On the other hand, we will see that aMI has a good short-run performance.

We define aMI as follows. At any time n, aMI chooses his favorite only among those non-MI-players who are *not* putative deceivers at time n. aMI switches to a new favorite N if either aMI's old favorite O has become a putative deceiver or O's success rate has dropped by more than ϵ below the maximal success rate of the players which are not putative deceivers (at the given time). Note that even the object-inductivist OI may become a putative deceiver, namely when a 'demonic' streams of events deceives OI whenever OI is the favorite of a meta-inductivist. In this case OI predicts as if OI were a systematic deceiver. Hence it is possible that the class of players which are not putative deceivers at a given time becomes empty; in such a case aMI predicts by a random guess.

A *putative non-deceiver* is defined as an ultimate non-deceiver who does *not* switch his deception status for more than p times. Thus the putative non-deceivers are a subclass of the ultimate non-deceivers. Theorem (4.1) tells us that aMI predicts approximatively optimal in regard to putative non-deceivers. The difference to Theorem (3.1) is that in Theorem (4.1), the prediction games may contain also ultimate deceivers. aMI has a good *short run* performance, because at any time she chooses her favorites only among non-deceivers: as Theorem (4.2) tells us, aMI's success rate is never more than $\epsilon_d + \frac{m}{n}$ below the weighted average of the non-MI-players' success rates at the last time when they were aMI's favorite. Since the non-MI-players have an ϵ-approximately optimal success rate while being favorite, this is a good short run result; moreover, the term $\frac{m}{n}$ quickly vanishes for $n >> m$.

Theorem 4 *For every prediction game $((e), \{P_1, \ldots, P_m, \text{aMI}\})$, the following holds:*

(4.1) (Long run:) aMI ϵ-approximates the maximal success rate in the subclass $Q \subseteq \{P_1, \ldots, P_m\}$ of putative non-deceivers (from below):

$$\lim_{n \to \infty} (max(\{suc_n(P_i) : P_i \in Q\}) - suc_n(\text{aMI})) \leq \epsilon.$$

(4.2) (Short run:) Let n_i be the latest time $\leq n$ such that $P_i = fav_{n_i}(\text{aMI})$. Then:

$$suc_n(\text{aMI}) \geq \left(\sum_{1 \leq i \leq m} suc_{n_i}(P_i) \cdot f_n(P_i) \right) - \epsilon_d + \frac{m}{n}.$$

Proof

(Theorem (4.1)) Let P be the class of non-MI-players who switch their deception-status for more than p times, and $Q := \{P_1, \ldots, P_m\} - P$ the class of non-MI-players who switch their deception status for not more than p times. Let the *critical* time c be defined as the earliest time after which (i) every player in P has switched his deception status for more than p times, and (ii) every player in Q has performed the last switch of his deception status. It follows that after time c every non-MI-player P will either be a putative deceiver forever, or P will never be a deceiver again, in which case P satisfies the definition of a putative non-deceiver. So for all times $n \geqslant c$ aMI favors only non-MI-players who are not deceivers at time n. Since the time c plays the same role as the ultimate time u in Theorem (3.2), Theorem (4.1) follows from Theorem (3.2).

(Theorem (4.2)) The absolute success a_n(aMI) equals the absolute success achieved by all non-MI-players during their favorite-phases for times until time n, which gives us for the success rates (a) $suc_n(\text{aMI}) = \sum_{1 \leqslant i \leqslant m} f_n(P_i) \cdot suc_n(P_i|\text{aMI})$ (compare the paragraph below Eq.4 in the proof of Theorem (3.1), Appendix). Since time n_i until time n, the aMI-conditional success rate of a player P_i is frozen, i.e. (b): $suc_n(P_i|\text{aMI}) = suc_{n_i}(P_i|\text{aMI})$. At time $n_i - 1$, P_i was aMI's favorite and, hence, was not recorded as a deceiver; whence $suc_{n_i-1}(P_i|\text{aMI}) \geqslant suc_{n_i-1}(P_i) - \epsilon$ and hence (c): $suc_{n_i}(P_i|\text{aMI}) \geqslant suc_{n_i}(P_i) - \epsilon_d - \frac{1}{n_i}$. From (a), (b), (c) and (d): $\frac{1}{n_i} \leqslant \frac{1}{n}$, we obtain the claim of Theorem (4.2) by a simple substitution of terms. ♠

Up to now we have discovered a variety of one-favorite meta-inductive strategies whose performance got successively improved. But none of these meta-inductivists predicts universally optimal because of the possibility of systematic deceivers. Thus, for one-favorite meta-inductivists our project of justifying meta-induction in a not-circular way does not fully go through. Are there meta-inductivists strategies which are indeed universally optimal? In the next section we show that these meta-inductive strategies do indeed exist: they are found among the class of *weighted average* meta-inductivists.

5 Weighted Average Meta-Induction for Real-Valued and for Binary Events

A weighted average meta-inductivist predicts a weighted average of the predictions of the non-MI-players, weighted by their 'attractiveness'. The weighted average of several predictions of zeros and ones will be a real value in the interval $[0,1]$. Therefore this method cannot directly be applied to *binary* prediction games, because in these games every player has to predict either zero or one. But weighted average meta-induction can be applied to *real-valued* prediction games. In this form, the method of weighted average prediction has been developed in the mathematical theory of *universal prediction* which was mentioned in Section 2 (cf. [1]). The results in this literature have not been related to the problem of induction, but as explained in Section 2, the problem setting is similar to my prediction games. The results established in the previous section are not covered by the theorems established in this area; rather, for the binary and non-probabilistic setting this research provides a merely negative result (cf.[1, p.67]).

So let us consider a real-valued prediction game whose task consists in predicting the values of a real-valued and normalized random variable X. Hence, both the events e_n of the sequence (e) and the predictions of the players $pred_n(P)$ are values in the interval $[0,1]$. We speak here of a $[0,1]$-prediction game. The *loss* of a player P achieved by his prediction for time n is naturally defined as $l_n(P) := |e_n - pred_n(P)|$ (this is a value between 0 and 1), and his score is defined

Meta-Induction: A Game-Theoretical Approach ...   257

as $score_n(P) := 1 - l_n(P)$. The notions of absolute success and success rate are defined from scores as before.

The *weighted average meta-inductivist* is abbreviated as wMI and defined as follows. For every non-MI-player P we define $at_n(P) := suc_n(P) - suc_n(\text{wMI})$ as P's *attractivity* (as a favorite) at time n. Let $PP(n)$ be the set of all non-MI-players with *positive* attractivity at time n. wMI's prediction for time 1 is set to $\frac{1}{2}$, and for all times > 1 it is defined as follows:

$$(\forall n \in \mathbb{N}^+ :) \quad pred_{n+1}(\text{wMI}) := \frac{\sum_{P \in PP(n)} at_n(P) \cdot pred_{n+1}(P)}{\sum_{P \in PP(n)} at_n(P)}.$$

In words: wMI's prediction for the next round is the attractivity-weighted average of the attractive players' predictions for the next round. Should it happen that $PP(n) = \emptyset$, $pred_{n+1}(\text{wMI})$ is set to $\frac{1}{2}$.

Informally explained, the reason why wMI cannot be deceived is the following. A non-MI-player who tries to deceive wMI would be one who starts to predict incorrectly as soon as his attractivity for wMI is higher than a certain threshold. The success rates of such wMI-adversaries must oscillate around each other. But wMI does not favor just one of them (who will predict incorrectly), but wMI predicts according to an attractivity-weighted average of correctly and incorrectly predicting adversaries, and therefore wMI's long-run succcess must approximate the maximal long-run success of his adversaries. A 'demonic' event sequence (which minimizes wMI's score at each time) does not change this result, because a low success rate of wMI goes hand in hand with a low success rates of the attractive non-MI-players.

Theorem (5.2) establishes that wMI is indeed a *universally* long-run optimal prediction strategy, even in the strict (and not approoximative) sense. Also wMI's short-run performace is excellent, as Theorem (5.1) reveals: the short-run loss $\sqrt{\frac{m}{n}}$ is under complete control because the number of non-MI-players is known, and it approaches zero for times $n \gg m$.

Theorem 5 *For every real-valued prediction game $((e), \{P_1, \ldots, P_m, \text{wMI}\})$:*

(5.1) *(Short run:)* $\forall n \in \mathbb{N}^+: suc_n(\text{wMI}) \geq maxsuc_n - \sqrt{\frac{m}{n}}$.
(5.2) *(Long run:)* $suc_n(wMI)$ *(strictly) approximates the non-MI-players' maximal success:*
$$\lim_{n \to \infty}(maxsuc_n - suc_n(\text{wMI})) = 0.$$

Proof We identify wMI with the polynomially weighted average forecaster F described on p.12 of [1, ch.2] with its parameter p set to 2. Our non-MI-players $\{P_1, \ldots, P_m\}$ are identified with the N 'experts' (i.e., $N = m$). The predictions of F are for the special case $p = 2$ defined as follows:

$$pred_{n+1}(F) = \frac{\sum_{P \in PP(n)} (L_n(F) - L_n(P)) \cdot pred_{n+1}(P)}{\sum_{P \in PP(n)} L_n(F) - L_n(P)} \quad (8)$$

In Eq.8, $L_n(P) := \sum_{1 \leq i \leq n} l_i(P)$ is the *cumulative* loss of a player P until time n, which is (by our definitions) equal to $n - a_n(P)$. Hence, $a_n(P) = n - L_n(P)$. So Eq.8 implies:

$$pred_{n+1}(F) = \frac{\sum_{P\in PP(n)}(a_n(P) - a_n(F)) \cdot pred_{n+1}(P)}{\sum_{P\in PP(n)} a_n(P) - a_n(F)}$$

$$= \frac{n \cdot \sum_{P\in PP(n)}(suc_n(P) - suc_n(F)) \cdot pred_{n+1}(P)}{n \cdot \sum_{P\in PP(n)} suc_n(P) - suc_n(F)}$$

$$= \frac{\sum_{P\in PP(n)} at_n(P) \cdot pred_{n+1}(P)}{\sum_{P\in PP(n)} at_n(F))} \qquad (9)$$

So, F's predictions coincide with wMI's predictions, and hence, $suc_n(F) = suc_n(\text{wMI})$. We abbreviate our loss function $l_n(P) := |e_n - pred_n(P)|$ as $l(p, e) := |e - p|$ where $e = e_n \in [0, 1]$ and $p = pred_n(P) \in [0, 1]$. We prove that $l(p, e)$ is *convex* in p, which means that for all values of e and for every $\gamma \in [0, 1]$ and every $a < b \in [0, 1]$ the following holds: $l((\gamma \cdot a + (1-\gamma) \cdot b), e) \leq \gamma \cdot l(a, e) + (1-\gamma) \cdot l(b, e)$. We have to distinguish four cases. Case 1, $a < b \leq e$: We must show that $e - \gamma \cdot a - (1-\gamma) \cdot b \leq \gamma \cdot e - \gamma \cdot a + (1-\gamma) \cdot e - (1-\gamma) \cdot b$. Simplification of this inequality leads to $e \leq e$. Case 2, $e \leq a < b$: We must show that $-e + \gamma \cdot a + (1-\gamma) \cdot b \leq -\gamma \cdot e + \gamma \cdot a - (1-\gamma) \cdot e + (1-\gamma) \cdot b$. Simplification leads to $-e \leq -e$. Case 3, $a \leq e \leq b$ and $e \geq \gamma \cdot a + (1-\gamma) \cdot b$. We must show that $e - \gamma \cdot a - (1-\gamma) \cdot b \leq \gamma \cdot e - \gamma \cdot a + (1-\gamma) \cdot b - (1-\gamma) \cdot e$. Simplification leads to $e \leq b$, which is true by the assumption of the case. Case 4, $a \leq e \leq b$ and $e \leq \gamma \cdot a + (1-\gamma) \cdot b$. We must show that $\gamma \cdot a + (1-\gamma) \cdot b - e \leq \gamma \cdot e - \gamma \cdot a + (1-\gamma) \cdot b - (1-\gamma) \cdot e$. Simplification leads to $a \leq e$, which is true by the assumption of the case.

Because the loss function $l_n(P)$ is convex in $pred_n(P)$, corollary 2.1 in [1, p.12f] establishes for F's predictions for all times n:

$$L_n(F) - min(\{L_n(P_i) : 1 \leq i \leq m\}) \leq \sqrt{m \cdot n}, \qquad (10)$$

hence

$$a_n(F) \geq max(\{a_n(P_i) : 1 \leq i \leq m\}) - \sqrt{m \cdot n}, \qquad (11)$$

and so (by dividing through n)

$$suc_n(F) = suc_n(wMI) \geq max(\{suc_n(P_i) : 1 \leq i \leq m\}) - \sqrt{\frac{m}{n}} \qquad (12)$$

Eq.12 is the claim of Theorem (5.1); and Theorem (5.2) is an immediate consequence of it. ♠

For binary prediction of binary events the loss function $|e_n - pred_n(P)|$ is not convex in $pred_n(P)$. Since this is required in the proof of Theorem 5, this theorem does not apply to binary prediction games. This can be seen by the following example mentioned in [1, p.67]: assume a meta-inductivist with two non-MI-players, one of them constantly predicting 1 and the other constantly predicting 0, and the event-sequence is 'demonic', i.e. constantly produces the opposite of the meta-inductivist's predictions. Then whatever the meta-inductivist predicts, his succes rate will be constantly zero, while the maximal success rate of the two non-MI-players must always be ≥ 0.5 (because their mean success rate must be 0.5). We can summarize this negative result as follows: a meta-inductive strategy which predicts long-run optimal for an *individual* player in arbitrary *binary* prediction games *does not exist*. Together with Theorem 5 this is a deep result, insofar it shows that a 'real-valued' nature is more friendly to the inductivist than a 'digitalized' (discrete) nature.

Nevertheless I have found a way to apply Theorem 5 indirectly also to the prediction of binary events, namely by means of assuming a *collective* of k meta-inductivists, abbreviated as $cwMI_1, \ldots, cwMI_k$, and by considering their *mean success rate* ('cwMI$_i$' stands short for 'collective weighted-average meta-inductivist no.i'). I treat a prediction game with binary events and binary predictions of the non-MI-players as a subcase of a real-valued game, and regard wMI's prediction $pred_n(\text{wMI})$ (defined as above) as an *ideal* real-valued prediction which is approximated by the mean value of the k binary predictions of the collective of cwMI-meta-inductivists as follows: for $1 \leq i \leq k$, cwMI$_i$ predicts 1 if $i \leq [p \cdot k]$, and else 0, where $[x]$ is the integer-rounding of the real number x. Thus, $[p_n \cdot k]$ cwMI's predict 1 and $k - [p_n \cdot k]$ cwMI's predict 0. In this way, one obtains a universal optimality result for the mean success rate of collective of cwMI's, abbreviated as $\overline{suc_n}(\text{cwMI})$, which stated in Theorem 6. Compared to Theorem 5, the mean success rate of cwMI is decreased by the additional term $\frac{1}{2 \cdot k}$ which reflects the maximal loss due to approximation of the ideal prediction by k binary predictions. This additional loss can be made arbitrarily small by increasing the number of meta-inductivists.

Theorem 6 *For every binary prediction game* $((e), \{P_1 \ldots, P_m, cwMI_1 \ldots, cwMI_k\})$:

(6.1) *(Short run):* $\forall n \in \mathbb{N}^+: \overline{suc_n}(\text{cwMI}) \geq maxsuc_n - \sqrt{\frac{m}{n}} - \frac{1}{2 \cdot k}$.

(6.2) *(Long run:)* $\overline{suc_n}(\text{cwMI})$ $\frac{1}{2 \cdot k}$*-approximates the non-MI- players' maximal success:*

$$lim_{n \to \infty}(maxsuc_n - \overline{suc_n}(\text{cwMI})) \leq \frac{1}{2 \cdot k}.$$

Proof At every time i ($1 \leq i \leq n$) the actually achieved mean score is $\frac{[p_n \cdot k]}{k}$ if $e_i = 1$ and $1 - \frac{[p_n \cdot k]}{k}$ if $e_i = 0$; while the ideally achieved score is p_n if $e_i = 1$, and $1 - p_n$ if $e_i = 0$. Since $|[p_n \cdot k] - p_n \cdot k| \leq \frac{1}{2}$, it follows that at every time, the actually achieved mean score deviates from the ideally achieved score by at most $\frac{1}{2 \cdot k}$. Hence we obtain for the mean success rate: $\overline{suc_n}(\text{cwMI}) \geq suc_n(wMI) - \frac{1}{2 \cdot k}$. From this and Theorem 5, the claims of Theorem 6 follow. ♠

The relation of this collective weighted average meta-induction to the situation of an individual meta-inductivist in binary prediction games is the following. The cwMI-adversaries may conspire against a particular individual, say against cwMI$_3$, and constantly deceive cwMI$_3$ (or alternatively, a 'demonic' event sequence may constantly deceive cwMI$_3$). But the cwMI-adversaries cannot deceive the other cwMI's at the same time, and their anti-cwMI$_3$-conspiration will not affect the optimality result for the cwMI's mean success.[3]

---

[3]This fact is related to a result of [1, chap.4.1-2] for the binary prediction game in which it is assumed that at every time n wMI predicts according to a probability distribution $P_n(x)$.

The collective actions of the cwMI's may be based on rational agreement. But it is sufficient to assume that the frequency of cwMI's who favor a non-MI-player P is proportional to P's attractivity (provided this is positive). In the latter perspective, Theorem 6 describes the predictive success of a population of cwMI's in a typical evolutionary setting. Under the assumption that the members of this population *share* their mean success, this provides a basis for interactive utility effects in the narrow game-theoretical sense.

Let me finally emphasize that Theorems 5 and 6 do not only hold for our special and most natural loss function $l_n(P) := |e_n - pred_n(P)|$, but for *every* loss function $l(pred_n(P), e_n)$ which is *convex* in $pred_n(P)$ (as explained in the proof of Theorem 5). Also of interest is the fact that in prediction games satisfying the conditions of Theorem 1, the strategy wMI will soon coincide with the simple MI-strategy: for after some time, only the best player will have positive attractivity, whence wMI and hence all of the cwMI$_i$'s will predict as if they would favor this best player forever.

6 Epistemological Conclusions

Tab.I summarizes my results on the long-run optimality of meta-induction. In order to express *differentiated* results, I relativize the notion of optimality to a class of strategies Σ w.r.t. which xMI is optimal, and to a class of worlds W in which the strategies in Σ are played. The use of this twofold relativization to 'players' and 'worlds' becomes clear when we compare eMI with aMI. Since Theorems 1–4 can be generalized to real-valued events (which I mention without proof), the distinction between real-valued and binary events affects only wMI and cwMI.

xMI-strategy	kind of optimality	strategies in Σ must be xMI-accessible and:	worlds in W contain finitely many xMI-accessible players satisfying:
MI (th.1)	strict	convergent	all players convergent, $\exists!$ best player
eMI (th.2)	ϵ-appr.	convergent	all players convergent
eMI (th.3)	ϵ-appr.	ult. non-deceivers	all players ult. non-deceivers
aMI (th.4)	ϵ-appr.	put. non-deceivers	no condition
wMI (th.5)	strict.	arbitrary	real-valued events
wMI (th.6)	coll. $\frac{1}{2 \cdot k}$-approx.	arbitrary	binary events

Table I: Optimality of xMI w.r.t. class of strategies Σ and class of possible worlds W.

As I have mentioned in Section 1, the optimality of a prediction strategy is not the strongest kind of best-alternative justification of it – a stronger justification

---

with $P_n(1) = p_n$. It follows that the expectation value of wMI's success rate behaves like that of wMI in the real-valued prediction game. The crucial assumption behind this theorem is that the predictions of the non-MI-players at time n, though they may depend on wMI's probability distribution $P_n(x)$, do *not* depend on the *actual* prediction which wMI chooses at time n (cf. p.69). Otherwise, wMI could be permanently deceived.

is the demonstration of *dominance*: a strategy is called (ϵ-approximatively) *dominant* w.r.t. a strategy class Σ and a world class W iff it is (ϵ-approximatively) optimal and no other strategy in Σ is (ϵ-approximatively) optimal w.r.t. Σ and W (cf. [14, p.10]). We cannot show that the meta-inductivists of Tab.I are generally dominant, because there exists refined versions of them which may improve them. The most important technique of refining inductive methods is by *conditionalization*. Conditionalized inductive methods exploit correlations in non-random worlds which obtain between the events e_n and prior events, internal or external to the sequence (e), with help of Reichenbach's principle of the narrowest reference class ([9, §72]). Assume $\{R_1,\ldots,R_r\}$ is a partition of the events in the (say q) past time units, described in terms of nomological predicates, such that the given player has reliable information about which cell of the partition was realized before the given time, and the cells are statistically relevant for the event E (i.e., $f_n(E|R_i) \neq f_n(E|R_j)$ for $j \neq i$). Then a *conditionalized OI-strategy* applies the straight- & maximum-rule to this partition as follows: given that R_i is realized at time n, predict E if $f_n(E|R_i) \geqslant 1/2$, otherwise predict $\neg E$. Provided that all involved frequencies converge to a limit, one can prove that conditionalizing to reference partitions may only improve the success, compared to the simple OI-strategy (for a proof in utility-theoretic terms cf. Good [3, chap.17]). The *conditionalized meta-inductivist* conditionalizes the success frequencies of the other players to the cells of this partition. For example, the conditionalized MI favors the first-best player P in the list of non-MI-players whose conditional success rate $suc_n(P|R_i)$ is maximal, where R_i is the cell which is realized at time n; and the conditionalized wMI computes the attractivities of the non-MI-players in terms of their conditional success rates $suc_n(P|R_i)$.

While a simple xMI approximates the maximal success always from below, the success rate of a conditionalized xMI may be even strictly greater than the success rates of the non-MI-players. This fact *does not affect* the optimality of the simple xMI (in the respective class of worlds), because we assume that refined meta-inductivistic techniques, *if they are accessible*, are among the methods of the alternative players (hence with a 'non-xMI-player' we mean a 'non-simple-xMI-player'). However, this observation shows that a simple meta-inductivist can *improve* his results by getting access to refined meta-inductivist (or object-inductivist) techniques. Moreover, a good meta-inductivist will not only observe the other strategies, but also try to understand and *internalize* them, because a meta-inductivist who can simulate a prediction strategy does no longer depend on the presence of other players who play that strategy.

Refined meta-inductive strategies are not the only reason for why the meta-inductivists of Tab.I are not dominant (w.r.t. the respective classes of strategies and worlds). Another reason are *mixed* strategies, which use non-inductive clairvoyance strategies only in 'para-normal' worlds in which these strategies are superior, while they use object-inductive strategies in 'normal' worlds. Also these mixed strategies are not dominated by meta-induction. We can establish dominance of meta-induction only w.r.t. *purely* non-inductive strategies, which are defined as prediction strategies which play non-inductive in at least one 'normal' world, i.e. in a world in which the *only* reason for non-accidental predictive success is the exploitation of inductive uniformity by object-induction. Hence for every purely non-inductive prediction strategy there exists at least one ('normal') world in which its predictive success in the long run is significantly worse than the predictive success of all (simple or refined) object-inductivistic strategies. It follows that each meta-inductivist of Tab.I is dominant w.r.t. all purely non-inductive prediction strategies in the respective class of possible worlds.

This result about dominance may be critized to be an almost trivial consequence of my definition of a 'purely non-inductive' strategy. But I have argued in Section 1 that already the *optimality argument* is a sufficiently strong justification of meta-induction, insofar meta-induction (be it unconditionalized or conditionalized) is the *only* prediction strategy for which optimality can be *rationally demonstrated*. Although there could exist a mixed induction-clairvoyant strategy of the described kind which is equally successful as meta-induction in all possible worlds, it would be impossible to prove this because we do not know an algorithmic definition of such a strategy.

Weighted average meta-induction has turned out as a universally optimal prediction method. One may criticize that for binary prediction games this optimality is only guaranteed for the mean success of a collective of meta-inductivists, but I think that *collective optimality* is nevertheless an important kind of universal optimality. As a further epistemological consequence, the universal optimality of weighted average meta-induction implies an important correction of a central claim of Gigerenzer and the ABC research group: their claim that all prediction methods, whether complex or simple, are *ecological* in the sense that their success is restricted to certain environmental conditions. This claim is falsified by Theorems 5 and 6 which have established weighted average meta-induction as a universally optimal prediction strategy. In conclusion, I think the achieved optimality results on meta-induction are strong enough to show that a non-circular justification of induction can be successful.

Appendix: Proof of Theorem 3

Proof

(Proof of Theorem (3.1)) By induction on switch-points. For each s_n the following holds:

$$suc_{s_n}(fav_{s_n+1}(\text{eMI})) > suc_{s_n}(fav_{s_n}(\text{eMI})) + \epsilon, \qquad (13)$$

where $fav_{s_n}(\text{eMI}))$ is the old (actual) and $fav_{s_n+1}(\text{eMI}))$ the new favorite. For $n := 1$, (1) gives the induction start, because $X_1 = 1$, and if P is eMI's first favorite, then $f_{s_1}(P) = 1$ and $suc_{s_1}(P|\text{eMI}) = suc_{s_1}(P)$. Now assume the induction hypotheses,

$$suc_{s_{n-1}}(fav_{s_{n-1}+1}(\text{eMI})) > \sum_{1 \leqslant i \leqslant m} f_{s_{n-1}}(P_i) \cdot suc_{s_{n-1}}(P_i|\text{eMI}) + X_{n-1} \cdot \epsilon, \qquad (14)$$

which in terms of absolute successes means:

$$a_{s_{n-1}}(fav_{s_{n-1}+1}(\text{eMI})) =$$
$$s_{n-1} \cdot suc_{s_{n-1}}(fav_{s_{n-1}+1}(\text{eMI})) >$$
$$> s_{n-1} \cdot \left(\sum_{1 \leqslant i \leqslant m} f_{s_{n-1}}(P_i) \cdot suc_{s_{n-1}}(P_i|\text{eMI}) \right) + s_{n-1} \cdot X_{n-1} \cdot \epsilon.$$
(15)

Let α be the scores earned by $fav_{s_{n-1}}$ between $s_{n-1}+1$ and s_n. eMI's favorite does not change during this interval; hence $fav_{s_{n-1}+1} = fav_{s_n}$. From this fact and (3) we obtain for the absolute success rate of fav_{s_n} at time s_n:

$$a_{s_n}(fav_{s_n}(\text{eMI})) =$$
$$s_n \cdot (suc_{s_n}(fav_{s_n}(\text{eMI}))) >$$
$$s_{n-1} \cdot \left(\sum_{1 \leqslant i \leqslant m} f_{s_{n-1}}(P_i) \cdot suc_{s_{n-1}}(P_i|\text{eMI}) \right) + \alpha + s_{n-1} \cdot X_{n-1} \cdot \epsilon.$$
(16)

Observe that for each P_i and time n, $n \cdot f_n(P_i) \cdot suc_n(P_i|\text{eMI}) = n \cdot \frac{n_i}{n} \cdot \frac{a_n(P_i|\text{eMI})}{n_i} = a_n(P_i|\text{eMI})$ (where n_i = number of times for which P_i was eMI's favorite). Therefore, the term $s_{n-1} \cdot \sum_{1 \leqslant i \leqslant m} f_{s_{n-1}}(P_i) \cdot suc_{s_{n-1}}(P_i|\text{eMI})$ is the absolute success earned by eMI's favorites until switch time s_{n-1}. Adding the score α to it gives the absolute success earned by eMI's favorites until switch time s_n, i.e. the sum $s_n \cdot \left(\sum_{1 \leqslant i \leqslant m} f_{s_n}(P_i) \cdot suc_{s_n}(P_i|\text{eMI}) \right)$. So from Eq.16 we obtain

$$a_{s_n}(fav_{s_n}(\text{eMI})) =$$
$$s_n \cdot suc_{s_n}(fav_{s_n}(\text{eMI})) >$$
$$s_n \cdot \left(\sum_{1 \leqslant i \leqslant m} f_{s_n}(P_i) \cdot suc_{s_n}(P_i|\text{eMI}) \right) + s_{n-1} \cdot X_{n-1} \cdot \epsilon.$$
(17)

From Eq.13 and Eq.17 we obtain for the new favorite at switch point s_n:

$$a_{s_n}(fav_{s_n+1}(\text{eMI})) =$$
$$s_n \cdot suc_{s_n}(fav_{s_n+1}(\text{eMI})) >$$
$$s_n \cdot \left(\sum_{1 \leqslant i \leqslant m} f_{s_n}(P_i) \cdot suc_{s_n}(P_i|\text{eMI}) \right) + s_{n-1} \cdot X_{n-1} \cdot \epsilon + s_n \cdot \epsilon.$$
(18)

By definition $s_{n-1} \cdot X_{n-1} \cdot \epsilon + s_n \cdot \epsilon = (s_1 + \ldots s_{n-1}) \cdot \epsilon + s_n \cdot \epsilon = (s_1 + \ldots + s_n) \cdot \epsilon = s_n \cdot X_n \cdot \epsilon$. Therefore Eq.18 gives us

$$s_n \cdot suc_{s_n}(fav_{s_n+1}(\text{eMI})) > \\ s_n \cdot \left(\sum_{1 \leq i \leq m} f_{s_n}(P_i) \cdot suc_{s_n}(P_i|\text{eMI}) \right) + s_n \cdot X_n \cdot \epsilon, \quad (19)$$

and by dividing Eq.19 on both sides through s_n we obtain the claim of Theorem (3.1) for switch time s_n.

(Proof of Theorem (3.2)) Let $u := max(\{u_{P_i} : 1 \leq i \leq m\})$. At no time $n \geq u$ some non-MI-player is a deceiver. We prove that the number of switches of eMI's favorite after time u is finite. It follows that after some time $s^* \geq u$, eMI will put on some P_i forever. This implies that eMI strictly approximate P_i's success and ϵ-approximates the maximal success of the ultimate non-deceivers (from below).

For *reductio ad absurdum*, assume that there exist players who switch from a non-favorite to a favorite (of eMI) *infinitely* many times. By a suitable re-indexing, let $P_1, \ldots, P_k$ be *all* of these players. Because of $k \leq m$ and $X_n \geq 1$, Theorem (3.1) entails for every switch point s_n:

$$\sum_{1 \leq i \leq k} f_{s_n}(P_i) \cdot suc_{s_n}(P_i|\text{eMI}) < suc_{s_n}(fav_{s_n+1}(\text{eMI})) - \epsilon. \quad (20)$$

By our assumptions, there must exist a time point $w \geq u$ after which eMI's favorite is always one of the $P_1, \ldots, P_k$. We consider now switch points s_n lying past w, $s_n \geq w$:

In Eq.20, the frequencies $f_{s_n}(P_i)$, for $1 \leq i \leq k$, do no longer add up to one, but only to $\gamma_{s_n} := 1 - \frac{w}{s_n} \cdot \delta_w$, where δ_w abbreviates the relative frequency of times until time w at which players different from $P_1, \ldots, P_k$ were favorites. Note that $lim_{n \to \infty}(\gamma_{s_n}) = 1$, since $lim_{n \to \infty}(\frac{w}{s_n}) = 0$. We define the *re-normalized* frequencies $f^*_{s_n}(P_i) := \frac{f_{s_n}(P_i)}{\gamma_{s_n}}$, which add up to one (for $1 \leq i \leq k$). Using this definition we obtain from Eq.20 that for all $s_n \geq w$:

$$\gamma_{s_n} \cdot \left(\sum_{1 \leq i \leq k} f^*_{s_n}(P_i) \cdot suc_{s_n}(P_i|\text{eMI}) \right) < suc_{s_n}(fav_{s_n+1}(\text{eMI})) - \epsilon. \quad (21)$$

Using the equation $\Sigma = \gamma \cdot \Sigma + (1-\gamma)) \cdot \Sigma \leq \gamma \cdot \Sigma + (1-\gamma)$ (for $\Sigma < 1$), we obtain from Eq.21 that for all $s_n \geq w$:

$$\sum_{1 \leq i \leq k} f^*_{s_n}(P_i) \cdot suc_{s_n}(P_i|\text{eMI}) < suc_{s_n}(fav_{s_n+1}(\text{eMI})) - \epsilon + (1 - \gamma_{s_n}). \quad (22)$$

Let $\delta := \epsilon - \epsilon_d$. We pass to the *first* switch point lying past w at which the term $(1 - \gamma_{s_n})$ (which converges to zero) has become smaller than $\frac{\delta}{2}$. We call this *distinguished* switch point σ_1. Because $\epsilon := \epsilon_d + \delta$, Eq.22 implies:

For all $s_n \geqslant \sigma_1$:

$$\sum_{1 \leqslant i \leqslant k} f^*_{s_n}(P_i) \cdot suc_{s_n}(P_i|\text{eMI}) < suc_{s_n}(fav_{s_n+1}(\text{eMI})) - \epsilon_d - \frac{\delta}{2}. \quad (23)$$

The sum term in Eq.23 is a weighted average of the $suc_{s_n}(P_i|\text{eMI})$'s for $1 \leqslant i \leqslant k$. Let $P_{min(\sigma_1)}$ be the (first-best) player among $\{P_1, \ldots, P_k\}$ which has a *minimal* eMI-conditional success rate at time σ_1. By the laws about weighted averages Eq.23 implies:
For all $s_n \geqslant \sigma_1$:

$$suc_{s_n}(P_{min(\sigma_1)}|\text{eMI}) < suc_{s_n}(fav_{s_n+1}(\text{eMI})) - \epsilon_d - \frac{\delta}{2}. \quad (24)$$

We now construct what we call a *min-max-hypercycle*: we pass from σ_1 to the next distinguished switch point, call it σ_2, at which the player $P_{min(\sigma_1)}$ becomes favorite once again; this switch point must exist because by assumption each player in $\{P_1, \ldots, P_k\}$ switches his favorite-status infinitely many often. The eMI-conditional success rate of $P_{min(\sigma_1)}$ is frozen between σ_1 and σ_2; and $P_{min(\sigma_1)}$ is not a deceiver at time σ_2; so it must hold:

$$suc_{\sigma_1}(P_{min(\sigma_1)}|\text{eMI}) = $$
$$suc_{\sigma_2}(P_{min(\sigma_1)}|\text{eMI})) \geqslant suc_{\sigma_2}(P_{min(\sigma_1)}) - \epsilon_d. \quad (25)$$

However (by our construction), $P_{min(\sigma_1)} = fav_{\sigma_2+1}(\text{eMI})$, and so, Eq.24 implies:

$$suc_{\sigma_2}(P_{min(\sigma_2)}|\text{eMI})) < suc_{\sigma_2}(P_{min(\sigma_1)}) - \epsilon_d - \frac{\delta}{2}. \quad (26)$$

It follows from Eq.25+26 that the player $P_{min(\sigma_2)}$ who has a minimal eMI-conditional success rate at time σ_2 must be *different* from $P_{min(\sigma_1)}$, and his eMI-conditional success rate at σ_2, compared to $P_{min(\sigma_2)}$'s at σ_1, *has declined* by more than $\frac{\delta}{2}$:

$$suc_{\sigma_2}(P_{min(\sigma_2)}|\text{eMI}) < suc_{\sigma_1}(P_{min(\sigma_1)}|\text{eMI}) - \frac{\delta}{2}. \quad (27)$$

We use the distinguished switch point σ_2 and the player $P_{min(\sigma_2)}$ to construct the next min-max-hypercycle, i.e., we pass to the next switch point σ_3 at which $P_{min(\sigma_2)}$ becomes favorite again, obtaining again a different player $P_{min(\sigma_3)}$ whose eMI-conditional success rate has dropped by at least $\frac{\delta}{2}$ below that of $P_{min(\sigma_2)}$, and so on. Since every min-max-hypercylce enforces a decline of the minimum of the eMI-conditional success rates by $\frac{\delta}{2}$, and since success rates cannot drop below zero, there must come a distinguished switch point σ^* after which no further hypercycle is possible, which *contradicts* our assumptions that all players in $\{P_1, \ldots, P_k\}$ switch their favorite status infinitely often. ♠

Bibliography

[1] Cesa-Bianchi, N., Lugosi, G. *Prediction, Learning, and Games*. Cambridge: Cambridge University Press, 2006.

[2] Gigerenzer, G., et al. *Simple Heuristics That Make Us Smart*. Oxford: Oxford Univ. Press, 1999.

[3] Good, I. J. *Good Thinking*. Minneapolis: Univ. of Minnesota Press, 1983.

[4] Greeno, J. Evaluating of statistical hypotheses. In Salmon, W. (Ed.), *Statistical Explanation and Statistical Relevance*, 89–104. Pittsburgh: Open Court, 1971.

[5] Hogarth, R. M., Karelaia, N. '"take-the-best" and other simple strategies'. *Theory and Decision*, 61:205–249, 2006.

[6] Kading, D. Concerning mr. feigl's 'vindication of induction. *Philosophy of Science*, 27:405–407, 1960.

[7] Kelly, K. *The Logic of Reliable Inquiry*. New York: Oxford Univ. Press, 1996.

[8] Reichenbach, H. *Experience and Prediction*. Chicago: Univ. of Chicago Press, 1938.

[9] Reichenbach, H. *The Theory of Probability*. Santa Mateo: Univ. of California Press, 1949.

[10] Rescher, N. *Induction*. Pittsburgh: Univ. of Pittsburgh Press, 1980.

[11] Salmon, W. Should we attempt to justify induction? *Philosophical Studies*, 8(3):45–47, 1957.

[12] Salmon, W. The pragmatic justification of induction. In Swinburne, R. (Ed.), *The Justification of Induction*, 85–97. Oxford: Oxford University Press, 1974.

[13] Skyrms, B. *Choice and Chance*. Encinco: Dickenson, 1975.

[14] Weibull, J. *Evolutionary Game Theory*. Cambridge/Mass: MIT Press, 1995.

How Particle Physics Cut Nature At Its Joints [1]

Oliver Schulte

Simon Fraser University

oschulte@cs.sfu.ca

ABSTRACT. This paper presents an epistemological analysis of the search for new conservation laws in particle physics. Discovering conservation laws has posed various challenges concerning the underdetermination of theory by evidence, to which physicists have found various responses. These responses include an appeal to a plenitude principle, a maxim for inductive inference, looking for a parsimonious system of generalizations, and unifying particle ontology and particle dynamics. The connection between conservation laws and ontological categories is a major theme in my analysis: While there are infinitely many conservation law theories that are empirically equivalent to the laws physicists adopted for the fundamental Standard Model of particle physics, I show that the standard family laws are the only ones that determine and are determined by the simplest division of particles into families.

1 Introduction: Conservation Laws and Underdetermination

The underdetermination of belief by evidence is a central topic in epistemology, as it is the point of departure for skeptical arguments against the possibility of knowledge. One aim of the philosophy of science is to study underdetermination as it arises in scientific practice. Such case studies refine our understanding by revealing different kinds of underdetermination. In addition, the epistemological problems that scientists face are rich and complex, and since scientists often confront them with success, the practice of science promises to teach us much about effective strategies for responding to the challenge of underdetermination. Moreover, a systematic philosophical analysis of the relationship between theory and evidence in a scientific domain can clarify the structure and the strength of the logic we find in practice.

In this paper I consider how physicists have resolved various problems of underdetermination that arise in the search for a theory of reactions among elementary particles. One type of underdetermination is *global underdetermination*, which arises when even complete knowledge of all observational facts does not determine the answer to the scientific question under investigation. Global underdetermination is a concern in particle physics in at least two ways. First, a

---

[1] The author is with Department of Philosophy and School of Computing Science, Simon Fraser University, Vancouver-Burnaby, B.C., V5A 1S6, Canada.

possible reaction among elementary particles may never materialize, in which case it will never be observed, and the observational facts may not determine whether the reaction is possible or not. Second, for a given set of (additive) conservation laws there are infinitely many different laws that are consistent with exactly the same processes among elementary particles. Thus even complete knowledge of all reactions ever to occur does not determine a unique set of laws without further considerations.

Local underdetermination arises when finite data samples do not determine the answer to a scientific question, even if complete knowledge of the infinitely many observational facts would. Local underdetermination is closely related to the classic problem of induction. An example of local underdetermination in particle physics is the issue of whether a reaction that has not been observed yet will be observed in a future experiment. Since conservation principles entail assertions about what reactions can and cannot be observed, the local underdetermination of possible reactions means that finite data may not determine which conservation principles make correct predictions.

The final type of underdetermination that I consider in this paper is the underdetermination of *particle ontology* by empirical phenomena. One of the goals of particle physics is to find a taxonomy for the particle zoo that relates properties of particles to empirical phenomena. The main result of this paper is that data about observed reactions among elementary particles determine an essentially unique simplest combination of particle classes and conservation laws. This finding supports and illustrates David Lewis' observation that "laws and natural properties get discovered together" [17, p.368].

This paper is not a historical account of the development of particle conservation laws from the 60s through the 80s. Instead I focus on some of the *epistemological* issues that arose in the development of conservation laws. With regard to historical and current practice, my study aims to be realistic, but within a limited scope. It is realistic in that I consider laws of a form found in actual particle theory—additive conservation laws—and I analyze actual reaction data from particle accelerators. It is limited in various respects, such as: (1) I leave aside other types of conservation laws (e.g., discrete spacetime symmetries [34, 35]). (2) I consider the formulation of conservation laws as a topic on its own, rather than in combination with the development of quark theory and the fundamental Standard Model of particle physics. (3) I do not analyze attempts to derive inductively inferred conservation laws from more fundamental principles—what Earman calls the search for "second-order laws" [4]. Thus the view of additive conservation laws in this paper is closer to the phenomenological approach of the 1950s and 1960s than to current theories about conservation laws. As far as the history of particle theory is concerned, my analysis is best viewed as pertaining to an early historical period of particle research. Despite these limitations, my discussion captures enough aspects of particle theory to raise and illuminate many philosophical questions about the underdetermination of theory and ontology by evidence. Also, the fact that the epistemological analysis matches the physicists' inferences, as I show below, indicates that my discussion captures much of the logic that guided the discovery of selection rules in particle research.

The paper is organized as follows. In Section 2 I outline the project of discovering new conservation laws in particle physics, and the principles that have guided this project as formulated by prominent practitioners such as Nobel Laureates Feynman and Cooper. Section 3 analyzes the logical structure of additive conservation laws—what empirical predictions can they express, and under what circumstances do different sets of laws make the same predictions? Following the general principles of scientific inference examined in Section 2, I formulate the *maximally strict inference rule*: posit conservation principles that explain the nonoccurrence of as many unobserved particle reactions as possible. Section 4 examines the status of the maximally strict inference rule. This rule can be justified in terms of a general inductive epistemology whose precept is that good inference rules serve the aims of inquiry (in the present analysis, the aims of true and stable belief). The main finding of this section is that given the particle reaction data that are currently available, physicists adopted conservation laws that follow the maximally strict inference rule. As there are infinitely many alternative law sets that are empirically equivalent to those physicists adopted, the question arises what is special about the standard laws. Section 5 answers this question by showing that the standard laws correspond to *a simple division of elementary particles into disjoint families*, and that they are essentially the only laws to do so. The enterprise of particle physics raises some of the great themes of the philosophy of science: natural laws, natural kinds, and simplicity. In the final section I comment on David Lewis' prominent views on these topics in the light of what my case study shows about conservation laws in particle physics. More details about the results of my analysis may be found in [27] and [30].

2 Conservation Laws and Underdetermination in Particle Physics

I use the term "global underdetermination" to refer to a situation with two possible worlds in which our experience is exactly the same [13]. This scenario is familiar in philosophy. For instance, Descartes' Meditations described two worlds in which our experience is the same, one in which an evil demon produces our illusions, and another in which our perceptions reflect reality. In particle physics, the possibility of global underdetermination arises in a more pedestrian manner. Consider a possible physical process such as $p \rightarrow \pi^0 + e^+$, the decay of a proton into a pion and a positron. Is it possible that the laws of physics permit this process, yet we never observe it? If yes, then global underdetermination arises: there are two physically possible worlds, one in which $p \rightarrow \pi^0 + e^+$ is a possible process, another in which the process is forbidden, yet our total evidence in each may be the same. This possibility may seem unlikely, and indeed particle physicists have explicitly ruled it out.

> There is an unwritten precept in modern physics, often facetiously referred to as Gell-Mann's totalitarian principle, which states that 'anything which is not prohibited is compulsory'. Guided by this sort of argument we have made a number of remarkable discoveries from neutrinos to radio galaxies. [1]

So the scenario I sketched above cannot happen: If $p \rightarrow \pi^0 + e^+$ is not forbidden, then by Gell-Mann's Totalitarian principle, the process must eventually occur. Kenneth Ford states a succinct version of the Totalitarian principle that emphasizes the role of conservation laws: "everything that *can* happen without violating a conservation law *does* happen" [7, p.82], Ford's emphasis. The next section elaborates on the importance of conservation laws in the search for the laws of particle dynamics.

If the concerns of global underdetermination are met, the total findings of an unbounded course of inquiry would settle all questions about what particle reactions are possible. However, at a given stage of inquiry, physicists will have explored but a finite amount of phenomena and gathered but a finite number of observations. Thus the challenge remains of how to generalize from the evidence available in the short run. As there are an infinite number of particle theories logically consistent with a finite amount of data, we face a problem of local underdetermination.

As Richard Kane astutely observes, the Totalitarian Principle not only defuses global underdetermination, but also carries implications for the direction of research and theory in particle physics [11]:

> What is interesting is that, in committing themselves to plenitude in this restricted form, modern physicists are committing themselves to the principle that what never occurs must have a sufficient reason or explanation for its never occurring.

Thus the Totalitarian Principle implies an explanatory strategy: to focus on reactions that *do not occur* as the primary explananda. If we recall Ford's conservation law version of the Totalitarian Principle, the strategy is to find quantities that are conserved in reactions known to be possible, but not conserved in problematic reactions that fail to be observed. Nobel Laureate Leon Cooper testifies to the importance of conservation principles, in particular selection rules (a type of conservation law that I will discuss in detail presently) [2, p.458].

> In the analysis of events among these new particles, where the forces are unknown and the dynamical analysis, if they were known, is almost impossibly difficult, one has tried by observing what does not happen to find selection rules, quantum numbers, and thus the symmetries of the interactions that are relevant.

These sources make clear that the emphasis on conservation principles has been a great help to physicists in their search for a theory of partice reactions. Much of this paper is concerned with working out exactly how. The assumption that conservation principles are adequate for describing particle dynamics rules out many logically possible particle worlds and thus amounts to a substantive empirical background assumption that limits skeptical possibilities. To see why that is so, we need to study the logical structure and predictive import of conservation principles in some detail.

3 The Logic of Additive Conservation Laws

This section introduces additive conservation principles and discusses their status in the fundamental Standard Model of Particle Physics and recent extensions of the Standard Model. An analysis follows of how the focus on additive conservation principles can resolve local underdetermination.

3.1 Conservation Laws in Particle Physics

A particle is an object that obeys the rules of quantum mechanics for a point with well-defined mass and charge [19, chap.1.1]. Physicists refer to particles that are neither atoms nor nuclei as 'elementary' particles.[2] Several kinds of conservation principles in particle physics describe what reactions among elementary particles are possible: General conservation principles such as conservation of energy, momentum and electric charge, discrete space-time symmetries such as parity and CPT, and so-called numeric or additive conservation principles [34, 35]. My analysis considers the last kind of conservation principle, also known as a **selection rule** [19, p.36]. A selection rule introduces a quantity and assigns a value for that quantity to each known elementary particle. Tab.I lists five such quantities, namely electric charge, baryon number, electron number, muon number and tau number; it specifies what value each of the 22 listed common particles has for a given quantity [34]. These quantities are the same for each kind of

	Particle	Charge(C)	Baryon#(B)	Tau#(T)	Electron#(E)	Muon#(M)
1	Σ^-	-1	1	0	0	0
2	$\overline{\Sigma}^+$	1	-1	0	0	0
3	n	0	1	0	0	0
4	$\overline{n}$	0	-1	0	0	0
5	p	1	1	0	0	0
6	$\overline{p}$	-1	-1	0	0	0
7	π^+	1	0	0	0	0
8	π^-	-1	0	0	0	0
9	π^0	0	0	0	0	0
10	γ	0	0	0	0	0
11	τ^-	-1	0	1	0	0
12	τ^+	1	0	-1	0	0
13	ν_τ	0	0	1	0	0
14	$\overline{\nu}_\tau$	0	0	-1	0	0
15	μ^-	-1	0	0	0	1
16	μ^+	1	0	0	0	-1
17	ν_μ	0	0	0	0	1
18	$\overline{\nu}_\mu$	0	0	0	0	-1
19	e^-	-1	0	0	1	0
20	e^+	1	0	0	-1	0
21	ν_e	0	0	0	1	0
22	$\overline{\nu}_e$	0	0	0	-1	0

Table I: Some Common Particles and Quantum Number Assignments

particle in all reactions (unlike, say, momentum). From now on, I will use the term "quantity" to refer to such process-invariant, or time-invariant, properties of particles. In this paper I consider only conservation principles that involve process-invariant quantities. To fix some notation, the fact that particle p carries x units of quantity q will be denoted by $q(p) = x$. For instance, if C denotes electric charge, we have $C(e^-) = -1$.

---

[2] except for the proton, which is considered an elementary particle although it is the nucleus of the hydrogen atom.

A reaction conserves a quantity just in case the total sum of the quantity over the reagents is the same as the total sum over the products. For example, the reaction $p + p \to p + p + \pi^0$ conserves Baryon number, since the Baryon total of the reagents is $2 \times Baryon\#(p) = 2 \times 1$, and the Baryon total of the products is $2 \times Baryon\#(p) + Baryon\#(\pi^0) = 2 \times 1 + 0$.

The conservation of the quantities electric charge, baryon number, tau number, electron number and muon number that are shown in Tab.I is part of the *Standard Model* that takes the quarks as building blocks of particles and was elaborated by particle physicists in decades of research [2, 8, 18, 34]. For brevity, I will sometimes abbreviate these quantities as C, B, T, E, M.

A comment is in order regarding the current status of these conservation laws. Since the recent discovery that neutrinos have nonzero mass (1998-2000), there has been extensive activity in extending the Standard Model of particle physics on the basis of new experimental data [33]. Briefly, the nonzero mass of neutrinos permits quantum-mechanical effects that violate traditional conservation laws such as muon number. For instance, in "neutrino oscillation" a muon neutrino turns into a tau neutrino, which clearly violates muon number conservation. Particle theory implies that such violations occur very rarely, but nonetheless there seems to be sufficient evidence that they do occur in nature. The picture that seems to be emerging is that some of the conservation laws of the Standard Model such as muon or baryon number conservation are partial symmetries that hold only "approximately", that is with very rare exceptions. There are complex current debates in particle theory about the best way to extend the Standard Model to account for these new phenomena. My aim in this paper is not to describe possible extensions of the Standard Model. Instead, I will focus on the pre-1998 Standard Model and the particle phenomena that it covers, where the quantities listed in Tab.I are universally conserved. It is possible to carry out the following analysis for a model with partial symmetries and approximate conservation laws. The results would be essentially the same, and the extra complications do not lead to new epistemological insights.

3.2 Selection Rules and Local Underdetermination

Let us see how the use of conservation principles can resolve local underdetermination. Suppose that other resources from physical theory, known conservation principles, etc., do not suffice to answer the question of how many pions can be produced in a collision of two protons. Suppose further that the answer must lie with conservation principles governing collisions of two protons. Now if we observe the reaction $p + p \to p + p + \pi^0$, we infer that *whatever* conservation principles govern collisions of two protons, the pion π^0 must carry 0 of any conserved quantity, because clearly the two protons on the left and on the right put the same weight into the conservation balance. But if the pion π^0 carries 0 of every conserved quantity, then two protons may produce *any* number of pions without violating a conservation law. Thus after observing one reaction such as $p + p \to p + p + \pi^0$ that produces pions, we can deduce that any number of pions can be produced in a collision of two protons—which is what current particle theory tells us [7, p.82].

So if one reaction is possible (such as $p+p \to p+p+\pi^0$), then selection rules must permit other reactions as well (such as $p + p \to p + p + \pi^0 + \pi^0$). Thus focusing on conservation principles allows us to infer that certain unobserved reactions are possible. For a given set of observed reactions R there will be a set R' of reactions that are entailed by the observed data R in the sense that any set of selection rules consistent with the reactions R must also be consistent with the reactions R'. I shall refer to the reactions R' entailed by a data set R as **the least generalization of** R.

If our goal is to explain the nonoccurrence of unobserved reactions, we would seek a set of selection rules that rule out as many unobserved reactions as possible. Let us say that such a set of selection rules is **maximally strict** for the observed reactions R. In other words, a set of selection rules Q is maximally strict for observed reactions R just in case the reactions that conserve all quantities in Q form the least generalization of R. The next proposition asserts that, for any set of observed reactions R, there are infinitely many maximally strict sets of conservation laws.

Proposition 1 *Let R be any set of observed reactions. Then there are infinitely many maximally strict sets of conservation laws, each of which allows exactly the least generalization of R.*

Here are the intuitive reasons why the proposition holds. The key observation is that we can think of particle reactions that are allowed by a given set of selection rules as forming a linear space. For example, given two reactions r and r' that are consistent with a given set of selection rules Q, we can "add" r and r' to obtain the reaction $r + r'$ that is also consistent with Q. To illustrate, both the neutron decay $n \to p + e^- + \bar{\nu}_e$ and muon decay $\mu^+ \to e^+ + \nu_e + \bar{\nu}_\mu$ conserve all standard quantities. Hence so does their sum, which is the collision $n + \mu^+ \to p + e^- + \bar{\nu}_e + e^+ + \nu_e + \bar{\nu}_\mu$. The least generalization of a given set of reactions R is the least linear space containing R; that is, the least generalization comprises the reactions that can be generated as linear combinations of reactions in R.

How can we find a set of selection rules that is consistent with exactly the least generalization of the reactions R? The answer is to note that the set of quantities that are conserved in the reactions R forms another linear space (technically, this linear space is the orthogonal complement of R). For example, given the standard model quantities Baryon Number and Electric Charge, we can define their sum Baryon-Charge to be the quantity that assigns to every particle the sum of its Baryon Number and its Electric Charge. To illustrate, the Baryon-Charge of the proton p is $1 + 1 = 2$. It is not hard to see that Baryon-Charge is conserved in every reaction that conserves Baryon Number and Electric Charge. In other words, Baryon-Charge is *redundant* given Baryon# and Electric Charge. In general, a linear combination of conserved quantities will also be conserved, but does not add any more constraints on particle reactions.

A maximally strict set of conservation laws is a set that contains as many irredundant laws as possible. Each law adds a constraint on how nature may behave, so the more irredundant laws a theory contains, the stricter it is. Given that linear combinations of quantities are redundant, what we seek is a maximal set of linearly independent quantities; in linear algebra terms, a *basis* for the space of quantities conserved in the reactions R. Fig.1 illustrates these facts.

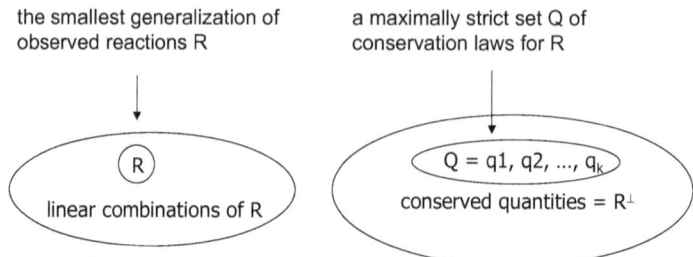

Figure 1: The least generalization of observed reactions R is to predict that all and only linear combinations of the data R are possible. The set of quantities that are conserved in all the reactions R forms a linear space, the orthogonal complement $R^\perp$. Any set of conservation laws Q that corresponds to a basis for $R^\perp$ allows precisely the least possible generalization of the observed reactions R.

It is easy to see that the space of conserved quantities has infinitely many bases. For example, we can multiply Baryon Number by 2 to form the quantity $2B$ defined by $2B(p) = 2 \times B(p)$. For instance, the proton carries two units of $2B$ since it carries one of B. Now we can replace the conservation of Baryon Number B by the conservation of $2B$ and obtain a predictively equivalent theory that is consistent with exactly the same reactions. More formally, the set of quantities $\{2B, M, E, T\}$ is predictively equivalent to the set $\{B, M, E, T\}$. Similarly, we could replace B by $3B, 4B$, etc. Formal statements and proofs of these results may be found in [27] and [30].

Given the emphasis that physicists have placed on explaining what does not occur (cf. Section 2), our analysis so far suggests the following inductive rule for generalizing from observed data to a set of conservation laws: *always adopt a set of conservation laws that are maximally strict*. I refer to this rule as the **maximally strict rule** (MSR for short).

There are two basic questions we may ask about the maximally strict rule. 1) the epistemologist's question: what sort of justification is there for the maximally strict rule, for example in terms of more general epistemic principles? 2) the naturalist's question: do maximally strict theories match what we find in scientific practice? I discuss these questions in the next section.

4 Inferring Conservation Laws: Methodological Analysis and Comparison With Practice

This section examines the status of the maximally strict inference rule, first from the point of view of general inductive methodology, second in terms of agreement with scientific practice.

4.1 Means-Ends Justification of the Maximally Strict Inference Rule

In a previous article [27] I discussed the status of the maximally strict rule from the point of view of an inductive epistemology based on means-ends analysis.

I review here the main results of the investigation of the MSR rule. The basic tenet of means-ends analysis is that good inference methods are those that attain the goals of inquiry. Means-ends epistemology examines a number of standards of empirical success, such as reliable convergence to the truth, fast convergence, and stable convergence [10, 13, 26]. To keep matters simple, let us consider two aims: *reliable and stable convergence* to a theory that makes correct predictions. The notion of reliable convergence to a correct theory stems from a Peircean vision of empirical success in which science may err in the short run, but corrects itself and eventually settles on the truth. Reichenbach's well-known vindication of induction is in this spirit: he showed that the posits of his straight rule come arbitrarily close to the true limiting relative frequency of an event, if that limit exists [24, 25]. Putnam generalized Reichenbach's idea to develop a general theory of inductive inference [20, 22], which has grown into a mathematical subject known as Formal Learning Theory [10, 13, 28].

Assume that some set of additive conservation laws is predictively equivalent for the totality of all reactions ever to be observed. Then the maximally strict rule is guaranteed to eventually arrive at a predictively adequate theory, and thus satisfies the Peirce-Reichenbach-Putnam ideal of empirical inquiry [27, 30]. A venerable philosophical tradition supports the idea that stable belief is a significant epistemic good. Since Plato's Meno, philosophers have been familiar with the idea that stable true belief is better than unstable true belief. Epistemologists such as Sklar have advocated principles of "epistemic conservatism" [32]. Kuhn tells us that a major reason for conservatism in paradigm debates is the cost of changing scientific beliefs [15]. In this spirit, means-ends epistemologists, starting with a seminal paper of Putnam's [21], have examined inductive methods that minimize the number of times that they change their theories before settling on their final conjecture [13, 26].

It turns out that the means-ends criteria of reliable convergence to the right answer and minimizing theory changes single out the maximally strict rule. I record this fact in the following proposition.

Proposition 2 *Assume that there is a set of selection rules that correctly predicts which processes among n known particles occur. The maximally strict rule is the only inference rule that satisfies the following two aims:*

1. *The rule is guaranteed to eventually settle on a set of conservation laws that correctly predicts which reactions can and cannot occur; and*
2. *the rule changes its predictions at most n times.*

The demonstration is in [27, Section.7]. This result warrants a few comments. (1) The maximally strict rule changes its mind only when its most recent conservation law theory is falsified (as neutrino oscillation falsifies the conservation of muon number, Sec. 3), so it is also the case that the MSR is the only rule whose predictions are falsified at most n times (under the assumptions of the proposition). (2) The proposition illustrates the philosophically important point that criteria of inductive success can strongly constrain empirical conjectures in the short run. Elsewhere I have shown that the same inductive goals select the generalization "all emeralds are green" over all "emeralds are grue" in a Goodmanian Riddle of Induction [26].

Thus it appears that means-ends analysis matches scientists' methodological practice, in the sense that means-ends analysis directs inductive inquiry to adopt maximally strict conservation law theories, which corresponds to the principle of explaining as many unobserved reactions as possible. In the next sections I examine how the inferences of the maximally strict rule compare with the theories physicists have developed in response to the currently available data. There are several steps in this project. First, to collect from the literature a suitable set of known reactions for applying the maximally strict method. Second, to compute a maximally strict set of conservation principles. Third, to determine whether the laws C, B, M, E, T shown in Tab.I form a maximally strict set for the observed reactions.

4.2 Collecting Particle Reaction Data

To facilitate computational analysis, we need to supply reaction data in electronic format. The most readily available kind of reaction data are decays of particles, since they are listed with each particle. For instance, decay modes are listed in the Annual Review of Particle Physics, an authoritative publication that summarizes the current state of knowledge in the field every year [5]. The data set includes a decay mode for each of the 182 particles that has one. This list comprises all particles except for $\gamma, e^-, e^+, p, \bar{p}, \bar{\nu}_e, \nu_\mu, \bar{\nu}_\mu, \nu_\tau, \bar{\nu}_\tau$. For example, for the upsilon1S particle whose symbol is $\Upsilon(1S)$, I listed the decay mode $\Upsilon(1S) \to \mu^+ + \mu^-$. In all, my data set includes 205 observed reactions, 199 of which are decays listed in the Review. Typically, the data include the most probable decay listed in the Review of Particle Physics. The additional reactions are important processes listed in textbooks. The complete particle database is posted in Excel format at [29]; more discussion and justification for focusing on decay modes may be found in [27, 30, 31].

4.3 Applying the Maximally Strict Inference Rule to Current Particle Data

After translating current particle data into vector format, we can run a program that produces a maximally strict set of conservation principles for this data.[3] As we saw in Section 3, finding a maximally strict theory is equivalent to finding a basis for the orthogonal complement of the given reaction data. Although for a human this is a formidable task, fortunately it is a well-studied problem in computational linear algebra [12], and efficient programs for solving it are available that can analyze even an extensive set of reaction data in minutes. The next step is to compare the set of conservation principles produced by the program with the Standard Model laws C, B, M, E, T. With the aid of the computer, it is simple to check that the two sets of laws span the same linear space and hence are empirically equivalent, that is, consistent with exactly the same reactions. This computation establishes that the combined laws C, B, M, E, T form a maximally strict theory for the current data.

Finding 1 *Let D be the particle reaction data described in Section 4.2. The combination of laws asserting the conservation of electric charge, baryon number,*

---

[3] A trace of a program run is posted at [29].

tau number, electron number and muon number, forms a maximally strict set of conservation laws for the reaction data D.

The finding answers the naturalist's question in the affirmative: The predictions made by the laws we find in the Standard Model are exactly those mandated by the maximally strict inference rule. According to Proposition 1, there are infinitely many sets of conservation laws that are empirically equivalent to the standard set—any basis spanning the same linear space will do. So even if local underdetermination is resolved by the directive to make maximally strict predictions, global underdetermination arises because there are many sets of laws that make the same (maximally strict) predictions. Among the infinitely many selection rules that make the same predictions as the set C, B, M, E, T, is there something special about this set?

5 Underdetermination and Particle Families

Naturally this question occupied particle physicists. Feynman thought the answer would lie in the fact that quantities such as Baryon number would turn out to relate to physical phenomena other than as a conserved quantity: "If charge is the source of a field, and baryon number does the same things in other respects it ought to be the source of a field too. Too bad that so far it does not seem to be, it is possible, but we do not know enough to be sure"[6, p.67]. Omnes devotes a chapter to deriving selection rules in his graduate text on particle physics. He gives "once and for all" a procedure for assigning these numbers [19, chap.2]. He does not address the question of whether his procedure uniquely determines a set of quantum numbers, although he comments that various aspects of the number assignments are arbitrary. Williams remarks, commenting specifically on the conservation of lepton number (the sum $M+E+T$): "this lepton number conservation is arbitrary and has no basis in more fundamental ideas" [34, p.285].

One of the main considerations that led physicists to these laws is that they correspond to ontological categories that find an independent interpretation and support in the Standard Model that takes the quarks as building blocks of particles. The term *family conservation law* for the conservation of baryon, electron, muon, and tau number reflects the connection between selection rules and particle ontology [34]. The idea that natural laws have a special connection to ontology has a long standing in philosophy; often the idea is expressed by the thesis that natural laws refer to fundamental ontological categories or natural kinds [14, p.6]. In this section and the next I pursue the connection between conservation laws and ontology to see how it affects the underdetermination of conservation principles.

The four quantities Baryon#, Electron#, Muon#, Tau# illustrated in Tab. I have the following interpretation in terms of particle families. First, the particle world is divided into Baryons and non-Baryons. Every non-Baryon receives Baryon# 0. Each Baryon that is regular matter is assigned Baryon# 1 (e.g., the proton p); each Baryon that is anti-matter is assigned Baryon# -1, and usually its symbolic name is marked by an overline, like for instance the antiproton $\bar{p}$. Similarly Electron# represents the Electron family, which comprises the particles

$e^-, e^+, \nu_e, \overline{\nu}_e$: all other particles receive Electron# 0, the two electron type particles e^- and ν_e Electron# 1, and the antiparticles e^+ and $\overline{\nu}_e$ receive Electron# -1, as illustrated in Tab.I. Muon Number similarly represents the muon particle family $\mu, \overline{\mu}, \nu_\mu, \overline{\nu}_\mu$ and Tau# the tau particle family $\tau, \overline{\tau}, \nu_\tau, \overline{\nu}_\tau$.

This section examines the underdetermination of particle taxonomy by reaction data. I establish two new theorems in linear algebra showing that a division of particles into disjoint families is, under quite general conditions, uniquely determined by reaction data. Then I apply these results to arrive at the main finding of the case study in this paper: the available reaction data did uniquely determine the Standard Model particle families comprising the baryons and the three lepton generations tau, electron and muon.

5.1 Particle Reaction Data Determine Particle Families In Principle

The partition of particles into baryons, and the muon, electron and tau types is not necesssary for a predictively adequate conservation theory: it is easy to produce theories that are equivalent to the standard laws but do not incorporate these distinctions. For a very simple example, suppose we form a new quantity q defined as the vector sum of baryon number and muon number, such that each particle p carries an amount of q the sum of its baryon and muon numbers (in symbols, $q(p) = B(p) + M(p)$). Now replace baryon number by q, which results in the law set $\{C, q, M, E, T\}$. This set is empirically equivalent to the standard rules, in that a reaction r conserves all quantities just in case r conserves all quantities in the set; but no conservation law in the new set corresponds to the family of baryons. A skeptic with nominalist leanings (sometimes called a "conventionalist" [3, section.4]) may point to such examples as evidence that ontological categories such as "baryon" have no reality in nature, but are imposed by us to explain the particle reaction phenomena, while equally good (i.e., predictively equivalent) explanations are available that do not employ these categories.

There is a reply to the nominalist: what is better about the standard family laws compared to the gerrymandered ones is that they divide the particle world into disjoint categories. Laws based on disjoint categories have several virtues. (1) They satisfy the intuition that the laws of nature should be based on fundamental distinctions among the objects in the world. (2) They can be interpreted as asserting that particles of one kind do not transform into particles of another, which seems physically meaningful. (3) They define ontological categories that may point to and receive support from independent ontologies. For example, in the quark model of particles the baryons turn out to be exactly the particles composed of three quarks.

Fair enough, says the nominalist. But surely we can imagine an alternative set of disjoint categories, different from the baryon, muon etc. categories, with corresponding conservation laws that are predictively equivalent to the standard selection rules? I will show that the answer is no: What the nominalist says he can imagine is in fact logically impossible. Say that a particle p **carries** a quantity q if $q(p) \neq 0$. A set of quantum numbers $\{q_1, \ldots, q_j\}$ **forms a family set** if no particle carries two quantities; formally, if $q_i(p) = 0$ whenever $q_j(p) \neq 0$, for all $i \neq j$. Given a family set of quantities, the carriers of the quantities form disjoint particle families. The standard quantities {Baryon#, Muon#, Electron#, Tau#}

form a family set. For example, the proton carries 1 unit of baryon number, but 0 of the muon, electron and tau numbers. In the debate with the nominalist, the issue is how many family sets in addition to the standard particle families there might be. The surprising answer is that if there is any predictively adequate family set, then all predictively adequate family law sets define the *same* particle ontology.

Theorem 1 *Let Q, Q' be any two predictively equivalent combinations of laws, with associated quantities. If both Q and Q' are family sets, then they define the same particle families as carriers of their quantities. It follows that if Q' is any family set of laws that is predictively equivalent to the conservation of baryon number, tau number, electron number and muon number, then Q' defines the same particle families as these laws (namely the baryons and the three lepton generations tau, electron and muon).*

The proof is in [30, 31]. The basic argument runs as follows. Consider two conservation theories, Q, Q' that are predictively equivalent (i.e., consistent with exactly the same reactions) such that Q and Q' are family sets. The structure of vector spaces requires that it must be possible to translate Q into Q'. However, the proof shows that if one family set Q can be translated into another Q', then the particle families corresponding to Q must be the same as those for Q'.

The theorem addresses both global and local underdetermination. For global underdetermination, consider the set of all reactions ever to occur in the course of nature; denote this set by *Total*. If this set is predicted correctly by a family law set Q, in the sense that Q permits exactly the reactions in *Total*, then all other family law sets Q' that permit exactly the reactions in *Total* define the same particle families. In other words, the total reaction phenomena *Total* determine a unique set of particle families. For local underdetermination, suppose we have particle accelerator data D, and there is some family law set Q that is maximally strict for D. Then the proposition implies that all other maximally strict family conservation theories for D define the same families. In other words, the reaction data D and the maximally strict rule together determine the division of particles into disjoint families, without recourse to other considerations or a more general ontological model (such as the Standard Model).

The theorem establishes a subtle interplay between the contingent and the necessary. It need not be the case that there is any partition of particles into families that corresponds to empirically adequate conservation laws. But given that there is such a partition, it is unique. In terms of the joints metaphor, there need not have been any empirically adequate way of carving up nature, but given that there is one, the cuts are uniquely determined.

5.2 Particle Reaction Data Determine the Standard Particle Families in Practice

Our results so far show how the logical structure of additive conservation laws resolves the underdetermination of theory by evidence in general; let us now consider how they apply to our case study. Consider the reaction data that were consistent with the conservation of Charge, Baryon#, Tau#, Electron#, and Muon# and led to the adoption of these conservation laws; this data includes the data set D described in Section 4.2. Theorem 1 suggests that if there is a maximally strict family set of laws for the reaction data D, then the corresponding particle families would be uniquely determined. But the theorem does not directly apply to the law set Charge, Baryon#, Tau#, Electron#, and Muon#, because this is *not* a family law set: the carriers of electric charge include, for instance, many baryons, such as the proton.

However, as Feynman pointed out (cf. Section 5), charge is different from the other additive conservation laws in that it is the source of a field and relates to other dynamic phenomena. Thus we can view the charge of a particle as being determined independently. Moreover, the law of the conservation of electric charge was well established in physics long before the study of subatomic particles. So for the project of finding new conserved quantities for the subatomic world, the conservation of the C quantity can be taken as given. The reason why this observation is important is that Theorem 1 can be generalized to show that *if Q is any set of laws, predictively equivalent to Charge, Baryon#, Tau#, Electron#, and Muon#, and of the form (electric charge + a set of family laws), then Q defines the same particle families as baryon number, tau number, electron number and muon number (namely the baryons and the three lepton generations tau, electron and muon).* A fuller desription of this result with proofs may be found in [30, 31].

The next finding records that if we take the conservation of electric charge as our starting point, the actual reaction data described previously uniquely determine the $\{B, M, E, T\}$ families.

Finding 2 *Let D be the particle reaction data described in Section 4.2. Let Q be any family set of conservation laws such that the combination of Q and electric charge is maximally strict for the reaction data D from particle physics. Then Q defines the same particle families as the Standard Model laws, namely baryons and the three lepton generations tau, electron and muon.*

Figure 2 illustrates this result.

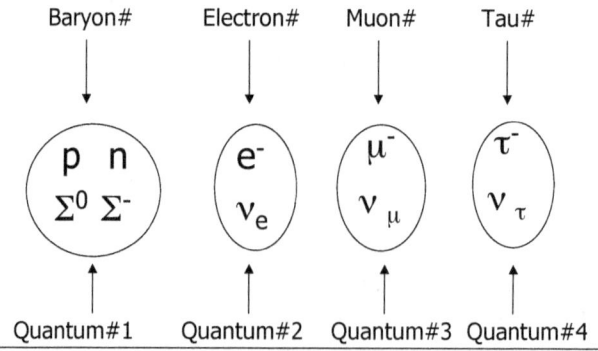

Figure 2: Given the law of conservation of electric charge, *any* maximally strict set of laws that partitions the particle world into disjoint categories employs the same categories as the standard family conservation laws.

The finding holds for the following reasons. From Finding 1 we know that the laws asserting the conservation of electric charge, baryon number, tau number, electron number and muon number, form a maximally strict set of conservation laws for the reactions D. So if Q plus electric charge is maximally strict for the reaction data D, then Q plus electric charge makes exactly the same predictions as the set of laws corresponding to the conservation of electric charge, baryon number, tau number, electron number and muon number. Then the generalization of Theorem 1 described above implies that the two sets of laws must respectively define the same particle families.

From the point of view of local underdetermination and inductive inference, Finding 2 gives strong guidance to finding a maximally strict conservation theory: it suggests to start with electric charge and its conservation as given, and then search for family laws that 1) correspond to disjoint ontological categories, and 2) in combination with electric charge yield a maximally strict theory. It turns out that the parsimony of a set of selection rules guides an inference method towards a simple ontology: A computer program that searches for a maximally parsimonious and strict set of selection rules rediscovers exactly the rules posited by physicists [30, 31]. "Parsimony" in this context means rules that assign quantities whose absolute values are as small as possible (e.g., 0 rather than -1, or -1 rather than 3).

I conclude my case study by comparing my algorithmic derivation of selection rules and particle families with the historical progress of particle theory. One difference with the historical development is that often ontological categories were epistemically prior to reaction data. For example, if one arranges elementary particles by mass, there is a large gap between the lightest baryon (the proton) and the heaviest non-baryon particle (baryon number was also called "heavy particle number"). So the category of baryons suggests itself independently of reaction data; the baryon number quantity can be derived from this category, and then its conservation checked against the available reaction data. This is the kind of procedure that Feynman referred to as a "quick way of guessing at the laws of nature" [6, p.67]. In contrast, in the perspective of my analysis particle families are epistemically posterior to reaction data, because they are discovered by analyzing the reaction data, not posited first and then checked against the data. The historical reliance on considerations and measurements in addition to reaction data, such as the masses of particles, would suggest that particle families cannot be discovered from reaction data alone, but must be derived from other sources. The surprising and novel result of my analysis is that this is not so; this new insight is what motivates developing a method that begins with reaction data to find particle families.

6 Discussion: Induction, Laws, Simplicity, and Natural Kinds in Particle Physics

My analysis has touched on some of the great themes in the philosophy of science as they pertain to conservation laws in particle physics. In this section I take a broader perspective and place my findings in the context of more general discussions in the philosophy of science. To recapitulate, we found that the Standard Model of particle physics features the set of conservation laws that yield the best generalizations about particle dynamics and the simplest particle taxonomy with the smallest number of particle families. The operative notion of ontological simplicity in my analysis has basically two aspects: first, *number*—to employ as few ontological categories as possible (in our case four not counting charge), and second, *overlap*—to find as many particle families as possible that are disjoint, that is, categories whose boundaries do not overlap.

Lewis on Laws, Kinds and Simplicity. David Lewis uses the term "natural property" for the properties that "physicists discover in the course of discovering laws" [17, p.365]. As the distinction between natural kinds and natural properties is not important for my purposes, I will mostly follow Lewis in referring to natural properties. . Many philosophers have noted connections between induction, laws and natural kinds or properties. Quine asked "what tends to confirm an induction?" [23, p.155] and held that "the answer lies in similarity" [23, p.157], "which is immediately definable in terms of kind; for, things are similar when they are two of a kind". Goodman's thesis that valid generalizations are the law-like ones is well-known. He also suggests that law-likeness involves natural kinds [9]. A connection between natural laws and natural kinds has been widely accepted by philosphers after Goodman. Kornblith encapsulates the view as follows [14, p.6].

> On the account of science which began to emerge from these authors [Putnam, Boyd, Field and a "host of others"], it is the business of science to discover the real causal structure of the world; what this means, in a word, is the discovery of natural kinds and the causal relationships among them.

In his book "Counterfactuals", David Lewis' account of natural laws is, roughly, that a natural law is a member of the system of true generalizations that makes an ideal trade-off between simplicity and information content [16]. Our case study suggests a promising synthesis of Lewis' analysis with the traditional connection between laws and natural kinds: if a key ingredient of the simplicity of a system of generalizations is ontological simplicity, then Lewis' account of laws makes a connection with ontology. Let me elaborate this idea.

Suppose we add to Lewis' account of laws the thesis that an important part of the simplicity of a system of generalizations is ontological simplicity. Specifically, consider the following two theses:

1. A law is a member of the simplest system of empirically adequate generalizations, where ontological simplicity is an important aspect of simplicity.
2. A natural property/kind is one that appears in the ontologically simplest system of empirically adequate generalizations.

Our case study illustrates both theses. In particular for 2), we saw that the standard particle families are the ones that correspond to the ontologically simplest maximally strict set of conservation laws. The two theses imply a connection between laws and natural properties or kinds. For by Thesis 1), laws will tend to to be ontologically simple empirically adequate generalizations. And by Thesis 2), ontologically simple empirically adequate generalizations involve natural properties. If the maximally simple system S of empirically adequate generalizations is also the ontologically simplest one, as in our case study, then it follows from 1) and 2) that the properties appearing in S are also natural properties. Figure 3 illustrates these connections.

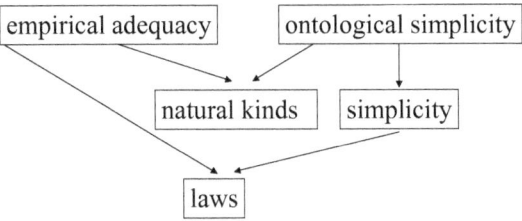

Figure 3: To illustrate the thesis that natural kinds are those that appear in the ontologically simplest empirically adequate system of generalizations. Ontological simplicity is a component of simplicity, which connects Lewis' simplicity-based account of natural laws with ontology. Our case study shows that the four Standard Model particle families Baryon, Electron, Muon, and Tau correspond to the ontologically simplest system of conservation laws for the given particle reaction data.

In a later article "New Work for a Theory of Universals", Lewis adds ontological considerations to his account of laws: for a given axiom set representing a system of generalizations, he stipulates that the primitive vocabulary of the axioms must refer only to natural properties [17, p.368]. Lewis' suggestion seems to assume a notion of natural property that is independent of empirical adequacy and of simplicity, and uses this independent notion to restrict the range of systems of generalizations that may count as laws. My proposal in contrast defines both the concept of law and that of natural property in terms of empirical adequacy, simplicity, and ontological simplicity.

On both views, "laws will tend to be regularities involving natural properties" and "the scientific investigation of laws and of natural properties is a package deal" [17, p.368]. Both views also agree that "in putting forward as comprehensive theories that recognise only a limited range of natural properties, physics proposes inventories of the natural properties instantiated in our world" [17, p.364]. My approach has an important advantage for understanding scientific discovery: Lewis' account does not address how scientists should go about finding natural properties (what Daly terms the "epistemological question" [3, Sec.2]), whereas on my proposal scientists can find natural properties by seeking a predictively adequate theory that is ontologically simple, as shown in Section 5.2 and [30]. In our case study, the difference is that while Lewis' account explains why the conservation of baryon, muon, electron and tau number count as laws of nature given the natural particle families comprising baryons, muon-type particles, electron-type particles and tau-type particles, his account does not entail that physicists may go about *discovering* these categories by pursuing ontologically simple laws as they have done.

7 Conclusion

Many physicists and philosophers have emphasized the importance of conservation laws. In his address as president of the PSA, John Earman said that

> Philosophers of science have barely scratched the surface of the topic of laws, symmetries, and symmetry breaking. What I find most attractive about this topic is that it brings into fruitful interaction issues from metaphysics, from mathematics and physics, from the philosophy of scientific methodology, and from foundations of physics. [4, p.1240].

My case study is an example of the rich interaction that Earman describes. I examined some of the newer conservation laws that particle physicists introduced for the realm of subatomic particles. I considered additive conservation laws that assign an amount of some quantity q to each particle, and are satisfied in a reaction if the sum total of q among the reagents is the same as the sum total among the products. An example of a classical additive conservation law is the conservation of electric charge; a well-known new conservation law is the conservation of baryon number. The search for conservation laws has shown how the available evidence can help science find the truth despite concerns about the underdetermination of theory by evidence. A plenitude principle—Gell-Mann's "totalitarian principle"—asserts that the course of inquiry is sufficient to eventually bring into actuality all particle interactions that are physically possible, at least given enough experimental attention and resources. Conversely, if a process fails to be realized experimentally, there should be a conservation law that explains its nonoccurrence. This leads to a principle for inductive inference: When given a list of reaction data, find conservation laws that explain the absence of as many unobserved processes as possible. The question arises whether the laws that physicists actually adopted conform to this principle. With the aid of a computer program, I confirmed that the answer is yes: a standard set of additive conservation laws rules out as many unobserved reactions as possible. However, my analysis also shows that there are infinitely many other sets of predictively equivalent conservation. Thus particle physics faces a problem of global underdetermination: how to select a theory from a set of alternatives that make exactly the same predictions. In the case of conservation laws, it is possible to resolve the underdetermination by connecting laws with ontology: the standard set of laws correspond to a grouping of elementary particles into disjoint categories (i.e., the baryon, muon, electron and tau families). A nominalist might expect that these ontological categories themselves would be underdetermined by the data, but an analysis of the logic of additive conservation laws shows that this is not so: Any set of laws that (1) explains the absence of as many unobserved processes as possible, and (2) employs disjoint ontological categories, must agree with the ontology of the standard set of laws. In this case, two problems are easier than one, because seeking a theory that reconciles ontology and dynamics constrains the alternatives more than considering ontology or dynamics in isolation.

We have encountered a number of standard themes from the philosophy of science in the course of this study: a plenitude principle, global underdetermination, the problem of induction, laws, simplicity and natural kinds. All these issues arise in practice, and physicists deal with them, albeit often implicitly as part of their research work rather than explicitly formulating the challenges as philosophical issues. Sometimes the solution is to appeal to a general principle, such as a plenitude principle, sometimes to pursue the implications of context-specific assumptions, such as the logic of additive conservation principles. It is by studying the interplay of general principles and specific assumptions that we can see how scientists resolve the underdetermination of theory by evidence.

Acknowledgments

This research was supported by grants from the Social Sciences and Humanities Research Council of Canada (SSHRC) and from the Natural Sciences and Engineering Research Council of Canada (NSERC). I am indebted to Matthew Strassler, Manuella Vincter and the late Robert Coleman for helpful discussion of the physics of conservation laws, and to Alex Rueger and Bernard Linsky for comments on the philosophical and methdological issues involved. Mark Drew collaborated on the computational search for the most parsimonious set of conserved quantities. Alexandre Korolev, Leo Chen and Greg Dostatni provided valuable assistance with the particle and reaction databases and with the programs that infer a maximally strict set of conservation laws.

Bibliography

[1] Bilaniuk, O.-M., Sudarshan, G. E. C. Particles beyond the light barrier. *Physics Today*, 22:43–52, 1969.

[2] Cooper, L. *Physics: Structure and Meaning*. Hanover: University Press of New England, 1970.

[3] Daly, C. Natural kinds. In Craig, E. (Ed.), *Routledge Encyclopaedia of Philosophy*. London: Routledge, 1998.

[4] Earman, J. Laws, symmetry, and symmetry breaking: Invariance, conservation principles, and objectivity. *Philosophy of Science*, 71:1227–1241, 2004.

[5] Eidelman, S., The Particle Data Group. The Review of Particle Physics. *Phys. Lett. B*, 592(1), 2004.

[6] Feynman, R. *The Character of Physical Law*. Cambridge, Mass.: MIT Press, 19th edn., 1965. Reprinted in 1990.

[7] Ford, K. W. *The World of Elementary Particles*. New York: Blaisdell, 1963.

[8] Gell-Mann, M., Ne'eman, Y. *The eightfold way*. New York: W.A. Benjamin, 1964.

[9] Goodman, N. *Fact, Fiction and Forecast*. Cambridge, MA: Harvard University Press, 1983.

[10] Jain, S., Osherson, D., Royer, J. S., Sharma, A. *Systems That Learn*. M.I.T. Press, 2 edn., 1999.

[11] Kane, R. Principles of reason. *Erkenntnis*, 24:115–136, 1986.

[12] Kannan, R., Bachem, A. Polynomial algorithms for computing the smith and hermite normal forms of an integer matrix. *SIAM Journal of Computing*, 8(4):499–507, 1979.

[13] Kelly, K. T. *The Logic of Reliable Inquiry*. Oxford: Oxford University Press, 1996.

[14] Kornblith, H. *Inductive Inference and Its Natural Ground*. Cambridge, Mass.: MIT Press, 1993.

[15] Kuhn, T. *The Structure of Scientific Revolutions*. Chicago: University of Chicago Press, 1970.

[16] Lewis, D. *Counterfactuals*. Cambridge, Mass.: Harvard University Press, 1973.

[17] Lewis, D. New work for a theory of universals. *Australasian Journal of Philosophy*, 61(4):343–377, 1983.

[18] Ne'eman, Y., Kirsh, Y. *The Particle Hunters*. Cambridge: Cambridge University Press, 1983.

[19] Omnes, R. *Introduction to Particle Physics*. London, New York: Wiley Interscience, 1971.

[20] Putnam, H. 'degree of confirmation' and inductive logic. In Schilpp, A. (Ed.), *The Philosophy of Rudolf Carnap*. La Salle, Ill.: Open Court, 1963.

[21] Putnam, H. Trial and error predicates and a solution to a problem of mostowksi. *Journal of Symbolic Logic*, 30:49–57, 1965.

[22] Putnam, H. Probability and confirmation. In *Mathematics, Matter and Method, Philosophical Papers*, vol. 1. London: Cambridge University Press, 1975.

[23] Quine, W. V. Natural kinds. In *Ontological Relativity and Other Essays*, 114–138. New York and London: Columbia University Press, 1969.

[24] Reichenbach, H. *The Theory of Probability*. London: Cambridge University Press, 1949.

[25] Salmon, W. Hans reichenbach's vindication of induction. *Erkenntnis*, 35:99–122, 1991.

[26] Schulte, O. Means-ends epistemology. *The British Journal for the Philosophy of Science*, 79(1):141–147, 1996.

[27] Schulte, O. Inferring conservation laws in particle physics: A case study in the problem of induction. *The British Journal for the Philosophy of Science*, 51:771–806, 2000.

[28] Schulte, O. Formal learning theory. In Zalta, E. N. (Ed.), *The Stanford Encyclopedia of Philosophy*. 2006. URL http://plato.stanford.edu/archives/spr2006/entries/learning-formal/.

[29] Schulte, O. Data sets and programs for particle reaction analysis, 2007. URL http://www.cs.sfu.ca/~oschulte/particles/.

[30] Schulte, O. The co-discovery of conservation laws and particle families. *Studies in the History and Philosophy of Modern Physics*, 39(2):288–314, 2008.

[31] Schulte, O., Drew, M. S. Algorithmic derivation of additive selection rules and particle families from reaction data. HEP Preprint Archive, http://arxiv.org/abs/hep-ph/0602011, 2007.

[32] Sklar, L. Methodological conservatism. *Philosophical Review*, LXXXIV:374–400, 1975.

[33] Tipler, P., Llewellyn, R. *Modern Physics*. W. H. Freeman, 4 edn., 2002.

[34] Williams, W. S. C. *Nuclear and Particle Phyics*. New York: Oxford University Press, 1997.

[35] Wolfenstein, L., Trippe, T. G. Tests of conservation laws. In *The Review of Particle Physics* [5].

Redoing the Foundations of Decision Theory [1]

Lawrence E. Blume
Cornell University
leb19@cornell.edu

David A. Easley
Cornell University
dae3@cornell.edu

Joseph Y. Halpern
Cornell University
halpern@cs.cornell.edu

ABSTRACT. In almost all current approaches to decision making, it is assumed that a decision problem is described by a set of states and set of outcomes, and the decision maker (DM) has preferences over a rather rich set of *acts*, which are functions from states to outcomes. However, most interesting decision problems do not come with a state space and an outcome space. Indeed, in complex problems it is often far from clear what the state and outcome spaces would be. We present an alternate foundation for decision making, in which the primitive objects of choice are syntactic *programs*. A program can be given semantics as a function from states to outcomes, but does not necessarily have to be described this way. A representation theorem is proved in the spirit of standard representation theorems, showing that if the DM's preference relation on programs satisfies appropriate axioms, then there exist a set S of states, a set O of outcomes, a way of viewing program as functions from S to O, a probability on S, and a utility function on O, such that the DM prefers program a to program b if and only if the expected utility of a is higher than that of b. Thus, the state space and outcome space are subjective, just like the probability and utility; they are not part of the description of the problem. A number of benefits of this approach are discussed.

1 Introduction

In almost all current approaches to decision making under uncertainty, it is assumed that a decision problem is described by a set of states and set of outcomes, and the decision maker (DM) has a preference relation on a rather rich set of *acts*, which are functions from states to outcomes. The standard representation theorems of decision theory give conditions under which the preference relation can be represented by a utility function on outcomes and numerical representation of beliefs on states. For example, Savage [18] shows that if a DM's preference order satisfies certain axioms, then the DM's preference relation can be represented by a probability Pr on the state space and a utility function mapping outcomes to the reals such that she prefers act a to act b iff the expected utility of a (with

---

[1] This is a slight variant of a paper that originally appeared in the *Proceedings of the Tenth International Conference on Principles of Knowledge Representation and Reasoning (KR 2006)*, 2006, pp. 14–24. Copyright © 2006, American Association for Artificial Intelligence (www.aaai.org). All rights reserved. The paper is supported in part by NSF under grants CTC-0208535, ITR-0325453, and IIS-0534064, by ONR under grants N00014-00-1-03-41 and N00014-01-10-511, and by the DoD Multidisciplinary University Research Initiative (MURI) program administered by the ONR under grant N00014-01-1-0795.

respect to Pr) is greater than that of b. Moreover, the probability measure is unique and the utility function is unique up to affine transformations. Similar representations of preference can be given with respect to other representations of uncertainty (see, for example, [6, 19]).

Most interesting decision problems do not come with a state space and acts specified as functions on these states. Instead they are typically problems of the sort "Should I buy 100 shares of IBM or leave the money in the bank?" or "Should I attack Iraq or continue to negotiate?". To apply standard decision theory, the DM must first formulate the problem in terms of states and outcomes. But in complex decision problems, the state space and outcome space are often difficult to formulate. For example, what is the state space and outcome space in trying to decide whether to attack Iraq? And even if a DM could formulate the problem in terms of states and outcome, there is no reason to believe that someone else trying to model the problem would do it in the same way. For example, reasonable people might disagree about what facts of the world are relevant to the pricing of IBM stock. As is well known [12], preferences can be quite sensitive to the exact formulation. To make matters worse, a modeler may have access to information not available to the DM, and therefore incorrectly construe the decision problem from the DM's point of view.

Case-based decision theory (CBDT) [7] deals with this problem by dispensing with the state space altogether and considering instead *cases*, which are triples of the form (p, a, r), where p is a problem, a is an action and r is a result. Actions in the CBDT framework are the analogue of acts, but rather than being functions from states to outcomes, actions are primitive objects in CBDT, and are typically described in English.

We take a different view of acts here. choose among some simple actions: "do x" or "do y". But they also perform various tests on the world and make choices contingent upon the outcome of these tests: "If the stock broker says t, do x; otherwise do y." We formalize this view by taking the objects of choice to be (syntactic) programs in a programming language. We then show that if the DM's preference relation on programs satisfies appropriate axioms, we, the modelers, can impute a state space, an outcome space, and an interpretation of programs as functions from states to outcomes such that the (induced) preferences on these functions has a subjective expected utility (SEU) representation, similar in spirit to that of Savage. Just as probability and utility are derived notions in the standard approach, and are tools that we can use to analyze and predict decisions, so too are the state space and outcome space in our framework.

This formulation of decision problems has several advantages over more traditional formulations. First, we can theorize (if we want) about only the actual observable choices available to the DM, without having states and outcomes, and without needing to view acts as functions from states to outcomes. Indeed, in the full paper, we show that we can test whether a DM's behavior is consistent with SEU despite not having states and outcomes. The second advantage is more subtle but potentially quite profound. Representation theorems are just that; they merely provide an alternative description of a preference order in terms of numerical scales. Decision theorists make no pretense that these representations have anything to do with the cognitive processes by which individuals make choices. But to the extent that the programming language models the language of the DM, we have the ability to interpret the effects of cognitive limitations having to do with the language in terms of the representation. For instance, there may be limitations on the space of acts because some sequence of tests is too computationally costly to verify. We can also take into account a DM's inability to recognize that two programs represent the same function. Finally, the approach lets us take into account the fact that different DMs use different languages to describe the same phenomena.

To understand where the state space and outcome space are coming from, first note that in our framework we need to ask two basic questions: (1) what is the programming language and (2) what is the semantics of a program; that is, what does a program *mean*. In this paper, to introduce the basic framework, we focus on a rather simple programming language, where the only construct is **if** ... **then** ... **else**. That is, if a and b are programs and t is a tests, then **if** t **then** a **else** b is also a program. For example, we could have a program such as **if** the moon is in the seventh house **then** buy 100 shares of IBM **else** sell 100 shares of Microsoft. Notice that once we consider **if** ... **then** ... **else** programs, we also need to define a language of *tests*. For some of our results we also allow a randomization construct: if a_1 and a_2 are programs, then so is $ra_1 + (1-r)a_2$. Intuitively, according to this program, the DM tosses a coin which lands heads with probability r. If it lands heads, the DM performs a_1; if it lands tails, the DM performs a_2. While people do not typically seem to use randomization, adding it allows us to connect our results to work in the more standard setting with states and outcomes that uses randomization, such as that of Anscombe and Aumann [1].

There are many different ways to give semantics to programs (see, for example, [20]). Here we consider what is called *input-output semantics*: a program is interpreted as a *Savage act*, i.e., a function from states to outcomes. That is, given a state space S and a outcome space O, there is a function ρ_{SO} that associates with a program a a function $\rho_{SO}(a)$ from S to O. We prove Savage-like representation theorems in this setting. We show that if the DM's preference order on programs satisfies certain axioms, then there is a state space S, an outcome space O, a probability distribution on S, a utility function on O, and a function ρ_{SO} mapping programs a to functions $\rho(a)$ from S to O such that the DM prefers act a to act b iff the expected utility of $\rho_{SO}(a)$ is at least as high as that of $\rho_{SO}(b)$. Again, we stress that the state space S and the outcome space O are subjective, just like the probability and utility; ρ_{SO} depends on S and O. If in our representation we choose a different state space S' and/or outcome space O', there is a correspondingly different function $\rho_{S'O'}$.

Besides proving that a DM's preferences can be represented by a probability and utility, Savage proves that the probability is unique and the utility is unique up to affine transformations. We cannot hope to prove such a strong uniqueness result, since for us the state space and outcome space is subjective and certainly not unique. We can show that, for the language with randomization, if acts are totally ordered (an assumption that Savage makes but we do *not* make in general), the expected utility of acts is unique up to affine transformations. However, without randomization, as we show by example, we cannot hope to get a uniqueness result. The set of acts we consider is finite, and is simply not rich enough to determine the expected utility, even up to affine transformations. To get uniqueness in the spirit of Savage, we seem to require not only randomization but a fixed outcome space.

The rest of this paper is organized as follows. In Section 2, we give the syntax and semantics of the simple programming languages we consider here, and define the notion of program equivalence. In Section 3, we state our representation theorems and discuss the (very few) postulates we needed to prove them. The key postulate turns out to be an analogue of the *cancellation* axiom [13]. We conclude in Section 4. Most proofs are left to the full paper.

2 The Programming Language

Before we describe programs, it is helpful to define the language of *tests*. For simplicity, we consider a simple propositional logic of tests. We start with some set T_0 of *primitive tests*. We can think of these primitive propositions as representing statements such as "the price/earnings ratio of IBM is 5", "the moon is in the seventh house" or "the economy will be strong next year". We then close off under conjunction and negation. Thus, if t_1 and t_2 are tests, then so are $t_1 \wedge t_2$ and $\neg t_1$. Let T be the set of tests that can be formed from the primitive tests in T_0 in this way. The language of tests is thus just basic propositional logic. Tests can be viewed as statements that the DM considers relevant to her decision problem. They are a part of her specification of the problem, just as states are part of the specification in the Savage and Anscombe-Aumann frameworks.

We can now describe two programming languages. In both cases we start with a set A_0 of primitive programs. We can think of the programs in $\mathcal{A}$, as representing such primitive actions as "buy 100 shares of IBM" or "attack Iraq". (Of course, it is up to the DM to decide what counts as primitive, both for tests and programs. It may well be that "attack Iraq" is rather complex, formed from much simpler actions.) For the first language, we simply close off the primitive programs in $\mathcal{A}$, under **if** ... **then** ... **else**. Thus, if a_1 and a_2 are programs and t is a test, then **if** t **then** a_1 **else** a_2 is a program. Let $\mathcal{A}_{\mathcal{A}_,,\mathcal{T},}$ consist of all programs that can be formed in this way. (We omit the subscripts $\mathcal{A}_,$ and T_0 if they are either clear from context or not relevant to the discussion.) Note that $\mathcal{A}$ allows nesting, so that we can have a program such as **if** t_1 **then** a_1 **else** (**if** t_2 **then** a_2 **else** a_3).

For the second language, we close off the primitive programs in $\mathcal{A}$, under both **if** ... **then** ... **else** and randomization, so that if a_1 and a_2 are programs and $r \in [0,1]$, then $ra_1 + (1-r)a_2$ is a program. We allow arbitrary nesting of randomization and **if** ... **then** ... **else**. Let $\mathcal{A}^+_{\mathcal{A}_,,\mathcal{T},}$ consist of all programs that can be formed in this way. Again, we omit subscripts when doing so does not result in loss of clarity.

We next give semantics to these languages. That is, given a state space S and an outcome space O, we associate with each program a function from S to O. The first step is to give semantics to the tests. Let π_S^0 be a *test interpretation*, that is, a function associating with each primitive test a subset of S. Intuitively, $\pi_S^0(t)$ is the set of states where t is true. We then extend π_S^0 in the obvious way to a function $\pi_S : T \to 2^S$ by induction on structure:

- $\pi_S(t_1 \wedge t_2) = \pi_S(t_1) \cap \pi_S(t_2)$
- $\pi_S(\neg t) = S - \pi_S(t)$.

A *program interpretation* assigns to each program a a (Savage) act, that is, a function $f_a : S \to O$. Let $\rho_{SO}^0 : A_0 \to O^S$ be a test interpretation for primitive programs, which assigns to each $a_o \in A$ a function from $S \to O$. We extend ρ_{SO}^0 to a function mapping all programs in A to functions from S to O by induction on structure, by defining

$$\rho_{SO}(\text{if } t \text{ then } a_1 \text{ else } a_2)(s) = \begin{cases} \rho_{SO}(a_1)(s) & \text{if } s \in \pi_S(t) \\ \rho_{SO}(a_2)(s) & \text{if } s \notin \pi_S(t). \end{cases} \quad (1)$$

To give semantics to the language $\mathcal{A}^+$, given S, O, and π_S, we want to associate with each act a a function from S to probability measures on O. Let $\Delta(O)$ denote the set of probability measures on O. Now, given a function $\rho_{SO}^0 : \mathcal{A}_I \to \cdot(\mathcal{O})^S$, we can extend it by induction on structure to all of $\mathcal{A}^+$ in the obvious way.[2] For **if** ... **then** ... **else** programs we use (1); to deal with randomization, define

$$\rho_{SO}(ra_1 + (1-r)a_2))(s) = r\rho_{SO}(a_1)(s) + (1-r)\rho_{SO}(a_2)(s).$$

That is, the distribution $\rho_{SO}(ra_1 + (1-r)a_2))(s)$ is the obvious convex combination of the distributions $\rho_{SO}(a_1)(s)$ and $\rho_{SO}(a_2)(s)$.

Note that tests and programs are amenable to two different interpretations. First, programs may not necessarily be implementable (just as acts may not be implementable in the standard Savage setting). The DM may not be able to run a primitive program and may not be able to tell whether a test in a program is actually true in a given setting. However, as long as the DM in some sense "understands" the tests and programs, she might be able to determine her preferences among them. A second interpretation is that the DM has preferences only on implementable programs, that is, ones where all primitive programs can be run and the DM can determine the truth of all tests involved. Our results hold with either interpretation.

3 The Representation Theorems

We assume that the decision maker actually has two preference orders, $\succeq$ and $\succ$, on choices. Intuitively, $a \succeq b$ means that a is at least as good as b from the DM's point of view; $a \succ b$ means that a is strictly better than b. If $\succeq$ is a total order, then $\succ$ is definable in terms of $\succeq$: $a \succ b$ iff $a \succeq b$ and $b \not\succeq a$. However, in the case of partial orders (which we allow here), $\succ$ is not necessarily definable in terms of $\succeq$. Indeed while it could be the case that $a \succeq b$ and $b \not\succeq a$ implies $a \succ b$, it could also happen that the DM could say, "a is surely at least as good as b to me, but I cannot distinguish whether it is better than b, or merely as good." In this case, $a \succeq b$, $b \not\succeq a$, and $a \not\succ b$. As usual, we write $a \sim b$ if $a \succeq b$ and $b \succeq a$.

We prove various representation theorems, depending on whether there is randomization in the language, and whether the set of outcomes is taken to be fixed. One assumption we will need in every language is the irreflexivity of $\succ$.

A1. $a \not\succ a$ for all a.

---
[2] We could, as before, take ρ_{SO}^0 to be a function from $\mathcal{A}$ to O^S, rather than from $\mathcal{A}$ to $\Delta(O)^S$. This would not affect our results. However, if we view general programs as functions from S to distributions over O, it seems reasonable to view primitive programs in this way as well.

For some of our results, we also assume that the preference relation is complete. In our setting, this is expressed as follows.

A2. For all a and b, either $a \succ b$ or $b \succeq a$.

A2 says that all alternatives are ranked, and if the DM is not indifferent between two alternatives, then one must be strictly preferred to the other.

The engine of our analysis is the *cancellation* postulate. Although simple versions of it have appeared in the literature (e.g. [13]) it is nonetheless not well known, and so before turning to our framework we briefly explore some of its implications in more familiar settings.

3.1 The Cancellation Postulate for Choices, Savage Acts, and AA Acts

In describing the cancellation postulate, we use the notion of a *multiset*. A multiset is just a set with repetitions allowed. Thus, $\{\{a\}\}$, $\{\{a, a\}\}$, and $\{\{a, a, a\}\}$ are distinct multisets. (Note that we use the $\{\{\cdot\}\}$ to denote multisets.) Two multisets are equal if they have the same elements with the same number of occurrences. Thus $\{\{a, a, b, b, b\}\} = \{\{b, a, b, a, b\}\}$, but $\{\{a, b, a\}\} \neq \{\{a, b, \}\}$. Let C be a set of choices (not necessarily programs) among which the DM has to choose, and let $(\succeq, \succ)$ be a pair of preference relations on C.

> **Cancellation on C:** If $\langle a_1, \ldots, a_n \rangle$ and $\langle b_1, \ldots, b_n \rangle$ are sequences of elements of C such that $\{\{a_1, \ldots a_n\}\} = \{\{b_1, \ldots, b_n\}\}$, then
>
> 1. If $a_i \succeq b_i$ or $a_i \succ b_i$ for $i \leq n-1$, then $b_n \succeq a_n$; and
> 2. if, in addition, $a_k \succ b_k$ for some $k \leq n-1$, then $b_n \succ a_n$.

Roughly speaking, cancellation says that if two collections of choices are identical, then it is impossible to prefer each choice in the first collection to the corresponding choice in the second collection. In this setting, cancellation is essentially equivalent to reflexivity and transitivity.

Proposition 1 *A pair $(\succeq, \succ)$ of preference relations on a choice set C satisfies cancellation iff*

(a) $\succeq$ *is reflexive,*

(b) $\succ \subset \succeq$, *and*

(c) $(\succeq, \succ)$ *is transitive; that is, if $a \succeq b$ and $b \succeq c$, then $a \succeq c$, and if either $a \succ b$ or $b \succ c$, then $a \succ c$.*

Proof First suppose that cancellation holds. To see that $\succeq$ is reflexive, take $A = B = \{\{a\}\}$. The hypothesis of the cancellatiom axiom holds, so we must have $a \succeq a$. To see that $\succ$ is a subset of $\succeq$, suppose that $a \succ b$. Take $a_1 = b_2 = a$, $a_2 = b_1 = b, A = \{\{a_1, a_2\}\}$, and $B = \{\{b_1, b_2\}\}$. Since $a \succ b$, by cancellation, we must have $b \succeq a$. Thus, $\succ \subseteq \succeq$. Finally, to see that cancellation implies transitivity, consider the pair of multisets $\{\{a, b, c\}\}$ and $\{\{b, c, a\}\}$. If $a \succeq b$ and $b \succeq c$, then cancellation implies $a \succeq c$, and if one of the first two relations is strict, then $a \succ c$. We defer the proof of the converse to the full paper. ♠

In the Savage framework we strengthen the cancellation postulate:

Cancellation for Savage acts: If $\langle a_1, \ldots, a_n \rangle$ and $\langle b_1, \ldots, b_n \rangle$ are two sequences of Savage acts defined on a state space S such that for each state $s \in S$, $\{\{a_1(s), \ldots, a_n(s)\}\} = \{\{b_1(s), \ldots, b_n(s)\}\}$, then

1. if $a_i \succeq b_i$ or $a_i \succ b_i$ for $i \leq n-1$, then $b_n \succeq a_n$; and
2. if, in addition, $a_i \succ b_i$ for some $i \leq n-1$, then $b_n \succ a_n$.

Cancellation for Savage acts is a powerful assumption because equality of the multisets is required only "pointwise". Were we to require the equality of multisets for sequences of functions, then the characterization of Proposition 1 would apply. But we require only that for each state s, the multisets of outcomes generated by the two sequence of acts are equal; there need be no function in the second sequence equal to any function in the first sequence. Nevertheless, the conclusion still seems reasonable, because the two collections of acts deliver the same bundle of outcomes in each state.

In addition to the conditions in Proposition 1, Savage cancellation directly implies *event independence*, a condition at the heart of all representation theorems (and can be used to derive the Sure Thing Principle). If $T \subseteq S$, let $a_T b$ be the Savage act that agrees with a on T and with b on $S - T$; that is $a_T b(s) = a(s)$ if $s \in T$ and $a_T b(s) = b(s)$ if $s \notin T$. We say that $(\succeq, \succ)$ satisfies event independence if for all acts a, b, c, and c' and subsets T of the state space S, if $a_T c \succeq b_T c$, then $a_T c' \succeq b_T c'$, and similarly with $\succeq$ replaced by $\succ$.

Proposition 2 *If $(\succeq, \succ)$ satisfies the cancellation postulate for Savage acts then $(\succeq, \succ)$ satisfies event independence.*

Proof Take
$$\langle a_1, a_2 \rangle = \langle a_T c, b_T c' \rangle$$
and take
$$\langle b_1, b_2 \rangle = \langle b_T c, a_T c' \rangle.$$
Note that for each state $s \in T$,
$$\{\{a_T c(s), b_T c'(s)\}\} = \{\{a(s), b(s)\}\} = \{\{b_T c(s), a_T c'(s)\}\},$$
and for each state $s \notin T$,
$$\{\{a_T c(s), b_T c'(s)\}\} = \{\{c(s), c'(s)\}\} = \{\{b_T c(s), a_T c'(s)\}\}.$$
Thus, again we can apply cancellation. ♠

Proposition 1 provides an axiomatic characterization of cancellation for choices. Is there a similar characterization for cancellation for Savage acts? For example, is cancellation equivalent to the combination of irreflexivity of $\succ$, the fact that $\succ$ is a subset of $\succeq$, the transitivity of $\succ$ and $\succeq$, and event independence? As the following example shows, it is not.

Example 1 *Suppose that $S = \{s_1, s_2\}$, $O = \{o_1, o_2, o_3\}$. Let (o, o') be an abbreviation for the Savage act a such that $a(s_1) = o$ and $a(s_2) = o'$. Clearly there are nine possible acts. Suppose that $\succ$ is the total order on these acts characterized by*
$$(o_1, o_1) \succ (o_1, o_2) \succ (o_2, o_1) \succ (o_2, o_2) \succ (o_3, o_1)$$
$$\succ (o_1, o_3) \succ (o_2, o_3) \succ (o_3, o_2) \succ (o_3, o_3);$$

let $\succeq$ be the reflexive closure of $\succ$. By construction, $\succ$ is irreflexive, $\succ$ is a subset of $\succeq$, $\succ$ and $\succeq$ are transitive. To see that $(\succeq, \succ)$ satisfies event independence, note that

- $(x, o_1) \succ (x, o_2) \succ (x, o_3)$ for $x \in \{o_1, o_2, o_3\}$;
- $(o_1, y) \succ (o_2, y) \succ (o_3, y)$ for $y \in \{o_1, o_2, o_3\}$.

However, cancellation for Savage acts does not hold, since $(o_1, o_2) \succ (o_2, o_1)$, $(o_2, o_3) \succ (o_3, o_2)$, and $(o_3, o_1) \succ (o_1, o_3)$.

We do not know if cancellation for Savage acts has a simple characterization. However, once we allow randomization, we can get an elegant characterization of cancellation. As usual, define an *AA act* (for Anscombe-Aumann) to be a function from states to lotteries over outcomes. We can define an analogue of cancellation for AA acts. Note that since an AA act is a function from states to lotteries over outcomes, if a and b are acts, $a + b$ is a well-defined function on states: $(a + b)(s)(o) = a(s)(o) + b(s)(o)$. While $(a + b)(s)$ is not a lottery on outcomes (since its range is $[0, 2]$, not $[0, 1]$), it can be viewed as an "unnormalized lottery".

Cancellation for AA acts: If $\langle a_1, \ldots, a_n \rangle$ and $\langle b_1, \ldots, b_n \rangle$ are two sequences of AA acts such that $\sum_{i=1}^n a_i = \sum_{i=1}^n b_i$, then

1. if $a_i \succeq b_i$ or $a_i \succ b_i$ for $i \leq n - 1$, then $b_n \succeq a_n$; and
2. if, in addition, $a_i \succ b_i$ for some $i \leq n - 1$, then $b_n \succ a_n$.

Note that cancellation for AA acts can be viewed as a generalization of cancellation for Savage acts. In the case of Savage acts, the probabilities are just point masses, so the fact that $(\sum_{i=1}^n a_i)(s) = (\sum_{i=1}^n b_i)(s)$ for all states s says that the multisets of outcomes must be equal for all states s.

For AA acts, there is a standard probabilistic independence axiom. We say that the preference orders $(\succeq, \succ)$ satisfy *AA act independence* if for all AA acts a, b, and c, and all $r \in (0, 1]$, we have $a \succeq b$ iff $ra + (1 - r)c \succeq rb + (1 - r)c$ and $a \succ b$ iff $ra + (1 - r)c \succ rb + (1 - r)c$. We say that $(\succeq, \succ)$ satisfies *rational AA act independence* if they satisfy AA act independence for all rational $r \in (0, 1]$.

Proposition 3 *If $(\succeq, \succ)$ satisfies $\mathbf{A2}$ and cancellation for AA acts, then $(\succeq, \succ)$ satisfies rational AA act independence.*

Proof Suppose that $a \succeq b$ and $r = m/n$. Let $a_1 = \cdots = a_m = a$ and $a_{m+1} = \cdots = a_{m+n} = rb + (1-r)c$; let $b_1 = \cdots = b_m = b$ and $b_{m+1} = \cdots = b_{m+n} = ra + (1-r)c$. It is easy to check that $\sum_{i=1}^{m+n} a_i = \sum_{i=1}^{m+n} b_i$. If $rb + (1-r)c \succ ra + (1-r)c$, then we get a contradiction to cancellation for AA actions. Thus, by $\mathbf{A2}$, we must have $ra + (1-r)c \succeq rb + (1-r)c$. The same collection of programs can be used for the converse implication as well as for the result with $\succ$. ♠

We can now characterize cancellation for AA acts for total orders.

Theorem 1 *If $(\succeq, \succ)$ is a pair of preference relations satisfies $\mathbf{A2}$, then $(\succeq, \succ)$ satisfies cancellation for AA acts iff $\succ$ is irreflexive, $\succ \subseteq \succeq$, $(\succeq, \succ)$ is transitive, and $(\succeq, \succ)$ satisfies rational AA act independence.*

Proof The fact that extended cancellation implies that $(\succeq, \succ)$ has the required properties follows from Propositions 1, and 3. For the converse, suppose that $\langle a_1, \ldots, a_n \rangle$ and $\langle b_1, \ldots, b_n \rangle$ are sequences of AA acts such that $a_1 + \cdots + a_n = b_1 + \cdots + b_n$ and $a_i \succeq b_i$ for $i = 1, \ldots, n-1$. By way of contradiction, suppose that $a_1 \succ b_n$. Let $c = \frac{1}{n-1}(a_2 + \cdots + a_n)$. Since $a_1 \succeq b_1$, by rational AA act independence, we get that

$$\frac{1}{n}(a_1 + \cdots + a_n) = \frac{1}{n}a_1 + \frac{n-1}{n}c \succeq \frac{1}{n}b_1 + \frac{n-1}{n}c$$

$$= \frac{1}{n}(b_1 + a_2 + \cdots + a_n).$$

By induction (using transitivity) and the fact that $a_n \succ b_n$, it follows that $\frac{1}{n}(a_1 + \cdots + a_n) \succ \frac{1}{n}(b_1 + \cdots + b_{n-1} + b_n)$. But this contradicts the assumption that $a_1 + \cdots + a_n = b_1 + \cdots + b_n$. Since it is not the case that $a_n \succ b_n$, by **A2**, we must have $b_n \succeq a_n$, as desired. If $a_i \succ b_i$ for some $i \in \{1, \ldots, n-k\}$, a similar argument shows that $b_n \succ a_n$. ♠

It seems that **A2** plays a critical role in the proof of Proposition 3 and hence Theorem 1. It turns out that by strengthening cancellation appropriately, we can avoid this use of **A2**. We first state the strengthened cancellation postulate for Savage acts, since we shall need it later, and then for AA acts.

> **Extended cancellation for Savage acts:** If $\langle a_1, \ldots, a_n \rangle$ and $\langle b_1, \ldots, b_n \rangle$ are two sequences of Savage acts defined on a state space S such that for each state $s \in S$ we have $\{\{a_1(s), \ldots, a_n(s)\}\} = \{\{b_1(s), \ldots, b_n(s)\}\}$, then
>
> 1. if there exists some $k < n$ such that $a_i \succeq b_i$ or $a_i \succ b_i$ for $i \leq k$, $a_{k+1} = \ldots = a_n$, and $b_{k+1} = \ldots = b_n$, then $b_n \succeq a_n$; and
> 2. if, in addition, $a_i \succ b_i$ for some $i \leq k$, then $b_n \succ a_n$.

Clearly, cancellation for Savage acts is just the special case of extended cancellation where $k = n - 1$. The intuition behind the extended cancellation postulate is identical to that for the basic cancellation postulate; moreover, it is easy to see that the existence of an SEU representation for $\succeq$ implies extended cancellation. There is an obvious analogue of extended cancellation for AA acts:

> **Extended cancellation for AA acts:** If $\langle a_1, \ldots, a_n \rangle$ and $\langle b_1, \ldots, b_n \rangle$ are two sequences of AA acts such that $\sum_{i=1}^{n} a_i = \sum_{i=1}^{n} b_i$, then
>
> 1. if there exists some $k < n$ such that $a_i \succeq b_i$ or $a_i \succ b_i$ for $i \leq k$, $a_{k+1} = \ldots = a_n$, and $b_{k+1} = \ldots = b_n$, then $b_n \succeq a_n$; and
> 2. if, in addition, $a_i \succ b_i$ for some $i \leq k$, then $b_n \succ a_n$.

Extended cancellation is just what we need to remove the need for **A2** in Proposition 3.

Proposition 4 *If $(\succeq, \succ)$ satisfies extended cancellation for AA acts, then $(\succeq, \succ)$ satisfies rational AA act independence.*

Proof The proof is identical to that of Proposition 3, except that we can show that $ra+(1-r)c \succeq rb+(1-r)c$ immediately using extended cancellation, without invoking **A2**. We leave details to the reader. ♠

We can now get the desired characterization of cancellation for AA acts.

Theorem 2 $(\succeq, \succ)$ *satisfies extended cancellation for AA acts iff $\succ$ is irreflexive, $\succ \subseteq \succeq$, $(\succeq, \succ)$ is transitive, and $(\succeq, \succ)$ satisfies rational AA act independence.*

Proof The "if" direction is immediate from Propositions 1, and 4. For the converse, we proceed much as in the proof of Theorem 1. Suppose that $\langle a_1, \ldots, a_n \rangle$ and $\langle b_1, \ldots, b_n \rangle$ are sequences of AA acts such that $a_1 + \cdots + a_n = b_1 + \cdots + b_n$, $a_i \succeq b_i$ for $i = 1, \ldots, n - k$, $a_{k+1} = \ldots = a_n$, and $b_{k+1} = \ldots = b_n$. Then from transitivity and rational AA act independence we get that

$$\frac{1}{n}(a_1 + \cdots + a_n) \succeq \frac{1}{n}(b_1 + \cdots + b_k + a_{k+1} + \cdots + a_n)$$
$$= \frac{1}{n}(b_1 + \cdots + b_k) + \frac{n-k}{n}a_n.$$

Since $b_{k+1} = \ldots = b_n$ and $a_1 + \cdots + a_n = b_1 + \cdots + b_n$, we have that

$$\frac{1}{n}(b_1 + \cdots + b_k) + \frac{n-k}{n}(b_n) = \frac{1}{n}(b_1 + \cdots + b_n)$$
$$= \frac{1}{n}(a_1 + \cdots + a_n).$$

Thus, by transitivity,

$$\frac{1}{n}(b_1 + \cdots + b_k) + \frac{n-k}{n}(b_n) \succeq \frac{1}{n}(b_1 + \cdots + b_k) + \frac{n-k}{n}(a_n).$$

By rational AA act independence, it follows that $a_n \succeq b_n$. Moreover, the same argument shows that $a_n \succ b_n$ if $a_i \succ b_i$ for some $i \leq k$. ♠

3.2 The Cancellation Postulate for Programs

We use cancellation to get a representation theorem for preference orders on programs. However, the definition of the cancellation postulates for Savage acts and AA acts make heavy use of states. We now show how we can get an analogue of this postulate for programs.

Definition 1 *Given a set $T_0 = \{t_1, \ldots, t_n\}$ of primitive tests, an atom over T_0 is a hypothesis of the form $t'_1 \wedge \ldots \wedge t'_n$, where t'_i is either t_i or $\neg t_i$.*

An atom δ can be identified with the truth assignment v_δ to the primitive tests such that $v_\delta(t_i) = \text{true}$ iff t_i appears unnegated in δ. If there are n primitive tests in T_0, there are 2^n atoms. Let $At(T_0)$ denote the set of atoms over T_0. It is easy to see that, for all tests $t \in T$ and atoms $\delta \in Atoms(T_0)$, either $\pi_S(\delta) \subseteq \pi_S(t)$ for all state spaces S and interpretations π_S or $\pi_S(\delta) \cap \pi_S(t) = \emptyset$ for all state spaces

S and interpretations π_S. (The formal proof is by induction on the structure of t.) We write $\delta \Rightarrow t$ if the former is the case.

A program in $\mathcal{A}$ can be identified with a function from atoms to primitive programs in an obvious way. For example, if a_1, a_2, and a_3 are primitive programs and $T_0 = \{t_1, t_2\}$, then the program $a = $ **if** t_1 **then** a_1 **else** (**if** t_2 **then** a_2 **else** a_3) can be identified with the function f_a such that

- $f_a(t_1 \wedge t_2) = f_a(t_1 \wedge \neg t_2) = a_1$;
- $f_a(\neg t_1 \wedge t_2) = a_2$; and
- $f_a(\neg t_1 \wedge \neg t_2) = a_3$.

Formally, we define f_a by induction on the structure of programs. If $a \in \mathcal{A}_I$, then f_a is the constant function a, and

$$f_{\text{if } t \text{ then } a \text{ else } b}(\delta) = \begin{cases} f_a(\delta) & \text{if } \delta \Rightarrow t \\ f_b(\delta) & \text{otherwise.} \end{cases}$$

The cancellation postulate that we use for the language $\mathcal{A}_I$ is simply extended cancellation for Savage acts, with atoms playing the role of states:

A3. If $\langle a_1, \ldots, a_n \rangle$ and $\langle b_1, \ldots, b_n \rangle$ are two sequences of programs in $\mathcal{A}_{\mathcal{A}_I, \mathcal{T}_I}$, and for each atom δ over T_0,

$$\{\{f_{a_1}(\delta), \ldots, f_{a_n}(\delta)\}\} = \{\{f_{b_1}(\delta), \ldots, f_{b_n}(\delta)\}\},$$

then

1. if there exists some $k < n$ such that $a_i \succeq b_i$ or $a_i \succ b_i$ for $i \leq k$, $a_{k+1} = \ldots = a_n$, and $b_{k+1} = \ldots = b_n$, then $b_n \succeq a_n$; and
2. if, in addition, $a_i \succ b_i$ for some $i \leq k$, then $b_n \succ a_n$.

Of course, we can prove analogues of Propositions 1 and 2 using **A3**. **A3** has another consequence when choices are programs: a DM must be indifferent between equivalent programs, where programs a and b are *equivalent* if, no matter what interpretation is used, they are interpreted as the same function. For example, **if** t **then** a **else** b is equivalent to **if** $\neg t$ **then** b **else** a; no matter what the test t and programs a and b are, these two programs have the same input-output semantics. Similarly, if t and t' are equivalent tests, then **if** t **then** a **else** b is equivalent to **if** t' **then** a **else** b.

Definition 2 *Programs a and b are* equivalent, *denoted $a \equiv b$, if, for all S, O, π_S, and ρ_{SO}, we have $\rho_{SO}(a) = \rho_{SO}(b)$.*

The general problem of checking whether two programs are equivalent is at least as hard as checking whether two propositional formulas are equivalent, and so is co-NP-hard. It is not hard to show that it is in fact co-NP-complete. Nevertheless, it is a consequence of cancellation that a DM must be indifferent between two equivalent programs.

Proposition 5 *Suppose that $(\succeq, \succ)$ satisfies* **A3**. *Then $a \equiv b$ implies $a \sim b$.*

Proof It is easy to check that if $a \equiv b$ then $f_a = f_b$. Let S be $At(T_0)$, the set of atoms, let O be $\mathcal{A}_I$, the set of primitive programs, and define $\rho_{SO}^0(c)$ to be the constant function c for a primitive program c. It is easy to see that $\rho_{SO}(c) = f_c$ for all programs c. Since if $a \equiv b$, then $\rho_{SO}(a) = \rho_{SO}(b)$, we must have $f_a = f_b$. Now apply **A3** with $a_1 = a$ and $b_1 = b$ to get $b \succeq a$, and then reverse the roles of a and b. ♠

To get a representation theorem for $\mathcal{A}^+$, we use the cancellation postulate for AA acts, again replacing states by atoms. The idea now is that we can identify each program a with a function f_a mapping atoms into distributions over primitive programs. For example, if t is the only test, then the program $a = \frac{1}{2}a_1 + \frac{1}{2}(\text{if } t \text{ then } a_2 \text{ else } a_3)$ can be identified with the function f_a such that

- $f_a(t)(a_1) = 1/2$; $f_a(t)(a_2) = 1/2$
- $f_a(\neg t)(a_1) = 1/2$; $f_a(\neg t)(a_2) = 1/2$.

Formally, we just extend the definition of f_a given in the previous section by defining
$$f_{ra_1+(1-r)a_2}(t) = rf_{a_1}(t) + (1-r)f_{a_2}(t).$$
We then get the obvious analogue of extended cancellation for AA acts:

A3′. If $\langle a_1, \ldots, a_n \rangle$ and $\langle b_1, \ldots, b_n \rangle$ are two sequences of acts in $\mathcal{A}^+_{\mathcal{A}_I,\mathcal{T}}$ such that $f_{a_1} + \ldots + f_{a_n} = f_{b_1} + \ldots + f_{b_n}$, then if $a_i \succeq b_i$ for $i = 1, \ldots, k$, where $k < n$, and $a_{k+1} = \ldots = a_n$, and $b_{k+1} = \ldots = b_n$, then $b_n \succeq a_n$. Moreover, if $a_i \succ b_i$ for some $i \in \{1, \ldots, n-k\}$, then $a_n \succ b_n$.

Again, **A3′** can be viewed as a generalization of **A3**.

Theorem 2 shows that we can replace **A3′** by **A1** and postulates that say that (a) $\succeq \subseteq \succ$, (b) $(\succeq, \succ)$ is transitive, (c) $(\succeq, \succ)$ satisfies rational AA act independence, and (d) $a \equiv b$ implies $a \sim b$.

3.3 A Representation Theorem for $\mathcal{A}$

We are now ready to state our first representation theorem. In the theorem, if f is a Savage act mapping states S to outcomes O and u is a utility function on O (so that $u : O \to \mathbb{R}$), then $u_f : S \to \mathbb{R}$ is the function defined by taking $u_f(s) = u(f(s))$. If Pr is a probability measure on S, then $E_{\text{Pr}}(u_f)$ is the expectation of u_f with respect to Pr.

Theorem 3 *If $(\succeq, \succ)$ are preference orders on acts in $\mathcal{A}_{\mathcal{A}_I,\mathcal{T}}$ satisfying **A1** and **A3**, then there exist a finite set S of states, a set $\mathcal{P}$ of probability measures on S, a finite set O of outcomes, a set $\mathcal{U}$ of utility functions on O, a set $\mathcal{V} \subseteq \mathcal{P} \times \mathcal{U}$, a test interpretation π_S^0, and a program interpretation ρ_{SO}^0 such that $a \succeq b$ iff $E_{\text{Pr}}(u_{\rho_{SO}(a)}) \leq E_{\text{Pr}}(u_{\rho_{SO}(b)})$ for all $(\text{Pr}, u) \in \mathcal{V}$ and $a \succ b$ iff for $E_{\text{Pr}}(u_{\rho_{SO}(a)}) < E_{\text{Pr}}(u_{\rho_{SO}(b)})$ for all $(\text{Pr}, u) \in \mathcal{V}$. Either $\mathcal{P}$ or $\mathcal{V}$ can be taken to be a singleton. Moreover, if **A2** also holds, then $\mathcal{V}$ can be taken to be a singleton and S can be taken to be $At(T_0)$.*

Note that, in Theorem 3, there are no uniqueness requirements on $\mathcal{P}$ or $\mathcal{U}$. In part, this is because the state space and outcome space are not unique. But even if **A2** holds, so that the state space can be taken to be the set of atoms, the probability and the utility are far from unique, as the following example shows.

Example 2 *Take $\mathcal{A}_I = \{\dashv, \sqcup\}$ and $T_0 = \{t\}$. Suppose that $\succeq$ is such that*

$$a \succ \text{if } t \text{ then } a \text{ else } b \succ \text{if } t \text{ then } b \text{ else } a \succ b.$$

*It is not hard to check that every program in $\mathcal{A}$ is equivalent to one of these four, so **A2** holds, and we can take the state space to be $S^* = \{t, \neg t\}$. Let $O^* = \{o_1, o_2\}$, and define $\rho^0_{S^* O^*}$ so that $\rho^0_{S^* O^*}(a)$ is the constant function o_1 and $\rho^0_{S^* O^*}(b)$ is the constant function o_2. Now define $\pi^0_{S^*}$ in the obvious way, so that $\pi^0_{S^*}(t) = \{t\}$ and $\pi^0_{S^*}(\neg t) = \{\neg t\}$. We can represent the preference order by using any probability measure Pr^* such that $\text{Pr}^*(s_1) > \text{Pr}^*(s_2)$ and utility function u^* such that $u^*(o_1) > u^*(o_2)$.*

As Example 2 shows, the problem really is that the set of actions is not rich enough to determine the probability and utility. By way of contrast, Savage's postulates ensure that the state space is infinite and that there are at least two outcomes. Since the acts are all functions from states to outcomes, there must be uncountably many acts in Savage's framework.

The next example shows that if the order is partial and we want the representation to involve just a single utility function, then we cannot take the state space to be the set of atoms.

Example 3 *Suppose that $T_0 = \emptyset$, and, as in Example 2, $\mathcal{A}_I$ (and hence $\mathcal{A}$) consists of the two primitive programs a and b, which are incomparable. In this case, the smallest state space we can use has cardinality at least 2. For if $|S| = 1$, then there is only one possible probability measure on S, so a and b cannot be incomparable. Since there is only one atom when there are no primitive propositions, we cannot take the state space to be the set of atoms. (There is nothing special about taking $T_0 = \emptyset$ here; similar examples can be constructed for arbitrary choices of T_0.) It is also immediate that there is no representation where the outcomes space has only one element. There is a representation where the state and outcome space have cardinality 2: let $S = \{s_1, s_2\}$ and $O = \{o_1, o_2\}$; define ρ^0_{SO} so that $\rho^0_{SO}(a)(s_i) = o_i$ and $\rho^0_{SO}(b)(s_i) = o_{i \oplus 1}$, for $i = 1, 2$ (where $\oplus$ represents addition mod 2); let $\mathcal{P}$ be any set of probability measures that includes measures Pr_1 and Pr_2 such that $\text{Pr}_1(s_1) > \text{Pr}_1(s_2)$ and $\text{Pr}_2(s_2) > \text{Pr}_2(s_1)$; let $\mathcal{U}$ be any set of utility functions that includes a utility function such that $u(o_1) \neq u(o_2)$. It is easy to see that this choice gives us a representation of the preference order that makes a and b incomparable. This suggests that there are no interesting uniqueness requirements satisfied by $\mathcal{P}$ and $\mathcal{U}$.*

We remark that it follows from the proof of Theorem 3 that, even if the order is partial, there is a representation involving a single probability measure (and possibly many utility functions) such that the state space is $At(T_0)$.

3.4 A Representation Theorem for $\mathcal{A}^+$

A3' does not suffice to get a representation theorem for the richer language $\mathcal{A}^+$. As we have observed **A3'** gives us rational AA act independence. To get a representation theorem, we need to have act independence even for acts with real coefficients as well. Moreover, we need a standard Archimedean property. The following two postulates do the job.

A4. If $r \in (0, 1]$, then $a \succeq b$ iff $ra + (1 - r)c \succeq rb + (1 - r)c$ and $a \succ b$ iff $ra + (1 - r)c \succ rb + (1 - r)c$.

A5. If $a \succ b \succ c$ then there exist $r, r' \in [0, 1]$ such that $a \succ ra + (1 - r)c \succ b \succ r'a + (1 - r')c \succ c$.

We now get the following analogue of Theorem 3. With randomization, if **A2** holds, we get some degree of uniqueness. Although the state space, outcome space, probability, and utility are not unique, expected utilities are unique up to affine transformations. To make our discussion of uniqueness clearer, say that $(S, O, \mathcal{P}, \pi'_S, \rho'_{SO}, \sqcap)$ *is a representation of* $(\succeq, \succ)$ if it satisfies the conditions of Theorem 3. Note that if a is a program, then $\rho_{SO}(a)$ is an AA act. As usual, we take $E_{\Pr}(u_{\rho_{SO}(a)})$, the expected utility of this AA act with respect to Pr, to be

$$\sum_{s \in S} \sum_{o \in O} \Pr(s) u(o) (\rho_{SO}(a))(s)(o).$$

Theorem 4 *If* $(\succeq, \succ)$ *are preference orders on acts in* $\mathcal{A}^+_{\mathcal{A}_I, \mathcal{T}_I}$ *satisfying* **A1**, **A3'**, **A4**, *and* **A5**, *then there exist a finite set S of states, a set $\mathcal{P}$ of probability measures on S, a finite set O of outcomes, a set $\mathcal{U}$ of utility functions on O, a set $\mathcal{V} \subseteq \mathcal{P} \times \mathcal{U}$, a test interpretation π^0_S, and a program interpretation ρ^0_{SO} such that* $a \succeq b$ *iff* $E_{\Pr}(u_{\rho_{SO}(a)}) \geq E_{\Pr}(u_{\rho_{SO}(b)})$ *for all* $(\Pr, u) \in \mathcal{V}$ *and* $a \succ b$ *iff* $E_{\Pr}(u_{\rho_{SO}(a)}) > E_{\Pr}(u_{\rho_{SO}(b)})$. *Either $\mathcal{P}$ or $\mathcal{V}$ can be taken to be a singleton. Moreover, if* **A2** *also holds, then $\mathcal{V}$ can be taken to be a singleton, S can be taken to be* $At(\mathcal{T}_0)$, *and if* $(S, O, \Pr, \pi^0_S, \rho^0_{SO}, u)$ *and* $(S', O', \Pr', \pi^0_{S'}, \rho^0_{S'O'}, u')$ *both represent* $(\succeq, \succ)$, *then there exist constants α and β such that for all acts* $a \in \mathcal{A}^+_{\mathcal{A}_I, \mathcal{T}_I}$, $E_{\Pr}(u_{\rho_{SO}(a)}) = \alpha E_{\Pr'}(u'_{\rho_{S'O'}(a)}) + \beta$.

If **A2** holds, then Theorem 4 says that the expected utility is unique up to affine transformation, but makes no uniqueness claims for either the probability of the utility. This is not surprising, given that, in general, the probability and utility will be over quite different spaces. But even if two representations use the same state and outcome spaces, not much can be said, as the following example shows.

Example 4 *Suppose that* $\mathcal{A}_I = \{\neg, \sqcup\}$, $\mathcal{T}_0 = \{t\}$, *and* $(\succeq, \succ)$ *is a pair of orders satisfying* **A1**, **A2**, **A3'**, **A4**, *and* **A5** *such that* $a \succ b$ *and* **if** t **then** a **else** $b \sim \frac{1}{2}a + \frac{1}{2}b$. *It is easy to see that these constraints completely determine* $(\succeq, \succ)$. *Let* $S = At(\mathcal{T}_0)$, $O = \{a_t, a_{\neg t}, b_t, b_{\neg t}\}$, $\pi^0_S(t) = \{t\}$, $\pi^0_S(\neg t) = \{\neg t\}$, *and* $\rho_{SO}(c)(s) = c_s$ *for* $c \in \{a, b\}$ *and* $s \in \{t, \neg t\}$. *There are many representations of* $(\succeq, \succ)$ *with this state and outcome space: for example, we could take* $\Pr_1(t) = 1/2$, $u_1(a_t) = u_1(a_{\neg t}) = 1$, *and* $u_1(b_t) = u_1(b_{\neg t}) = 0$; *or we could take* $\Pr_2(t) = 3/4$, $u_2(a_t) = 4$, $u_2(a_{\neg t}) = 8$, $u_2(b_t) = 0$, *and* $u_2(b_{\neg t}) = 4$. *We leave it to the reader to check that these choices both lead to representations, which give the same expected utility to acts.*

As in the case of the language $\mathcal{A}$, we cannot in general take the state space to be the set of atoms. Specifically, if $\mathcal{A}_I$ consists of two primitive programs, and we take all programs in $\mathcal{A}_I^+$ to be incomparable, then the same argument as in Example 3 shows that we cannot take S to be $At(T_0)$, and there are no interesting uniqueness requirements that we can place on the set of probability measures or the utility function.

3.5 Objective Outcomes

In many applications where it may seem reasonable to consider the outcome space to be objective, rather than subjective. For example, if we are considering decisions involving trading in securities, we can take the outcome space to be dollar amounts. Given a fixed, finite set O of outcomes, we consider the set of acts that result when, in addition to the primitive acts in $\mathcal{A}_I$, we assume that there is a special act a_o for each outcome $o \in O$. Call the resulting language $\mathcal{A}_{\mathcal{A}_I,T_I,\mathcal{O}}$ or $\mathcal{A}^+_{\mathcal{A}_I,T_I,\mathcal{O}}$, depending on whether we allow randomization. We then define ρ_{SO} so that $\rho_{SO}(a_o)$ is interpreted as the constant function o for each outcome o.

With an objective outcome space O, we have not been able to prove a representation theorem for $\mathcal{A}$, but we can get a representation theorem for $\mathcal{A}^+$ that is quite close to that of Savage and Anscombe and Aumann. We need a postulate that, roughly speaking, ensures that all outcomes o are treated the same way in all states. This postulate is the obvious analogue of one of Savage's postulates. Given a test t, we write $a \succeq_t b$ if for some (and hence all, given cancellation) acts c, we have

if t then a else $c \succeq$ if t then b else c.

We say that t is *null* if, for all a, b, and c, we have

if t then a else $c \sim$ if t then b else c.

Define a *generalized outcome act* to be a program of the form $ra_{o_1} + (1-r)a_{o_2}$, where o_1 and o_2 are outcomes.

A6. If t is not null and a_1 and a_2 are generalized outcomes, then $a_1 \succeq a_2$ iff $a_1 \succeq_t a_2$ and similarly with $\succeq$ replaced by $\succ$.

A7. There exist outcomes o_0 and o_1 such that $a_{o_1} \succ a_{o_0}$.

Theorem 5 *If $(\succeq, \succ)$ are preference orders on acts in $\mathcal{A}^+_{\mathcal{A}_I,T_I,\mathcal{O}}$ satisfying* **A1**, **A3'**, *and* **A4-A7**, *then there exist a set S of states, a set $\mathcal{P}$ of probability measures on S, a set $\mathcal{U}$ of utility functions on O, a set $\mathcal{V} \subseteq \mathcal{P} \times \mathcal{U}$, a test interpretation π^0_S, and a program interpretation ρ^0_{SO} such that $a \succeq b$ iff for all $(\Pr, u) \in \mathcal{V}$, $E_{\Pr}(u_{\rho_{SO}(a)}) \geq E_{\Pr}(u_{\rho_{SO}(b)})$ and $a \succ b$ iff for all $(\Pr, u) \in \mathcal{V}$, $E_{\Pr}(u_{\rho_{SO}(a)}) > E_{\Pr}(u_{\rho_{SO}(b)})$. If* **A2** *holds, then we can take $\mathcal{V}$ to be a singleton, the probability is uniquely determined, and the utility function is determined up to affine transformations. That is, if $(S, \Pr, \pi^0_S, \rho^0_{SO}, u)$ and $(S', \Pr', \pi^0_{S'}, \rho^0_{S'O}, u')$ are two representations of the preference order, then, for all tests t, $\Pr(\pi_S(t)) = \Pr'(\pi_{S'}(t))$, and there exist α and β such that $u' = \alpha u + \beta$.*

Note that in Theorem 5, even if acts are totally ordered, we cannot take the state space to consist only of atoms as the following example shows.

Example 5 *Suppose that there are two outcomes: o_1 ($\$1,000$) and o_0 ($\$0$), and one primitive act a: buying 10 shares of IBM. Act a intuitively can return somewhere between $\$0$ and $\$1,000$; thus, $o_1 \succ a \succ o_0$. There are no tests. If there were a representation with only one state, say s, in state s, a must return either $\$1,000$ or $\$0$. Whichever it is, we cannot represent the preference order.*

Our earlier representation theorems always involved a single utility function. As the following example shows, we can use neither a single utility function nor a single probability measure here. Moreover, there cannot be a representation where the set of probability-utility pairs has the form $\mathcal{P} \times \mathcal{U}$.

Example 6 *Suppose that $O = \{o_1, o_0, o\}$, $\mathcal{A}_I = \emptyset$, and $T_0 = \{t\}$. Let u_1 be a utility function such that $u_1(o_1) = 1$, $u_1(o_0) = 0$, and $u_1(o) = 3/4$; let u_2 be a utility function such that $u_2(o_1) = 1$, $u_2(o_0) = 0$, and $u_2(o) = 1/4$; let Pr_1 be a probability measure on $S = At(T_0)$ such that $\text{Pr}_1(t) = 1/4$, and let Pr_2 be a probability measure on S such that $\text{Pr}_2(t) = 3/4$. Consider the preference order on $\mathcal{A}^+_{\mathcal{A}_I, T_I, \mathcal{O}}$ generated from $\mathcal{V} = \{(\text{Pr}_\infty, \sqcap_\infty), (\text{Pr}_\in, \sqcap_\in)\}$, taking $\pi_S(t) = \{t\}$. It is easy to see that this preference order has the following properties:*

- $a_{o_1} \succ a_o \succ a_{o_0}$;
- a_o and $\frac{1}{2}a_{o_1} + \frac{1}{2}a_{o_0}$ are incomparable;
- (**if** t **then** a_{o_1} **else** a_{o_0}) and $\frac{1}{2}a_{o_1} + \frac{1}{2}a_{o_0}$ are incomparable; and
- (**if** t **then** a_{o_1} **else** a_{o_0}) $\sim a_o$.

Consider a representation of this order. It is easy to see that to ensure that a_o is not comparable to $\frac{1}{2}a_{o_1} + \frac{1}{2}a_{o_0}$, the representation must have utility functions u_1 and u_2 such that $u_1(o) > \frac{1}{2}(u_1(o_1) + u_1(o_0))$ and $u_2(o) > \frac{1}{2}(u_2(o_1) + u_2(o_0))$. To ensure that **if** t **then** a_{o_1} **else** a_{o_0} and $\frac{1}{2}a_{o_1} + \frac{1}{2}a_{o_0}$ are incomparable, there must be two probability measures Pr_1 and Pr_2 in the representation such that $\text{Pr}_1(\pi_S(t)) < 1/2$ and $\text{Pr}_2(\pi_S(t)) > 1/2$. Finally, to ensure that (**if** t **then** a_o **else** a_{o_0}) $\sim a_o$, we cannot have both (Pr_1, u_1) and (Pr_1, u_2) in $\mathcal{V}$.

4 Conclusions

Most critiques of Bayesian decision making have left two assumptions unquestioned: that beliefs may be represented with a single number, and that all possible states and outcomes are known beforehand. The work presented here directly addresses these concerns. We have shown that by viewing acts as programs rather than as functions from states to outcomes, we can prove results much in the spirit of the well-known representation theorems of Savage and Anscombe and Aumann; the main difference is that the state space and outcome space are now objective, rather than being given as part of the decision problem. So what does all this buy us? To the extent that we can prove only such representation theorems, the new framework does not buy us much (although we still claim that thinking of acts as programs is typically more natural than thinking of them as functions—rather than thinking about a state and outcome space, a DM need just think of what he/she can do). We have proved these results just to show how our approach relates to the standard approaches. We believe that the real benefit of our approach will be realized when we move beyond the limited setting we have considered in this paper. We list some possibilities here:

- Once we move to game-theoretic settings with more than one agent, we can allow different agents to use different languages. For example, when trying to decide whether to buy 100 shares of IBM, one agent can consider quantitative issues like price/earnings ratio, while another might consult astrological tables. The agent who uses astrology might not understand price/earnings ratios (the notion is simply not in his vocabulary) and, similarly, the agent who uses quantitative methods might not understand what it means for the moon to be in the seventh house. Nevertheless, they can trade, as long as they both have available primitive actions like "buy 100 shares of IBM" and "sell 100 shares of IBM". Agreeing to disagree results [2], which say that agents with a common prior must have a common posterior (they cannot agree to disagree) no longer hold. Not only is there no common prior, once the state space is subjective, the agents do not even have a common state space on which to put a common prior. Moreover, notions of unawareness [3, 8, 10, 17] come to the fore. Agents need to reason about the fact that other agents may be aware of tests of which they are unaware. See [4, 5, 9] for preliminary work taking awareness into account in games.

- We focus here on static, one-shot decisions. Dealing with decision-making over time becomes more subtle. Learning becomes not just a matter of conditioning, but also learning about new notions (i.e., expanding the language of tests). Note that we can think of learning unanticipated events as a combination of learning about a new notion and then conditioning. This framework lends itself naturally to vocabulary expansion—it just amounts to expanding the possible set of programs.

- We have considered only a very simple programming language with input-output semantics. Interesting new issues arise once we consider richer programming languages. For example, suppose that we allow concatenation of programs, so that if a and b are programs, then so is $a; b$. Intuitively, $a; b$ means "do a, and then do b". We can still use input-output semantics for this richer programming language, but it might also be of interest to consider a different semantics, where we associate with a program the sequence of states it goes through before reaching the outcome. This "path semantics" makes finer distinctions than input-output semantics; two programs that have the same input-output behavior might follow quite different paths to get to an outcome, starting at the same initial state. The framework thus lets us explore how different choices of semantics for programs affect an agent's preference order.

- As we have seen (Proposition 5), cancellation forces a DM to be indifferent between two equivalent programs. But since testing program equivalence is co-NP-complete, it is unreasonable to expect that agents will necessarily be indifferent between two equivalent programs. It would be very interesting to consider weakenings of the cancellation axiom that do not force a DM to be indifferent between equivalent programs. Such considerations are impossible in the Savage setting, where acts are functions.

Besides exploring these avenues of research, we would like to understand better the connection between our work and the work on *predictive state representations* [11, 15]. A predictive state representation is a way of representing dynamical systems that tries to move away from a state space as part of the problem description but, rather, constructs the state space in terms of tests and actions (which are taken to be primitive). While the technical details and motivation are quite different from our work, there are clearly similarities in spirit. It would be interesting to see if these similarities go deeper.

Bibliography

[1] Anscombe, F. J., Aumann, R. J. A definition of subjective probability. *Annals of Mathematical Statistics*, 34:199–205, 1963.

[2] Aumann, R. J. Agreeing to disagree. *Annals of Statistics*, 4(6):1236–1239, 1976.

[3] Fagin, R., Halpern, J. Y. Belief, awareness, and limited reasoning. *Artificial, Intelligence*, 34:39–76, 1988.

[4] Feinberg, Y. Subjective reasoning—games with unawareness technical report resarch paper series #1875. Tech. rep., 2004. Stanford Graduate School of Business.

[5] Feinberg, Y. Games with incomplete awareness technical report resarch paper series #1875. Tech. rep., 2005. Stanford Graduate School of Business.

[6] Gilboa, I., Schmeidler, D. Maxmin expected utility with a non-unique prior. *Journal of Mathematical Economics*, 18:141–153, 1989.

[7] Gilboa, I., Schmeidler, D. *A Theory of Case-Based Decisions*. Cambridge, Mass.: MIT Press, 2001.

[8] Halpern, J. Y., Rêgo, L. C. Interactive unawareness revisited. In *Theoretical Aspects of Rationality and Knowledge: Proc. Tenth Conference (TARK 2005)*, 78–91. 2005.

[9] Halpern, J. Y., Rêgo, L. C. Extensive games with possibly unaware players. In *Proc. Fifth International Joint Conference on Autonomous Agents and Multiagent Systems*. 2006.

[10] Heifetz, A., Meier, M., Schipper, B. Multi-person unawareness. In *Theoretical Aspects of Rationality and Knowledge: Proc. Ninth Conference (TARK 2003)*, 148–158. 2003. An extended version with the title "Interactive unawareness" will appear in the *Journal of Economic Theory*.

[11] Jaeger, M. Observable operator models for discrete stochastic time series. *Neural Computation*, 12:1371–1398, 2000.

[12] Kahneman, D., Slovic, P., Tversky, A. (Eds.). *Judgment Under Uncertainty: Heuristics and Biases*. Cambridge/New York: Cambridge University Press, 1982.

[13] Krantz, D. H., Luce, R. D., Suppes, P., Tversky, A. *Foundations of Measurementm Vol 1: Additive and Polynomial Representations*. New York: Academic Press, 1971.

[14] Kreps, D. *Notes on the Theory of Choice*. Boulder, Colo.: Westview Press, 1988.

[15] Littman, M. L., Sutton, R. S., Singh, S. Predictive representations of state. In *Advances in Neural Information Processing Systems 14 (NIPS 2001)*, 817–823. 2001.

[16] McNeill, B. J., Pauker, S. J., Sox, J. H. C., Tversky, A. On the elicitation of preferences for alternative therapies. *New England Journal of Medicine*, 306:1259–1262, 1982.

[17] Modica, S., Rustichini, A. Unawareness and partitional information structures. *Games and Economic Behavior*, 27(2):265–298, 1999.

[18] Savage, L. J. *Foundations of Statistics*. New York: Wiley, 1954.

[19] Schmeidler, D. Subjective probability and expected utility without additivity. *Econometrica*, 57:571–587, 1989.

[20] Winskel, G. *The Formal Semantics of Programming Languages*. Cambridge, Mass: MIT Press, 1997.

The Philosophy of Computer Simulation [1]

Kevin B. Korb
Monash University
kbkorb@gmail.com

Steven Mascaro
Monash University
sm@voracity.org

ABSTRACT. The growth in the application of computer simulation across the sciences, and especially the application of artificial life techniques (agent-based and individual-based modeling) to evolutionary biology and psychology, the social sciences, epidemiology and ecology, raises many philosophical questions. One basic question is: How can we learn about reality by simulating it? Attempts to answer that question revolve around the relation between experimental procedure and simulation. We consider the relation between simulation and real-world experiment and find it to be the identity relation.

1 The Scope and Limits of Computer Simulation

Computers have rapidly become the primary intellectual tool deployed by humans. This is natural. One of the first things students of computer science learn is that computers are *universal:* within the range of computable functions, there is simply *nothing* that computers cannot do. Every normal programming language is, in fact, a universal Turing machine, as can be proved easily by programming a simple universal Turing machine in that language. This universality, coupled with the rapid expansion of computational power, means that computers support almost everything that occurs in developed economies. It also means that computers can be applied to nearly any intellectual task, if not as an independent source of innovation, a use waiting at least upon some considerable new developments in artificial intelligence, then as a helpmate and support. Computers can be, and have been, applied to further research in biology, chemistry, microphysics, macrophysics, economics, sociology, art, music and philosophy. Computer simulation has become a reliable and regular contributor to investigation in each of these fields of endeavor, and probably every science. This expanding reach of computation has led to extreme reactions, including those who see computation as essentially inferior to human inference, and these uses of computation as epistemologically suspect, and those who see no bounds to computer application and who call for a new epistemology to underwrite these activities.

Here we shall attempt to develop some ground between the more extreme reactions to scientific computer simulations. In particular, we find considering the

---

[1] The authors are with Clayton School of Information Technology, Monash University, Clayton, Victoria, Australia.

relation between computer simulations and scientific experiments to be interesting and fruitful. Many within the new epistemology camp have been suggesting that simulating "lies between" theorizing and experimenting [24, 38, 43]; if simulation is part theory and part experiment, then the old stories of how we learn about theory from experiment can hardly apply. However, we do not agree with this "in-betweenness" theory; rather, we suggest that the old stories about the growth of scientific knowledge, whether right or wrong about science before the computer, are equally right or wrong about current science.

1.1 What Computers Can't Do

There is general agreement that (ordinary) computers cannot compute non-computable functions, e.g., solving Turing's Halting Problem or computing the Busy Beaver numbers. To generate solutions to such problems computers would need to have access to infinite precision real numbers, for example, which is something no finite digital machine can manage. There is not any consensus about what this restriction really means, however. Some, such as Penrose [32], seem to think this implies that computers are significantly inferior in potential computational ability to analog computers, such as humans. But for such a potential to be manifested, one must find an analog means of taking advantage of infinite precision real numbers, which presupposes overcoming quantum limits and pervasive low-level thermal noise. Since such limits have been operative throughout the entire evolutionary history of humanity, and since human mental capacities have certainly evolved largely for their adaptive value, it follows that human mentation as it currently exists has no more ability to break the barrier of non-computability than does the humble desktop computer. If there is potential to break through that barrier, we can hardly expect unassisted evolution to find it. We find fully satisfactory Turing's original answer to a similar complaint put to the possibility of machine intelligence — that any formal system is constrained to a proper subset of the truth by having a Gödel sentence true of it: who is to say we are any different? Computers are limited. But only a fool can fail to see the many severe limitations of humans.

If we find something that humans *can* do, then we have a *prima facie* case that computers can do it too. If we are too stupid to figure out how to get them to do it, that is a problem not attributable to them.

2 What Is Simulation?

So, what are simulations? The PC game "The Sims" is a simulation: it simulates the life and times of various characters who worry about getting jobs and cleaning toilets. Aircraft and naval piloting simulators simulate conditions involved in normal and abnormal maneuvers of aircraft and ships. And Second Life simulates a large range of human and non-human activities. Despite many commentators on the philosophy of simulation taking these sorts of cases seriously (e.g., [17, 25, 26]), in all of these simulations a human user plays an essential and central role, which is not to the point in simulation science. It is also not to the point that in ordinary language these processes are called simulations; that usage is simply emphasizing that humans are being put into other-than-real-world situations.

Such simulations are not in general being used to expand our scientific knowledge, and so they do not raise the epistemological questions we wish to engage here. The simulations of interest to us here are those in which the *entire* simulation occurs within a computer, as a computer process. Indeed, we shall argue that the simulations of interest here are computer processes which simulate *other processes* — whether chemical, ecological, astrophysical or from whatever other scientific study.

2.1 A Definition of Simulation

A commonly used definition is:

Definition 1 *A* **computer simulation** *is "the use of a computer to solve an equation that we cannot solve analytically."* (Frigg and Reiss, 2008)

See also, for example, [23, 26, 33, 42].[2] A comment of Reddy's [36, p.162] might be confused with this kind of definition: "Simulation is a tool that is used to study the behaviour of complex systems which are mathematically intractable." That, however, would be a confusion of an accidental with an essential property: we use tools where they are useful, and not where they can be used but are unhelpful.

Definition 1 itself, however, includes both too much and too little.[3] Whether "we" can or cannot solve an equation analytically is surely immaterial. For one thing, that would render the term absurdly relative to the individual; for example, many programs which for us would be simulations would not count as simulations for a John von Neumann. For another, as new analytic techniques become available, what once counted as a simulation may not any longer.

We do not want a concept of simulation which is relative to time, place or individual calculational ability; we want a concept which is secured by a methodological role within science. But focusing on the positive side of the definition, things only get worse.

It's true that computer simulation began with the work of von Neumann, Metropolis and others working out ways of computing solutions to equations required for the development of the hydrogen bomb. This lead to such procedures as "Metropolis sampling", Monte Carlo (MC) integration and Monte Carlo methods in general. Monte Carlo integration is a method of numerically solving an unanalysable (or difficult to analyse) integral;[4] it does so by averaging pseudo-randomly selected values of the function in question. One can think of it as throwing darts at a board where the curve is drawn and using the frequency of darts under the curve as an estimate of its area. MC integration contrasts with numerical quadrature, which sums the areas of rectangles bounding portions of the curve. Nobody talks of the latter as simulation, but it solves equations just as well as the MC approach (at least given moderately well-behaved curves and low dimensionality). However, under the definition above quadrature counts as simulation. And, if MC integration were to be counted as simulation, we can't

---

[2] It is worth noting that Humphreys has retracted this view, finding the arguments of Hartmann [21] persuasive [25, p.108].

[3] This is a point originally made by Hartmann [21].

[4] Note, however, that unanalysability is not a part of anyone's *definition* of Monte Carlo methods; it's just that analysable integrals are analysed instead!

see any reason to deny the application to numerical methods generally, since they are all about solving things with computers that we cannot solve in our heads.

However, we think it is far preferrable to deny that equation solving is simulation and reserve that term for (computer) processes which mimic relevant features of a dynamic physical process under study (which is Hartmann's definition [21, p.83]; see also [34, 44]). Racynski and Bargiela [35] have recently put this nicely in their first sentence: "To put it simply, computer simulation is a process of making a computer [process] behave like a cow, an airplane, a battlefield, a social system, a terrorist, [an] HIV virus, a growing tree, ... or any other thing."

2.2 Dynamic versus Static

Frigg and Reiss [17] have objected to the idea that simulations are inherently dynamic, being processes that model other processes. It does not matter, according to them, that the computer process *takes* time, so long as it *represents* time: "[A]ll that matters is that the computer provides states that come with a time index.... If ... we have a computer that can calculate all the states in no time at all, surely we don't feel we lose anything."[5]

It is, of course, true that if the computer process encodes a representation of all the time steps of the target process, whether simultaneously or not, then it contains all the *information* that a simulation would carry or convey. However, it hardly follows that it *is* a simulation. For example, we might have function state(cond,i) which returns the simulated process's i-th state, given initial condtions cond. This function contains all the information contained in the simulation; indeed, by iteration it could be used to run the simulation. However, we can equally well use it to run all sorts of processes which are *not* the simulation, for example, the states indexed by the Fibonacci sequence.

For a more homely example, a similar point can be made about a feature film sliced into individual frames and put in an album: the album is not the feature film. The album contains all and only the information within the movie. But a movie moves, an album does not; and the album will remain not-movie until someone splices it back together. The methodologically relevant point is that one can poke a computer process and *then* see what happens. That is at least part of the point when people note that simulations embody aspects of experimentation. But if the process is already completed, one cannot poke it. There is no experimental side to things, even if the computer program incorporates all the information of the original simulation. Since the information is in there, presumably there is some way of extracting the same information as one would from experimenting with a simulation; but it would not be by some intervention which mimics experimentation. One might well say, along with Frigg and Reiss [17], that since all the information is there, none of this matters. But keeping *some* connection with ordinary language and ordinary semantics is necessary, and calling a photo album a movie is just silly.[6]

---

[5]We should like to point out that, despite our differences on some particular issues, and especially the definition of simulation, Frigg and Reiss [17] present a parallel argument to our own, in particular advancing our shared claim that the epistemology of simulation is the epistemology of experimentation.

[6]For a final analogy, you might consider Hans Moravec's proposal for life-extension: downloading the information content of your brain into a disk and "waiting" for technological devel-

2.3 Artificial Life Simulations

A final, and we hope decisive, objection to Definition 1 is that there have arisen very large regions of simulation research which are not plausibly described as equation solving at all, covering at least the vast bulk of agent-based modeling in artificial life and social simulation (which is our primary area of simulation research) and individual-based modeling in ecology [19]. Although some equations will inevitably describe some characteristics of such simulations, it is at most *unusual* for the solution of equations to be the motivating factor in such investigations. The motivation is more typically the investigation of high-level properties of the system which emerge from an explicitly defined lower level of simulation. Some example motivations are:

- Demonstrating a feasible mechanism for the Baldwin effect in evolution [22]
- Showing that flocking behavior can result from independent decision-making throughout a flock [37]
- Determining the minimal space requirements for beech forests to survive in isolated patches [19, section 1.2.2 & section 6.8.3]
- Finding conditions supporting or underminining the main postulated mechanisms for the evolution of dimorphic parental investments in offspring [27]
- Investigating the effectiveness of different possible public health interventions in response to a smallpox epidemic [12]

If the philosophy of simulation is not to be left behind by the science of simulation, Definition 1 must be abandoned. Therefore, we shall adopt Hartmann's definition, but rendering it more explicit.

2.4 Another Definition of Simulation

One immediate benefit of Hartmann's definition is that it rules out the virtual reality scenarios directly: since the human user (trainee) is a necessary ingredient, these are not computer simulations. We will nevertheless now drop the word "computer" and talk about simulation most generally, as this will help us understand the relation between the epistemology of simulation and the epistemology of experiment. Our proposed semi-formal rendition of Hartmann's mimicking account is:

Definition 2 *S is a* **simulation** *of P if and only if*

1. *P is a physical process or process type*
2. *S is a physical process or process type*
3. *S and P are both correctly described by a dynamical theory T containing (for S; parenthetically described for P):*
   - *an ontology of objects O_S (O_P) and types of objects $\Psi_i(x)$ ($\Phi_i(x)$)*
   - *relations between objects $\Psi_i(x_1, \ldots, x_n)$ ($\Phi_i(x_1, \ldots, x_n)$); hence, there are states of the system, s*

---

opment to support "your" reanimation. We suggest the incredulity this idea induces in most people is simply rational.

The Philosophy of Computer Simulation 311

- *dynamical laws of development (possibly stochastic):*
  $f_S(s) = s'$ $(f_P(s) = s')$

In other words, both the simulation S and the target of the simulation P are physical processes with a common dynamical theory T. Computer simulations are then easily defined as:

Definition 3 *S is a **computer simulation** of P if and only if*

1. *it is a simulation of P*
2. *and it is a computer process or process type.*

An immediate objection to Definitions 2 and 3 might occur to you. It is symmetric: according to this, we could just as well use the sun to simulate our astrophysical programs as vice versa! This clearly won't do; our simulations, whether computer based or not, are surely *intrinsically* simpler than their targets of study. To use a metaphor of Giere's [18] (borrowed from Borges [5]), if we were to construct a map of the earth on a 1:1 scale, it's true that we could more accurately measure distances using this very fine resolution map than using cruder maps, but, obviously, all the other advantages of maps would be lost. (Borges' characters start shifting domicile from reality to map!) The problems with symmetry are both practical and theoretical. Theoretically, whatever one's view may be about the nature of scientific explanation and theories, it's entirely clear that they some how *summarize* features of the world. Computer scientists would say they *compress* information about the world. In short, they are shorter than any direct, exhaustive description of their objects.

All of this is well taken, but it doesn't follow that we need to acknowledge the point formally, within our definition. Simulations are typically constrained by both a lack of understanding of fine details of the objects of our simulations and by a lack of time to wait for the implications of fine details to filter through our simulations. Frequently, however, crude simulations are made less crude, as advances are made in both our understanding of the physical systems and in our computational capacities. If we were somehow to extend this advance in resolution power indefinitely, we might begin to approach the 1:1 scale contemplated by Giere. Admittedly, going all the way would be pointless. However, that doesn't mean that by going all the way we would no longer be dealing with a simulation; a pointless simulation remains a simulation.

3 Homomorphic Simulation

So, for practical and theoretical reasons, we require that our simulations *not* be as detailed as the processes we simulate. Instead we require that:[7]

Proposition 1 *There should exist a **homomorphism** h from P to S.*

Definition 4 *A **homomorphism** h from P to S is a mapping $h: P \to S$ such that*

1. *For every object $x \in O_P$, $h(x) \in O_S$.*

---

[7] For a similar account, see Norton and Suppe [30], although their account is somewhat cluttered with a variety of idealized, averaged and approximate models.

2. *For every relation Φ, $\Phi(x_1, \ldots, x_n)$ is true of P iff $h(\Phi) = \Psi$ and $\Psi(h(x_1), \ldots, h(x_n))$ is true of S*
3. *For every state transition function f in P, $f(s) = s'$ iff $f_h(h(s)) = h(s')$ (or, for stochastic laws, the distributions over states should be identical)*

The application of homomorphisms to simulation, to be sure, should be taken with a grain of salt. That is, it is an ideal and one which we are unlikely actually to reach with non-trivial simulations. It is frequently noted in the literature that our simulations often diverge from reality in small ways and sometimes in large ways. Nearly every simulation diverges in at least this way: digital computational processes cannot exactly simulate continuous time, whereas real systems at least appear to develop in continuous time; thus, these systems support relations ("in-between times") that have no counterpart in their simulations. Nevertheless, at least for most problems, time can be discretized to a fineness where this difference does not matter. The epistemological problem is to sort out when the divergences do matter to inferences about the real systems.

Our central epistemological proposal is that simulations can be tested for adequacy by testing whether a homomorphism between the real and the virtual system holds. In the simulation literature this is called "validation".

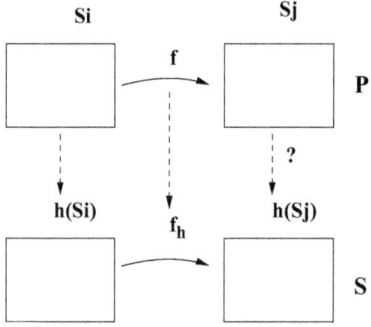

Figure 1: Testing for homomorphism.

3.1 Testing for Homomorphism (Validation)

By observing, or arranging for, the physical system P in state s_i and its subsequent transition to state s_j, we are enabled to test whether the simulation undertakes in homomorphic initial conditions the like transition (or vice versa). In Figure 1 this corresponds to checking that simulation S shows a transition from $h(s_i)$ to $h(s_j)$. Validation can be thought of as parallel to confirmation. Instead of confirming how well a theory represents reality, we are confirming how well a simulation maps reality. As such, validation comes in degrees, as there will be more or less severe tests possible for the adequacy of the mapping. Indeed, we would assert that the degrees come in the form of prior and posterior probabilities of the existence of a homomorphism, exactly as with ordinary confirmation theory, were we to allow ourselves the diversion into more traditional issues in the philosophy of scientific method.

Grimm and Railsback [19, chap 9] present a likely account of how validation might proceed. They suggest first testing low-level submodels which describe non-emergent phenomena in the simulation and only subsequently looking at higher-level systems, including properties of the simulation that emerge from interactions between submodels. At the higher levels we are conducting simulated versions of controlled experiments [19, p.316]: "We pose alternative theories for the individual behavior as the hypotheses to be tested, implement each hypothesis in the [simulation], identify some patterns as the 'currency' [standard] for evaluating the hypotheses, and then conduct simulations that determine which hypotheses fail to reproduce the patterns." For example, the beech forest simulation was designed to reproduce both the horizontal mosaic pattern of tree stands and the vertical pattern of tree cover. But subsequently unplanned for patterns in the simulation were discovered and put to good use [19, p.7; our emphasis]:

> [The simulation] was so rich in structure and mechanism that it also produced independent predictions regarding aspects of the forest not considered at all during model development and testing. These predictions were about the age structure of the canopy, spatial aspects of this age structure, and the spatial distribution of very old and large trees. All these predictions were in good agreeement with observations, considerably increasing the model's credibility. *The use of multiple patterns to design the model obviously led to a model that was structurally realistic.*

Other than the fact that this procedure is dealing with a computer simulation rather than directly with an ecological theory, there is no interesting methodological difference between this and standard theory testing. A rich, multi-patterned simulation offers a variety of opportunities for testing its conformity to the target process. And just as in standard confirmation theory [14, pp.123–129], the more varied the predictions of a simulation, or the submodels used to make them, that are tested and confirmed against reality, the greater our confidence that the simulation indeed maps that reality. With a sufficient variety of such tests, testing diverse transitions under diverse conditions, we may well be able to conclude that the simulation is, or is not, homomorphic, either approximately or exactly.

The existence of an approximate homomorphism is *crucial*: it underwrites the relevance of the simulation for the system being simulated and, in particular, its use both for explaining events in the real world and in predicting them.

As noted, homomorphisms may exist at a variety of levels of resolution. The level of resolution of the homomorphic simulation depends upon two major points:

1. How well do we (think we) understand P? How detailed a theory do we have to test?
2. Pragmatic constraints upon our simulation (e.g., how much time can we spend waiting).

The levels of potential simulation may lie upon one-another Shrek-like, as in an onion (see Figure 2):

At the top-level is a simulation with such a small ontology that nothing useful can be simulated. Below the actual simulation are more detailed potential simulations which are unused for such reasons as: the theory describing such detail

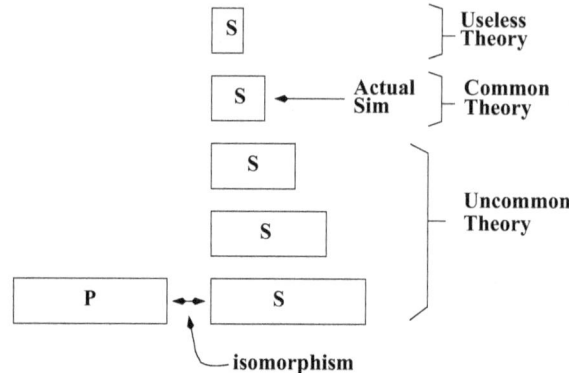

Figure 2: The simulation onion. (Smaller layers indicate fewer details, but greater generality.)

has not been invented; the simulations at that level of detail are impractical; the level of detail describes events of no interest to us.[8]

4 Simulations as Experiments

Many people have been attracted to the idea that simulations have no empirical side to them and, in particular, that they are basically revved-up thought experiments. Oreskes and others claim that simulations cannot be used to acquire *any* empirical knowledge about the world, directly or indirectly (e.g., [2, 9, 31]). Rather, simulations are limited to extending our understanding of the theories being simulated, by exploring their deductive consequences. Di Paolo, Noble and Bullock [9], for example, examine the Hinton and Nowlan [22] simulation study of the Baldwin effect — the acceleration of genetic evolution via learning by individuals in a population [4]. Prior to that study, the Baldwin effect had been given little attention; it sounded too much like a Lamarckian process and the mechanism for fixing learned behavior genetically was not understood. The simulation of Hinton and Nowlan [22] changed that by producing a plain, easily inspected mechanism, which demonstrably exhibited the Baldwin effect. Di Paolo et al. [9] argue that this is essentially a theoretical, deductive use of simulation, making plain what was implicit in the theory. While it is clear that simulations can be used to explore the deductive consequences of theories, it is not clear that that is the only role they may have in empirical science. Nor is it clear that Hinton and Nowlan's *mechanism* was in any sense implicit in Baldwin's theory. We now proceed to argue that they have potentially every role that experiments may have in empirical science.

---

[8]In other words, this kind of account is very far from requiring the "perfect mimesis" of isomorphism that Winsberg [43, p.116] claims is implied.

The Philosophy of Computer Simulation 315

4.1 A Comparison with Real Experiments

To further our claim that simulation studies share epistemology with traditional scientific experiments, we can consider Allan Franklin's experimental strategies [15]. Franklin emphasizes that his strategies are neither exclusive of other strategies nor exhaustive. Nevertheless, they provide a good indication of what happens in physical experiments; we annotate the list with reference to simulation studies ([15, p.104]):[9]

1. Experimental checks and calibration, in which apparatus reproduces the known phenomena
2. Reproducing artifacts that are known in advance to be present
   Regarding 1 and 2, reproducing known phenomena is a standard check of adequacy in simulation studies.
3. Intervention, in which the experimenter manipulates the object under observation
   The relative ease of manipulating simulations is one of their key advantages in experimental studies.
4. Independent confirmation using different experiments
   In simulation research there is always an opportunity to test very different kinds of initial conditions, and sometimes an opportunity to test the operation of distinct subprocesses [19, 20]. Replication of simulation results using distinct simulations is also a possibility (e.g., [3, 11]).
5. Elimination of plausible sources of error and alternative explanations of the result
   These are activities integral to both verifying and validating simulations.
6. Using the results themselves to argue for their validity
   By this Franklin meant that an experiment may create results which are highly unlikely to be artifacts of the measurement process or experimental procedure and so by themselves support the claim that they reflect an external reality. Similarly, simulation results may likewise be determined to be highly unlikely to be due to bugs, not just because of steps in the verification process, but also because of the results themselves.
7. Using an independently well confirmed theory of the phenomena to explain the results
   This is one leg of our triangle of Figure 3 below.
8. Using an apparatus based on well confirmed theory
   The apparatus here is the simulation, and associated software; verification is part of the process of justifying the claim that it is based on well confirmed theory.
9. Using statistical arguments
   It has been frequently remarked that the use of, or rather the need for, automated data analysis and data visualization techniques is a striking feature of simulation research. Epstein and Axtell [13], for example, employ a variety of graphics to good effect.

Clearly, at a phenomenological level, simulation research is very akin to traditional experimental research. But this does not demonstrate that at an "epistemological level" they are again alike.

---
[9]We have corrected Franklin's "corroboration" with "confirmation".

4.2 The Epistemology of Simulation

There are two acknowledged steps to justifying claims that a simulation is informative of the real world:

Verification: Determine whether the simulation correctly implements the theory being investigated,

- e.g., by performing design verification, debugging the simulation, and consistency checks.

Validation: Determine whether the simulation as implemented conforms to the target process.[10]

- This is testing for the existence of a homomorphism, by comparing simulation results with the target process and vice versa.

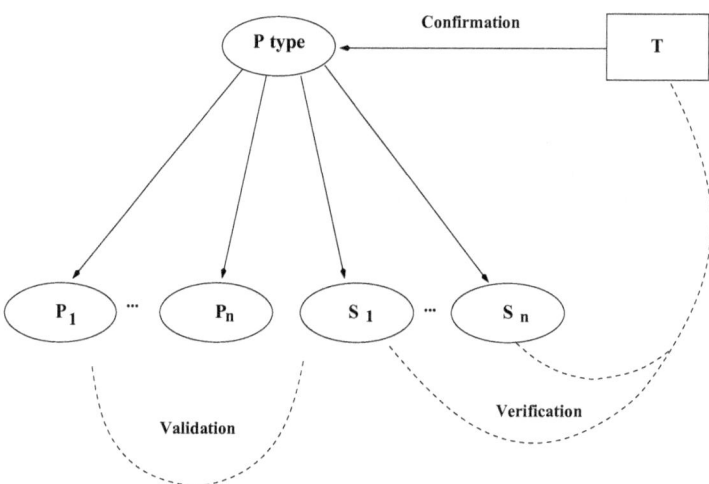

Figure 3: The verification, validation and confirmation triangle.

These steps are portrayed graphically in Figure 3. A theory T has been developed for the type of physical process P.[11] A range of token processes, $P_1, \ldots, P_n$ instantiate that type. And a simulation has been developed for those processes and/or the process type, leading to simulation processes $S_1, \ldots, S_m$. We can think of the computer program used to launch the simulation processes as a process type S, not depicted in the figure. This situation presents us with a triangle with three legs of possible justificatory test for the relevance of simulation S to its target P: whether the theory T represents the reality P (confirmation); whether S properly implements T (verification); whether S corresponds to P (validation). Each test can be carried out independently of the others. Once any two tests have

---

[10] From a history of the philosophy of science perspective, these uses of "verify" and "validate" are backwards. However, this usage comes from the software engineering tradition.

[11] Incidentally, we are favorably inclined towards the semantic interpretation of scientific theories [39], but nothing in our account hangs directly upon that.

The Philosophy of Computer Simulation 317

been conducted, and assuming their outcomes are not in dispute, then the third test becomes irrelevant, since we then know everything there is to know about the relations between T, P and S. This explains a range of observations previously made about simulation. Given verification, S can be manipulated to investigate the implications of T (this is the "S as thought experiment" above). Again, given verification, determining that there is a homomorphism between P and S is tantamount to confirmation; given validation, S can be used in exploratory theorizing as well as predicting the consequences of intervening in P systems; given confirmation, failures of correspondence between P and S indicate verification failures. When any two justificatory steps have been successful, the S_k are just as much instantiations of P as are the P_i: they all provide supervenience bases for that type of physical process.

4.3 Experiments as Simulations

An obvious rebuttal to the claim of epistemological sameness between experiments and simulations is: "Unlike simulations, when you're testing the real-world at least you know what you're testing is *real*! You can't be testing the wrong thing!" Morgan [29], for example, claims there is an inferential gap between simulation and reality that doesn't exist between real-world experiments and real-world target systems. This is a seductive thought, but it is wrong. The inferential gap is always there in any scientific study. It is nearly an everyday occurrence to hear about some medical study in which the experimental groups turn out to be unrepresentative of some target population. And it is an old joke that experimental psychology has accumulated a large body of evidence about the psychology of college students.

We previously observed that, following our Definition 2, the targets of our computer simulations — the physical processes in the world — might be construed as simulations (but not *computer* simulations) of our computer processes, even if that is not pragmatic. The general point is that some (non-computer) physical processes may be used to simulate others. For example, scale models built with clay and water are used to assess water flow and tidal action; or, wind tunnels are used to assess aerodynamic flows. Again, of course, the simpler and smaller processes are generally said to be simulations of the more complex and larger processes which are the targets of investigation. Which is the simulation is, in the first instance, driven by which process is wanting to be understood. But other factors play a role, including accessibility, ease of intervening in the process, and the ethics of intervening in the process. If there is a single process, and not a process *type*, of interest, and if it is accessible, etc., then there may be no recourse to a simulation: a simple intervention in the target process may be attempted. There is then no inferential gap, because the studied system is also the target system. Engineering applications and most treatments by medical doctors are of this type. If our interest spans an entire type of process, and if we are not looking for a specific outcome other than learning about that type, then a direct examination of all instances of the type is unlikely to be possible. In that case, we shall have to simulate the process type of interest with another that is more accessible. For example, we might use a sample of adult humans in Australia today to simulate the category of all adult humans across all of time and space. This is the common practice in medical research. Similarly, we might be interested in physical conditions immediately after the Big Bang, but have little prospect of directly measuring them; instead, we might simulate such conditions using high-intensity collisions of subatomic particles. Again, Galileo famously tested his telescopes on terrestrial objects to gain support for his inferences concerning celestial objects [6, p.72]. While there are many complications and nuances to each of these stories, in all these cases and all such experiments we are using one physical process, or type of process, to simulate another. And, in all such experiments, there is the same potential for things to go wrong, for the experiment to be uninformative, because the experimental subject fails to simulate the target subject, because the homomorphism between reality and experimental process fails.

In short, "studying proximate systems as stand-ins for target systems of interest ... pervades all science" [17]. The epistemology of computer simulation is the same as the epistemology of experimentation for the simple reason that all experiments are simulations and computer simulation experiments are just a special type.

4.4 Special Epistemology

Regardless of our arguments and proposed interpretation of the epistemology of simulation, various philosophers of science have claimed that computer simulation is a new methodology demanding a new epistemology (e.g., [41, 43], [25, p.54]). Computer simulation is certainly a new methodology, involving new tools and techniques. Expertise in experimental physics is hardly interchangeable with expertise in physics simulation. But what additional reasons have been advanced for demanding a new epistemology? Some of the features of simulation said to require new epistemological thinking are:

- *Visualization.* Coupling computer simulations with visualization, including animations, is very common; indeed, computer simulation is being used specifically to create artistic animations (e.g., [28]). Humphreys [25, pp.111–114] seems to think that the use of visualization is a defining characteristic of simulation, which is clearly going too far. For example, in our discussion below of our simulation of parental investments, we have not found it necessary to include any graphics or imagery, nor would our argument be weakened had we never produced any. Regardless, Frigg and Reiss [17] are surely right that the importance of visualization in coping with massive amounts of data is a property that simulation shares with experimentation.
- *Approximation.* All computer simulations, short of simulations isomorphic to the target process, of course, are approximate. But this again is no unique property of computer simulation. There are many examples of the use of physical experiments which are known to distort the properties of target systems. Wind tunnels are used to investigate aerodynamics, however their walls introduce "unnatural" turbulence which can affect the process under study [30]. Some such distortion is true of any scale model and, more generally, of any physical system not strictly identical to the target system.
- *Discretization.* Something which may well be distinctive about computer simulations is that they are housed within discrete von Neumann machines.

We've already mentioned that time must be represented as a sequence of time steps, whereas real processes do not step through time. More generally, any state variable must be represented with some finite degree of precision, implying misrepresentation of real numbers. However, once again, this point of potential epistemological concern generalizes to experimental apparatus quite generally: although in a real-world experiment both the experimental and target processes may well both be continuous processes, the experimenter will have no way of taking advantage of that (in either manipulation or observation) beyond some finite degree of error.

- *Calibration.* Typically simulations have various parameters that need to be calibrated so as to reproduce known phenomena of the real system. For example, in a simulation of the evolution of group selection the strength of altruistic behavior needed to be adjusted to produce a stable population of altruists [1]. This might suggest that you can get whatever result you want by recalibrating your simulation. It's implausible that science constructs reality according to its wishes, despite the more extreme views of social constructivists, but it's far more plausible that computer scientists can construct virtual reality according to their wishes. We accept that there is some danger here; the flexibility of universal computation can cover many faults of theory, if allowed to do so. However, we again assert the parallelism between simulation study and experimental study: given the validation of structural properties of a simulation, the calibration of parameters of the simulation can only push the results so far and not infinitely far. Such calibration serves the identical purpose with calibration in physical experiment, that of finding the settings which support previously observed measurements of a target system under given initial conditions, and so supporting the claim that measurements under new conditions will be informative.

5 Example Simulation: The Evolution of Parental Investment

We would like to illustrate some of the issues that have arisen with an artificial life simulation. Such simulations have received considerably less attention in the philosophical literature than Monte Carlo estimation, yet they are becoming much more prominent in the work of biologists and social scientists, usually under the rubric "agent-based" or "individual-based" modeling. Here we discuss an example of evolutionary ALife simulation, one of a number we have used to investigate issues in the theory of evolution which have been contentious and which resist any easy recourse to ordinary experimental test [27]. Because of the contentiousness of these issues and because of the complexity of the real systems being hypothesized about, it is not likely that these simulations can soon resolve the problems addressed. However, it should be clear that there is potentially a rich field of patterns to validate such homomorphisms as we can create.

Our simulation is an agent-based evolutionary ALife simulation. This means that there are agents with a phenotype (behavior) arising from an interaction between a heritable genotype and their environment. The agents' behavioral repertoire includes reproduction, when suitable mates are present and when their level of health is sufficient. Offspring are created by chromosomal crossover and mutation and receive an initial donation of health from each parent. The health of agents is a function of their foraging abilities, their levels of activity and how much health they donate to offspring. When an agent's health drops to zero, the agent dies. The agent will die regardless when it reaches its maximum life span.

We used this simulation to investigate various hypotheses proposed to explain the evolution of dimorphic sex-linked traits and, in particular, why parental investments differ between males and females, with females characteristically (but not universally) investing more in their offspring than males [40]. In order to do this, the amount of health investment in offspring was a heritable, sex-linked trait in our simulation; we also simulated variable gestation terms. (For further details of our simulation see [27].)

5.1 Hypotheses

We designed simulation experiments to test three widely discussed explanatory hypotheses for the evolution of differing parental investments:

1. *Concorde hypothesis:* The sex that makes the greater initial investment has the more to lose and is thus the sex more likely to evolve further investment [40]. Thus, differential investments may arise by chance and then be fixed by subsequent evolution.
2. *Desertion hypothesis:* The sex which has the first chance to desert an offspring (leaving it with the other parent) will do so [8]. For example, in many fish it is the male who looks after the offspring; since males must wait for females to spawn their eggs before fertilizing them, females have the first opportunity to desert. Since fertilization occurs inside the female in mammals, males have the first opportunity to desert and, indeed, generally invest less.
3. *Paternal uncertainty hypothesis:* The sex which is less certain of being the parent of an offspring will invest less in its (apparent) offspring, particularly in species where females go through a gestation period [40].

5.2 Experiments

5.2.1 Concorde Hypothesis

Dawkins and Carlisle [8] attacked Trivers' first hypothesis above, suggesting it involves fallacious reasoning of the sort used to defend continued spending on a project based on sunk costs rather than on future potential. They used the then topical example of continuing government spending on the Concorde supersonic airliner, "justified" by prior extravagant waste on the project. It seems unlikely that the forces of evolution, unlike the forces of government, would succumb to such fallacious reasoning, rather than responding to the better supported principle of evolving traits to maximize expected fitness. As we agreed with Dawkins and Carlisle's reasoning, we fully expected our simulation of the Concorde hypothesis to fail to establish or sustain dimorphic parental investments.

It's clear that our basic simulation must favor some combined level of investment from both parents, since investment is necessary for offspring to achieve a level of health required for reproduction and since over-investment will be punished by an inability for the parents to further reproduce. However, the basic simulation offered no advantage to one sex or the other for differential investment, as is clear from the random walk in the relative size of sex-linked investments that results from running it.

In order to test the Concorde hypothesis, the simulation needed to be set up with the initial condition that a randomly selected sex starts with a higher investment than the other. Otherwise, the basic simulation was unchanged. The result was that all parental investments rapidly converged on 1/2 the level of combined investment optimal for the basic simulation. As the simulation incorporates all and only the basic ingredients of evolution, two sexes and parental investments, this seems to be a clear experimental refutation of Trivers' conjecture; alternatively, assuming the falsity of that conjecture was not in doubt, this is a clear experimental support for the adequacy of our basic simulation.

5.2.2 Desertion Hypothesis

To test the desertion hypothesis we allowed parents to invest health for an evolvable period after birth, contingent upon their maintaining contact with their offspring. The child needed a minimum total investment period; if one parent quit investing before that minimum, the other parent was forced to make up the difference as required by the offspring. Furthermore, females had a fixed minimum investment period in addition to an evolvable variable period, making up their total investment period. Males only had an evolvable variable period, allowing them an opportunity to desert first. Thus, if the hypothesis is correct, then average investment periods should lengthen for females and shrink for males, when they are initialized to be roughly equal. In most of our simulation runs there was a clear diminution of investment periods for males, although they did not drop to zero; female investment periods were reliably sustained by evolution well above the fixed minimum. Under some circumstances male and female investments would not diverge, such as when the females could not make up the difference or when agent mobility was reduced, so the opportunity to desert failed to arise. In summary, our simulation appears to have supported Dawkins and Carlisle's hypothesis that an earlier opportunity to desert combined with a fixed minimum amount of parental investment will result in sexual dimorphism.

5.2.3 Paternal Uncertainty Hypothesis

We tested the paternal uncertainty hypothesis by fixing the probability of paternity as a parameter of the simulation. Mothers always invested in their own offspring. However, males were chosen by the mother to invest in her offspring, according to the probability of paternity. In other words, if the paternity probability was set at 1, then the female always chose the true father for investment; if 0, then the female never chose the true father; and otherwise she chose males randomly, but with a probability of paternity fixed by the simulation parameter.

We're sure the reader can anticipate our results at this point. In simulations where the paternity parameter was 0, health investments by "fathers" were largely altruistic, and the investment level evolved downwards.[12] On the other

---

[12]Since mobility is a factor in our simulations, there was in fact kin selection pressure in favor

hand, the parameter 1 resulted in maintaining high levels of investment and intermediate parameters resulted in intermediate levels of investment evolving.

5.3 Parental Investments

None of our experimental work with parental investment is particularly profound. We implemented a basic, straightforward simulation framework suitable for testing theories about parental investment, and then we implemented straightforward, clear mechanisms for each hypothesis, enitrely in accord with Grimm and Railsback's account of simulation as an experimental inquiry. The result is we have two candidate explanatory mechanisms and one which is unviable. There are, of course, other conceivable explanations for the evolution of dimorphic investments, which can be tested likewise. And moving on from demonstrating viable mechanisms for evolution to *asserting* the correctness of one or another hypothesis as an explanation for any actual case of dimorphic investments will require, minimally, investigations establishing the existence of a corresponding (homomorphic) mechanism in the real system and the non-existence of other processes which would negate or overwhelm it. So, what we have done with our simulations here is modest: we have made clear which of these three mechanisms is *capable* of being incorporated in legitimate explanations, in circumstances reasonably close to those we have simulated.

6 Conclusion

The experimentation with computer simulations that has become a prominent feature in the sciences is more than experimentation in name only. It is full-blooded experimentation. It carries problems, techniques and methods which are clearly new, such as debugging methods. It carries with it problems, techniques and methods which are old, as well, such as figuring out which statistics to capture to obtain an informative view of what is happening. None of the issues raised in this paper to this point actually identify any interesting limitation on computer simulation or the need for any new epistemology. The difficulties with sorting out the epistemology of experimental science are not yet adequately resolved; but there is no reason to believe that that epistemology won't have rich enough resources to accommodate what scientists are today doing with their computers.

The limits of computer simulation are, thus far, the limits of Turing computation. We know some of what lies beyond those limits, what has been called hypercomputations [7], computations which, for example, infinitary machines can deal with. Despite the pessimism of Dreyfus, Penrose and Humphreys [10, 16, 32], and many others, however, we have been given no reason to believe that human mental capacities are beyond the capabilities of Turing computability. In consequence, the prospects for the "ultimate" simulation, that of the human brain — completing the practical goal of producing a genuine artificial intelligence, are very real, if also rather distant.

---

of sustaining some level of investment. That is, even though the designated fathers were never the true fathers, they were more likely to be near relatives than unselected (because distant) agents.

Bibliography

[1] Appalanaidu, C. Group selection: An investigation into the potential for the evolution of virtual ecosystems. Clayton School of IT, Monash University, 2007. Honours Thesis.

[2] Axelrod, R. *The Complexity of Cooperation: Agent-Based Model of Competition and Collaboration.* Princeton, NJ: Princeton University Press, 1997. Suggests that agent-based modelling (not specifically simulation) is used to substitute for deduction, when formal mathematics can not generate appropriate conclusions and is a kind of intuitive technique.

[3] Axtell, R., Axelrod, R., Epstein, J. M., Cohen, M. D. Aligning simulation models: A case study and results. *Computational and Mathematical Organization Theory*, 1:123–141, 1996.

[4] Baldwin, M. J. A new factor in evolution. *The American Naturalist*, 30:441–451, 1896.

[5] Borges, J. *Historia Universal de la Infamia.* Buenos Aires: Emece, 1954.

[6] Chalmers, A. F. *What is this thing called science?* St. Lucia, Qld: University of Queensland Press, 1982.

[7] Copeland, B. J. Hypercomputation. *Minds and Machines*, 12:461–502, 2002.

[8] Dawkins, R., Carlisle, T. Parental investment, mate desertion and a fallacy. *Nature*, 262:131–133, 1976.

[9] DiPaolo, E. A., Noble, J., Bullock, S. Simulation models as opaque thought experiments. In Bedau, M. A., McCaskill, J. S., Packard, N. H., Rasmussen, S. (Eds.), *Artificial Life VII: Proceedings of the Seventh International Conference on Artificial Life*, 497–506. Cambridge, MA: MIT Press, 2000.

[10] Dreyfus, H. L. *What Computers Still Can't Do: A Critique of Artificial Reason.* Cambridge, Mass: MIT Press, 3rd edn., 1992.

[11] Edmonds, B., Hales, D. Computational simulation as theoretical experiment. *Journal of Mathematical Sociology*, 29:209–232, 2005.

[12] Eidelson, B. M., Lustick, I. Vir-pox: An agent-based analysis of smallpox preparedness and response policy. *Journal of Artificial Societies and Social Simulations*, 7(3), 2004.

[13] Epstein, J. M., Axtell, R. *Growing Artificial Societies: Social Science from the Bottom Up.* Cambridge: MIT Press, 1996.

[14] Franklin, A. *The Neglect of Experiment.* Cambridge University Press, 1986.

[15] Franklin, A. *Experiment, right or wrong.* New York: Cambridge University Press, 1990.

[16] Freedman, D., Humphreys, P. Are there algorithms that discover causal structure? *Synthese*, 121:29–54, 1999.

[17] Frigg, R., Reiss, J. A critical look at the philosophy of simulation. *Synthese*, 2008. Forthcoming.

[18] Giere, R. Using models to represent reality. In Magnani, L., Nersessian, N., Thagard, P. (Eds.), *Model-Based Reasoning in Scientific Discovery*, 41–57. New York: Kluwer, 1999.

[19] Grimm, V., Railsback, S. *Individual-based Modelling and Ecology.* Princeton: Princeton University Press, 2005.

[20] Grimm, V., Revilla, E., Berger, U., Jeltsch, F., Mooij, W. M., Railsback, S. F., Thulke, H., Weiner, J., Wiegand, T., DeAngelis, D. L. Pattern-oriented modeling of agent-based complex systems: Lessons from ecology. *Science*, 310:987–991, 2005.

[21] Hartmann, S. The world as a process: Simulation in the natural and social sciences. In Hegselmann, R., Müller, U., Troitzsch, K. (Eds.), *Modelling and Simulation in the Social Sciences from the Philosophy of Science Point of View*, 77–100. Kluwer, 1996.

[22] Hinton, G. E., Nowlan, S. J. How learning can guide evolution. *Complex Systems*, 1:495–502, 1987.

[23] Humphreys, P. Computer simulations. In *Philosophy of Science Association 1990*, vol. 2, 497–506. 1991.

[24] Humphreys, P. Numerical experimentation. In Humphreys, P. (Ed.), *Patrick Suppes: Scientific Philosopher*, vol. 2. Dordrecht: Kluwer, 1993.

[25] Humphreys, P. *Extending Ourselves: Computational Science, Empiricism, and Scientific Method*. Oxford: Oxford University Press, 2004.

[26] Kueppers, G., Lenhard, J., Shinn, T. Computer simulation: Practice, epistemtology, and social dynamics. In Lenhard, J., Kueppers, G., Shinn, T. (Eds.), *Simulation: Pragmatic Construction of Reality*. Dordrecht: Springer, 2006.

[27] Mascaro, S., Korb, K. B., Nicholson, A. E. An alife investigation on the origins of dimorphic parental investments. In Abbass, H. A., Bossamaier, T., Wiles, J. (Eds.), *Proceedings of the Australian Conference on Artificial Life*, 171–185. 2005.

[28] McCormack, J. A developmental model for generative media. In Capcarrere, M. S., Freitas, A. A., Bentley, P. J., Johnson, C. G., Timmis, J. (Eds.), *Proceedings of the 8th European Conf. on Advances in Artificial Life*, 88–97. 2005.

[29] Morgan, M. Model experiments and models in experiments. In Magnani, L., Nersessian, N. J. (Eds.), *Model Based Reasoning: Science, Technology, Values*, 41–58. Springer, 2002.

[30] Norton, S. D., Suppe, F. Why atmospheric modelling is good science. In Miller, C., Edwards, P. (Eds.), *Changing the Atmosphere: Expert Knowledge and Environmental Governance*, 67–105. Cambridge, MA: MIT Press, 2001.

[31] Oreskes, N., Shrader-Frechette, K., Belitz, K. Verification, validation and confirmation of numerical models in the earth sciences. *Science*, 263(5147):641–646, February 1994. 2003-02-17 A fairly strong position against simulation contributing to knowledge.

[32] Penrose, R. *The Emperor's New Mind : Concerning Computers, Minds, and the Laws of Physics*. Oxford: Oxford University, 2nd edn., 1999.

[33] Pritsker, A. A. B. Compilation of definitions of simulation. *Simulation*, 33:61–63, 1979.

[34] Pritsker, A. A. B. *Introduction to Simulation and SLAMII*. John Wiley & Sons, 1984.

[35] Racynski, S., Bargiela, A. *Modeling And Simulation: Computer Science of Illusion*. Research Studies Pr, 2007.

[36] Reddy, R. Epistemology of knowledge-based systems. *Simulation*, 48:161–170, 1987.

[37] Reynolds, C. W. Flocks, herds, and schools: A distributed behavioral model. *Computer Graphics*, 21:15–34, 1987.

[38] Rohrlich, F. Causal discovery via MML. In *PSA 1990*, vol. II, 507–518. 1991.

[39] Suppe, F. *The Structure of scientific theories*. Urbana: University of Illinois Press, 1977.

[40] Trivers, R. L. Parental investment and sexual selection. In Campbell, B. (Ed.), *Sexual Selection and the Descent of Man*, 136–179. London: Heinemann, 1972.

[41] Winsberg, E. Sanctioning models: The epistemology of simulation. *Science in*

Context, 12(2):275–292, 1999.

[42] Winsberg, E. Simulations, models, and theories: Complex physical systems and their representations. In *Proceedings of the 2000 Biennial Meetings of the Philosophy of Science Association (Supplement to Philosophy of Science)*, vol. 68, S442–S454. 2001.

[43] Winsberg, E. Simulated experiments: methodology for a virtual world. *Philosophy of Science*, 70(1):105–125, 2003.

[44] Zeigler, B. *Theory of Modeling and Simulation*. New York: Wiley-Interscience, 1976.

Updating Classical Mereology [1]

Thomas Mormann

University of the Basque Country UPV/EHU

ylxmomot@sf.ehu.es

ABSTRACT. In this paper it is argued that classical Boolean mereology cannot deal with problems concerning structured wholes and structured parts. Elementary examples of structured mereological systems are provided by Boolean algebras, groups, similarity structures and topological structures. In general mereological systems turn to be non-Boolean. Classical Boolean mereology is to be considered only as a very special case. A truly general mereology as a general theory of parthood has to take into account the various kinds of structures present in the mereological wholes under consideration.

This idea can be rendered precise in the framework of category theory. More precisely, it can be shown that every category comes along with its own specific mereology. Depending on the category's structure the category-relative mereology more or less deviates from classical Boolean mereology.

1 Classical Mereology

Classical mereology is the theory of classical mereological systems. I take a classical mereological system to be a complete Boolean algebra. In this way, formally, classical mereology may be considered as an elementary part of the theory of Boolean algebras, since philosophers usually ignore the more advanced parts of the theory of Boolean algebras, and leave it to the mathematicians.[2]

Standard examples of classical mereological systems in this sense are the power sets of sets. Take for instance the set $X = \{a, b, c\}$. Then the resulting mereological system can be depicted by the following familiar lattice diagram in Fig.1:

In other words, the *elements* of the Boolean algebra PX are the parts of X. Not all complete Boolean algebras are power sets PX, however. A more general class of classical Boolean systems is provided by the class of Boolean algebras

---

[1] The author is with the Department of Logic and Philosophy of Science, University of the Basque Country UPV/EHU, Donostia-San Sebastian, Spain. Email:ylxmomot@sf.ehu.es

[2] Many mereologists prefer to exclude the bottom element 0 of a Boolean algebra. According to them, there is no "empty part". Then they conceive a classical mereological system as a Boolean algebra minus 0. Others have qualms even with the existence of a top element 1, often called the universe that comprises all parts of mereological system, still others only admit certain kinds of fusions, in particular they consider fusions of infinitely many mereological individuals as unpalatable. In this paper I don't want to discuss problems of this kind. Since this paper is intended to be a contribution to the area of formal methodology I will stick to the mathematically more convenient presentation of the theory. Hence from now on I consider a classical mereological system to be a complete Boolean algebra with bottom 0 and top 1.

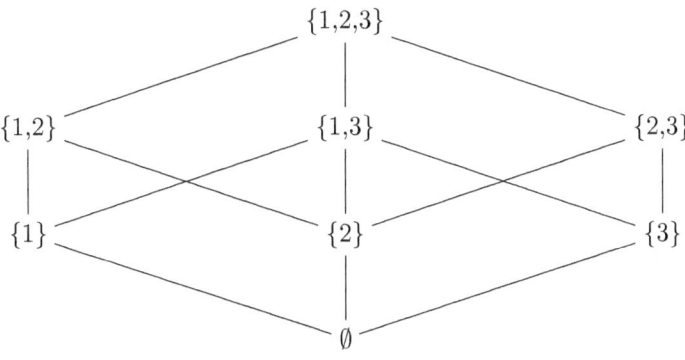

Figure 1:

O^*X of regular open subsets of topological spaces X.[3] Notwithstanding this fact, power sets PX may still be considered as the most important class of classical (complete) mereological systems, in particular since David Lewis in *Parts of Classes* vigorously advocated the thesis that sets and their subsets are to be conceived as mereological systems. More precisely, Lewis argued that mereology and set theory are intimately related in that the former is to be conceived as the elementary "innocent" part of the latter (cf. [7, chap.1]). Conceiving classical Boolean mereology as the elementary part of set theory he self-confidently asserted: "I myself take mereology to be perfectly understood, unproblematic, and certain." (ibidem, 75). For philosophers like him, mereology is just a tool that by itself is philosophically not very interesting. This is a rather bold claim. In the following I'd like to argue that it is wrong. Mereology is more complex and more interesting than partisans of classical mereology might have imagined. Asserting that mereology is to be understood formally as the theory of Boolean algebras is to claim that the mereological concepts of part and whole are fully captured by the conceptual framework of Boolean algebras. In other words, everything concerning parts and wholes can be expressed in terms of complete Boolean algebras. This is a rather strong claim. In this paper I'll argue that it is not only strong but untenable.

Instead of casting mereology in the narrow framework of Boolean algebras I propose to conceive the notions of part, whole, and their relatives as *context dependent*. It is not plausible to assume that "part" and "whole" mean always the same, in particular it is not plausible that the parts of a whole always form a complete Boolean algebra. Take, for instance, the body of a living being. Then one may consider its heart as a part of the whole body. It seems, however, strange to say that the "body minus the heart" is just another part of the body that serves as the "complement" of the "heart-part". But this is required from the perspective of Boolean mereology. Thus, classical mereology is not very good in dealing with structured wholes and structured parts. But most wholes to be met in the world are structured wholes in some way or other.

Traditional mereology bluntly assumes that "structured wholes" are of no concern for mereology in its genuine sense. Mereology is assumed thus abstract

---

[3] Actually, due to Stone's representation theorem, all complete Boolean algebras can be represented in this way.

that it can safely ignore any non-mereological aspects of the systems it is dealing with. Thereby the ken of mereology is severely restricted, and not much is left for it. The problem of the relation between structure and mereology not only concerns "difficult" entities such as "organic wholes", it pops up also for artifacts. If the carburetor of my car is part of the car, is the "car minus the carburetor" another "complementary" part of the car? From the point of view of a car mechanic the latter "part" of the car hardly makes sense. Or, take another more formal example. Take Boolean algebras and consider them as candidates of mereological investigations asking the question

What are the parts of a Boolean algebra?

In analogy to Lewis's claim that the parts of a set are its subsets it does not seem too far-fetched to contend that the parts of a Boolean algebra are its Boolean subalgebras. As we shall see in a moment, this opens the gate to a wealth of non-classical, i.e. non-Boolean mereologies, since, as it is well-known, the *Boolean subalgebras* of a Boolean algebra in general do *not* form a Boolean mereological system. Thus, the mereology of Boolean algebras does not fit into the framework of classical Boolean mereology. This is already shown by the algebra of Boolean subalgebras of the power set $P(\{a,b,c\})$ of a set with three elements a, b, and c. It looks as follows:

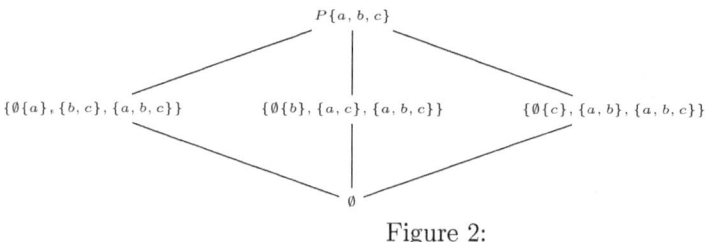

Figure 2:

This lattice, often called the "diamond" (cf. [3, p.132]), is clearly not a Boolean lattice, it is not even a distributive lattice. Thus, conceiving a Boolean algebra B as a structured whole whose structured parts are its Boolean subalgebras leads us outside the ken of classical Boolean mereology. This phenomenon is, of course, not restricted to Boolean algebras. Virtually all non-trivially structured entities lead to a non-Boolean structural mereologies. To put it bluntly, Boolean mereological systems are not the rule but rather the exception. In general, mereological systems are non-Boolean systems. I take the failure of classical Boolean mereology to cope with problems of structure as a good reason to consider the project of revising and updating classical mereology.

The outline of this paper is as follows: In Section 2 we will deal with an elementary example of a structural mereology, to wit, the structural mereology of groups. The mereology of groups, it will be argued, may be considered as a paradigmatic example of a structural mereology that considerably differs from standard Boolean mereology. In Section 3 we outline the general format of structural mereologies in the framework of the mathematical theory of categories. The upshot will be that every category **C** comes along with its own mereology. Depending on the kind of **C**, this **C**-mereology turns out to be more or less similar to the classical Boolean mereology of sets.[4] As is to be expected the classical Boolean mereology of sets can be identified with the mereology of the category **SET** of sets. As another example of a structural mereology, which is rather similar but still different from set-theoretical mereology, in Section 4 we consider the mereology of similarity structures, i.e., sets endowed with a reflexive and symmetric similarity relation. A similarity structure may be considered as a rather simple kind of a spatial structure that allows to introduce the notion of neighborhood. Another example of a structural mereology related to spatial structures is discussed in Section 5, namely the structural mereology of spaces dealing with topologically well behaved parts of topological spaces. In Section 6 it is argued that the general category-theoretical perspective on mereology may also shed new light on a classical mereological problem that has been considered already by Plato and Aristotle, namely, the problem whether the whole is "more" than its parts. The paper concludes with some general remarks on the prospects of a generalized mereology.

2 Structural Mereology I: Groups

The objects of the world are rarely blobs lacking any structure, rather they are structured in some way or other. One may even doubt that the concept of an object without any structure makes sense at all. Hence let us take as our starting point the general assumption that the objects of the world we are dealing with are structured objects, "structured" to be understood in a broad sense that need not be specified for the moment. Then, given a structured whole W, it is reasonable to ask for its *structured* parts, not just for its parts.

In the first section we already mentioned the case of Boolean algebras and their parts. If one considers the Boolean subalgebras of a Boolean algebra as its parts, the resulting mereological system of subalgebras is in general *not* a Boolean algebra. For didactical reasons, instead of Boolean algebras, I propose to consider the even more elementary case of groups and their structural parts. Recall that a group G is a set endowed with an associative multiplication $G \times G \xrightarrow{m} G$ such that there is a unique neutral element $e \in G$ satisfying $m(a,e) = m(a,e) = e$, and for all $a \in G$ there is an $a^* \in G$ with $m(a,a^*) = m(a^*,a) = e$. The element a^* is called the inverse of a, and the neutral element e is often called the unit of G. For $a, b \in G$ the product $m(a,b)$ is often denoted by $a \cdot b$ or simply by ab. A group is called abelian or commutative if $m(a,b) = m(b,a)$. A subgroup H of G is a subset of G that is a group under the multiplication of G, i.e., it contains e, and with $a, b \in H$ also a^{-1} and $m(a,b)$ are elements of H.

Groups abound in mathematics, physics, and elsewhere. Let us just mention the group of integers $\mathbf{Z}$, the groups $\mathbf{Z}_n$, of natural numbers modulo n, $n \in \mathbf{N}$, the real numbers $\mathbf{R}$, and symmetry groups such as the Lie groups $O(m), SU(m)$ endowed with their standard (matrix) multiplications.

There are several candidates for the office of the structural parts of a group G, but certainly the most straight-forward choice is to take the subgroups of G

---

[4] For more detailed accounts of this category-relative mereology (although not under this name) from a mathematical perspective the reader may consult the presentations to be found in Lawvere and Schanuel's *Conceptual Mathematics* [5](elementary), and Lawvere and Rosebrugh's *Set Theory for Mathematicians* [4] (more advanced).

as its structural parts.[5] In line with our experiences with parts of sets one may naively conjecture that the more elements a group has the more complicated its subgroup structure tends to be. This is not always true, however. Take the groups $\mathbf{Z_p}$ of integers modulo p, $p \in \mathbf{N}$ and prime. Since there are arbitrarily large prime numbers p there are groups of arbitrary large order whose only parts are the trivial group E having only one element and $\mathbf{Z}_p$ itself. Mereologically more interesting is the group $\mathbf{Z}$ of integers (with standard addition + as group operation) that has infinitely many subgroups. They are all of the form $n\mathbf{Z} := \{\cdots, -2n, -n, 0, n, 2n, \cdots\}$ for $n \in \mathbf{N}$.

In order to keep matters as simple as possible, in the rest of this paper we will consider only finite groups having therefore finitely many subgroups. To deal with these groups it is expedient to characterize them by generators and relations. More precisely, we will deal with the following groups ($z^k = z \times \cdots \times z$ (k times)):

$$\begin{aligned}
\mathbf{Z}_m &:= \{e, z, z^2, , z^{m-1}; z^m = e, m \in \mathbf{N}\} \\
K &:= \{e, x, y, xy; x^2 = y^2 = e \text{ and } xy = yx\} \\
&= Z_2 \times Z_2 \text{ (Klein group)} \\
S_3 &:= \{e, x, y, y^2, xy, xy^2; x^2 = y^3 = e \text{ and } xy = y^2x\}
\end{aligned}$$

In analogy to the Boolean lattice Fig.1, which describes the parts of the set $\{1, 2, 3\}$, the structural parts of these groups may be conspicuously exhibited by lattices $PART(G)$. For $G = Z_3, Z_4, Z_6, K$, and S_3 the lattices $PART(G)$ can be depicted diagrammatically as follows:

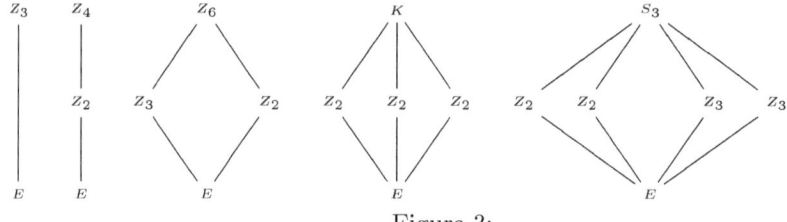

Figure 3:

More explicitly, these lattices are to be read as follows: If, according to the recipe given above, the group $\mathbf{Z}_4$ can be considered as the set $\{e, z, z^2, z^3\}$ its subgroup $\mathbf{Z}_2$ is to be conceived of as the set $\{e, z^2\}$ endowed with the canonical multiplication inherited from $\mathbf{Z}_4$. Similarly, if $\mathbf{Z}_6$ is given by set $\{e, z, z^2, z^3, z^4, z^5\}$ the subgroup $\mathbf{Z}_2$ is given by $\{e, z^3\}$ and the subgroup $\mathbf{Z}_3$ is given by $\{e, z^2, z^4\}$ endowed with the multiplication inherited from $\mathbf{Z}_6$. A bit more interesting is the case of the Klein group K. Its three subgroups are given by $\{e, x\}$, $\{e, y\}$, and $\{e, xy\}$, respectively. Following these lines the reader may calculate $PART(S_3)$ for himself.

As is directly observed only $PART(\mathbf{Z}_3)$ and $PART(\mathbf{Z}_6)$ are Boolean.[6] Thus the structural mereology of groups in general is not Boolean. Already these

---

[5] Another plausible choice would be to consider only *normal* subgroups of G as parts of G. From this mereological perspective simple groups (= groups having no normal subgroups) have a trivial mereological structure.

[6] As is easily shown that lattice $PART(\mathbf{Z}_m)$ of structural parts of $\mathbf{Z}_m$ is Boolean iff the natural number m is square-free.

elementary examples show that in general the structural mereology of groups is not Boolean. Different types of subgroup lattices $PART(G)$ may be distinguished and related to the types of the groups G. For a comprehensive account of this perspective on group theory see [10]. For instance, groups of prime order such as $\mathbf{Z}_3$ have a trivial subgroup lattice, since there only part is the trivial group E. The group $\mathbf{Z}_4$ is a bit more complex having a single proper part $\mathbf{Z}_2$. Evidently the parthood structure of $\mathbf{Z}_4$ is non-Boolean since its only non-trivial proper part $\mathbf{Z}_2$ has no complementary counter-part such that the fusion of it and $\mathbf{Z}_2$ would yield the whole $\mathbf{Z}_4$.

More interesting are the cases of the Klein group K that may be described as the Cartesian product $\mathbf{Z}_2 \times \mathbf{Z}_2$ of two copies of the group $\mathbf{Z}_2$ and the symmetric group $\mathbf{S}_3$ that can be characterized as the group of permutations of three elements 1, 2, and 3. Naively one might have guessed that the subgroup $\mathbf{Z}_2$ appears exactly twice in K and once in $\mathbf{S}_3$ due to the fact that K has four elements and $\mathbf{S}_3$ has six elements. The lattices $PART(K)$ and $PART(\mathbf{S}_3)$ refute this guess. The lattice $PART(K)$ is the "diamond" $\mathbf{M}_3$ in which the structural part $\mathbf{Z}_2$ of K appears three times, and $PART(\mathbf{S}_3)$ is even more complicated, exhibiting three copies of $\mathbf{Z}_2$ and one of $\mathbf{Z}_3$. Nevertheless, this lattice structure is not to be interpreted extensionally in the sense that this K is construed of three copies of $\mathbf{Z}_2$, since the group K has only four elements. Rather, one and the same group $\mathbf{Z}_2$ appears three times in the part structure of the Klein group K. Still more complicated is the case of the symmetric group $\mathbf{S}_3$.

In *Against Structural Universals* [6] David Lewis pointed out that some intricate metaphysical problems are lurking here. Does it really make sense to say that, say, one and the same part $\mathbf{Z}_2$ appears three times in one and the same whole, or that the same parts can form different mereological wholes, depending on how they are assembled? Lewis flatly denied that this is possible. He contended that a composition of this kind cannot claim to be mereological at all. According to him the talk that one and the same part appears repeatedly in one and the same whole is simply unintelligible.[7] I don't want to discuss Lewis's examples in this paper, rather I am content to show that his objections can be defused for groups and similar structures.

The appropriate general framework for dealing with this kind of questions concerning structural mereology has turned out to be the mathematical theory of categories. A category **C** may be described as a local mathematical universe of discourse dealing with a certain kind of objects, called the **C**-objects, and their relations called **C**-morphisms. Thereby one obtains, say, the category **TOP** of topological spaces, the category **GROUP** of groups, and countless categories in mathematics, physics, computer science, and other areas (cf. [4, 5, 8]).

Before we deal with mereological problems on the general level of categories, I'd like to deal with the mereology of a concrete and easily accessible specific category, to wit, the category of groups. Then the general case may be more easily grasped.

As has become evident in the evolution of modern mathematics, groups and other structures do not live in isolation. Rather, group theory does not only study groups in themselves, an essential part of group theory is the study of

---

[7] Lewis did not deal with groups but with "structural universals" such as "methane" or "butane" that allegedly were composed of more primitive universals such as "hydrogene" and "carbon" as is indicated by their chemical formulas CH_4 and C_4H_8, respectively.

relations between groups. The most important relations are homomorphisms (or structure-preserving mappings) $H \xrightarrow{f} G$ defined as follows:

Definition 1 *Let H and K be groups. A homomorphism h with domain H and codomain K is a set-theoretical map $H \xrightarrow{h} K$ satisfying the conditions:*

$$f(ab) = f(a)f(b) \qquad (1a)$$
$$f(e) = e \qquad (1b)$$
$$f(a^*) = f(a)^* \qquad (1c)$$

The set $h(H) := \{h(x); x \in H\}$ is a subset of G and called the image of H in G. As is easily shown the image $h(H)$ of H is a subgroup of G. Between any two groups H and K there always exists the trivial homomorphism $H \xrightarrow{t} K$ mapping all elements a of H to $t(a) = e$. Given two groups H and K it is usually difficult to find a non-trivial homomorphism $H \xrightarrow{f} K$. It may even happen that none exists. For instance, there is no non-trivial homomorphism between groups whose order is prime to each other, e.g. $\mathbf{Z}_2$ and $\mathbf{Z}_3$. For the following we need to distinguish between different types of homomorphisms:

Definition 2 *A group homomorphism $H \xrightarrow{h} K$ is a monomorphism if and only if $h(x) = h(y)$ entails $x = y$. A group homomorphism $H \xrightarrow{h} K$ is an epimorphism if and only if for every $y \in K$ there is at least one $x \in H$ such that $h(x) = y$. A homomorphism is an isomorphism if and only if it is a monomorphism and an epimorphism.*

Definition 3 *Let $H \xrightarrow{h} K$ and $H' \xrightarrow{h'} K$ be two monomorphisms with the same target K. The monomorphisms h and h' are called equivalent if and only if there is an isomorphism $H \xrightarrow{s} H'$ such that $h = h' \cdot s$, i.e., for all $a \in H$ one has $h(a) = h'(s(a))$.*

It is easily proved that Definition 3 defines indeed an equivalence relation between monomorphisms. The equivalence class of a monomorphism $H \xrightarrow{h} G$ may be denoted by $[h]$. In order not to overburden notion we will often blur the distinction between h and $[h]$, i.e., we will talk of a monomorphism $H \xrightarrow{h} G$ even if we really mean the equivalence class $[h]$ of h. Now we are ready to formulate the central notion of this section, to wit, the concept "part of a group":

Definition 4 *Let G be a group. A part of G is an equivalence class of monomorphisms $H \xrightarrow{h} G$ according to the equivalence relation Definition 3. If $H \xrightarrow{h} G$ represents a part of G, then H is called the type of h (or $[h]$). If it is not necessary to refer explicitly to h, we may simply call H a part-type of G.*

Assume that $H \xrightarrow{h} G$ and $H' \xrightarrow{h'} G$ are equivalent monomorphisms, i.e., define the same part of G. Then the images $h(H)$ and $h'(H')$ coincide. Since $h(H) = h'(H')$ is a subgroup of G one can identify a part of G with a subgroup of G, to wit the image $h(H)$ or $h'(H')$. Since we aim at a general clarification of the concept of part, which is not confined to group theory, it is nevertheless expedient to stick the clumsier terminology of parts as (equivalence classes of) monomorphisms $H \xrightarrow{h} G$, even if for groups a more elegant terminology is available.

If $H \xrightarrow{h} G$ is a part of G it is natural to express this by saying that h is a way how the part-type H is involved in G. Or, still differently, we may say a part of G is a part-type-in-a-way. Distinguishing between parts and part-types allows us to speak meaningfully that one part may appear in different ways. This will be shown by the following elementary examples. More precisely, we will show that one and the same group H – conceived as a part-type of an other group G - may give rise to different parts of G, i.e., there may exist non-equivalent monomorphisms $H \xrightarrow{h} G$ and $H \xrightarrow{g} G$. Take, for instance the Klein group K, and denote the generator of the group $\mathbf{Z}_2$ by z. Then three non-equivalent monomorphism $\mathbf{Z}_2 \xrightarrow{h_i} K$, $i = 1, 2, 3$, are defined by

$$h_1(z) = x$$
$$h_2(z) = y$$
$$h_3(z) = xy$$

According to Definition 3, these homomorphisms define *different* parts of K all of which are of the same type $\mathbf{Z}_2$. It goes without saying that this phenomenon is not restricted to the Klein group but occurs for many groups. The group of permutations S_3 provides a particularly interesting example since it reveals that in the case of groups a structured whole is not necessarily determined by its part-types alone. It might happen that two different structured wholes have the same part types. This is quite in line with our mereological intuitions in that often the same parts may be assembled in different ways so that they form different wholes. Take the groups $\mathbf{Z}_6$ and S_3. Although they both have six elements they are non-isomorphic, since the former is commutative and the latter not. A classical theorem of group theory tells us that they can only have parts of type $\mathbf{Z}_2$ or $\mathbf{Z}_3$. Indeed, the only parts of Z_6 are given by the monomorphisms $\mathbf{Z}_2 \xrightarrow{h} Z_6$ and $\mathbf{Z}_3 \xrightarrow{k} Z_6$ defined by $h(z) = x^3$ and $k(z) = x^2$. On the other hand, S_3 has a more complicated parthood structure since there are three non-equivalent ways $\mathbf{Z}_2 \xrightarrow{g_i} S_3$, $i = 1, 2, 3$ for $\mathbf{Z}_2$ to become a subgroup of S_3:

$$g_1(w) = x$$
$$g_2(w) = xy$$
$$g_3(w) = xy^2$$

On the other hand there is only one way $\mathbf{Z}_3 \longrightarrow S_3$ of how $\mathbf{Z}_3$ be a part of S_3, namely by mapping its generator z onto y. In sum, it is possible that the very same components $\mathbf{Z}_2$ and $\mathbf{Z}_3$ can be assembled in two different ways. This is in no way paradoxical if one bases one's considerations on the sophisticated definition of parthood, which takes into account the way, how the components are assembled, to wit, the monomorphisms by which the "abstract" groups $\mathbf{Z}_2$ and $\mathbf{Z}_3$ are embedded into Z_6 and S_3. Thereby it is revealed that there is only one way that renders $\mathbf{Z}_2$ and $\mathbf{Z}_3$ parts of Z_6, while in the case of S_3 the group $\mathbf{Z}_2$ gives rise to three different parts, while for $\mathbf{Z}_3$ there is still only one way to become part of S_3.

Pace Lewis, then, according to Definition 4, for groups the talk of "a part many times over" *does* make sense. One and the same group H may be part of a larger group G in many different ways, namely, provided there are different, non-equivalent monomorphisms that embed H in G. There is nothing mysterious about it.

In order to convince a Lewisian skeptic it may be more expedient, not to rely on groups and their structural parts but to show that the same kind of argument goes through for Lewis's preferred kind of mereological systems, to wit, set-theoretical ones. In this case, instead of group-theoretical monomorphisms we simply deal with set-theoretical monomorphisms, i.e. 1-1-set-theoretical functions $Y \xrightarrow{m} X$. Following Definition 3 a part of the set X is defined to be an equivalence class of set theoretical monomorphisms $A \xrightarrow{m} X$. Then m defines a subset of X, namely $m(A) \subseteq X$. An equivalent monomorphism $A' \xrightarrow{m'} X$ yields the same subset $m(A) = m'(A')$. Hence, every part of X defines a unique subset of X in the ordinary sense. On the other hand, if A is a subset of X in the ordinary sense, then the inclusion of A in X yields a canonical monomorphism $A \xrightarrow{i} X$. The equivalence class of this monomorphism defines a unique part of X in the sense of Definition 4. Hence, the notions of subsets and parts of X coincide. Denoting the class of subsets of X by PX we get that PX is a lattice with respect to the order relation $\leq$ that is just the familiar set-theoretical inclusion. More precisely, PX is a Boolean lattice with bottom element $\emptyset$ and top element X.

The essential difference between the lattices PX of subsets of X and the lattices $PART(G)$ of group parts is that the PX are *Boolean* lattices while the subgroup lattices $PART(G)$ in general are not Boolean.[8] The lack of Booleaness for the lattices $PART(G)$ reflects the extra structure present in groups G but not present in arbitrary sets X. If G is a group, not all subsets $A \subseteq G$ are G-parts but only those that are compatible with the group structure, i.e., those that are subgroups. In other words, only a selected subclass of subsets of G qualifies as group parts. Formally, then, the generalization of traditional mereology to a wealth of structural mereologies amounts to giving up the requirement that mereological lattices have to be Boolean lattices. Instead, we subscribe to a more liberal account that allows for other types of lattices as well.

Definition 4 is, however, only the beginning of a full-fledged theory of structural mereology for groups. Up to now, we have only defined the notions of structural parts and part type but have said nothing why the class $PART(G)$ of structural parts G is actually a lattice. For groups, this is intuitively more or less

---

[8] Among the parthood lattices displayed above, only that of $\mathbf{Z}_3$ and Z_6 are Boolean. In general, lattices of subgroups are not even distributive as is shown by the examples K and S_3. For a comprehensive treatment of subgroup lattices see [10].

clear since the subgroups of G are subsets of the set of elements of G endowed with a group multiplication inherited from that of G. Thereby it is easily seen that the set $PART(G)$ of structural parts inherits a lattice structure from the power set PG of G.

Nevertheless, in view of the general category-theoretical account of structural mereology to be developed in the next section, it is desirable to show that this lattice structure can also be obtained from the new definition of part as laid down in Definition 4. That is to say, we'd like to formulate the basic mereological notion such as overlapping, disjointness, fusion in terms of monomorphisms. This can indeed be done, as is shown by the following definition, which provides the base for a full-fledged structural mereology of groups:

Definition 5 *Let $H \xrightarrow{h} G$ and $H' \xrightarrow{h'} G$ represent two parts of G. If there is a monomorphism $H \xrightarrow{p} H'$ such that $h = h' \cdot p$ this is denoted by $h \leq h'$, or, by an abuse of language, simply by $H \leq H'$. Then, again committing an abuse of language, H is called a smaller part of G than H'.*

The definition of $\leq$ is reasonable in the sense that it depends only on the equivalence classes of h and h', not on the representing monomorphisms h and h'. Then it is easily seen that $\leq$ is an order structure on the parts of G, i.e., $\leq$ is reflexive, transitive, and anti-symmetric. Even more can be proved:

Proposition 1 *Let G be a group and denote the set of parts of G as defined in Definition 5 by $PART(G)$. Then $PART(G)$ endowed with $\leq$ has the structure of a lattice, i.e., $(PART(G), \leq)$ is an order structure with minimal element E, maximal element G and every finite subset of $PART(G)$ has an infimum and a supremum with respect to $\leq$. As usual, the infimum $inf(H, K)$ is denoted by $H \wedge K$, and the supremum $sup(H, K)$ is denoted by $H \vee K$.*

This lattice structure on $PART(G)$ enables us to speak of inclusion, composition, overlapping, and disjointness of G-parts much in the same way as for the parts of classical mereological systems such as the system PX of subsets of a set X. For instance, $H \leq H'$ is to be interpreted as that H is included in H', and $H \wedge H' \neq E$ means that H and H' overlap nontrivially. It may be expedient to spell out in some detail what is the infimum $H \wedge K$ and the supremum $H \vee K$, respectively. By definition H and K – as parts of G – are defined by some monomorphisms $H \xrightarrow{h} G$ and $K \xrightarrow{g} G$, respectively. Then $h(H) \cap g(K) \subseteq G$ is a subgroup of G, independent of the representing monomorphisms h and g. Hence the inclusion $h(H) \cap g(K) \xrightarrow{i} G$ defines a well defined part of G in the sense of Definition 4. This part is denoted by $H \wedge K$. Suppressing the various monomorphisms involved one may simply say that $H \wedge K$ is just the intersection of H and K. Thus, the construction of the infimum $H \wedge K$ of the G-parts H and K is not essentially different from the construction of the infimum of two sets X and Y in set theory, which is just the intersection $X \cap Y$.

Things become more interesting when we consider the supremum of $H \vee K$. In this case $H \vee K$ is not the set-theoretical union of H and K, since this is usually not a subgroup of G. Rather, $H \vee K$ may be described as the subgroup of G generated by H and K. More precisely, the following holds: Let H and K – as parts of G – be represented by the monomorphisms $H \xrightarrow{h} G$ and $K \xrightarrow{g} G$, respectively. Then there is a smallest subgroup of G that contains $h(G)$ and $g(K)$. This subgroup may be denoted by $H \vee K \subseteq G$ and defines a part of G in the sense of Definition 5. In other words, just as in the case of standard mereology, for G-parts one can form the infimum (intersection, overlapping) and the supremum (fusion, composition) in such a way, that these operations render the set $PART(G)$ of G-parts a lattice. Although the composition $H \vee K$ of H and K is not just the set-theoretical union but something different. Nevertheless, the "non-mereological" composition $H \vee K$ makes perfect sense, since it is a well-defined subgroup of G. *Pace* Lewis, then, group theory is an example of a domain which provides an honest notion of a "sui generis composition".

It should be clear that our approach not only works for groups but for many other structures as well. The details will be described precisely in the next section. What is going on may be informally described as follows. According to Fig.2 a group G is a "structured set" in that it is a "set plus group structure". Other types of structured sets may be defined analogously. Usually it is not difficult to define appropriate structure-preserving homomorphisms between these structures. Thereby for each type of structured sets one may set up a specific notion of parthood and composition mimicking the definitions Definition 2–5. Thereby we obtain for each structure S a structure-specific mereology encapsulated in the notion structure specific lattice $PART(S)$ of structured parts of S. This program will be carried out in detail in the next section in terms of category theory. The category-theoretical generalization of mereology reveals that the essential structure of generalized or relativised mereology that does not depend on the specific features of group theory or set theory. Moreover, in the category theoretical framework it can be shown that the traditional Boolean structure of set-theoretical mereology is only a special case of the general approach of structural mereology. In other words, standard Boolean mereology, which Lewis took as the only feasible one, and other structural mereologies are on an equal footing. All of them are special cases of a general theory of parthood and composition.

3   A Category-theoretical Framework for General Mereology

The mereologically interesting point is that every category **C** comes along with its own specific **C**-mereology that deals with the **C**-parts of its **C**-objects. Instead of describing how this works in precise abstract terms, let us be content to state that the structural mereologies of Boolean algebras and groups we just mentioned correspond exactly to the category-theoretical mereologies of the categories **BOOLE** of Boolean algebras, and the category **GROUP** of groups. As it should be, the mereology of sets favored by Lewis, is nothing but the mereology of the category **SET** of sets. The examples of the categories **BOOLE** and **GROUP** show that most categories have mereologies that are not classical. From a general category-theoretical point of view, then, there is no reason for mereologists to restrict their attention to Boolean mereology that is only a small facet of the whole range of mereological possibilities. Mereology in general is structural mereology, and it is just a "Boolean prejudice" to ignore non-Boolean mereological systems.

Updating Classical Mereology

After having established that one can set up a structural mereology for groups, encapsulated in the concept a group-part, let us now consider how this recipe can be generalized to all kind of structures. This is done by showing that nothing depends on the specific features of groups in the refined definition of parthood as given in Definition 5. Rather, conceiving the theory of groups and group homomorphisms as one category among many, for every category **C** one may set up its own specific "**C**-mereology".

Definition 6 *A category C is given by the following ingredients:*

1. *A collection of things $A, B, D, \cdots$ called **C**-objects.*
2. *A collection of things $f, g, h, \cdots$ called **C**-morphisms.*
3. *An operation that assigns to each **C**-morphism f a **C**-object $dom(f)$ (the domain of f), and an other operation that assigns to f a **C**-object $cod(f)$ (the codomain or target of f). Thus, morphisms may be displayed as $A \xrightarrow{f} B$ whereby it is assumed that A is the $dom(f)$ and B is $cod(f)$.*
4. *An operation assigning to each pair (f, g) of **C**-morphisms with $dom(g) = cod(f)$ a **C**-morphism $(g \cdot f)$ with $dom(g \cdot f) = dom(f)$ and $cod(g \cdot f) = cod(g)$ such that the following law of associativity is satisfied. Given the configuration $A \xrightarrow{f} B \xrightarrow{g} D \xrightarrow{h} E$ one has $h \cdot (g \cdot f) = (h \cdot g) \cdot f$.*
5. *An assignment to each **C**-object A a **C**-morphism $A - id_A \rightarrow A$ such that for any **C**-morphisms $A \xrightarrow{f} B$ and $B \xrightarrow{g} A$ one has $id_A \cdot g = g$ and $f \cdot id_A = f$.*

Intuitively, the category-theoretical notion of a morphism intends to capture the essential features of the idea of a set-theoretical function. More precisely the requirements Definition 63–5 generalize the essential aspects of the concatenation of set-theoretical functions. Thus, it is an easy exercise to show that there is a category **SET** whose objects are sets, and whose morphisms are set-theoretical maps such that $g \cdot f$ is the familiar set-theoretical concatenation of set-theoretical functions. **SET** is not, however, the only category. Categories abound in mathematics and elsewhere. Large lists of categories occurring in mathematics, computer science and other sciences can be found in any textbook on category theory (cf. [4, 5, 8]). The category **GROUP** of groups is defined as having as objects groups and as morphisms homomorphisms in the sense of group theory. Analogously, categories of manifolds, vector spaces, rings, fields and other structures may be defined. It should be noted, however, that there are categories of a quite different kind than **GROUP** or **SET** whose morphisms are not set-theoretical mappings at all. For instance, a lattice or any other ordered structure may be conceived as a category. All of them have specific notions of parthood and composition.

In order to set up a general category-theoretical analogue of Proposition 1 of group-theoretical parthood, we need purely category-theoretical characterizations of the concepts of monomorphism, epimorphism and isomorphism that do not presuppose the notions of set, structure, and elementhood. This is achieved by the following definition:

Definition 7 *Let **C** be a category.*

1. A **C**-morphism $B \xrightarrow{m} D$ is a **C**-monomorphism if and only if for all **C**-morphisms $A \xrightarrow{f,g} B$ the identity $m \cdot f = m \cdot g$ entails $f = g$.
2. A morphism $D \xrightarrow{e} E$ is a C-epimorphism if and only if for all morphisms $E \xrightarrow{s,t} F$ the identity $s \cdot e = t \cdot e$ entails $s = t$.
3. A **C**-morphism $X \xrightarrow{h} Z$ is a **C**-isomorphism if and only if it is a **C**-monomorphism and a C-epimorphism.

Informally, monomorphisms are morphisms that can be cancelled from the left, and epimorphisms can be cancelled from the right. As is easily checked set-theoretical and group-theoretical homomorphisms are iso/epi/mono/morphisms in the ordinary sense if and only if they are iso/epi/mono/morphisms in the sense of 7. Thus, 7 is the "correct" category-theoretical generalization of the original, more restricted versions of these notions. The point is that 7 does not refer to set-theoretical notions such as set and element. Observing that the equivalence relation 4 is already formulated in purely category-theoretical terms the desired category-theoretical generalization of the concept of parthood is at hands:

Definition 8 Let X be an object of a category **C**, and $Z \xrightarrow{f} X$ a **C**-monomorphism. A monomorphism $Z' \xrightarrow{f'} X$ is equivalent to $Z \xrightarrow{f} X$ if and only if there is a **C**-isomorphism $Z \xrightarrow{j} Z'$ such that $f' \cdot j = f$. Then a **C**-part[9] of X is defined as an equivalence class of **C**-monomorphisms $Z \xrightarrow{f} X$.

Analogously as for the category **GROUP** for any category **C** one may define the relation $\leq$ between **C**-parts of X. Then, under some mild conditions on **C**, for every **C**-object X one obtains an order structure, or, under somewhat stronger conditions, even a *lattice* of its **C**-parts (cf. [4]). In this way, every category comes along with its own ready-made notions of **C**-mereology.

Thus, 8 and the ensuing theory of **C**-parts and **C**-composition answers Lewis's question of what is the general notion of composition and parthood: The general theory of parthood and composition is the theory of the category-relative concepts of **C**-parthood and **C**-composition, **C** an arbitrary category. It should be noted that this theory is not a philosopher's fancy invention but rather is an established mathematical enterprise built up over the last decades.

Before we leave this sketch of a general theory of parthood let us prove that for the category **SET** of sets the definition 8 yields what it should, namely, the standard mereological theory of parthood and composition. This is seen as follows. Let X be a set and $A \subseteq X$ a subset. Then the inclusion of A in X yields a canonical monomorphism $A \xrightarrow{i} X$. The equivalence class of this monomorphism defines a **SET**-part of the **SET**-object X in the sense of 8. On the other hand, every monomorphism $B \xrightarrow{m} X$ defines a subset of X, namely $m(B) \subseteq X$. It is clear that equivalent monomorphism $B \xrightarrow{m} X$ and $B' \xrightarrow{m'} X$

---

[9] In category theory, a **C**-part of a **C**-object X is often called a subobject of X. The family of subobjects is denoted by $Sub(X)$.

yield the same subset $m(B) = m'(B')$. Thus, every **SET**-part of X in the sense of 7 defines a unique subset of X and vice versa. Therefore the set PX of subsets of X and the set of **SET**-parts of X coincide. This evidences that the definition 8 is the correct generalization of the familiar mereological notion of a subset to arbitrary categories. This means that our definition of structural parthood is indeed continuous with the original mereological definition.

4 Structural Mereology II: Similarity Structures

The concept of similarity enjoys a mixed reputation in philosophy. On the one hand, authors such as Goodman and Quine considered similarity or resemblance as highly suspicious and finally philosophical useless. Their verdict was that "similarity was a quack ...". On the other hand, there have been philosophers such as Carnap who, at least for some time, considered the relation of similarity as a sufficient base for carrying out "the logical constitution of the world" (cf. [2]). Although the detractors of similarity have dominated the debate on the philosophical dignity of this concept there are some signs for a change of the tide. After all, similarity plays an important role in many realms of scientific and common sense argumentation.

The aim of this section is to show how mereology can be contextualized in such a way that it takes into account the concept of similarity. For this purpose we start from a rather weak concept of similarity inspired by Carnap's notion of "recollection of similarity" (cf. [2]). There similarity is conceived as a binary relation $\sim$ defined on some domain S of objects, and it is assumed that $\sim$ is reflexive and symmetric, i.e., every object x of S is similar to itself ($x \sim x$) and if x is similar to y then y is also similar to x ($x \sim y \Rightarrow y \sim x$). It is not assumed that the similarity relation $\sim$ is transitive, i.e. from $x \sim y$ and $y \sim z$ one cannot infer that $x \sim z$. Apparently, this is a quite weak, almost trivial notion of similarity. Goodman and Quine have argued at length that this notion of similarity is too weak to be interesting. In the following I'd like to argue that their assessment was wrong.

To get started let us fix notation as follows: Let $(S, \sim)$ be a similarity structure, i.e. a set S endowed with a binary similarity relation $\sim$.[10] In the usual way, one may define various categories of similarity structures, depending on what kinds of similarity structures and similarity-preserving morphisms are admitted. In the following we will restrict our attention to one such category, denoted by **SIM**.[11] As every category **SIM** comes along with a ready-made specific **SIM**-mereology. Indeed, **SIM**-mereological systems are complete Heyting-algebras that can be conceived as generalizations of the classical Boolean systems PX that arise from **SET**-objects X. This is not too surprising since the category **SET** of sets may be conceived as a subcategory of **SIM**, namely, the subcategory of similarity structures $(X, \sim)$ endowed with a trivial similarity relation $\sim$ defined by $x \sim y := x = y$. The **SIM**-mereology provide an elementary and easily

---

[10] Similarity structures appear in the literature under many names: tolerance spaces, coherence spaces and others. These names emphasize the broadly spatial character of these structures.

[11] **SIM** has some interesting categorial properties. Among other things it can be shown that it is almost a topos, namely a quasi-topos.

accessible example of a structural mereology that, although similar to standard **SET**-mereology, differs in some aspects from it.

Definition 9 *The category* **SIM** *of similarity structures is defined as follows:*
    SIM-*Objects:*   Similarity structures $(S, \sim)$.

    SIM-*Morphisms:*   Set-theoretical maps $S \xrightarrow{f} T$ that are similarity-preserving in the sense that $x \sim y \Rightarrow f(x) \sim f(y)$.

Since identity maps $S \xrightarrow{id} S$ are obviously **SIM**-morphisms and the concatenations of **SIM**-morphisms are **SIM**-morphisms again, 9 defines a category. The parts of a **SIM**-object $(S, \sim)$ are defined in the usual way as equivalence classes of **SIM**-monomorphism (see 8). Denote the set of all **SIM**-parts of $(S, \sim)$ by $PART(S, \sim)$. Then $PART(S, \sim)$ is endowed with an order structure $\leq$ as explained in the previous section. Then the following proposition can be proved:

Proposition 2 *Let* $(S, \sim)$ *be a similarity structure. Then the lattice* $PART(S, \sim)$ *of structural parts of* $(S, \sim)$ *is a complete Heyting algebra. If the similarity relation* $\sim$ *coincides with identity is trivial, then* $PART(S, \sim)$ *is just the power set* PS.

Similarity structures abound. For instance, a natural intuitive interpretation of similarity conceives similarity as some sort of nearness, i.e., two things are considered as similar, if they are near to each other in some sense to be specified. Given a element x of a similarity structure $(S, \sim)$ one may therefore consider the set $S(x) := y; x \sim y$ as a kind of neighborhood of x. Thereby a similarity structure $(S, \sim)$ obtains a kind of rudimentary spatial structure. A concrete example of such a spatial similarity structure can be obtained as follows: Let (X, d) be a metrical space, i.e., a set endowed with a non-negative real-valued function $X \, x \, X \xrightarrow{d} R$. Let $\varepsilon > 0$. Then X is rendered a similarity structure by stipulating $x \sim y := d(x, y) < \varepsilon$.

A quite different, more algebraic kind of similarity structure can be defined for Boolean algebras and similar structures. If $(B, \leq)$ is a Boolean algebra with bottom element 0, the set $B - -\{0\}$ is rendered a similarity structure by stipulating $x \sim y := infimum(x, y) \neq 0$. These two kinds of similarity structures in no way exhaust the possibilities. Similarity is a very flexible structure that allows to define similarity structures almost ad libitum.

5 Spatial Mereology

Let us conclude the series of examples of structural mereologies by an example that has met considerable attention in recent years, to with, the mereology of space. To be as specific as possible let us concentrate on the most prominent case, to wit, Euclidean space E. It should be clear, however, that the following considerations apply to a much larger variety of spaces. Conceiving E as a mereological system requires to give a reasonable answer to the mereologically fundamental question "What are the parts of Euclidean space?" The standard Euclidean answer is "The parts of Euclidean space E are sets of Euclidean points". This is, however, not a very convincing answer. It amounts to conceive E simply as a set of points, thereby completely ignoring its genuine spatial structure. In other words, the point of a genuinely spatial mereology is somehow to take into account the spatial structure of E. There are several ways to achieve this. One is to describe the Euclidean space as a topological space. Recall that a topological space (X, O^*X) is defined as a set X for which a set $O^*X \subseteq PE$ is singled out.[12] The elements of O^*X are called the open sets of the topological space X and form a complete Heyting algebra with bottom element $\emptyset$ and top element X. The Euclidean space E carries a canonical topological structure OE inherited from its canonical metrical structure. Then a reasonable answer, at least prima facie, to the fundamental mereological question "What are the parts of E?" would be "The parts of Euclidean space are the open subsets of E, i.e. the elements of OE." Considering only the elements of OE as spatial parts of E, and not just any contrived subset of E amounts to the requirement that genuine spatial parts have to be structurally "nice" or "natural", or in other words, the notion of parthood in the case of E has to take into account the spatial structure of E. This is done by denying that certain "wild" subsets of E deserve the predicate of "being a spatial part of E". For this maneuver, there is, of course, a price to pay. The resulting mereology, which takes as parts of E only the elements of OE, and not all elements of PX, is no longer Boolean, but solely Heyting. This entails that the elements of OE no longer have Boolean complements. All this can be cast in the category-theoretical framework of Section 3 introducing the category **TOP** whose objects are topological spaces and whose morphisms are continuous maps.

It should be noted that the class of open subsets O^*X of a topological space X is in no way the only reasonable choice for genuine spatial parts of X. Since even open sets may often look rather unwieldy and unnatural, one may prefer to restrict the ken of proper spatial parts of a space X still further. Accordingly, some authors have proposed to consider only regular open subsets of X as its spatial parts. This class, denoted by O^*X, is a proper subclass of O^*X. Indeed, the move from O^*X to O^*X has some advantages. For instance, O^*X is a complete Boolean algebra – and not only a Heyting algebra. On the other hand, new difficulties arise. For instance, while (under some mild restrictions) a topological space (X, O^*X) is determined (up to isomorphism) by the Heyting algebra O^*X of its parts, this is no longer true for O^*X. There may be topologically different spaces X and Y having isomorphic algebras O^*X and O^*Y. This can be interpreted as the fact that the spatial mereology of a space no longer fully determines it. Accepting only regular open subsets $A \in O^*X$ of X as spatial parts entails that some information on X is lost in that the Boolean lattice O^*X alone does not uniquely determine the point set X. Hence, the full topological structure of (X, O^*X) escapes a mereological description in terms of O^*X. In other words, topological mereology is inevitably non-Boolean mereology.

6 Platonic versus Aristotelian Mereology

The category-theoretical generalization of classical mereology has not only the virtue to offer a unifying framework for a large variety of structural mereologies.

---

[12] For a succinct account of the basic notions of topology the reader may consult [3, chap.10]

It may also be used to shed new light on a classical metaphysical problem that already occupied Plato and Aristotle, namely, the time-honored problem whether a whole is identical with its parts. The pertinent texts are *Socrates' Dream* in the *Theaetetus* and *Aristotle's Metaphysics Z 17*, respectively (cf. [9, chap.4, 60ff]). As is well known, Plato held that the whole is identical with its parts. Accordingly, he claimed that the syllable "SO" is identical to the letters "S" and "O". Then the difficulty arises that the letters "S" and "O" may not only be assembled to form the syllable "SO" but also to the syllable "OS". This may be considered as analogous to the group-theoretical fact that the groups $\mathbf{Z}_2$ and $\mathbf{Z}_3$ may be assembled in two different ways, namely, one way to yield the abelian cyclic group $\mathbf{Z}_6$ and another way to yield the non-abelian group $\mathbf{S}_3$.

In contrast to Plato, Aristotle maintained that the whole and its parts are different. The syllable, said Aristotle is not just the letters "S" and "O" but something else, too, since when the syllable is "dissolved, the whole, i.e. the syllable, no longer exists, but the elements of the syllable exist." Hence, Aristotle concluded, the syllable consists of the elements plus a further item, which is of a completely different type than the elements namely its substance. As Scaltsas points out, the classical dispute between a Platonic and an Aristotelian account to mereology finds a certain rehearsal in the dispute between Armstrong, Lewis and other contemporary philosophers on the possibility of non-mereological composition of structural universals (cf. [1, 6]). Structural mereology in the sense of category-relative mereological as outlined in this paper may not directly contribute to this problem, but at least it may widen the horizon of mereologists and thereby help indirectly to better understand some classical problems of traditional mereology. For reasons of space I cannot further elaborate this point, but it seems that the account of structural mereology presented in this paper favors an Aristotelian stance in mereological matters.

7 Concluding Remarks

In this paper I have argued that classical Boolean mereology needs to be contextualized or customized in the sense that it has to take into account the various kinds of structures that are present in the objects of our mereological investigations. This kind of customized mereology defies the confines of standard Boolean mereology. Elementary examples of structured mereological systems in this sense are provided by Boolean algebras, groups, similarity structures, topological structures, and many other structures as well. These examples show that in general, the lattices of structural parts of structured wholes are not Boolean. These considerations can be rendered precise in the framework of category theory according to which every category **C** comes along with its own specific **C**-mereology. Classical Boolean mereology turns out to be closely related to the category **SET** of lattices. In sum, then, I'd to contend that mereology is far from being "perfectly understood, and unproblematic" part of formal methodology, as the late David Lewis maintained. Rather, it is to be considered as an open field of research that offers a wealth of interesting philosophical, logical, and mathematical problems.

Bibliography

[1] Armstrong, D. M. *Universals and Scientific Realism: A Theory of Universals.* Cambridge: Cambridge University Press, 1980.

[2] Carnap, R. *Der logische Aufbau der Welt, Berlin, Weltkreis Verlag.* 1928. Translated by George, R. as *The Logical Structure of the World*, Berkeley and Los Angeles:University of California Press, 1967.

[3] Davey, B. A., Priestley, H. A. *Introduction to Lattices and Order.* Cambridge: Cambridge University Press, 1990.

[4] Lawvere, F., Rosebrugh, R. *Sets for Mathematicians.* Berlin and New York: Springer, 2003.

[5] Lawvere, F., Schanuel, S. *Conceptual Mathematics: A First Introduction to Categories.* Cambridge: Cambridge University Press, 1997.

[6] Lewis, D. Against structural universals. *Australasian Journal of Philosophy*, 64:25–46, 1986.

[7] Lewis, D. *Parts of Classes.* Oxford: Basil Blackwell, 1991.

[8] Mac Lane, S. *Categories for the Working Mathematician.* New York: Springer, 1998.

[9] Scaltsas, T. *Substances and Universals in Aristotle's Metaphysics.* Ithaca and London: Cornell University Press, 1994.

[10] Schmidt, R. Subgroup lattices of groups. In *Expositions in Mathematics*, vol. 14. Berlin: de Gruyter, 1994.

An Informational Interpretation of Monadology [1]

Soshichi Uchii

Kyoto University, Emeritus

uchii@mac.com

ABSTRACT. In this paper, I will try to exploit the implication of Leibniz's statement in *Monadology* (1714) that "there is a kind of self-sufficiency which makes them [monads] sources of their own internal actions, or incorporeal automata, as it were" (*Monadology*, sect.18). Leibniz's monads are simple substances, with no shape, no magnitude; but they are supposed to produce the phenomena resulting from their activities, which for us humans look as the whole world, the nature. The activities of a monad are characterized by mental terms, *perceptions* (internal states) and *appetites* (which change the internal state). By means of perceptions, a monad becomes a "perpetual living mirror of the universe"; it can receive the information of other monads and it can send its own, in turn, to others. The communication and interconnection thus produced result in the physical and the psychical phenomena observed by us, humans. According to Leibniz, all monads are governed by the *teleological* law given by the God, and the world of phenomena are governed by the *causal and mechanical* law. Leibniz argues that there is a *pre-established* harmony among the monads so that this double character is no problem.

Now, I will propose an informational interpretation of monadology, which regards the monads as an automaton governed by the God's program and arranged appropriately; and I will argue that Leibniz's scenario can be defended in terms of this interpretation. The crucial part of this interpretation is that the God's program and the monads' activities are related with the phenomenal world by means of a *coding* by God. This interpretation is also defended on the textual basis, with a special reference to Leibniz's distinction between *primitive* and *derivative* forces. Drawing on R. M. Adams's careful reading of Leibniz's texts (*Leibniz: Determinist, Theist, Idealist*, [1]), I will argue that his rendering is quite in conformity with my interpretation, although he does not seem to be aware of the notion of coding.

1 The Basic Features of Monadology

In this paper, I should like to present an interpretation of Leibniz's monadology in terms of the concepts of the informatics, the theory of information. And this is meant to illustrate a way to utilize our contemporary tools for interpreting a great historical figure of philosophy of science, and to make his valuable insights alive again in our days. Since we are now in China, let me quote from Confucius:

---

[1] I wish to thank Dr. M. Matsuou for reading and commenting on the two earlier versions of this paper. His comments and questioning were greatly helpful to clarifying several obscure points in my interpretation.

An Interpretation of Monadology

"Cherishing the old and thereby deducing something new, you may become a teacher" (*Analects* 2-11). My research was made in the spirit of this saying, but I will venture to add a new phrase, as far as the philosophy of science is concerned, in the conclusion.

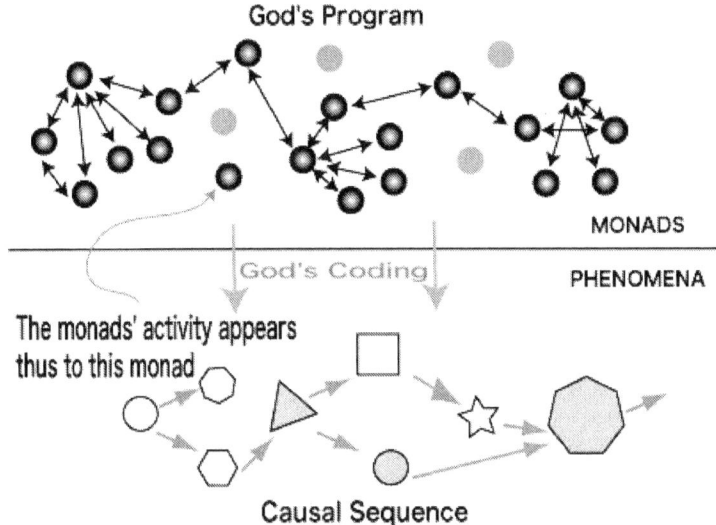

Figure 1: Diagram of Monadology

It is well known that Leibniz likened a monad to an automaton. In the Section 18 of *Monadology*, he said, "there is a kind of self-sufficiency which makes them [monads] sources of their own internal actions, or incorporeal automata, as it were." I am going to take this statement seriously, and I will claim the word "automata" should be understood roughly in the same sense as our modern sense in the informatics. I will further claim:

1. my interpretation makes easier to unify teleology and causality in Leibniz, and
2. it can clarify the relationship between the monads and the phenomena, in conformity with Leibniz's text; and finally,
3. it can illustrate the informational aspect of Leibniz's philosophy of science.

But first let us review the basic features of *Monadology*.

Monads are, according to Leibniz, the ultimate individuals that are supposed to produce all phenomena of our world, physical and mental. They are simple entities with no shape, no magnitude. But they have internal states (called *perceptions*), and the capacity (called *appetite*) to change their internal states. By means of these perceptions, each monad can reflect the states of other monads, and by this interaction, the monads produce the physical and the mental phenomena altogether. It should also be remembered that the monads are organized into groups; in particular, conscious beings, such as animals or humans, have a central monad (called *soul*), and other monads in the group, roughly speaking, constitute its *body*; the soul somehow acts as a Central Processing Unit (CPU) in the group, and the whole group forms an *organic* body.

I understand the "phenomena" are higher order perceptions of some organic body (such as humans), i.e., conscious perceptions produced by a multitude of perceptions (and other activities of monads), as is indicated in the figure 1.

It may be interesting to see that this idea of organic body composed of monads finds a specific example in the 20th century: John von Neumann's cellular automata. Two-dimensional infinite space is filled with the same unit automata (each is a cell automaton), connected with the four neighbors around it, and any finite complex automaton can be constructed in this space (see Fig.2).

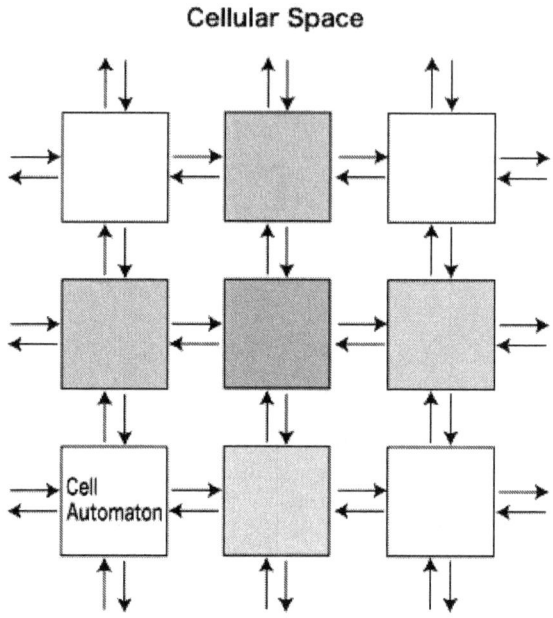

Figure 2: Cellular Space

In particular, a universal Turing machine can be constructed, and, given a tape—again composed of cell automata—with coded information of its own structure, it can reproduce the same machine as itself, within the space (see Fig.3). Each cell is an automaton, but a higher order automaton, including its own CPU composed of a number of cell automata, is constructible. Its basic idea goes back to Leibniz, although von Neumann may have had no direct connection with Leibniz.

Now, back to monadology. According to Leibniz, the monads are governed by the teleological law (final cause) imposed by God, whereas the phenomena (which appear to us, humans) are governed by the (*efficient*) causal law. Leibniz argues that there is no inconsistency between these two aspects, because God created the monads and their phenomena with a Pre-established Harmony; that is, the two aspects fit together nicely according to God's design. But many people may question this claim: "Why, and how is this harmony possible?" One great virtue of my informational interpretation is that we can easily answer this question.

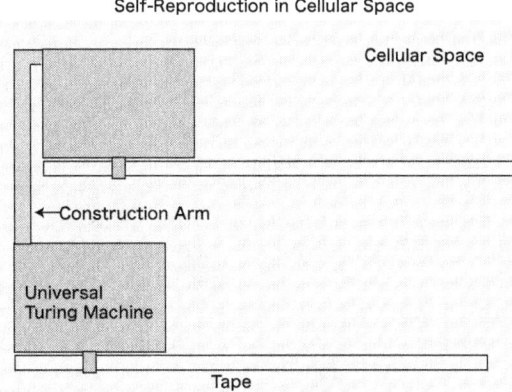

Figure 3: Self-Reproducing Automata

2 Outline of my Informational Interpretation

Leibniz's monads are primarily the bearers of information (what Leibniz called perception), and they are programmed by God to change their internal states (informational contents). Now, it is clear that God's program is teleological, since His program is meant to fulfill some purpose. Even our programs for computers are teleological, since they are meant to do some job. However, a program must be related with something else, external to the program itself.

Take a program for the Turing machine, for instance; say, a program for computing the successor function. It is clear that this program is teleological, since it has a definite purpose to fulfill. Supposing the tape of the Turing machine has its content in binary code (either a mark or blank), a program is a series of instructions to handle the tape and its content. But in order for this program to do the intended job, we have to establish the correspondence between the configurations on the tape and the numbers (which are independent of the tape or configurations of mark). This correspondence is achieved by means of coding. Thus the program makes sense only in combination with the coding. This is indeed an elementary point the informatics teaches us. All right. But if we want to build a physical device for the same computation as the Turing machine with this program, what are we going to do? We have to design such a device in conformity with the laws of physics, and the computation of this device must be executed causally, in our actual world. Thus, in a word, the teleological program must be executed causally, in the actual, physical world. In this case, no one would deny the compatibility of teleology and causality. Then, roughly the same holds, for Leibniz's monadology. This is the crux of my Informational Interpretation of monadology. The monads and their perceptions can be handled by informatics, in terms of teleology; but the activities of the monads are supposed to be responsible for the world of phenomena, which we humans take as the world of physical and mental events. The world of phenomena is regulated by causal laws, and God's program must be executed in this world of phenomena. Thus, God has to supply a coding for the correspondence between the activities of monads and the world of phenomena. The phenomena, as I understand, are higher order, conscious perceptions produced in some organic body; hence the problem is, how to bridge between lower perceptions and higher perceptions.

Here, some complications come in, because, according to Leibniz, the phenomena are produced as a result of the activities of the monads. The phenomena are, so to speak, how the monads' activities appear to the mind of an observer. But I suppose God can choose the way the monads' activities look to an observer, and that is determined by God's coding. Our coding for the Turing machine presupposes two independent entities: there are natural numbers, on the one hand, and configurations on the tape, on the other hand. But Leibniz's God creates the monads, and He determines the coding for the phenomena, how the activities of the monads look to organic bodies with consciousness. Thus God is the system designer of the world of monads, together with the phenomena produced by their activities. God's program is teleological, but the laws of phenomena, determined by the monads' activities and God's coding, are causal and mechanical, which are subject of our scientific knowledge.

Further, we may ask: Can we ever know God's program, or at least a portion of it, and if we can, by what means? I think Leibniz's answer is "Yes", and I would venture to say that he thinks scientific research of phenomena is an indispensable means for that. This was clearly indicated in his correspondence with Christian Wolff, in terms of the relationship between the primitive forces and the derivative forces (for this, I owe a great deal to [1, chap.13]). I will come back to this point later, when I try to clarify the relation between the informational content and Leibniz's notion of forces.

Thus, according to my Informational Interpretation, the whole structure of Leibniz's monadology looks like this: (1) First, from God's side: He creates the monads and programs their activities (teleological law). These activities proceed, and they are projected, so to speak, onto each monad's internal screen; how these activities appear to monads is coded by God, so that the phenomena are produced according to the monads' activities *together with* this coding. The result is that the phenomena proceed according to the causal law. (2) Second, from our (man's) side: we notice that the phenomena seem to occur according to some regularity, and we try to extract the causal laws underlying the phenomena; this is scientific activity, including the common sense. But if we further wish to know the ultimate law of the phenomena, there still is a higher level of knowledge, i.e., decoding the phenomena and their law. If we are lucky enough, we may know some portion of God's program. This is quite analogous to the physicist's search for the ultimate theory of the world; but Leibniz goes further than this, because he wants to explain the mental phenomena too.

3 Leibniz's Classification of Forces

So far, I have merely presented an outline of my interpretation. Now I have to state the reasons why this interpretation looks good to me. A good clue is Leibniz's classification of forces.

In *A Specimen of Dynamics* (1695), Leibniz classifies the forces into four categories: The primitive force, active and passive, and the derivative force, active and passive. The *primitive active* force is said to be "inherent in all corporeal substance as such, since it is contrary to the nature of things that there should be any body which is wholly at rest"; and also that it is "none other than the first entelechy" and "corresponds to the *soul* or *substantial* form". But by this explanation alone, we cannot figure out what it is. As regards the *derivative active force*, it is said to be "as it were the limitation of primitive force brought about by the collision of bodies with each other". Again, by this explanation alone, it is hard to understand. A similar explanation of passive force (primitive and derivative) follows, but I will skip it. What is important here is that Leibniz is trying to get at the mechanical notion of force applicable in physics, and his specific discussion in this paper is wholly restricted to the derivative force, in such phenomena as motion and collision.

From other sources, we can infer that the primitive force is attributed to the level of substance, and the derivative force to the level of phenomena, including physics. All the same, the relation between the primitive forces and the derivative forces is quite obscure at this stage.

4 Leibniz's definition of "Action/Passion" in *Monadology*

With this unresolved problem in view, let us now turn to *Monadology*, Leibniz's final form of metaphysics. Surprisingly, in this brief writing, the word "force" never appears. Instead, the interactions between monads are now characterized in terms of perceptual concepts. That is, a monad is *active* vis-à-vis another, in so far as the former has *perfection*; and "perfection" means "perceptions are *distinct*". Passivity (or Passion) is the reverse of this, and a monad is passive in so far as it has confused perceptions. Thus, clearly, Leibniz now takes an informational approach for the characterization of the monads' activities. For, the words "distinct" and "obscure" are primarily concerned with the informational content of a perception: a clear perception in one monad can explain what happens in another monad.

Then, what is the relation between this notion of activity and Leibniz's previous notion of forces? A clear statement is found in Leibniz's correspondence with de Volder, around 1704. Leibniz says that it is obvious that "primitive forces can be nothing but the internal strivings of simple substances, strivings by means of which they pass from perception to perception in accordance with a certain law of their nature" [2, p.181]. In short, Leibniz now regards perceptions as basic, and the primitive forces can be reconstructed in terms of the transition of perceptions. This is the main reason why I call Leibniz's approach informational.

5 R. M. Adams' Clarification

Having ascertained the basic stance of Leibniz in *Monadology*, let us come back to the question of the relationship between primitive and derivative force. Robert M. Adams pursued this question, and he has shown, convincingly on my opinion, that the derivative forces have a mixed character, covering both physical force and intramonadic force. Recall that in *A Specimen of Dynamics* (1695), the derivative forces were placed in the physical world, the world of phenomena. But if derivative force is a *modification* of primitive force that is in a monad (substance), how can it be in the physical world? How are they related? These are the unresolved problems.

Adams' clarification is illuminating. As we have already seen, the primitive force is identified with "the internal strivings of simple substances, strivings by means of which they pass from perception to perception in accordance with a certain law of their nature". The primitive force thus characterized should be understood as a *comprehensive* force *persisting* together with the monad itself. But the "operation" of this force takes place consecutively, one operation at an occasion, another at another occasion, etc. In this way, each operation is restricted to an occasion. Thus each operation of the primitive force should be regarded as the *derivative* force, a *modification* of the primitive force. This is Adams' interpretation. The point is that, while the primitive force is eternal, its modification or derivative force is restricted to each occasion for the operation.

Thus Adams claims that the operation of the primitive force thus characterized is not the present perceptions of a monad, but the present *appetites*, since the operation connects the present perceptions to the next perceptions. That is exactly what appetites are supposed to do.

The same point can be more easily seen, in terms of a transition function of an automaton. An automaton changes its internal state according to a transition function; this function is a totality, covering all possible combinations of internal states and inputs. But for its operation, this function needs specific data such as the present state, or the present input; given these, the transition function determines the next internal state. This contrast between the transition function (totality) itself and its operation at a given instant, is quite analogous to the distinction between the primitive force and its operation, derivative force.

Thus, the derivative force as intramonadic force has been clarified. In my own terminology, the derivative force signifies specific operations of a monad's program (given by God); it is synchronized with the operations of all other monads (again by God's programming), so that one may concentrate on the given monad alone, and hence the derivative force as well as the primitive can be regarded as internal; anyway, other monads are reflected in each monad.

But, then, how can we explain the derivative force as physical force? Many bodies physically interact with each other, in such phenomena as collisions, for instance. Thus it seems that an internal state of such a body is clearly insufficient for explaining such phenomena. How is it possible that physical forces such as inertia or kinetic energy of several bodies concur to produce a mechanical phenomenon?

Here, again, Adams' analysis [1, pp.383–386] is illuminating. Drawing on the Leibniz-Wolff correspondence during 1710-11, Adams reconstructs Leibniz's answer to Wolff, as regards the role of the primitive force in physical phenomena such as collisions. Leibniz's answer is basically this: even in such physical phenomena as collisions, no force is transmitted from one body to another; each body is moved only by its own derivative force, which is internal to it. How is this possible, according to Leibniz's monadology?

An organic body composed of monads has within itself a *representation* of all the external circumstances that affect it. This means that, if this organic body is to collide with another, its internal representation includes this circumstance, and the correspondence of its motion with the other's is assured by a Pre-established Harmony. Leibniz says, "the Entelechy itself is modified corresponding to these mechanical or derivative forces"; and this means the primitive force is internally modified in a monad, thereby becoming internal derivative forces, and these in turn correspond to the mechanical forces. Thus, derivative physical forces acting in such a collision can be all reduced, in a way, to the derivative forces internal to each monad. This is, in essence, Adams' rendering. I agree that this is quite in conformity with Leibniz's text.

However, there is still a missing link in this rendering. Physical forces are at the level of phenomena, whereas the primitive force and the internal derivative forces are within each monad. What is it that bridges the two sides? Adams says it is "expression" or "representation". But still, the nature of such "representation" is unspecified. That is why I added a qualifying phrase "in a way", when it is said that "physical forces can be all reduced to the derivative forces". The nature of this "reduction" is not clarified yet, and this is the only complaint I can make of Adams' rendering.

6 Representation via Coding

At last, I can begin to state my own message. As we have already seen, the realm of monads and the realm of phenomena are quite different. The former is reality, the latter mere appearance. In order to connect these two different realms, coding is necessary, just as we connect the realm of natural numbers with the realm of possible configurations of marks on a tape. The phenomena, i.e. the appearances of the activities of the monads to some monads, are realized, so to speak, by the "software" of God, God's programming. Then, of course, some coding is presupposed; and only this can achieve the correspondence between the reality and the phenomena; for that correspondence cannot be causality in the usual sense (efficient cause), but a sort of teleological mapping of one realm to another, which is nothing but coding in the general sense.

Thus the concepts of the informatics provide a great deal of help for spelling out the exact nature of the "reduction" or the "correspondence" which Leibniz and Adams were trying to get at. And, as I have argued in the outline of my interpretation (section 2), the notion of coding enables us to see clearly the compatibility of teleology and causality.

7 Science as Decoding

The last question I wish to address is the role of science in monadology: if the reality is coded in the phenomena, what is our scientific activity supposed to do, according to Leibniz's monadology? This problem was also touched upon in the Leibniz-Wolff correspondence during 1710-11. Again, my interpretation heavily depends on Adams' rendering, but I believe mine emphasizes a neglected aspect of monadology.

Let us follow more of the Leibniz-Wolff correspondence during 1710-11, with the help of Adams' rendering. Despite Leibniz's explanation of derivative forces, Wolff could not understand the relationship between intramonadic forces and physical forces. This is quite understandable, because Wolff could not appeal to the help of the concepts of informatics! Pressed by Wolff's request, Leibniz now discloses his opinion as regards the means by which we may know how the primitive force is modified in monads, say, when a heavy body is falling and accelerated. His answer is surprisingly simple: "the modification of the primitive force that is in the Monad itself cannot be explained better than by expounding how the derivative force is changed in the phenomena." That is to say, since the physical force is an expression of the intramonadic derivative force, the best way to detect how the primitive force is modified in the monad is to study the manner of change of the physical force. Although Adams himself quotes this passage as a ground for saying that the physical force and the intramonadic derivative force are *identical* [1, p.386], I diverge from Adams here.

According to my Informational Interpretation, the physical force and the intramonadic derivative force cannot be identical, since they belong to different realms. As I have repeated again and again in this paper, the relation between them is that of coding, i.e., correspondence between two entirely *different* entities. Therefore, even if we agree with Leibniz that the study of physical phenomena is the best way to know that the primitive force is modified, the exact manner how it is modified is still to be known. How should we proceed to this deeper knowledge? My Informational Interpretation naturally suggests the following: the subject of this deeper knowledge is God's program together with His coding; and the basic method for attaining such knowledge should be *decoding*, i.e., reconstructing the original message from the coded message. Thus, from physical phenomena, we have to extract regularities (laws), and from these regularities we aim at a unification of them, and then we have to figure out the underlying coding; if we can succeed in this, then we may be able to recover some portion of the original program.

I believe Leibniz's view of science was close to this picture. And the science in the last and this century came closer to this picture. For, the method of informatics became indispensable in many fields of science. For instance, the study of quantum gravity suggests that there is a discrete unit of physical information: continuous mathematics breaks down at the level of Planck length, and the minimal area which can contain one bit of information is known to be 4 square Planck length. Again, in the field of molecular biology, the method of informatics is now a common sense. Thus, whether or not we aim at God's program, an informational approach in science is becoming more and more important.

But this is not my final message. What I wish to emphasize in this conclusion is that applying new concepts to old philosophy may well be fruitful. If my argument in this paper was successful to some extent, you may have realized this already. So, let us go back to the words of Confucius I have quoted in the beginning. "Cherishing the old and thereby deducing something new, you may become a teacher", and I may add, "Interpreting the old in terms of something new, you may revive the old".

Bibliography

[1] Adams, R. M. *Leibniz: Determinist, Theist, Idealist*. Oxford: Oxford University Press, 1994.

[2] Ariew, R., Garber, D. (Eds.). *G. W. Leibniz, Philosophical Essays*. Hackett, 1989.

[3] Gerhardt, G. I. (Ed.). *Briefwechsel zwischen Leibniz und Christian Wolff*. Halle: H. W. Schmidt, 1860.

[4] Gerhardt, G., I. (Ed.). *Die Philosophischen Schriften von G. W. Leibniz*, vol. 2. Berlin, 1879.
[5] Jolley, N. (Ed.). *The Cambridge Companion to Leibniz*. Cambridge University Press, 1995.
[6] Loemker, L. E. (Ed.). *G. W. Lebniz Philosophical Papers and Letters*. Kluwer, 2nd edn., 1969.
[7] Rescher, N. *G. W. Leibniz's Monadology*. University of Pittsburgh Press, 1991.
[8] Woolhouse R, S., Francks, R. (Eds.). *G. W. Leibniz Philosophical Texts*. Oxford University Press, 1998.

Descartes' attempt, in the *Regulae*, to base the certainty of algebra on mental vision – A conjectural reconstruction [1]

Henk J. M. Bos

Aarhus University, Utrecht University

h.j.m.bos@uu.nl

ABSTRACT. At the end of the extant text of his unfinished *Regulae* Descartes explained how the basic algebraic operations could be applied in geometry. In that connection he referred to a technique he needed for solving a particular geometrical problem. The problem was the "application of areas" and the technique can be identified as the use of the "gnomon figure" as explained and employed in the first two books of Euclid's *Elements*. This makes it possible to reconstruct, tentatively, the argument that should have followed immediately, as well as the kind of reasoning which Descartes explored in this context. The conjectural reconstruction should be of interest in connection with Descartes' early ideas about the certainty of mathematical reasoning and the nature of clear and distinct intuition as a means to attain such certainty.

1 A reconstruction

In this article I present a conjectural reconstruction of a mental activity which, I assume, Descartes undertook while composing the *Regulae*. The activity consisted in using the two mental faculties of *imagination* and *intellect*, which he defined in the same text, to understand the algebraic operations addition ($+$), subtraction ($-$), multiplication ($\times$), division ($\div$), and square root extraction ($\sqrt{\ }$) when applied to geometrical line segments. The aim of the activity was to decide whether these operations could be understood with certainty, and thereby be accepted as reliably providing certain knowledge. According to Descartes' criterion for certain knowledge, also explained in the *Regulae*, this would be the case if, by the imagination, one could form images of the working of these operations, and if the intellect, contemplating these images, would see that they were fully *clear* and *distinct*. Descartes was convinced that every human being was in principle able personally to perform such a mental activity.

I base my conjectural reconstruction on the assumption that Descartes himself actually undertook this mental activity in the case of the algebraic operations and that we (I) can reconstruct it on the basis of what he explained about this process of understanding in general and what we know about his knowledge of mathematics, and especially algebra, at the time.

---

[1] The author is professor emeritus in the Department of Mathematics of Utrecht University (The Netherlands), and honorary professor in the Department of Science Studies of Aarhus University (Denmark).

I feel a certain apprehension about the reconstruction. It is somewhat unusual because it is not about a result which some mathematician from the past had (in private notes, for instance) and which is now lost. Nor is it about a concept which a mathematician might have formed or contemplated and which is now forgotten. It is about a formal mental activity which Descartes himself undertook and suggested his reader to undertake.

This is unusual terrain for me and I therefore the more welcome the opportunity to present the conjecture to colleagues, first at the congress and now in its *Proceedings*.[2]

2 The *Regulae*

Descartes wrote his *Regulae ad directionem ingenii* [6] (*Rules for the direction of the mind*) during the 1620's. He never completed it and the extant text was not published during his life time; it exerted no influence on the understanding of Cartesian philosophy and mathematics in the 17th century. It is, however, a crucial source for understanding Descartes' earliest thinking. It documents his attempts to compose a practical philosophical treatise on how to obtain certain knowledge by methodical reasoning.

The treatise was planned to be in three parts, each containing 12 succinctly formulated rules of reasoning, each rule followed by an extensive explanation. The algebraic operations are discussed in Rules 17 and 18. These are the last rules with explanation, the next three are stated only, and then the extant text breaks off. In the preceding rules Descartes developed his practical method and his criteria for attaining certain knowledge. These he modeled on arithmetic and geometry, which, he wrote, were the only sciences that thus far had provided "certain and indubitable cognition." [6, p. 362, p. 10][3] Mathematics, in which he had received a thorough training while he was a student at the Jesuit college in La Flèche, was a strong inspiration for Descartes. The geometrical component of this inspiration is unsurprising because of the role of visual imagination in Descartes' model of understanding. The crucial role of algebra in the *Regulae* is perhaps less easily connected to the context of knowledge and certainty. It underscores Descartes' familiarity with the then current mathematical interests among which algebra was one of the first. Moreover, Descartes' use of algebra in the *Regulae* is not disjoint from geometry; on the contrary, he combines the two in his method of reasoning, and the combination of algebra and geometry was a much debated item among mathematicians at the time.

Thus Descartes's attempt to base the certainty of the algebraic operations on mental vision was closely linked to its contemporary mathematical context, on which I draw heavily in my reconstruction.

---

[2] I am thankful to Kirsti Andersen for her in-depth and very helpful comments on drafts of the present article.
[3] All quotations from [6] are from the English translation in [10, vol.1], to which the second of the page references refers.

3 Merging algebra and geometry

3.1 Obstacles

During the decades around 1600, between the Renaissance absorption of classical Greek mathematics in Western Europe and the great early modern inventions of analytic geometry and the calculus, the most important developments in mathematics concerned algebra. One of these developments was what may be termed the *merging of algebra and (pure) geometry*, leading to the beginnings of analytic geometry in the writings of Fermat and Descartes.

The possibility and the usefulness of this merging was not self-evident; many mathematicians saw conceptual and technical obstacles for the use of algebraic methods in pure geometry. Below I discuss the three main objections against the merging and the possibilities to overcome them. There were indeed also incentives to overcome the obstacles; these I discuss in the next section.

It should be noted that the obstacles and incentives I have in mind here concerned the merging of algebra and *pure* geometry. Within practical geometry, such as used by astronomers, surveyors, and builders, the obstacles presented no particular difficulty. Indeed numerical calculations had been routinely and successfully applied to geometrical problems from times well before the rise of Greek mathematics. And by 1600 there was an active interest among mathematical practitioners not only in using algebra in their calculations but also in developing new techniques of manipulating and solving equations.

3.2 Transfer of methods

Algebra had been developed by medieval Arabic mathematicians in the context of numerical calculations and was seen primarily as a collection of methods for solving problems about numbers. It belonged therefore to arithmetic, which was defined as the science of *numbers*. Numbers were primarily the (positive) integer numbers and, by extension to fractions, the rational numbers. Although at the time mathematicians experimented with irrationals such as the square roots of non-square whole numbers, these were generally seen as exceptional and of problematic status, witness such expressions for them as 'surd' or 'ineffable' numbers.

Geometry, in the sense of the pure geometry of Euclid and Apollonius, was the science of *magnitude*, in particular the geometrical magnitudes such as line segments, rectilinear plane and solid figures and some curved lines and figures. These magnitudes were not numbers. Number was discrete quantity, geometrical magnitude was continuous quantity. Pure mathematics was the science of abstract quantity, and the dichotomy of quantity split the field in two essentially disjoint sciences, arithmetic and geometry. They differed because the things they studied were different and, according to the generally accepted Aristotelian conception of sciences, scientific methods should conform to the subject matter they were applied to, and should therefore not be transferred from the one science to the other.

Apart from this Aristotelian objection in principle against transfer of methods, there was also a basis in mathematical fact: Numbers and continuous magnitudes, both taken in the classical and early modern senses of the terms, are fundamentally different and their difference presents conceptual obstacles against the merging of algebra and geometry. These obstacles can be grouped under two headings: *numerical approximation versus geometrical exactness*, and *arithmetical versus geometrical multiplication*.

3.3 Numerical approximation versus geometrical exactness

In practical geometry, numbers were used for measurement. By measuring with respect to a unit length and its square and cube, distances, areas and volumes are given numbers expressing their size. These numbers were rational. The positional and dimensional relations between the measured objects are effectively represented by algebraic relations between the numbers expressing their measures; in particular, (in the case of the foot as unit length) the area of a rectangle in square feet is equal to the product of the lengths in feet of its two sides. Although the numbers used in practical geometry were rational only, the difference between commensurable and incommensurable line segments presented no difficulties. Measuring with numbers implied imprecision and approximation anyway; the fact that for two incommensurable lines the exact measures could not be both rational did not matter in practice.

But approximation and imprecision, however acceptable in practice, were inadmissible in pure geometry, which had to be precise and exact and in which incommensurability and irrational ratios were crucial theoretical issues which did not admit the fuzziness of approximations. Thus there was good reason to believe that numbers, in the then common meaning of rational numbers, could never be used effectively and reliably in pure geometry.

3.4 Arithmetical versus geometrical multiplication

The other obstacle concerned the operation of multiplication. The definition of multiplication (as well as of its inverse, division) in arithmetic is essentially dependent on the fact that numbers denote the multitude of a set of objects. In multiplying 3 and 7, 3×7, we take 7 three times (or 3 seven times) using the fact that three (or seven) can indicate a multitude of 'time's. We can extend the meaning of multiplication to cases in which one of the factors is not a number; for instance we can take some line segment in geometry and multiply it by 3 by joining three copies of it. But we cannot multiply, in the sense indicated by the term itself (making multiples), when neither factor is a number. In particular we cannot multiply geometrical figures in this sense: if a and b are line segments, for instance, 'a times' has no meaning, hence neither has 'a times b'.

Yet there is an obvious candidate in geometry to take over at least some of the roles of the arithmetical multiplication; it is the operation (construction) which forms, out of two given line segments a and b, a rectangle with sides equal to a and b. The affinity [4] between multiplying numbers and forming rectangles was well known in classical as well as early modern mathematics, but mathematicians were also aware that the two operations were not the same. The main difference

---

[4]Clavius used the term 'affinitas' for the relation between rectangles and products of two numbers in his edition of Euclid's *Elements*: "Habet autem comprehensio haec parallelogrammi rectanguli sub duabus rectis lineis angulum rectam continentibus magnam affinitatem cum multiplicatione unius numeri in alterum." [3, vol.1, p. 82].

was that the product of two numbers is again a number, whereas the rectangle formed from two line segments is not a line segment but a rectangle, that is, a figure of higher dimension. In geometry, therefore, multiplication of more than three line segments was impossible as it would lead to an inconceivable four-dimensional figure. In arithmetic there was no such restriction. The change of dimension also implied that, if a geometrical analogon of division (seen as the inverse of multiplication) was to be introduced, the dividend and the divisor had to have different dimensions; thus a rectangle could be divided by a line segment (the result being a line segment) but not vice versa.

3.5 Three approaches

The arguments above suggest that there were essentially three possible approaches to adjust the incomplete analogy between multiplication of numbers and forming rectangles out of line segments, each with its particular advantages and disadvantages. The first was to identify each line segment with a number, its length, measured with respect to some unit, and then define the geometrical product of two line segments, a and b with lengths α and β, as the line segment with length equal to the arithmetical product $\alpha\beta$ of the numbers α and β. Any such convention, however, would give rational numbers as lengths, which meant introducing approximation and giving up exactness in many situations. This was the approach taken in practical geometry, and we have seen that it was not acceptable in pure geometry.

The second approach was to redefine geometrical multiplication in such a way that the product of two line segments would again be a line segment. This required the introduction, in analogy to the measuring procedures of practical geometry, of a unit line segment in pure geometry. Descartes chose this approach in his *Géométrie* [5, p. 98], defining the product ab of two line segments a and b with respect to a unit segment u as the "fourth proportional" of u, a and b, that is, the line segment c satisfying the proportionality $u : a = b : c$. Earlier mathematicians had used a unit length in geometry, but not with the express intention of creating an arithmetical structure for non-numerical objects. In the *Regulae* Descartes, significantly, used another definition based on rectangle formation rather than on proportionality. I return to that definition, and its difference from the later one, in Section 5.2.

As the history of analytic geometry shows, Descartes' approach in the *Géométrie* proved to be the most successful approach. Yet its adoption may now seem more obvious than it was. In pure geometry there is no cogent argument for singling out one line segment rather than another as unit. But the multiplication operation essentially depends on this choice; the product of two line segments a and b, defined with respect to a unit segment u_1 is different from their product with respect to another unit u_2. Without the benefit of hindsight it can hardly have seemed promising to introduce an operation in geometry that only applied to line segments, eliminated differences of dimension, and whose results depended on an arbitrary choice. [1, pp. 298–299]

The third approach was one that avoided the choice of a unit and accepted the dimensional implications of multiplication as rectangle formation. It consisted in the introduction of a system of abstract magnitudes with dimensions, corresponding to geometrical magnitudes for the first three dimensions and extended fully abstractly to higher dimensions. Within this system the algebraic operations were introduced axiomatically and chosen such that for the first three dimensions they coincided with the ones based on rectangle formation as multiplication. This was the approach of Viète (and taken over in simplified form by Fermat in his version of analytic geometry). Viète's system had no unit length and required homogeneity in the sense that only terms of the same dimension could be summed or subtracted. [1, pp. 145–154]

4 Incentives for merging

4.1 Reasoning with unknowns

Considering the above mentioned obstacles against the merging of algebra and (pure) geometry, and the complexity of both Descartes' and Viète's redefinitions of the algebraic operations for geometrical use, the question arises what expectations stimulated the attempts to merge algebra and geometry when they appeared to be so fundamentally incompatible. There were two main incentives, a technical one, the handling of unknowns and indeterminates, and a philosophical one, the idea of a *mathesis generalis*.

The distinctive feature of algebra was the introduction of an unknown number and its powers. With these unknowns, for which special symbols had been introduced (later to be replaced by the notation x, x^2, x^3, ..., suggested by Descartes), one could translate the data of a problem about numbers, together with the required property of its yet unknown solution, into an equation. The numerical value of the unknown could then be found by solving the equation. By 1600, methods were known for solving linear, quadratic, cubic and quartic equations (in one unknown).

The unknown numbers were the wonder of algebra. While at present the unknown x is familiar to almost everyone as a quintessentially mathematical object, it is difficult to realize the bewildering novelty of the new numbers featuring in algebra, not only the unknowns that later became our x and its powers, but also roots of non-square numbers, and roots of sums of unknown and known numbers, all of which led to equations that, at least if their degree was smaller than 5, could be solved.[5]

In geometry there was another novelty, of classical Greek origin but rediscovered comparatively late, and made known through the publication, (in Latin translation) in 1588 of the *Collectio* [12] written by the late classical Greek geometer Pappus of Alexandria. It contained intriguing references to a method, called *analysis* and used by Apollonius and Archimedes. Although Pappus' explanations were often confusing, it was clear that the method used unknowns in solving geometrical problems in which some unknown point or line with a particular property had to be found within a given (i.e. known) figure. Apparently the method relied on a formalized reasoning about unknown and given elements of figures. It prescribed to find a sequence of relations between these elements, linking the element to be constructed to the given elements. These relations then indicated a way along which that element could be constructed.[6]

---

[5] Cf. Clavius' explanation of these three varieties [2, pp. 6 sqq].
[6] On the early modern reception of this analysis see [1, chap. 5, pp. 95–117].

This was reasoning, not calculating, with unknowns, but the analogy with algebra was remarkable, the more so because the relations found in the geometrical analysis often had the form of equalities involving unknown line segments, and these equalities only lacked a notation for becoming equations. As a result mathematicians became convinced that algebra had something to do with the geometrical analysis reported by Pappus. Some (Descartes among them) thought that the ancients knew algebra but kept it secret. The similarity, whether as ancient secret knowledge, or as early modern insight, was noted and provided a stimulus to explore the possibilities of merging algebra and geometry despite the obstacles mentioned above.

4.2 *Mathesis universalis*

'*Mathesis universalis*'[7] was a term current among early modern mathematicians. It indicated, in mostly wishful discourse, a kind of mathematics generally applicable to all things quantitative. One of its origins was in the commentary on Book I of Euclid's *Elements* by the fifth-century neo-Platonist philosopher and mathematician Proclus, in which he speculated about a general science consisting of the mathematical theories that apply to both discrete and continuous magnitudes, in particular proportion theory [13, pp. 15–17]. Mathematicians and philosophers elaborated on this idea and some used the term '*mathesis universalis*' in that connection. It was a search for a kind of mathematics – universal mathematics – spanning the divide between arithmetic and geometry and combining the two by a theory, or a set of rules, valid for any kind of quantity, discrete as well as continuous, numerical, geometrical, or arising in other situations where multitudes or magnitudes could be counted or mutually compared. In this connection, algebra was seen as a promising, if as yet inadequate, version of such a universal mathematics.

Clearly, any attempt to create a universal mathematics had to provide precise definitions of the algebraic operations as applied to quantities in general, or at least to both numbers and geometrical quantities. The period we are discussing witnessed a number of such attempts, one by Viète, one[8] by van Roomen, and two by Descartes.

In the 1590's Viète published a series of treatises intended to form what he called "a work on the restored mathematical analysis or the new algebra."[9] It contained a fully consistent universal algebra of abstract quantities, with the algebraic operations introduced not by definitions but by axioms describing their formal rules. Here he introduced, as mentioned above, the abstract extension of geometry to include objects of all dimensions larger than three, which were necessary to salvage the dimensional aspect of multiplication in geometry. Viète saw arithmetic and geometry as two different representations of this general algebra; the actual workings of the operations addition, multiplication, etc. were different but they followed the same rules as the ones Viète codified in his new algebra. His work may be considered as the most complete elaboration of the idea of a *mathesis universalis*. [10]

---

[7] Cf. [4], the notes by Marion in [9, pp. 156–158, 160–164, 302–309], and [16].
[8] or two, see note 11 below.
[9] Translated from the title of the first treatise of the series, cf. [17].
[10] Cf. [1, chap. 8, pp. 145–158].

Van Roomen published in 1593 a book in which he sketched a *mathesis universalis* [14];[11] it contains long lists of axioms which should define the operations of algebra in this extended domain. His system was much less precisely formulated than Viète's and drew little interest from contemporary mathematicians.

As we'll see in detail below, Descartes' *Regulae* contained a sketch of a *mathesis universalis* in the context of a philosophical treatise. The mathematical aspects are hardly worked out, but it does contain an explicit definition of multiplication of line segments through which Descartes wanted to safeguard the applicability of algebra in geometry. Later he directed the inspiration he received from the idea of a mathesis universalis more exclusively to mathematics. In the *Géométrie* [5] of 1637 we find the results of that activity in the form of a treatise on the application of algebraic methods in geometry. It contains the above mentioned definition of multiplication of line segments by means of a proportionality involving a unit segment. The *Géométrie* became the founding text of analytic geometry. Yet it was sufficiently linked to his earlier ideas that we may recognize it as the second and final version of his mathesis universalis.

5 The algebra of line segments in the *Regulae*

5.1 Mathematics in the *Regulae*

In the *Regulae* Descartes repeatedly stressed his inspiration by mathematics. He stated that only in pure mathematics – arithmetic and geometry – fully certain knowledge and understanding could be achieved.[12] He embraced the *mathesis universalis* idea and believed it to be the only base for reasoning with certainty. He realized that an effective *mathesis universalis* would need the analytic methods of algebra.

As we have seen, these ideas and expectations were alive in mathematics at his time, although not generally accepted. What makes Descartes' case exceptional is, primarily, that he extended his expectations for a *mathesis universalis* from a general science of quantities to a general method of reasoning applicable to all matters, philosophical and otherwise, about which indubitable knowledge can at all be reached. Moreover, he intended not to rely on the mere authority of mathematics by taking its results as absolutely certain; rather he worked out a mental technique for testing the certainty of reasoning in mathematics (and other fields) and thus to achieve true knowledge independently.

To attain broad applicability of his method while keeping mathematics as the basis, Descartes had to argue that a very general class of subjects could be interpreted as essentially mathematical. The argument is part of the material dealt with in the first 16 rules, in which Descartes built up his philosophy of knowledge and cognition, and explained the possibilities and restrictions of attaining certainty. For brevity I summarize the argument by quoting rules 2 and 14. Rule 2 restricts the matter to which the rules are applicable:

---

[11] Summarized in [15]. Recently Paul Bockstaele identified two copies of a text by van Roomen containing a reworked version of this sketch, printed c. 1602 but apparently never effectively published; a publication is forthcoming in *Archive for History of Exact Science*.

[12] See note 3.

> We should attend only to those objects of which our minds seem capable of having certain and indubitable cognition [6, p. 10, p. 362].

Rule 14 completes the process of reducing questions to a form in which they can effectively be understood and answered with certainty:

> The problem should be re-expressed in terms of the real extension of bodies and should be pictured in our imagination entirely by means of bare figures. Thus it will be perceived much more distinctly by our intellect. [6, p. 56, p. 438]

I shall return below to the roles of the intellect and the imagination mentioned in this rule. For now I note that Descartes takes extension to be "whatever has length, breadth and depth, [6, p. 438, p. 59]" and thus arrives at geometry as the science in terms of which all questions open to philosophical scrutiny can be dealt with.

Descartes shared with other proponents of the *mathesis universalis* idea the conviction that algebra was essential for its success. With geometry as quintessential science, then, he could not avoid dealing with he obstacles mentioned above for merging algebra and geometry. In particular he had to propose a clear redefinition of multiplication for line segments and corresponding definitions of division and (square- and higher-order) root extraction. In the next sections I show how he approached these matters; then I return to what he wrote, and what he may have thought, about the certainty of the algebraic operations.

5.2 Multiplication, the unit segment

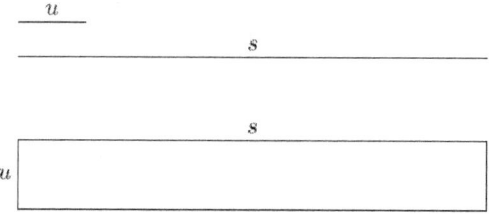

Figure 1: Equivalence of line segments and rectangles, $s \simeq s \times u$.

As explained above, the arithmetical definition of multiplication required at least one of the factors to be a (rational) number, and could therefore not be applied to line segments. The obvious candidate for a redefinition was the forming of rectangles, but that required a solution for the dimension problem: if the product was a rectangle, either one could not multiply more than three factors, or one had to accept objects of higher dimension than 3; neither was acceptable to Descartes. He realized that the product rectangle had somehow to be reduced to a product segment. He did so, in Rule 18 by introducing a unit segment u and defining an equivalence relation between line segments and rectangles (see Figure 1) [6, pp. 465–467, pp. 73–75]:[13]

---

[13] For Descartes' own formulation of the equivalence see the quotation from [6, p. 468, p. 76] in Section 5.3: "we always conceive the lines as rectangles, one side of which is the length which we adopted as our unit."

Any line segment s is equivalent to a rectangle with sides s and u.

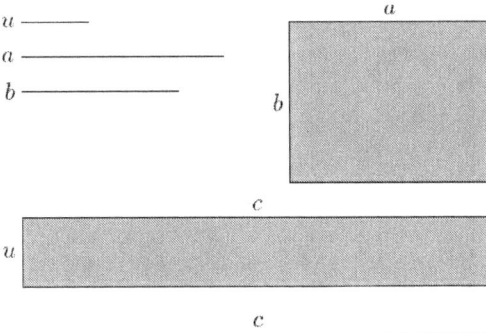

Figure 2: Equivalent rectangles in Descartes' definition of multiplication.

He then multiplied segments as follows (see Figure 2): Given two line segments a and b, and a unit segment u, form the rectangle with sides a and b, this is the product of a and b in rectangular form. Now transform this rectangle into another one, equal in area and with one side equal to the unit u. (This of course raises the question, to which I return in Section 5.3, how Descartes imagined that this new rectangle could be made.) The other side of the rectangle, c in the figure, will be the product of a and b in line segment form.

For ease of the modern reader I introduce some notation (not used by Descartes): × stands for the geometrical product in rectangle form, = is the equality sign, which, for rectangles, means equality in area (not necessarily in shape), and ≃ denotes the equivalence illustrated in Figure 2. Then the definition of the product in line segment form can be represented as:

$$a \times b = u \times c \simeq c.$$

At this point I should note the similarities and differences between the above definition of multiplication and the one which Descartes gave, later, in the *Géométrie*. There Descartes did not refer to rectangles but, as mentioned above in Section 3.5, defined the multiplication by means of a proportionality: the product ab of two line segments a and b was the fourth proportional of u, a and b [5, pp. 297–299]. The two definitions define the same line segment. The proof that that is so is based on Prop. 16 of Book VI of the *Elements* which links proportionalities and rectangles; it asserts that the proportionality $a : b = c : d$ holds if and only if the rectangles $a \times b$ and $c \times d$ are equal. [14] Yet, as will become clear below, there is a crucial difference between the two definitions with respect to visual evidence. Of the two equalities in Prop. VI-16, the one, an equality of rectangles, is much easier visualizable than the other, which is a proportionality, that is, an equality of ratios.

---

[14] One might say that, in stead of the above equivalence $u \times c \simeq c$ between a rectangle and a line segment, Descartes used the equivalence $c : u \simeq c$ in the *Géométrie*.

5.3 The missing argument: transformation of rectangles

As noted, there is an argument missing in the reasoning about multiplication so far: How to transform a rectangle such as $a \times b$ into to a rectangle of equal area but having one of its sides equal to u? If we can perform this transformation we actually have the other side c; if not, we have only an implicit definition of the product c but not a means to attain or achieve it, and thus we have no effective multiplication, let alone an effective algebra.

At the end of Rule 18 Descartes takes up this issue explicitly and announces an explanation:

> It is therefore important to explain here how every rectangle can be transformed into a line, and conversely how a line or even a rectangle can be transformed into another rectangle, one side of which is specified. Geometers can do this very easily, provided they recognize that in comparing lines with some rectangle (as we are now doing) we always conceive the lines as rectangles, one side of which is the length which we adopted as our unit. In this way the whole business is reduced to the following problem: given a rectangle, to construct upon a given side another rectangle equal to it. The merest beginner in geometry is of course perfectly familiar with this; nevertheless I want to make the point, in case it should seem that I have omitted something. [6, p. 468, p. 76]

The explanation, however, if Descartes wrote it down, is lost; the extant remaining text of the *Regulae* consists only of the three rules 19–21 which prescribe in general terms how a problem should be translated into an algebraic equation. These rules lack further explanation, so the above quotation is the last of the explanatory texts in the unfinished treatise.

Which procedure of transforming rectangles did Descartes intend to explain at this point? His text is clear enough to identify it; it is the 'application of areas' in the case of rectangles.

5.4 The application of areas

The *application of areas* is one of the key results in the first two books of Euclid's *Elements*.[15] It is a procedure for solving the following general problem:

> Given a line segment a, an angle α and any rectilinear figure F, construct a parallelogram equal in area to F and with one side equal to a and one angle equal to α.

The procedure is explained and proved in Props I-44, 45; it is the basis of the last proposition of the second book, II-14, which gives the quadrature of any rectilinear area, that is, its transformation into a square of equal area.

The application of areas procedure is based on the use of a particular configuration called the *gnomon*, to which I return below. This figure made it possible

---

[15] On the history and the importance of the result see Heath's note to Prop. I-44 in [11, vol. 1, pp. 342–346].

to construct the required parallelogram and to prove the correctness of the construction without the use of the theories of similar figures and proportionality, which are explained later on in Books V and VI. There the application of areas recurs, but now based on theorem VI-16, mentioned above, which states that, if a, b, c, d are line segments, then the proportionality $a : b = c : d$ holds if and only if the rectangles $a \times d$ and $b \times c$ are equal in area. Books V and VI were generally recognized as very difficult in Descartes' time, definitely not stuff for "mere beginners"; in contrast the gnomon figure and its properties were seen as quintessentially simple and clear.[16] I assume therefore that Descartes intended to explain the application of rectangular areas by means of the gnomon figure.

5.5 The gnomon

In the rectangular case, the *gnomon figure* (see Figure 3) consists of a rectangle with one of its diagonals; through a point on the diagonal two lines are drawn parallel to the sides of the rectangle. These lines, together with the diagonal,

 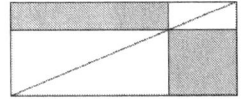 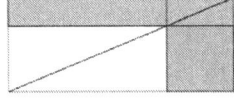

Figure 3: The gomon figure, its main property and the gnomon shape.

divide the rectangle in six parts, four triangles and two rectangles. The configuration has the *gnomon property*: the two rectangular parts are equal in area. The property is proved in *Elements* I-43 for parallelograms (hence also for rectangles); it makes the gnomon figure very effective in proving the equality of rectangles. Euclid used the gnomon figure in Book II to prove various equalities between parts of squares and rectangles that arise if lines parallel to the sides are drawn. The name gnomon is used in Greek mathematical texts to indicate a figure like the one shaded in the third diagram in Figure 3.[17] The Euclidean proof of the gnomon property is as follows:

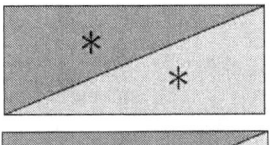

The diagonal divides the rectangle in two large triangles above and below the diagonal; these triangles are equal.

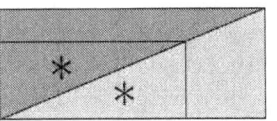

For the same reason the two triangles in the lower left corner of the figure are equal.

---

[16] In his comments to the definition of the gnomon in his edition of the *Elements* (book II, Def. 2), Clavius praised the simplicity and the strength of the propositions in Book II which, he claimed, provided the basis of "those admirable rules of algebra.": "Nam ex nonnullis harum propositionum demonstrantur regulae illae admirabiles Algebrae, quibus vix credo in disciplinis humanis praestantius aliquid reperiri, quippe cum miracula quaedam numerorum (ut ita dicam) eruant tam abstrusa, ac recondita, ut facultas illa omnem captum humanum superare videatur, tanta nihilominus facilitate, atque voluptate, ut facilius vedeatur esse nihil." [3, vol. 1, p. 83].

[17] The gnomon figure shows the difference between two similar rectangles (or parallelograms); in particular in the case of squares it represents the difference of two squares and thereby plays a key role in book II of the *Elements* which concerns equalities of areas analogous to algebraic equalities such as $a^2 - b^2 = (a-b)^2 + 2(a-b)b = (a-b)(a+b)$ (cf. Prop. II-4).

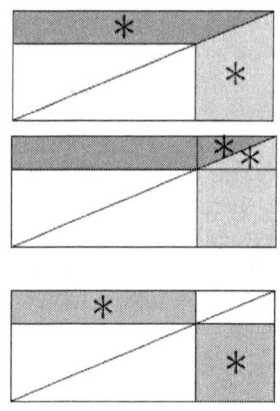

If these triangles are removed, the areas remaining in the larger triangles are equal, because if equal parts are taken away from equal figures the remaining figures are also equal.

Similarly the two triangles in the upper right corner are equal, and if these are also removed,

again the remaining parts of the two large triangles are the same. These remaining parts are precisely the two rectangles in the gnomon figure, so these are equal and the gnomon property is proved.

The proof relies upon the third of the five axioms or "common notions" in Book I of Euclid's *Elements*: "If equals are subtracted from equals, the remainders are equal." [11, vol. 1, p. 155] I have formulated the gnomon property and its proof, and added the corresponding sequence of figures, to exhibit the power of the argument, especially in combination with figures, to convince of a geometrical truth that is not at all obvious at first sight. One cannot simply 'see' from the gnomon figure that the two rectangles are equal. But contemplation of the figure and modifying it judiciously can provide conviction and certainty about a geometrical truth such as the gnomon property.

6 Descartes on certainty and mental vision

From its position in the general argument of the *Regulae* it is clear that, at least before he decided to abandon the text, Descartes considered the method he was going to explain at the end of Rule 18 as an example of geometrical reasoning providing the kind of certainty his philosophy was to make available outside mathematics as well. What were his criteria for such certainty? Like in Section 5.1, I shall have to be brief here.

Descartes based his requirements for true understanding on a metaphysical and physiological model of the working of the human mind. The model distinguishes various mental faculties involved in attaining knowledge. The central faculty is the *intellect*, which has two main instruments for acquiring knowledge, an immediate one, *intuition*, by which the intellect directly contemplates a subject, and an indirect one, *deduction*, by which it makes inferences from earlier acquired pieces of knowledge to conclude new knowledge. For both activities one can train one's mind to reach a state of awareness and clarity which enables the intellect to recognize with certainty that the insights gained by intuition or deduction are true. The test for this is a direct mental experience in which the intuition or deduction is recognized as *clear* and *distinct*. As deduction always involves earlier attained truths, all knowledge is ultimately built on simple and direct intuitions. One of the ways intuition works is a kind of mental vision, called the *imagination*. Although not all the results of intuition are gained in that way, imagination plays a crucial role, especially in dealing with problems reduced to geometrical form.

Descartes saw imagination as a process taking place in the brain when the immaterial intellect, by a kind of mental vision, views a material bodily organ best described as a kind of screen. The intellect can project figures on the imagination, change them at will, contemplate the resulting configurations, and thus by direct intuition gain understanding about them. Simple geometrical truths in particular could be gained in this way. Descartes used the term imagination both in a broad sense as the process in which the intellect contemplates by mental vision the screen-like object, and in a restricted sense as particularly the screen itself; I'll follow his usage in that respect.

I find it striking how precisely the proof of the gnomon property discussed in Section 5.5 fits Descartes' description of the interplay of imagination and intellect in acquiring certain knowledge through intuition. Twice the figure in the imagination visualizes the taking away of equal triangles from equal subfigures of the gnomon figure. Intuition convinces the intellect that the two halves of the rectangle shown at the beginning are equal and that after each of the two removals the remaining figures are equal because of the third Euclidean axiom about removing equal parts from equal things. I assume that this reasoning exemplifies how, according to Descartes, the intellect achieves certain knowledge through clear and distinct intuition of figures in the imagination. Indeed I find it likely that this Euclidean proof of the gnomon property was one of the mathematical examples on which Descartes based his model of the working of the human mind.[18] It appears, then, safe to assume that the next subject planned for rule 18 was to ascertain that the operations of the algebra of line segments could be clearly and distinctly understood, by providing proofs well-suited to mental vision and employing the application of areas procedure based on the gnomon property.

This leads me to the reconstruction of what Descartes would have mentally 'seen' when applying his mind to the algebraic operations in geometry according to the method of intuition by means of the imagination.

7 The Reconstruction

7.1 Imagining division a/b

I start with the elementary operations addition, subtraction, multiplication and division as applied to line segments. Earlier in rule 18, Descartes had already

---

[18] One of the few explicit mathematical examples of intuition in the *Regulae* occurs in Rule 4 [6, p. 369, pp. 14–15], it is an arithmetical one:

> The self-evidence and certainty of intuition is required not only for apprehending single propositions, but also for any train of reasoning whatever. Take for example, the inference that 2 plus 2 equals 3 plus 1: not only must we intuitively perceive that 2 plus 2 make 4, and that 3 plus 1 make 4, but also that the original proposition follows necessarily from the other two.

Descartes here points to one step in the argument that should not be omitted in intuiting the truth of the proposition, namely the one concluding from $2+2 = 4$ and $1+3 = 4$ that $2+4 = 3+1$. That inference is based on the first Euclidean axiom "Things which are equal to the same thing ar also equal to one another" [11, vol. 1, p. 155]. I note that the geometrical reasoning I propose above has the same structure as the arithmetical one from Rule 4: Two instances of directly intuited equality, combined with one of the Euclidean axioms. The difference between the two is that in the arithmetical example the intuition of the first two equalities is less explicitly spatial.

explained, with simple diagrams, how the first two can be visualized in the imagination: the two line segments are posited along a horizontal line with the one's right end point coinciding with the other's left end point; thus combined they form a new line segment: their sum. Subtracting one segment from another (which should be larger) is similarly visualized. Multiplication and division are rather straightforward once one decides on the gnomon; I give the division as example: Given two line segments a and b, I have to find a line segment c such that $c = a/b$.

Imagine the two line segments a and b joined together horizontally as if they were added.

I know that $a/b = c$ means that $a = b \times c$. But the line segment a can only be equal to the rectangle $b \times c$ if a is taken in its equivalent form as rectangle $a \times u$, in which u is the unit length: $a \times u = b \times c$. I can now follow the application of areas procedure of *Elements* I-44 to find a rectangle equal to $a \times u$ and having one of its sides equal to b.

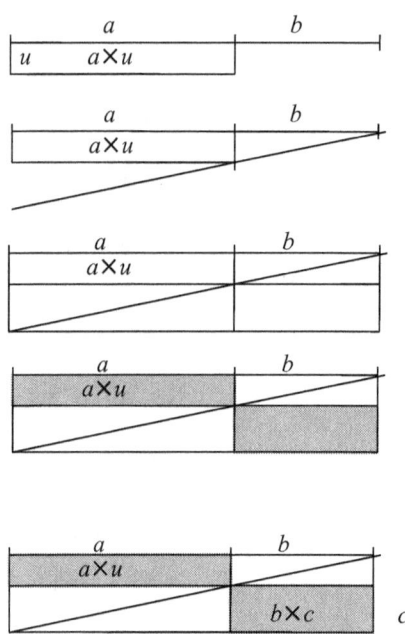

The unit u is given as well as a, so I can locate the rectangle $a \times u$ below a; it will be part of the gnomon figure I'm going to use.

I can now also locate the diagonal because it passes through the end point of b and the down-right corner of $a \times u$.

I can now locate the down-left corner of the large rectangle and complete the gnomon figure.

Knowing the gnomon property I may now deduce that the rectangle in the down-right corner is equal to the rectangle $a \times u$.

But I see that one of the sides of that rectangle is b and so the other side does exactly what the c I am looking for should do, namely $a \times u = b \times c$; so I have found c: it is the vertical side of the down-right rectangle.

I have used the term "locate" for what in geometry would be "construct" in order to avoid the assumption that for Descartes the action in the imagination would simply be identical with Euclidean construction. Yet the similarity with construction suggests to note here which operations have to be allowed (i.e. judged to be fully clear and distinct to the intellect) for the performance of a division as above to be acceptable. They are: locating a given (i.e. previously located) line segment at the endpoint of another line segment, either in the same direction or at a right angle, locating the points of intersection of located lines, locating lines through two given points, and prolonging line segments beyond their

Descartes on mental vision and the certainty of algebra

end points. It seems to me that none of these operations, which the intellect here is supposed to perform by projecting figures on the screen-like imagination, is beyond Descartes' requirements for clear and distinct intuition.

The reconstructed imagination of division does more than visualizing the successive steps leading from the given line segments a, b and u, to the one to be found, c. The resulting image, being a gnomon figure, also provides the material for the intellect to intuit, in the way explained above in Section 5.5, that the two rectangles $a \times u$ and $b \times c$ are equal and that therefore c is truly the quotient of a and b. The image carries the solution as well as its proof.

7.2 Imagining root extraction $\sqrt{a}$

Descartes was well aware that addition, subtraction, multiplication and division were not sufficient to employ the power of algebra, one would at least also need root extraction. It is therefore likely that he tried to find a procedure for root extraction as convincing as the one above for division. The first case to check would have been the square root:

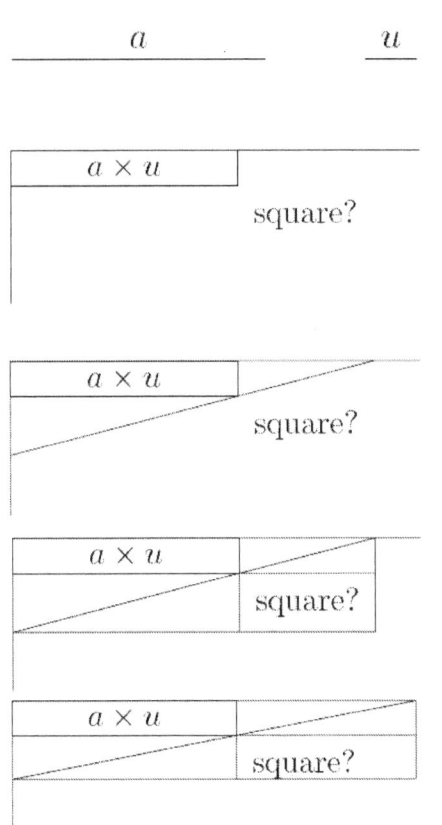

Given a line segment a and a unit segment u, I now have to locate a line segment, call it d, equal to $\sqrt{a}$. Thus $d^2 = a$, which, as earlier, I interpret as $d \times d = a \times u$.

This suggests to consider the rectangle $a \times u$ and try locating an appropriate gnomon figure in which it fits. So I extend the left and upper sides of $a \times u$ and note that I would like to find a square located somewhere down-right and forming a gnomon figure together with $a \times u$.

But I have no givens on which to base such a figure. Let me try nevertheless and draw any line through the down-right corner of the rectangle and locate its intersections with the prolongations of the sides.

And let me complete the gnomon. I did not find a square.

So let me try once more. This one is worse. The diagonal should be steeper than the one I tried first.

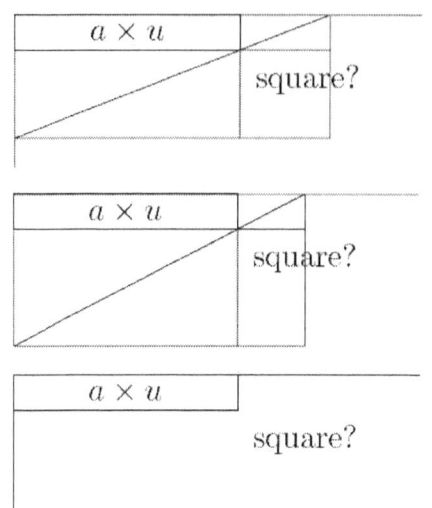

This looks better, but is it a square? How can I decide, by mental vision alone, whether a rectangle is a square?

Anyway, this diagonal is too steep.

It seems I can only try and err or hit a square by accident and without guarantee. The square root appears unattainable by a procedure as clear and distinct as the one used for division. I give up.

It appears that if, as I conjecture, Descartes has tried to find a procedure for square root extraction suitable for intuition by means of the imagination, he would have found that it could not be done along the lines suggested by the gnomon-based application of areas. What alternatives could he then have found within contemporary geometry, which meant in Euclid's *Elements*?

7.3 Alternatives for imagining root extraction

Descartes was no doubt aware that square root extraction in the geometrical sense is not beyond the Euclidean means of construction. Indeed the construction is explained twice in the *Elements* (II–14 and VI–13, 16), it is as follows:

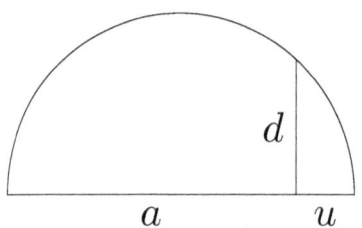

Join the line segment a and the unit segment u; draw a semicircle on the resulting line; from the point where a and u are joined, draw a perpendicular line, mark its intersection with the semicircle, call the resulting segment d; then the square $d \times d$ will be equal to the rectangle $a \times u$.

Evidently an imagination of square root extraction based on this construction would involve a new element as compared to the imagination of the first four algebraic operations, namely the location of a semicircle on a given line segment and the location of its intersection with a given line. Moreover, while the images for the four elementary operations, as illustrated in the case of division, carried both the result and its proof, the figure of the semicircle and the segments a, u and d does not as directly provide the arguments why the square $d \times d$ is equal to the rectangle $a \times u$. This becomes clear if we look at the two – different – proofs in the *Elements*. The proof in Prop. II-14 is based on the gnomon theorem I-43, the theorem of Pythagoras I-47, and Prop. II-5 which provides a gnomon-based argument. The latter argument could be visualized, which would require a considerable extension of the figure (Cf. the figure at Prop. II-5 in [11, vol. 1, p. 382] which should be appended under the base of the semicircle). Also Pythagoras' theorem could be proved by a visually convincing proof. But the combination of the two (in one intricate figure) would hardly serve a direct intuition of the equality of the square $d \times d$ and the rectangle $a \times u$; the full proof of II-14 consists of a chain of deductions and is therefore much less direct than the argument supporting division.

The construction by means of the semicircle recurs in Book VI of the *Elements*; it is used in Prop. 13 to find a "mean proportional" of two given line segments. (A magnitude M is called the mean proportional between two magnitudes A and B of the same kind as M if $A : M = M : B$.)

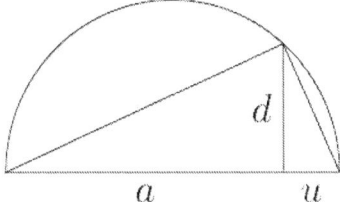

Euclid's proof in VI-13 is based on the theories of proportion and similarity of figures: The triangles with bases a and u respectively are similar, hence $a : d = d : u$, that is, d is the mean proportional of a and u.

Prop. VI-16 (see Section 5.4) then yields that the square $d \times d$ is equal to the rectangle $a \times u$, that is, in the context of the *Regulae*, that $d = \sqrt{a}$.

As noted at the beginning of Section 5.5 the theories of similarity and proportions are intricate, and in particular the (Eudoxean) definition of equality of ratios is hardly amenable to visual confirmation.

Thus the alternatives held little promise for clear and distinct visualizations. Moreover, the occurrence of the circle in the construction of the mean proportional would have made Descartes extra aware that his model of visual perception, depending on the creation by the intellect of figures on the imagination, would require a practice analogous to geometrical construction. And not only construction by straight lines and circles ('ruler and compass') as in the *Elements*, but also higher-order constructions by special curves or instruments. Indeed the next root to be considered, the cubic root $\sqrt[3]{}$ was known not to be constructible by straight lines and circles.

It appears, then, from the exercise in imagination in Section 7.2 and the above survey of alternatives, that any attempt to base the certainty of square root extraction on clear and distinct imagination would induce strong doubts about basing geometrical certainty of other than the simplest operations on intuition of imagined figures, especially in comparison with the appealing visual conviction of the gnomon-based diagrams.

8 Conjecture

I have assumed in my reconstruction that Descartes' description of clear and distinct mental perception in the *Regulae* was based on personal experience and practice. The alternative would be to consider the arguments on perception as mere (and disingenuous) rhetoric to legitimize his new philosophy, which I find difficult to match with his writing. I also take at face value his repeated assertions that he first found in mathematics the kind of certainty he was aiming at.

Let me now formulate my conjecture on the basis of the above arguments.

It is this: Probably quite early in his mathematical studies Descartes learned about the application of areas by means of the gnomon figure and realized its visually convincing quality. By the time he shaped his ideas about clear and distinct mental perception he returned to this piece of mathematics, and other relatively simple ones,[19] and used them in his personal practical exploration of the workings of inner perception. Then during the 1620's, while sketching and composing the *Regulae*, he elaborated most of his ideas about attaining certainty on the assumption that mathematics provides not only examples of methodical achievement of certainty but also equally certain general methods for problem solving, especially by means of equations. Also early in his career he realized that a successful *mathesis universalis* would require a well-based interpretation of multiplication and division as applied in geometry, but he did not explore root extraction far enough to realize the difficulties involved in visualizing algebraic operations beyond the elementary four.[20] Then, when the *Regulae* project was well under way, he came to the point where he had to explain and demonstrate the algebraic operations as applied in geometry. Now he could not postpone root extraction because his system would require equation solving, of which root extraction is a special case, namely solving the special equations $x^2 = a$, $x^3 = a$, etc. In that connection he was confronted with the difficulties sketched in Sections 7.2 and 7.3, and realized that further attempts to base the certainty of algebra on mental vision would fail.

I note that the conjecture concerns the process leading to Descartes' realization of unavoidable failure, and the role of the gnomon arguments in that process, rather than the realization itself, which has been noticed before, in particular by Schuster.[21]

What, then, are results of this exercise in reconstructing early Cartesian mental vision? If the conjecture formulated above in this section is granted we have some new mathematical examples illustrating Descartes' early conception of intuition by means of imagination. The first is the proof of the gnomon property (section 5.5), extant in Euclid's *Elements*, and which, independently of the conjecture, would qualify as a likely inspiration for Descartes. Then there are the visualizations of multiplication and division of line segments (sections 5.2 and 7.1), credible as examples of clear and distinct intuition and less trivial than the examples of addition and subtraction which Descartes does give in Rule 18. And finally the case of square root extraction (sections 7.2 and 7.3) offers a plausible example of an attempted intuition which Descartes would have rejected as insufficiently clear and distinct.

---

[19] Such as the one mentioned in note 18.

[20] The opinion of Clavius on the strength of the propositions in book II of the *Elements* (see note 16) may have played a role here.

[21] In his detailed and very informative study [16] Schuster comes to the conclusion that three areas of difficulties contributed to Descartes' decision to abandon the writing of the *Regulae* (pp. 73-79). The last of these related to the "intuitive grounding in the imagination" (p. 73) of mathematical objects and operations. He sees Descartes' efforts in that area primarily as legitimatory. He writes: "[Descartes] probably realized that the solution of quadratic and higher-order equations, or the extraction of square- and higher-order roots, would elude the excessively simple manipulations of lines and rectangles demanded by his legitimatory aims and doctrine." (p. 78).

Bibliography

[1] Bos, Henk, J. M. *Redefining geometrical exactness : Descartes' transformation of the early modern concept of construction,* (Sources and studies in the history of mathematics and physical sciences). New York: Springer, 2001.

[2] Clavius, Christoph. *Algebra.* Rome, 1608.

[3] Clavius, Christoph. *Opera Mathematica (4 vols).* Mainz, 1611–1612.

[4] Crapulli, Giovanni. *Mathesis universalis; genesi di una idea nel XVI secolo.* Rome: Edizioni dell'Ateneo, 1969.

[5] Descartes, René. La geometrie. In *Discours de la méthode pour bien conduire sa raison et chercher la verité dans les sciences; plus la dioptrique les meteores et la geometrie qui sont des essais de cete methode,* (Leiden, 1637), 297–413. Standard edn. In [8], vol. 6, 367–485; Translated (English) in [7].

[6] Descartes, René. Regulae ad directionem ingenii. In [8], vol. 10, 356–469. Translated (French) in [9]; (English) in [10], vol. 1, 7–78; page references to editions in [8] vol. 10 and [10], vol. 1, respectively.

[7] Descartes, René. *The geometry of René Descartes,* Smith, D. E., Latham, M. L.(Ed.). New York: Dover, with facs. of orig. edn., 1954.

[8] Descartes, René. *Oeuvres de Descartes.* In Charles, Adam, Paul, Tannery (Eds.), *nouvelle présentation 11 vols,* vol. 11. Paris: Vrin, 1964–1974. The earlier edition is from 1897–1913.

[9] Descartes, René. *Règles utiles et claires pour la direction de l'esprit et la recherche de la vérité; traduction selon le lexique Cartésien, et annotation conceptuelle.* Martinus Nijhoff: The Hague, 1977. tr. annot. Jean-Luc Marion, math. annot. P. Costabel.

[10] Descartes, René. *The philosophical writings of Descartes,* Cottingham, J., Soothoff, R., Murdoch, D., and Kenny, Anthony (Eds.) *3 vols.* Cambridge: Cambridge University Press, 1985–1991.

[11] Euclid. *The thirteen books of Euclid's Elements,* 3 vols (tr. intr. comm. Thomas L. Heath). New York: Dover, reprint of 1926 edn., 1956.

[12] Pappus. *Pappi Alexandrini mathematicae collectiones a Federico Commandino Urbinate in latinum conversae at commentariis illustratae.* Pesaro, 1588.

[13] Proclus. *A commentary on the first book of Euclid's Elements* (tr. intr. Glenn R. Morrow, pref. Ian Mueller). Princeton: Princeton University Press, reprint of 1970 edn., 1992.

[14] Roomen, Adriaan van (Adrianus Romanus). *In Archimedis circuli dimensionem expositio et analysis: Apologia pro Archimede.* Würzburg, 1597.

[15] Roomen, Adriaan van (Adrianus Romanus). *Universae mathesis idea, qua mathematicae universim sumptae natura, praestantia, usus et distributio brevissime proponuntur.* Würzburg, 1602.

[16] Schuster, John, A. Descartes' mathesis universalis: 1619–28. In Gaukroger, S. W. (Ed.), *Descartes, philosophy, mathematics and physics,* 41–96. Totowa, Brighton: Barnes and Noble, Harvester, 1980.

[17] Viète, François. *In artem analyticen isagoge: seorsim excussa ab opere restitutae mathematicae analyseos seu algebra nova.* Tours, 1591.

C

PHILOSOPHICAL ISSUES OF PARTICULAR SCIENCES

Structuralism, Fictionalism, and Applied Mathematics[1]

Mary Leng
University of Liverpool
mcleng@liv.ac.uk

ABSTRACT. A natural and attractive view of the applications of mathematics in empirical science is that expressed by *ante rem* structuralist, Stewart Shapiro, according to whom "mathematics is to reality as pattern is to patterned" [6, p. 248]. Viewing mathematics as the science of abstract patterns, *ante rem* structuralists such as Shapiro have the means for an explanation of the applicability of mathematics which proceeds by considering the various ways in which our theories about 'freestanding' mathematical patterns can be used to provide information about the patterns instantiated in physical reality.

But does the view that the world is mathematically structured require a belief in the existence of abstract structures? This paper considers three ways in which mathematical structures are appealed to in the context of our scientific theories, and considers how a fictionalist might view these appeals to mathematical structure. While the first of these three kinds of applications of mathematics does not require us to posit abstract mathematical structures that are related in various ways to nonmathematical objects, the second and third applications do require the supposition of ideal models and/or pure mathematical structures over and above the physical reality our theories aim to describe. In such cases, since a fictionalist cannot accept the existence of these ideal models/mathematical structures, to the extent that such applications are indispensible to our empirical scientific theories, a fictionalist must be anti-realist about such theories, believing not that they are true, but only that they are *good* representations of the physical world, in that "the physical world holds up its end of the "empirical-science bargain"" [1, p. 134].

In this respect, then, it seems that *ante rem* structuralism has some advantages over fictionalism. Through its realism about mathematical structures, *ante rem* structuralism can preserve the assumption that our scientific theories are (at least approximately) true, and can do so without appeal to the less than crystal clear notion of the "physical world" being the way it would have to be for our theories to be true. This paper argues that these advantages are illusory. Given structuralist metaphysics, the mere claim that our scientific theories are true is far too weak to account for the applicability of those theories. In order to account for applicability, the structuralist will also have to hold that "the physical world holds up its end of the "empirical-science bargain"". If it is this claim that is doing the work in accounting for the applicability of our scientific theories, it is unclear what is gained by the additional claim that our empirical theories are true.

---

[1] The author is with Department of Philosophy, University of Liverpool.

1 Introduction

Our scientific theories are mathematical through and through. It seems that we cannot even state their laws without talking, not just about physical systems and their properties, but about their relations to mathematical objects. Why should this be? Why should a body of truths about abstract, nonspatiotemporal things turn out to be indispensable to our best theories of the physical world? Mathematical structuralism provides the beginnings of an attractive answer to this question. According to structuralism, mathematics is the science of patterns, and, to quote one prominent structuralist, Stewart Shapiro, "mathematics is to reality as pattern is to patterned" [6, p. 248]. The physical world is structured mathematically, so it is no wonder that mathematics, as the general theory of structure, is relevant in understanding that world.

Stewart Shapiro is an *ante rem* structuralist. That is, he holds that "freestanding" abstract mathematical structures exist over and above any physical instantiations. The structuralist view of applications slots in well to this metaphysical outlook, since it allows one to be straightforwardly realist about both sides of the pattern/patterned relation. However, as Shapiro himself notes [6, p. 248], one does not have to adopt the structuralist metaphysics to adopt something like this view of applications. Traditional Platonists, who do not reify mathematical *structures* but do believe in structured systems of mathematical *objects* can deal with applications in much the same way, albeit with a slight difference of emphasis. Rather than holding that freestanding mathematical universals are sometimes instantiated in particular systems of physical objects, a traditional Platonist can hold that the structures exemplified in physical systems are modelled in systems of mathematical objects, expressing this in terms of the existence of isomorphisms between systems of physical objects and systems of mathematical objects.

But does the view that mathematics applies to physical reality because the physical world is mathematically structured require a belief in *any* abstract mathematical reality, whether that be a reality of structures and their positions, or of more traditional mathematical objects? I would like to probe the structuralist account of mathematics a little more carefully in order to get clearer on its commitments, with a view to answering this question. In the process, I will argue that a version of the account is available to mathematical fictionalists, who do not accept the existence of abstract mathematical objects *or* of abstract mathematical structures. Furthermore, while one may think that the *ante rem* structuralist's use of this account has some advantages over this fictionalist account, in that it can be stated more straightforwardly and allows us to remain wholesale realists about our scientific theories, I will argue that these advantages are illusory.

2 Kinds of Applications of Mathematical Structure

The slogan that mathematics is to reality as pattern is to patterned is an attractive one, but needs some filling out. Shapiro points to two closely related senses in which mathematical patterns can relate to physical systems, by direct exemplification and by modelling:

> At least some applications consist of incorporating mathematical structures into physical theories, so that physical systems exemplify mathematical structures. What is almost the same thing, in some theories, the structures of physical systems are modeled or described in terms of mathematical structures. [6, p. 243]

The Platonist account mentioned above aims to incorporate all applications into the modelling account, avoiding talk of exemplification altogether. On the other hand, as this quote from Shapiro makes clear, even though the structuralist account can make some use of the notion of a mathematical structure being exemplified in physical systems, at least some applications will require us to talk of the *relations* between abstract mathematical structures and physical systems. As we will see, this aspect of the structuralist account of applications causes some difficulty for fictionalists, as it makes it hard for a fictionalist to be a scientific realist. But before we turn to this point, it is worth looking a bit more closely at kinds of applications by instantiation and by modelling, and consider the question of just what, according to the structuralist account, is being modelled.

2.1 Type I: Direct and Approximate Exemplification

The most basic type of application is via direct exemplification. Shapiro takes the example of the Kariera system, a class structure at work in a tribe discovered by Lévi-Strauss:

> A tribe is divided into four classes, and there is a certain function for determining the class of a child from the classes of his or her parents. There is an "identity class" in the sense that when a member of this class mates with any member of the tribe, the offspring are members of the other's class. If two members of the same class mate, then their offspring are of the identity class. [6, p. 249]

It turns out that, taking the four classes as objects together with the rules for determining the class of a child as a function on those four classes, the structure exemplified is that of the Klein group. Theorems about the Klein group thus receive an interpretation in terms of the class system, and can allow us to infer properties of that system (for example, as Shapiro points out, the fact that the Klein group is Abelian means that if a child's parents are from class A and class B, we do not need to know which parent is from which class in order to determine the class of the child).

Now this kind of application is relatively easy for a fictionalist, who does not believe in abstract mathematical objects or structures, to make sense of, since it does not require us to presuppose that there are any such things. When a mathematical theory is directly exemplified, its nonlogical terminology can be interpreted in such a way that its axioms assert ordinary truths about nonmathematical objects and their relations. One view of mathematical axioms suggests that we think of axioms such as the axioms of group theory as *implicit definitions* of their primitive terms: a group $\langle G, +, 0 \rangle$ is any system of objects together with a binary relation satisfying certain properties and a distinguished identity element. A Klein group is any such system which satisfies some further restrictions. In the case of the Kariera system, the claim that the Klein group structure is instantiated in this system is just the claim that, when we interpret classes as objects, the rules for determining the class of a child as a function on those classes, and a particular class as the identity, the defining features of the Klein group are true of the Kariera system. And since, as a matter of logical consequence, theorems about Klein groups hold true of any system satisfying the definition, such theorems will also be true, suitably interpreted, about the Kariera class system. This application requires no appeal to abstract mathematical objects or structures. All that is required is for us to find interpretations of the primitives G, $+$ and 0 as ordinary nonmathematical objects and relations, such that the axioms for group theory are true in that interpretation.

Unfortunately, such neat applications are not always readily available. Indeed, more often than not, even in applications by instantiation, the axioms of the mathematical theories we apply are not strictly speaking true of their nonmathematical interpretation. Take, for example, Euclidean geometry: we apply this theory to physical space with a high degree of success, interpreting the theory's assumptions about points and straight lines as truths about physical points and straight lines. But in fact these assumptions are not true of physical points and straight lines: according to our current best science, ordinary physical space is not Euclidean. Or consider our mathematical theory of fluid dynamics. The 'fluids' of this theory are continuous substances, while fluids in the real world are not continuous. Yet the mathematical theory is highly successful when we interpret its technical term 'fluid' as applying to fluids in the usual sense.

There are two ways in which such applications can be understood. One way is to continue with the 'application by instantiation' idea, holding in this case not that the axioms of the theory are *true* of the physical system in question, but that they are nevertheless *approximately true*. In this case, the account of applications is not substantially different from the account given for group theory above: we do not need to suppose that there are any abstract mathematical objects our structures, only that certain statements are approximately true when interpreted as talking about concrete things. Such an account is fairly plausible for the application of Euclidean geometry, where locally we know that the Euclidean axioms are close enough to the truth that the approximation makes no detectable difference. But the example of fluid dynamics is somewhat more complex: do we really wish to say that it is even *approximately* true that fluids are continuous?

In the case of fluid dynamics, a more natural understanding of such an application holds that real fluids behave in important respects *like* the continuous fluids of our mathematical models. Thus, in a fairly standard fluid dynamics textbook we are asked to assume the following 'continuum hypothesis' (!):

> ...that the macroscopic behaviour of fluids is the same *as if* they were perfectly continuous in structure [2, p. 4, my italics].

One way of understanding this *as if* claim is as a direct comparison between actual fluids and the ideal fluids of some ideal, abstract model of the pure mathematical theory of fluid dynamics. And in such a case, in order to make such a comparison it seems that we have to suppose, not that our mathematical theory is true when interpreted about some specific concrete system, but rather, that that theory is true of a system of abstract objects, some ideal model. While such applications appear relatively straightforward for *ante rem* structuralists (who hold that mathematical theories describe abstract structures and are automatically true of the positions in those structures, considered as objects), fictionalists will have more difficulty in making sense of these cases while still refusing to accept the existence of *abstracta*.

2.2 Type II: Structural Embeddings

In Type I applications, even when our mathematical theories are not perfectly instantiated in a physical system, we do at least have a resemblance between all parts of our ideal mathematical models and the concrete systems they are used to describe. But in many applications of mathematics, there are parts of our abstract mathematical theories that correspond to nothing in the 'concrete' physical realm. Shapiro points to applications of complex analysis and set theory in this regard:

> A case in point is complex analysis. The very name "imaginary number" suggests that no straightforward physical interpretation of these mathematical structure positions is forthcoming. Yet complex analysis is useful in physics and engineering.[2] A related phenomenon is the use of a physically uninterpreted theory, such as set theory, to solve problems that are undecidable in a weaker theory, such as arithmetic. ... [S]ome of these results have applications in recursive function theory and thus in computability. One would be hard put to find a physical exemplification of the set-theoretic hierarchy, let alone an exemplification that relates to computability. [6, pp. 249–50]

Here we have mathematical structures being applied even when the systems to which they are applied do not even approximately exemplify those structures. What can the structuralist view of applications say about these?

In fact, when we think in terms of structure, such applications should not be too problematic, as can be seen if we think first about applications of mathematics to mathematics itself. Why is set theory helpful in solving problems about arithmetic? Because the structure of the natural numbers can be modelled in the sets. Why is complex analysis helpful in learning about real-valued functions? Because the real numbers are embedded in the complex plane. If one structure is embedded in a larger one, then we may find ourselves appealing to the larger structure to discover truths about the embedded structure. And in such cases, if the embedded structure also receives a concrete exemplification, we may find ourselves learning truths about the concrete exemplification by appeal to mathematical structures that are not themselves exemplified.

So such applications of unexemplified mathematical theories to concrete physical reality are not surprising, so long as their applicability is due to their ability to tell us about theories that are (at least approximately) physically exemplified. This of course leaves some applications potentially mysterious (as Shapiro points out [6, p. 247], we cannot neatly explain all uses of mathematics in quantum mechanics by means of this kind of account, simply because we have so little idea

---

[2]Shapiro is rather too quick in assuming that there could be no straightforward applications of complex numbers via direct exemplification. In fact, the use of complex numbers in quantum theory to measure wave phase can be given a straightforward 'modelling' interpretation. (I am grateful to McLarty, Colin for stressing this point.) However, so long as complex numbers are used in *some* contexts where they do not directly represent any physical phenomenon, Shapiro's point, that not all applications can be via direct exemplification, stands.

of what the underlying 'physical reality' we are describing is like). But insofar as we do have an understanding of what it is that we are applying our mathematics *to*, we should be able to make use of the notion of structural embeddings in order to explain the applicability of wider mathematical structures.

Again, though, for a fictionalist who doesn't believe that there are 'wider mathematical structures' over and above those structures that are physically instantiated, something more will need to be said about such applications.

2.3 Type III: Applications to Non-Physical Systems

In our discussion of applications so far, we have assumed that the ultimate 'targets' of our mathematical models are systems of concrete physical objects. However, many applications of mathematics do not have such obvious targets. Many of our physical theories, including classical Hamiltonian mechanics and quantum mechanics, are so-called 'phase space' theories. Such theories use vectors to represent, not actual systems of physical objects (as when vectors are used to represent space-time points), but *possible states* of a physical system. In these cases, mathematics may still be to reality as patterned is to patterned, but the reality that is patterned is not a reality of actual, physical objects. If we want to look for a system that exemplified the mathematical structure in such cases, it looks like it would have to be a system of *possible* objects, ways that one and the same physical system could have been.

Do structuralists who wish to adopt the pattern/patterned account of applications have to be realists about *possibilia* in order to account for these cases? Shapiro is unclear on this. On the one hand, he emphasizes that in such applications

> the theorist describes a class of mathematical objects or structures and claims that this class represents the structures of all possible systems of a certain sort. *Relations among the objects or structures represent relations among the possible objects.* [6, p. 250, my italics.]

Talk of relations among possible objects certainly suggests a commitment to the existence of possibilia which are modelled by our mathematical theories. On the other hand, though, in what follows he seems to suggest that the move to structure allows us to avoid realism about possibilia themselves:

> Classical mechanics entails that there is at least a continuum of possible configurations of physical objects. There is even a continuum of possible pairs of point masses. *We do not have to reify the "possibilities"; we speak of their structure instead.* [6, p. 250]

This is rather confusing: Shapiro claims that mathematics is to reality as pattern is to patterned, but when the patterned system appears to be a system of *possibilia*, he seems to want to retreat and say we need only accept the existence of the pattern itself. But perhaps an explanation of this attitude can be gleaned from another comment he makes about such applications. When a mathematical theory applies to a class of 'possible systems', Shapiro tells us that

> theorems about the class of structures will correspond to facts about the possible systems—about what is and what is not possible. [6, p. 250]

Now, while talk of 'facts about possible systems' suggests the reification of such systems and their objects (possibilia), the alternative talk of facts 'about what is and what is not possible' invites another reading. If one wishes to be realist about modality without being realist about possible worlds and their contents, one will accept that there are (modal) facts about what is and is not possible, without holding these to be reducible to facts about what is true in systems of really existing *possibilia*. One might in such a case think that talk of possible systems of possible objects is a helpful way of representing facts about what is and is not possible, even if one does not think that there are any such possible objects. The reason that speaking *as if* there are systems of such objects is helpful is that the structure one supposes such systems to have respects the structure of the modal facts one is aiming to model.

This, of course, is extremely speculative – I do not pretend to know what Shapiro had in mind in his own tantalizing comments on possibilities in this discussion, but rather suggest this as a reading consistent with some of his comments. There is certainly no need for a structuralist such as Shapiro to take this option: there is nothing in structuralism that would speak against adopting a standard reductive account of modality. On the other hand, fictionalists who do not accept abstract objects are likely to be equally squeamish about possibilia (while still accepting primitive, unreduced modal facts: as far as I know, no fictionalist has tried to do without modality). So if it is possible for the structuralist to bypass commitment to possibilia while making use of mathematical theories that appear to describe the structure of possible systems of objects, then this will be a welcome option for fictionalists too.

3 Fictionalism and Type I-III Applications

Of the types of applications of mathematics considered here, only the most basic type I applications (via the interpretation of our mathematical axioms as truths or approximate truths about ordinary systems of concrete nonmathematical objects) are easily dealt with by fictionalists. At least on the face of it, all other cases seem to require that we suppose that our mathematical theories correctly describe systems of problematic objects (whether they be abstract 'ideal' counterparts of physical objects, such as ideal fluids (Type I); abstract purely mathematical objects (such as transfinite sets or complex numbers) (Type II); or systems of possible objects (Type III, setting aside for now the proposed anti-reductionist dodge). How one deals with these cases as a fictionalist will depend on whether one wishes to remain a realist about our scientific theories, in the sense that one holds those theories to be true or approximately true.

3.1 Realist Fictionalism

Hartry Field is best known as having attempted a defense of realist fictionalism. As a scientific realist, Field wants to hold that our best scientific theories are at least approximately true. But if we take the applications we have considered at face-value, as representative of ordinary scientific theorizing, it looks like our scientific theorizing involves us in (a) comparing physical systems to mathematically-described systems of ideal objects; (b) embedding mathematically-described physical systems in more complex mathematical structures and learning about those systems by consideration of their relation to those more complex structures; and (c) providing mathematical models of systems of possible objects. If one does not believe in any of the systems of ideal, mathematical, or possible objects posited in the context of our ordinary scientific theorizing, how can one claim to be a realist about the scientific theories that make use of these systems in helping to understand the physical world?

Field's answer in *Science without Numbers* is to agree that he cannot be a realist about *those* theories, but to hold that these are not our ultimate theories of spatiotemporal reality. Rather, he argues, we can express our theories of the underlying physical systems that our mathematically-stated theories are normally used to describe directly, in non-mathematical terms, and it is these theories whose truth we are committed to in doing science. Furthermore, given these descriptions, he argues that we can *prove* that the physical systems we believe in are correctly represented by the systems of mathematical objects we use to model them. That is, if Field is correct in his assumption that his nominalization project can by carried out, then *given the suppositions our mathematical theories make about the structured systems of mathematical objects they concern*, it will be possible to show that, if there *were* such objects, they would indeed be related to the physical objects of our nonmathematical theories in the way our mathematically-stated empirical theories describe.

Field's nominalization project has, therefore, two aims. First, to provide attractive nonmathematical versions of our usual scientific theories, that a scientific realist can believe to be (approximately) true without thereby becoming committed to the existence of mathematical objects. And second, to use those theories to *explain* why the mathematically-stated versions of our scientific theories are so useful. Such an explanation would hold that, since we have shown that, *if there were* mathematical objects satisfying the hypotheses of our mathematical theories, then they *would be* related to nonmathematical objects in the ways our mathematically-stated theories describe, then if we indulge in the pretense that there *are* such objects, and work out the consequences of that assumption, we will be able to uncover truths about the nonmathematical objects that we really believe to exist.[3]

Field's project is most plausible for Type I and II applications. In such applications, we learn about systems of nonmathematical objects by relating them to the objects of mathematical models, and inquire into the nature of these mathematical models by describing their essential structure and perhaps embedding that structure into wider models. So long as we can describe the original systems of nonmathematical objects in nonmathematical terms (itself no mean feat, since many of our ordinary concepts are infused with mathematics), then the prospects for an explanation of the applicability of the mathematical versions of our scientific theories in terms of these nonmathematical descriptions are reasonable. But Type III applications are more problematic for Field, as critics of Field's nominalization programme have been quick to point out. If the system

---

[3] I have presented Field's project in non-technical (and arguably somewhat loaded) terms for my own rhetorical purposes, as will become clear. His own presentation, in terms of representation theorems and the semantic conservativeness of mathematics over nonmathematical theories, is presented in [3], and is well known, so I make no apologies for applying my own slant on the essence of the project here.

being 'modelled' by the mathematics is not itself a system of nominalistically acceptable objects, then there should be no solace in an explanation of the applicability of the mathematical 'pattern' that proceeds by means of a detailed nonmathematical description of the equally problematic 'patterned' reality. As Malament, David puts it, at best in applying Field's strategy to a phase-space theory we will be able to

> reformulate the theory so that its subject matter is the set of "possible dynamical states" (of particular physical systems) and various relations into which they enter. But this is no victory at all! Even a generous nominalist like Field cannot feel entitled to quantify over *possible dynamical states*. [4, p. 533]

Perhaps Field could try to deal with such applications in a similar manner to our proposal for the structuralist who does not believe in a realm of possible objects. That is, Field could take the basic realm to be described by our mathematical theory to be a realm of irreducible modal facts, which are nevertheless modelled well if we suppose that there is a realm of possible dynamical states that are related to these modal facts. Field's scientific realism requires him to do more than the structuralist does at this point, though: he is required to give a theory which yields the modal facts he accepts as consequences, and does so without quantifying over either possible dynamical states *or* mathematical objects. But arguably our main access to those facts is through thinking in terms of possible dynamical states. It is, then, unclear whether we would be able to come up with an attractive theory which yielded the modal facts we wanted without detouring via consideration of a realm of possibilia. In fact, this problem is quite general, arising even for attempted nominalizations of more straightforward physical subject matters. Can we be confident that, for any of the more complex uses of mathematics to describe a nonmathematical subject matter, we will be able to come up with attractive theories of the subject matter in question that do not make use of mathematical concepts? Field's sample nominalization replaces talk of real number quantities applied to measure properties of physical objects (for example, the lengths of physical line segments) in favour of comparison predicates: 'is congruent (e.g., in respect of length) to' and 'is greater (e.g., in respect of length) than'. But not all applications of mathematics involve the use of quantities as measurements in this way. It is just a hope that we can find appropriate non-mathematical substitutes for all of the mathematical concepts we find ourselves applying in describing the nonmathematical world.

3.2 Instrumentalist Fictionalism

We said that Field's project had two aims: first, to preserve his scientific realism while remaining a nominalist, by providing attractive nonmathematical versions of our usual scientific theories, and second, to use those theories to explain why we should expect our ordinary mathematically stated theories to be useful instruments in finding out about the nonmathematical realm, even if we don't believe those theories to be true. Instrumentalist fictionalists[4] focus on Field's second aim. In his (sketched) nominalization of Newtonian gravitational theory, Field

---

[4] Amongst whose number I count myself and, in some incarnations, Balaguer, Mark [1, Part II] and Joseph Melia [5].

has a plausible explanation of why one mathematically-stated scientific theory manages to be good without being true – i.e., that it is correct in the way it represents the nonmathematical objects it talks about. Instrumentalist fictionalists can seize on this explanation in providing an (inductive) argument for the instrumental success of other mathematically-stated theories. We can explain why it is good to indulge in the pretense that there are real numbers related to nonmathematical objects in the way our mathematical version of gravitational theory describes, on the grounds that that theory is correct in the picture it paints of the nonmathematical objects it concerns (something we can *prove* once we have our nominalistic theory together with appropriate representation theorems in hand). Why not, then, think that other mathematically-stated scientific theories are successful for the same reason: that is, even in cases where we do not have a nominalistic version of the 'nonmathematical' content of our theory to hand, we can suppose that the reason such a theory is predictively successful is not that it is *true* in all of its parts, but that it provides a good representation of the systems of nonmathematical objects it is ultimately concerned with describing. The appropriate attitude to such theories would not, then, be to *believe* that they are literally true, but rather, to indulge in the pretense that they are true for the representative advantages they bring.

Fictionalists of this sort cannot be straightforwardly realist about those scientific theories that they cannot nominalize. Insofar as their use of such theories involves them in an element of belief, the belief must be in the truth of the *nominalistic content* of our theories, i.e., in the correctness of the picture they paint of the *nonmathematical* realm. Balaguer expresses the attitude as follows:

> Now, of course, the actual mixed sentences of empirical science would not be true, but the point is that in moving from empirical science to its nominalistic content—that is, to the claim that the physical world holds up its end of the "empirical-science bargain"—we do not lose any important part of our picture of the physical world. [1, p. 134]

Unfortunately, in the absence of nominalized versions of our scientific theories, this picture is something we may only be able to access by means of the 'pretense' that there are mathematical objects. We have to indulge in what Joseph Melia calls the "way of the weasel", using the framework of mathematics to paint a picture of the nonmathematical world, but having done so, withholding belief from the mathematical parts of the framework. Indeed, Melia thinks that this is quite a plausible interpretation of the attitude taken by many ordinary scientists:

> Whilst almost all scientists will admit that they must quantify over numbers in order to formulate their scientific theories, almost all will go on to deny that there are such things as mathematical objects. Philosophers typically represent these scientists as engaging in doublethink—denying by night what they believe by day. But it is surely uncharitable to regard so many scientists as hypocrites! Surely it is more charitable to think that we must have misinterpreted them. But look at the kind of things they say: "The force between two massive objects is *proportional* to the *product* of the masses *divided* by the square of the distance"; "There is a one-to-one differentiable *function* from the points of space-time onto *quadruples of real numbers*" —

how can we have misinterpreted them? By thinking that any theorist who presents a theory of the world must do so by asserting a set of sentences, each one believed by the theorist. This is our mistake. As soon as we allow theorists to take away details that were added before, to subtract parts of their earlier discourse, the theorists no longer appear to believe contradictory things. The mathematics is the necessary scaffolding upon which the bridge must be built. But once the bridge has been built, the scaffolding can be removed. It is surely more charitable to take scientists to be weasels rather than inconsistent hypocrites. [5, p. 469]

Plausible though this interpretation may be, instrumentalist fictionalists must admit that it is a disadvantage that they have to rely on unanalyzed notions such as that of "the physical world holding up its end of the "empirical-science bargain"", or of mathematical "scaffolding", in stating what it is to believe the nominalistic content of our scientific theories. Surely, in this respect, standard realist structuralism has an advantage over fictionalism?

4 Structuralism and Applied Mathematics

Let us suppose that Field's nominalization project cannot be satisfactorily completed: that we cannot give attractive nonmathematical versions of our ordinary mathematically-stated scientific theories, and must instead adopt the hypothesis that there are mathematical objects in presenting our best scientific theories. In that case, the only option for fictionalists is to reject scientific realism and adopt instrumentalist fictionalism, together with its belief in the correctness of the elusive 'nominalistic content' of our ordinary scientific theories. *Ante rem* structuralists, on the other hand, appear to do better: they believe in freestanding mathematical structures, and hence in mathematical objects interpreted as pure positions in those structures. So (setting worries about the realms of *possibilia* presupposed by some Type III applications aside), they can account for those applications of mathematics that are not via direct instantiation in terms of relations between nonmathematical objects and the positions in mathematical structures. So they can be realist about our scientific theories, holding simply that they are true, and thus avoiding the instrumentalist fictionalist's elusive claim that they are correct in their nominalistic content.

Actually, things are not so simple, as a look at structuralist metaphysics will make clear. For *ante rem* structuralists, the truth of a mathematically stated theory comes cheap: Shapiro hypothesizes that any coherent theory correctly describes some structure.[5] Insofar as it is interpreted as talking about the positions in that structure, then, any coherent theory is automatically true. Take, for example, a simple scientific theory which describes an isomorphism between the objects in some physical system and the positions in some freestanding mathematical structure. This theory itself can be viewed as a mathematical theory (standardly, in the language of set theory with urelements, the isomorphism the theory asserts to exist will just be some set of ordered pairs whose first members are urelements and whose second members are pure sets). If it is any good at all,

---

[5] Shapiro's 'coherence' is a primitive (unreducible) modal analogue of semantic consistency.

this theory will be coherent, and hence will describe some mathematical structure, about whose positions it asserts the truth. But content of the scientific realist's claim that our best scientific theories are true or approximately true cannot be that they are true in this sense, since *any* coherent scientific theory so-interpreted will be equally true. What, then, is the claim that structuralists who wish to be scientific realists will have to make about our ordinary, mathematically-stated scientific theories?

Clearly, truth of a 'freestanding' mathematical structure is not enough. Rather, the theory will also have to remain true when its urelements are interpreted as the nonmathematical objects the theory intends to model, with relations the theory induces on these urelements being mirrored in relations that do hold between those objects. In other words, the theory won't just be true of some positions in some abstract mathematical structure: *the nonmathematical world will also have to be as the theory describes it as being*. Might we express this account by saying that the nonmathematical world holds up its side of the "empirical-science bargain"? It is not clear to me what else the structuralist can be saying here. If I am right, then although *ante rem* structuralists can be scientific realists, they must express this realism as holding not only that our scientific theories are true, but also that they are true in their nonmathematical content. And if so, it is not clear (to me at least) what would be wrong with simply holding on to the latter claim as one's account of the applicability of empirical science while ditching the former.

5 Conclusion

Of course, structuralists may well have more to say about the sense in which they take our mathematically-stated scientific theories to be not just true, but true in their picture of the nonmathematical objects they concern. It remains to be seen whether a more detailed account is available, and indeed whether such an account is adaptable to instrumentalist-fictionalist purposes. But for now at least, I think that the jury should remain out on whether *ante rem* structuralism offers any advantage over instrumentalist fictionalism. For if the notion of a scientific theory's being correct in its picture of the nonmathematical realm is required by both accounts, it is unclear what is added by the additional claim that that theory is also true.

Bibliography

[1] Balaguer, M. *Platonism and Anti-Platonism in Mathematics*. Oxford: Oxford University Press, 1998.

[2] Batchelor, G. K. *An Introduction to Fluid Dynamics*. Cambridge: Cambridge University Press, 1967.

[3] Field, H. *Science Without Numbers: A Defence of Nominalism*. Princeton, NJ: Princeton University Press, 1980.

[4] Malament, D. Review of *Science without Numbers: A Defense of Nominalism*. *Journal of Philosophy*, 79(9):523–34, 1982.

[5] Melia, J. Weaseling away the indispensability argument. *Mind*, 109(435):455–79, 2000.

[6] Shapiro, S. *Philosophy of Mathematics: Structure and Ontology*. Oxford: Oxford University Press, 1997.

What's it take to interpret a physical theory? [1]

Laura Ruetsche

University of Michigan

ruetsche@umich.edu

ABSTRACT. A theory of "ordinary QM" associates with a physical system an observable algebra $\mathfrak{B}(\mathcal{H})$, the collection of bounded operators on a separable Hilbert space $\mathcal{H}$. The *maximal beable approach* is a scheme for interpreting ordinary QM that uses projection operators minimal in $\mathfrak{B}(\mathcal{H})$'s projection lattice—these stand in one-to-one correspondence with pure states on $\mathfrak{B}(\mathcal{H})$— to characterize possible conditions of a quantum system. The scheme is widespread: most familiar interpretations of QM are instances of it. But, I argue, the scheme is not readily extendible to the setting of QM_∞, whose theories treat systems with infinitely many degrees of freedom, such as quantum fields. What impedes the maximal beable approach is the absence of minimal projection operators from the observable algebras typical of QM_∞. I argue that this leaves the maximal beable approach without an informative way to characterize quantum possibilities, or to explicate quantum probabilities.

1 Introduction

What's it take to interpret a physical theory? The question is ambiguous. In one sense it asks what it takes *to be* an interpretation of a physical theory; that is, what *qualifies* an intellectual offering as an interpretation of a theory. In another sense, it asks what it takes *to construct* an interpretation of a physical theory; that is, what resources are fair game for articulating an offering satisfying the qualifying criteria. I believe that a fairly widespread (although often implicit) answer to the question in its first sense is:

1. *An interpretation of a physical theory is a characterization of the worlds possible according to that theory.*

And I believe that a fairly widespread (although even more often implicit) answer to the question in its second sense is:

2. *To construct an interpretation of a physical theory, one may draw upon that theory's own formal apparatus, as well as ones own metaphysical scruples.*

---

[1] The author was with the Department of Philosophy, University of Pittsburgh. Now the author is with the Department of Philosophy, University of Michigan. Email: ruetsche@umich.edu

An example of drawing on a theory's formal apparatus in the service of its interpretation is setting up a correspondence between worlds possible according to the General Theory of Relativity and triples (M, g_{ab}, T_{ab}) satisfying the field equations, or between worlds possible according to the General Theory of Relativity and equivalence classes under diffeomorphism of such triples. An example of drawing on ones metaphysical scruples in the service of interpretation is opting on the basis of Leibniz shift arguments for the correspondence between possible worlds and equivalence classes.

My motivating concern is with the adequacy of the foregoing picture to the rough and tumble of how physical theories do what they ought to do, which is (among other things) unify and explain phenomena. But my particular agenda here is quite narrow. It's to explicate one expression of the picture framed by the tenets 1 and 2, and to chronicle the difficulties it faces. The expression is the *maximal beable approach* to interpreting quantum theories, devised for the setting, familiar from philosophical expositions such as Redhead [19] or Hughes [14], of what I'll call *ordinary quantum mechanics*, wherein the observables pertaining to a quantum system are the self-adjoint elements of $\mathfrak{B}(\mathcal{H})$, the collection of bounded operators acting on a separable Hilbert space $\mathcal{H}$, aka a Type I von Neumann algebra. The maximal beable approach struggles when it encounters quantum theories of a sort less precedented in the philosophical literature, but typical of a setting I'll call QM_∞, encompassing quantum field theories and the thermodynamic limit of quantum statistical mechanics. In general, the observables pertaining to a QM_∞ system belong to von Neumann algebras of types more general, and more surprising, than the Type I algebra $\mathfrak{B}(\mathcal{H})$. I aim here to indicate why this is so and how it spells trouble for the maximal beable approach.

The paper is organized as follows. Section 2 presents a series of prominent interpretations of ordinary QM as instances of the maximal beable approach. Section 3 begins with a rudimentary review of lattice theory, then deploys the terms introduced to recharacterize the maximal beable approach in ordinary QM. The crux of the recharacterization is that the maximal beable approach hinges on the presence in $\mathfrak{B}(\mathcal{H})$ of minimal projection operators, that is, atoms in its projection lattice. Section 4 documents the absence of minimal projection operators from a variety of observable algebras encountered in QM_∞. There I contend that this absence leaves the maximal beable approach unable to say either what quantum probabilities are probabilities for or what the values of those probabilities are. A closing section takes stock.

2 Ordinary QM and its interpretation

2.1 Ordinary QM

What I'll call a theory of ordinary QM associates with a quantum system the collection $\mathfrak{B}(\mathcal{H})$ of bounded operators acting on a separable Hilbert space $\mathcal{H}$; identifies observables pertaining to the system with self-adjoint elements of $\mathfrak{B}(\mathcal{H})$; and identifies possible states of the system with normed, positive, countably additive linear functionals $\omega : \mathfrak{B}(\mathcal{H}) \to \mathbb{C}$. Taking the real number $\omega(A)$ as the *expectation value*—that is, long run experimental average—of the observable associated with the self-adjoint element $A \in \mathfrak{B}(\mathcal{H})$ endows the theory with empirical content. For Hilbert spaces of dimension greater than 2, Gleason's theorem sets states on

$\mathfrak{B}(\mathcal{H})$ in one-to-one correspondence with density operators (trace class operators of trace one) in $\mathfrak{B}(\mathcal{H})$. Where W is the density operator implementing the state ω, the prescription $\omega(A) = Tr(WA)$ for all $A \in \mathfrak{B}(\mathcal{H})$ establishes the correspondence. Schrödinger's equation equips the kinematical structure just specified with dynamics. An isolated system in initial state $W(0)$, with Hamiltonian (energy) operator H, evolves in a time t to the state $W(t) = e^{-iHt}W(0)e^{iHt}$.[2] The evolution operators e^{-iHt} are unitary and form a strongly continuous family.

The associations and identifications just laid out are in the first instance. Interpretations of ordinary QM can and do complicate or revise them. An interpretation of ordinary QM undertakes to say what the world would have to be like, in order for a theory of the structure just laid out to be empirically adequate. Thus an interpretation characterizes a space of possible worlds as worlds of which ordinary QM is empirically adequate.[3]

2.2 The Maximal Beable Approach

An interpretation of a quantum theory should answer three related questions about the probabilities that theory assigns. First,

Question 1 *What are quantum probabilities probabilities for?*

The worlds possible according to a quantum theory are the recipients of its probability assignments. Thus, to answer Question 1 is to characterize the worlds possible according to a quantum theory, that is, to begin to interpret it. Second,

Question 2 *What is the nature of these probabilities?*

If an answer to Question 1 identifies the worlds possible according to a quantum theory, an answer to Question 2 tells us how to understand the probabilities that quantum theory assigns those worlds.

An interpretation of a quantum theory should also disclose that theory as *empirical*, that is, as committed to specific testable predictions about the course of nature. Hence the third question

Question 3 *What values do these probabilities have?*

An answer to Question 3—which typically will either invoke the Born Rule directly [textbook, modal, hidden variable] or reconstitute it from other principles [Many Worlds]—equips the quantum theory with articulate empirical content, in the form of explicit statistical predictions. To be sure, it is the job of the theory itself to generate statistical predictions, but it is the job of the theory's interpretation to explicate those predictions as predictions about the possibilities the interpretation allies with the theory. A natural way for an interpretation of a theory of ordinary QM to proceed is to take each state W on $\mathfrak{B}(\mathcal{H})$ to correspond to a set of possible configurations of the system associated with $\mathfrak{B}(\mathcal{H})$, that is, worlds (consisting of that system) possible according to the theory. (The

---

[2]Setting factors of Planck's constant to 1.
[3]Why not 'true', a more common formulation? To make rooms for positions, recognized as interpretive, that alter the substance of the physics just presented, for instance by revising Schrödinger dynamics or by enriching the observable set with magnitudes, such as position and momentum in Bohmian mechanics, without correlate in $\mathfrak{B}(\mathcal{H})$.

W/world correspondence is allowed to be one-many rather than required to be one-one to accommodate approaches which interpret quantum probabilities in terms of statistical distributions.) A noble interpretation will attempt to characterize possible worlds without appeal to problematic and ambiguous notions such as 'measurement,' 'subjectivity,' 'classicality,' and so on. Familiar with such appeals from various contemporary explications of Copenhagen orthodoxy (for a sample, see Section I of Wheeler and Zurek [23]), John Bell conjures an interpretive strategy in the form of an alternate future for quantum mechanics:

> ...it is interesting to speculate on the possibility that a future theory will not be *intrinsically* ambiguous and approximate. Such a theory could not be fundamentally about 'measurements,' for that would again imply incompleteness of the system and unanalyzed interventions from outside. Rather it should again become possible to say of a system not that such and such may be *observed* to be so, but that such and such *be* so. The theory would be not be about '*observ*ables' but about '*be*ables'. ([4], p.41)

Drawing upon the interpretive resource of the theoretical apparatus, suppose that the properties in whose terms a quantum system associated with $\mathfrak{B}(\mathcal{H})$ is characterized correspond to self-adjoint elements of $\mathfrak{B}(\mathcal{H})$, or more precisely to the assignment of eigenvalues to those elements. The option of taking W to describe an ensemble of systems, each of which exhibits a determinate eigenvalue for each observable pertaining to it, is exercised at (what most commentators' metaphysical scruples reckon to be[4]) considerable cost, if the dimension of $\mathcal{H}$ is greater than 2. Bell imagines another option:

> Could not one just promote *some* of the observables of the present quantum theory to the status of beables? The beables would then be represented by linear operators in the state space. The values which they are allowed to *be* would be eigenvalues of those operators. For the general state the probability of a beable *being* a particular value would be calculated just as was formerly calculated the probability of *observing* that value. ([4], p.41)

The project Bell envisions is the *maximal beable approach* to interpreting quantum theories: to first approximation, the approach of identifying the largest set of quantum observables pertaining to a system that can (subject to constraints arising from metaphysical scruples) enjoy determinate values simultaneously. Determinate value assignments to this set indicate what worlds are possible according to the quantum theory; that is, they answer Question 1.

Bell has the maximal beable approach aspire to *explicate* quantum probabilities. Successfully executing the approach, we come to understand not only what quantum probabilities are probabilities for (specific patterns of beable instantiation) but also how to calculate the values of those probabilities, and so answer Question 3. This sort of explication is intregral to understanding QM as an empirical theory — a theory whose empirical commitments take the form of *characterized* possible worlds assigned *explicit* probabilities.

---

[4]Famously, Bell's metaphysical scruples are exceptional; see, for instance, [3], p.8-9

Clifton and Halvorson translate the interpretive approach Bell conjures into the formal apparatus of operator theory. Their articulation casts $\mathfrak{B}(\mathcal{H})$ as a C^* algebra, so we will begin with an introduction to that notion.

Definition 1 *An **algebra** $\mathfrak{A}$ over the field $\mathbb{C}$ of complex numbers is a set of elements $(A, B, ...)$ that is (Ai) closed under a commutative, associative operation $+$ of binary addition; (Aii) closed with respect to a binary multiplication operation $\cdot$, which is associative and distributive with respect to addition, but not necessarily commutative; and (Aiii) closed with respect to multiplication by complex numbers. A $*$-algebra is an algebra closed under an involution $* : \mathfrak{A} \to \mathfrak{A}$.[5] A C^* algebra is a $*$ algebra equipped with a norm $\| \ \|$[6] that is complete in the topology induced by that norm.*

Simply put, an algebra is a closed linear vector space (Ai and Aii) equipped with a multiplication operation (Aiii). $\mathfrak{B}(\mathcal{H})$ is an example. A subset of $\mathfrak{B}(\mathcal{H})$ satisfying (Ai)-(Aiii)—the linear closure of the spectral resolution of a non-degenerate self adjoint element of $\mathfrak{B}(\mathcal{H})$ for example—is a *subalgebra* of $\mathfrak{B}(\mathcal{H})$. Like $\mathfrak{B}(\mathcal{H})$ all the algebras considered here are $*$ algebras. $\mathfrak{B}(\mathcal{H})$ is also a C^* algebra, with the Hilbert space operator norm providing the C^* norm.

Three more definitions preface Clifton and Halvorson's articulation of the maximal beable approach.

Definition 2 *A **state** ω on a C^* algebra $\mathfrak{A}$ is a normed, positive, linear functional $\omega : \mathfrak{A} \to \mathbb{C}$.*

This is just a state in the sense familiar from ordinary quantum mechanics, generalized to algebras that needn't take the form $\mathfrak{B}(\mathcal{H})$, and relieved of the (seemingly natural!) requirement of countable additivity. The next definition unpacks the intuitive notion of possessing a determinate value:

Definition 3 *A state ω on a subalgebra $\mathfrak{D}$ of a C^* algebra $\mathfrak{A}$ is **dispersion free** iff $\omega(A^2) = (\omega(A))^2$ for each self adjoint $A \in \mathfrak{D}$.*

Thus a system in a dispersion free state ω on $\mathfrak{D}$ possesses a determinate value for every self-adjoint element of $\mathfrak{D}$. These elements are beables for a system in ω. (NB a dispersion-free state needn't be countably additive, even if $\mathfrak{D}$ contains enough projection operators for a requirement of countable additivity to make sense.)

When ω is dispersion-free on $\mathfrak{D}$, every observable in $\mathfrak{D}$ possesses a determinate value, moreover one which ω predicts with certainty. The next definition is introduced to accommodate the possibility that a state corresponds to a straightforward statistical mixture over possible patterns of beable instantiation.

Definition 4 *A state ω on a subalgebra $\mathfrak{D}$ of a C^* algebra $\mathfrak{A}$ is a **mixture of dispersion-free states** on $\mathfrak{D}$ iff there exists a probability measure μ_ω on the space Λ of dispersion free states ω_λ on $\mathfrak{D}$ such that for all $A \in \mathfrak{D}$*

$$\omega(A) = \int_\Lambda \omega_\lambda(A) d\mu_\omega(\lambda) \tag{1}$$

---

[5] satisfying:
$$(A^*)^* = A, \quad (A+B)^* = A^* + B^*, \quad (cA)^* = \bar{c}A^*, \quad (AB)^* = B^*A^*$$
for all $A, B \in \mathfrak{A}$ and all complex c (where the overbar denotes complex conjugation).

[6] satisfying $\|A^*A\| = \|A\|^2$ and $\|AB\| < \|A\|\|B\|$ for all $A, B \in \mathfrak{A}$.

Now, for Clifton and Halvorson, to take the maximal beable approach to interpreting QM is to proceed as follows

> Given a state on an algebra of observables, characterize those subalgebras of 'beables,' that are maximal with respect to the property that the state's restriction to the subalgebra is a mixture of dispersion-free states. Such *maximal beable subalgebras* could then represent maximal sets of observables with simultaneously determinate values distributed in accordance with the state's expectation values. ([13], p.2442; cf. [8], p.117,119)

Underlying the maximal beable approach are metaphysical scruples about the nature of properties (to be instantiated is to take a value that is not disjunctive or interval-valued or otherwise fuzzy, but dispersion-free) and the desirability of plenitude (the imperative to find *maximal* beable algebras: if an interpretation can add to the list of properties instantiated in a quantum world, it should).

Maximal beable subalgebras, so defined, exist for any state ([13], p.2447). In the case of *faithful states*, maximal beable subalgebras are readily characterized.

Definition 5 *A state ω on $\mathfrak{A}$ is* **faithful** *iff $\omega(A) = 0$ implies $A = 0$ for all $A \in \mathfrak{A}$.*

Fact 1 *If ω is a faithful state on $\mathfrak{A}$ then it can be represented as a mixture of dispersion free states on a subalgebra $\mathfrak{D}$ of $\mathfrak{A}$ iff $\mathfrak{D}$ is abelian (that is, pairwise commuting). (Cf. [9], p.172)*

The maximal beable approach to interpreting a theory of ordinary QM sets the generic C* algebra $\mathfrak{A}$ in the foregoing characterizations equal to $\mathfrak{B}(\mathcal{H})$. It follows from Fact 1 that maximal beable subalgebra of $\mathfrak{B}(\mathcal{H})$ for a faithful state is a maximal abelian subalgebra of $\mathfrak{B}(\mathcal{H})$, that is, an abelian subalgebra of $\mathfrak{B}(\mathcal{H})$ not properly contained in any other abelian subalgebra of $\mathfrak{B}(\mathcal{H})$. There is a simple recipe for generating a maximal abelian subalgebra of $\mathfrak{B}(\mathcal{H})$: start with a complete set $\{E_i\}$ of orthogonal one-dimensional projection operators in $\mathfrak{B}(\mathcal{H})$ and close in the weak topology. The result will be an algebra consisting of every element of $\mathfrak{B}(\mathcal{H})$ that has $\{E_i\}$ as a spectral resolution.[7] Call this the maximal abelian subalgebra of $\mathfrak{B}(\mathcal{H})$ *generated by* $\{E_i\}$.

Most physically interesting states in QM_∞ are faithful, so what follows will deal exclusively with faithful states. Bub 1997 gives a comprehensive account of the maximal beable approach in the presence of non-faithful states.

2.3 Variations on the maximal beable approach

The maximal beable approach is a template for interpreting QM wherein a possible world is a 'maximal set of co-obtaining properties' (Bub 1997, [8], p.18), with a property understood as the assignment of an eigenvalue to a self-adjoint operator. Different extant variations on the maximal beable approach can be individuated by how they complete that template—that is, by how they identify maximal beable subalgebras—and by the commentary on quantum probability they append to it.

---
[7]See [6], Section 3.2 for an argument.

Collapse

Collapse interpretations conjoin the *eigenstate-eigenvector link*, according to which a system in a state $|\psi\rangle$ has a determinate value for A if and only if $|\psi\rangle$ is an eigenstate of A, in which case the value is the associated eigenvalue, to the ignoble *postulate of measurement collapse*, according to which an A measurement performed on a system in a state $|\psi\rangle$ instantaneously and discontinuously collapses $|\psi\rangle$ to the A eigenstate corresponding to the outcome observed, with the Born rule probabilities prescribed by $|\psi\rangle$ furnishing a probability distribution over candidate collapses.

Now, assume non-degenerate A, with eigenstates $\{|\alpha_i\rangle\}$, is measured on an object in initial state $|\psi\rangle = \sum_i c_i |\alpha_i\rangle$. According to the postulate of measurement collapse, the density operator $W = \sum_i |c_i|^2 E_{\alpha_i}$ (where E_{α_i} is the projection operator for the subspace spanned by $|\alpha_i\rangle$) describes the post-measurement object state. W is faithful if none of the c_is is 0. The complete orthogonal set of projections E_{α_i}, as well as every $A \in \mathfrak{B}(\mathcal{H})$ having that set as a spectral resolution, is a maximal abelian subalgebra, call it $\mathfrak{D}_A$, of $\mathfrak{B}(\mathcal{H})$. It's also the set of observables that have determinate values after measurement, according to the eigenstate-eigenvalue link.[8] Thus a collapse interpretation instantiates the maximal beable approach. Its commentary on quantum probability answers (Question 1)-(Question 3) as follows:

(Question 1): Each projection operator E_{α_i} encodes a possible condition of the system described by W via the recipe that the value of A in $\mathfrak{D}_A$ is the eigenvalue associated with E_{α_i}. (NB This is the eigenvector/eigenvalue link again.)

(Question 2): Quantum probabilities are epistemic; they reflect our ignorance of the endpoint of collapse.

(Question 3): The probability of a collapse to the condition encoded by E_{α_i} is $Tr(WE_{\alpha_i})$. Notice that because the E_{α_i} are orthogonal, these probabilites are non-interfering.

"Bohm"ian interpretations

A "Bohmian" interpretation dictates a prefered determinate observable. For simplicity, suppose that it's discrete and non-degenerate and call it R. (Bohm, of course, supposes R to be position, an operator with a continuous spectrum. Technical difficulties attendant upon this complication—difficulties not with articulating Bohmian mechanics but with shoe-horning it into a framework where properties correspond to self-adjoint operators on a separable Hilbert space— motivate the shudder quotes around "Bohm" in the heading.) R's eigenprojections E_{r_i} correspond to pairwise orthogonal one-dimensional subspaces of $\mathcal{H}$; they generate a maximal abelian subalgebra $\mathfrak{D}_R$ of $\mathfrak{B}(\mathcal{H})$. This is the beable subalgebra, according to "Bohm." Notably, it is independent of W if W is faithful. Possible conditions of a system described by W are coded by R's eigenprojections E_{r_i}; as before, the eigenstate-eigenvalue link is the decoder (Question 1). The probability that the system is in the condition coded by E_{r_i} is the Born Rule probability $Tr(WE_{r_i})$ (Question 3); this probability is epistemic (Question 2).

---

[8]Or, to be more careful, the intersection, over the collection of possible collapses, of the sets that have determinate values.

Modal interpretations

Same song, slightly different verse. Suppose W is faithful and non-degenerate. Then $\mathfrak{D}_W$, the maximal abelian subalgebra of $\mathfrak{B}(\mathcal{H})$ generated by W's spectral projections E_i, is the beable algebra; possible conditions of the system ("value states") stand in one-to-one correspondence with W's spectral projections, with the eigenstate-eigenvalue link dictating what's true of a system in value state E_i (Question 1); the probability that the system is in the condition coded by E_i is the Born Rule probability $Tr(WE_i)$ (Question 3); this probability is epistemic (Question 2).

Relative state formulations

Same song, different register. For the sake of exposition, suppose that the universe admits a preferred decomposition into subsystems, and that the Hilbert space for each subsystem has a prefered basis (dictating the character of the 'worlds' corresponding to different branches of the universal wave function). Let W be the reduced state of some prefered subsystem induced by the universal wave function, and let $\{E_i\}$ be the complete set of eigenprojections onto the preferred basis for that subsystem. Then proceed as with the modal interpretation, only with this epicycle: the E_i's keep track not of value states (mutually exclusive possible conditions of the system) but of *worlds*, of which there are many, and to none of which the system is confined simpliciter. In the world kept track of by E_i, what's true of the system is given by applying the eigenstate-eigenvector link to E_i (Question 1). The profligate metaphysics precludes a straightforward epistemic interpretation of quantum probabilities; there are other options, for instance, that Born rule probabilities encapsulate the degrees of belief of a rational observer in a universe truly described by a relative state formulation ([22]; for another approach to relative state formulation probabilities, see [2]).

3 Von Neumann algebras and their projection lattices

Here I introduce some apparatus for re-describing what I've just reviewed, in order to notice a possible problem: in the case that the von Neumann algebra pertaining to a system fails to be a Type I factor, the formal pivot on which all the foregoing interpretations hinged—the encoding one-dimensional projections E_i (and their analogs), decoded by the eigenstate-eigenvalue link to answer Question 1 and plugged into the trace prescription to answer Question 3—can go missing.

3.1 Von Neumann algebras

Given an algebra $\mathfrak{D}$ of bounded operators on a Hilbert space $\mathcal{H}$, its *commutant* $\mathfrak{D}'$ is the set of all bounded operators on $\mathcal{H}$ that commute with every element of $\mathfrak{D}$. So, for example, the commutant of $\mathfrak{B}(\mathcal{H})$ consists of scalar multiples of the identity operator. $\mathfrak{D}$'s *double commutant* $\mathfrak{D}''$ is $\mathfrak{D}'$'s commutant. Every element of $\mathfrak{B}(\mathcal{H})$ commutes with the identity operator, and so with every element of

$\mathfrak{B}(\mathcal{H})'$. This makes $\mathfrak{B}(\mathcal{H})$ its own double commutant. It also qualifies $\mathfrak{B}(\mathcal{H})$ as a von Neumann algebra.[9]

Definition 6 *A **von Neumann algebra** $\mathfrak{M}$ is a *-algebra of bounded operators such that $\mathfrak{M} = \mathfrak{M}''$.*

$\mathfrak{B}(\mathcal{H})$ is moreover a von Neumann factor:

Definition 7 *A von Neumann algebra $\mathfrak{M}$ is a **factor** if and only if $\mathfrak{M} \cap \mathfrak{M}'$ contains only multiplies of the identity.*

For the sake of simplicity, we will confine our attention here to factor algebras.

Every von Neumann algebra $\mathfrak{M}$ is a subalgebra of $\mathfrak{B}(\mathcal{H})$ for some $\mathcal{H}$, but it doesn't follow that every $\mathfrak{M}$ has projection operators corresponding to every subspace of the Hilbert space on which it acts. Let $\mathcal{P}(\mathfrak{M})$ be the set of projection operators in the von Neumann algebra $\mathfrak{M}$. It is a consequence of von Neumann's double commutant theorem that $\mathfrak{M} = \mathcal{P}(\mathfrak{M})''$. A typology of von Neumann algebras originating with Murray and von Neumann [17] is based on the character of the projections $\mathcal{P}(\mathfrak{M})$ does contains. The characters that interest us most are infinite projections and minimal projections.

Infinite, finite, and minimal projections

The *range* of a projection E in von Neumann algebra $\mathfrak{M}$ acting on a Hilbert space $\mathcal{H}$ is the linear span of $\{|\psi\rangle \in \mathcal{H} : E|\psi\rangle = |\psi\rangle\}$. Thus the range of E is a closed subspace of $\mathcal{H}$ (cf. [15], Prop.2.5.1). Two projections E and F in $\mathfrak{M}$ are *equivalent* (written $E \sim F$) just in case their ranges are isometrically embeddable into one another, *by an isometry that is an element of* $\mathfrak{M}$. Equivalence so construed is manifestly relative to $\mathfrak{M}$. When E's range is a subspace of F's range (written $E \leq F$), E is a *subprojection* of F. Equivalent criteria are that $FE = EF = E$ and that $|E|\psi\rangle| \leq |F|\psi\rangle|$ for all $|\psi\rangle \in \mathcal{H}$. We use the subprojection relation to define the relation *weaker than* (written $\preceq$), which imposes a partial order on projections in a von Neumann algebra: E is *weaker than* F if and only if E is equivalent to a subprojection of F. Because $\preceq$ is a partial order, $E \preceq F$ and $F \preceq E$ together imply that $E \sim F$.

A projection $E \in \mathfrak{M}$ is *infinite* if and only if there's some projection $E_0 \in \mathfrak{M}$ such that $E_0 < E$ and $E \sim E_0$. In this case, E_0's range is both a proper subset of, and isometrically embeddable, into E's range. $E \in \mathfrak{M}$ is *finite* if and only if it is not infinite.

A non-zero projection $E \in \mathfrak{M}$ is *minimal* if and only if E's only subprojections are 0 and E itself. It follows that minimal projections are finite. The minimal projections of the factor $\mathfrak{B}(\mathcal{H})$ are the one-dimensional ones.

A Classification of Von Neumann Algebras

The Murray-von Neumann classification applies in the first instance to von Neumann algebras which are factors; on such algebras, the weaker than relation $\preceq$ imposes a total order (see [16], Prop.6.2.6).

---

[9]Topologically characterized, a *von Neumann algebra* $\mathfrak{M}$ is a *-algebra of bounded operators that is strong operator-closed in its action on some Hilbert space. Von Neumann showed that the strong and weak closures of a self-adjoint algebra $\mathfrak{D}$ of bounded Hilbert space operators coincide—and coincide as well with $\mathfrak{D}$'s double commutant. For more on von Neumann algebras and operator topologies, consult Kadison and Ringrose ([15], Ch.5; [16], Ch.6).

Type I Type I factors contain minimal projections, which are therefore also finite.

The algebras $\mathfrak{B}(\mathcal{H})$ of bounded operators on a separable Hilbert space—that is, the observable algebras of ordinary QM—are Type I factors, and each Type I factor is isomorphic to some $\mathfrak{B}(\mathcal{H})$. Type II and III factors may be less familiar.

Type II Type II factors contain no minimal projections, but do contain (non-zero) finite projections.

Indeed, in a sense that can be made precise ([21], Section 1.3), Type II factors have projections whose ranges are subspaces of *fractional* dimension.

Type III Type III factors have no (non-zero) finite projections and so no minimal projections. All their projections are infinite and therefore equivalent (cf. [16], Corr. 6.3.5).

For concrete examples of factors of Types II and III, see Sunder [21].

3.2 The lattice of projections

The set $\mathcal{P}(\mathfrak{M})$ is partially ordered by the relation $\leq$ of subspace inclusion, defined in terms of the Hilbert space on which $\mathfrak{M}$ acts. Thus $E \leq F$ if E's range is a (not necessarily proper) subspace of F's range. (The following will employ expressions like "E" to refer both to projections and to the closed subspaces that are their ranges.) This partial order enables us to define for each pair of elements $E, F \in \mathcal{P}(\mathfrak{M})$, their *greatest lower bound* (aka *meet*) $E \cap F$ as the projection whose range is largest closed subspace of $\mathcal{H}$ that is contained in both E and F; and their *least upper bound* (aka *join*) $E \cup F$ as the projection whose range is smallest closed subspace of $\mathcal{H}$ that contains both E and F. Thus $E \cap F$ is just the projection whose range is the intersection of E's range and F's, and $E \cup F$ is just the projection whose range is the linear span of E's range and F's. A *lattice* is a partially ordered set every pair of elements of which has both a least upper bound and a greatest lower bound. Thus the foregoing definitions render $\mathcal{P}(\mathfrak{M})$ a lattice. Indeed, $\mathcal{P}(\mathfrak{M})$ is an orthocomplemented lattice.

Definition 8 (orthocomplemented lattice) *A lattice S has a zero element 0 s.t. $0 \leq a$ for all $a \in S$ and a unit element 1 s.t. $a \leq 1$ for all $a \in S$. The zero operator (the projection operator for the null subspace) is the zero element of $\mathcal{P}(\mathfrak{M})$ and the identity operator I is the unit element. The* **complement** *of an element a of a lattice S is an element $a' \in S$ such that $a' \cup a = 1$. A lattice is complemented if each of its elements has a complement. It's* **orthocomplemented** *if these complements obey*

$$a'' = a \qquad a \leq b' \text{ if and only if } b \leq a' \tag{2}$$

The complement $E^\perp$ of $E \in \mathcal{P}(\mathfrak{M})$ supplied by the projection $I - E$, whose range is the orthogonal complement of E's range.

$\mathcal{P}(\mathfrak{M})$ is not necessarily a distributive lattice.

Definition 9 *All a, b, c in a **distributive lattice** S satisfy the distributive law*

$$a \cup (b \cap c) = (a \cup b) \cap (a \cup c) \qquad (3)$$
$$a \cap (b \cup c) = (a \cap b) \cup (a \cap c)$$

Let $\mathfrak{M} = \mathfrak{B}(\mathcal{H})$, and let A, B, and C be projections whose ranges are subspaces spanned by the vectors $\mathcal{H} \ni |\alpha>$, $|\beta>$ and $|\gamma> = |\alpha> + |\beta>$, where $|\alpha>$ and $|\beta>$ are orthogonal. It is easy to verify that (3) breaks down. However, in the special case that $\mathfrak{M}$ is abelian, $\mathcal{P}(\mathfrak{M})$ is distributive. Indeed, it's a *Boolean lattice* (aka a *Boolean algebra*), that is, is a distributive complemented lattice. The simplest Boolean lattice is the set $\{0, 1\}$, where each element is the other's complement and meet and join correspond to set-theoretic intersection and union respectively. Call this lattice B_2. Notice that B_2's elements can be put into one-one correspondence with the truth-values *false* (0) and *true* (1).

A *Boolean (or two-valued) homomorphism* between Boolean lattices B and B_2 is a map $h : B \to B_2$ preserving Boolean operations.[10] Construing B as a lattice of propositions, we can construe lattice operations — join ($\cup$), meet ($\cap$), and complement (′)—as logical operations — disjunction (v), conjunction (&), and negation ($\sim$), respectively. Given this construal, a two-valued homomorphism $h : B \to B_2$ on a Boolean lattice B is a *truth valuation on B respecting the classical truth tables* for disjunction, conjunction, and negation.

When a Boolean lattice B is finite— that is, it has finitely many elements— there's a simple recipe for obtaining its two-valued homomorphisms. An element a of a lattice S is an *atom* if and only if S contains no non-zero elements "smaller" than a.[11] We encountered the notion of "atom" in a different guise earlier in this section: the minimal projections in von Neumann algebra $\mathfrak{M}$ (if it has any) are the atoms of the lattice $\mathcal{P}(\mathfrak{M})$. An atom a in a Boolean lattice B generates a two-valued homomorphism h_a on B as follows:

$$h_a(b) = 1 \quad \text{if } a \leq b \qquad (5)$$
$$h_a(b) = 0 \quad \text{otherwise}$$

If B is finite *all* its two-valued homomorphisms are determined in this way ([5], Cor.5.3).

3.3 The maximal beable approach revisited

We can use the notions just reviewed to recharacterize ordinary QM instantiations of the maximal beable approach. Given an ordinary QM system in a faithful

---

[10]to wit,

$$\begin{aligned} h(a \cup b) &= h(a) \cup h(b) \\ h(a \cap b) &= h(a) \cap h(b) \\ h(a') &= h(a)' \\ h(0) &= 0 \\ h(1) &= 1 \end{aligned} \qquad (4)$$

(NB the second and third (equivalently the second and fourth) are sufficient to define a Boolean homomorphism; the remaining properties are consequences.)

[11]That is, for all $b \in S$, $b \leq a$ implies $b = a$ or $b = 0$.

state ω on $\mathfrak{B}(\mathcal{H})$, familiar interpretations use a maximal beable subalgebra $\mathfrak{D}$ of $\mathfrak{B}(\mathcal{H})$ to identify a maximal abelian subalgebra $\mathcal{P}(\mathfrak{D})$ of $\mathcal{P}(\mathfrak{B}(\mathcal{H}))$. As we have seen, different interpretations identify $\mathcal{P}(\mathfrak{D})$ in different ways. But they all share the strategy of using atoms in $\mathcal{P}(\mathfrak{D})$—minimal projection operators in the Type I von Neumann algebra $\mathfrak{B}(\mathcal{H})$—to code worlds possible for the system in state ω. Each world possible corresponds to a two-valued homomorphism on $\mathcal{P}(\mathfrak{D})$ determined by its coding atom. The classical semantic structure of such worlds is an attractive feature of the maximal beable approach. The state ω defines a probability distribution over these homomorphisms: where W is the density operator in $\mathfrak{B}(\mathcal{H})$ implementing ω, the probability distribution is determined by plugging the atoms generating the homomorphisms into the trace prescription. Another attractive feature of the maximal beable approach is that the probabilities obtained by restricting ω to $\mathcal{P}(\mathfrak{D})$ will be non-interfering, because $\mathcal{P}(\mathfrak{D})$ is abelian. They have the character of classical probabilities, and can bear either epistemic or subjective interpretations just as well as those probabilities can.

Of course, there is nothing to stop us from extending the maximal beable approach beyond ordinary QM by replacing $\mathfrak{B}(\mathcal{H})$ with an *arbitary* von Neumann algebra $\mathfrak{M}$. The maximal beable recipe becomes

> **Maximal beable recipe** *Given a system in a faithful state ω on a von Neuman algebra $\mathfrak{M}$, identify a maximal abelian subalgebra $\mathcal{P}(\mathfrak{D})$ of $\mathcal{P}(\mathfrak{M})$. Two-valued homomorphisms on $\mathcal{P}(\mathfrak{D})$ correspond to worlds possible for the system; ω defines a probability distribution over those.*

The next section sets out inducements to pursue such an extension, in the form of physically significant algebras of quantum observables not isomorphic to $\mathfrak{B}(\mathcal{H})$.

4 QM$_\infty$

4.1 Physical setting in which Type III factors arise

Von Neumann algebras not isomorphic to $\mathfrak{B}(\mathcal{H})$ for some separable $\mathcal{H}$ are commonplace in QM$_\infty$. Consider, for instance, axiomatic approaches to local relativistic Quantum Field Theory, which associate with each open bounded region O of Minkowski spacetime a von Neumann algebra $\mathfrak{M}(O)$ of observables pertaining (in some sense) to that region. These local algebras are typically Type III. For example, in the Minkowski vacuum state for the mass $m \geq 0$ Klein-Gordon field, if O is a region with non-empty spacelike complement, the standard axioms imply that $\mathfrak{M}(O)$ is a Type III factor ([1]).

For another example, consider the thermodynamic limit of quantum statistical mechanics (reached by letting the number of systems one considers and the volume they occupy go to ∞ while keeping their density finite). Equilibrium states at finite temperatures correspond to Type III factors for a wide variety of physically interesting systems: Bose and Fermi gases, the Einstein crystal, the BCS model (see [11], p.139-140; [7], Corr.5.3.36). At temperatures at which phase transitions occur (if there are any for the systems in question), equilibrium states correspond to direct sums/integrals of Type III factors.

Type II factor states also abound in QM$_\infty$, as do non-factor algebras of Types II and III. But the examples of Type III factors just provided are enough to provoke the question: can strategies and techniques for interpreting QM evolved in the environment of Type I factors be adapted to the environment of QM_∞?

4.2 Atomlessness

There is an impediment to adapting familiar variations on the maximal beable approach to the Type III factor algebras typical of QM_∞. The impediment is that if $\mathfrak{M}$ is a type III factor, its maximal abelian subalgebras—the very things the maximal beable approach would cast as maximal beable algebras—lack minimal projections—the very things familiar variations on the maximal beable approach use to characterize the worlds possible according to a quantum theory, and to assign probabilities to those worlds.

Let's approach this impediment by way of some examples. First, an atomless Boolean algebra. Consider the collection of subsets of $\mathbb{R}$ of the form $[x, \infty)$, partially ordered by set theoretic inclusion, with set theoretic union, intersection, and complementation supplying the Boolean operations of $\cup$, $\cap$, and $\sim$. This Boolean algebra is atomless: for any element $X = [x, \infty)$, there's an element $Y = [y, \infty)$, $y > x$, such that $Y < X$.

Next, an atomless maximal abelian von Neumann algebra: Let $\mathcal{H}$ be the separable Hilbert space L_2 of square integrable functions on the unit interval $[0, 1]$ equipped with the Lebesgue measure. Where S is a borel subset of $[0, 1]$, let χ_S be the operator on L_2 corresponding to multiplication by the characteristic function for f. Notice that if S is a set of measure 0, then χ_S is the zero operator on L_2, and that if T and V differ by a set of measure 0, χ_T and χ_V are the same operator on L_2. The collection $\{\chi_S\} = \mathfrak{D}_Q$ (with addition and multiplication defined pointwise) is a maximal abelian von Neumann algebra acting on $\mathcal{H}$ ([16], Example 5.1.6; see also [12]).[12] The following consideration suggest (truly ([16], Lemma 8.6.8) that $\mathfrak{D}_Q$ lacks atoms. For each measurable subset X of $[0, 1]$, the characteristic function χ_X is a projection in $\mathfrak{D}_Q$. $\chi_Y < \chi_X$ iff Y is a measurable subset of X. But every measurable subset of X itself has a measurable proper subset. Thus no projection in $\mathfrak{D}_Q$ is minimal. It turns out that every atomless maximal abelian von Neumann algebra is isomorphic to $\mathfrak{D}_Q$ ([16], p.665 ff.).

To see that such atomlessness afflicts any maximal abelian subalgebra of a Type III factor algebra, suppose that $\mathfrak{M}$ is such an algebra, and that $\mathfrak{D}$ is a maximal abelian subalgebra of $\mathfrak{M}$. And suppose, for reductio, that E is an atom in $\mathcal{P}(\mathfrak{D})$. Because $\mathfrak{M}$ contains no minimal projections, there exists $F \in \mathfrak{M}$ such that $F < E$. Because E is an atom in $\mathcal{P}(\mathfrak{D})$, $F \notin \mathcal{P}(\mathfrak{D})$. But then $\mathcal{P}(\mathfrak{D})$ is not a maximal abelian subalgebra of $\mathcal{P}(\mathfrak{M})$, which is our contradiction.
Proof $\mathcal{P}(\mathfrak{D})$ is non-maximal because there is an element of $\mathcal{P}(\mathfrak{M})$ in $\mathcal{P}(\mathfrak{D})$'s commutant but not in $\mathcal{P}(\mathfrak{D})$. That element is F. Because $\mathcal{P}(\mathfrak{D})$, its hypothesized atom E is either orthogonal to or a subprojection of every other element G of $\mathcal{P}(\mathfrak{D})$. That is, for any $G \in \mathcal{P}(\mathfrak{D})$, either (i) $EG = E$, or (ii) $EG = 0$. In case (i), $F < G$, because $F < E$ by hypothesis, and subspace inclusion is transitive. $F < G$ implies $FG = GF = F$. So F commutes with G. In case (ii), $FG = FEG = 0$, and F commutes with G. Thus F lies in $\mathcal{P}(\mathfrak{D})$'s commutant but not in $\mathcal{P}(\mathfrak{D})$. ♠

---

[12] To be more precise, $\mathfrak{D}_Q$ consists of equivalence classes of characteristic functions for Borel subsets of $[0, 1]$, where χ_T and χ_V belong to the same equivalence class if and only if T and V differ by a set of measure 0.

We conclude that any maximal abelian subalgebra of the projection lattice of a Type III factor algebra is atomless.

Familiar variations on the maximal beable approach appeal to atoms in the projection lattice of a maximal abelian subalgebra of $\mathfrak{B}(\mathcal{H})$ to answer Question 1-3. When, as the generalization to QM_∞ demands, a Type III factor $\mathfrak{M}$ is substituted for $\mathfrak{B}(\mathcal{H})$ as the observable algebra of interest, the atomlessness of $\mathfrak{M}$'s maximal abelian subalgebras and their projection lattices impedes the adaptation of these variations to the more general environment.

4.3 Ultrafilters to the Rescue?

Atomless though they may be, maximal abelian subalgebras of $\mathcal{P}(\mathfrak{M})$ for arbitrary $\mathfrak{M}$ admit *ultrafilters*. Here I'll indicate why this is so, explain how it's being so might lend hope to the maximal beable approach to QM_∞, then temper that hope.

Ultrafilters on maximal abelian subalgebras of $\mathcal{P}(\mathfrak{M})$ for arbitrary $\mathfrak{M}$

Definition 10 (Ultrafilter) *A **filter** on a lattice S is a non-empty proper subset F of S such that for all $a, b \in S$*

1. *if a and b are both elements of F, then so is their meet $a \cap b$;*
2. *if a is an element of F and $a \leq b$, then b is an element of F;*
3. $0 \notin F$.

*As subsets of S, filters can be ordered by inclusion. An **ultrafilter** on S is a filter on S that's not a proper subset of any other filter on S.*

An example of an ultrafilter is the subset F_a of B mapped to 1 by the homomorphism h_a defined in (5):

$$F_a = \{b \in B \text{ such that } a \leq b\} \tag{6}$$

F_a's distinctions do not end there. For F_a is also a *principal ultrafilter*, which is just an ultrafilter on a lattice generated (in the manner of (6)) by an atom of that lattice.

We care about ultrafilters because an ultrafilter on a Boolean lattice amounts to a truth valuation:

Fact 2 *Each ultrafilter F of a Boolean lattice B generates a two-valued homomorphism $h : B \to B_2$ via*

$$\begin{aligned} h(a) &= 1 \text{ if } a \in F \\ h(a) &= 0 \text{ if } a \notin F \end{aligned} \tag{7}$$

One way to demonstrate the existence of an ultrafilter is

Fact 3 (Ultrafilter Extension Theorem) *Any subset S of a Boolean lattice possessing the finite meet property[13] is contained in some ultrafilter (Bell and Machover 1977, Cor. 3.8).*

---

[13] That if $x_1, ..., x_n \in S$ then $x_1 \cap ... \cap x_n \neq 0$.

The proof of the Ultrafilter Extension Theorem invokes Zorn's lemma, which is equivalent to the axiom of choice. Because a general Boolean lattice needn't have atoms, its ultrafilters needn't be principal. The Ultrafilter Extension Theorem may tell us that they exist, but not what they look like. This reticence will be treated in more detail presently.

No matter what the type of a von Neumann algebra $\mathfrak{M}$, if $\mathfrak{M}$ admits a faithful state, $\mathcal{P}(\mathfrak{M})$ will have maximal abelian subalgebras $\mathcal{P}(\mathfrak{D})$ (where $\mathfrak{D}$ is a maximal abelian subalgebra of $\mathfrak{M}$), because maximal beable subalgebras always exist, and coincide with maximal abelian subalgebras in the presence of faithful states. In virtue of Fact 1, a faithful state on $\mathfrak{M}$ is a dispersion-free state on the abelian algebra $\mathfrak{D}$. The projection lattice of $\mathcal{P}(\mathfrak{D})$ is also a Boolean lattice admitting ultrafilters, because each dispersion-free state on $\mathfrak{D}$ induces an ultrafilter (and so a two-valued homomorphism) on $\mathcal{P}(\mathfrak{D})$. (This follows from the fact that if ω is dispersion free on $\mathfrak{D}$, $\omega(A)$ lies in A's spectrum for each $A \in \mathfrak{D}$. The elements of $\mathcal{P}(\mathfrak{D})$ mapped to 1 by dispersion-free ω thus constitute an ultrafilter for $\mathcal{P}(\mathfrak{D})$.)

Hope?

These ultrafilters won't be principal (i.e. generated by atoms), but so what? Recall the recipe for the maximal beable approach spelled out in Section 3.3. Atoms are invoked nowhere in that recipe. Although familiar variations on the maximal beable approach code facts and mediate probability assignments by atoms, the master recipe explicitly calls only for maximal abelian subalgebras (the beables) and two-valued homomorphisms on their projection lattices (the possible worlds, obtained as consistent eigenvaluations on those beables). And these ingredients, we've just seen, *will* be available in the more general setting. So even a general von Neumann algebra harbors possible worlds in the maximal beable approach's favored sense of maximal sets of co-obtaining properties.

What's more, for a state ω on a general von Neumann algebra $\mathfrak{M}$, we can still express ω (a la equation (1)) as a mixture of dispersion free states on a maximal abelian subalgebra $\mathfrak{D}$ of $\mathfrak{M}$. That is, for all $A \in \mathfrak{D}$, there will be a probability measure μ_ω on the space Λ of dispersion free states ω_λ on $\mathfrak{D}$ such that

$$\omega(A) = \int_\Lambda \omega_\lambda(A) d\mu_\omega(\lambda)$$

And this is all the maximal beable approach requires for a quantum state to define a probability distribution over possible worlds in its favored sense. Regarding the maximal beable approach in these general terms suggests that, despite the proclivities of variations developed for ordinary QM, the maximal beable approach does not presuppose that $\mathcal{P}(\mathfrak{M})$ contains atoms.

Tempered

But upon closer examination, the promise of non-principle ultrafilters rings hollow. Consider the states ω_λ in the expression above, and the ultrafilters they define. What is a possible world coded by such an ultrafilter *like*? This is a fair question; in fact, it's the question (Question 1) a maximal beable approach to a system in a state ω on $\mathfrak{M}$ must answer to lend content to the theory interpreted. It is also a thorny question, as Hans Halvorson makes clear: "Although we 'know' that there are ultrafilters (i.e. pure states[14]) on [atomless $\mathfrak{D}$)], we do not know this because someone has constructed an example of such an ultrafilter... We are told that there is some pure state ω on $\mathfrak{D}$, but we are not given a recipe for determining the value $\omega(A)$ for an arbitrary element $A \in \mathfrak{D}$" (Halvorson 2001, 41). The rub is that the ultrafilters in question are non-constructable. Their existence is demonstrated by appeal to the ultrafilter extension theorem, which by way of presupposing Zorn's Lemma, presupposes the axiom of choice. To show that these ultrafilters exist, one finds a family of elements of $\mathcal{P}(\mathfrak{D})$ satisfying the finite meet property, then invokes the ultrafilter extension theorem to conclude that this family belongs to an an ultrafilter on $\mathcal{P}(\mathfrak{D})$.[15]

The ineffability of ultrafilters on non-atomic $\mathcal{P}(\mathfrak{D})$ suggests that we have no handle, analogous to the one supplied in the Type I case by applying the eigenstate-eigenvalue link to the atom generating a principle ulatrafilter, on how to decode the facts these ultrafilters encode. It also suggests that we have no handle, analogous to the one supplied in the Type I case by applying the trace prescription to the system state and the atom generating a principal ultrafilter, on how to assign those facts probabilities. Confronted with the non-atomic von Neumann algebras of QM_∞, then, the maximal beable approach shirks two key interpretive tasks. First, it fails to explicitly characterize the worlds possible according to the theory. It tells us that there exist homomorphisms defined by pure states ω_λ corresponding to these possible worlds, but it doesn't identify those states or lend content to those possible worlds. Second, it fails to explicate the probabilities the theory assigns to these worlds. Again, we know that a probability distribution μ_λ exists, but ignorant of the identities of states ω_λ, we're ignorant as well of what probabilities ω assigns them. Thus the maximal beable approach fails to equip QM_∞ with empirical content, in the form of specific probabilities assignments to explicitly characterized worlds possible for systems described by non-atomic von Neumann algebras.

5 What now?

Taking the maximal beable approach to QM_∞, I have suggested, will leave us in difficulty. But why? Here is a non-comprehensive list of possible culprits.

At its most general, the maximal beable approach makes a resource of the formal apparatus of von Neumann algebras and their self-adjoint elements. In particular, the maximal beable approach assimilates a physical property to a determinate eigenvalue assignment to a self-adjoint element of the von Neumann algebra of observables for the system whose property it is. Perhaps this is the wrong formal apparatus. Perhaps the salient algebra has been misidentified. Perhaps C* algebras (cf. [20]) or universal enveloping von Neumann algebras (cf. [18]) will prove more fruitful frameworks. Or perhaps the aspiration to explicate the metaphysical commodity *property* so directly in terms of formal apparatus is misguided. (I think that the discipline inherent in the aspiration respects the contribution of physics to philosophy of physics. I don't think such respect is

---

[14]Cf. [[15], Prop. 4.4.1] If ω is dispersion free on abelian $\mathfrak{D}$, then ω is a pure state of $\mathfrak{D}$.
[15]For the sake of space, I am setting to one side the alarming fact that if $\mathcal{P}(\mathfrak{D})$ lacks atoms, then pure states on $\mathfrak{D}$ fail to be countably additive. For a more complete discussion of the non-constructability of ultrafilters on nonatomic algebras, see Halvorson (2001), which also investigates ways to avoid the difficulties this non-constructability creates.

a *mandatory* component of philosophical reflection on physics. But I think it's nice that it's a component of some reflections.)

The maximal beable approach also rests on a set of metaphysical scruples, for instance, that a possible world is a *maximal* set of co-obtaining properties. Perhaps these scruples are to blame for the approach's inadequacy to QM_∞. Rob Clifton's [9] strategy for interpreting the exotic von Neumann algebras encountered in QFT dispenses with the maximality scruple. To oversimplify, given a state ω on a von Neumann algebra $\mathfrak{M}$, Clifton identifies a beable algebra $\mathfrak{D}$ that isn't a maximal abelian subalgebra of $\mathfrak{M}$ but *is* the largest abelian subalgebra of $\mathfrak{M}$ characterizable, by means Clifton deems admissible, in terms of ω and $\mathfrak{M}$. In other words, $\mathfrak{D}$ can be embedded in larger abelian subalgebras of $\mathfrak{M}$, but ω doesn't tell us which ones. The merits of refraining from plenitude deserve further debate. Some demerits of Clifton's particular strategy for refraining have been discussed elsewhere ([9], [10]): many observable algebras admit a dense set of states for which Clifton's beable algebra is trivial, in the sense that it contains only multiples of the identity operator!

The suggestion I find interesting is that the maximal beable approach goes wrong by allowing itself only the resources of formal apparatus and metaphysical scruples. However, the strategy of modifying the sets of *those* resources exploited by the maximal beable approach opens up many promising avenues for developing interpretations of QM_∞. Thus the considerations aired in this essay aren't even the shadow of an argument that those resources are inadequate to the task of interpreting QM_∞.

About that task, this discussion holds one lesson at least. Minimal projections, and interpretive strategies they underwrite, are not a perfectly general feature of the sorts of von Neumann algebras that arise in physical applications. It follows that assumptions widespread in the semantics of ordinary QM— assumptions that pure states are both fact encoders and probability bearers — are upset by QM_∞. The maximal beable approach can be freed of these assumptions and extended to QM_∞. But due to the non-constructability of ultrafilters on non-atomic abelian von Neumann algebras, the extension shirks the main tasks of quantum semantics: the *characterization* of worlds possible according to a quantum theory, and the *explication* of the probabilities that theory assigns those worlds. It appears that we can't interpret QM_∞ by *simply* extending or even adapting our favorite semantics for ordinary QM to infinite quantum systems.

Bibliography

[1] Araki, H. Type of von neumann algebra associated with free field. *Progress in Theoretetical Physics*, 32:956–965, 1964.

[2] Barrett, J. *The Quantum Mechanics of Minds and Worlds*. Oxford: Oxford University Press, 1999.

[3] Bell, J. On the problem of hidden variables in quantum mechanics, reprinted. In *Speakable and unspeakable in quantum mechanics*, 1–13. Cambridge: Cambridge University Press, 1987[1966].

[4] Bell, J. Subject and object, reprinted. In *Speakable and unspeakable in quantum mechanics*, 40–44. Cambridge: Cambridge University Press, 1987[1973].

[5] Bell, J. L., Machover, M. *A Course in Mathematical Logic*. North Holland: Amsterdam, 1977.

[6] Beltrametti, E., Cassinelli, C. *The Logic of Quantum Mechanics*. MA: Addison-Wesley, 1981.

[7] Bratteli, O., Robinson, D. *Operator Algebras and Quantum Statisical Mechanics II*. Berlin: Springer-Verlag, 2 edn., 1997.

[8] Bub, J. *Interpretting the Quantum World*. Cambridge: Cambridge University Press, 1997.

[9] Clifton, R. The modal interpretation of algebraic quantum field theory. *Physics Letters A*, 271:167–177, 2000.

[10] Earman, J., Ruetsche, L. Relativistic invariance and modal interpretations. *Philosophy of Science*, 72:557–583, 2005.

[11] Emch, G. *Algebraic methods in statistical mechanics and quantum field theory*. New York: Wiley, 1972.

[12] Halvorson, H. On the nature of continuous physical quantities in classical and quantum mechanics. *Journal of Philosophical Logic*, 30:27–50, 2001.

[13] Halvorson, H., Clifton, R. Maximal beable subalgebras of quantum mechanical observables. *International Journal of Theoretical Physics*, 38:2441–2484, 1999.

[14] Hughes, R. *The Structure and Interpretation of Quantum Mechanics*. Cambridge, MA: Harvard University Press, 1989.

[15] Kadison, R., Ringrose, J. *Fundamentals of the Theory of Operator Algebras*, vol. 1. New York: Academic Press, 1997.

[16] Kadison, R., Ringrose, J. *Fundamentals of the Theory of Operator Algebras*, vol. 2. New York: Academic Press, 1997.

[17] Murray, F., von Neumann, J. On rings of operators. *Annals of Mathematics*, 37:116–229, 1936.

[18] Primas, H. *Chemistry, Quantum Mechanics, and Reductionism*. New York: Springer-Verlag, 1983.

[19] Redhead, M. *Incompleteness, nonlocality, and realism*. Oxford: Oxford University Press, 1988.

[20] Segal, I. E. The mathematical meaning of operationalism in quantum mechanics. In Henkin, L., Suppes, P., Tarski, A. (Eds.), *Studies in Logic and the Foundatios of Mathematics*, 341–352. North-Holland: Amsterdam, 1959.

[21] Sunder, V. *An Invitation to von Neumann Algebras*. Berlin: Springer-Verlag, 1987.

[22] Wallace, D. Everettian rationality: defending Deutsch's approach to probability in the Everett interpretation. *Studies in the History and Philosophy of Modern Physics*, 34:415–438, 2003.

[23] Wheeler, J., Zurek, W. H. *Quantum theory and measurement*. Princeton University Press, 1983.

Causal models: A Philosophical Definition Built on Statistical Concepts[1]

Tianjiao Chu

University of Pittsburgh

tic19@pitt.edu

ABSTRACT. In this paper I propose a new definition of causal models that is mathematically rigorous and philosophically sound. This new definition will help to clarify what is implied by a causal model, and make explicit the assumptions required for the derivation of a causal model. Under the new definition, a causal model is interpreted as a statistical model accompanied with a decision rule that guides its application to different statistical populations. That a causal model is by definition to be applied to multiple statistical populations differentiates it from other statistical models, which are intended as descriptions of a single statistical population. This new definition of causal models does not employ, either explicitly or implicitly, the concepts of causation, intervention/manipulation, counterfactuals, etc. On the contrary, the concepts of causation, intervention/manipulation, and counterfactuals will be understood in term of causal models. The new definition can accommodate both deterministic and inherently indeterministic causal models.

> The only immediate utility of all sciences, is to teach us, how to control and regulate future events by their causes. ... [W]e may define a cause to be an object, followed by another, and where all the objects similar to the first are followed by objects similar to the second. Or in other words where, if the first object had not been, the second never had existed.
>
> Hume, *An Enquiry Concerning Human Understanding*, §7

Correlation is not causation – so is every student of statistics told. But what exactly is meant by this statement? Correlation is well defined, as the ratio of the covariance of two random variables to the product of their standard deviations. The concept of causation, however, is much vague and lacks mathematical rigority: Philosophers have debated over the meaning of causation for the last few centuries without reaching a consensus, while statisticians are often less than enthusiastic in providing a theory of causation.

Despite the absence of a precise definition of causation, in the last two decades, remarkable progress has been made in the field of causal inference [3, 5]. Among the major results of casual inference are the algorithms of deriving causal relations among a set of variables from their correlation matrix. The success of the study

---

[1] The author is with Department of Obstetrics, Gynecology, & Reproductive Sciences, University of Pittsburgh, Pittsburgh, PA 15213, USA.

of causal inference makes it more urgent to clarify the concept of causation, which would help us understand the results of causal inference and their implications. One of the most notable recent efforts to answer what is causation is [6]. Inspired by the works in causal inference, Woodward provides a philosophical definition of causation based on a slightly modified concept of intervention. Unfortunately, his definition is circular in that the concept of intervention itself has to be defined in term of causation.[2]

The main purpose of this paper is to provide a conceptual foundation for the study of causal inference. Focusing strictly on the implication of causal claims from the statistical perspective, I propose a rigorous and non-circular definition of causal models that does not rely on the concepts of intervention or causation. After the causal model being defined, definition of causation then is a straightforward derivation. The definition of causation presented in this paper is not intended to cover all or most of the major usages of the word "causation." However, the new definitions of causal model and causation should help the philosophers' search for more satisfactory theories of causation.

The development of a rigorous definition of causal models has to be guided by our intuition and experiences with causal models, such as those described by Hume in the quoted text at the beginning of this paper. I shall talk frequently about, from the statistical perspective, what a causal model should look like and how to use a causal model, even before presenting a full definition of causal models. It is hoped that the formal definition of causal models presented in this paper will match these experiences and intuition.

1 What should we expect from a causal model

One of the major goals of statistical study is to make prediction. From the statistical perspective, causal models may be understood mainly as a tool to help us predict observables **Y** in some statistical populations from knowledge of other observables **X**. Of course, it is meaningless to talk about predicting **Y** from **X** unless the information of **X** is available earlier than that of **Y**. That is, we need to make sure that **X** precedes **Y** informatically in the following sense:

Definition 1 (Informatical Precedence) *Consider a statistical population S, where **X** and **Y** are two sets of distinct observables.*

(i) *The observable $Y_j \in \mathbf{Y}$ is preceded informatically by the observable $X_i \in \mathbf{X}$ in S if, regardless of when the information of Y_j is available, in principle the information of X_i could always be available earlier.*

(ii) *The observables **Y** are preceded informatically by the observable $X_i \in \mathbf{X}$ in S if:*

- *There is an observable $Y_j \in \mathbf{Y}$ such that Y_j is preceded by X_i informatically.*

---

[2][6] claims that this is not a vicious circularity, because in practice, such as in randomized experiments, we may claim that an action α, e.g., the randomization of the treatment, is an intervention on a variable X with respect to another variable Y even if we do not know whether X is a cause of Y. However, this argument is valid only for the justification of using intervention to *discover* the specific causal relation between X and Y, not for the justification of using intervention to *define* the very concept of causation.

- X_i is *NOT preceded informatically by any observable in* **Y**.

(iii) *The observables* **Y** *are preceded informatically by the observables* **X** *in S if* **Y** *is preceded informatically by each observable* $X_i \in$ **X**.

Then, as a prediction tool, a causal model M about observables **X** and **Y** should enable us to predict, for a collection $\mathbf{S}_a$ of statistical populations, the conditional distribution of **Y** in each statistical population $S_i \in \mathbf{S}_a$ from the knowledge of **X** in S_i, where **X** precedes **Y** informatically in all statistical populations in $\mathbf{S}_a$.[3]

If the predicted conditional distribution of **Y** is degenerated, we can check the correctness of the prediction for a statistical population S made by M by comparing the observed value of **Y** in S with the predicted value. If the predicted conditional distribution of **Y** is non-degenerated, we can either evaluate the correctness of the prediction using the likelihood of getting the observed value in S were the predicted distribution correct, or estimate the posterior of the correctness of the predicted distribution.

Another goal of statistical study is to summarize our observations. In this sense, from the statistical perspective, causal models could also be understood as an explanation tool. Suppose the value of the observables **Y** in a population S are already known, we could say that a causal model M explains the distribution of **Y** in S if the predicted conditional distribution of **Y** by M from the knowledge of **X** is correct when checked against the known value of **Y**.

Throughout the paper, M is used to denote a causal model, S a population to which we may apply the causal model M, **Y** the observables to be predicted by M, and **X** the observables used by M to predict **Y**.

The class of statistical populations **S** to which a causal model M may potentially be employed to make a prediction or explanation is called the domain of M.[4] For many causal models, the associated domain is self-evident. For example, the domain of a causal model about the smoking and the risk of heart disease for adult human male could include any collection of adult human males. In general, a complete specification of the domain of a causal model M requires substantial knowledge about the subject matter that M concerns, and different users of the same causal model may have different understanding on which population should be included. Throughout this paper, whenever a causal model is mentioned, it is assumed that its domain is already specified and agreed upon. Two necessary conditions for a class of population to be the domain of a causal model are given at the end of next section.

The complete specification of a causal model M consists of two components. Let **S** be the domain of M. The first component of M is a mathematical structure G, called the *invariant structure*, that specifies how to predict the distribution of observables **Y** in a population $S \in \mathbf{S}$ from knowledge of the observables **X**. We say that the invariant structure G of M is imposed on S if M is used to predict **Y** in S from **X**. Imposing G on S does not necessarily produce the correct prediction. A population S is said to be predictable by M, or equivalently, predictable by G, if the prediction obtained by imposing G on S is true.

The second component of M is a binary decision rule D, called the admissibility rule, which accepts a population S in the domain of M only if S is predictable

---

[3]Note that the predicted conditional distribution of **Y** is itself a function of **X** only, hence the prediction can be made before the information of **Y** is available.

[4]Note that the domain of M may include a lot of populations for which M cannot make correct predictions.

by M. S is said to be admissible if it is accepted by D. Note that D is conservative in the sense that it is desirable, but not required, for D to accept all populations predictable by M

When applying a causal model $M = (G, D)$ to a population S to predict the observables $\mathbf{Y}$, first D is used to determine if S is admissible. Here only information about observables preceding $\mathbf{Y}$ informatically in S can be made available to D. Next, if S is accepted by D, the invariant structure G is imposed on S to predict the distribution of $\mathbf{Y}$ in S based on the knowledge of the predictors $\mathbf{X}$.

A causal model M is correct if for any population S accepted by its admissibility rule D, the prediction of S by M is true. D is correct relative to M if every population S accepted by D is predictable by M. G is correct relative to M if imposing G on any population S accepted by D of M always produces the correct prediction.

Invariant structures will be formally defined in the next section. In section 3, I shall argue for the necessity of the admissibility rules. Section 4 is a discussion about how to find an appropriate admissibility rule for a causal model. The paper ends with new definitions of causation and intervention in term of causal models.

2 Invariant structures

It is often argued that one of the major advantage of causal models is that they are stable, and invariant under "manipulation" or "intervention" [3, 5, 6]. The invariant structure component of a causal model M represents exactly what is invariant in all the populations predictable by M. In this section, I shall first define the invariant structures for causal models that predict only a single observable. The definition of this type of invariant structures then is used to define invariant structures in causal models that predict multiple observables.

2.1 Atomic invariant structures

Atomic causal models are causal models that predict only a single observable. The invariant structures of an atomic causal model is called an *atomic invariant structure*.

Definition 2 (Atomic Invariant Structure) *G is an atomic invariant structure with exogenous variables $\mathbf{X} = \{X_1, \cdots, X_k\}$ and endogenous variable Y if*

$$G = \Big(\mathbf{X} \cup \{Y\}, \{\mathcal{X}, \mathcal{Y}\}, [(T, f_Y)], (\Omega, \mathscr{F}, P)\Big),$$

where

(i) $\mathbf{X} = \{X_1, \cdots, X_k\}$ *and Y are real valued variables.*
(ii) $(\Omega, \mathscr{F}, P)$ *is the probability space associated with G*
(iii) $\mathcal{X}$ *and $\mathcal{Y}$ are Borel sets, with $\mathcal{X} = \mathcal{X}_1 \times \cdots \times \mathcal{X}_k$ where $\mathcal{X}_i \in \mathscr{R}$, and $\mathcal{Y} \in \mathscr{R}$.*
(iv) $[(T, f_Y)]$ *is an equivalence class that includes (T, f_Y), where:*

- $T = T(\omega)$ *is a measurable mapping from $(\Omega, \mathscr{F})$ to a measurable space $(\mathcal{T}, \mathscr{T})$*

- $f_Y(\mathbf{x}, t) = f_Y(x_1, \cdots, x_k, t)$ is a measurable mapping from $\mathcal{X} \times \mathcal{T}$ to $\mathcal{Y}$.

- (T, f_Y) and (T', f'_Y) are equivalent if and only if for each $\mathbf{x} \in \mathcal{X}$, $f_Y(\mathbf{x}, T(\omega))$ and $f'_Y(\mathbf{x}, T'(\omega))$ have exactly the same distribution.

While an atomic invariant structure could have an empty set of exogenous variables, i.e., $\mathbf{X} = \emptyset$, for simplicity, throughout this paper, unless specifically noted, we shall always assume that $\mathbf{X}$ is non-empty.

$\mathcal{X}_i$ and $\mathcal{Y}$ are respectively the range of the exogenous variable X_i and the endogenous variable Y.[5] For any $(T', f'_Y) \in [(T, f_Y)]$, f'_Y is called a version of the predictive function of G, T' is called the uncertainty factor corresponding to f'_Y.

Note that the same symbol is used to denote both a variable in an invariant structure G of a causal model M, and an observable in a population S in the domain $\mathbf{S}$ of M. This naming convention effectively constitutes a mapping rule that associates a variable in an invariant structure with an observable of a population. The endogenous variable Y in G represents the predictee Y of the populations in the domain of M, the exogenous variables $\mathbf{X}$ in G correspond to the predictors $\mathbf{X}$ of the populations, and the equivalence class $[(T, f_Y)]$ determines the predicted distribution of the predictee Y conditional on the knowledge of the predictors $\mathbf{X}$ obtained by imposing G on S: Given that the value of predictors $\mathbf{X}$ equals $\mathbf{x}$, the predicted distribution of predictee Y in S is precisely the distribution of $f'_Y(\mathbf{x}, T')$ on $\mathcal{Y}$, where (T', f'_Y) can be any element of $[(T, f_Y)]$ in G.

The definition of invariant structures looks almost the same as some previous definitions of causal models in the literature of causal inference. For example, an atomic invariant structure differs from the functional causal model [3] mainly in a single aspect: Instead of treating T as an unmeasured exogenous variable to explain the uncertainty of the prediction made by the causal model, in this paper an equivalence class $[(T, f_Y)]$ is provided to represent the uncertainty of the prediction. T by itself does not represent any observable of any population, nor is it identifiable from any query about any population.

The reason why T is treated as purely a random quantity defined on $(\Omega, \mathscr{F}, P)$ is that, on the one hand, the functional representation $f_Y(\mathbf{x}, T(\omega))$ of the conditional distribution is very convenient when we later talk about "partialization." On the other hand, in this paper, the definition of causal model is required to be able to represent genuinely indeterministic causal relations, such as those implied by the quantum theory, where uncertainty cannot be explained away by some hidden variables. A mathematical formalization of causal models that presumes implicit understanding of the concept of causation, such as [3], may be allowed to exclude some perfectly legitimate scientific causal models for the sake of simplicity. But such an exclusion is unacceptable for a definition of causal models intended to serve as the foundation based on which the very concept of causation is derived.

---

[5]Strictly speaking $\mathcal{X}$ can be any Borel set in $\mathscr{R}^k$, not necessarily the Cartesian product of k linear Borel sets. However, by requiring $\mathcal{X}$ to be the product of $\mathcal{X}_i$'s, the notation will be simpler when we talk about ranges of exogenous variables in causal models with multiple predictees.

2.2 Composite invariant structures

To define the invariant structures of a *composite causal model*, which is a causal model that predicts multiple observables, we need first to define a partially ordered set of atomic structures called the *dependent set of atomic invariant structures*.

Definition 3 (Dependent Set of Atomic Invariant Structures) *Given m atomic invariant structures $G_1, \cdots, G_m$, the partially ordered set $(\{G_1, \cdots, G_m\}, <)$ is called a set of dependent atomic invariant structures if the following constraints are satisfied:*

(i) *The same probability space $(\Omega, \mathscr{F}, P)$ is associated with all atomic structures G_i for $i = 1, \cdots, m$.*
(ii) *Any two distinct atomic invariant structures G_i and G_j cannot share the same variable as their endogenous variable.*
(iii) *If a variable V appear in two atomic invariant structures G_i and G_j, it must have the same range in G_i and G_j.*
(iv) *The endogenous variable of G_i could appear as an exogenous variable in another atomic structure G_j if and only if $G_i < G_j$.*

Let Y_i and $\mathbf{X}_i$ be the endogenous and exogenous variables in G_i respectively. Variables in $\mathbf{Y} = \{Y_1, \cdots, Y_m\}$ and in $\mathbf{X} = \bigcup_{i=1}^{m} \mathbf{X}_i \setminus \mathbf{Y}$ are called respectively the endogenous and exogenous variables of the dependent set of atomic invariant structures $(\{G_1, \cdots, G_m\}, <)$. Let f_i be a version of predictive function of atomic invariant structure G_i, T_i the corresponding uncertainty factor, $\mathcal{X}$, $\mathcal{Y}$ and $\mathcal{T}$ ranges of $\mathbf{X}$, $\mathbf{Y}$, and $\mathbf{T} = \{T_1, \cdots, T_m\}$ respectively. We can construct recursively a m-dimensional function $\mathbf{f} = (f^1, \cdots, f^k)$ from $\mathcal{X} \times \mathcal{T}$ to $\mathcal{Y}$, called the predictive function of $(\{G_1, \cdots, G_m\}, <)$ corresponding to $\mathbf{T}$, such that for each possible value $\mathbf{x}$ of $\mathbf{X}$ in $\mathcal{X}$:

- $f^1(\mathbf{x}, \mathbf{T}) = f_1(\mathbf{x}_1, T_1)$, where $\mathbf{x}_1 = (x_{1_1}, \cdots, x_{1_{k_1}})$ are corresponding entries in $\mathbf{x}$ for $\mathbf{X}_1$, and $\mathbf{X}_1 \subset \mathbf{X}$.
- $f^i(\mathbf{x}, \mathbf{T}) = f_i(\mathbf{x}_i, T_i)$, where $\mathbf{x}_i = (x_{i_1}, \cdots, x_{i_{k_i}})$. x_{i_j} is the corresponding entry in $\mathbf{x}$ for X_{i_j} if $X_{i_j} \in \mathbf{X}$. $x_{i_j} = f^k(\mathbf{x}, \mathbf{T})$ if $X_{i_j} = Y_k$ is an endogenous variable in G_k for some $k < i$.

We can define an equivalence relation on the set of pairs $(\mathbf{T}, \mathbf{f})$ of corresponding predictive function and uncertainty factor of a dependent set of atomic invariant structures such that $(\mathbf{T}, \mathbf{f})$ and $(\mathbf{T}', \mathbf{f}')$ are equivalent if and only if for each $\mathbf{x} \in \mathcal{X}$, $\mathbf{f}(\mathbf{x}, \mathbf{T}(\omega))$ and $\mathbf{f}'(\mathbf{x}, \mathbf{T}'(\omega))$ have exactly the same distribution. Obviously, for each $\mathbf{x} \in \mathcal{X}$, the equivalence class $[(\mathbf{T}, \mathbf{f})]$ uniquely determines a distribution on $\mathcal{Y}$, i.e., the distribution of $\mathbf{f}'(\mathbf{x}, \mathbf{T}'(\omega))$, where $(\mathbf{T}', \mathbf{f}')$ can be any element in $[(\mathbf{T}, \mathbf{f})]$.

The invariant structure of a composite causal model, called *composite invariant structure*, is basically a dependent set of atomic invariant structures associated with an equivalence class of pairs of uncertainty factor and predictive function.

Definition 4 (Composite Invariant Structure) *Let*

$$(\{G_1, \cdots, G_m\}, <)$$

be a dependent set of atomic invariant structures, and $[(\mathbf{T}, \mathbf{f})]$ *an equivalence class of corresponding uncertainty factors and predictive functions of* $(\{G_1, \cdots, G_m\}, <)$. *Then*

$$G = \Big((\{G_1, \cdots, G_m\}, <), [(\mathbf{T}, \mathbf{f})]\Big)$$

is called a composite invariant structure.

The endogenous and exogenous variables of $(\{G_1, \cdots, G_m\}, <)$ are also called the endogenous and exogenous variables of the invariant structure G. The elements in $[(\mathbf{T}, \mathbf{f})]$ are also called corresponding uncertainty factors and predictive functions of G.

The interpretation of a composite invariant structure G as a component of a composite causal model M is almost the same as the interpretation of an atomic invariant structure. The endogenous variable $\mathbf{Y}$ in G represents the multiple observables $\mathbf{Y}$ of the populations M is to predict, where the exogenous variables $\mathbf{X}$ in G correspond to the observables $\mathbf{X}$ based on which $\mathbf{Y}$ is predicted. $[(\mathbf{T}, \mathbf{f})]$ determines the predicted distribution of the observables $\mathbf{Y}$ given the knowledge of the observables $\mathbf{X}$: Conditional on the observables $\mathbf{X}$ being $\mathbf{x}$, the distribution of the observables $\mathbf{Y}$ is precisely the distribution of $\mathbf{f}'(\mathbf{x}, \mathbf{T}')$ in G, where $(\mathbf{T}', \mathbf{f}')$ can be any element in $[(\mathbf{T}, \mathbf{f})]$.

2.3 Two necessary conditions for the domain of a causal model

Let us first define the *ancestor* relation on a dependent set of atomic invariant structures $(\{G_1, \cdots, G_m\}, <)$, following a similar definition by [5].

Definition 5 (Ancestor) *Let V be any variable in $\mathbf{X} \cup \mathbf{Y}$, i.e., V could be either an endogenous or an exogenous variable in some G_i. The set of ancestors $A(V)$ of V in G is defined recursively:*

(i) $V \in A(V)$.
(ii) *If there is an atomic invariant structure G_i in G such that V is the endogenous variable of G_i, then $\mathbf{X}_i \subset A(V)$, where $\mathbf{X}_i$ is the set of exogenous variables in G_i.*
(iii) *If there is an atomic invariant structure G_j in G with endogenous variable Y_j, and $Y_j \in A(V)$, then $\mathbf{X}_j \subset A(V)$, where $\mathbf{X}_j$ is the set of exogenous variables in G_j.*

Let M be a casual model with exogenous variables $\mathbf{X}$ and endogenous variables $\mathbf{Y}$. Below are two necessary conditions for a class $\mathbf{S}$ of populations to be the domain of M:

Formal Applicability Condition *M must be formally applicable to each population S in $\mathbf{S}$ in the sense that:*

   (i) *Each of the endogenous and exogenous variables in M represent a distinct observable of S.*
   (ii) *The observable in S represented by $Y_i \in \mathbf{Y}$ is preceded informatically by all observables in S represented by $A(Y_i) \setminus \{Y_i\}$, where $A(Y_i)$ is the set of ancestors of Y_i in M.*

(iii) None of the observables in S represented by the exogenous variables $\mathbf{X}$ in M is preceded informatically by any observables represented by the endogenous variables $\mathbf{Y}$ in M.

Conditionalization Condition: $\mathbf{S}$ *must be closed under conditionalization in the sense that if* $S \in \mathbf{S}$, *and* $\mathbf{Z}$ *is some arbitrary set of observables preceding the predictees* $\mathbf{Y}$ *informatically in* S, *then* $S' = \{s : s \in S, \ \mathbf{Z}(s) \in A\}$, *i.e., the subset of* S *obtained by taking all elements of* S *where the value of* $\mathbf{Z}$ *is in set* A, *is also a population in* $\mathbf{S}$.

3 Admissible populations

The second component of a causal model M is the admissibility rule, which determines if a population S is predictable by M. In this section I shall show why this component is needed, and why we could not simply impose G of M on each population in its domain.

Let M_C be an atomic causal model for smoking and cancer which states that high smoking level (the predictor) increases the likelihood of cancer occurrence (the predictee), $\mathbf{S}$ be the domain of M_C. Let

$$G = \Big(\{X,Y\}, \{\mathcal{X}, \mathcal{Y}\}, [(T, f)], (\Omega, \mathscr{F}, P)\Big) \tag{1}$$

be the invariant structure of M_C, where X and Y represent respectively smoking level and cancer occurrence. Besides smoking, there could be many other observables preceding cancer occurrence informatically. Suppose there is a set of observables called the complementary risk factors such that for all populations in the domain of M_C, smoking level and the complementary risk factors contain all the potentially attainable information relevant to cancer occurrence available before the observation of cancer occurrence. In other words, cancer occurrence is independent of all other preceding observables given smoking level and the complementary risk factors. Suppose that there is an random quantity T' defined on the probability space of G, and a measurable mapping g, such that in all populations in $\mathbf{S}$, given smoking level equals x and the complementary factors equal $\mathbf{z}$, the conditional distribution of cancer occurrence is precisely the distribution of $g(x, \mathbf{z}, T')$.

The above assumptions imply the existence of a causal model M'_C with invariant structure

$$G' = \Big(\{\{X\} \cup \mathbf{Z}, Y\}, \{\mathcal{X} \times \mathcal{Z}, \mathcal{Y}\}, [(T', g(x, \mathbf{z}, T'))], (\Omega, \mathscr{F}, P)\Big) \tag{2}$$

such that each population in $\mathbf{S}$ is predictable by M'_C. Note that G' of M'_C and G of M_C share the same probability space. The assumed existence of M'_C, which will be called as the *domain assumption* of M_C, is based on the *causality postulate* that will be introduced in the next section.

For any population S in $\mathbf{S}$, we can construct a *probabilistic model for S based on* G' in the following way:

Let $(\Omega, \mathscr{F}, P)$ in G' be the probability space of the probabilistic model. Pick up a pair of corresponding uncertain factor and predictive function, say, $(T'(\omega), g)$, from the equivalence class $[(T', g(x, \mathbf{z}, T'))]$ of G'. Let $X(\omega)$ and $\mathbf{Z}(\omega)$

be random mappings independent of $T'(\omega)$ such that the joint distribution of X and $\mathbf{Z}$ on $\mathcal{X} \times \mathcal{Z}$ is exactly the same as the joint distribution of the smoking level and the complementary risk factors in S, where $\mathcal{X} \times \mathcal{Z}$ is the range of smoking level and the complementary risk factors in all populations in $\mathbf{S}$.[6]

Let $Y(\omega) = g(X, \mathbf{Z}, T')(\omega)$. The conditional distribution of Y given X and $\mathbf{Z}$ in the model is exactly the same as the distribution of cancer occurrence given smoking level and the complementary risk factors in S predicted by M'_C:

$$\begin{aligned}
P\Big(Y \in A, (X, \mathbf{Z}) \in B\Big) \\
= \int_B \int P\Big(g(X, \mathbf{Z}, T') \in A | X, \mathbf{Z}, T'\Big) d\mu_{T'|X,Z} d\mu_{X,Z} \\
= \int_B \int I_A[g(X, \mathbf{Z}, T')] d\mu_{T'|X,Z} d\mu_{X,Z} \\
= \int_B \int I_A[g(X, \mathbf{Z}, T')] d\mu_{T'} d\mu_{X,Z} \quad (3)
\end{aligned}$$

where A is any measurable subset of $\mathcal{Y}$, B any measurable subset of $\mathcal{X} \times \mathcal{Z}$. Note that $\int I_A[g(X, \mathbf{Z}, T')] d\mu_{T'}$ is precisely the conditional probability of the cancer occurrence in S falling into set A given certain smoking level and the complementary risk factors, as implied by M'_C.

Let $S_0 \in \mathbf{S}$ be a population where smoking level and the complementary risk factors are independent. Now $X(\omega)$, $\mathbf{Z}(\omega)$, and $T'(\omega)$ are jointly independent in the probabilistic model for S_0 based on G'. The conditional probability that cancer occurrence in S_0 falling into set A given certain smoking level is:

$$\begin{aligned}
\mu_{Y|X}(A) &= P(Y \in A | X) \\
&= \int \int I_A[g(X, \mathbf{Z}, T')] d\mu_{T'} d\mu_{Z|X} \\
&= \int \int I_A[g(X, \mathbf{Z}, T')] d\mu_{T'} d\mu_Z \quad \text{w.p.1} \quad (4)
\end{aligned}$$

Suppose that S_0 is predictable by M_C, M_C must predict that the conditional distribution of cancer occurrence given smoking level x is $\mu_{Y|X}(x)$ as specified in Equation 4. Thus, for any $(T', f') \in [(T, f)]$ in G of M_C, the distribution of $f'(x, T')$ in G must be given by the measure $\mu_{Y|X}(x)$ on $\mathcal{Y}$ in the probabilistic model for S_0 based on G'. In particular, given that G' and the probabilistic model for S_0 share the same probability space with G, $(\mathbf{Z} \cup \{T'\}, g)$ must be an element of $[(T, f)]$ in G,.

Let us now consider four new populations in $\mathbf{S}$ produced by four different actions taken on S_0, and examine if they are predictable by G.

The first new population S_1 is obtained by taking all elements in S_0 whose value of smoking level falls into an interval $(x_1, x_2]$. Obviously, smoking level and the complementary risk factors are still independent in S_1. A probabilistic model

---

[6]The independence between $T'(\omega)$ and $\{X(\omega), \mathbf{Z}(\omega)\}$ in the probabilistic model is required because $T'(\omega)$ is simply a random quantity in G', where $X(\omega)$ and $\mathbf{Z}(\omega)$ also represent observables in S.

for S_1 based on G' with the same choice of (T', g) is similar to the one defined above for S_0, except that the random variable $X(\omega)$ representing smoking level has a different distribution in the model for S_1. Equation 4 also holds for S_1. Because g, μ_Z and $\mu_{T'}$ are the same in the models for S_0 and S_1, the conditional distribution of cancer occurrence in S_1 given smoking level equal x is the same as the conditional distribution of cancer occurrence in S_0 given smoking level equal x. Clearly, S_1 is predictable by G.

It turns out that the admissibility of S_1 is not by coincidence. The following theorem give sufficient conditions for new populations generated from a known predictable population to be predictable:

Theorem 1 Conditionalization Theorem
Let S be any population in the domain $\mathbf{S}$ of a causal model M.

(i) Any subpopulation of S obtained by conditionalization on its predictor observables $\mathbf{X}$ is predictable by M if and only if S is predictable by M.

(ii) Let S be any population in the domain $\mathbf{S}$ of a causal model M. Suppose there is another observable Z in S such that conditional on the predictors $\mathbf{X}$, Z and the predictees $\mathbf{Y}$ are independent. Then any subpopulation of S obtained by conditionalization on Z is predictable by M if and only if S is predictable by M.

Proof To show the part (ii), let B be any subset of the range of Z, A any subset of the range of $\mathbf{Y}$. Given that conditional on $\mathbf{X}$, Z and $\mathbf{Y}$ are independent, we have, with probability 1,

$$P(\mathbf{Y} \in A | \mathbf{X}, Z \in B) = P(\mathbf{Y} \in A | \mathbf{X})$$

Thus, if S is predictable by M, the subpopulation S_Z obtained by conditionalization on Z being in B is also predictable by M.

Part (i) follows from part (ii) directly when we realize that we can set $Z = I_A[\mathbf{X}]$, where A is any set of values of $\mathbf{X}$. Z and $\mathbf{Y}$ are independent given $\mathbf{X}$. Thus, if S is predictable by M, according to part (ii) of the theorem, the subpopulation S_X obtained by conditionalization on $\mathbf{X}$ being in A is also predictable by M. ♠

The conditionalizations described in Theorem 1 are called *conservative conditionalizations*, because they preserve the admissibility of the original population.

Suppose after learning M_C, the government of S_0 decides to discourage people from smoking by banning smoking in public places. This produces another new population S_2. Assuming that the ban is effective in reducing smoking level, but smoking level remains independent of the complementary risk factors in S_2, and the distribution of the complementary risk factors in S_2 is the same as in S_0. By the same argument as for S_1, it can be shown that S_2 is also predictable by G.

We have seen two new populations where M_C can make correct predictions. It is also easy to imagine populations where M's predictions are incorrect. For example, suppose that because of a nuclear plant accident, people in population S_0 are exposed to higher dose of harmful radiation, which is a cancer risk factor. The exposure to harmful radiation effectively produced a new population S_3. It is reasonable to assume that the exposure level and smoking level are still independent in S_3. However the application of M_C to S_3 may generate an incorrect prediction of distribution of cancer occurrence given a certain smoking level: In S_3, because of the exposure to harmful radiation, for any given level of smoking, people would be more likely to have cancer than in S_0.

Alternatively, suppose a drug is discovered to curb the addiction to smoking, but this drug is only effective for people with a certain gene mutation, and that gene mutation, unfortunately, is a cancer risk factor. The introduction of this drug effectively produces yet another new population S_4. Again the application of M_C to S_4 will produce an incorrect prediction: In S_4, because the smokers are less likely to have that gene mutation, conditional on a higher level of smoking, people would be less likely to have cancer than in S_0.

A closer examination of S_3 and S_4 reveals why they are not predictable by M_C. In S_3, the increase of exposure to harmful radiation means that the distribution of the complementary risk factors is changed. That is, in a probabilistic model for S_3 based on G', the Equation 4 still holds, but the random vector $\mathbf{Z}$ has a different distribution than in the model for S_0 based on G'. [7] Thus the conditional distribution of cancer occurrence in S_3 given smoking level, represented by $\mu_{Y|X}$ in Equation 4, is no longer the same as in S_0.

In S_4, carrying a certain gene mutation is another member of the complementary risk factors. The introduction of new drug, while not changing the marginal distribution of the complementary risk factors, makes the complementary risk factors and smoking level dependent in S_4. This time, Equation 4 no longer holds in a probabilistic model for S_4 based on G'. The conditional distribution of Y given X in the model for S_4 thus is different from $\mu_{Y|X}$ of Equation 4 in the probabilistic model for S_0 based on G'.

Now it is clear that even if a population S_0 is predictable by the causal model M, other populations that differ only slightly from S_0 may well be not predictable. This illustrates the need of providing a decision rule to the causal model M to check if a population S is predictable before making prediction of S with M.

4 Admissibility rules and universal causal models

The admissibility rule is needed in order to make sure that when the invariant structure of a causal model M is imposed on a population, it will produce the correct prediction. While most admissibility rules are tied to specific causal models and difficult to describe, there are a few types of admissibility rules that be presented concisely.

Definition 6 (Universal Admissibility Rule) *D is a universal admissibility rule for a causal model M if D accepts all populations in the domain of M.*

Definition 7 (Empty Admissibility Rule) *D is an empty admissibility rule for a causal model M if D accepts no population in the domain of M.*

Definition 8 (Ad hoc Admissibility Rule) *D is an ad hoc admissibility rule for a causal model M if D accepts only a set of populations, as well as their*

---
[7] We can construct probabilistic models for S_3 and S_4 based on G' because both are predictable by G'.

Causal models: A Philosophical Definition ...

subpopulations obtained through conservative conditionalization, that are already *tested to be predictable by M.*

A causal model is called a universal causal model if it contains a universal admissibility rule, or an empty casual model if it contains an empty admissibility rule, or an ad hoc casual model if it contains an ad hoc admissibility rule. Universal causal models can be further divided into the following two categories:

Definition 9 (Deterministic Causal Model) *A causal model $M = (G, D)$, where G is an invariant structure and D an admissibility rule, is deterministic if:*

(i) Let $\mathcal{X}$ be the range of the exogenous variables $\mathbf{X}$, $[(\mathbf{T}, \mathbf{f})]$ the equivalence class of pairs of corresponding uncertain factor and predictive function in G. For all $\mathbf{x} \in \mathcal{X}$, there is a $\mathbf{y}$ in the range $\mathcal{Y}$ of the endogenous variables $\mathbf{Y}$ such that $P(\mathbf{f}(\mathbf{x}, T(\omega)) = \mathbf{y}) = 1$ in G.

(ii) D is a universal admissibility rule.

Definition 10 (Genuinely Indeterministic Causal Model) *A causal model $M = (G, D)$, where G is an invariant structure and D an admissibility rule, is genuinely indeterministic if:*

(i) Let $\mathcal{X}$ be the range of the exogenous variables $\mathbf{X}$, $\mathcal{Y}$ the range of the endogenous variables $\mathbf{Y}$, $[(\mathbf{T}, \mathbf{f})]$ the equivalence class of pairs of corresponding uncertain factor and predictive function in G. There is a subset $C \subset \mathcal{X}$ such that:

- *There is at least one population S in the domain of M such that in a probabilistic model for S based on G, $P(\mathbf{X} \in C) > 0$.*
- *For each $\mathbf{x} \in C$, there are two disjoint sets $A, B \subset \mathcal{Y}$ such that $P(\mathbf{f}(\mathbf{x}, T(\omega)) \in A) > 0$ and $P(\mathbf{f}(\mathbf{x}, T(\omega)) \in B) > 0$ in G.*

(ii) D is a universal admissibility rule.

The deterministic and genuinely indeterministic causal models are also *complete causal models*.

Definition 11 (Complete Causal Model) *A causal model M is complete if for any population S in the domain of M, the predictees $\mathbf{Y}$ are independent of any other observables preceding $\mathbf{Y}$ informatically, conditional on the predictors $\mathbf{X}$.*

Theorem 2 Complete Causal Model Theorem
Each universal causal model is a complete causal model.

Proof
Obviously a deterministic model M must be complete, because in any population S in the domain of M, the value of the predictees $\mathbf{Y}$ is uniquely determined by the value of the predictors $\mathbf{X}$.

To show that the genuinely indeterministic models are also complete, let S be a population in the domain of a genuinely indeterministic causal model M such that there is an offending observable Z in S preceding the predictees $\mathbf{Y}$, and $\mathbf{Y}$ and Z are dependent given the predictors $\mathbf{X}$ in S. In a probabilistic model for S based on the invariant structure of M, $\mathbf{Y}(\omega)$ and $Z(\omega)$ are dependent conditional on $\mathbf{X}(\omega)$. There must be sets $A \in \mathscr{Y}$, $B \subset \mathscr{Z}$, and $C \subset \mathscr{X}$, where $\mathscr{Y}$, $\mathscr{Z}$ and $\mathscr{X}$ are the σ-fields associated with the ranges of $\mathbf{Y}$, Z, and $\mathbf{X}$ respectively, that satisfy the following conditions:

(i) $P(\mathbf{X} \in C) > 0$,
(ii) $P(\mathbf{Y} \in A, Z \in B | \mathbf{X} = \mathbf{x}) \neq P(\mathbf{Y} \in A | \mathbf{X} = \mathbf{x}) P(Z \in B | \mathbf{X} = \mathbf{x})$ for each $\mathbf{x} \in C$
(iii) $P(\mathbf{Y} \in A | \mathbf{X} = \mathbf{x}) > 0$ and $P(Z \in B | \mathbf{X} = \mathbf{x}) > 0$ for each $\mathbf{x} \in C$

Note that condition (iii) follows from condition (ii) and that with probability 1, $P(\mathbf{Y} \in A | \mathbf{X}) P(Z \in B | \mathbf{X}) = 0$ implies $P(\mathbf{Y} \in A, Z \in B | \mathbf{X}) = 0$.

It then follows that, for each $\mathbf{x} \in C$,

$$P(\mathbf{Y} \in A | \mathbf{X} = \mathbf{x}, Z \in B) = \frac{P(\mathbf{Y} \in A, Z \in B | \mathbf{X} = \mathbf{x})}{P(Z \in B | \mathbf{X} = \mathbf{x})}$$
$$\neq P(\mathbf{Y} \in A | \mathbf{X} = \mathbf{x}) \qquad (5)$$

Let S_Z be the subpopulation of S such that for each element in S_Z, the value of the observable Z falls into B. Because Z precedes the predictees $\mathbf{Y}$ informatically, and the domain of a causal model is closed under conditionalization on observables preceding the predictees, S_Z must belong to the domain of M. However, according to Equation 5, the conditional distribution of the predictees $\mathbf{Y}$ given the predictors $\mathbf{X}$ in S_Z will not be the same as it in S. Given that S is predictable, S_Z cannot be predictable, which contradicts the assumption that M is a genuinely indeterministic causal model with a universal admissibility rule. ♠

For those models that are neither deterministic nor genuinely indeterministic, we would like to choose an admissibility rule that maximizes the number of predictable populations accepted by the admissibility rule without accepting populations that are not predictable. Theorem 1 shows that if a population is predictable, then its subpopulations obtained by conservative conditionalization is also predictable. But they provide no guidance when a new population is not a subpopulation of a known predictable population.

It turns out that with an assumption called the *causality postulate*, using a procedure called "partialization," it is often possible to derive an admissibility rule to accept new populations that are not subpopulation of known predictable populations:

Causality Postulate: *Let* **S** *be the domain of a causal model M with endogenous variables* $\mathbf{Y}$ *and exogenous variables* $\mathbf{X}$. *There exists a complete causal model M', i.e., a causal model that is either deterministic or genuinely indeterministic, with endogenous variable* $\mathbf{Y}$ *and exogenous variables* $\mathbf{X} \cup \mathbf{Z}$ *and associated with domain* **S'** $\supseteq$ **S**.

Definition 12 (Partialization) *Consider an invariant structure*

$$G' = \Big((\{G'_1, \cdots, G'_m\}, <), [(T', g(\mathbf{x}, \mathbf{z}, T'))]\Big) \qquad (6)$$

with exogenous variables $\mathbf{X} \cup \mathbf{Z}$. We can derive from G' a new invariant structure

$$G = \Big((\{G_1, \cdots, G_m\}, <), [(\mathbf{Z} \cup \{T'\}, g)] \Big) \tag{7}$$

with exogenous variables $\mathbf{X}$ through the following steps:

- Pick up any element, say, $(T'(\omega), g(x, \mathbf{z}, T'(\omega))$ from $[(T', g(x, \mathbf{z}, T'))]$ of G',
- Assign a random mapping $\mathbf{Z}(\omega)$ to a subset $\mathbf{Z}$ of the exogenous variables in G' such that $\mathbf{Z}(\omega)$ and $T'(\omega)$ are independent
- Replace the $[(T', g(x, \mathbf{z}, T'))]$ of G' with a new equivalence class $[(\mathbf{Z} \cup \{T'\}, g)]$
- Remove the exogenous variables $\mathbf{Z}$ from G'

The above procedure is called partialization, and G is called a partialization of G' obtained by assigning a measurable mapping $\mathbf{Z}(\omega)$ to $\mathbf{Z}$.

Partialization can be applied to any invariant structure with at least two exogenous variables. If we allow an invariant structure to have no exogenous variables, the partialization procedure can be applied to invariant structures with only one exogenous variable.

The following theorem shows the close relation between an invariant structure and its partializations:

Theorem 3 Partialization Theorem

Let M' be a causal model with invariant structure G' as specified by Equation 6. Let S be predictable by G', and observables $\mathbf{X}$ and $\mathbf{Z}$ independent in S. Then:

(i) *If S is also predictable by an invariant structure G, where G has the same endogenous variables and probability space as G', and the exogenous variables $\mathbf{X}$ in G is a subset of the exogenous variables in G', then G must be a partialization of G' obtained by assigning $\mathbf{Z}(\omega)$ to $\mathbf{Z}$ of G', where the distribution of $\mathbf{Z}(\omega)$ in G is the same as the observables $\mathbf{Z}$ in S.*

(ii) *If G is an invariant structure as specified by Equation 7, i.e., a partialization of G' obtained by assigning $\mathbf{Z}(\omega)$ to $\mathbf{Z}$ of G', where the distribution of $\mathbf{Z}(\omega)$ in G is the same as the observables $\mathbf{Z}$ in S, then S is predictable by G.*

Proof

The part (i) of Theorem 3 was proved during our discussion in the previous section about the population S_0 where smoking is independent of other cancer risk factors. To show part (ii), it is easy to see that in a probabilistic model for S based on G',

$$P(\mathbf{Y} \in A, \mathbf{X} \in B) = \int_B \int I_A[g(\mathbf{X}, \mathbf{Z}, T')] d\mu_{Z,T'|X} d\mu_X$$

Given that $\mathbf{X}$ is independent of $\mathbf{Z}$ and T',

$$P(\mathbf{Y} \in A, \mathbf{X} \in B) = \int_B \int I_A[g(\mathbf{X}, \mathbf{Z}, T')] d\mu_{Z,T'} d\mu_X$$

which is exactly what is predicted by G. Thus, S is predictable by M. ♠

The following corollary shows how to predict if a new population is predictable:

Corollary 1 Admissibility Rule Derivation
Consider a causal model M with domain $\mathbf{S}$, invariant structure G, exogenous variables $\mathbf{X}$, and endogenous variables $\mathbf{Y}$. Let $S \in \mathbf{S}$ be predictable by M, $S' \in \mathbf{S}$ another population that is NOT a subpopulation of S. Suppose there is a causal model M' with domain $\mathbf{S}$, invariant structure G', exogenous variables $\mathbf{X} \cup \mathbf{Z}$, and endogenous variables $\mathbf{Y}$, such that both S and S' are predictable by G'. Suppose that observables $\mathbf{X}$ and $\mathbf{Z}$ are independent in S, then S' is predictable by M if observables $\mathbf{X}$ and $\mathbf{Z}$ are also independent in S', and the distribution of observables $\mathbf{Z}$ in S' is the same as in S.

Note that the causal M' mentioned in Corollary 1 does not have to be a complete model, though it is guaranteed by the causality postulate that there is at least a complete causal model that could be used as M'. The application of Corollary 1 does not require full knowledge about either the causal model M' or the distribution of $\mathbf{Z}$ in S and S'. All needed is that the new population S' is similar to S in the sense that the observables $\mathbf{Z}$ have the same distribution and are independent of $\mathbf{X}$. This requirement is not so difficult to fulfill, especially when S and S' are closely related. For example, if S' is obtained by taking some action on S, as long as we believe that the action does not change either the distribution of the observables $\mathbf{Z}$ nor the independence between $\mathbf{Z}$ and $\mathbf{X}$, S' should be predictable if S is.

A toy example of the application of Corollary 1 is the argument for showing that S_2, the population where smoking is banned in public places, is predictable by the causal model M_C. The discussions of populations S_3 and S_4 illustrates that new populations are unlikely to be predictable when the preconditions of the above guidance are not satisfied.

5 Causes and intervention

Based on the definition of causal models, other related concepts can be easily defined.

5.1 Causes and direct causes

In line with the previous definitions of causal relations in the literature of causal inference, in this paper, the causal relation between two observables X and Y is defined relative to a causal model M that contains both X and Y.

Definition 13 (Cause and Direct Cause) *Let $M = (G, D)$ be a correct nonempty composite causal model associated with domain $\mathbf{S}$. Let*

$$G = \Big((\{G_1, \cdots, G_m\}, <), [(\mathbf{T}, \mathbf{f})] \Big)$$

It is easy to see that the casual model $M_i = (G_i, D)$ is a correct non-empty atomic causal model associated with domain $\mathbf{S}$. Let X_{ij} be one of the exogenous variables $\mathbf{X}_i$ in G_i, and Y_i the endogenous variable in G_i. X_{ij} is called a direct cause of Y_i in M if there is no atomic invariant structure G'_i with exogenous variables $\mathbf{X}_i \setminus \{X_{ij}\}$ and endogenous variable Y_i such that the atomic causal model $M'_i = (G'_i, D)$ associated with domain $\mathbf{S}$ is also a correct causal model.

Moreover, if there is a sequence of variables $\{V_1, \cdots, V_j\}$ in G such that, for $i = 1, \cdots, j-1$, V_i is a direct cause of V_{i+1} in M, then V_1 is said to be a cause of V_j in M.

If X is a cause of Y in a causal model M, we say X and Y are causally related in M. Not all causal relations are interesting. In particular, a causal relation implied by an ad hoc causal model M is likely to be trivial. An ad hoc causal model M can be derived in the following way:

- Pick up a population S
- Estimate the conditional distribution of the predictees $\mathbf{Y}$ given the predictors $\mathbf{X}$ in S
- Represent the estimated conditional distribution by a pair of uncertain factor $\mathbf{T}$ and predictive function $\mathbf{f}$.
- Define the ad hoc admissibility rule D which accepts only S and its subpopulations obtained through conservative conditionalization.

The causal relation between $\mathbf{X}$ and $\mathbf{Y}$ implied by the above ad hoc causal model M is of little significance, because M lacks the power of prediction or explanation.

5.2 Intervention

We can also define intervention or manipulation based on the new definition of causal models, mostly following the definition of intervention or manipulation in the literature, e.g., in [5] or [1]. For simplicity, I shall only show how to define an intervention on an endogenous variable.

Definition 14 (Intervention) *Consider a correct casual model $M = (G, D)$ with equivalence class of uncertain factors and predictive functions $[(\mathbf{T}, \mathbf{f})]$. Let $\mathbf{X}_i$ and Y_i be the exogenous and endogenous variables of an atomic invariant structure G_i in G, and $[(T_i, f_i)]$ the equivalence class of G_i. Create a new atomic invariant structure G'_i by adding a binary exogenous variable W_i to G_i. Let the equivalence class of G'_i be $[(T_i, f'_i)]$ such that the distribution of $f'_i(T_i, \mathbf{x}_i, w_i) = f_i(T_i, \mathbf{x}_i)$ when $w_i = 0$, and $f'_i(T_i, \mathbf{x}_i, w_i) = g_i(T_i) \neq f_i(T_i, \mathbf{x}_i)$ for some function g_i of T_i when $w_i = 1$. Create a new invariant structure G' from G by replacing G_i of G with G'_i, and adjust the equivalence class of G' accordingly to $[(\mathbf{T}, \mathbf{f}')]$.*

For each population in the domain of M, let $W_i = 1$ if an action α_i is already taken, and $W_i = 0$ otherwise. Let S be a population predictable by M, and $W_i = 0$ in S. If by taking the action α_i on S we produce a new population S' such that S' is predictable by G', not predictable by G, the action α_i is called an intervention *on the observable Y_i of the population S, and G' the extended invariant structure for α_i.*

It is often difficult to *predict* if an action would be indeed an intervention. Corollary 1 may help here. Assume there is another causal model with invariant structure G'' such that the extended invariant structure G' for α_i is a partialization of G'' by assigning a random mapping to some exogenous variables $\mathbf{Z}$ in G'', and both S and S' are predictable by G''. If it is reasonable to believe that $\mathbf{X} \cup \{W_j\}$ are independent of $\mathbf{Z}$ in S (recall that W_j is the binary variable representing if action α_j is taken), and that α_j does not change the distribution of $\mathbf{Z}$, nor makes it dependent with $\mathbf{X} \cup \{W_j\}$ in S', then α_i is an intervention on S.

Interventions could be very useful in causal discovery. In particular, in controlled experiments, if we ignore the placebo effect, the randomization of treatment can be thought roughly as an intervention. However, a causal model learned from an intervened population in general should not accept populations other than those produced by similar intervention. For example, without further assumptions about the joint independence between predictors and unmeasured observables potentially affecting the predictees, the casual models learned from experiments with randomized treatment may not be able to make correct predictions for populations produced by different ways of applying the treatment.

For illustration, suppose we want to investigate the effect of two competing drugs d_1 and d_2 on a certain type of disease. A large pool of patient representative of the population S_0 are given randomly either d_1 or d_2. Let $M = (G, D)$ be the causal model learned from the experiment, S_1 the population produced by prescribing d_1 to all patients in S_0, and S_2 the population where all patients taking d_2. Clearly, both S_1 and S_2 are predictable by M.

Now suppose there is a binary observable Z unmeasured in the experiment, such that taking the drug d_1 is beneficial when $Z = z_1$, harmful when $Z = z_2$, while taking d_2 is harmful when $Z = z_1$, beneficial when $Z = z_2$. If $P(Z = z1) > 0.5$ in S_0, the model M learned from the experiment would imply that d_1 is generally beneficial, and d_2 generally harmful. Suppose that a new population S_3 is produced by prescribing the drugs to patients in S_0 in a more flexible way, and that the patients whose Z value is z_2 are more likely to take d_1, while the patients whose Z value is z_1 more likely to take d_2. Obviously S_3 is not predictable by M, because people taking d_1 in S_3 generally will be harmed, while people taking d_2 generally will benefit from the drug.

6 Discussion

At the end of this paper, I would like to comment on the implications of the new definitions of causal models and causation.

First of all, it is clear that based on the new definition of causal models, what the causal inference algorithms infer from the data are invariant structures. To get full causal models, we may still need expert's knowledge to find an appropriate admissibility rule. On the other hand, it can be shown that the basic assumptions behind some causal inference algorithms make it possible to derive admissibility rule with the help of Corollary 1, hence less reliant on expert's knowledge.

For example, the assumptions behind the PC algorithm [5] are basically:

(i) There is a complete (deterministic) causal model M'' such that the population S where the observation data is drawn is predictable by M''.
(ii) All the predictors $\mathbf{X}$ and $\mathbf{Z}$ of M'' are jointly independent in S.

(iii) Only the values of **X** and the predictees **Y** are measured for the population S.

Suppose G is an invariant structure learned from S using the PC algorithm. How can we find the decision rule D that accepts not only S, but also some other populations so that the causal model $M = (G, D)$ is not an ad hoc model? Because of the independence among observables in **X** and the independence between **X** and the unmeasured observables **Z**, by Corollary 1, an action α_i on some set of predictor $\mathbf{X}_i \subseteq \mathbf{X}$ of S would produce a new population S' predictable by M as long as the action does not change the distribution of **Z** nor makes **X** and **Z** dependent. In other words, if we are confident that the action α_i's impact on S is relatively "local," we could add the population S' produced by α_i to the list of populations predictable by M.

Secondly, the definitions of causal models and causation presented in this paper are intended to be neutral to different scientific world of views. It allows causal models to used to predict the statistical populations not only from the world of classic physics and the world of special or general theory of relativity, but also from the world of quantum mechanics. Indeed, the genuinely indeterministic causal models are designed to be applied to quantum mechanic systems. Of course, we should be careful when specifying the invariant structure of a causal model for a quantum mechanic system: Two observables of the same quantum state cannot be defined as random variables on the same probability space unless they commute. It is recommended that, when develop a causal model for quantum systems, the principles of consistent histories should be followed ([2]).

Thirdly, although in this paper causal models are considered mainly as a prediction tool, this does not prevent them from performing the task demanded by Hume, i.e., "to teach us, how to control and regulate future events by their causes." Causal models, by the new definition, do not tell directly how to manipulate one observable to achieve desired affect on the other observable. But they can make predictions on the new populations produced by various actions, provided that these populations are predictable. Based on their predictions, we can choose the actions that bring out the most desirable populations.

Finally, it should be pointed out that the fact that causal models are mainly a prediction tool leads to its dependence on the informatical precedence relation between the predictors and the predictees, as shown in the Formal Applicability Condition for a class of populations to be the domain of a causal model. This dependence limits the range of the application of causal models: Causal models are not suitable to describe the relation between observables where the informatical precedence relation is not well defined. This limitation may partly explain why, according to Russell, "in advanced sciences such as gravitational astronomy, the word 'cause' never occurs..." [4].

Bibliography

[1] Dawid, A. P. Influence diagrams for causal modelling and inference. *International Statistical Review*, 70:161–189, 2002.

[2] Griffiths, R. B. *Consistent Quantum Theory*. Cambridge, UK: Cambridge University Press, 2003.

[3] Pearl, J. *Causality*. Cambridge, UK: Cambridge University Press, 2000.

[4] Russell, B. On the notion of cause. In *Proceedings of the Aristotelean Society*, 1–26. 1913. Reprinted from Russell, B. 1917, Mysticism and Logic, London: Unwin, pp.171–196.

[5] Spirtes, P., Glymour, C., Scheines, R. *Causation, Prediction, and Search*. New York, NY: MIT Press, second edn., 2000.

[6] Woodward, J. *Making Things Happen: a Theory of Causal Explanation*. Oxford, UK: Oxford University Press, 2003.

Logics in a New Frame of Cognitive Science On Cognitive Logic, its Objects, Methods and Systems [1]

Shushan Cai

Tsinghua University

sscai@tsinghua.edu.cn

ABSTRACT. The central place of rationalism in philosophy in the last century has been challenged by cognitive science since the 1970s. This paper proposes a new framework for logic in cognitive science, consisting of six areas: philosophical logic, logic of language, mental logic or logic of mind, logic in the cultures and evolutions, logic in artificial intelligence and logic in the brain-nervous systems. Cognitive logic is founded in natural languages, it attends to experience, rather than to formalization and deductivism, and it reflects the regularities of cognition. Cognitive logic has given up the idea (held by logicians since the age of ancient Greece) that logic is the laws of thought. Instead, it emphasizes the role of humans, the users of languages, in its theory of languages, as well as in its methods of researchs. Cognitive logic has combined language, logic and cognition as a whole system with the human and his brain and mind as its core. Cognitive logic concerns individual physiology, mental states, language and culture just as cognitive science does. So, cognitive logic is a new research program or theory that will recombine all the disciplines of modern logic into six groups and show more clearly what cognitive science will tell us in the new century.

1 Introduction

Cognitive science was constituted by six related disciplines: philosophy, psychology, linguistics, computer science, anthropology, and neuroscience. We have six core branches of cognitive science that is: (1) philosophy of mind or philosophy of cognitive science; (2) cognitive psychology; (3) cognitive linguistics or language and cognition; (4) artificial intelligence; (5) cognitive anthropology or culture evolution and cognition; (6) cognitive neuroscience. We have eleven more sub-fields of cognitive science if we let the six core branches interact with each other.

These six supporting disciplines, six branches and eleven sub-fields compose the system of cognitive science shown by Fig.1:[12].

Then we have a new system of logics in the framework of cognitive science, which together we call cognitive logic.

I use "cognitive logic" to refer to a new system of logics in the framework of cognitive science as well as in the background of modern formal logics, namely,

---

[1] Professor Cai is with Center for Psychology and Cognitive Science, Tsinghua University, Beijing, 100084 China.

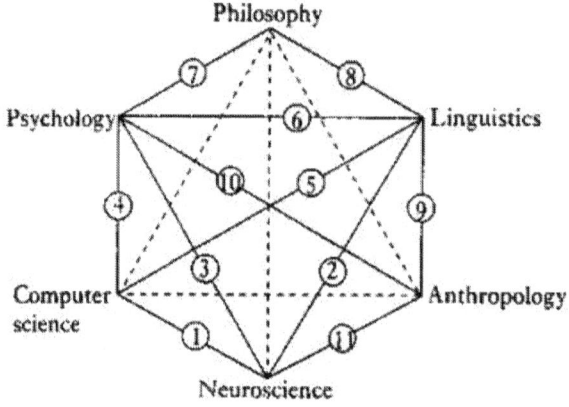

Figure 1: the cognitive hexagon

a system as a result from the interaction of modern logics with cognitive science. Let us just put Fig.1 into the background of modern logics (the square in Fig.2). Let us imagine a mapping from the framework of cognitive science onto the background of modern logics; then we can get a new framework for logics like Fig.3

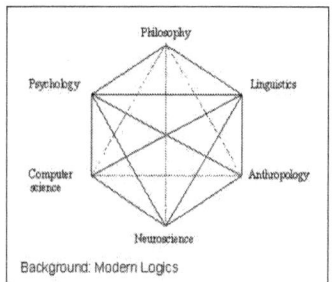
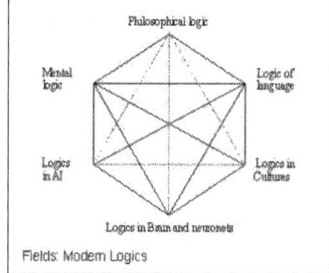

Figure 2: the frame of cognitive science mapping onto the background of modern logics

Figure 3: cognitive logic i.e. logics in a new frame of cognitive science

2 The Linguistic Foundation of Cognitive Logic

Logic is a system about the mode of knowing and the constitution of reasoning, based on a special language and special methods. So, only different formal languages or different methods or models of reasoning differentiate logical systems from one another.

We know that the foundation of every logic system is a kind of language. For example, the linguistic foundation of first order logic in first order language, i.e. an alphabet $x_1, x_2, x_3, \cdots$; $f_1, f_2, f_3, \cdots$; $P1, P2, P3, \cdots$; $\forall$; $\neg, \rightarrow$ and some formal rules, call them L_1. Cognitive logic is a theory about cognitive models and reasoning systems that is based on a special language. But what is that?

In my opinion, the foundation of cognitive logic is the so called 'return to natural language', by which I refer to the kind of language analysis that comes from natural language, developed as a part of those belonged to UG and analyzed by formal syntax (Noam Chomsky, 1957) and semantics [7], and then returns to natural language. Such languages are matchlessly abundant, including all the knowledge about mankind's languages.

This kind of return, from formal language to natural language, started from L. Wittgenstein, and was matured by the development of philosophers of the school of ordinary language, especially by figures such as N. Chomsky, R. Montague, J. L. Austin and J. R. Searle, who investigated natural languages with formalized methods. Thereby, the three main branches of modern linguistics come into being: syntax, semantics and pragmatics.

We have cognitive linguistics within in the framework of cognitive science, which includes these fruits in modern linguistics and philosophy of language, and which constitute the foundation of cognitive logic.

Cognitive linguistics, I think, experienced two main stages. The first one, we call rationalist cognitive linguistics, presented by N. Chomsky. The second, we call empiricist cognitive linguistics presented by G. Lakoff.

Chomsky described cognitive linguistics like this: 'All organisms have special subsystems that lead them to deal with their environment in a specific way. Some of these subsystems are called "mental" or "cognitive", ... the development of cognitive system, like others, is influenced by the environment, but the general course is genetically determined. ... The evidence is strong that among the human cognitive systems is a "faculty of language"(FL), to borrow a traditional term: some subsystem of (mostly) the brain. The evidence is also overwhelming that apart from severe pathology, FL is close to uniform for humans: it is a genuine species property. The "initial state" of FL is determined by the common human genetic endowment.'[17, F15]. Chomsky thought that FL appears to be biologically isolated in critical respects, hence a species property in a stronger sense than just being a common human possession. According to his theory, a person is equipped to engage in the 'creative use of language' if he has 'internalized languages' (I-language). Chomsky viewed the study of language as part of biology, taking its place just as the study of the visual system. Such a system can be studied at various levels. In the case of cognitive systems, these are sometimes called the "psychological" and "physiological" levels. He wrote: 'what is learned at the "psychological level" commonly provides guidelines for the inquiry into neural mechanisms; and reciprocally, insights into neural mechanisms can inform the psychological inquiries that seek to reveal the properties of the organism in different terms. ... We have no idea what the outcome maybe of today's efforts to unify the psychological and physiological levels of scientific inquiry into cognitive capacities of organisms, human language included.'[17, F15] Chomsky summed up and looked forward: 'The past half century of the study of language has been rich and rewarding, and the prospects for moving forward seem exciting, not only within linguistics narrowly conceived but also in new directions, even including the long-standing hopes for unification of linguistics and the brain sciences, a tantalizing prospect, perhaps now at the horizon.'[17, F15]

By the end of 20 century, two excellent theorists, George Lakoff and Mark Johnson brought forward a new blueprint for cognitive science. In their epoch-making book, *Philosophy in the Flesh*[6], they gave three important judgments about mind: The mind is inherently embodied; thought is mostly unconscious; abstract concepts are largely metaphorical. Their critical understanding of reason is: Reason is not disembodied, but arises from the nature of our brains, bodies, and bodily experience; reason is evolutionary; Reason is not 'universal' in the transcendent sense; reason is not completely conscious, but mostly unconscious; reason is not purely literal, but largely metaphorical and imaginative; reason is not dispassionate, but emotionally engaged. Therefore, they concluded: "...there is no Cartesian dualistic person with a mind separate from and independent of the body; there exists no Kantian radically autonomous person with absolute freedom and a transcendent reason that correctly dictates what is and isn't moral; the utilitarian person for whom rationality is economic rationality – the maximization of utility – does not exist; the phenomenological person who through phenomenological introspection alone can discover everything there is to know about the mind and the nature of experience is a fiction; there is no poststructuralist person; there exists no Fregean person; there is no such thing as a computational person and there is no Chomskyan person.[6, pp.5-7]."

N. Chomsky and his challenger, G. Lakoff, the two most important philosophers of our century, shoot at the same target – mind and cognition – although their weapons are different: rationalism for Chomsky, experientialism for Lakoff.

The foundation and the goal of contemporary linguistics have been changed, I think. In the last century, we came from Bloomfield's experiential linguistics, went to Chomsky's rationalist linguistics, then, we came back to Lakoff's experiential linguistics. Remember that this is not simple return, we have received many new fruits from linguistics: cognitive linguistics, neurolinguistics, computational linguistics, psycholinguistics, anthropological linguistics, etc.

Now, we can give a definition to cognitive logic as follows:

Definition 1 *cognitive logic is a logical system, its foundation is cognitive linguistics and its subjects are cognitive models and rules.*

This definition shows the difference between cognitive logic and other logical systems. First, the difference in linguistic foundations: Cognitive logic with its foundation as cognitive linguistics, different from traditional logic with its foundation as natural language, and from mathematical logic and other modern logics with their foundations in formal languages. Second, the difference in research objects: cognitive logic with cognitive models and rules as its object, different from any other logics, especially epistemic logic with its object as epistemology, cognitive logic will contain all other logics related to cognitions.

3 The Methods of Cognitive Logic

Cognitive logic is supported by modern linguistics and logics, therefore, the methods used in modern linguistics and logics are also the methods in cognitive logic.

Syntax, semantics and pragmatics are three main approaches in modern linguistics. In [9], the three terms in question were defined as follows: pragmatics as the study of 'the relation of signs to interpreters', semantics as the study of 'the relations of signs to the objects to which the signs are applicable', syntax as the study of 'the formal relations of signs to one another'.[9, p.6]. [4, p.9] gave more details about these three terms:

> If we are analyzing a language, then we are concerned, of course, with expressions. But we need not necessarily also deal with speakers and designata. Although these factors are present whenever language is used, we may abstract from one or both of them in what we intend to say about the language in question. Accordingly, we distinguish three fields of investigation of languages. If in an investigation explicit reference is made to the speaker, or, to put it in more general terms, to the user of a language, then we assign it to the field of pragmatics. If we abstract from the user of the language and analyze only the expressions and their designata, we are in the field of semantics. And if, finally, we abstract from the designata also and analyze only the relations between the expressions, we are in (logical) syntax. The whole science of language, consisting of the three parts mentioned, is called semiotic.

We can view the development from syntax, semantics to pragmatics as an expansionary process according to Morris and Carnap. And, we can see this process exactly from Chomsky's syntax, Montague's semantics and Austin's speech act theory. We can introduce some important conclusions in this process:

First, cognitive logic receives more and more rich content from the development of modern linguistics. Chomsky's syntax, formulated in the 1950s, has developed more and more content from Chomsky himself and by his colleagues or his students, including systems known as TG, UG, GB, P&P, MP, etc. Montague's intensional semantics, formed in 1970's, put more content in the meaning of natural language, his famous sentence is "there is no important theoretical difference between natural languages and the artificial languages of logicians."[7, p.222] Austin put forward his famous theory of speech acts, according to which factors such as speaker, hearer, time, place and context were considered in understanding the function of the natural languages. In this process, many logical methods were used in the researches of linguistics, such as formal mathematical methods utilized in Chomsky's syntax and in Montague's semantics, as well as set theoretic methods used in John R. Searle and D. Vanderveken's work [15] about illocutionary logic.

Second, the philosophers turned their attention to mind after the last century's research of languages. In the beginning of the 20th century, the philosophers paid attention to formal languages. They thought that artificial languages are better than natural languages because of their accuracy and rigor. But in 1931, a shock given by Gödel in his famous theorems awakened philosophers from their dreams about ideal languages. The consistency of any (consistent) axiomatizable theory adequate to represent arithmetic cannot be proved by the means available in that theory, and in any such theory there remain unprovable truths of arithmetic. (As it stands, the sentence is false, since if T is a consistent axiomatizable such theory and you add Con_T (an arithmetical sentence saying that T is consistent), you get an axiomatizable extension of T in the same language that proves the consistency of T (though not that of $T + Con_T$)) After Gödel's results, philosophers and linguists paid attention to natural languages again and they thought natural languages would be richer and better than artificial ones.

Chomsky created his theory formally, and he was viewed as one of the leaders of first generation of cognitive science for his rationalist linguistics. Almost every philosopher of language, such as Chomsky, Searle, etc. turned their attention to mind and cognition after recognizing the disparity of formal and natural languages. In fact, we have two 'language turns' in the last two centuries. First, we turned from natural language to formal language in the period from the end of the 19th century to the middle of the 20th century. We call it the "formalization turn", in which G. Frege and B. Russell were the two pioneers, and Gödel, the pinnacle. Second, we turned from formal language back to natural language in the period from the middle of the 20th century to the end of that century. We call it the "language turn" or "naturalization turn", in which N. Chomsky, R. Montague and J. L. Austin were the pioneers, and also leaders in the study of syntax, semantics and pragmatics respectively. Wittgenstein was a great figure during those two language turns. Now, by the end of last century and the beginning of this century, we have another great turn, which we call as "mind turn" or "cognitive turn". That is, we turned from language to mind or cognition searching for the secret of cognition, viz. how does our brain produce mind and how does it work? We call it the final secret of God because there will be no secret left after this work is completed. Of course, I don't think my God will let the descendants of Adam and Eve, sinful mankind, know His last secret. But Man, who was bodacious in eating the fruit from the tree of knowledge of good and bad, will pry into Gods final secret. N. Chomsky and G. Lakoff are the two great figures in this new revolution, I think, but I really don't know who will ultimately eat a new fruit from the tree of knowledge.

Third, the cognition turn reflects an important change in the process of investigation of languages, that is, the turn from rationalism to experientialism. The best realization of rationalism in modern linguistics is Chomsky's transformational grammar, in which methods that originated from Leibniz and were developed by Frege and Russell were used, viz. we could get all the well formed sentences just by using a few initial symbols and a few deductive rules.[10, pp.43–60] Then Montague reached the peak of perfection of formalization using only initial symbols, e and t, to build his theory of formal semantics.[8, pp.247–270] But this process of full formalization met its troubles in the investigation of computer programs because the various contexts of utterance (speaker, hearer, time, place and other context factors) are very difficult to analyze formally. In 1990s, G. Lakoff put forward his second-generation of cognitive science, the embodied view of mind. He said: [6, p.78]

> In short, second-generation cognitive science is in every respect a cognitive science of the embodied mind. Its findings reveal the central role of our embodied understanding in all aspects of meaning and in the structure and content of our thought. Meaning has to do with the ways in which we function meaningfully in the world and make sense of it via bodily and imaginative structures. This stands in contrast with the first-generation view that meaning is only an abstract relation among symbols (in one view) or between symbols and states of affairs in the world (in another view), having nothing to do with how our understanding is tied to the body.
>
> What we are calling "first-generation" versus "second-generation"

cognitive science has nothing to do with the age of any individual or when one happened to enter the field. The distinction could just as well be called "disembodied" versus "embodied" or "assuming tenets of formalist analytic philosophy" versus "not assuming tenets of formalist analytic philosophy". The distinction is one of philosophical and methodological assumptions.

Logic, from Aristotle to Frege, Russell, even Gödel; and from Kant to Husserl, Heidegger, even Descartes; from Chomsky to Montague and even Austin, had been viewed as a universal frame of thought having nothing to do with the person who used the language and logic. But in Lakoff's philosophy in the flesh, conceptual systems are pluralistic, not monolithic; the structure of concepts includes prototypes of various sorts, most concepts are not characterized by necessary and sufficient conditions and each type of prototype uses a distinct form of reasoning; reason is embodied in that our fundamental forms of inference arise from sensorimotor and other body-based forms of inference; reason is imaginative in that bodily inference forms are mapped onto abstract modes of inference by metaphor.[6, p.76]

Now, it is time to consider and characterize again the features of logic in the new frame of cognitive science.

4 The System of Cognitive Logic

4.1 Philosophical logic

Philosophical logic is the most mature and most fruitful of the six branches of cognitive logic, I think. The famous references includes D. Gabbay and F. Guenthner's work during 1980s and 2000s, Handbook of Philosophical Logic, which have 11 volumes in its second edition in 2001. You can also consult Lou Goble 2001, D. Lewis 1998, A. C. Grayling 1997, M. Sainsbury 1991, S. Wolfram 1989, R. Simonds 1988, G. H. von Wright 1983, U. Monnich 1981, J. W. Davis 1969, N. Rescher 1968, P. F. Strawson 1967. So, I omit my descriptions here.

Philosophical logic is a discipline developed from philosophical modern logics. "Philosophical logic comprises the sorts of logic that hold greatest interest for philosophers. Philosophical logic develops formal systems and structures to be applied to the analysis of concepts and arguments that are central to philosophical inquiry."[5, p.1] The traditional concepts of philosophical logic include necessity, knowledge, obligation, existence, time and reasoning itself. The objects of philosophical logic include also the basic concepts of logic itself, such as conditionals, negation, quantifiers, truth, logical consequence and so on. These core concepts are investigated and analyzed by using the methods of classical logics, namely, first order logic and higher order logic, as well as modal logic, deontic logic, epistemic logic, temporal logic, intuitionistic logic, free logic, relevant logic, many-valued logic, probability logic, nonmonotonic logic, etc. Further, philosophical logic investigates natural language, the foundation of logic, the relation between logic and natural language, as well as the relation between formal language and natural language.

4.2 Mental logic or logic of mind

Mental logic is the most interesting and attractive field of cognitive logics, I think. It grew from the crossfertilization of logic and psychology (especially cognitive psychology). L. J. Rips 1986 gave many examples in psychology to support this new discipline.[14, p.279] P. C. Wason, 1966 developed a famous game to show how a person would be influenced by his mind when he was tested in a simple inference. In this best-known test, the subjects were shown cards that had numbers on one side and letters on the other. They were told a rule such as *"if a card has an A on one side, then it has 4 on the other."* The subjects were shown four cards like Fig.4 and they would indicate exactly which card must be turned over to determine whether this rule held or not. As we expect, most people knew that they must turn the A over to check whether it had 4 on the other side. This situation can be interpreted as that most people know how to use modus ponens correctly. But it is surprising that a great many people refuse to check the 7 in spite of the fact that checking it is as important as checking the A. This shows that great many people don't know about appreciate of modus tollens. However, some people were puzzled about this task and they turned over the B or 4 even though these two cards had nothing to do with the determining the truth of the rule.

Figure 4: Cards in Wason's selection task

This experiment is interesting because it suggests that people approach this kind of reasoning task with representations and computations quite different from those used in formal logic. Johnson-Laird and Byrne 1991 argue that deductive reasoning is carried out neither by formal logical rules nor by content-specific rules or schemas, but by mental models. People turned over the A because it is represented, but they refused to turn over the 7 because it is not represented in Wason's task.[16, pp.36–37]

Unfortunately, we have these puzzles not only in Wason's task but also in other ordinary situations. Another example similar to Wason's task is so called bar-and-age task. In this test, we have a rule like this: *'if a person is in the bar, then he or she is over 21.'* And we have four cards like this: IN-BAR, NOT-IN-BAR, 23, 18. People's selections were better in this test than in the abstract letter-and-number test as most people recognized that it is necessary not only to turn over the IN-BAR car to check the age, but also to turn over the 18 card to check the state on the other side. Cheng and Holyoak 1985 argued that people approach these tasks not with mental logic, but with pragmatic reasoning schemas. Another psychological experiment shows that people may not abide by principles of probability theory when they reason inductively. Subjects were told in this experiment that Frank likes to read a lot of serious literature, attend foreign movies, and discuss world politics. The subjects were then asked to estimate the probability that Frank is a college-educated, that Frank is a carpenter, and that Frank is a college-educated carpenter. As we expect, most people in experiment knew that Frank would more probably be college-educated than be a carpenter. But people often violate probability theory by judging it to be more likely that Frank is college-educated carpenter than that he is a carpenter.[16, pp. 36–37] Kahneman, Slovic, and Tversky 1982 gave numerous other instances to show that people's inductive reasoning appears to be based on something other than formal rules of probability theory.

Traditionally, logical reasoning can be divided into three classes: deductions, inductions and analogies. Each of them includes several logical models. For example, implications (including modus ponens and modus tollens), negations, disjunctions, conjunctions are deductive models; full or not full enumerations are inductive models; simulations are analogical models. Now, we can consider the relationship between psychological models and logical models. First, I divide psychological models into four parts: innate models or genetic models that are from inheritance not from learning or from practice; acquired models that are acquired by learning; experiential models that are known by exhausting practice; mental models that are known by inexhaustive practice, that is, just by several tries.

Then, I set up a function mapping psychological models into logical models, that is, logical model = f (psychological model). For example, we have negation and modus ponens as innate models, modus tollens, disjunctions and conjunctions as acquired models, full enumerations as experiential models, not full enumerations and simulations as mental models, etc. I call this function PL-function. My suggestions are as in the following table.

psychological models	logical models	Names	Classes
Innate (Genetic) Models	if A then not $\neg A$ or A is opposite to $\neg A$	negation	
	if $A \to B$ and A then B	modus ponens	deductions
Acquired Models	if $A \to B$ and $\neg B$ then $\neg A$	modus tollens	
	If A or B then $A \vee B$	disjunctions	
	if $A \vee B$ and $\neg A$ then B[2]		
	if A and B then $A \wedge B$	conjunctions	
	if $A \wedge B$ then A and B		
Experiential Models	if $S_1,\ldots,S_n$ is P then S is P	full enumerations	inductions
Mental Models	if $S_1,\ldots,S_i,\ldots$ is P then S is P	not full enumerations	
	if S_1 is P then S_2 is P as $S_1 \sim S_2$	simulations	analogies

Table I: psychological model vs. logical models

From the PL-function, we know that psychological models are original, and logical models are just the expressions of psychological models in thinking. Now that psychologists have accepted logic as a part of psychology,[1, pp.7–22] will not logicians do a similar thing to psychology to make it as a part of logic?

---
[2]We have other evidences to support that this model is also innate

4.3 Logic of Language

We also call it logic and language.[18] It may be the most important branch for modern logic because it deals with more about the foundation of logic and language. As I mentioned above, we have had two important language turns in the last two centuries. First, we turned from natural language to artificial language; second, we turned back from artificial language to natural language.

During these times, we developed many logical systems, such as classical logics (first order logic and higher order logic), extensions of classical logic (modal logic, deontic logic, epistemic logic, temporal logic, etc.), alternatives to classical logic (intuitionistic logic, free logic, relevant logic, many-valued logics, nonmonotonic logic, probability logic, etc.) We call these three groups of logics "Basic Logics" or "Foundational Logics" that is the foundation of modern logics. Using these basic logics, we can investigate philosophy, language and linguistics, and artificial intelligence; then we get other new systems of logic, known as philosophical logic, logic of language, and logics in AI, etc.

The logic of language grows from three roots, linguistics, philosophy and logic. First, its three main branches, syntax, semantics, and pragmatics, come directly from linguistics, especially from semiotics. Second, its important contents come from philosophy, especially from philosophy of language, such as truth, meaning, speech acts, reference and descriptions, names and demonstratives, propositional attitudes, possible worlds and situations, metaphor, private language. Third, its methods of inference and of semantics come from modern logics including classical logic, extensions of classical logic and alternatives to classical logic.

But what is the difference between the logic of language and its three roots? The logic of language is neither linguistics, nor philosophy, but logic itself. And, logic of language is different from other logics such as first-order logic, higher-order logic, modal logic, many-valued logics, and so on. These logics are only tools we use to build the theories or systems of the logic of language. The main difference between the logic of language and its three roots is that a logical system including syntactic methods and semantic models will be given when we analyze an object that we list in linguistics or in philosophy of language. For example, when we discuss speech acts, the most important topic in pragmatics, we are in the fields of linguistics or philosophy. But we are in the field of logic of language exactly if we analyze the concepts and sentences of speech acts by building a logical system and prove its theorems and discuss the relations between this system and other logical systems. Thus, John R. Searle's and Daniel Vanderveken's 1985, *Foundation of Illocutionary Logic*,[15] belongs to logic of language, not to linguistics or philosophy. So does my work 1998, *Speech Acts and Illocutionary Logic*. In this book, I built the formal systems of illocutionary logic including the propositional illocutionary logic system PF, the quantificational illocutionary logic system QF and the modal illocutionary logic system MF. I proved all theorems of these systems and gave the semantics for every system. Finally, I proved the soundness, consistency and completeness of them, and I also proved the difference between illocutionary logic and classical propositional logic, first-order logic, second-order logic, orthodox modal logic, intuitionistic logic, etc. In the last chapter of this book, I specially recommended the application of my systems in computer science, especially in AI.[2]

The last question is whether the logic of language is necessary. It is said that we do not need logic of language because we have logic as well as linguistics and that is enough. I argue against this erroneous point of view by putting the logic of language into the framework of cognitive logic. You can see clearly that the logic of language is as necessary and important as philosophical logic, mental logic, logics in AI, and so on.

4.4 Logics in Culture and evolution

That is anthropological logic. We call it the logics of culture and evolution because we investigate in it the views of culture and evolution of mankind.

In a famous book, *Not by Genes Alone: How Culture Transformed Human Evolution*, Richerson, Peter J. and Robert Boyd told us that, 'Culture changes the nature of human evolution in fundamental ways.' It is true, they argued, because culture is a necessity part of the design problem for human psychology, culture is an ultimate cause of human behavior, and culture makes us odd. They provide important points of view such as 'Cultural differences account for much human variation;' 'Technology is culture, not environment;' 'Variation in the social environment is not enough to explain human variation (Cultures can 'reappear' after long suppression);' 'Little behavioral variation between groups is genetic;' 'Much culture is not evoked;' 'Culture adaptations evolve by the accumulation of small variations;' 'The magnitude of human variation is explained by culture;' 'Culture is (mostly) information in brains; "Cultural evolution is Darwinian;' 'Culture evolution have forces: random forces, Decision-making forces, Natural selection;' 'Population thinking is useful even if cultural variants aren't much like genes;' then they concluded that we need some kind of evolutionary models because good models are like good tools: they are known to do a certain job reasonably well. The following is a so-called Darwinian analysis of cultural evolution:[13]

- draw up a model of the life history of individuals;
- fit an individual-level model of the cultural (and genetic, if relevant) transmission processes to the life history;
- decide which cultural (and genetic) variants to consider;
- fit an individual-level model of the ecological effects to the life history and to the variants;
- scale up by embedding the individual-level processes in a population; and
- extend over time by iterating the one-generation model generation after generation.

In fact, there are different cognitive models between various nations, languages, or cultures. For example, different nations have various psychological reactions to the same color; different nations or those of different cultural backgrounds have various explanations for the stimulus of same symbol. As for the metaphor of concepts, different nations get various psychological hints even from the same concept. Let us take Chinese characters as examples.

Chinese characters are a special pictography, unique in the world. The shape, sound (and tone), and meaning are combined in each Chinese character, i.e. each Chinese character has a shape (a single independent graph), a sound (a single syllable that can be divided into four tones), a meaning (an independent concept). So, using Chinese character, we can do nice handwritings like pictures, fair poems in serious rule and form like music, and elegant couplet in cultural meanings like aphorism. Furthermore, we need not any space between Chinese characters in a sentence, even any punctuations between sentences in a paper, because of the independence of Chinese character as a single graph. The syntax of Chinese, especially ancient Chinese, is functioned only by the order of the characters, not by the auxiliary word as in English and other phonetic systems. So, we can use Chinese to write palindrome in sentences or even in poems, and to make metaphors in vague sentences. This is the special cognitive base of the language and the culture of Chinese.

The methodological conclusion in our analysis is that the differences in cultures come from their foundations, the languages. For example, the Chinese culture or Chinese cultures (including all the cultures in Asia based on Chinese) are experimentalist because Chinese is a kind of language that is suitable to individual experience and to psychological and emotional perception. On the other hand, western cultures are rationalist because English and other phonetic languages are suitable to rational thinking. It is not surprising that the western philosophers have been looking for some kinds of axiom systems in their theories from Euclid, Aristotle, Descartes, Frege, Russell, to Chomsky, even Lakoff. But the Eastern philosophers never do things like these! In logics, western philosophy manifests itself in reason and deduction, but eastern philosophy in experiment and induction.

In the age of the second-generation of cognition, the so called "picture-reading age", maybe Chinese is the more suitable language for computer. It is said that we need only about 6000 icons enough to do all things in computers, as known; these are just the number of basic Chinese characters![3]

4.5 Logics in AI

Using modus ponens, we can get β from the premises $\alpha \to \beta$ and α, but get nothing from $\alpha \to \beta$ and β. So, in many logics including all the classical logics, first-order logic and second-order logic, and other deductive logics, we cannot do anything in the premises $\alpha \to \beta$ and β. For example, if someone did not study hard, he would not get $A+$ in his examination. Now, Bill has not gotten $A+$ in his examination, – can we infer that he did not study hard? No, obviously we cannot because there are some other possible reasons that might interfere with an excellent score in his examination: maybe he was not so bright; maybe he was inferior in his approach; maybe he was in poor health the day of examination and so on.

But in Prolog, a famous language in AI, we can guess α from $\alpha \to \beta$ and β. This is so called 'trace' or 'backing-up-and-trying-again' process that can be used in AI. The Prolog, program in Logic, is based on this logic. We can use this system to make various expert systems. The Prolog program for Bill's study is as in the following,

```
domains
```

---

[3] We have about 3000 Chinese characters called "the first class of Chinese characters", and other about 3000, called "the second class of Chinese characters".

Logics in a New Frame of Cognitive Science

```
  person, state, score = symbol
predicates
  get(person, score)
  study(person, state)
  be(person, state)
clauses
  study(bill, hard).
  be(bill, not-so-bright).
  be(bill, not-so-better-in-approach).
  be(bill, poor-in-health).
  get(Person, A+):-study(Person, hard),
                   be(Person, bright),
                   be(Person, better-in-approach),
                   be(Person, good-in-health).
```

In this program, if we input

```
goal
  get(bill, A+).
```

The answer in the screen is "No" because just one of the four conditions for the goal, get (bill, A+) that means "bill got A+ in his examination", has been satisfied. In this program, if we input

```
goal
  get(bill, A+).

goal
  study(bill, hard).
```

The answer in the screen is "Yes" because it is a fact, i.e. a clause in the Prolog program. And, if we input

```
goal
  study(Who, hard).
```

The answer in the screen is "bill".

The so-called "geek computer" can actually solve many problems for people in medical treatment, games, sale systems, on so on. John R. Searle thought in his famous Chinese Room that there is no strong AI, that is, there is no computer as intelligent as human or even more intelligent than human because such a machine would be just like a man who did not know anything about Chinese but can translate any Chinese sentence into another language based on a dictionary in his room.[11]

Logics in AI are the best developed of our all six branches of cognitive logic. But, according to the philosopher's standard, [3, pp.18–22] it has many difficulties and problems.

4.6 Logic in the Brain and Neuro-nets

> Nothing is currently known about the neurological plausibility of formal logic. Our understanding of the brain is far too limited to say

whether neurons use formal representations or accomplish something like modus ponens. Metaphorically, every synaptic connection between neurons looks like a miniature inference of this kind: if neuron 1 fires, then neuron 2 fires. ... But it is obvious that single neurons do not represent whole propositions, and how groups of neurons perform inferences is unknown.[16, p.38]

[16] discussed the two connectionist models of mind and brain, a simple local network and a distributed processing network. Fig.5 and 6 are these two models:

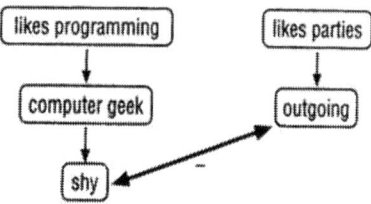

Figure 5: Simple local network with excitatory links (thin lines) and an inhibitory link (thick line with minus sign).From [16, p.110]

Figure 6: A visual analogy for a distributed processing net work.From Wet Mind: The New Cognitive Neuroscience, by Stephen M. Kosslyn and Olivier Koenig. Copyright 1992 by Stephen M. Kosslyn and Olivier Koenig. Reedited from [16, p.111]

Thagard [16] discussed possible applications of these models in problem solving, planning, decision, explanation, learning, language and knowledge. But he thought that these models are much more neurologically plausible than logical ones. He said: 'Connectionist models are much more neurologically plausible than logical ones, but we cannot rule out the possibility that one of the things that human neural networks do is perform inferences in accord with formal logic. Such performance certainly occurs in humans trained in formal logic, even if logic does not come naturally to people in general.'[16, p.39]

We can divide these six disciplines or fields of so-called cognitive logic into three groups. The first group includes philosophical logic, the logic of language and logics in AI. We call this group well developed fields as we can find many works, books and other references in it. The second group is mental logic. I call it a developing field because we can find a few, but not so many works and books in it, but it is a new and attractive field. The third group includes logics in culture and evolution, and logic in brains and neuro-nets. I do not call it a developing field because we don't even know what this logic is and how it works, we will have a lot of work to do on it.

It is said that cognitive science will work and will bring a revolution in our life this century. And so will cognitive logic, I think.

Bibliography

[1] Bonatti, L. Why it took so long to bake the mental-logic cake: Historical analysis of the recipe and its ingredients. In Braine, M. D. S., O'Brien, D. P. (Eds.), *Mental Logic*. Mahwah, New Jersey: Lawrence Erlbaum Associates, 1998.

[2] Cai, S. *Speech Acts and Illocutionary Logic*. Beijing: China Social Sciences Publishing House, 1998.

[3] Cai, S. How to understand ai from a philosopher's view. *Studies in Dialectics of Nature.*, 17(11), 2001.

[4] Carnap, R. *Introduction to Semantics*. Cambridge, MA: Harvard University Press, 1942.

[5] Goble, L. (Ed.). *The Blackwell guide to Philosophical Logic*. Blachwell Publishers, 2001.

[6] Lakoff, G., Johnson, M. *Philosophy in the Flesh: The Embodied Mind and Its Challenge to Western Thought*. 1999. Basic Books, A Member of the Perseus Books Group.

[7] Montague, R. Universal grammar. In Thomason, R. H. (Ed.), *Formal Philosophy: Selected Papers of Richard Montague,* 1974. New Haven and London: Yale University Press, 1970.

[8] Montague, R. The proper treatment of quantification in ordinary (english, reprinted). In Thomason, R. H. (Ed.), *Formal Philosophy: Selected Paper of Richard Montague.* New Haven and London: Yale University Press, 1974.

[9] Morris, C. *Foundations of the Theory of Signs. International Encyclopedia of Unified Science*, vol. 1. Chicago: University of Chicago Press, 1938.

[10] Ouhalla, J. *Introducing Transformational Grammar: From Principles and Parameters to Minimalism*. Edward Arnold Limited, 2nd edn., 1999.

[11] Preston, J., Bishop, M. (Eds.). *Views into the Chinese Room: New Essays on Searle and Artificial Intelligence*. Oxford: Clarendon Press, 2002.

[12] Pylyshyn, Z. Information science: its roots and relations as viewed from the perspective of cognitive science. In *Machlup and Mansfield*, 76. 1984.

[13] Richerson, P. J., Boyd, R. *Not by Genes Alone: How Culture Transformed Human Evolution*. Chicago and London: University of Chicago Press, 2005.

[14] Rips, L. J. Mental muddles. In Brand, M., Harnish, R. M. (Eds.), *The Representation of Knowledge and Belief*. Tucson, Ariz: University of Arizona Press, 1986.

[15] Searle, J. R., Vanderveken, D. *Foundation of Illocutionary Logic*. Cambridge: Cambridge University Press, 1985.

[16] Thagard, P. *Mind: Introduction to Cognitive Science*. The MIT Press, 1996.

[17] Ungerer, F., Schmid, H. J. *A Introduction to Cognitive Linguistics*. Beijing: Foreign Language Teaching and Research Press, 2001. Preface by Chomsky.

[18] van Benthem, J. F. A. K., ter Meulen, A. G. B. (Eds.). *Handbook of Logic and Language*. Amsterdam, New York: Elsevier, 1997. Cambridge, Mass.: MIT Press.

Minimal Rationality vs Optimized Brain-Wiring:[1]

Christopher Cherniak
University of Maryland
cherniak@umd.edu

ABSTRACT. One critique of standard rationality idealizations is that they require the agent to be an Ideal Logician. – For instance, if finitely represented, this deductive competence would violate Church's Undecidability Theorem. Such an apercu motivates bounded-resource models of the agent, where rationality instead falls between nothing and perfection. In turn, this resource-realistic perspective focuses attention on bounded-resource models down at the neural hardware level – on the limited neuroanatomical connectivity available. However, deployment of some brain-wiring turns out to be optimized to the limits of detectability (e.g., layout of interconnected *C. elegans* ganglia, cat cerebral cortex areas, etc.). Furthermore, simple physical processes – without e.g., DNA involvement – seem to suffice to generate some of this complex structure. Prima facie, finely minimized brain-wiring phenomena appear in tension with the minimal rationality models that drove their discovery. This clash pertains to the Minimalist program in linguistics – the question whether language processing resembles optimal neural hardware.

1 Preface

The emphasis here is on combinatorial network optimization – that is, minimization of connection costs among interconnected components in a system. The picture will be that such wiring minimization can be observed at various levels of nervous systems, invertebrate and vertebrate, from placement of the entire brain in the body down to the subcellular level of neuron arbor geometry. In some cases, the minimization appears either perfect, or as good as can be detected with current methods – a predictive success story. In addition, these instances of optimized neuroanatomy include candidates for some of the most complex biological structures known to be derivable "for free, directly from physics" – i.e., purely from simple physical energy minimization processes. Such a "Physics suffices" picture for some biological self-organization directs attention to innate structure via non-genomic mechanisms, an underlying leitmotif of the Minimalist program in Chomskian linguistics.[20]

---

[1] The author is with the Committee for Philosophy and the Sciences, Department of Philosophy, University of Maryland, College Park, MD 20742 USA.

Real-World Concept	Formal Concept	Problem-Examples
⇐[C-T thesis]⇒		
"Absolutely Impossible" (?)	Absolutely uncomputable	Arithmetic Full pred calc
"Practically Unfeasible" (?)	ND-exponential time [$t = E(i)$]	WS1S Presburger arith Monadic pred calc
	NP-time NP-complete	Steiner tree Component placement Travelling salesman Propositional calc
	. . . . .	
"Practically Feasible" (?)	Polynomial time [$t = P(i)$]	Min spanning tree Nim Multiplication

Table I: A computational intractability hierarchy. (Conceptual / schematic cartoon.) Standard rationality models require an Ideal Logician, who can perform absolutely uncomputable inferences, as well as NP-complete ones

2 Rationality

Standard conceptions of rationality derive from models of the rational agent in microeconomic, game, and decision theory. The underlying idealization is that the agent, given its belief-desire system, optimizes its choices. To achieve this perfection of appropriate decisions in turn would require vast inferential insight.

While such logical omniscience might properly characterize a Supreme Being, prima facie it seems to clash with the most basic law of human psychology: We are finite entities. Among the more extreme departures from reality, for such ideal agents, portions of the deductive sciences would be trivial. (E.g., the role of the discovery of the set-theoretic paradoxes in the development of logic over the past century then ceases to be intelligible.) For a computational theory of mind, the agent's deductive competence must be represented as a finite algorithm; this ideal agent would in fact have to violate Church's Undecidability Theorem for first-order logic.[3]

Furthermore, formally correct and complete inference procedures are typically computationally complex, with surprisingly small-scale problem instances sometimes requiring cosmically unfeasible time and memory resources. To an extent, this practical intractability parallels classical absolute unsolvability. (Tab.I, Fig.1)

Since most of us on good days are not computationally hogtied or rationality-paralyzed, we must use instead quick and dirty heuristics – i.e., trade off al-

```
Test of truth-functional consistency [SAT]  of Belief Set
    by truth-table method,  by ideal computer   I:

I checks each line of truth-table in 1 "supercycle":
    the time a lightray takes to traverse dia of proton.
[Speed of light = 299,726 km/sec,  proton = 10⁻¹³ cm dia.]
So,  supercycle = 2.9 x 10⁻²³ sec.

Maximum number cycles available  in history of Universe:
    2 x 10¹⁰ years x 365 days x 24 hrs x 60 min x 60 sec
    x 2.9 x 10²³ cycles  <  2 x 10⁴¹ cycles total.

What is largest Belief Set  I could test?
    Number lines in truth table = 2ⁿ,
    where n = number of logically independent props.

2¹³⁸ atoms   require more than 3 x 10⁴¹ lines.

So:  at 1 truth-table line checked / supercycle,
    I cannot evaluate the truth-table for even 138
    independent propositions during interval from
    BigBang to present.
```

Figure 1: Cosmic cost of simple truth-functional consistency tests. SAT is in-principle computable, but NP-complete. – The exponential growth of the cost-function is not humanly intuitive.[4, p.403]

gorithm correctness and / or completeness for speedup, etc. In turn, is this deployment itself anything like finely optimized, rather than just moderately well-managed? That does not seem likely, nor advisable. One can get trapped in paralyzing regresses if limited resources are expended on perfecting decisions in turn about how to use the resources, etc., etc. (E.g., Ryle[25].)

One response to problems with the traditional idealizations is a via media strategy. Acknowledging that nothing could count as an agent or person that conformed to no rationality constraints, one pauses to ask: Must one leap to a conclusion that the agent has to be ideally rational? Is rationality all-or-nothing, or is there some golden mean ranging between unattainable perfect unity of mind and utter disintegration of personhood? The normative and empirical rationality models of Simon [26] are among the earliest of this less stringent type: The central principle is that, rather than optimizing or maximizing, the agent only "satisfices" its expected utility. Similarly for agent-constitutive rationality [2]: "No rationality, no agent."

Such a resource-realistic rationality framework possesses systematicity, in that it links together independent research programs in cognitive psychology and in computer science. The psychological studies focussed on people's strikingly ubiquitous use of reasoning procedures that are formally incorrect and/or incomplete heuristics (e.g., Kahneman et al.[23]). Another field that developed contemporaneously with, but separately from, non-ideal rationality accounts and experimental studies of reasoning heuristics is computational complexity theory in computer science[22]: As mentioned above, the basic insight is that formally correct and complete inference procedures appear to be intrinsically intractable. So, if formally adequate procedures entail practical paralysis, then the above heuristics may constitute the ultimate tradeoff of perfection for speed and usability.

A bounded-resource paradigm for rationality also of course meshes with the basic scientific worldview of humans' place in Nature. Like other creatures in the natural order, we are only finite.[1]

3 From Rationality to Neuroanatomy

The philosophical framework here extrapolates from the above rationality critique: Tacit unbounded-resource assumptions, that human cognitive resources are virtually without limit, pervade all levels of mind / brain science. (Fig.2) – Such "impossibility engines" turn up even at the most concrete hardware level, neuroanatomy. For example, some anatomical descriptions of cortex connectivity resources are not even quantitatively consistent.[5] It is as if neural connections were virtually infinitely thin wires, with effectively instantaneous signal transmission.

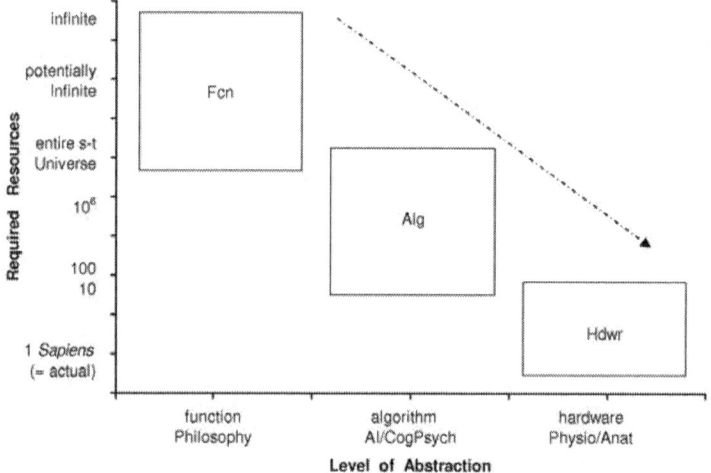

Figure 2: Required resources as a function of level of abstraction of model in cognitive neuroscience. (Conceptual cartoon.) "1 Sapiens" ≈ resources of 1 *H. sapiens*. The less abstract the model, the less overestimation of resources presupposed; but, even at the most concrete level, overestimation persists.[6, p.239]

For instance, for connectionist models of massively parallel and interconnected computation in the brain, how much brain wiring in fact is available? And, how well is it employed; how well do neurons optimize even just local connectivity?

Long-range connections in the brain are a critically constrained resource, hence there seems strong selective pressure to optimize finely their deployment. The "formalism of scarcity" of interconnections is network optimization theory, which characterizes efficient use of limited connection resources. The field matured in the 1970's for microcircuit design, typically to minimize the total length of wire needed to make a given set of connections among components. When this simple "Save wire" idea is treated as a generative principle for nervous system organization, it turns out to have applicability: To an extent, "Instant brain structure – just add wire-minimization." The main caveat is that in general network optimization problems are easy to state, but enormously computationally costly to solve exactly. The ones reviewed here are "NP-hard", each conjectured to require computation time on the order of brute-force search of all possible solutions, hence often intractable. The discussion here focuses upon component placement optimization.

4 Component Placement Optimization

A key problem in microcircuit design is component placement optimization: Given a system of interconnected components, find the positioning of the components on a 2-d surface that minimizes total connection cost (e.g., wirelength). This concept seems to account for aspects of neuroanatomy at multiple hierarchical levels.

"Why the brain is in the head" is a 1-component placement problem. That is, given the positions of receptors and muscles, positioning the brain as far forward in the body axis as possible minimizes total nerve connection costs to and from the brain, because more sensory and motor connections go to the anterior than to the posterior of the body. This seems to hold for the vertebrate series (e.g., humans), and also for invertebrates with sufficient cephalization to possess a main nervous system concentration (e.g., nematodes).[8, 10]

Multiple-component problems generally require exponentially exploding costs for exact solutions; for an n-component system, all $n!$ alternative layouts must be searched. One neural wiring optimization result is for placement of the 11 ganglionic components of the nervous system of the roundworm *Caenorhabditis elegans*, with about 1,000 interconnections. (See Fig.3.) This nervous system is the first to be completely mapped, which enables fair approximation of wirelengths of connections. (See Fig.4.) When all $\sim$ 40 million alternative possible ganglion layouts are generated, the actual layout turns out in fact to be the minimum wirelength one.[8] Some optimization mechanisms provide convergent support for this finding: A simple genetic algorithm, with wirecost as fitness-measure, will rapidly and robustly converge upon the actual optimal layout.[14] Also, a force-directed placement ("mesh of springs") algorithm, with each connection approximated as a microspring acting between components, attains the actual layout as a minimum-energy state, with relatively little trapping in local minima.[14] (See Fig.5.)

(There is also statistical evidence that this "brain as microchip" wire-minimization framework applies in the worm down to the level of clustering of individual neurons into ganglionic groups, and even to cell body positioning within ganglia to reduce connection costs.[8])

Finally, the wiring-minimization approach can be applied to placement of functional areas of mammalian cerebral cortex. Since wirelengths of intrinsic cortex connections are difficult to derive, an alternative approach is to examine instead a simpler topological measure of connection cost, conformance of a layout

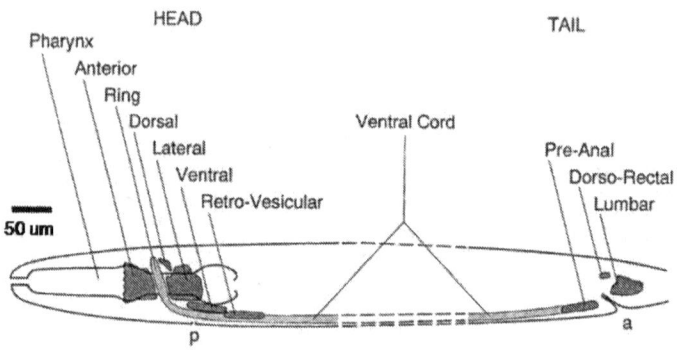

Figure 3: *C. elegans* ganglion components: their body locations and schematized shapes.

to a wire-saving heuristic Adjacency Rule: If components a and b are connected, then a and b are adjacent. Exhaustive search of all possible layouts is still required to identify the cheapest ones. A promising calibration of this approach is that the minimum wirecost actual layout of the nematode ganglia is among the layouts with fewest violations of this adjacency rule. For 17 core visual areas of macaque cortex, the actual layout of this subsystem ranks in the top 10^{-7} layouts best fitting this adjacency-costing; for 15 visual areas of cat cortex, the actual layout ranks in the top 10^{-6} of all layouts.[15]

In general, a Size Law seems to apply to cases like macaque, cat (and worm) with local-global tradeoffs: The larger proportion of a total system the evaluated subsystem is, the better its optimization. (We have observed this Size Law trend recently also for rat olfactory cortex and for rat amygdala.[16]) For the largest systems studied (visual, auditory, plus somatosensory areas of cat cortex), there is evidence of optimization approaching limits of current detectability by brute-force sampling techniques. (See Fig.6.) A similar Size Law pattern also appears to hold for Steiner subtree optimization of neuron arbor topologies. [13]

5 Optimization: Mechanisms and Functional Roles

The innateness hypothesis is typically expressed in the DNA era as a thesis that some cognitive structure is encoded in the genome. In contrast, an idea of "non-genomic nativism"[11] can be explored, that some biological structure is inborn, yet not genome-dependent; instead, it arises directly from simple physical processes. – Not only, then, is the organism's *tabula rasa* in fact not blank, it is "pre-formatted" by the natural order: a significant proportion of structural information is pre-inscribed via physical and mathematical law.

Noam Chomsky describes a strong Minimalist thesis, that "a principled account" of language is possible: "If that thesis were true, language would be something like a snowflake, taking the form it does by virtue of natural law, ..." [17–20] The snowflake reference recalls D'Arcy Wentworth Thompson's *On*

Minimal Rationality vs Optimized Brain-Wiring

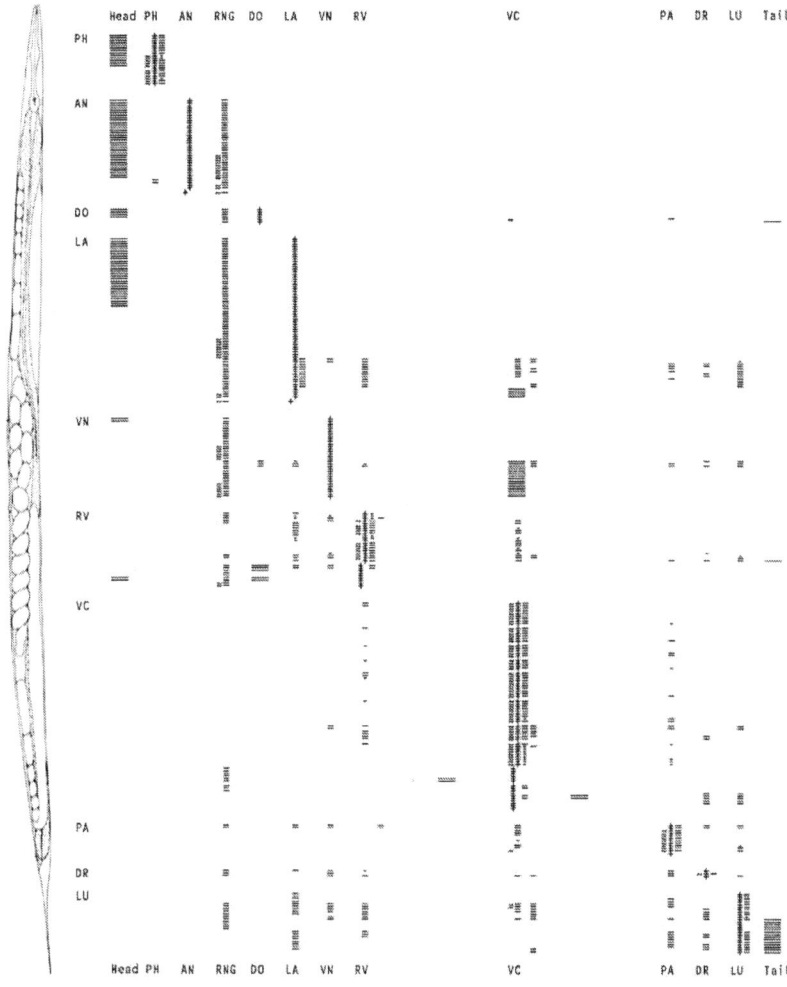

Figure 4: Complete ganglion-level connectivity map for *C. elegans* nervous system. (This may be the first depiction of approximately complete connectivity of a nervous system down to synapse level.) Each horizontal microline represents one of the 302 neurons. This actual ganglion layout requires the least total connection length of all 39,916,800 alternative orderings.[8]

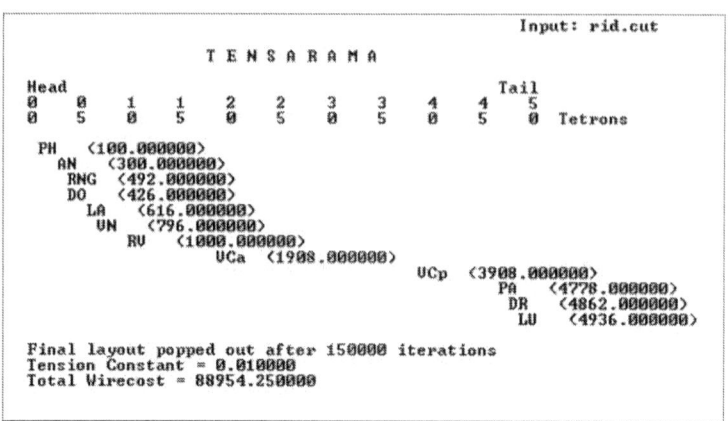

Figure 5: Tensarama, a force-directed placement algorithm for optimizing layout of *C. elegans* ganglia. This "mesh of springs" vector-mechanical energy-minimization simulation represents each of the worm's ∼ 1,000 connections (not shown) acting upon the movable ganglia PH, AN, etc. The key feature of Tensarama performance for the actual worm connectivity matrix is its low susceptibility to local minima traps. [14] – Unlike Tensarama performance for small modifications of the actual connectivity matrix, and unlike such force-directed placement algorithms in general for circuit design. Here Tensarama is trapped in a slightly sub-optimal layout, by a "killer" connectivity matrix that differs from the actual matrix by only *one* less connection.

Growth and Form (1917/1961)[27], where the paradigmatic example of mathematical form in nature was the hexagonal packing array, of which snow crystals are an instance. However, even the thousand pages of the unabridged 1917 edition of Thompson's opus contained few neural examples. Similarly, Alan Turing's study[28] of biological morphogenesis via chemical diffusion processes began an inquiry that needs to continue. In effect, we have seen here how far this type of idea presently can be seen to extend for biological structure at the concrete hardware level of neuroanatomy. The key concept linking the physics and the anatomy is optimization of brain wiring.

The picture here has been of limited connections deployed very well – a predictive success story. The significance of ultra-fine neural optimization remains an open issue. Another question is where it does in fact occur, and how good it is. Tradeoffs of local optimality for better cost minimization of a total system are one way in which global optimization can be obscured.

The high levels of connection optimization in the nervous system seem unlike levels of optimization common elsewhere in organisms. Optimization to nearly absolute physical limits also can be observed in human visual and auditory sensory amplitude sensitivities, and in silk moth olfactory sensitivity to pheromones[14] – i.e., at the very boundary of the neural with its environment. Why should the neural realm sometimes demand optimization, rather than the more familiar biological satisficing?

Mechanisms of neural optimization are best understood against the background sketched earlier, that the key problems of network optimization theory are

Minimal Rationality vs Optimized Brain-Wiring

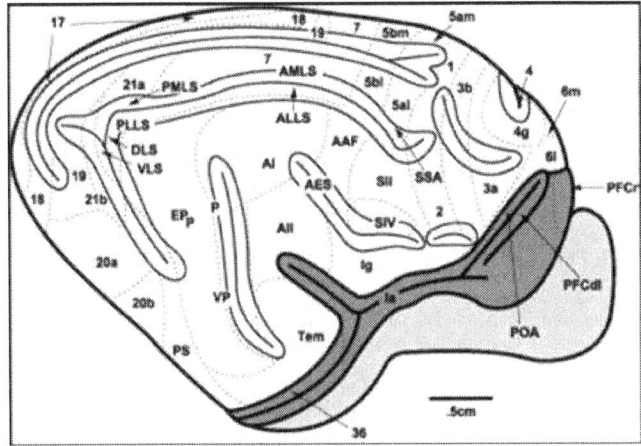

Figure 6: Cerebral cortex of cat. (Lateral aspect; rostral is to right.) Placement of 39 interconnected functional areas of visual, auditory, and somatosensory systems (in white). Exhaustive search of samples of alternative layouts suggests this actual layout ranks at least in the top 100 billionth of all possible layouts with respect to adjacency-cost of its interconnections.[15] – "Best of all possible brains"?

NP-complete, hence exact solutions in general are computationally intractable. For example, blind trial and error exhaustive search for the minimum-wiring layout of a 50 component system (such as all areas of a mammalian cerebral cortex hemisphere), even at a physically unrealistic rate of one layout / picosecond, would still consume more than the age of the Universe.[9] Thus, to avoid these Cosmos-crushing costs, even terrestrial evolution instead must exploit "quick and dirty" approximation / probabilistic heuristics.

One such possible strategy discernible above is optimization "for free, directly from physics". That is, as some structures develop, physical principles cause them automatically to be optimized. There is some evidence for neuron arbor optimization via fluid dynamics, as well as for nematode ganglion layout optimization via "mesh of springs" force-directed placement simulation. For the neural optimization examples above, some of this structure from physics depends in turn on exploiting anomalies of the computational order.[12] For component placement optimization, there is the chicken-egg question of whether components begin in particular loci and make connections, or instead start with their interconnections and then adjust their positions, or some mix of both causal directions. It is worth noting that both a force-directed placement algorithm for ganglion layout, and also genetic algorithms for layout of ganglia and of cortex areas, suggest that simple "connections → placement" optimization processes can suffice.

If the brain had unbounded connection resources, there would be no pressure to refine employment of wiring. So, to begin with, the very fact of neural finitude appears to drive "Save wire" fine-grained minimization of connections. Another part of the functional role of such optimization may be the picture here of "physics → optimization → neural structure". Optimization may be the means to anatomy. The human brain is often characterized as the most complex structure known in the Universe. Perhaps the harmony of neuroanatomy and physics provides an economical means of self-organizing complex structure generation, to ease brain structure transmissibility through the "genomic bottleneck"[4, 7] – the limited information carrying-capacity of the genome. This constitutes a thesis of non-genomic nativism, that significant innate complex biological structure is not encoded in DNA, but instead derives from basic physical principles.[7, 11]

6 Levels of Abstraction for Optimization Models

Marr's [24] "function / algorithm / hardware" levels in cognitive science were preceded by a similar set of levels in top-down methodology of computer engineering (e.g., Dijkstra, late 1960's). This in turn was reflected in the emergence of functionalism / computational psychology in 1960's philosophy of mind. And Chomsky [17] had distinguished between abstract competence and realistic performance models. These levels seem a good way to relate hardware optimization findings to main results of cognitive science at the computational (algorithm, software) level.

An uncomfortable instance of the levels approach has emerged for my research program in recent years. It concerns (in)consistency of optimal neural hardware observations with less idealized rationality models (at the function level). My own critique was that standard rationality idealizations – drawn from microeconomic, game, and decision theory – can benefit from a more realistic approach, particularly regarding limited available cognitive resources.

So, on the one hand, there is a tension or un-consilience – if that is the right euphemism – with the "best of all possible brains" neural hardware results. For what irony it is worth, as noted, the research program in computational neuroanatomy optimization in fact stemmed directly from my own prior philosophical critiques of classical rationality idealizations. – "We don't have God's brain." Our resources at all three levels are bounded. So, neural hardware optimization does not seem to entail rationality optimization. In the other direction, rationality optimization would require much more than brain optimization; indeed, hardware optimization could be dispensed with, if the ideal agent had unbounded resources. Efficiency is then not needed. (As noted, even absolute unsolvability results like Church's Theorem, etc. in fact no longer apply for such an ideal agent.)

However, on the other hand, such suboptimal rationality of our cognitive system does not ipso facto exclude optimality at other levels higher than hardware. – For instance, computational subsystems. Chomsky has focused on syntax, pointedly eschewing even semantics. (E.g., [17].) Fodor extended the Chomskian program beyond syntax; but even he explicitly excludes the central cognitive system from such an approach.[21, p.101] So, the way still seems open for a Minimalist program; the hardware optimization findings might yet point to an approach. The conclusion remains inconclusive.

Bibliography

[1] Cherniak, C. *Pragmatism and Realism*. B. Litt. thesis, University of Oxford, 1973.

[2] Cherniak, C. Minimal rationality. *Mind*, 90:161–183, 1981.

[3] Cherniak, C. *Minimal Rationality.* Cambridge: MIT Press, 1986.
[4] Cherniak, C. Undebuggability and cognitive science. *Communications of Assoc for Computing Machinery,* 31:402–412, 1988.
[5] Cherniak, C. The bounded brain: Toward quantitative neuroanatomy. *Journal of Cognitive Neuroscience,* 2:58–68, 1990.
[6] Cherniak, C. Meta-neuroanatomy: The myth of the unbounded mind/brain. In Agazzi, E., Cordero, A. (Eds.), *Philosophy and the Origin and Evolution of the Universe,* 219–252. Boston: Kluwer, 1991.
[7] Cherniak, C. Local optimization of neuron arbors. *Biological Cybernetics,* 66:503–510, 1992.
[8] Cherniak, C. Component placement optimization in the brain. *Journal of Neuroscience,* 14:2418–2427, 1994.
[9] Cherniak, C. Philosophy and computational neuroanatomy. *Philosophical Studies,* 73:89–107, 1994.
[10] Cherniak, C. Neural component placement. *Trends in Neurosciences,* 18:522–527, 1995.
[11] Cherniak, C. Innateness and brain-wiring optimization: Non-genomic nativism. In Zilhao, A. (Ed.), *Cognition, Evolution, and Rationality,* 103–112. Londo: Routledge, 2005.
[12] Cherniak, C. Neuroanatomy and cosmology. In Bickle, J. (Ed.), *Oxford Handbook of Neuroscience and Philosophy.* New York: Oxford, 2009.
[13] Cherniak, C., Changizi, M., Kang, D. Large-scale optimization of neuron arbors. *Physical Review E,* 59:6001–6009, 1999.
[14] Cherniak, C., Mokhtarzada, Z., Nodelman, U. Optimal-wiring models of neuroanatomy. In Ascoli, G. (Ed.), *Computational Neuroanatomy: Principles and Methods,* 71–82. Totowa, NJ: Humana, 2002.
[15] Cherniak, C., Mokhtarzada, Z., Rodriguez-Esteban, R., Changizi, B. Global optimization of cerebral cortex layout. *Proceedings National Academy of Sciences,* 101:1081–1086, 2004.
[16] Cherniak, C., Rodriguez-Esteban, R. Information processing limits on generating neuroanatomy: Layout optimization of olfactory cortex and amygdala of rat. *Journal of Biological Physics,* 2009. In press.
[17] Chomsky, N. *Aspects of the Theory of Syntax.* Cambridge: MIT Press, 1965.
[18] Chomsky, N. Beyond explanatory adequacy. In Belletti, A. (Ed.), *Structures and Beyond: The Cartography of Syntactic Structures,* 2. New York: Oxford, 2004.
[19] Chomsky, N. Three factors in language design. *Linguistic Inquiry,* 36:1–22, 2005.
[20] Chomsky, N. Opening remarks. In Piattelli-Palmarini, M., Uriagereka, J., Salaburu, P. (Eds.), *Of Minds and Language.* New York: Oxford, 2009.
[21] Fodor, J. *Modularity of Mind.* Cambridge: MIT Press, 1983.
[22] Garey, M., Johnson, D. *Computers and Intractability.* San Francisco: WH Freeman, 1979.
[23] Kahneman, D., Slovic, P., Tversky, A. (Eds.). *Judgment under Uncertainty: Heuristics and Biases.* New York: Cambridge, 1982.
[24] Marr, D. *Vision.* San Francisco: WH Freeman, 1982.
[25] Ryle, G. *Concept of Mind.* Chicago: University of Chicago, 1949.
[26] Simon, H. *Models of Bounded Rationality.* Cambridge: MIT Press, 1982.
[27] Thompson, D. *On Growth and Form.* New York: Cambridge, 1917.

[28] Turing, A. Chemical basis of morphogenesis. *Phil Trans Roy Soc Lond B*, 237:37–72, 1952.

Logic in the study of autism: reasoning with rules and exceptions [1]

Michiel van Lambalgen

University of Amsterdam

M.vanLambalgen@uva.nl

ABSTRACT. Autism is a psychiatric disorder characterised by (sometimes very severe) impairments in verbal and social communication, and in executive tasks. The aetiology of autism is controversial. A prominent theory holds that autism is a consequence of a 'theory of mind' deficit, as manifested in significantly worse than normal performance in 'false belief' tasks. An apparently very different theory maintains that autism is a consequence of defects in executive function, and again there are (nonverbal) tasks which support this interpretation. Clearly the aetiology has implications for treatment. A logical analysis of theories and diagnostic tasks has turned out to be very helpful here. It so happens that both in the false belief tasks and in the tasks supporting executive dysfunction closed world reasoning (CWR) plays a large role. CWR counsels to take those propositions as false which one has no reason to suppose true. Once it has been observed that CWR is important in existing tasks, it becomes possible to devise an experimental paradigm in which CWR is tested more directly. In fact an existing task in the psychology of the reasoning (of healthy adults), the 'suppression task', filled the bill. The logical analysis alluded to predicted that autists would behave very differently on this task compared to normals. An experiment on 28 autists and 28 matched controls indeed showed this to be the case. On a more theoretical level, what is interesting is that CWR seems to correspond to a neural architecture in which inhibitory interneurons play an important role; and recent neurological evidence indicates that these are compromised in autism.

1 Prologue: a role for logic in cognitive science

The reader is not to be blamed if she finds the conjunction of the words 'logic' and 'autism' in our title improbable. Logic and psychiatric disorders seem to lie at opposite ends of the spectrum of rationality. To dispel such doubts, we preface our exposition of an experiment on reasoning in autists with a discussion of the place of logic in cognitive science generally, and from this discussion it will hopefully become clear why logic may have a role to play in explaining behavioural patterns in psychiatric disorders. The reader who does not need to be persuaded of this can skip directly to Section 2.

---

[1]ILLC/Department of Philosophy, University of Amsterdam. Many ideas in this article were developed in close collaboration with Keith Stenning (Edinburgh). The experiment on reasoning in autists was executed by Judith Pijnacker (F.C. Donders Center for Cognitive Neuroimaging, Nijmegen).

Arguably, logic played an important role in the genesis of cognitive science, but its importance for this field has progressively diminished, as a consequence of many different pressures. We list some of these, in arbitrary order. They are formulated as generally perceived; it will later become apparent why we believe they are wide off the mark.

1. experiments with reasoning tasks, such as the famous Wason selection task [37], show that logical form is not a determinant of reasoning
2. logic cannot deal with vagueness and prototypicality
3. logic cannot deal with uncertainty and must be replaced by probability theory, which is after all *the* calculus of uncertainty
4. what we know about the neocortex suggests that the computations executed by the brain are very different from logical computations
5. the computational complexity of logic is too high for logic to be a realistic model of cognition

To set the stage, we begin by discussing the selection task and its aftermath. Wason's original task was presented by means of a form as depicted in Fig.1. In order to appreciate the tremendous difficulty posed by this task, the reader – who has probably seen the task before – should realise that this is all the information provided to the subjects.

> Below is depicted a set of four cards, of which you can see only the exposed face but not the hidden back. On each card, there is a number on one of its sides and a letter on the other.
>
> Also below there is a rule which applies only to the four cards. Your task is to decide which if any of these four cards you *must* turn in order to decide if the rule is true. Don't turn unnecessary cards. Tick the cards you want to turn.
>
> **Rule:** *If there is a vowel on one side, then there is an even number on the other side.*
> **Cards:**
>

Figure 1: Wason's selection task

This experiment has been replicated many times. If one formulates the rule *If there is a vowel on one side, then there is an even number on the other side.* as an implication $p \to q$, then the observed pattern of results is typically given as in Tab.I. Wason believed there to be only one 'logically correct' answer, namely

p	p,q	$p,\neg q$	$p,q,\neg q$	misc.
35%	45%	5%	7%	8%

Table I: Typical scores in the selection task

$p, \neg q$, and concluded that according to this standard, the vast majority of reasoners are irrational. In addition, the selection task was used to argue that most adults do not reach what Piaget [23] considered the pinnacle of cognitive development, the formal operational stage, which is basically mastery of classical propositional logic. Not only do subjects typically fail to master the *modus tollens* inference supposed to be at work here, they do not even have a workable concept of logical form to guide their reasoning. For if Wason's 'abstract' rule *If there is a vowel on one side, then there is an even number on the other side.* is replaced by the 'concrete' rule *if you want to drink alcohol, you have to be over 18*, performance suddenly jumps to more acceptable levels, around 75%. These two reasoning tasks – 'abstract' and 'concrete' – have the same logical form. If logical form determines performance in a task, performance on these tasks must be comparable. The data show that performance differs wildly, whence logical form is irrelevant to reasoning. A tacit inference followed: if logic plays no role in reasoning, why would it play a role in other domains of cognition? There is much to say about this line of argument (see [29, 32]), but here we continue with the other listed doubts about the cognitive relevance of logic.

The argument from uncertainty contrasts logic, which is supposed to be dealing in certain truth, with real life, where

> ...it is, in fact, rational by the highest standards to take proper account of the probability of one's premises in deductive reasoning (Stevenson and Over [34, p.615]).

> ...performing inferences from statements treated as absolutely certain is uncommon in ordinary reasoning. We are mainly interested in what subjects will infer from statements in ordinary discourse that they may not believe with certainty and may even have serious doubts about [34, p.621].

It is a moot point whether the implied contrast is at all valid (see [33] for discussion), but for our present purposes it suffices to note that logic is widely perceived to fail on this score. A whole research program, 'Rational Analysis' as practised by Oaksford and Chater [21], is built on this perception.

In the current scientific climate, objection 4 is probably the most formidable. Neural computation is parallel and distributed (hence fast, as opposed to logic: 5), and has no need for symbols, which are an obstacle to 'graceful degradation' (see 2) anyway. Moreover, neural networks have a built-in capacity to learn, whereas equipping symbolic systems with learning capability is laborious.

In the face of such forceful objections, it seems a daunting task to defend a continuing role for logic in cognitive science. Still, this is what we shall do. We shall argue, using Marr's model of explanation in cognitive science [19, chap.1], that the criticisms directed against logic fall foul of a confusion of levels of explanation. More importantly, we show that logic, rightly conceived, can still make a positive contribution to explanation and prediction in cognitive science. The case we focus on is autism. This may seem stony ground indeed: why would there be a role for *logic* in autism? This disorder, about which more below, has been variously characterised as due to a deficit in 'theory of mind', as defective central processing, and as executive dysfunction, to mention but a few proposed aetiologies. On the surface of it, logic has little to do with any of these.

Considering the place of logic in cognitive explanation will clarify these matters. We must first note that so far we have been talking about logic in the singular, as is common in cognitive science, which proceeds mostly as if classical logic were the only game in town. But to appreciate logic's place in cognitive science, a broader conception of logic is necessary. This involves acknowledging the need for many different logics, varying in syntax, semantics and definition of validity [32]. In fact, the impression that logic is useless in cognitive science is for the most part due to the uncritical identification of logic with classical logic, or in any case with a system of rules.[2]

This need for many logics arises because at a suitably abstract level cognitive functions, i.e. information processing tasks, can be usefully described in *a* logic, not necessarily classical, and not necessarily given by a system of rules. To see why this can be so, consider David Marr's three levels of cognitive inquiry:

1. identification of the information processing task as an input–output function;[3] this competence model expresses what the information processing task considered is supposed to do
2. specification of an algorithm which computes that function
3. neural implementation of the algorithm specified[4]

As an example of how a competence model can be specified, consider (a particular view of) language comprehension. Here, the input consists of natural language discourse, and the output consists in discourse models, or more technically, 'discourse representation structures' (DRS) [10, 15]. Both input and output can be specified in terms of a formal language. The formal structure of the input language is simple: concatenation of sentences. For the output we need a language describing the various ways in which DRSs can be combined. The specific 'competence' aspect is given by the norm relating input and output: if the input discourse is true in a model $\mathcal{A}$, the output DRS can be embedded in $\mathcal{A}$. As regards the second level, there are algorithms transforming natural language discourse into a corresponding DRS – corresponding in the sense that the competence model is satisfied. A neural implementation does not exist yet – but see below.

To return to the competence level, observe that the norm relating input and output can be viewed as a definition of validity of 'arguments' $S_1, \ldots, S_n \vdash K$, where the S_i are sentences and K is (a sentence describing) a DRS. Thus we have all the ingredients of a logic here: syntax, semantics, validity. But why is this a useful way of looking at the competence level?

One useful feature of such a fully specified competence model is that it allows a correctness proof for the algorithm of the second level. Equally importantly, in setting up the competence model formally one may uncover assumptions which restrict its scope unduly. An example is in order here. The formal model for discourse comprehension sketched earlier has as a consequence that, as the discourse grows, the set of discourse models forms a monotonic chain. Assume that the given discourse grows from $S_1, \ldots, S_n$ to $S_1, \ldots, S_n, S_{n+1}$, and that the semantic force of the comma is conjunction, so that '$S_1, \ldots, S_n, S_{n+1}$ is true on $\mathcal{A}$' implies '$S_1, \ldots, S_n$ is true on $\mathcal{A}$. Let K_n be the discourse model corresponding

---

[2]In logic itself this conception was finally laid to rest with the advent of 'model theoretic logics', for which see [4].

[3]Marr calls this the 'computational level', but this term is infelicitous in view of what the next level does. We prefer the term 'competence model' instead.

[4]This tripartite scheme of explanation should be read neither bottom-up nor top-down, but as a set of constraints; for example, the algorithm may be determined both by the competence model and by what is known about neural computation.

to $S_1,\ldots,S_n$, and K_{n+1} that for $S_1,\ldots,S_n,S_{n+1}$. Then both K_n and K_{n+1} are embeddable in $\mathcal{A}$, and since the universe of K_n is a subset of that of K_{n+1}, it follows that K_n must be a substructure of K_{n+1}. Thus the sequence of discourse models is monotonic.

As soon as one realises this, one sees that the proposed competence model lacks cognitive plausibility. Consider the simple discourse

> Some real estate agents are crooks. In fact all of them are.

Gricean maxims tell you that after the first sentence the discourse model verifies 'Some real estate agents are not crooks', a statement that is flatly denied by the second sentence. So here the progression of discourse models cannot be monotonic. While this example is from pragmatics, the following is squarely inside semantics, and concerns the meaning of the English progressive [1].

> The door of the living-room was closed. Inside the radio played classical music. The girl was writing letters [∗] when her friend spilled coffee on the tablecloth/paper.

Behavioural data show that if the sentence ends with 'tablecloth', 75% of subjects conclude that in the end a letter was written, whereas if the final word is 'paper', this percentage drops to 37%. The most parsimonious explanation for this is that in the discourse model of the discourse up to the point ∗ a finished letter is introduced, so that after seeing 'paper', but not after seeing 'tablecloth', that entity must be removed again.[5] Therefore this sequence of discourse models is non-monotonic. It is argued in [1–3] that a fairly well-established discourse processing principle, the 'principle of immediacy', which says that

> every source of information that constrains the interpretation of an utterance (syntax, prosody, word-level semantics, prior discourse, world knowledge, knowledge about the speaker, gestures, etc.) can in principle do so immediately [i.e. before sentence boundary], [9]

forces a non-monotonic progression of discourse models. Again, this shows that the earlier competence model, in particular its definition of validity, is not cognitively plausible. In the face of such difficulties, a common tactic in cognitive science is to drop all reference to a competence model and be content with an algorithm instead. The downside of this tactic is that one ends up with a bunch of specialised algorithms, and that commonalities between tasks may disappear from view; more on this below, in our discussion of autism. So what has to replace the discarded competence model? The definition of validity cannot be direct comparison with the real world, but it may come from internal considerations on what is the 'best' notion of non-monotonicity from a cognitive point of view.

To get some idea of what this could be, we consider a very general cognitive function, one that is involved every time a task involves top-down processing: executive function. 'Executive function' is an umbrella term for processes responsible for higher-level action control that are necessary for maintaining a goal and achieving it in possibly adverse circumstances. There is no unanimity on how to partition executive function into meaningful sub-components, but we

---

[5]There exist EEG data consistent with this view [2].

may take executive function to be composed of planning, initiation, inhibition, coordination and control of action sequences, leading toward a goal held in working memory. It will be seen that at the competence level executive function lends itself very well to a characterisation in logical terms. Indeed, at a logical level, the operation of executive function can be described as non-monotonic conditional reasoning and aberrations thereof.[6] We propose that the non-monotonic logic that works well for executive function also works well for more specialised tasks such as discourse processing, thus replacing the notion of validity derived from comparison with the world by an internal criterion.

We are now finally in a position to explain why logic can contribute to explanation and prediction in the case of psychiatric disorders [31], in particular autism. There exist a number of experimental tasks on which autistic performance is markedly different from that of normal controls. These tasks have been taken to corroborate the theory on the basis of which they were developed. The most famous example is the 'false belief task' which is viewed as supporting the theory that holds that autism is a deficit in (high level, perhaps domain-specific) 'theory of mind', and as refuting the theory which says that autism is due to (low level, domain-general) executive dysfunction. This picture is challenged when following the methodology advocated here. We may devise a competence model of the false belief task, formulated in logical terms, and also competence models for other tasks, including those of a more executive flavour. It will turn out that these competence models are almost (but not quite) identical. The common component involves executive function, and dysfunction predicts deviant autistic performance in these tasks. The interim conclusion to be drawn from this is that the experiments do not provide support for a sharp distinction between theory of mind deficit and executive dysfunction. Furthermore, the competence model also predicts deviant autistic performance in a domain hitherto hardly investigated in autists: conditional reasoning. Thus, the advantages of having a formally specified competence model are twofold: (i) the competence model of a task describes what is essential in that task (ii) the competence model, together with auxiliary assumptions, predicts novel phenomena.

After these preliminaries, it is time to make good on our promises. We begin with a more precise logical characterisation of executive function.

2 Executive function and logic

As mentioned in the previous section, 'executive function' is a pointer to processes responsible for higher-level action control that are necessary for maintaining a goal and achieving it in possibly adverse circumstances. In the following we abstract from the co-ordination and control component of executive function, and concentrate on goal maintenance, planning and (contextually determined) inhibition.

By definition, planning consists in the construction of a *sequence* of actions which will achieve a given goal, taking into account properties of the world and

---

[6]This may seem surprising, especially given the prevalent conception of reasoning as a conscious and somewhat laborious activity starting from explicitly given premises. How then can fast and largely automatic executive function be profitably described by a logic? It will turn out, however, that the logics most useful in this context, variants of closed world reasoning, do allow fast and automatic processing.

the agent, and also events that might occur in the world. The relevant properties include stable causal relationships obtaining in the world, and also what might be termed 'inertia', in analogy with Newton's first law. If a property has been caused to hold by the occurrence of an event, we expect that the property persists until it is terminated by another event. This is the inertial aspect of causality: a property does not cease to hold (or come to hold) spontaneously, without identifiable cause. Such inertia is a prerequisite for successful action in the world; and we will have to find a formal way to express it. It does however not suffice for successful planning.

The problem is that in the definition of planning, 'will achieve' definitely cannot mean: '*provably* achieves in classical logic', because of the notorious frame problem: it is impossible to take into account all eventualities whose occurrence might be relevant to the success of the plan, but classical logic forces one to consider all models of the premises, including those that contain farfetched possibilities. Therefore the question arises: how to characterise formally what makes a good plan?

A reasonable informal suggestion is: the plan works to the best of one's present knowledge. More formally, this idea can be reformulated semantically as: the plan achieves the goal in a 'minimal model' of reality; where a minimal model is characterised by the property that, very roughly speaking, every proposition is false which you have no reason to assume to be true. In particular, in the minimal model no events occur which are not forced to occur by the data, and only explicitly mentioned causal influences are represented in the model. This makes planning a form of non-monotonic reasoning.

We thus postulate that the logical idea underlying planning is *closed world reasoning*: the principle which says that every proposition which is not forced to hold by the data available can be assumed to be false. This can apply to propositions about occurrences of events as well as those expressing causal relationships. One may identify a number of areas to which closed world reasoning is applicable, each time in slightly different form:

1. lists: train schedules, airline databases, ...
2. diagnostic reasoning and abduction
3. unknown preconditions
4. causal and counterfactual reasoning
5. attribution of beliefs and intentions[7]
6. construction of discourse models, in particular event structures from verb tenses[8]

It is of some interest that several psychiatric disorders come with disturbances in one or more forms of reasoning from this list. Children with Attention-Deficit Hyperactivity Disorder tend to have specific difficulties with ordering events in a narrative. Autists have difficulties with at least 3, 4 and 5. They also have a special relationship with lists, in the sense that they feel lost without lists, such as timetables to organise daily activities; they have great difficulty accommodating unexpected changes to the timetable, and try to avoid situations such as holidays

---

[7]It may not be obvious that this is a planning problem at all, but Section 3.2 will make clear why this is so.
[8]It is explained in [36] what this has to do with closed world reasoning.

in which rigid schedules are not applicable. One may view this as an extreme version of closed world reasoning, sometimes even applied in inappropriate circumstances. But before one concludes from this that autists are good at closed world reasoning to the point of over-applying it, one must carefully distinguish several components of closed world reasoning. On the one hand, there is the inference from *given* premises to a conclusion. In [32], Chapter 8 it is shown that such inferences can be executed fast by suitable neural networks. In a wider sense, non-monotonic reasoning also involves 'pre-processing' the given situation or discourse, that is, *encoding* the law-like features of a situation in a particular type of premises. Laws and regularities always allow exceptions, and skill at 'exception handling' is required – which involves identifying and encoding the relevant exceptions, and knowing when 'enough is enough'. Autists appear to do worse than normals on this last aspect, although they behave normally with respect to the non-monotonic inferences occurring after the encoding stage.

We have thus identified closed world reasoning as an important component of executive function. A good formal representation of closed world reasoning as relevant to planning is the *Event Calculus* as formulated in logic programming with negation as failure[9] [36], but due to space constraints we rely on an ad hoc formalism here, which brings out the intimate connection between inhibition and closed world reasoning.

In the logical model of executive function proposed here, inhibition is represented through the special logical form of causal properties of actions, where the link from action to effect is mediated by a slot labelled $\neg ab$:

$$A \wedge \neg ab \rightarrow E \tag{1}$$

This conditional is read as 'if A and *nothing abnormal is the case*, then E', where the expression '*nothing abnormal is the case*' is governed by closed world reasoning. For instance, if there is nothing known about a possible abnormality, i.e. if the causal system is closed, one concludes $\neg ab$, hence from A it follows that E. If however there is information of the form $C \rightarrow ab$, i.e. if there is a context C which constitutes an abnormality, and C is the case, then the link from A to E is inhibited. In the neural model of closed world reasoning proposed in [32], Chapter 8, ab corresponds to an (artificial) neuron situated between the neurons for A and E, such that C is connected to ab via an inhibitory link; and this is the general way of incorporating contextual influences. Defects in the inhibitory neurons would thus lead to deficient context processing, as we see in autism. In [32], Chapter 9 we present some recent evidence indicating that in the brain of autists inhibition is compromised at the neurological level, among other reasons because inhibitory interneurons are underdeveloped. We will briefly discuss this topic in Section 6.

3 Non-monotonicity in autism: rules and exceptions

Autism is a clinical syndrome first described by Leo Kanner in the 1940s, often first diagnosed in children around 2–3 years of age as a deficit in their affective relations and communication. The full range of symptoms includes

---

[9]This is a formalisation of closed world reasoning.

- poor or unusual social interaction skills (e.g. the autistic child typically refuses eye-contact and is indifferent or hostile to demonstrations of affection)
- delayed development of difficulties in both verbal and non-verbal communication (the latter comprises for example gestures, pointing, showing)
- the presence of repetitive behaviours (sometimes self-harming), and an insistence on sameness.

Autistic children do not engage spontaneously in make-believe play and show little interest in the competitive social hierarchy, and in personal possessions. Autism comes in all severities – from complete lack of language and severe retardation, to mild forms continuous with the 'normal' range of personalities and IQs.

More than any other psychiatric disorder, autism has captured the imagination of the practitioners of cognitive science, because, at least according to some theories, it holds the promise of revealing the essence of what makes us human. This holds especially for the school which views autism as a deficit in 'theory of mind', the ability to represent someone else's feelings and beliefs. Some go so far as to claim that in this respect autists are like our evolutionary ancestors, given that chimpanzees have much less 'theory of mind' than humans.

3.1 Theory of mind and reasoning

A famous experiment, the 'false belief' task [22], investigates how autistic subjects reason about other people's belief. The standard design of the experiment is as follows. A child and a doll (Maxi) are in a room together with the experimenter. Maxi and child witness a bar of chocolate being placed in a box. Then Maxi is brought out of the room. The child sees the experimenter move the chocolate from the box to a drawer. Maxi is brought back in. The experimenter asks the child: 'Where does Maxi think the chocolate is?' The answers to this question reveal an interesting cut-off point, and a difference between autists and normally developing children. Before the age of about 4 years , the normally developing child responds where the child knows the chocolate to be (i.e. the drawer); after that age, the child responds where Maxi must falsely believe the chocolate to be (i.e. the box). By contrast, autists go on answering 'in the drawer' for a long time.

This experiment has been repeated many times, in many variations, with fairly robust results. The outcomes of these experiments have been argued to support the 'theory of mind deficit' hypothesis on the cause of autism. Proposed by Leslie in 1987 [16], it holds that human beings have evolved a special 'module' devoted specifically to reasoning about other people's minds. As such, this module would provide a cognitive underpinning for empathy. In normals the module would constitute the difference between humans and their ancestors – indeed, chimpanzees seem to be able to do much less in the way of mind-reading. In autists, this module would be delayed or impaired, thus explaining abnormalities in communication and also in the acquisition of language, if it is indeed true that the development of joint attention is crucial to language learning (as claimed for instance by Tomasello [35]).

This seems a very elegant explanation for an intractable phenomenon, and it has justly captured the public imagination. Upon closer examination the question arises whether it is really an explanation rather than a description of one class of symptoms. For instance, the notion of a 'module' is notoriously hazy. In this context it is obviously meant to be a piece of dedicated neural circuitry. In this way, it can do the double duty of differentiating us from our ancestors and being capable of being damaged in isolation. But it is precisely this isolation, or 'encapsulation' as Fodor called it, that is doubtful. One reason is that evolution does not generally proceed by adding new modules, but instead by tweaking old ones, and another is that much of the problem of functionally characterising human reasoning about minds is about interactions between modules. 'Theory of mind' requires language to formulate beliefs in and it also entails a considerable involvement of working memory, as can be seen in 'nested' forms of theory of mind, as in Dunbar's example

> Shakespeare intended us to realise that Othello believes that Iago knows that Desdemona is in love with Cassio.

However, as soon as one realises that a 'module' never operates in isolation, then the 'theory of mind deficit' hypothesis begins to lose its hold. We are now invited to look at the (possibly defective) interactions of the 'module' with other cognitive functions (language, working memory, ...), which leads to the possibility that defects in these functions may play a role in autism. And there is of course also the problem of what the 'module' would have to contain, given that for instance reasoning about other people's *desires* (as opposed to beliefs) is possible for both autists and non-human primates.[10]

3.2 Reasoning in the false belief task

It is tempting to view the false belief task as concerned with reasoning about belief, and hence to attempt a formalisation of the reasoning in some variant of multi-agent epistemic logic. However, a more fine-grained analysis is possible, which takes into account the way in which beliefs are formed and maintained. From this analysis it will become clear that the 'theory of mind' capacity is much less *sui generis* than commonly thought, and in fact intimately linked with executive function. This said, there may after all be some role for epistemic logic, and we will return to this issue at the end of our analysis.

We will now analyse attribution of belief as it occurs in the false belief task as consisting of three components

1. awareness of the causal relation between perception and belief, which can be stated in the form: 'if φ is true in scene S, and agent a sees S, then a comes to believe φ'.

---

[10] Apart from these theoretical problems, it is experimentally controversial at what stage 'theory of mind' abilities emerge. False-belief tasks were initially proposed as diagnosing a lack of these abilities in normal three-year-olds and their presence in normal four-year-olds [16]. Others have proposed that irrelevant linguistic demands of these tasks deceptively depress three-year-olds' performance. For example, in the 'Maxi' task, the child sees the doll see the chocolate placed in a box, and then the child but not the doll sees the chocolate moved to the drawer. Now if the child is asked 'Where will the doll look for the chocolate *first*?' (instead of 'Where will the doll look for the chocolate?') then children as young as two can sometimes solve the problem [27]. These arguments, if correct, push reasoning about intentions earlier in ontogeny, although the results have been disputed by Clements and Perner [6].

2. awareness of the inertial properties of belief: beliefs do not form spontaneously, but must be generated by limited number of causes, such as perception and inference
3. inhibition of response tendencies when necessary; more generally the involvement of executive function

An agent solving the task correctly first of all needs to have an awareness of the causal relation between perception and belief, component 1. Applied to the situation at hand, this means that Maxi comes to believe that the chocolate is in the box. An application of the principle of inertia (component 2 now yields that Maxi's belief concerning the location of the chocolate persists unless an event occurs which causes him to have a new belief, incompatible with the former. The story does not mention such an event, whence it is reasonable to assume – using closed world reasoning – that Maxi still believes that the chocolate is in the box when he returns to the experimenter's room. An explanation for performance in the false belief also needs to account for the incorrect answers given by children younger than 4 and autists. These subjects almost always answer 'in the drawer', when asked where Maxi believes the chocolate to be.

This is where component 3 comes in. Normal and autistic performance in the false belief task are both analysed as conditional reasoning with instances of the general executive function rule 1. The agent a is supposed to be governed by *response rules* of the type

$$B_a(\varphi) \wedge \neg ab_{a,\varphi} \to R_a(\varphi), \qquad (2)$$

in words

> If agent a Believes φ and nothing *ab*normal is the case, then he Reports φ

The key to understanding performance in the task is the competition between two different instances of 2

1. φ represents the actual location of the chocolate (known to the agent)
2. φ represents Maxi's belief about the location of the chocolate

After substitution, the two resulting response rules can be made to inhibit each other by suitable conditions on the abnormalities. Let p represent the actual location of the chocolate, then we have the following substitution instance of 2

$$B_a(p) \wedge \neg ab_{a,p} \to R_a(p) \qquad (3)$$

To model this, we borrow a notion from executive dysfunction theory, and hypothesise that the 'prepotent response' is always for the child to answer where it knows the chocolate to be. In some children, this response can be inhibited, in other children it cannot, for various reasons that we shall explore below.

This said, there remain intuitive considerations on the false belief task which suggest that some sort of modal principle of positive introspection is operative after all. An experiment by Clements and Perner [6] shows that normal 3 year olds may give the wrong verbal answer in the false belief task, while simultaneously looking at the right place. Hauser's interesting paper 'Knowing about knowing' [12] glosses this result by saying that these 3 year olds have (implicit) knowledge about the right response, but no knowledge of their knowledge, i.e. no explicit knowledge. This distinction can be represented by a slight change in the set-up. We keep the predicate $R_a(p)$ for 'agent a (verbally) reports her belief that p', but introduce a new predicate $A_a(p)$ with the intended meaning 'agent a *acts out* her belief that p', for example by looking. We then get two general response schemata instead of the one given as 3, namely

$$B_b(\varphi) \wedge \neg ab_{b,\varphi} \to A_b(\varphi) \qquad (4)$$

and

$$B_b(B_b(\varphi)) \wedge \neg ab_{b,\varphi} \to R_b(\varphi) \qquad (5)$$

Only positive introspection leads to congruent answers here. That is, the argument given above for the normal child older than 4 now applies to 'acting out' only, i.e. with R_b replaced by A_b everywhere; positive introspection is needed to give the corresponding verbal response The reader may well wonder why we identified B and BB with implicit and explicit belief, respectively. The reason is mainly that matters may be arranged such that $B(B(\varphi))$ implies $B(\varphi)$, but not conversely, so that, at least formally, BB represents a stronger form of belief.

3.3 Executive dysfunction and the box task

Russell's executive function deficit theory [25] takes as basic the observation that autists often exhibit severe perseveration. They go on carrying out some routine when the routine is no longer appropriate, and exhibit great difficulty in switching tasks when the context calls for this (that is, when switching is not governed by an explicit rule). This perseveration, also observed in certain kinds of patients with frontal cortex damage, would give rise to many of the symptoms of autism: obsessiveness, insensitivity to context, inappropriateness of behaviour, literalness of carrying out instructions. Task-switching is the brief of *executive function*, introduced in Section 2.

Perseveration is illustrated in a paradigmatic experiment designed by Hughes and Russell [14], the 'box task' (see Fig.2).

The task is to get the marble which is lying on the platform (the truncated pyramid) inside the box. However, when the subject puts her hand through the opening, a trapdoor in the platform opens and the marble drops out of reach. This is because there is an infrared light-beam behind the opening, which, when interrupted, activates the trapdoor-mechanism. The switch on the left side of the box deactivates the whole mechanism, so that to get the marble you have to flip the switch first. In the standard set-up, the subject is shown how manipulating the switch allows one to retrieve the marble after she has first been tricked by the trapdoor mechanism.

Even though this task is non-verbal, the pattern of results is strikingly similar to that exhibited in the false belief task: normally developing children master this task by about age 4, and before this age they keep reaching for the marble, even when the marble drops out of reach all the time. Autistic children go on failing this task for a long time. The performance on this task is conceptualised as follows. The natural, 'prepotent', plan is to reach directly for the marble, but this plan fails. The child then has to re-plan, taking into account the information

Logic in the study of autism: reasoning ...

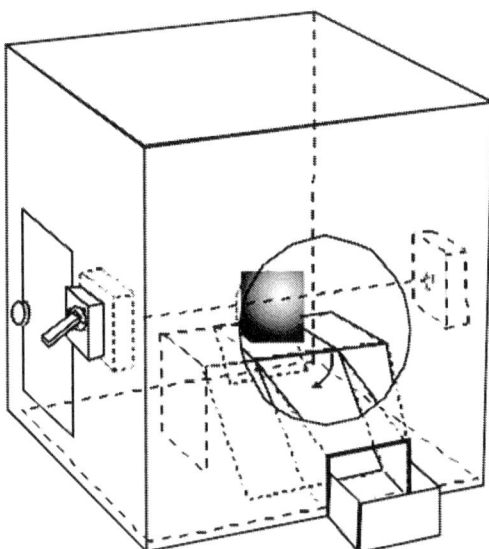

Figure 2: Russell's box task

about the switch. After age 4 the normally developing child can indeed integrate this information, that is, inhibit the pre-potent response and come up with a new plan. It is hypothesised that autists cannot inhibit this prepotent response because of a failure in executive function. But to add precision to this diagnosis we have to dig deeper.

It is important to note here, that the ability to plan and re-plan when the need arises due to changed context, is fundamental to human cognition, no less fundamental than 'theory of mind' abilities. Human beings (and other animals too) act, not on the basis of stimulus-response chains, but on the basis of (possibly distant) goals which they have set themselves. That goal, together with a world-model lead to a plan which suffices to reach the goal in the assumed circumstances. But it is impossible to enumerate *a priori* all events which might possibly form an obstacle in reaching the goal. It is therefore generally wise to keep open the possibility that one has overlooked a precondition, while at the same time not allowing this uncertainty to inhibit one's actions. It is perhaps this flexibility that autists are lacking. This point can be reformulated in logical terms. The autist's concept of a rule is one in which the consequent invariably follows the antecedent. By contrast, a normal subject's rule is more likely to be of the exception-tolerant variety. Indeed, Russell writes (following a suggestion by Donald Peterson)

> [T]aking what one might call a 'defeasibility stance' towards rules is an innate human endowment – and thus one that might be innately lacking ... [H]umans appear to possess a capacity – whatever that is – for abandoning one relatively entrenched rule for some novel ad hoc procedure. The claim can be made, therefore, that this capacity is lacking in autism, and it is this that gives rise to failures on 'frontal' tasks – not to mention the behavioural rigidity that individuals with the disorder show outside the laboratory [26, p.318].

Russell goes on to say that one way this theory might be tested is through the implication that "children with autism will fail to perform on tasks which require an appreciation of the defeasibility of rules such as 'sparrows can fly'." This is what we shall do; but to get started we first need a logical description of what goes on in the box task.

3.3.1 Closed world reasoning in the box task

For the formalisation we borrow some self-explanatory notation from the situation calculus. Let c be a variable over contexts, then the primitives are

- the predicate $do(a, c)$, meaning 'perform action a in context c'
- the function $result(a, c)$, which gives the new context after a has been performed in c.

The actions we need are g ('grab'), u ('switch up'), d ('switch down'). We furthermore need the following context-dependent properties:

- $possess(c)$: the child possesses the marble in c;
- $up(c)$: the switch is up in c (= correct position)
- $down(c)$: the switch is down in c (= wrong position).

The following equations give the rules appropriate for the box task

$$down(c) \land do(u, c) \neg ab'(c) \to up(result(u, c)) \qquad (6)$$
$$do(g, c) \land \neg ab(c) \to possess(result(g, c)) \qquad (7)$$

We first model the reasoning of the normal child > 4 yrs. Initially, closed world reasoning for $ab(c)$ gives $\neg ab(c)$, reducing the rule 7 to

$$do(g, c) \to possess(result(g, c)) \qquad (8)$$

which prompts the child to reach for the marble without further ado. After repeated failure, she reverts to the initial rule 7, and concludes that after all $ab(c)$. After the demonstration of the role of the switch, she forms the condition

$$down(c) \to ab(c) \qquad (9)$$

She then applies closed world reasoning for ab to 9, to get

$$down(c) \leftrightarrow ab(c) \qquad (10)$$

which transforms rule 7 to

$$do(g, c) \land up(c) \to possess(result(g, c)) \qquad (11)$$

Define context c_0 by putting $c = result(u, c_0)$ and apply closed world reasoning to rule 6, in the sense that $ab'(c)$ is set to $\bot$ due to lack of further information, and $\to$ is replaced by $\leftrightarrow$. Finally, we obtain the updated rule, which constitutes a new plan for action

$$down(c_0) \land do(u, c_0) \land c = result(u, c_0) \land do(g, c) \to possess(result(g, c)) \qquad (12)$$

As in the previous tasks, both the normal child younger than 4, and the autistic child are assumed to operate in effect with a rule of the form

$$do(g,c) \to possess(result(g,c)) \tag{13}$$

which cannot be updated, only replaced *in toto* by a new rule such as 12.

It is tempting to speculate on the computational complexities of both these procedures. Russell wrote that 'humans appear to possess a capacity – whatever that is – for abandoning one relatively entrenched rule for some novel ad hoc procedure [26, p.318]'. The preceding considerations suggest that 'abandoning one relatively entrenched rule' may indeed be costly, but that normal humans get around this by representing the rule in such a way that it can be easily updated. It is instructive to look at the computation that the normal child older than 4 is hypothesised to be performing. The only costly step appears to be the synthesis of the rule 9; the rest is straightforward logic programming, hence efficient. Rule 7 is never abandoned; a new rule is derived without having to ditch 7 first.

To close this discussion, we compare the false belief task to the box task from the point of view of the formal analysis. The tasks are similar in that for successful solution one must start from rules of the form $A \wedge \neg ab \to E$, identify conditions which constitute an abnormality, and apply closed world reasoning; and also that in both cases a failure of ab to exercise its inhibitory function leads to the inability to inhibit the prepotent response. A difference is that in the false belief task, one needs a 'theory' relating ab to sensory, or inferred, information, whereas it suffices to operate with rules for actions in the box task.

4 A task with the same formal structure: the suppression task

When considered formally, the tasks mentioned have a logical structure in common, besides showing undeniable differences. The common logical structure is closed world reasoning applied to possible exceptions. The question is whether the formal analogies between the tasks are indicative of a single cognitive function exercised in these tasks. We claim there is: it is executive function, conceptualised as reasoning with exception-tolerant, mutually inhibiting conditionals. Here one shouldn't think of reasoning as a necessarily conscious activity. Whether reasoning becomes conscious or not has to do with the complexity of the logic. Closed world reasoning as formulated in logic programming has an efficient neural (implementation [13]; and [32, chap.8]), and may therefore be executed largely unconsciously. Classical reasoning has no efficient neural implementation, and for this reason it may be more laborious.

It therefore becomes of interest to see whether one can devise a task which investigates the form of reasoning involved in executive function. Surprisingly, a task with the required properties has been around for some time, although it was not treated as such: Byrne's 'suppression task' [5].

If one presents a subject with the following premises:

1. (a) *If she has an essay to write she will study late in the library.*
   (b) *She has an essay to write.*

roughly 90% of subjects[11] draw the conclusion 'She will study late in the library'. Next suppose one adds the premise

2. *If the library is open, she will study late in the library.*

and one asks again: what follows? In this case, only 60% concludes 'She will study late in the library'. This known as the 'suppression' of *modus ponens*.

However, if instead of the above, the premise

3. *If she has a textbook to read, she will study late in the library*

is added, then the percentage of 'She will study late in the library'-conclusions is around 95%.

In this type of experiment one investigates not only *modus ponens* (MP), but also *modus tollens* (MT), and the 'fallacies' *affirmation of the consequent* (AC), and *denial of the antecedent* (DA), with respect to both types of added premises, 2 and 3. The results are that MT is suppressed in the presence of a premise of the form 2 (but not 3), and that both AC and DA are suppressed in the presence of a premise of the form 3 (but not 2).

4.1 A formal analysis

Byrne viewed the suppression effect mainly as showing that subjects are not guided by the rules of classical logic, but instead let their inferences be determined by semantic content. We believe a more informative account of the suppression effect can be given, also establishing its relevance outside the reasoning domain, namely as showing that (normal) subjects are capable of flexible management of rules in context. For instance, normal subjects generally allow rules to have exceptions (and actions to have unknown preconditions), and they are quite good at exception-handling. This capacity involves some form of closed world reasoning. To take our paradigmatic example, in

'If Marian has an essay to write she will study late in the library.
Marian has an essay to write.'

no exception is made salient, therefore the subject can draw the *modus ponens* inference: 'She will study late in the library'. The addition of premise 2

'If the library is open, Marian will study late in the library'

makes salient a possibly disabling condition in first rule, namely the library's being shut. But since no other disabling conditions are mentioned, it is assumed that there aren't any.

Speaking informally, we represent conditionals such as

If Marian has an essay to write she will study late in the library.

as *defaults* of the form

If Marian has an essay to write, *and nothing abnormal is the case*, she will study late in the library.

---

[11] The figures we use come from the experiment reported in [7], since the experiments reported in this study have more statistical power than those of [5].

As in our formalisation of the box task, the italicised phrase introduces an overt marker for a possible abnormality or unknown precondition, which can be given concrete semantic content by other material given by the discourse. The claim is that this is a natural thing for a subject to do, because most rules indeed have exceptions, or unstated preconditions. The task at hand is to turn this intuition into a formal model.

Formally, we write a conditional as

$$p \wedge \neg ab \rightarrow q,$$

where ab is again a proposition letter representing an unspecified abnormality. As before, the logic governing ab is closed world reasoning.

In general, one may give ab concrete content by adding implications of the form

$$s \rightarrow ab,$$

which express that the eventuality denoted by s constitutes an abnormality. Now suppose that there are n such implications in all, i.e., we have the implications

$$s_1 \rightarrow ab, \ldots, s_n \rightarrow ab.$$

In the absence of further implications beyond the n mentioned, we want to conclude that we have listed *all* abnormalities. This can be done by *defining*[12] ab as

$$ab \leftrightarrow \bigvee_{i \leq n} s_i.$$

Two special cases are of particular interest. If $n = 1$, i.e. if we only have the implication $s \rightarrow ab$, the definition yields $ab \leftrightarrow s$. Furthermore, for the case $n = 0$, the definition entails that ab is false, i.e. $\neg mathitab$. That is, if there is no information about the abnormality ab, we assume it does not occur.

These formal stipulations will help us explain the logic behind the suppression task. We do two illustrative cases; for the full treatment we refer to [30].

Modus ponens Consider again

> If Marian has an essay to write she will study late in the library.
> Marian has an essay to write.

Formally, this becomes

$$p \wedge \neg ab \rightarrow q; \; p.$$

Closed world reasoning yields $\neg ab$, which suffices to draw the conclusion q. Therefore modus ponens also follows in this non-classical context, once closed world reasoning is applied. Failure to apply modus ponens may then be evidence of a resistance to apply closed world reasoning to the abnormality.

The situation becomes slightly more complicated in the case of a further type 3 premise:

> If Marian has an essay to write she will study late in the library.
> If Marian has an exam she studies late in the library.
> Marian has an essay to write.

---

[12]This form of definition is technically known as the *completion*.

There are now two conditional premises, each with its own disabling abnormality. The formalisation thus becomes

$$p \wedge \neg ab \to q;\ r \wedge \neg ab' \to q;\ p.$$

Since the discourse does not provide information either about ab or about ab', they are both set to false, that is, we have $\neg ab$ and $\neg ab'$. The discourse thus becomes equivalent to

$$p \vee r \to q;\ p,$$

which again justifies the conclusion q.

Real complications arise in the case of a premise of type 2:

> If Marian has an essay to write she will study late in the library.
> If the library is open Marian studies late in the library.
> Marian has an essay to write.

Again there are two conditional premises, each with its own disabling abnormality, but in this case there is interaction, because the antecedent of the second conditional highlights a possible precondition. The formalisation is therefore not

$$p \wedge \neg ab \to q;\ r \wedge \neg ab' \to q;\ p,$$

as it was in the previous case, but rather

$$p \wedge \neg ab \to q;\ r \wedge \neg ab' \to q;\ \underline{\neg r \to ab};\ p,$$

where the added underlined implication reflects the assumption that the second conditional has made an abnormality for the first conditional salient. Closed world reasoning applied to this implication yields $ab \leftrightarrow \neg r$, and if we then substitute r for $\neg ab$ in the first conditional we get

$$p \wedge r \to q,$$

to which modus ponens can no longer be applied. The conclusion from this formal exercise is that suppression of modus ponens can be explained as an instance of closed world reasoning. This is definitely *not* to say that subjects *should* choose this underlying formal representation. It is very well possible to stick to the classical interpretation of the conditional, not containing a marker for a possible exception, in which case modus ponens should not be suppressed – indeed this is a plausible hypothesis to explain what autists appear to be doing.

Denial of the antecedent Classical. fallacies and their suppression can be explained similarly. As an example we treat denial of the antecedent, in the case of the premises

> If Marian has an essay to write she will study late in the library.
> Marian does not have an essay to write.

The premises can be formalised as

$$p \wedge \neg ab \to q;\ \neg p,$$

and since there is no information about ab, by closed world reasoning one may assume $\neg ab$. This particular inference involves more closed world reasoning however: one also has to assume that, in the absence of further information, $p \wedge \neg ab$ is the *only* reason to conclude q, so that we have in effect

$$q \leftrightarrow p \wedge \neg ab.$$

Given $\neg p$, it indeed follows from this that $\neg q$.[13]

Suppose we now add a further conditional premise of type 3, to get

> If Marian has an essay to write she will study late in the library.
> If Marian has an exam she studies late in the library.
> Marian does not have an essay to write.

The formalisation is

$$p \wedge \neg ab \rightarrow q;\ r \wedge \neg ab' \rightarrow q;\ \neg p.$$

Closed world reasoning yields $\neg ab$ and $\neg ab'$, which reduces the formalised premises to

$$p \rightarrow q;\ r \rightarrow q;\ \neg p.$$

Closed world reasoning applied to the two implications $p \rightarrow q$ and $r \rightarrow q$ yields

$$q \leftrightarrow p \vee r,$$

from which given only $\neg p$ nothing follows. The addition of the second conditional premise may thus lead to a suppression of DA inferences. The reader may verify that nothing happens in the case of an extra premise of the form 2.

It is of some importance for our discussion of the autism data to distinguish the two forms of closed world reasoning that play a role here. On the one hand there is the closed world reasoning applied to abnormalities or exceptions, which takes the form: 'assume only those exceptions occur which are explicitly listed'. On the other hand there is the closed world reasoning applied to rules, which takes the form of diagnostic reasoning: 'if B has occurred and the only known rules with B as consequent are $A_1 \rightarrow B, \ldots, A_n \rightarrow B$, then assume one of $A_1, \ldots, A_n$ has occurred'. These forms of closed world reasoning are in principle independent, and in our autist population we indeed see a dissociation between the two.

5 Autists' performance in the suppression task

In order to test these hypotheses, formulated generally as

> 4. Autists can apply closed world reasoning, but have a decreased ability in handling exceptions to rules.

---

[13] Thus from the point of view of closed world reasoning, denial of the antecedent is not a fallacy at all.

an experiment was carried out at the F.C. Donders Center for Cognitive Neuroscience Nijmegen, in collaboration with the Psychiatry Department of the Radboud University Nijmegen Medical Centre.[14] The full results will be published in Pijnacker et al. [24]; here we give a synopsis.

The materials consisted of the inference patterns MP, MT, DA, AC in the standard two-premise form, in the extended form with an additional premise (of type 2) and with an alternative premise (of type 3). There were 120 reasoning problems in all. Subjects were 28 high-functioning autists (i.e. with $IQ > 85$), and 28 controls matched for IQ, age, gender and handedness. The problems were presented to the subjects on a computer. After the question was presented (e.g. 'Does she study late in the library?'), the subjects could answer 'yes', 'no', or 'maybe', by pressing the corresponding keys. Reaction times were measured, but these will not be discussed here.[15]

The prediction was that autists show less suppression on MP and MT with additional premise, but behave like the normal controls otherwise. Table II presents the data.

In the two-premise inferences, both groups endorsed MP and MT at equally high rates, and did not differ significantly in the number of yes responses for MP and no responses for MT. Endorsement of AC and DA (i.e. a yes response for AC and no response for DA) was at a lower rate in both groups. There was a trend for the autistic group to endorse AC and DA less often than the control group ($p = 0.054$).

Suppression of MP and MT in the case of an additional premise is indeed significantly rarer in autists than in normals ($p = 0.025$). By contrast, there is no significant difference between the groups for additional premises in the inferences DA and AC, and no significant difference for alternative premises in all conditions.

These results are thus consistent with the hypothesis that autists are specifically impaired in the processing involved in the incorporation of exceptions. The autists have no trouble incorporating an extra conditional premise in the discourse, as is shown by the results for AC and DA with alternative premise. Also the impact of alternative premises in a closed world context is quite well understood. It is specifically the exception-highlighting character of the additional premise that causes the problems. We have some qualitative data illustrating the peculiarity of the responses here; see Section 7.

6 Inhibition, closed world reasoning, and the neural substrate

We certainly do not claim that autism, say, is completely characterised by some peculiarities in reasoning (either explicit or implicit) *vis à vis* normal controls. What is of most importance is that the formal structure tells us something about

---

[14]The experiment was a larger-scale replication of an earlier pilot experiment performed by Heleen Smid on 6 subjects [28]. The results were so striking it was decided to go for a replication which could yield statistically significant results.

[15]This experiment formed part of a larger experimental program, investigating pragmatics in autism. Interestingly, in standard Gricean pragmatic inferences (e.g. from 'Some real estate agents are crooks' to 'Some real estate agents are not crooks') no differences between the autists and the normal controls could be found.

argument pattern	base	+ alternative	+ additional
MPc	96.1%	97.5%	51.1%
MPa	89.6%	92.9%	71.0%
MTc	92.8%	95.0%	45.0%
MTa	79.6%	90.3%	62.1%
DAc	69.1%	10.4%	33.6%
DAa	48.0%	15.7%	28.9%
ACc	67.1%	9.7%	35.7%
ACa	45.0%	12.2%	28.1%

Table II: Results on suppression task in autists (n=28) and matched controls (n=28). The letters 'c' and 'a' refer to controls and autists, respectively.

underlying neural peculiarities. In the case of autism we saw that the analysis of the relevant diagnostic tasks highlighted the role of closed world reasoning about unknown preconditions, which at the level of neural implementation was seen to be intimately related to inhibition; and we could then appeal to data showing that for various reasons inhibition is compromised in autism. The full story is given in Stenning and van Lambalgen [32, chap.9], but here is the rough outline.

In the neural implementation of closed world reasoning given in [32], Chapter 8 the effect of $\neg ab$ in the basic executive function rule $A \wedge \neg ab \to E$ is modelled by an inhibitory neuron which ensures that no activation can flow from neuron A to neuron E. In this context, it is of some interest that there are known peculiarities with inhibition in the autistic brain.

- The growth pattern of autistic brain is different from normal growth. It has often been reported that autists have rather large heads, typically about 15–40% having *macrocephaly*, defined as a head circumference greater than two standard deviations about the mean [20]. The temporal profile of head growth is also different: a growth spurt in the first two years, followed by much decreased growth (compared to normals). The inhibitory interneurons are the latest neurons to mature, so that the autistic brain may have relatively underdeveloped inhibitory neurons.

- An important component of brain development is pruning, the cutting-away of synapses that are little used. In normals, pruning affects excitatory synapses more than inhibitory synapses (see Luciana [18], p.161), but this is not so in autists [17]. Thus the autistic brain may end up having an imbalance between excitation and inhibition, and C. Frith [8] has proposed this as one source of deviant autistic cognition.

Of course, many gaps have to be filled before these connections can be made tight, but it is at least interesting that several indicators point in the same direction.

7 Conclusion

The logical analysis of standard tasks diagnostic for autism identified a common element in these tasks, namely closed world reasoning with exceptions. It was hypothesised that autists have a specific difficulty with this form of reasoning, and this hypothesis was put to the test using the suppression task whose main component is reasoning about exceptions. Logical analysis allowed one to make a very precise prediction: less suppression in the case of MP and MT with additional premise, no difference from normals otherwise. This prediction is much more precise than what one would get from processing accounts of autism such as Weak Central Coherence theory [11], which would predict difficulties with discourse integration *per se*, not the observed differentiation.

The logical analysis also shows that aetiologies like 'theory of mind deficit' and 'executive dysfunction' may not be mutually exclusive after all. If the analysis given here is correct, 'theory of mind' calls upon executive function to do much of the work, and the 'theory' component added is not concerned specifically with other minds, but with the relation between beliefs and perception, inference, and other information sources. Thus, there is little reason to speculate about a theory of mind module [16].

Lastly, a logical analysis is not in principle inimical to considerations of neural implementation. It depends very much on the result of the analysis: if it is some form of closed world reasoning, a neural network implementation can be found. This is of course not yet an implementation in neuronal circuitry as it occurs in the brain, but it is at least suggestive. The link between reasoning with exceptions and inhibition in the neural network model suggests looking for peculiarities with inhibition in the autistic brain, and there indeed turned out to be evidence pointing in that direction.

Appendix:
what autists say when doing the suppression task

We now present some conversations with our subjects while engaged in the suppression task.[16] The subjects were presented with either two or three premises, and were asked whether another sentence was true, false or undecided. We then proceeded to ask them for a justification of their answer.

Excerpts from dialogues: MP We recall the argument:

> If Marian has an essay to write she will study late in the library.
> (*) If the library stays open then Marian will study late in the library.
> Marian has an essay to write.
> Does Marian study late in the library?

Here is subject C, first engaged in the two-premise case, i.e. without (*):

> C: But that's what it says!
> E: What?
> C: If she has an essay then she studies.
> E: So your answer is 'yes'?
> C: Yes.

The same subject engaged in the three-premise argument:

---

[16] The data are taken from the earlier experiment by Heleen Smid [28].

C. *Yes*, she studies late in the library.
E. Ehh, why?
C. Because she *has to write* an essay.

Clearly the possible exception highlighted by the second conditional is not integrated; the emphasis shows that the first conditional completely overrides the second.

Excerpts from dialogues: MT In this case the argument pattern is

> If Marian has an essay to write she will study late in the library.
> (*) If the library stays open then Marian will study late in the library.
> Marian will not study late in the library.
> Does Marian have an essay?

Here is again subject C:

> C. No, she has ... oh no, wait a minute ... this is a bit strange isn't it?
> E. Why?
> C. Well, it says here: if she *has to write* an essay ... And I'm asked whether she has to write an essay?
> E. Mmm.
> C. I don't think so.

This is probably evidence of the general difficulty of MT, but note that the second conditional does not enter into the deliberations. In the dialogue, E. then prompts C. to look at second conditional, but this has no effect: C. sticks to his choice.

Here is a good example of the way in which a subject (in this case B) can be impervious to the suggestions of the experimenter. The dialogue refers to the argument with three premises; we give a rather long abstract to show the flavour of the conversations.

> B: No. Because if she had to make an essay, she would study in the library.
> E: Hmhm.
> B: And she doesn't do this, so she doesn't have an essay.
> E: Yes.
> B: And this means ... (inaudible)
> E: (laughs) But suppose she has an essay, but the library is closed?
> B: Ah, that's also possible.
> E: Well, I'm only asking.
> B: Well, according to these two sentences that's not possible, I think.
> E: How do you mean? B: Ehm, yes she just studies late in the library if she has an essay.
> E: Hmhm.
> B: And it does not say 'if it's open, or closed ...'
> E: OK.
> B: So according to these sentences, I know it sounds weird, but ....
> E: Yes, I
> B: I know it sounds rather autistic what I'm saying now (laughs).

E: (laughs) B: Eh yes.
E: So it is like you said? Or perhaps that she
B: Yes, perhaps the library closes earlier?
E: You may say what you want! You don't have to try to think of what should be the correct answer!
B: OK, no, then I'll stick to my first answer.
E: OK, yes.
B: (laughs) I know it's not like this, but (laughs).
E: Well, that's not clear. It's possible to say different things about reasoning here, and what you say is certainly not incorrect.

In the above we have seen examples of how our autistic subjects refuse to integrate the information about exceptions provided by (*). The next extracts show that this need not be because they are incapable of integrating a second conditional premise, or of applying closed world reasoning. We consider the 'fallacies' DA and AC, which can be valid if seen as a consequence of closed world reasoning, and which can be suppressed by supplying a suitable second conditional premise, e.g. (†) below.

Excerpts from dialogues: AC The argument is

If Marian has an essay to write she will study late in the library.
(†) If Marian has an exam, she will study late in the library.
Marian studies late in the library.
Does Marian have an essay?

Here is subject C, in the two-premise argument without (†).

C: Yes.
E: Why?
C: It says in the first sentence 'if she has an essay then she does that [study late etc.] ... But Marian is just a name, it might as well be Fred.

Now consider the three-premise case.

C: Mmm. Again 'if', isn't it?
E: Yes.
C: If Marian has an essay, she studies late in the library'...
E: Yes.
C: If Marian has an essay, she studies late in the library ...
E: Hmhm.
C: Marian studies late in the library. Does Marian have an essay?
E: Hmhm.
C: No.
E: Hmhm.
C: It does say she has to make an essay.
E: Hmhm. But she studies late in the library, can you conclude from this that she has to make an essay?
C: No you can't, because she could also have an exam.

We see in this example that C correctly judges the import of (†): after having applied closed world reasoning to the two-premise case, he notices that it is powerless in this case.

Bibliography

[1] Baggio, G., van Lambalgen, M. Processing consequences of the imperfective paradox. *Journal of Semantics*, 24(3), 2007.

[2] Baggio, G., van Lambalgen, M., Hagoort, P. Computing and re-computing discourse models: an ERP study. *Submitted*, 2007.

[3] Baggio, G., van Lambalgen, M., Hagoort, P. Principles of semantic unification. *Submitted*, 2007.

[4] Barwise, J. Model-theoretic logics: Background and aims. In Barwise, J., Feferman, S. (Eds.), *Model-Theoretic Logics*, chap. I. New York: Springer-Verlag, 1985.

[5] Byrne, R. Suppressing valid inferences with conditionals. *Cognition*, 31:61–83, 1989.

[6] Clements, W., Perner, J. Implicit understanding of belief. *Cognitive Development*, 9:377–395, 1994.

[7] Dieussaert, K., Schaeken, W., Schroyen, W., d'Ydewalle, G. Strategies during complex conditional inferences. *Thinking and reasoning*, 6(2):125–161, 2000.

[8] Frith, C. What do imaging studies tell us about the neural basis of autism? In Bock, J., Goode, G. (Eds.), *Autism : neural basis and treatment possibilities, Novartis Foundation symposium*, 149–166. Wiley, 2003.

[9] Hagoort, P., Berkum, J. v. Beyond the sentence given. *Philosophical Transactions of the Royal Society B: Biological Sciences*, 362(1481):801–811, 2007.

[10] Hamm, F., Kamp, H.and van Lambalgen, M. There is no opposition between formal and cognitive semantics. *Theoretical Linguistics*, 32(1):1–40, pp.41–115 have comments on this target article., 2006.

[11] Happé, F. *Autism: an introduction to psychological theory*. London: UCL Press, 1994.

[12] Hauser, M. Knowing about knowing: Dissociations between perception and action systems over evolution and during development. *Annals of the New York Academy of Sciences*, 1:1–25, 2003.

[13] Hoelldobler, S., Kalinke, Y. Towards a massively parallel computational model of logic programming. In *Proceedings ECAI94 Workshop on combining symbolic and connectionist processing*, 68–77. ECAI, 1994.

[14] Hughes, C., Russell, J. Autistic children's difficulty with disengagement from an object: its implications for theories of autism. *Developmental Psychology*, 29:498–510, 19993.

[15] Kamp, H., Reyle, U. From discourse to logic, introduction to modeltheoretic semantics of natural language, formal logic and discourse representation theory, part 1. In *Studies in Linguistics and Philosophy*, vol. 42. Dordrecht: Kluwer Academic Publishers, 1993.

[16] Leslie, A. Pretence and representation: the origins of a 'theory of mind'. *Psychological Review*, 94:412–26, 1987.

[17] Levy, J. Connectionist models of over-specific learning in autism. In Bowman, H., Labiouse, C. (Eds.), *Connectionist Models of Cognition and Perception*, vol. II, 115–126. World Scientific, 2004.

[18] Luciana, M. The neural and functional development of human prefrontal cortex. In Johnson, M. H., Haan, M. D. (Eds.), *The cognitive neuroscience of development*, chap. 7, 157–179. Hove: Psychology Press, 2003.

[19] Marr, D. *Vision: A Computational investigation into the human representation and processing of visual information*. San Fransisco: W.H. Freeman, 1982.

[20] Miles, J. H. L. T. T., Hillman, R. Head circumference is an independent clinical finding associated with autism. *American Journal of Medical Genetics*, 95:339–350, 2000.

[21] Oaksford, M., Chater, N. *Bayesian rationality*. Oxford: Oxford University Press, 2007.

[22] Perner, J., Leekham, S., Wimmer, H. Three-year olds' difficulty with false belief: the case for a conceptual deficit. *British Journal of Developmental Psychology*, 5:125–137, 1987.

[23] Piaget, J. *Logic and psychology*. Manchester: Manchester University Press, 1953.

[24] Pijnacker, J., Geurts, B., van Lambalgen, M., Buitelaar, J., Kan, C., Hagoort, P. Conditional reasoning in high-functioning adults with autism. *Submit*, 2007.

[25] Russell, J. (Ed.). *Autism as an executive disorder*. Oxford: Oxford University Press, 1997.

[26] Russell, J. Cognitive theories of autism. In Harrison, J., Owen, A. (Eds.), *Cognitive deficits in brain disorders*, 295–323. London: Dunitz, 2002.

[27] Siegal, M., Beattie, K. Where to look first for children's knowledge of false beliefs. *Cognition*, 38:1–12, 1991.

[28] Smid, H. Reasoning with rules and exceptions in autism, 2005. URL `http://staff.science.uva.nl/~michiell`.

[29] Stenning, K., van Lambalgen, M. A little logic goes a long way: basing experiment on semantic theory in the cognitive science of conditional reasoning. *Cognitive Science*, 28(4):481–530, 2004.

[30] Stenning, K., van Lambalgen, M. Semantic interpretation as reasoning in non-monotonic logic: the real rule-following. *Cognitive Science*, 29(6):919–960, 2005.

[31] Stenning, K., van Lambalgen, M. Logic in the study of psychiatric disorders: executive function and rule-following. *Topoi, Special issue on Logic and Cognitive Science*, 26(1):97–114, 2007.

[32] Stenning, K., van Lambalgen, M. *Human reasoning and cognitive science*. Cambridge, MA: MIT Press, 2008.

[33] Stenning, K., van Lambalgen, M. Logic in a noisy world. In Oaksford, M. (Ed.), *The psychology of conditionals*. Oxford: Oxford University Press, 2008.

[34] Stevenson, R., Over, D. Deduction from uncertain premises. *Quarterly Journal of Experimental Psychology A*, 48(3):613–643, 1995.

[35] Tomasello, M. *Constructing a language. A usage-based theory of language acquisition*. Boston: Harvard University Press, 2003.

[36] van Lambalgen, M., Hamm, F. *The proper treatment of events*. Oxford and Boston: Blackwell, 2004.

[37] Wason, P. C. Reasoning about a rule. *Quarterly Journal of Experimental Psychology*, 20:273–281, 1968.

Causal inference from time series: What can be learned from Granger causality?[1]

Michael Eichler

University of Maastricht

m.eichler@ke.unimaas.nl

ABSTRACT. In time series analysis, inference about cause-effect relationships among multiple time series is commonly based on the concept of Granger causality, which exploits temporal structure to achieve causal ordering of dependent variables. One major and well known problem in the application of Granger causality for the identification of causal relationships is the possible presence of latent variables that affect the measured components and thus lead to so-called spurious causalities. In this paper, we present a new graphical approach for describing and analysing Granger-causal relationships in multivariate time series that are possibly affected by latent variables. We show how such representations can be used for inductive causal learning from time series and discuss the underlying assumptions and their implications for causal learning.

1 Introduction

The notion of causality and the identification of new causal relationships play a central role in scientific research. In time series analysis, inference about cause-effect relationships is commonly based on the concept of Granger causality [13], which is defined in terms of predictability and exploits the direction of the flow of time to achieve a causal ordering of dependent variables. This concept of causality does not rely on the specification of a scientific model and thus is particularly suited for empirical investigations of cause-effect relationships. On the other hand, it is commonly known that Granger causality basically is a measure of association between variables and thus can lead to so-called spurious causalities if important relevant variables are not included in the analysis [17]. Since in most analyses involving time series data the presence of latent variables that affect the measured components cannot be ruled out, this raises the question whether and how the causal structure can be recovered from time series data.

Recent advances in the understanding of such latent variable structures were based on graphical models, which provide a general framework for describing and inferring causal relations [e.g. 21, 24]. For time series, this graphical approach for the discussion of causal relationships in systems that are affected by latent variables has been considered first by Eichler [5]. Based on previously

---

[1] The author is with Department of Quantitative Economics, University of Maastricht, P.O. Box 616, 6200 MD Maastricht, The Netherlands.

developed graphical representations of Granger-causal relationships in multivariate time series [4, 8], a new class of path diagrams for the representation of the interrelationships in multivariate time series with latent variables has been introduced. These general path diagrams allow a more complete encoding of the conditional independencies if the system is affected by latent variables. In Eichler [5], a multi-step procedure for the identification of such general path diagrams was proposed although no definite algorithm has been given.

In this paper, we discuss graphical representations for multivariate dynamic systems affected by latent variables in more detail. For this, we review in Section 2 the concept of Granger causality as originally introduced by Granger [13, 14] and in Section 3 related graphical representation for multivariate time series and their Markov properties. In Section 4, we discuss some properties of general path diagrams for systems affected by latent variables and introduce dynamic ancestral graphs. An approach for causal learning from time series data is presented in Section 5. More details and proofs for the results presented will be provided in a longer technical version of the paper. Section 6 contains some concluding remarks.

2 Causality in multivariate time series

While controlled experiments still provide the ideal framework for causal analysis, many complex phenomena such as information processing in the brain can only be studied from non-experimental or quasi-experimental data. It is therefore important to have an operational definition of causality that allows inference about cause-effect relationships also from observational studies. For multivariate time series, such a definition has been introduced by Granger [13, 14]. In this section, we review this concept and problems related to it.

2.1 Granger causality

Suppose that $X = \bigl(X(t)\bigr)_{t \in \mathbb{Z}}$ and $Y = \bigl(Y(t)\bigr)_{t \in \mathbb{Z}}$ are two stationary time series that are statistically dependent on each other. When is it justified to say that the one series X causes the other series Y? In order to come up with a general definition, Granger [13, 14] evokes the following two fundamental principles:

1. The effect does not precede its cause in time.
2. The causal series contains unique information about the series being caused that is not available otherwise.

The first principle of temporal precedence of causes is commonly accepted and has been also the basis for other probabilistic theories of causation [e.g. 11, 12, 29]. In contrast, the second principle is more subtle as it requires the separation of the special information provided by the former series X from any other possible information. To this end, Granger considers two information sets:

- $\mathcal{I}^*(t)$ - the set of all information in the universe up to time t;
- $\mathcal{I}^*_{-X}(t)$ - the same information set excluding the values of series X up to time t.

Here it is assumed that all variables in the universe are measured at equidistant points in time, namely $t \in \mathbb{Z}$. Now, if the series X causes series Y, we expect by the above principles that the probability distributions of $Y(t+1)$ conditionally on the two information sets $\mathcal{I}^*(t)$ and $\mathcal{I}^*_{-X}(t)$ differ.

Granger's definition of causality (1969, 1980). We say that the series X *causes* the series Y if

$$\mathbb{P}\big(Y(t+1) \in A | \mathcal{I}^*(t)\big) \neq \mathbb{P}\big(Y(t+1) \in A | \mathcal{I}^*_{-X}(t)\big)$$

for some measurable set $A \subseteq \mathbb{R}$ and all $t \in \mathbb{Z}$.

It is important to note that this concept of causality covers only direct causal relationships. For example, if X affects Y only via a third series Z, then $\mathcal{I}^*_{-X}(t)$ comprises the past values of Z and $Y(t+1)$ is independent from the past values of X given $\mathcal{I}^*_{-X}(t)$. Implicitly, the separation of the two information sets $\mathcal{I}^*(t)$ and $\mathcal{I}^*_{-X}(t)$ is based on the assumption that the universe considered is discretized not only in time but also in space.

Because of the metaphysical character of the information set $\mathcal{I}^*(t)$, the above definition cannot be used with actual data. In practice, only the background knowledge available at time t can be incorporated into an analysis. Therefore, the definition must be modified to become operational.

In the sequel, suppose that we have observed variables X_v, $v \in V$, and let $\boldsymbol{X}_V = (\boldsymbol{X}_V(t))_{t \in \mathbb{Z}}$ be the corresponding multivariate time series. Substituting the information set $\mathcal{I}^*(t)$ by the information set $\mathcal{I}_V(t)$ generated by the values of $\boldsymbol{X}_V$ up to time t and, similarly, $\mathcal{I}^*_{-X}(t)$ by $\mathcal{I}_{V \setminus \{b\}}(t)$, the information set generated by the values of $\boldsymbol{X}_{V \setminus \{b\}}$ up to time t, we obtain the following modified version of the above definition [14, 15].

Definition 1 *Let $a, b \in V$.*

(i) *The series X_b is said to be a* prima facie *cause of the series X_a with respect to $\mathcal{I}_V$ if*

$$\mathbb{P}\big(X_a(t+1) \in A | \mathcal{I}_V(t)\big) \neq \mathbb{P}\big(X_a(t+1) \in A | \mathcal{I}_{V \setminus \{b\}}(t)\big)$$

for some measurable set $A \subseteq \mathbb{R}$ and all $t \in \mathbb{Z}$.

(ii) *The series X_b does not cause the series X_a with respect to $\mathcal{I}_V$ if*

$$\mathbb{P}\big(X_a(t+1) \in A | \mathcal{I}_V(t)\big) = \mathbb{P}\big(X_a(t+1) \in A | \mathcal{I}_{V \setminus \{b\}}(t)\big)$$

for all measurable sets $A \subseteq \mathbb{R}$ and all $t \in \mathbb{Z}$.

The condition in (ii) is equivalent to that $X_a(t+1)$ and $\mathcal{I}_V(t)$ are independent conditionally on $\mathcal{I}_{V \setminus \{b\}}(t)$, abbreviated as $X_a(t+1) \perp\!\!\!\perp \mathcal{I}_V(t) | \mathcal{I}_{V \setminus \{b\}}(t)$. In that case, we say that X_b is *Granger-noncausal* for X_a with respect to $\mathcal{I}_V$; otherwise we say that X_b *Granger-causes* X_a with respect to $\mathcal{I}_V$. We note that besides this *strong* version of Granger causality the notions of *Granger causality in mean* [14, 15] and *linear Granger causality* [9, 16] exist.

It is clear from the general definition given above that Granger intended the information to be chosen as large as possible including all available and possibly relevant variables. Despite of this, most (econometric) textbooks [e.g. 22] introduce Granger causality only in the bivariate case. This has lead to some confusion about a multivariate definition of Granger causality [e.g. 18].

2.2 The problem of spurious causality

One major drawback of the above operational definition of causality is its dependence on the information set $\mathcal{I}_V$. If two or more variables are jointly affected by variables that are not included in the analysis and hence in the information set, this can induce conditional dependences among the observed variables that are wrongly interpreted as causal relationships. It therefore becomes necessary to find criteria to distinguish such spurious causalities from true cause-effect relationships.

A first step in this direction is the paper by Hsiao [17], who discussed causal patterns for vector time series of three variables. The general idea is that direct causes—described by Granger's general definition—persist regardless of the background information used for the analysis whereas indirect as well as spurious causes can be identified by either adding new variables to the analysis or removing already included variables from the information set.

Patterns of causality (Hsiao, 1982)

(a) X_1 is a *direct cause* of X_2 if X_1 Granger-causes X_2 with respect to $\mathcal{I}_{\{1,2\}}$ as well as with respect to $\mathcal{I}_{\{1,2,3\}}$.

(b) X_1 is an *indirect cause* of X_2 if
- X_1 Granger-causes X_2 with respect to $\mathcal{I}_{\{1,2\}}$ but not with respect to $\mathcal{I}_{\{1,2,3\}}$ and
- X_1 Granger-causes X_3 and X_3 Granger-causes X_2 both with respect to $\mathcal{I}_{\{1,2,3\}}$.

(c) X_1 is a *spurious cause of type II* for X_2 if
- X_1 Granger-causes X_2 with respect to $\mathcal{I}_{\{1,2\}}$ but not with respect to $\mathcal{I}_{\{1,2,3\}}$ and
- X_3 Granger-causes X_1 and X_2 with respect to $\mathcal{I}_{\{1,2,3\}}$.

(d) X_1 is a *spurious cause of type I* for X_2 if
- X_1 Granger-causes X_2 with respect to $\mathcal{I}_{\{1,2,3\}}$ but not with respect to $\mathcal{I}_{\{1,2\}}$ and
- X_1 Granger-causes X_3 and X_3 Granger-causes X_2 both with respect to $\mathcal{I}_{\{1,2,3\}}$.

This characterization of causal patterns could be generalized to higher dimensions although the formulation of the appropriate conditions seems technical. In the following sections, we discuss an alternative approach based on graphical representations of Granger-causal relationships.

3 Graphical representations for multivariate time series

Let $\boldsymbol{X}_V = \big(\boldsymbol{X}_V(t)\big)_{t \in \mathbb{Z}}$ with $\boldsymbol{X}_V(t) = \big(X_v(t)\big)_{v \in V}'$ be the vector time series of interest. For simplicity, we assume that $\boldsymbol{X}_V$ is a stationary Gaussian process with mean zero and covariances $\boldsymbol{\Gamma}(u) = \mathbb{E}\boldsymbol{X}_V(t)\boldsymbol{X}_V(t-u)'$. Throughout the paper, we make the following assumption.

Causal inference from time series

Assumption 1 *The spectral density matrix*

$$\mathbf{f}(\lambda) = \frac{1}{2\pi} \sum_{u=-\infty}^{\infty} \mathbf{\Gamma}(u)\, e^{-i\lambda u}$$

of $\mathbf{X}_V$ exists, and its eigenvalues are bounded and bounded away from zero uniformly for all $\lambda \in [-\pi, \pi]$.

Under this assumption, the process $\mathbf{X}_V$ has a mean-square convergent autoregressive representation

$$\mathbf{X}_V(t) = \sum_{u=1}^{\infty} \mathbf{\Phi}(u)\, \mathbf{X}_V(t-u) + \boldsymbol{\varepsilon}_V(t), \tag{1}$$

where $\mathbf{\Phi}(u)$ is a square summable sequence of $V \times V$ matrices and $(\boldsymbol{\varepsilon}_V(t))_{t\in\mathbb{Z}}$ is a Gaussian white noise process with non-singular covariance matrix $\mathbf{\Sigma}$.

3.1 Path diagrams and Granger causality

The autoregressive structure of the time series $\mathbf{X}_V$ can be visualized by a path diagram, in which the vertices correspond to the variables of $\mathbf{X}_V$ while the edges—arrows and lines—between vertices indicate non-zero coefficients in the autoregressive representation of $\mathbf{X}_V$ [6, 8].

Definition 2 *Let $\mathbf{X}_V$ be a stationary Gaussian process with autoregressive representation (1). Then the* path diagram *associated with $\mathbf{X}_V$ is the graph G with vertex set V and edge set such that for distinct $a, b \in V$*

  (i) $a \longrightarrow b$ *not in G if and only if $\Phi_{ba}(u) = 0$ for all $u \in \mathbb{N}$;*
  (ii) $a \,\text{---}\, b$ *not in G if and only if $\Sigma_{ab} = \Sigma_{ba} = 0$.*

Since path diagrams of this form may contain two types of edges, they will be referred to as mixed graphs. Furthermore, we note that, unlike in graphs commonly used for graphical modelling, two vertices a and b may be connected by up to three edges, namely $a \longrightarrow b$, $a \longleftarrow b$, and $a \,\text{---}\, b$. Similar path diagrams have been used to represent linear structural equation models [10, 19, 30][2].

As an example, consider a vector autoregressive process given by the equations

$$\begin{aligned}
X_1(t) &= \Phi_{11}\, X_1(t-1) + \Phi_{13}\, X_3(t-1) + \varepsilon_1(t), \\
X_2(t) &= \Phi_{22}\, X_2(t-1) + \Phi_{24}\, X_4(t-1) + \varepsilon_2(t), \\
X_3(t) &= \Phi_{33}\, X_3(t-1) + \Phi_{31}\, X_1(t-1) + \Phi_{32}\, X_2(t-1) + \varepsilon_3(t), \\
X_4(t) &= \Phi_{44}\, X_4(t-1) + \Phi_{43}\, X_3(t-1) + \Phi_{45}\, X_5(t-1) + \varepsilon_4(t), \\
X_5(t) &= \Phi_{55}\, X_5(t-1) + \Phi_{53}\, X_3(t-1) + \varepsilon_5(t).
\end{aligned} \tag{2}$$

Additionally, we assume that $\varepsilon_{\{1,2,3\}}(t)$, $\varepsilon_4(t)$, and $\varepsilon_5(t)$ are pairwise uncorrelated. Then the path diagram associated with this process is given by the graph in Fig. 1.

---

[2] In path diagrams for structural equation systems, correlated errors commonly are represented by bi-directed edges ($\leftrightarrow$) instead of dashed lines (---). Since in our approach directions are associated with temporal ordering, we prefer (dashed) undirected edges to indicate correlation between the error variables. Dashed edges with a similar connotation are used for covariance graphs [e.g. 3], whereas undirected edges (——) are commonly associated with nonzero entries in the inverse of the variance matrix [e.g. 20].

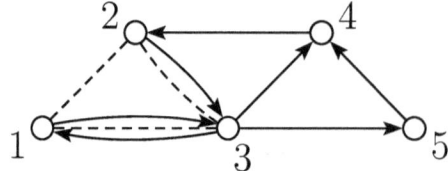

Figure 1: Path diagram associated with the process $\boldsymbol{X}_V$ in (2).

It is well known that the pairwise Granger-causal relationships among the variables of a process $\boldsymbol{X}_V$ are reflected in the autoregressive coefficients of the process and, thus, in the presence and absence of edges in the associated path diagram. More precisely, we have the following result.

Lemma 1 *Let G be the path diagram associated with a stationary Gaussian process $\boldsymbol{X}_V$ satisfying Assumption 1. Then for distinct $a, b \in V$*

(i) *$a \longrightarrow b$ not in G if and only if X_a is Granger-noncausal for X_b with respect to $\mathcal{I}_V$;*

(ii) *$a \text{ --- } b$ not in G if and only if X_a and X_b are contemporaneously independent with respect to $\mathcal{I}_V$, that is, $X_a(t+1) \perp\!\!\!\perp X_b(t+1) | \mathcal{I}_V(t)$ for all $t \in \mathbb{Z}$.*

Because of this result, the path diagram associated with a process $\boldsymbol{X}_V$ is also called the *Granger causality graph* of the process $\boldsymbol{X}_V$.

3.2 Markov properties

Under the assumptions imposed on $\boldsymbol{X}_V$, more general Granger-causal relationships than those in Lemma 1 can be derived from the path diagram associated with $\boldsymbol{X}_V$. This global Markov interpretation is based on a path-oriented concept of separating subsets of vertices in a mixed graph, which has been used previously to represent the Markov properties of linear structural equation systems [e.g. 19, 28]. Following [25] we will call this notion of separation in mixed graphs *m-separation*.

Let G be a mixed graph and $a, b \in V$. A *path* π in G is a sequence $e_1, \ldots, e_n$ of edges e_i in G with an associated sequence of nodes $v_0, \ldots, v_n$ such that e_i is an edge between v_{i-1} and v_i. The vertices v_0 and v_n are the *endpoints* while $v_1, \ldots, v_{n-1}$ are the *intermediate vertices* of the path. Notice that paths may be self-intersecting since we do not require that the vertices v_i are distinct.

An intermediate vertex c on a path π is said to be an *m-collider* on π if the edges preceding and succeeding c both have an arrowhead or a dashed tail at c (i.e. $\longrightarrow c \longleftarrow$, $\longrightarrow c \text{ ---}$, $\text{---} c \longleftarrow$, $\text{---} c \text{ ---}$); otherwise c is said to be an *m-noncollider* on π. A path π between a and b is said to be *m-connecting* given a set C if

(i) every m-noncollider on π is not in C and
(ii) every m-collider on π is in C;

otherwise we say that π is *m-blocked* given C. If all paths between a and b are m-blocked given C, then a and b are said to be *m-separated* given C. Similarly, two sets A and B are said to be m-separated given C if for every pair $a \in A$ and $b \in B$, a and b are m-separated given C.

With this notion of separation, it can be shown that path diagrams for multivariate time series have a similar Markov interpretation as path diagrams for linear structural equation systems [cf 19]. For disjoint subsets $A, B, C \subseteq V$, the subprocesses $\boldsymbol{X}_A$ and $\boldsymbol{X}_B$ are said to be *conditionally independent* given $\mathcal{I}_C$ if $\mathbb{E}(g(\boldsymbol{X}_A)|\mathcal{I}_{B\cup C}) = \mathbb{E}(g(\boldsymbol{X}_A)|\mathcal{I}_C)$ for all real-valued measurable functions $g(\cdot)$ on $\mathbb{R}^{A\times\mathbb{Z}}$, where $\mathcal{I}_{B\cup C}$ and $\mathcal{I}_B$ are the information sets (σ-algebras) generated by the complete series $\boldsymbol{X}_{B\cup C}$ and $\boldsymbol{X}_B$, respectively. In this case, we write $\boldsymbol{X}_A \perp\!\!\!\perp \boldsymbol{X}_B|\mathcal{I}_C$. Then separation in the path diagram can be translated into conditional independence among complete subprocesses of $\boldsymbol{X}_V$ [8].

Proposition 1 *Let $\boldsymbol{X}_V$ be a stationary Gaussian process that satisfies Assumption 1, and let G be its path diagram. Then, for all disjoint $A, B, C \subseteq V$, if A and B are m-separated given C then*

$$\boldsymbol{X}_A \perp\!\!\!\perp \boldsymbol{X}_B|\mathcal{I}_C.$$

Derivation of such conditional independence statements requires that all paths between two sets are m-blocked. For the derivation of Granger-causal relationships, it suffices to consider only a subset of these paths, namely those having an arrowhead at one endpoint. For a formal definition, we say that a path π between a and b is b-*pointing* if it has an arrowhead at the endpoint b; furthermore, a path between sets A and B is said to be B-*pointing* if it is b-pointing for some $b \in B$. Then, to establish Granger noncausality from $\boldsymbol{X}_A$ to $\boldsymbol{X}_B$, it suffices to consider only all B-pointing paths between A and B. Similarly, a graphical condition for contemporaneous independence can be obtained based on *bi-pointing* path, which have an arrowhead at both endpoints.

Definition 3 *A stationary Gaussian process $\boldsymbol{X}_V$ is Markov for a graph G if, for all disjoint subsets $A, B, C \subseteq V$, the following two conditions hold:*

(i) if every B-pointing path between A and B is m-blocked given $B \cup C$, then $\boldsymbol{X}_A$ is Granger-noncausal for $\boldsymbol{X}_B$ with respect to $\mathcal{I}_{A\cup B\cup C}$;

(ii) if the sets A and B are not connected by an undirected edge (---) and every bi-pointing path between A and B is m-blocked given $A \cup B \cup C$, then $\boldsymbol{X}_A$ and $\boldsymbol{X}_B$ are contemporaneously independent with respect to $\mathcal{I}_{A\cup B\cup C}$.

With this definition, it can be shown that path diagrams for Gaussian vector autoregressive processes can be interpreted in terms of such global Granger-causal relationships [cf 8].

Theorem 1 *Let $\boldsymbol{X}_V$ be a stationary Gaussian process satisfying Assumption 1, and let G be the associated path diagram. Then $\boldsymbol{X}_V$ is Markov for G.*

4 Representation of systems with latent variables

As pointed out above, the notion of Granger causality is based on the assumption that all relevant information is included in the analysis [13, 14]. The omission of important variables can lead to temporal correlations among the observed

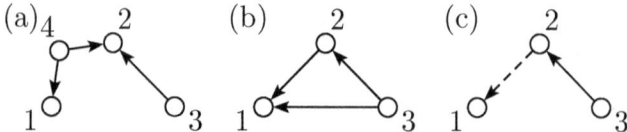

Figure 2: Graphical representations of the four-dimensional VAR(2) process in (3): (a) path diagram associated with $X_{\{1,2,3,4\}}$; (b) path diagram associated with $X_{\{1,2,3\}}$; (c) general path diagram for $X_{\{1,2,3\}}$.

components that are falsely detected as causal relationships. The detection of such so-called spurious causalities becomes a major problem when identifying the structure of systems that may be affected by latent variables.

Of particular interest will be spurious causalities of type I, where a Granger-causal relationship with respect to the complete process vanishes when only a subprocess is considered. Since the path diagrams from the previous section are defined, by Lemma 1, in terms of the pairwise Granger-causal relationships with respect to the complete process, they provide no means to distinguish such spurious causalities of type I from true causal relationships. To illustrate this remark, we consider the four-dimensional vector autoregressive process X_V with components

$$
\begin{aligned}
X_1(t) &= \alpha\, X_4(t-2) + \varepsilon_1(t), \\
X_2(t) &= \beta\, X_4(t-1) + \gamma\, X_3(t-1) + \varepsilon_2(t), \\
X_3(t) &= \varepsilon_3(t), \\
X_4(t) &= \varepsilon_4(t),
\end{aligned}
\tag{3}
$$

where $\varepsilon_i(t)$, $i = 1, \ldots, 4$ are uncorrelated white noise processes with mean zero and variance one. The true dynamic structure of the process is shown in Fig. 2(a). In this graph, the 1-pointing path $3 \to 2 \leftarrow 4 \to 1$ is m-connecting given $S = \{2\}$, but not given the empty set. By Theorem 1, we conclude that X_3 is Granger-noncausal for X_1 in a bivariate analysis, but not necessarily in an analysis based on $X_{\{1,2,3\}}$.

Now suppose that variable X_4 is latent. Simple derivations show [cf 5] that the autoregressive representation of $X_{\{1,2,3\}}$ is given by

$$
\begin{aligned}
X_1(t) &= \frac{\alpha\beta}{1+\beta^2} X_2(t-1) + \frac{\alpha\beta\gamma}{1+\beta^2} X_3(t-2) + \tilde{\varepsilon}_1(t), \\
X_2(t) &= \gamma\, X_3(t-1) + \tilde{\varepsilon}_2(t), \\
X_3(t) &= \varepsilon_3(t),
\end{aligned}
$$

where $\tilde{\varepsilon}_2(t) = \varepsilon_2(t) + \beta\, X_4(t-1)$ and

$$
\tilde{\varepsilon}_1(t) = \varepsilon_1(t) - \frac{\alpha\beta}{1+\beta^2} \varepsilon_2(t-1) + \frac{\alpha}{1+\beta^2} X_4(t-2).
$$

The path diagram associated with $X_{\{1,2,3\}}$ is depicted in Fig. 2(b). In contrast to the graph in Fig. 2(a), this path diagram contains an edge $3 \to 1$ and, thus, does not entail that X_3 is Granger-noncausal for X_1 in a bivariate analysis.

Table I: Creation of edges by marginalizing over i.

Subpath π in G	Associated edge e_π in $G^{\{i\}}$
$a \longrightarrow i \longrightarrow b$	$a \longrightarrow b$
$a \dashrightarrow i \longrightarrow b$	$a \dashrightarrow b$
$a \text{---} i \longrightarrow b$	$a \dashrightarrow b$
$a \longleftarrow i \longrightarrow b$	$a \leftarrow\!\!\dashrightarrow b$
$a \dashleftarrow i \longrightarrow b$	$a \leftarrow\!\!\dashrightarrow b$

As a response to such situations, two approaches have been considered in the literature. One approach suggests to include all latent variables explicitly as additional nodes in the graph [e.g. 8, 24]; this leads to models with hidden variables, which can be estimated, for example, by application of the EM algorithm [e.g. 2]. For a list of possible problems with this approach, we refer to Richardson and Spirtes [26, section 1].

The alternative approach focuses on the conditional independence relations among the observed variables; examples of this approach include linear structural equations with correlated errors [e.g. 19, 23] and maximal ancestral graphs [26]. In the time series setting, this approach has been discussed by Eichler [5], who considered path diagrams in which dashed edges represent associations due to latent variables. For the trivariate subprocess $\boldsymbol{X}_{\{1,2,3\}}$ in the above example, such a path diagram is depicted in Fig. 2(c).

Following this latter approach, we consider mixed graphs that may contain three types of edges, namely undirected edges (---), directed edges ($\longrightarrow$), and dashed directed edges ($\dashrightarrow$). For the sake of simplicity, we also use $a \leftarrow\!\!\dashrightarrow b$ as an abbreviation for the triple edge $a \overset{\leftarrow\text{---}}{\underset{\dashrightarrow}{}} b$. Unlike path diagrams for autoregressions, these graphs in general are not defined in terms of pairwise Granger-causal relationships, but only through the global Markov interpretation according to Definition 3. To this end, we simply extend the concept of m-separation introduced in the previous section by adapting the definition of m-noncolliders and m-colliders. Let π be a path in a mixed graph G. Then an intermediate vertex n is called an m-noncollider on π if at least one of the edges preceding and suceeding c on the path is a directed edge ($\longrightarrow$) and has its tail at c. Otherwise, c is called an m-collider on π. With this extension, we leave all other definition such as m-separation or pointing paths unchanged.

4.1 Marginalization

The main difference between the class of mixed graphs with directed ($\longrightarrow$) and undirected (---) edges and the more general class of mixed graphs that has been just introduced is that the latter class is closed under marginalization. This property makes it suitable for representing systems with latent variables.

Let G be a mixed graph and $i \in V$. For every subpath $\pi = <e_1, e_2>$ of length 2 between vertices $a, b \in V \setminus \{i\}$ such that i as an intermediate vertex and an m-noncollider on π, we define an edge e_π according to Tab. I.

Let $A^{\{i\}}$ the set of all such edges e_π. Furthermore, let $E^{\{i\}}$ be the subset of edges in E that have both endpoints in $V \setminus \{i\}$. Then we define $G^{\{i\}} = (V \setminus \{i\}, E^{\{i\}} \cup A^{\{i\}})$ as the graph obtained by marginalizing over $\{i\}$. Furthermore, for $L = \{i_1, \ldots, i_n\}$ we set $G^L = ((G^{\{i_1\}})^{\{i_2\}} \cdots)^{\{i_n\}}$, that is, we proceed iteratively by marginalizing over i_j, for $j = 1, \ldots, n$. Similarly as in Koster [19], it can be shown that the order of the vertices does not matter and that the graph G^L is indeed well defined.

We note that the graph G^L obtained by marginalizing over the set L in general contains self-loops. Simple considerations, however, show that G^L is Markov-equivalent to a graph $\tilde{G}^L$ with all subpaths of the form a --- b --→ b and a ←-- b --→ b replaced by $a ⇢ b$ and a ←→ b, respectively, and all self-loops deleted, that is, the graphs G^L and $\tilde{G}^L$ encode exactly the same Granger-causal and contemporaneous independence relations. It therefore suffices to consider mixed graphs without self-loops. We omit the details.

Now suppose that, for some subsets $A, B, C \subseteq V \backslash L$, π is an m-connecting path between A and B given S. Then all intermediate vertices on π that are in L must be m-noncolliders. Removing these vertices according to Table I, we obtain a path π' in G^L that is still m-connecting. Since the converse is also true, we obtain the following result.

Proposition 2 *Let G be a mixed graph, and $L \subseteq V$. Then it holds that, for all distinct $a, b \in V \backslash L$ and all $C \subseteq V \backslash L$, every path between a and b in G is m-blocked given C if and only if the same is true for the paths in G^L. Furthermore, the same equivalence holds for all pointing paths and for all bi-pointing paths.*

It follows that, if a process X_V is Markov for a graph G, the subprocess $X_{V \backslash L}$ is Markov for the smaller graph G^L, which encodes all relationships about $X_{V \backslash L}$ that are also encoded in G.

We note that insertion of edges according to Tab. I is sufficient but not always necessary for representing the relations in the subprocess $X_{V \backslash L}$. This applies in particular to the last two cases in Tab. I. For an example, we consider again process (3) with associated path diagram in Fig. 2(a). By Tab. I, the subpath suggests that X_1 Granger-causes X_2 (as does the path $1 \longleftarrow 4 \longrightarrow 2$ in the original path diagram), while in fact the structure can be represented by the graph in Fig. 2(c).

4.2 Markov equivalence and dynamic ancestral graphs

The operation of marginalization is of theoretical importance as it shows that the class of general path diagrams is rich enough to represent causal structures of systems with latent variables. It is of much less use for the identification of such structures as the relations determining the complete system are not available. Here, we face the problem that the set of Granger-causal relationships and contemporaneous independences that hold for the observed process does not uniquely determine a graphical representation within the class of general path diagrams. As an example, Fig. 3 displays two graphs that are Markov equivalent, that is, they encode the same set of Granger-causal and contemporaneous independence relations among the variables. Therefore, the corresponding graphical models—models that obey the conditional independence constraints imposed by the graph—are statistically indistinguishable.

In order to simplify the identification problem, the search for a suitable graphical representation can be restricted to a smaller class of graphs provided that this class contains at least one representative from every Markov equivalence

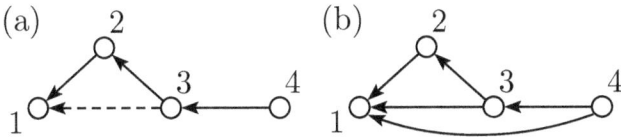

Figure 3: Two Markov equivalent graphs: (a) non-ancestral graph and (b) corresponding ancestral graph.

class. Following Richardson and Spirtes [26], we consider graphs that satisfy an ancestrality property. More precisely, we say that a vertex a is an ancestor of another vertex b if $a = b$ or there exists a directed path $a \to \cdots \to b$ in G. The set of ancestors of b is denoted by $\mathrm{an}(b)$. Then a graph G is called a dynamic ancestral graph if it satisfies the condition

$$a \in \mathrm{an}(b) \text{ then } a \dashrightarrow b \text{ not in } G \qquad (4)$$

for all distinct $a, b \in V$. We note that, in contrast to ancestral graphs [26], we do not require acyclicity (which is hidden in the time ordering) nor that edges are joined by at most one edge. The above condition (4) does imply, however, that two vertices a and b cannot be connected by the two edges $a \to b$ and $a \dashrightarrow b$ at the same time.

In order to obtain a Markov-equivalent ancestral graph for a general path diagram we have to substitute a dashed directed edge $a \dashrightarrow b$ by a directed edge $a \to b$ whenever a is an ancestor of b. Additionally, if G contains also the path $c \to a \dashrightarrow b$, an additional edge $c \to b$ needs to be inserted; similarly, the edges $c \dashrightarrow a$, $c \dashleftarrow a$, and $c \mathrel{-\!\!-\!\!-} a$ lead to the insertion of edges $c \dashleftarrow b$, $c \dashleftrightarrow b$, and $c \dashrightarrow b$, respectively. Iterating over all edges, we finally obtain a graph that satisfies condition (4) and is Markov-equivalent to the original graph.

Proposition 3 *Let G be a general path diagram. Then there exists a dynamic ancestral graph $\tilde{G}$ that is Markov-equivalent to G.*

From Tab. I and the way we have constructed dynamic ancestral graphs, it is clear that two vertices a and b are connected by a directed path $a \to \cdots \to b$ in a dynamic ancestral graph if and only if the same holds true in the path diagram associated with the underlying complete system. This leads us to the following general definition of causal effects in multivariate time series.

Definition 4 *A series X_a is said to have a causal effect—direct or indirect—on another series X_b if there exists some multivariate process $\boldsymbol{X}_V$ with $a, b \in V$ such that every graph in the Markov equivalence class of dynamic ancestral graphs for $\boldsymbol{X}_V$ contains a directed path $a \to \cdots \to b$.*

A further distinction between different causal patterns such as direct, indirect, or spurious causality is only possible with respect to a given information set $\mathcal{I}_V$ and requires to consider all graphs in the Markov equivalence class of general path diagrams for $\boldsymbol{X}_V$. We omit further details.

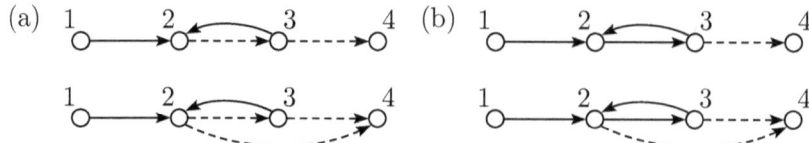

Figure 4: Inducing paths: (a) Dynamic ancestral graph with an inducing 4-pointing path between 2 and 4 and its corresponding Markov equivalent maximal dynamic ancestral graph. (b) Maximal dynamic ancestral graph with an inducing 4-pointing path between 2 and 4; the graph with additional edge 2 --→ 4 is not Markov equivalent.

5 Learning latent variable structures

There are two major approaches for learning causal structures: One approach utilizes constraint-based search algorithms such as the Fast Causal Inference (FCI) algorithm [27] while the other consists of score-based model selection. In the following, we briefly discuss the former approach for the identification of dynamic ancestral graphs. The latter approach has been investigated in Eichler [7].

The constraint-based search tries to find a graph that matches the empirically determined conditional independences. It usually consists of two steps:

1. identification of adjacencies of the graph;
2. identification of the type and the orientation of edges whenever possible.

In the case of ancestral graphs, the first step makes use of the fact that for every ancestral graph there exists a unique Markov-equivalent maximal ancestral graph (MAG), in which every missing edge corresponds to a conditional independence relation among the variables. Here an ancestral graph G is said to be maximal if addition of further edges would change the Markov equivalence class. MAGs are closely related to the concept of inducing paths; in fact, Richardson and Spirtes [26] used inducing paths to define MAGs and then showed the maximality property.

Definition 5 *In a (dynamic) ancestral graph, a path π between two vertices a and b is called an inducing path if every intermediate vertex on π is, firstly, a collider on π and, secondly, an ancestor of a or b.*

Figures 4(a,b) give two examples of dynamic ancestral graphs, in which 2 --- 3 --→ 4 resp. 2 —→ 3 --→ 4 are inducing 4-pointing paths. The graph in (b) shows that—unlike in the case of ordinary ancestral graphs—inducing paths may start with a tail at one of two vertices. As a consequence, insertion of an edge 2 --→ 4 or 2 —→ 4 changes the encoded Granger-causal relationships: while the upper graph implies that X_1 is Granger-noncausal for X_4 with respect to $\mathcal{I}_{\{1,2,3,4\}}$, this is not true for the lower graph. It follows that the method used for identifying adjacencies in ordinary MAGs does not apply to dynamic ancestral graphs.

The problem can be solved by observing that m-connecting pointing paths not only encode Granger-causal relationships but, depending on whether they start with $a \to c$ or $a \dashrightarrow c$, also a related type of conditional independences. More precisely, we have the following result.

Proposition 4 *Suppose that a and b are not connected by an edge $a \rightarrow b$ or $a \dashrightarrow b$ or by a b-pointing inducing path starting with $a \leftarrow c$, $a \leftdasharrow c$, or $a \text{---} c$. Then there exist disjoint subsets S_1, S_2 with $b \in S_1$ and $a \notin S_1 \cup S_2$ such that*

$$X_a(t-k) \perp\!\!\!\perp X_b(t+1) | \mathcal{I}_{S_1}(t) \vee \mathcal{I}_{S_2}(t-k) \vee \mathcal{I}_a(t-k-1)$$

for all $k \in \mathbb{N}$ and all $t \in \mathbb{Z}$.

The proof is based on the fact that inducing paths starting with an edge $a \rightarrow c$ or $a \dashrightarrow c$ only induce an association between $X_a(t-k)$ and $X_b(t+1)$ if one conditions on $X_c(t-k+1), \ldots, X_c(t)$. To block any other paths, we set S_2 to be the set of all intermediate vertices on all b-pointing inducing paths connecting a and b, and S_1 to be the set of all ancestors of a and b except a and S_2.

This leads us to the following algorithm for the identification of the Markov equivalence classes of dynamic ancestral graphs. Here, we use dotted directed edges $\dashrightarrow$ to indicate that the tail of the directed edge is (yet) undetermined.

Identification of adjacencies:

1. insert $a \text{---} b$ whenever X_a and X_b are not contemporaneously independent with respect to $\mathcal{I}_V$;
2. insert $a \dashrightarrow b$ whenever
    - X_a Granger-causes X_b with respect to $\mathcal{I}_S$ for all $S \subseteq V$ with $a, b \in S$ and
    - $X_a(t-k)$ and $X_b(t+1)$ are not conditionally independent given $\mathcal{I}_{S_1}(t) \vee \mathcal{I}_{S_2}(t-k) \vee \mathcal{I}_a(t-k-1)$ for some $k \in \mathbb{N}$, all $t \in \mathbb{Z}$, and all disjoint $S_1, S_2 \subseteq V$ with $b \in S_1$ and $a \notin S_1 \cup S_2$

Identification of tails:

1. *Colliders:*
   Suppose that G does not contain $a \dashrightarrow b$, $a \rightarrow b$, or $a \dashdashrightarrow b$. If $a \dashrightarrow c \dashrightarrow b$ and X_a is Granger-noncausal for X_b with respect to $\mathcal{I}_S$ for some set S with $c \notin S$, replace $c \dashrightarrow b$ by $c \dashdashrightarrow b$.
2. *Non-colliders:*
   Suppose that G does not contain $a \dashrightarrow b$, $a \rightarrow b$, or $a \dashdashrightarrow b$. If $a \dashrightarrow c \dashrightarrow b$ and X_a is Granger-noncausal for X_b with respect to $\mathcal{I}_S$ for some set S with $c \in S$, replace $c \dashrightarrow b$ by $c \rightarrow b$.
3. *Ancestors:*
   if $a \in \mathrm{an}(b)$ replace $a \dashrightarrow b$ by $a \rightarrow b$;
4. *Discriminating paths:*
   A fourth rule is based on the concept of discriminating paths. For details, we refer to Ali et al. [1].

We note that in contrast to the case of ordinary ancestral graphs only the tails of the dotted directed edges need to be identified. The positions of the arrow heads are determined by the time ordering of the Granger-causal relationships. The above algorithm probably can be complemented by further rules, see also Zhang and Spirtes [31].

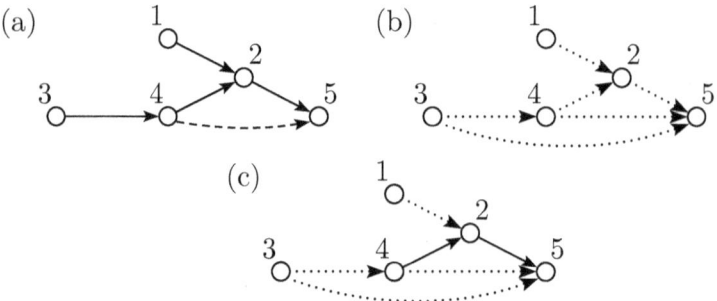

Figure 5: Identification of dynamic ancestral graphs: (a) underlying structure; (b) adjacencies; (c) identification of tails.

To illustrate the identification algorithm, we consider the graphs in Fig. 5. The original general path diagram is depicted in (a). Since 4 is an ancestor of 5, this graph is not ancestral. The adjacencies determined by the algorithm are shown in (b). Since X_1 does not Granger-cause X_5 with respect to $\mathcal{I}_V$, we find that 2 and 5 are connected by $2 \longrightarrow 5$. Similarly, X_3 is Granger-noncausal for X_2 with respect to $\mathcal{I}_{\{2,3,4\}}$, which implies that the graph contains also the edge $4 \longrightarrow 2$. No further tails can be identified; the final graph is given in (c).

6 Conclusion

The concept of Granger causality is widely used for inference about causal relationships from time series observations. One of the main problems in its application, however, is the possible presence of latent variables that affect the measured variables and thus can lead to so-called spurious causalities.

In this paper, we have presented a graphical approach for analysing cause-effect relationships in multivariate time series based on general path diagrams and dynamic ancestral graphs. In particular, we provided (a sketch of) an constraint-based search algorithm that allows identifying Markov equivalence classes of a dynamic ancestral graphs. The causal interpretation of the resulting graph is based on two fundamental assumptions. Firstly, the *causal Markov assumption* requires that all observed dependences are due to causal influences; this is a key assumption underlying all approaches to causal inference from observational data. Secondly, the faithfulness assumption states that the independences observed are structural and not due to cancellation of several causal influences. This assumption allows to detect spurious causes of type II when observing spurious causes of type I. Under these assumptions, directed edges can be interpreted as causal links although not as direct causes in the original sense of Granger. Due to identifiability up to Markov equivalence, only edges that are invariant in an equivalence class—and thus have their tail identified by the search algorithm—allow to be interpreted definitely as a cause if it is a directed edge and as a spurious cause if the edge is a dashed directed edge.

Bibliography

[1] Ali, R. A., Richardson, T. S., Spirtes, P. Markov equivalence for ancestral graphs. Tech. Rep. 466, Department of Statistics, University of Washington, 2004.

[2] Boyen, X., Friedman, N., Koller, D. Discovering the hidden structure of complex dynamic systems. In *Proceedings of the 15th Conference on Uncertainty in Artificial Intelligence*, 91–100. San Francisco: Morgan Kaufmann, 1999.

[3] Cox, D. R., Wermuth, N. *Multivariate Dependencies - Models, Analysis and Interpretation*. London: Chapman & Hall, 1996.

[4] Eichler, M. Graphical modelling of multivariate time series. Tech. rep., Universität Heidelberg, 2001. (arXiv:math.ST/0610654).

[5] Eichler, M. A graphical approach for evaluating effective connectivity in neural systems. *Philosophical Transactions of The Royal Society B*, 360:953–967, 2005.

[6] Eichler, M. Graphical modelling of dynamic relationships in multivariate time series. In Winterhalder, M., Schelter, B., Timmer, J. (Eds.), *Handbook of Time Series Analysis*, 335–372. Wiley-VCH, 2006.

[7] Eichler, M. Graphical modelling of multivariate time series with latent variables. Tech. rep., University of Heidelberg, 2006.

[8] Eichler, M. Granger causality and path diagrams for multivariate time series. *Journal of Econometrics*, 137:334–353, 2007.

[9] Florens, J. P., Mouchart, M. A linear theory for noncausality. *Econometrica*, 53:157–175, 1985.

[10] Goldberger, A. S. Structural equation models in the social sciences. *Econometrica*, 40:979–1001, 1972.

[11] Good, I. J. A causal calculus (I). *British Journal for the Philosophy of Science*, 11:305–318, 1961.

[12] Good, I. J. A causal calculus (II). *British Journal of the Philosophy of Science*, 12:43–51, 1962.

[13] Granger, C. W. J. Investigating causal relations by econometric models and cross-spectral methods. *Econometrica*, 37:424–438, 1969.

[14] Granger, C. W. J. Testing for causality, a personal viewpoint. *Journal of Economic Dynamics and Control*, 2:329–352, 1980.

[15] Granger, C. W. J. Some recent developments in a concept of causality. *Journal of Econometrics*, 39:199–211, 1988.

[16] Hosoya, Y. On the Granger condition for non-causality. *Econometrica*, 45:1735–1736, 1977.

[17] Hsiao, C. Autoregressive modeling and causal ordering of econometric variables. *Journal of Economic Dynamics and Control*, 4:243–259, 1982.

[18] Kamiński, M., Ding, M., Truccolo, W. A., Bressler, S. L. Evaluating causal relations in neural systems: Granger causality, directed transfer function and statistical assessment of significance. *Biological Cybernetics*, 85:145–157, 2001.

[19] Koster, J. T. A. On the validity of the Markov interpretation of path diagrams of Gaussian structural equations systems with correlated errors. *Scandinavian Journal of Statistics*, 26:413–431, 1999.

[20] Lauritzen, S. L. *Graphical Models*. Oxford: Oxford University Press, 1996.

[21] Lauritzen, S. L. Causal inference from graphical models. In Barndorff-Nielsen, O. E., Cox, D. R., Klüppelberg, C. (Eds.), *Complex stochastic systems*, 63–107. London: CRC Press, 2001.

[22] Lütkepohl, H. *Introduction to Multiple Time Series Analysis*. New York: Springer,

1993.

[23] Pearl, J. Causal diagrams for empirical research (with discussion). *Biometrika*, 82:669–710, 1995.

[24] Pearl, J. *Causality*. Cambridge, UK: Cambridge University Press, 2000.

[25] Richardson, T. Markov properties for acyclic directed mixed graphs. *Scandinavian Journal of Statistics*, 30:145–157, 2003.

[26] Richardson, T., Spirtes, P. Ancestral graph Markov models. *Annals of Statistics*, 30:962–1030, 2002.

[27] Spirtes, P., Glymour, C., Scheines, R. *Causation, Prediction, and Search*. Cambridge, MA: MIT Press, 2nd edn., 2001. With additional material by David Heckerman, Christopher Meek, Gregory F. Cooper and Thomas Richardson.

[28] Spirtes, P., Richardson, T. S., Meek, C., Scheines, R., Glymour, C. Using path diagrams as a structural equation modelling tool. *Soc. Methods Res.*, 27:182–225, 1998.

[29] Suppes, P. *A probabilistic theory of causality*. Amsterdam: North-Holland, 1970.

[30] Wright, S. The method of path coefficients. *Annals of Mathematical Statistics*, 5:161–215, 1934.

[31] Zhang, J., Spirtes, P. A characterization of Markov equivalence classes for ancestral graphical models. Tech. Rep. 168, Department of Philosophy, Carnegie Mellon University, 2005.

Probability and Structure in Econometric Models[1]

Kevin D. Hoover
Duke University
kd.hoover@duke.edu

ABSTRACT. The difficulty of conducting relevant experiments has long been regarded as the central challenge to learning about the economy from data. The standard solution, going back to Haavelmo's famous "The Probability Approach in Econometrics", involved two elements: first, it placed substantial weight on *a priori* theory as a source of structural information, reducing econometric estimates to measurements of causally articulated systems; second, it emphasized the need for an appropriate statistical model of the data. These elements are usually seen as tightly linked. I argue that they are, to a large extent, separable. Careful attention to the role of an empirically justified statistical model in underwriting probability explains puzzles not only in economics, but more generally with respect to recent criticisms of Reichenbach's principle of the common cause, which lies behind graph-theoretic causal search algorithms. And it provides an antidote to the pessimistic understanding of the possibilities for passive observation of causal structure in econometrics and related areas of Nancy Cartwright and others.

1 Econometrics and the Problem of Passive Observation

For nearly two centuries – at least since Mill [23, p.327] – philosophers have observed, and economists have lamented, the barriers to turning economics into an experimental science. At one point, the lack of scope for controlled experiments was seen as a serious barrier to the application of modern, probability-based statistics to economics. The situation was saved – or, at least, economists were comforted – with the publication of Trgyve Haavelmo's " The Probability Approach in Econometrics" [7] and the subsequent development of the theory of econometric identification by the Cowles Commission ([11, 21]; see Bouman's [3] excellent history of these developments).

There were two critical ideas in the new approach. The first is that statistical controls, accounting for covariates, could take the place of experimental controls

---

[1]The author is with Departments of Economics and Philosophy, Duke University, Durham, North Carolina, U.S.A. This paper is an abridgment of Hoover (2007). An earlier draft was presented to the 13th International Congress of Logic, Methodology, and the Philosophy of Science at Tsinghua University, Beijing, China 8-16 August 2007. I thank Julian Reiss for comments on that draft.

[24, chap.8, esp. pp.246–248]. Haavelmo proposed that an economic process could be partitioned into a deterministic and a random part. If the causal structure of the deterministic part were articulated fully and accurately enough, the random part would conform to the laws of probability.

The second critical idea was that the causal structure of the deterministic part of the economic process had to be articulated accurately. Haavelmo suggested that *a priori* economic theory could do the job. This second idea was the necessary prop of the first. The Cowles Commission, which took up Haavelmo's project, made the crucial discovery that causal structure is richer than, or (at least) distinct from, the probability structure [11, 21].

Take a very simple example. Suppose that A causes B, where A and B are two stochastic variables. Their relationship can be represented graphically as $A \to B$ and algebraically, with some additional structure, as:

$$A \Leftarrow \varepsilon_A \qquad (1)$$

$$B \Leftarrow \alpha A + \varepsilon_B \qquad (2)$$

where ε_A and ε_B are random error terms; for convenience we assume that each is distributed independent normal with mean zero and variances σ_A^2 and σ_B^2 (notated $\varepsilon_A \sim$ independent $N(0, \sigma_A^2)$ and $\varepsilon_B \sim$ independent $N(0, \sigma_B^2)$). Each is independent across successive draws and independent from each other (which implies that the covariance of ε_A and ε_B is zero ($cov(\varepsilon_A, \varepsilon_B) = 0$). The coefficient α is a fixed parameter. And the arrowhead on the equal sign turns it into an assignment operator, a reminder that the model incorporates the asymmetry of causation.

The causal structure of Eq.1 and Eq.2 determines its probability structure. Substituting Eq.1 into Eq.2 yields what econometricians refer to as reduced forms, which completely characterize the probability structure of the variables:

$$A = \varepsilon_A = E_A \qquad (3)$$

$$B = \alpha \varepsilon_A + \varepsilon_B = E_B \qquad (4)$$

E_A and E_B are themselves distributed $E_A \sim$ independent $N(0, \sigma_A^2)$ and $E_B \sim$ independent $N(0, \alpha^2 \sigma_A^2 + \sigma_B^2)$. But they are not independent of each other; in fact, $cov(E_A, E_B) = \alpha \sigma_A^2 = \Sigma \neq 0$.

The sense in which the causal structure is richer than the probabilistic structure is that the implication runs one-way: Eq.1 and Eq.2 imply Eq.3 and Eq.4, but not the other way round. In fact, if $B \to A$ with a causal structure analogous to Eq.1 and Eq.2, instead of $A \to B$, we can generate the reduced forms:

$$A = \varepsilon_A' + \beta \varepsilon_B' = E_A \qquad (5)$$

$$B = \varepsilon_B' = E_B \qquad (6)$$

The important thing is despite the difference in causal structures (reflected in the difference in the middle terms in the two sets of equations), both equations Eq.3 and Eq.4 and equations Eq.5 and Eq.6 define the same random terms, E_A and E_B. And these terms have the same interdependence in each case: $cov(E_A, E_B) = \beta \sigma_{B'}^2 = \Sigma$. They define the same probability distribution; they are observationally equivalent; or, in the argot of econometrics, they are not identified.

The observational equivalence of the two sets of equations means that we can work backwards to form estimates of the parameters only if we are willing to commit to a particular causal structure. If we believe that theory (or some other extra-statistical source) tells us that Eq.1 and Eq.2 constitute the correct causal structure, then we can use observations on A and B to estimate the parameter α. But what if, as economists often believe, causation is mutual $(A \leftrightarrow B)$? For example, suppose that the causal structure is

$$A \Leftarrow \beta' B + \varepsilon''_A \tag{7}$$

$$B \Leftarrow \alpha' A + \varepsilon''_B \tag{8}$$

Then there are infinite combinations of causal strengths connecting them, so that the equivalence class is itself infinite.[2] There is no way to recover estimates of α' or β' without further non-empirical assumptions (e.g., about their relative strengths). This is the classic *identification problem in econometrics*.

The classic solution is to imagine that A and B are subject to independent experimental control. Suppose

$$X \to A \leftrightarrow B \leftarrow Y \tag{9}$$

where X is a means of intervening on A independent of B or Y, and Y is a means of intervening on B independent of A or X. The Cowles Commission showed that, in such circumstances, unique estimates of α' or β' could be recovered. If we have independent reasons for thinking that the world is structured like Eq.9 and that X and Y can somehow be observed, then passive observations could replace controlled experiments.[3] In macroeconomics, the analogy with controlled experiments is the basis for techniques of causal inference based on patterns of invariance and noninvariance [12, chap.8–10]; while in microeconomics, it motivates the search for "natural experiments" [1]. The Cowles Commission itself and econometric orthodoxy in the second half of the 20th century downplayed experimental analogues, emphasizing instead the role of *a priori* economic theory in selecting the warranted causal structure such as Eq.9, in which X and Y are just additional observed variables, now christened *instrumental variables* (or just *instruments*).

The Cowles Commission's strategy opened an era of optimism about the possibilities for passive observation and the articulation of causal structure. Soon, however, pessimism set in: where does our confidence in the causal structure come from? Sims [31] famously stigmatized the assumed causal order as relying on "incredible" identifying restrictions (i.e., assumptions about which variables are *not* causally connected) – cf.[22]. For a while, some economists were resigned to using reduced forms only, but one can say so little about policy problems without a causal understanding that a whiff of *a priori* theory was soon reintroduced; and,

---

[2]The reduced forms are $A = \left(\dfrac{1}{1-\alpha'\beta'}\right)(\varepsilon''_A + \beta'\varepsilon''_B) = E_A$ and $B = \left(\dfrac{1}{1-\alpha'\beta'}\right)(\alpha'\varepsilon''_A + \varepsilon''_B) = E_B$. Once again, as indicated by the right-hand terms, they define the same probability distribution as Eq.3 and Eq.4 or Eq.7 and Eq.8.

[3]Scheines [29] provides a careful exposition not only of the logic of such inference from natural experiments, but also of the close analogy with the logic of controlled experiments.

for those who were still skeptical of *a priori* theory, natural experiments became the lodestone.[4]

Behind the alternating optimism and pessimism lies perhaps the biggest questions in empirical economics: how, and exactly what, can economists learn from passively data?

2 Probability Models: Function and Inference

In clarifying the identification problem, however, the Cowles Commission raises issues that go beyond economics. Cartwright [4] maintains that econometrics provides the clearest example of how probability should function in physical, as well as social, sciences. What impresses Cartwright are the detailed theoretical assumptions that inform theoretically identified econometric models. These correspond, in experimental contexts to experimental controls and shielding, which she argues allow "nature's capacities" to display themselves without conflating interference, just as the set of instrumental variables allow the strengths of causal connections to be measured in econometric models. Cartwright [5, p.173] emphasizes the stringency of the conditions needed to throw capacities into clear relief:

> My claim is that it takes hyperfine-tuning ... to get a probability. Once we review how probabilities are associated with very special kinds of models before they are linked to the world, both in probability theory itself and in empirical theories like physics and economics, we will no longer be tempted to suppose that just any situation can be described by some probability distribution or other. It takes a very special kind of situation with arrangements set just right – and not interfered with – before a probabilistic law can arise.

And while she is grateful to econometricians for clarifying the logic of the problem, she is pessimistic with respect to the project of applied econometrics, seeing econometricians as having themselves laid the groundwork for showing it to be a quite hopeless undertaking [5, chap.6–7].[5]

Cartwright's pessimism about econometrics is based in what it teaches her about the application of probability. First, probabilities are not simply there in the world to be invoked whenever it suits our inferential purposes. Rather, following Hacking [8], she argues that probabilities are ways of codifying the propensities of physical (and perhaps social) machinery to display behavior with frequencies that follow certain patterns. The propensity of a coin to display heads half the time and to provide no evidence of dependence between successive tosses (in short, to follow a binomial probability model with the key parameter set to $\frac{1}{2}$), requires that the coin, the flipping device, and the environment to be constituted in highly particular ways, and only then can we expect the probability model to apply. Physical experiments are examples of such *nomological machines*; while what an applied economic theory describes is a *socioeconomic machine*.

Second, Cartwright asks, how do probabilities attach to the world? Her answer is that probabilities attach to the world through models. It is the actual

---

[4] See [16, 18] for fuller accounts.
[5] Hoover [13, 14] shows that Cartwright's position is genuinely and unnecessarily, if only *implicitly*, pessimistic.

success of the binomial probability model in describing the frequencies generated in the coin-flipping nomological machine that warrants claims about probabilities – for example, claims about how often five heads in a row ought to be expected.

Two of the theses that I shall elaborate and defend are reactions or qualifications of the lessons that Cartwright draws from econometrics. First, her position that probabilities arise only in chance setups is correct up to a point. The generation of stable frequencies is a property of real-world (not just physical) systems appropriately configured. But consistent with Hacking [8, pp.24–25], it is not the frequencies directly but the frequencies on particular kinds of trials that exhibit chance behavior. The same data may be viewed on different kinds of trials: "there is nothing unusual about regarding one event under several aspects" [8, p.25]. Different probability models may be applied to the same data without conflict or contradiction.

And this connects to my second thesis: Cartwright is correct to stress the role of models, but their role is not merely to attach probabilities to the world, but to create probabilities. Without the models, there are no probabilities to discuss. This is not an anti-realist thesis. For I suppose that that some models are better than others when judged in relation to actual frequencies from a particular aspect or point of view and that different points of view may lead to different, but not contradictory, probability models. The upshot of my theses is that a (perhaps, *the*) central problem of econometrics is to establish appropriate probability models. While there are plenty of statistical tools devoted to specification testing and specification search, the logical role of probability models in econometric inference is a relatively neglected topic.[6]

I also want to defend a third thesis that is well illustrated by the equivalent probabilistic implications of the causal structures $A \to B$, $A \leftarrow B$ and $A \leftrightarrow B$ (see equations Eq.1-Eq.8 above): namely, probability models do not in general require causal presuppositions. In saying this, I do not wish to contradict another of Cartwright's [4] well-known principles that an output of causal knowledge is delivered only by inputs of causal knowledge. Rather I want to defend the weaker claim that, while prior causal knowledge may be useful and, sometimes at least, essential, some causal claims may be supported by facts about probability models that do not depend on assumptions about the truth of these very same causal claims.

To illustrate, consider a simple causal structure: $A \to C \leftarrow B$, where

$$A \Leftarrow \varepsilon_A \tag{10}$$
$$B \Leftarrow \varepsilon_B \tag{11}$$
$$C \Leftarrow \alpha A + \beta B + \varepsilon_C \tag{12}$$

and $\varepsilon_i \sim$ independent $N(0, \sigma_i^2)$, $i = A, B, C$ and $cov(\varepsilon_i, \varepsilon_j) = 0$ for all $i \neq j$. The arrangement is one with a *common effect* of two independent causes, sometimes known as an *unshielded collider* [36, p.10]. The causal structure itself cannot be directly observed, but we can observe realizations of the variables.

Now suppose that we want to infer the causal structure from the data. We begin by trying to establish a probability model of the data. A normal distribution is often a good place to start. There are standard statistical tests for normality. The joint normal model of three variables can be described by nine

---

[6]However see [10, esp. chs.1, 15], [19, 20, 34, 35].

parameters: three means (call them $\bar{A}, \bar{B}, \bar{C}$), three variances ($\sigma_A^2, \sigma_B^2, \sigma_C^2$), and three covariances or, equivalently, population correlations ($\rho_{AB}, \rho_{AC}, \rho_{BC}$).[7] A particularly simple model of the data assumes that the three variables are distributed independent normal with constant means and variances:

Model 1 $(A, B, C) \sim N(\bar{A}, \bar{B}, \bar{C}; \sigma_A^2, \sigma_B^2, \sigma_C^2; 0, 0, 0)$

where the last three arguments indicate that each of the population correlations is zero. Probabilistic independence can be defined formally as occurring when $P(X, Y) = P(X)P(Y)$. Informally, it means that the probability distribution of a variable is the same whatever realization is taken by another variable. Probabilistic independence implies that the corresponding population correlation is zero.

Each probability model sees the data from a point of view. Is Model 1 a good model? The answer, of course, depends in part on our purposes. If we, subscribe to an inferential scheme that requires judgments about probabilistic dependence (for example, the various causal-search algorithms in Spirtes and Pearl [26, 36]), then it is not a good model, since it *assumes* that there is no probabilistic dependence. A better model would be

Model 2 $(A, B, C) \sim N(\bar{A}, \bar{B}, \bar{C}; \sigma_A^2, \sigma_B^2, \sigma_C^2; \rho_{AB}, \rho_{AC}, \rho_{BC})$

There is no loss from taking this point of view, since Model 2 nests Model 1: if we decide that on our best estimates $\rho_{AB} = \rho_{AC} = \rho_{BC} = 0$, then Model 2 collapses to Model 1.

Whether a model is good also depends on its relationship to the data. Various statistical techniques allow us to estimate the parameters of Models 1 and 2 and whether one encompasses or nests the other, as well as to test their errors for normality, randomness and probabilistic independence against various alternatives. It is not to our purpose to discuss them in any detail.

The assertion that a probability model truly describes some portion of the world is a conjecture from which we can deduce that the model accounts for the co-occurrences of the data (both observed and yet-to-be-observed) except for some random residual. Conjectures about probability models, like all scientific conjectures, are accepted because they are supported by the right kind of data. They are never deductively certain, and they always remain open to doubt and criticism. Serious criticisms must be adjudicated in the light of the data and may lead to a reassessment of the appropriateness of a probability model. The crucial point is that the probability model is not a directly observable fact about the frequencies displayed by the data; rather it is a conjecture, the support for which depends on a complex of statistical inferences.

Returning to our illustration, suppose that we have obtained estimates for the parameters of Model 2, we have tested it for normality, and we have tested and rejected Model 1 as a special case. (These by no means exhausts all that the statistician might do to convince himself that Model 2 is a good model.) Now, here is a principle invoked in many causal search algorithms:

---

[7] The relationship of the population correlation to the covariance is, for example, $\rho_{AB} = \dfrac{cov(A, B)}{\sigma_A \sigma_B}$.

Principle of the Common Effect if X and Y are probabilistically independent conditional on some set of variables (possibly the null set) excluding Z, but are probabilistically dependent conditional on Z, then Z is the common effect of X and Y (or Z forms an unshielded collider on the path XZY).

I do not wish to defend this principle here, but instead consider the logic of its application.

We start with an estimate of Model 2. Suppose that a statistical test tells us that we cannot reject $\rho_{AB} = 0$ and that the correlation of A and B conditional on C does not equal zero ($\rho_{AB|C} \neq 0$). Then, we are working with a particularization of Model 2, call it

Model 2' $(A, B, C) \sim N(\hat{\bar{A}}_A^2, \hat{\bar{B}}_B^2, \hat{\bar{C}}_C^2; \hat{\sigma}_A^2, \hat{\sigma}_B^2, \hat{\sigma}_C^2; 0, \hat{\rho}_{AC}, \hat{\rho}_{BC})$

where, in the custom of econometricians, the "hats" indicate estimated values and the estimates $\hat{\rho}_{AC}$ and $\hat{\rho}_{BC}$ are constrained to fulfill the condition $\hat{\rho}_{AB|C} = 0$.[8]

Probabilistic dependence is a property of probability distributions and not of realized data. The important judgments here are $P(AB) = P(A)P(B)$ and $P(AB|C) \neq P(A|C)P(B|C)$. The crucial point is that these are deductive consequences of Model 2' and are not unmediated consequences of observed data. This is easily seen by noting that Model 1 is, for other purposes and from other points of view, a perfectly acceptable model of the data; and Model 1 does not imply $P(AB|C) \neq P(A|C)P(B|C)$. Given Model 2', we can *deduce* that the antecedents of the Principle of the Common Effect are fulfilled and, therefore, conclude that the data, through the mediation of Model 2', imply $A \rightarrow B \leftarrow C$, which we know by assumption is the causal structure that generated the data.

There are two points to take away from this illustration. The first is that an inference such as the one from the Principle of the Common Effect is a two-step process. Step 1 establishes the probability model through statistical inferences; step 2 deduces the probabilistic (in this case, causal) consequence from the inferential principle applied to the probability model. Commentators on causality frequently seem confused on the two-step nature of the inference because the parameters of common probability models frequently have easily calculated analogues among descriptive sample statistics. For example, Pearson's sample correlation coefficient r_{XY} is analogue to ρ_{XY}.[9] One cannot, however, work directly with the sample correlation coefficient or other descriptive statistics without reference to the probability model:

(a) parameterization is distribution-specific; some distributions may have no parameter closely related to ρ_{XY} in the normal distribution and, so, nothing to which r_{XY} can serve as an analogy;

---

[8] Which requires in particular that $(\hat{\rho}_{AC}^2 - \hat{\rho}_{AC}\hat{\rho}_{BC})\left(\sqrt{1 - \hat{\rho}_{AC}^2}\sqrt{1 - \hat{\rho}_{BC}^2}\right)^{-1} = 0$, so that, given $\hat{\rho}_{AB} \neq 0$, any nonzero values for both $\hat{\rho}_{AC}$ and $\hat{\rho}_{BC}$ are sufficient.

[9]
$$r_{XY} = \sum_{j=1}^{N}(X_j - \bar{X})(Y_j - \bar{Y}) / (\sum_{j=1}^{N}(X_j - \bar{X})^2 \sum_{j=1}^{N}(Y_j - \bar{Y})^2)^{\frac{1}{2}}$$

and

$$\rho_{XY} = E((X - E(X))(Y - E(Y))) / (E(X - E(X))^2 E(Y - E(Y))^2)^{\frac{1}{2}}$$

(b) even when the analogy holds, r_{XY} may not coincide with the best estimate of ρ_{XY}, since the parameters of a probability distribution are typically estimated jointly (e.g., by maximum likelihood methods) rather than individually;

(c) sample descriptive statistics are calculated without the aid of a probability distribution, and it is only through one or other distribution that they can have any bearing at all on probability – to act otherwise is to commit a category mistake.

The principal interest of most researchers with respect to causal search is in the second step of inferring causal structure from patterns of probabilistic dependence. They frequently take for granted that the problem of justifying a particular probability model from the data has been (or can easily be) solved.

The second point to take away from the illustration is that, in inferring the pattern of causal dependence from which causal order is itself inferred, we nowhere refer to the facts about causal structure that form the endpoint of our inferential chain (namely, the connection of A and B to their common effect C). That is not to say that we do not use causal knowledge at all. In restricting our model to three variables, we have implicitly judged that none of the other causal connections that our three variables has is structured in such a way as to interfere with the appropriateness of Models 2 or Model 2'. Such a judgment may, of course, be challenged, suggesting further investigation. That we cannot step out of a causal context notwithstanding, the key point is that we have begged no question.

People often intuitively think of probabilistic dependence as a causal notion. Hacking [8, p.20] provides one formulation:

> Two events are commonly said to be independent of each other if neither causes the other, and if knowledge that one occurred is no aid in discovering if the other occurred.

But Hacking also agrees to a second formulation: X and Y are independent if $P(XY) = P(X)P(Y)$. Not only is no assumption about causation cited in this second, *standard* formulation, the statistical tests of independence are based on patterns of co-occurrence without causal reference. Whether knowledge of one aids in discovering whether the other occurred depends importantly on what knowledge we have in mind and what we mean by "aids".

Generally, two watches give knowledge of each other: if I know that my watch says 2:39 PM, then it is a fair bet that my neighbor's watch is pretty close. My watch even gives me knowledge of what a watch in Australia is likely to read, knowing the difference in time zones between the east coast of the United States and, say, the west coast of Australia. And although we do not have a common reference point, I suspect that, if there were Martians and Martians had watches, then an hour passed on my watch would be an hour passed on a Martian's watch (suitably adjusting Martian units to our own). But generally, it is a well-supported conjecture that my watch is probabilistically independent of my neighbor's, the Australian's, and the Martian's watches;(cf.[37, section 2], [28, p.181]).

For example, take the standard time signal provided by the U.S. National Institute of Standards and Technology and the U.S. Naval Observatory as a reference time. Define the random variable A = the difference between the time on my watch and the reference time and B = the difference between the time on my neighbor's watch and the reference time. I maintain that typically we will find a well-supported probability model in which $P(AB) = P(A)P(B)$. More directly, if b is a particular realization of B, then we will find $P(A|B = b) = P(A)$. That watches convey knowledge about the likely behavior of other watches and clocks explains their widespread use. That watches are typically probabilistically independent of each other explains why, when the power has been cutoff, we can usefully look to our wristwatches to reset the clock on the microwave.[10]

Contrary to Hacking, it would be more accurate to say that

> *two random variables are independent of each other if the realization of one conveys no information about the distribution of the other.*[11]

Formulated this way, probabilistic independence does not invoke causal order conceptually, nor do statistical tests of independence presuppose causal order.

3 The Principle of the Common Cause

The importance of clarity with respect to the two-step inferential process – from data to probability model, from probability model to causal structure – is thrown into relief by recent discussions of the Principle of the Common Cause, a version of which lies at the heart of the graph-theoretic analysis of causal structure and related search algorithms. Hans Reichenbach [27, p.157] provides the original statement:

Principle of the Common Cause (Reichenbach) "If an improbable coincidence has occurred, there must exist a common cause."

Reise [28, p.184] gives a version, which he attributes to Hoover [15, p.548], that is clearer for the issues to hand:

Principle of the Common Cause (Hoover) "If variables X and Y are probabilistically dependent ..., then either X causes Y or Y causes X, or X and Y are the joint effects of a common cause."[12]

Reiss states the principle mainly to criticize it.

The background for Reiss's criticisms is Sober's [32, 33] putative counterexample.[13] In Sober's example, bread prices in England and sea levels in Venice are both rising and *ex hypothesi* not causally connected. In a sense that is less than

---

[10] I say "typically," because, for example, two old-fashioned electric clocks on the same circuit that use the cyclicality of the household electricity to control the speed of their motors may be probabilistically dependent after all.

[11] And two *events* are independent of each other if each is a realization of a mutually independent random variable.

[12] Reiss states he needs to modify my statement of the principle to make it consistent with his own paper. I agree that this formulation is better than my original formulation, which was specific to a particular context.

[13] Hoover [15] offers a detailed refutation of Sober's counterexample. While I remain convinced of its argument, at some key points the exposition apparently misled some readers about its essence. I hope to be clearer here.

perfectly clear, Sober maintains that bread prices and sea levels are correlated and, therefore, probabilistically dependent.

Reiss categorizes various reactions to Sober's counterexample as following one of two strategies: the first strategy argues that Sober's claim that there is a genuine probabilistic dependence between bread prices and sea levels is defective; the second proposes to defuse the counterexample by showing that it fails to apply to data when they are appropriately prepared.[14] Reiss treats these two strategies separately, but observes that they may be complementary. I would put the point more strongly: if "data preparation" (second strategy) is understood appropriately and if the first strategy is stated in its positive form (patterns of probabilistic dependence may support causal inference when genuine), then there is no legitimate way to separate the two strategies, for the second is part of establishing the *bona fides* of the probability model necessary for the first.

Although Sober does not offer any formal measures of correlation between bread prices and sea levels, he does provide some cooked data and notes that "higher than average sea levels tend to be associated with higher than average bread prices" [33, p.332,334]. Unlike Reiss, Sober [33, p.343] acknowledges the two-step inferential process involved in applying the Principle of the Common Cause; so, we are entitled to ask what the sample association of bread prices and sea levels says about probabilistic dependence. Note that on a common measure of sample association, Pearson's correlation coefficient (r), which was previously defined in footnote 8, bread prices (B) and sea levels (S) are highly correlated ($r_{BS} = 0.99$, where $-1 \leq r \leq 1$ and $|r| = 1$ indicates perfect correlation, whereas $r = 0$, indicates no correlation.). We cannot interpret this high correlation in terms of probability without a probability distribution. This is obvious, since it is exceedingly rare to find a correlation coefficient that is *exactly* zero; we must judge whether it is *effectively* zero or not relative to an assumed probability distribution.[15]

The stationary, multivariate normal distribution is the workhorse of statistics. It has the nice property that r is an analogue for its population parameter ρ, and that it can be shown that, as the sample size increases, the expected value of r converges to ρ. Roughly speaking, a distribution is stationary when its moments (mean, variance, and higher moments) are constant through time. But a stationary distribution is not a good model of Sober's data. A stationary distribution implies that a time-series will cross and re-cross its sample mean frequently. Sober's data cross their sample means only once. While this is a nice clue, there are also formal tests for non-stationarity. There are a number of non-stationary alternatives to the stationary, multivariate normal distribution – none of which display the correspondence between the sample correlation coefficient r and a fixed population parameter, like ρ.

One alternative is the random-walk in which the best expectation of the value of a time series at $t+1$ is its value at time t. If the data were generated by two probabilistically independent random walks, then r would be a worse-than-useless measure of probabilistic dependence; for the expectation of r as the sample size grows converges not to a single value but to a uniform distribution over the interval -1 to 1 [10, p.128]. This means that when the world is populated

---

[14]Reiss offers [15] as an example of the first strategy and [2, 36, 37] as examples of the second.

[15]Notice that this is true whether we accept classical statistical testing or a decision-theoretic approach.

by random walks that it is easy (and meaningless) to find high levels of sample correlation among some of them.

The paradigm random walk can be expressed as:
$$X_{t+1} = X_t + \varepsilon_t \tag{13}$$
or equivalently as
$$\Delta X_{t+1} = \varepsilon_t \tag{13'}$$
where ε_t is a stationary random error term (e.g., normal). The form (13') suggest to some commentators (e.g,. [6, 9, 25]) a quick fix. If we difference the non-stationary time series X, it becomes stationary, and the correlation coefficient between two such differenced, non-stationary time series is an indicator of probabilistic dependence. The problem with this approach is that, while the differences of nonstationary variables may be probabilistically dependent, so may the levels (even when the differences are not), and differencing the data eliminates the information about this relationship between the levels. Nonstationary variables that display probabilistic dependence in levels are said to be *cointegrated*. If the nonstationary random walk is sometimes illustrated by the path of drunk stumbling aimlessly as he leaves the bar, then cointegration is the situation in which the drunk has a faithful friend who follows at a discreet distance to make sure that he comes to no harm.[16]

Sober's counterexample "works" only because he trades on our implicitly judging probabilistic dependence against a probability model in which casual measures of association have a natural interpretation. But even at a casual level, it is obvious that a stationary probability model is not a good characterization of the data. And in any non-stationary model, the sample association of bread prices and sea levels is both natural and not indicative of probabilistic dependence. The situation is exactly like the association between time kept on two watches: rising bread prices in England give some indication of sea levels in Venice, but the distribution of sea levels is the same whether the current realization of bread prices is a rise or a fall.

The point is not that any particular non-stationary model fits Sober's counterexample. Rather it is that we must establish the probability model before we can make any judgment of probabilistic dependence. Sober may acknowledge the two-step process, but he fails to do the work – either statistically by testing the data or hypothetically by establishing the true distribution in his thought experiment – necessary to move from the first to the second step. What Reiss thinks of as data preparation is integral to establishing the probability model from which probabilistic dependence is ascertained. On the one hand, the probability model must be appropriate to the data; on the other hand, the probability model helps to guide the meaningful preparation of the data. This is crucial work for statistics or econometrics, though it is typically neglected in discussions of the Principle of the Common Cause.

In contrast to my analysis of Sober's counterexample, which is not so much a defense of the Principle of the Common Cause as a demonstration that it does not fail for Sober's particular reasons, Reiss offers a defense of the principle – a defense which falls into the category of "destroying-the-village-in-order-to-save-it." Reiss agrees with Cartwright and Hacking that probabilities (i.e., frequencies

---

[16] See [15, section 4–5], for a more technical exposition.

that are correctly described within the canons of a axiomatization of the behavior of random variables) arise only in well-constructed *chance setups*. At the same time, he objects to the two-step inferential process: he claims that my strategy "deprives the principle of much of its inferential power and to some extent betrays the motivation behind it" [28, p.185]. Reiss defends the principle as a sometimes useful heuristic, providing what he refers to as an "epistemological reading" as opposed to the "metaphysical reading" implicit in the two-step inferential process.

The contrast between an epistemological and a metaphysical reading is spurious. The two-step process is about inference (that is, classically epistemological) and says nothing about what causation actually is.

Reiss's strategy is explicitly analogous to Patrick Suppes' [38] well-known probabilistic analysis of causation. Suppes begins by defining *prima facie* cause as the case in which $P(A|B) > P(A)$. He then tries to catalogue the cases in which *prima facie* causes fails to correspond to actual cause and to suggest appropriate corrections. In parallel, Reiss takes the Principle of the Common Cause as providing a rule for inferring *prima facie* cause, and then catalogues a (partial) list of exceptions to the rule. Reiss's heuristic rule is not stated as a relationship of causal structure to probabilistic dependence, but as a relationship of sample association (or frequency) to causal structure, thus short-circuiting the first step of the two-step inferential process:

Principle of the Common Cause (Reiss) "The proposition $e =$ 'Random variables X and Y are (sample or empirically) correlated' is *prima facie* evidence for the hypothesis $h =$ 'X and Y are casually connected.' If all alternative hypotheses h_i^a (e.g., 'the correlation is due to sampling error,' 'the correlation is due to the data-generating process for X and Y being non-stationary,' 'X and Y are logically, conceptually, or mathematically related') can be ruled out, the e is genuine evidence for h." [28, p.193]

It is instructive to see how the two-step inferential process handles Reiss's exceptions. Reiss considers seven specific exceptions (non-stationary nonsense correlations, colliders, mixing, stationary nonsense correlation, homoplasies, non-statistical nonsense correlations, and laws of coexistence) and suggests that the list is actually open-ended. All six are easily treated using the two-step inferential process (see [17] for details). We have already discussed the case of non-stationary nonsense correlations and, in the interest of space, we consider only one other case here, that of *colliders*:

Reiss [28, pp.187–188] takes the causal configuration $A \to C \leftarrow B$ as an exception to the Principle of the Common Cause, because the correlation of A and B conditional on C does not indicate their causal connection (see equations Eq.10-Eq.12 above for the structure of the probability model). At first this seems clearly wrong: the Principle of the Common Cause begins with the claim that A and B are *correlated* (whereas the Principle of the Common Effect begins with them *uncorrelated*); hence the antecedent of the Principle of the Common Cause is not fulfilled, so it fails to provide a counterexample. What Reiss has in mind, however, is that the data on A and B may be collected in such a way that, without knowing it, we observe them only conditional on C, so that they appear to be unconditionally correlated.

A real example illustrates Reiss's concern. [30] Data on child molestation and exposure to child pornography was collected from prisoners in jail for possession of child pornography. Eighty-five percent admitted to having molested children. Consider three binary $(0,1)$ variables: $A = 1$ means viewed child pornography, $B = 1$ means molested children, and $C = 1$ means incarcerated for possession of child pornography. A concern immediately expressed by various critics of the study amounts to asserting the possibility that the variables form an unshielded collider and that A and B are correlated only because the mode of data collection implicitly conditions on C, the fear being that those who *both* view child pornography and molest children are more likely to be incarcerated for possession than those who merely view child pornography and that viewing it and child molestation may be unconditionally independent.

The intuition of the critics of the study can be interpreted with the two-step inferential process as the requirement that we get the right probability model, which means seriously entertaining the criticism and widening the scope of the data collection, so that the alternative hypothesis of an unshielded collider can be assessed. This strategy is suggested by a wider understanding of the world. But that is not an objection; there is nothing in the two-step process that suggests that a probability model is a black box for processing statistics without reference to their nature and provenance nor that only statistical criteria can be used to support a particular probability model. It is possible that sometimes we may make a mistake and do not notice accidental conditioning on an unobserved variable. Reiss persistently confuses the epistemic with the practical. It is nonsense to attack an inferential principle as metaphysical and not epistemic because we sometimes make errors in practice. The right response to accidental conditioning is to try to use all our knowledge to anticipate situations that give rise to such errors, to criticize research and to respond to criticism, and to test, test, test.

Reiss's remaining cases (mixing, stationary nonsense correlation, homoplasies, non-statistical nonsense correlations, laws of coexistence) are all answered in much the same way as his case of colliders (see [17]).

4 Cartwright's Pessimism Confounded

We have not offered a general defense of the Principle of the Common Cause. Instead, we have demonstrated that certain ways of attacking it illustrate the general proposition that we cannot neglect the need to provide a convincingly supported probability model if we wish to draw probabilistic conclusions from data. Reiss's "defense" of the Principle of the Common Cause goes wrong because he accepts simultaneously two premises: 1) Cartwright's view that only tightly controlled nomological (or socio-economic) machines generate frequencies that can be modeled probabilistically; and 2) that something like the Principle of the Common Cause is used – as a matter of fact – in actual research. Omitting the step of establishing a probability model, the conditions for which, Cartwright has argued, are too severe to be met in many practical contexts, is a fairly desperate move. Reiss's defense ends with the "irony" that, once his open-ended list of exceptions has been taken on board, we had better stick to experiments and eschew passive observation [28, p.194]. In the name of strengthening the practical applicability of the principle, Reiss kills it. His first premise is the mortal enemy of his second premise. In the end, he is left exactly where he started with Cartwright's view that the scope of probability models is very narrow and almost entirely restricted to experimental contexts.

Cartwright's position is deeply pessimistic with respect to economics and other sciences that rely on passive observation. Is it justified? Cartwright's argument points to a substantial disanalogy between experiments and passive observation. The disanalogy is real enough; it is the bane of empirical economics. But there is also a strong analogy between experiments and passive observation, which was central to Haavelmo's analysis sixty years ago.

Cartwright stresses that experiments generate probabilistically well behaved data (and, generally, display nature's capacities) only when they are arranged "just so". There is a danger of overstating the case. In a discussion of coin-flipping machines, Cartwright [5, p.166] says:

> Imagine ... that we flip the coin a number of times and record the outcomes, but that the situation of each flip is arbitrary. In this case we cannot expect any probability at all to emerge.

Approach this claim empirically: take any coin in general circulation and sit down anywhere and flip it how you will, recording heads and tails. Having done the experiment, I am sure that over the course of, say, 200 flips your data will conform – as judged by standard statistical tests – to a binomial probability model with a probability 1/2 for heads.

It matters that on a typical flip, the coin rotates at least once or twice. It matters that you do not wait to decide how to record the coin until you can see its resting position clearly. (For example, if a coin is leaning against the leg of table showing heads, and you decide after you see it that the rule for that particular flip will be to turn its visible side down before reading it, then the implicit preference for tails – if it persists in other such cases – may skew the results.) It does not matter whether you catch the coin in midair and turn it over on your wrist before reading it or let it fall to the floor or pick it out of the crack between the cushions on the sofa or fish it out from under the table. It does not matter whether the coin is new or worn or clean or dirty. The frequencies displayed by coins are very robust.

The point is not that every capacity nor every probability is similarly robust. It is quite difficult to construct a machine that will robustly deliver any probability other than zero for the frequency of a coin falling on its edge. Rather, the point is that it is a mistake to assert that a *very high* degree of control is an *a priori* requirement of frequencies conforming to stable probability models. Whether they do or do not is just something that we have to learn about the world in particular cases.

Even in controlled experiments, the range of factors that we attempt to control are frequently quite limited. Partly because we judge that certain factors are irrelevant and partly out of ignorance, many factors are left to nature's whims. And when we (or some other researcher) "replicates" our experiment, we cannot set every control in precisely the same way, and whimsical nature picks different values for factors that we have left uncontrolled. To paraphrase Heraclitus: you cannot perform the same experiment twice. At best each experiment is a model of its fellows. As with all models, we have to ask, is it a good or successful model? We learn from experience and from diagnostic tests whether we have successfully implemented controls of the right type and what we may neglect or ignore. Sometimes we find out later that we were wrong and that a neglected or overlooked factor is essential to the result, so that our experiment needs to be reinterpreted, redone, or set aside as uninformative.

The situation is not different in kind from what we face in formulating probability models for passively observed data. We may construct a chance setup and discover that it is reliable. Equally, we may observe the world and discover that there is a way of modeling it that reliably acts like a fabricated chance setup would. Hacking [8, p.1] introduces the notion of a chance setup and promptly illustrates it with a passively observed example: "the frequency of traffic accidents on foggy nights in a great city is pretty constant." The controls (day or night, foggy or clear, in the city or not) are of the same nature as the ones that find their ways into econometric models. The controls may be represented in coarse or finely delineated categories, as may the category of traffic accidents itself. Whether we need finer controls or more controls or controls of different kind and what kind of probability model should tie them together is a matter – exactly as it is for physically controlled experiments – for experience and testing to reveal.

Passive observation is, in many respects, at a disadvantage in comparison to active experimentation. That fact poses serious challenges for empirical economics. Nevertheless, that the inferential logic of passive observation is not of a radically different kind and that statistics provides many useful tools that help us to specify and test appropriate probability models is grounds for optimism.

Bibliography

[1] Angrist, J., Krueger, A. Instrumental variables and the search for identification: From supply and demand to natural experiments. *Journal of Economic Perspectives*, 15(4):69–85, 1999.

[2] Arntzentius, F. Reichenbach's common cause principle. In Zalta, E. N. (Ed.), *Stanford Encyclopedia of Philosophy*. Spring, 2005 edn., 2005. URL http://plato.stanford.edu/archives/spr2005/entries/physics-Rpcc/.

[3] Boumans, M. *The Problem of Passive Observations*. University of Amsterdam, 2007, unpublished manuscript.

[4] Cartwright, N. *Nature's Capacities and Their Measurement*. Oxford: Clarendon Press, 1989.

[5] Cartwright, N. *The Dappled World*. Cambridge: Cambridge University Press, 1999.

[6] Forster, M. Sober's principle of the common cause and the problem of comparing incomplete hypotheses. *Philosophy of Science*, 55(4):538–559, 1988.

[7] Haavelmo, T. The probability approach in econometrics. *Econometrica*, 12 (supplement), 1944. July.

[8] Hacking, I. *The Logic of Statistical Inference*. Cambridge: Cambridge University Press, 1965.

[9] Hausman, D. M., Woodward, J. Independence, invariance, and the causal Markov condition. *British Journal for the Philosophy of Science*, 50(4):521–583, 1999.

[10] Hendry, D. F. *Dynamic Econometrics*. Oxford: Oxford University Press, 1995.

[11] Hood, W. C., Koopmans, T. C. (Eds.). *Studies in Econometric Method*, Cowles Commission Monograph 14. New York: Wiley, 1953.

[12] Hoover, K. D. *Causality in Macroeconomics.* Cambridge: Cambridge University Press, 2001.

[13] Hoover, K. D. *The Methodology of Empirical Macroeconomics.* Cambridge: Cambridge University Press, 2001.

[14] Hoover, K. D. Econometrics and reality. In Mäki, U. (Ed.), *Fact and Fiction in Economics: Models, Realism, and Social Construction.* Cambridge: Cambridge University Press, 2002.

[15] Hoover, K. D. Nonstationary time series, cointegration, and the principle of the common cause. *British Journal for the Philosophy of Science,* 54(4):527–551, 2003.

[16] Hoover, K. D. The past as future: The marshallian approach to post-walrasian econometrics. In Colander, D. (Ed.), *Post Walrasian Macroeconomics: Beyond the Dyanamic Stochastic General Equilibrium Model,* 239–257. Cambridge: Cambridge University Press, 2006.

[17] Hoover, K. D. Probability and structure in econometric models. unabridged working paper, downloadable from, 2007. URL http://econ.duke.edu/~kdh9/Source%20Materials/Research/Probability%20and%20Structure%2019%20September%202007.pdf.

[18] Hoover, K. D. Economic theory and causal inference. In Mäki, U. (Ed.), *Handbook of the Philosophy of Economics.* Amsterdam: Elsevier/North-Holland, forthcoming. In Gabbay, D., Thagard, P. and Woodsthe, J. (Ed.): *Handbook of the Philosophy of Science.*

[19] Johansen, S. Confronting the economic model with the data. In Colander, D. (Ed.), *Post Walrasian Macroeconomics: Beyond the Dynamic Stochastic General Equilibrium Model,* 287–300. Cambridge: Cambridge University Press, 2006.

[20] Juselius, K. Models and relations in economics and econometrics. *Journal of Economic Methodology,* 6:259–290, 1999.

[21] Koopmans, T. C. *Statistical Inference in Dynamic Economic Models,* Cowles Commission Monograph 10. New York: Wiley, 1950.

[22] Liu, T. Underidentification, structural estimation, and forecasting. *Econometrica,* 28(4):855–865, 1960.

[23] Mill, J. S. On the definition of political economy. In Robson, J. M. (Ed.), *Essays on Some Unsettled Questions of Political Economy,* 309–340. Toronto: University of Toronto Press, 1844, reprinted in 1967. Collected Works of John Stuart Mill, vol.4, Essays on Economics and Society: 1824-1845.

[24] Morgan, M. S. *The History of Econometric Ideas.* Cambridge: Cambridge University Press, 1990.

[25] Papineau, D. Can we reduce causal direction to probabilities. In Hull, D., Okruhlik, K. (Eds.), *PSA 1992,* vol. 2, 238–242. East Lansing, MI: Philosophy of Science Association, 1992.

[26] Pearl, J. *Causality: Models, Reasoning, and Inference.* Cambridge: Cambridge University Press, 2000.

[27] Reichenbach, H. *The Direction of Time.* Berkeley: University of California Press, 1956.

[28] Reiss, J. Time series, nonsense correlations and the principle of the common cause. In Russo, F., Williamson, J. (Eds.), *Causality and Probability in the Sciences,* 179–196. London: College Publications, 2007.

[29] Scheines, R. The similarity of causal inference in experimental and non-experimental studies. *Philosophy of Science,* 72(5):927–940, 2005.

[30] Sher, J., Carey, B. Child-porn study raises question of behavior. *New York Times,*

2007. URL http://www.nytimes.com/2007/07/19/us/19sex.html?_r1. Online edition, 19 July.

[31] Sims, C. A. Macroeconomics and reality. *Econometrica*, 48(1):1–48, 1980.

[32] Sober, E. The principle of the common cause. In *From a Biological Point of View*, 158–174. Cambridge: Cambridge University Press, 1994.

[33] Sober, E. Venetian sea levels, british bread prices, and the principle of the common cause. *British Journal for the Philosophy of Science*, 52(2):331–346, 2001.

[34] Spanos, A. On theory testing in econometrics: Modeling with nonexperimental data. *Journal of Econometrics*, 67(1):189–226, 1995.

[35] Spanos, A. *Probability Theory and Statistical Inference : Econometric Modeling with Observational Data*. Cambridge: Cambridge University Press, 1999.

[36] Spirtes, P., Glymour, C., Scheines, R. *Causation, Prediction, and Search*. Cambridge, MA: MIT Press, 2nd edn., 2000.

[37] Steel, D. Making time stand still: A response to sober's counter-example to the principle of the common cause. *British Journal for the Philosophy of Science*, 54(2):309–317, 2003.

[38] Suppes, P. A probabilistic theory of causality. *Acta Philosophica Fennica*, Fasc. XXIV, 1970.

Variable Definition and Causal Inference[1]

Peter Spirtes
Carnegie-Mellon University
ps7z@andrew.cmu.edu

ABSTRACT. In the last several decades, a confluence of work in the social sciences, philosophy, statistics, and computer science has developed a theory of causal inference using directed graphs. This theory typically rests either explicitly or implicitly on two major assumptions:

Causal Markov Assumption: For a set of variables in which there are no hidden common causes, variables are independent of their non-effects conditional on their immediate causes.

Causal Faithfulness Assumption: There are no independencies other than those entailed by the Causal Markov Assumption.

A number of algorithms have been introduced into the literature that are asymptotically correct given these assumptions, together with various assumptions about how the data has been gathered. These algorithms do not generally address the problem of variables selection however. For example, are commonly used psychological traits such as extraversion, agreeableness, conscientiousness, etc. actually mixtures of different personality traits? In fMRI research, there are typically measurements of thousands of different tiny regions of the brain, which are then clustered into larger regions, and the causal relations among the larger regions are explored. Have the smaller regions been clustered into larger regions in the "right" way, or have functionally different regions of the brain been mixed together? In this paper I will consider the reasonableness of the basic assumptions, and in what ways causal inferences becomes more difficult when the set of random variables used to describe a given causal system is replaced by a different, but equivalent, set of random variables that serves as a redescription of the same causal system.

1 Introduction

In the last several decades, a confluence of work in the social sciences, philosophy, statistics, and computer science has developed a theory of causal inference using directed graphs. This theory typically rests either explicitly or implicitly on two major assumptions:

Causal Markov Assumption: For a set of variables in which there are no hidden common causes, variables are independent of their non-effects conditional on their immediate causes.

---

[1]The author is with Department of Philosophy, Carnegie-Mellon University. The author wishes to thank Clark Glymour and Richard Scheines for extensive discussions of the proper way to perform and interpret the translations of the causal assumptions and causal operations to new systems of random variables.

Causal Faithfulness Assumption: There are no independencies other than those entailed by the Causal Markov Assumption.

A number of algorithms have been introduced into the literature that are asymptotically correct given these assumptions, together with various assumptions about how the data has been gathered. These algorithms do not generally address the problem of variables selection however. For example, are commonly used psychological traits such as extraversion, agreeableness, conscientiousness, etc. actually mixtures of different personality traits? In fMRI research, there are typically measurements of thousands of different tiny regions of the brain, which are then clustered into larger regions, and the causal relations among the larger regions are explored. Have the smaller regions been clustered into larger regions in the "right" way, or have functionally different regions of the brain been mixed together? In this paper I will consider the reasonableness of the basic assumptions, and in what ways causal inferences becomes more difficult when the set of random variables used to describe a given causal system is replaced by a different, but equivalent, set of random variables that serves as a redescription of the same causal system.

This section will explain the basic setup for causal inference that I will consider. Although causal relations are often thought of as holding between events in space-time (e.g. turning a light switch, launching a ship), the causal laws I will consider here are laws that relate states to each other, rather than changes in states to each other. The usual laws of physics and many laws of biology are differential equations, and it is not entirely clear how well they fit into this framework, but they can be approximated by having the state of a property A at time t depend upon the state of A at time $t-1$. I am also not going to be concerned with actual causation here, although I will refer to specific individual at a specific time. So I am not interested in whether the aspirin was actually taken at time t, and if it was, whether it actually prevented a headache at $t+1$; I am merely interested in whether there is a causal law that make the probability of headache at $t+1$ different when aspirin is taken at t. The basic setup is similar to ones described in [8] and [10] and assumed in structural equation modeling.

1.1 Random Variables

Assume first that the relata of the type causal relation are the properties of an individual (or unit) at given times: e.g. taking aspirin at time t is a type cause of some (presumably diminished) strength of headache at $t+1$, for a given individual *Joe*. Also assume that for each unit, there is a set of properties at each time. To simplify matters, I will assume that time is discrete. Extending the basic setup to continuous time is non-trivial.

Each property will be represented by a real-valued random variable, and $A[t, u]$ will stand for the value of random variable A at time t for unit u. As an example, I will consider a population in which each unit in the population consists of a pair of particles, each of which has a position along a z-axis. I will call them *Position$_1$* and *Position$_2$*.

1.2 Structural Equations

Assume that for each property A of a unit U at a given time t, there is a causal law that relates the property to other properties at previous times. Further assume that these laws are a structural equation, that specifies the value of A as a function of other properties at previous times: $A[t, u] := f(\mathbf{V}[t-1, u], \cdots, \mathbf{V}[t-n, u])$, for some finite n.[2] The equations are "structural" and an assignment operator (":=") is used in the equation to indicate that the equation is not to be interpreted as merely stating the relationship between the actual values of $A[t, u]$ and $\mathbf{V}[t-1, u], \cdots, \mathbf{V}[t-n, u]$. The equation also has counterfactual implications: regardless of how $\mathbf{V}[t-1, u], \cdots, \mathbf{V}[t-n, u]$ were to receive their values (e.g. naturally occurring, experimental manipulation) and regardless of which values they received, the equation would be satisfied. I will restrict attention to structural equations that relate different variables for the same unit. If it is natural to consider equations that relate variables for different units, it is possible to aggregate units into larger units - e.g. individual people can be aggregated into families.[3] In the example, the equations are:

$$Position_1[u, t] := Position_1[u, t-1], \quad t > 0$$
$$Position_2[u, t] := Position_2[u, t-1], \quad t > 0$$

The assignment operator (":=") in the structural equations indicates that the equation holds regardless of how $Position_1[u, t-1]$ and $Position_2[u, t-1]$ obtain their values - either through the normal evolution of the system, or through being set by some intervention from outside the causal system.

1.3 Initial Conditions

In order to prevent an infinite regress, there is also an initial value for each variable. Suppose that there is a population of these units, and a probability distribution of the values of each random variable in the population. Another possibility is to start with a probability (propensity) distribution for each unit, but I will not pursue that possibility here. It is not necessary for what follows that each unit obey exactly the same set structural equations with the same parameters. Under certain assumptions, it is possible to allow units to have different parameter values, or even different equations. Thus the initial conditions that specify a value for each random variable at time 0 entails a joint distribution for the random variables at time 0. For example,

$$Position_1[t = 0] \sim N(0, 1)$$
$$Position_2[t = 0] \sim N(0, 1)$$

---

[2] Individual variables are in upper case italic, sets of variables are in upper case bold face, values of individual variables are in lower-case italic, and the values of sets of variables are in lower-case boldface. In some social sciences such as econometrics, time series equations allow a property A at time t to causally depend upon other properties at time t. This is called "contemporaneous causation" but its interpretation is controversial. I will take it as an approximation to the case where if the time index t is divided more finely, A at time t causally depends upon properties at earlier times than t, but times much closer to t than the original $t - 1$ is.

[3] There is also a second form of equation in common usage. $Pr(A[t, u]) := f(\mathbf{V}[t-1, u], \cdots, \mathbf{V}[t-n, u])$, where Pr is a countably additive propensity. This second form of equation in effect includes the structural equations as the special case where all of the propensities are zero or one. I will not use this formulation further in this paper. It would be an interesting project to translate the results of this paper formulated for structural equation models into the probabilistic equations.

Probability and Structure in Econometric Models

I($Position_1, Position_2$) (where I($\mathbf{A}, \mathbf{B}|\mathbf{C}$) means $\mathbf{A}$ is independent of $\mathbf{B}$ conditional on $\mathbf{C}$, and I($\mathbf{A}, \mathbf{B}$) means $\mathbf{A}$ is independent of $\mathbf{B}$). If the set of variables contains a single member, such as $Position_1$, I will write $Position_1$ rather than $\{Position_1\}$).

The structural equations across the population determine how the initial probability distribution evolves across time. Thus the value of the random variable for unit u at t is denoted $A[u, t]$, and the random variable itself is denoted as $A[t]$.

1.4 Actions and Manipulations

Also assume that it is possible to take actions that set the value of each random variable to a value that it would not necessarily take on in the course of the normal evolution of the system. After any such action, the system then evolves according to the structural equations, until another action is taken from outside to intervene on the causal system.

There are a variety of different kinds of such actions. Actions may either be specified at the level of a unit, or at the level of populations. In the former case, the most general kind of action I will consider is to set the value of each variable at time t to take on a value as a function of the actual values of the variables at t and some finite set of times earlier than t.[4] This is denoted by $Set(\mathbf{V}[u,t]) := f(\mathbf{V}[u,t], \cdots \mathbf{V}[u,t-n])$ for some positive integer n. A special case of an action sets the value of $A[u,t]$ to a constant: $Set(A[u,t] := x)$, where x is either a real number, or the special value null. If x is a real number, this corresponds to replacing the structural equation for $A[u,t]$ with $A[u,t] := x$. "null" corresponds to letting the variable value keep the propensity that it would take on in the normal course of the evolution of the system according to the structural equations; this corresponds to not replacing the structural equation for $A[u,t]$. $Set(A[t] := x)$ sets the value of $A[u,t]$ to x for all members u of the population.

The effect of an action on a causal system will generally depend upon how *each* variable in the system is set. I will adopt the convention that value of an action at any time is *null* unless it is specifically stated otherwise. Any action that sets the values of some subset of variables $\mathbf{B}$, and has a null effect on $\mathbf{V}\backslash\mathbf{B}$, is a *manipulation* of $\mathbf{B}$. If the value that $\mathbf{B}$ is set to is not a function of $\mathbf{V}$, then the manipulation is *unconditional*. Note that every action is a manipulation of some subset of variables, but whether or not an action is a manipulation of $\mathbf{B}$ depends upon what other variables are in $\mathbf{V}$. This is an important point I will return to later.

For example, the most general form of action that I will consider in the <$Position_1, Position_2$> system at the level of a unit is the following: $Set(Position_1[u,t]) := h(Position_1[u,t], \cdots, Position_1[u,t-n], Position_2[u,t], \cdots, Position_2[u,t-n])$ where n is some finite integer. There are a variety of interesting special cases of these general actions. For example, $Set(P(Position_1[u,t] := a), Position_2[u,t] = null)$ is an unconditional individual manipulation of $Position_1$ and a null manipulation of $Position_2$ because $Position_1$ is set to value a regardless of the values of $Position_1$ and $Position_2$ (unconditional), and no action is

---

[4] More realistically, the value of the variable is set at time $t+\varepsilon$, where ε is a small compared to the time interval between t and $t+1$.

take on *Position$_2$* (individual). A null manipulation of *Position$_2$*[u, t] does not change the current value of *Position$_2$*[u, t]. There might be several different ways of performing manipulations in an actual causal system. $Set(Position_2[u,t] := null)$ could be performed by simply not touching a given *Position* and leaving it as it is. It could also be performed by any method that does not affect the evolution of the system with respect to the variables in the system.

At the level of populations, the most general kind of action that I will consider is to set the joint distribution of the variables at time t as a function of the values of **V** at t and earlier times; this is denoted by $Set(P(\mathbf{V}[t]) := f(P(\mathbf{V}[t]), \cdots, P(\mathbf{V}[t-n])))$. An action that sets the joint distribution of a set of variables **B** and is null for the variables in $\mathbf{V}\backslash\mathbf{B}$ is a *manipulation* of **B**. If the distribution that **B** is set to is not a function of any variables in **V**, then the manipulation is *unconditional*. If $P'(\mathbf{V}[t])$ has support at a single point, it is a *deterministic* manipulation, and is abbreviated by $Set(\mathbf{V}[t] := \mathbf{v})$. Again, note that whether or not an action counts as a *manipulation* of $A[t]$ depends upon what other variable have been included in the causal system.

1.5 Causal Graphs

The causal graph corresponding to a set of variables can either be drawn for an individual unit, or for a population. For an individual unit, the causal graph G has a directed edge from $A[u, t-n]$ to $B[u, t]$ if and only if for some finite m there exists a pair of unconditional manipulations of $\mathbf{V}[u, t-m], \cdots \mathbf{V}[u, t-1]$ that differ only in the values assigned to $A[u, t-n]$ such that the resulting values of $B[u, t]$ are different. For a population, the causal graph G has a directed edge from $A[t-n]$ to $B[t]$ if and only if for some finite m there exists a pair of unconditional manipulations of $\mathbf{V}[t-m], \cdots \mathbf{V}[t-1]$ that differ only in the distributions assigned to $A[t-n]$ such that the resulting distributions of $B[t]$ are different. (Note that the existence of an edge in the population causal graph entails the existence of an edge in some individual causal graph, but not vice-versa. It is possible that a manipulation changes the value of every individual unit but leaves the distribution of properties in the population the same, e.g. by permuting the values among the units of the population.) If there is a directed edge from $A[t-n]$ to $B[t]$ then $A[t-n]$ is said to be a *direct cause* of $B[t]$. Because of the assumptions about the structural equations, each graph is acyclic, and there are no edges between variables at the same time. If there is an edge from $A[t-n]$ to $B[t]$ in the causal graph, then $A[t-n]$ is a *parent* of $B[t]$ and $B[t]$ is a *child* of $A[t-n]$. If there is a directed path from $A[t-n]$ to $B[t]$ in the *causal graph*, $A[t-n]$ is an *ancestor* of $B[t-n]$ and $B[t-n]$ is a *descendant* of $A[t-n]$. It also turns out to be convenient to consider each variable to be its own ancestor and descendant (but not its own parent or child.) For example, the causal graph for the $<Position_1, Position_2>$ causal system is: Note that

$Position_1[t=0] \rightarrow Position_1[t=1] \cdots \rightarrow Position_1[t=n] \cdots$
$Position_2[t=0] \rightarrow Position_2[t=1] \cdots \rightarrow Position_2[t=n] \cdots$

Figure 1:

$Position_1[t]$ is independent of $Position_2[t']$ for all t, t', and that according to the causal graph and the structural equations, manipulating $Position_1[t]$ affects $Position_1[t']$ for all $t' > t$, and manipulating $Position_2[t]$ affects $Position_2[t']$ for all $t' > t$. However, manipulating $Position_1[t]$ has no effect on $Position_2[t']$ for all t, t', and manipulating $Position_2[u,t]$ has no effect on $Position_1[u,t']$ for all t, t'.

The causal graph might seem redundant because the information it contains is a subset of the information in the structural equations. However, it plays an important role for several reasons. First, it is used by several algorithms for predicting the effects of manipulations (e.g. the Manipulation Theorem of [8], and the Causal Calculus of [4]). It is also a useful step in the inference of causal systems from data. First, the data, background knowledge, and partial experimental knowledge can be used to infer a causal graph, or more typically, a set of causal graphs. Then the causal graph, together with the data, background knowledge, and partial experimental knowledge can be used to estimate the parameter values in the structural equations. Some more details about this process will be presented below.

1.6 Weak Causal Markov Assumption

One of the (often implicit) fundamental assumptions used both in calculating the effects of manipulations, and in drawing inferences about causal graphs from sample data is the Causal Markov Assumption. I will not attempt to justify this assumption here, but several justifications are given in [4].

The version of the assumption I will use here can be stated in terms of the causal graph at the population level. If a probability distribution P and a DAG G have the relationship that each variable in P is independent of its non-descendants in G conditional on its parents in G, then P is said to be *Markov* to G. A set of variables **V** is said to be *causally sufficient* when every direct cause of two variables in **V** is also in **V** (i.e. there are no confounders of variables in **V** that have been left out of **V**.) The Causal Markov Assumption states that for a causally sufficient set of variables whose causal graph G is a directed acyclic graph (DAG), the population distribution is Markov to G. (Note that I am assuming that all causal graphs are acyclic because either they are indexed by time, or they are stripped of their time indices only when they can be represented by an acyclic graph. There are reversible causal systems or equilibrium causal systems that I am not considering in this section that it would be natural to represent with cyclic causal graphs.) Although the Causal Markov Assumption applies only to causally sufficient sets of variables, it is not the case that causal inference is impossible if the measured variables are not causally sufficient. See [4] for details.

For the case of DAGs, there is an equivalent formulation of the Causal Markov Assumption. *D–separation* is a graphical relation between three disjoint sets of variables in a DAG that is defined in the Appendix - the basic idea is that **A** is d-separated from **B** conditional on **C** when certain kinds of paths between members of **A** and members of **B** (that bear a relationship to **C**) do not exist in the graph. All of the conditional independence consequences of a distribution P being Markov to a DAG G are captured by d-separation: **A** is entailed to be independent of **B** conditional on **C** in a distribution P that is Markov to a DAG G if and only if **A** is d-separated from **B** conditional on **C** in G. Hence there is a formulation of the Causal Markov Assumption that is superficially a stronger assumption, but is actually an equivalent formulation. The Global Causal Markov Assumption states that for a causally sufficient set of variables, if the causal graph G for a population is a DAG, and **A** is d-separated from **B** conditional on **C** in G, then **A** is independent of **B** conditional on **C** in the population.

In the deterministic structural equation framework that I am using in this paper, the Causal Markov Assumption is entailed by a weaker assumption, the Causal Markov Assumption: For a causally sufficient set of variables whose causal graph is a DAG, if no member of **A** is a descendant of any member of **B**, and no member of **B** is a descendant of any member of **A**, and there is no third variable that is a direct cause of a member of **A** and **B**, then **A** and **B** are independent (i.e. causally unconnected variables are independent.)

For example, if the causal graph is the DAG depicted in Fig.1, then according to the Weak Causal Markov Assumption, $Position_1[t = i]$ and $Position_2[t = j]$ are independent conditional on $Position_1[t = i-1]$. The Global Causal Markov Assumption also entails that $Position_1[t = i]$ and $Position_2[t = j]$ are independent.

One problem with applying the Causal Markov Assumption to a causal graph that starts at an arbitrary point in time is that there seems to be no reason to assume that the initial condition that specifies the joint distribution at $t = 0$ should make the variables at $t = 0$ independent, particularly if they were generated by the same structural equations as the variables for $t > 0$. For example, if $P(A[t = 0], B[t = 0])$ is generated by applying the structural equations to earlier variables that are not represented in the causal system, and both are effects of some $C[t = -1]$ that is unrepresented in the causal system, then typically $A[t = 0]$ will be associated with $B[t = 0]$. This is not a violation of the Weak Causal Markov Assumption, because this is a case where the set of variables in the system is not causally sufficient (because it does not contain $C[t = -1]$).

There are typically two ways of dealing with this problem:

1. Assume the causal system is one in which the initial conditions have no effect (or approximately no effect) on the values of the system after a long enough waiting period. In that case, the Causal Markov Assumption can be applied to variables for large n. This is typically the case, for example, in linear systems in which noise affects each measured variable at every point in time.
2. Introduce a double-headed arrow between variables at $t = 0$ to represent any associations between pairs of variables at $t = 0$ due to a common cause in the past prior to $t = 0$. It is then easy to extend the definition of d-separation to graphs with double-headed arrows, and apply the Global Causal Markov Assumption to the set of variables even though they are not causally sufficient. (The way this is handled in [8] is slightly more complicated, but the complications make no difference to the issues discussed in this paper.)

1.6.1 A Special Case - Time-Index Free DAGs

In this paper, I will consider as a starting point a particularly simple kind of causal graph that is used quite commonly in structural equation models. The assumptions under which this particularly simple kind of causal graph is accurate are very strong and often unrealistic. Fortunately, they are not essential to causal inference, but they do make causal inference simple. Moreover, they will illustrate how even when very strong assumptions are made about causal systems, translating that causal system into the "wrong" variables can introduce many complications that make causal inference much more difficult.

The first major assumption is that there are no "hidden confounders" - that is the set of *measured* variables **V** in the causal system is *causally sufficient*.

The second major assumption is that there is a total ordering O of the variables in the causal DAG such that for each variable A and all t, there is no edge from a variable later than A in the ordering O to A. Suppose, for example, there is a causal system in which:

$A[0] \sim N(0,1)$, $B[0] \sim N(0,1)$, $C[0] \sim N(0,1)$, $A[0]$, $B[0]$, $C[0]$ jointly independent

$$A[t] := A[t-1], \quad t > 0$$
$$B[t] := A[t-1], \quad t > 0$$
$$C[t] := B[t-1], \quad t > 0$$

The causal graph is shown in Fig.2. In this case, the variables can be ordered

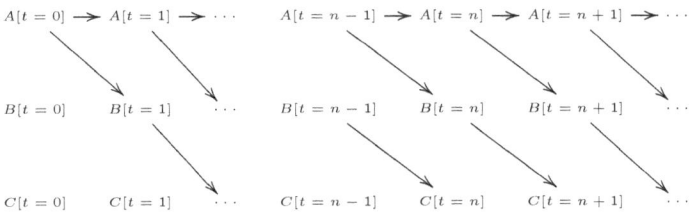

Figure 2:

as <A, B, C>, each temporally indexed A variable is directly caused only by other A variables (which are not later in the ordering <A, B, C>), each temporally indexed B variable is directly caused only by A variables (which are not later in the ordering <A, B, C>), and each temporally indexed C variable is directly caused only by B variables (which are not later in the ordering <A, B, C>). It is then possible to write time-index free structural equations as well.

However, the following causal graph is an example of a graph that violates the assumption: no ordering of the variables has the property that each variable is directly caused only by variables that precede it in the ordering: $Position_1[t]$ is directly caused by $Position_2[t-1]$, and $Position_2[t]$ is directly caused by $Position_1[t-1]$.

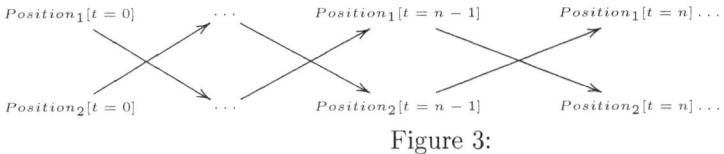

Figure 3:

The third major assumption is that that if $A[u, t-i]$ directly causes $B[u, t]$, then $A[u, t-j]$ does not directly cause $B[u, t]$ unless $i = j$.

Finally, I will assume that if the causal edge with the largest temporal gap (from a variable at $t - n$ to t) is n, the unit values are measured after having been stationary (i.e. had the same value) longer than n. The effect of these three assumptions (and the Causal Markov and Faithfulness Assumptions, described below) is that it is possible to make inferences about the structure of the causal graph from evidence that is all gathered at some time t, rather than having to observe the evolution of the system. This is plausible for systems in which the causes change very slowly, e.g. for an adults' height. If someone's height at 25 causes an effect, for most purposes it suffices to use their measured height at 45 as a surrogate for their height at 25.

For example, under the four assumptions listed, the causal graph in Fig.2 can be represented as $A \to B \to C$, and the causal Markov Assumption will still apply to this DAG and the joint population distribution as measured at a particular time. In the DAG in Fig.2, the Causal Markov Assumption entails that $C[t = 3]$ is independent of $A[t = 1]$ conditional on $B[t = 2]$. However, under the assumptions, $A[t = 3] = A[t = 1]$, and $B[t = 3] = B[t = 2]$. Hence $C[t = 3]$ is independent of $A[t = 3]$ conditional on $B[t = 3]$. If instead of representing the variables at different times and observing the evolution of the system, we instead simply represent the resulting stationary distribution, the Causal Markov Assumption will apply to the abbreviated DAG without the temporal indices.

It is also possible to write the structural equations without time-indices:

$$B := A$$
$$C := B$$

These equations hold at a given time when the system is in equilibrium. For a fixed unit, a manipulation of B to a particular value b should be interpreted as setting B to the value b at all times after some given time.

1.7 Causal Faithfulness Assumption

The Causal Faithfulness Assumption is the converse of the Causal Markov Assumption. It states that for a causally sufficient set of variables, if the causal graph G for a population is a DAG, **A** is d-separated from **B** conditional on **C** in G only if **A** is independent of **B** conditional on **C** in the population. This is the assumption made in [8], and as described below, it plays a key role in causal inference.

It has subsequently been shown that a weaker condition suffices for the existence of asymptotically correct causal inference algorithms for DAGs in the Gaussian and discrete cases. In the case of linear non-Gaussian variables it has been shown that under the assumption of no hidden common causes even this weaker assumption is not needed [6]. The Weak Causal Faithfulness Assumption states that for a causally sufficient set of variables, if the causal graph G for a population is a DAG, and there is a directed edge from A to B in the population causal graph G, then A and B are not independent conditional on any subset of the other variables. For example, consider the following DAG:

Suppose that $A := \varepsilon_A$, $B := x \times A + \varepsilon_B$, and $C := y \times B + z \times A + \varepsilon_C$, where x, y, and z are real constants. Assume all of the variables are Gaussian, and the ε's are independent. Then if the two paths from A to C cancel, i.e. if

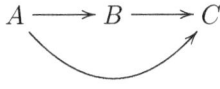

Figure 4:

$x \times y = -z$, then A and C will be independent, although they are not d-separated conditional on the empty set. The Weak Causal Faithfulness Assumption entails that A and C are not independent, because they are adjacent in the causal graph; hence in this case, the assumption entails $x \times y \neq -z$. I will not rehearse the arguments for this assumption here - see [8]. However, it has been shown that for several parametric families (linear Gaussian, discrete) violations of the Causal Faithfulness Assumption have Lebesgue measure 0 in the space of parameters.

This assumption plays a key role in causal inference for the following reason. Suppose that the true causal DAG is $A \rightarrow B \leftarrow C$. Any distribution that fits $A \rightarrow B \leftarrow C$ also fits the DAG in Fig.4, but only if the Weak Causal Faithfulness Assumption is violated. Without the Weak Causal Faithfulness Assumption, one would never be able to conclude that the DAG $A \rightarrow B \leftarrow C$ is the correct causal DAG, as opposed to the more complex DAG in Fig.4.

1.8 Consistency

Using the Causal Markov and Causal Faithfulness Assumptions, as well as distributional and sampling assumptions, it is possible to make inferences about what set of causal DAGs the true causal DAG belongs to. For example, if one makes the assumptions that allow for a DAG to be represented as a time-free DAG, then in the large sample limit, both the GES [1] and PC algorithms [8] identify a class of DAGs containing the true causal DAG.

The following algorithms are also relevant to the possibility of causal discovery of simple causal systems, after the variables have been translated, as explained in the section on translation. If one drops the assumptions that all confounders are measured, and allows for the possibility that sampling is affected by the measured variables, then in the large sample limit, the FCI algorithm identifies a class of DAGs containing the true causal DAG - the price that one pays for the dropped assumptions is increasing the computational complexity of the algorithm, and increasing the size of the set of DAGs containing the true DAG. Similarly if one makes the assumption that there are no hidden common causes, but there may be cycles, the Cyclic Causal Discovery Algorithm is provably reliable in the large sample limit [5]. Again, the price that one pays for allowing the possibility of cycles is that the algorithm is computationally more complex, and less informative. [2] describes an asymptotically correct algorithm for causal inference of vector autoregression models. At this time, there is no known algorithm for causal discovery given the possibility of both cycles and hidden common causes that is provably reliable in the large sample limit.

2 Translations

How does translating the description of a causal system into an alternative, but equivalent descriptions of the same causal system affect causal inference? It is standard to use random variables (i.e. functions from units to the reals) to represent the value that a property takes on in a unit in the population. However, there is not a 1-1 function mapping properties to random variables. On the one hand, the same property (e.g. height) can be represented by multiple random variables (e.g. height can be measured in inches, feet, meters, etc. which assign different numbers to each unit, and hence are different functions). On the other hand, multiple properties can be represented by the same function, if they happen to be coextensive in a population (e.g. "having a heart" and "having a kidney".)

The fact that height can be measured in inches or in feet obviously does not affect causal inference. The two different random variables *height in inches* and *height in feet* are related by an invertible function, i.e. *height in inches = height in feet* × *12*. Of course, there are transformations of *height in inches*, such as *height in inches squared*, which could lead to problems in causal inference, but these are not invertible functions.

Transformations of random variables that "mix" random variables together are more problematic. The pair of variables <*height, weight*> can be transformed into another pair of variables <*height + weight, height − weight*> such that there is a 1-1 function <*height + weight, height − weight*> = f(<*height, weight*>), and similarly <*height, weight*> = f^{-1}(<*height + weight, height − weight*>). Does the pair of variables *height + weight* and *height − weight* represent simply an odd way of talking about the properties *height* and *weight*, or do they represent somewhat odd properties in their own right? Nothing in this article depends on the answer to that question; I will refer to *height + weight* and *height − weight* as translations of height and weight.

Intuitively, there is an *important* difference between *height* and *weight* on the one hand, and *height + weight* and *height − weight* on the other hand. One might believe that each of *height* and *weight* represent "real" properties, while *height + weight* and *height − weight* do not individually represent "real" properties. I will consider various senses in which this might be the case later. However, it is clear that whatever can be said about the *pair* of properties height and weight can be said in the language of the *pair* of random variables <*height + weight, height−weight*>, and that whatever is said in the language of the random variables <*height + weight, height − weight*> can be translated into the language of <*height, weight*>. In that sense the pair of random variables <*height + weight, height − weight*> describe the same pair of properties that are described by <*height, weight*>.

There seems to be relatively little reason why someone would choose to frame the description of a causal system in terms of <*height+weight, height−weight*>. This is true particularly of many physical systems where we have well worked out and powerful theories. However, in many cases, particularly in the social sciences, it is often the case that it is not clear what the random variables ought to be, and which ones represent properties that are in some sense "real" or alternatively represent mixtures of "real" properties. For example, are commonly used psychological traits such as extraversion, agreeableness, conscientiousness, etc. actually mixtures of other "real" properties? In fMRI research, there are typically measurements of thousands of different tiny regions of the brain that are then clustered into larger regions, and the causal relations among the larger regions are explored. Have the smaller regions been clustered into larger regions in the "right" way, or have functionally different regions of the brain been mixed together?

A slightly different kind of example is related to the history of the relationship between cholesterol and heart disease. At one time it was thought that high levels of cholesterol caused heart attacks. Later, it discovered that there are two kinds of cholesterol, LDL cholesterol that causes heart attacks, and HDL cholesterol that prevents heart attacks. The original way of measuring cholesterol measured both LDL and HDL cholesterol - in other words it was the sum of the two. The measured variable was a kind of mixture of the underlying "real" properties. This is slightly different that what I have been discussing so far, because it is not possible to reconstruct the LDL and HDL cholesterol levels from the total cholesterol levels. So this is a case where both mixing of "real" properties occurred, but in addition information was thrown out. I will return to the question of this combination of problems later.

What is the difference between translations of random variables versus the structural equations relating random variables? Translations of random variables are always deterministic, instantaneous, and invertible; also it is not possible to independently set both the original random variable and the transformed variables. These properties are not always true of causal relations between random variables in the world, nor is it clear whether they are ever true. In the causal models that I consider here, they are never true, because the relationship of direct causation only exists between variables at different times - it is never instantaneous.

When the various pieces of a causal system - the structural equations, the causal graph, actions, the Causal Markov Assumption, etc. - are translated into the alternative representation, what does the resulting system look like, and how does it affect causal inference? I will consider each piece in turn, and illustrate it with the example depicted in Fig.1.

2.1 Translating Random Variables

First, the translation of the random variables is quite simple. By assumption there is an invertible function such that $\mathbf{V}' = g(\mathbf{V})$. For example:

$$\begin{aligned} C[u,t] &= 2 \times Position_1[u,t] + Position_2[u,t] \\ D[u,t] &= Position_1[u,t] + 2 \times Position_2[u,t] \end{aligned}$$

2.2 Translating Initial Conditions

The translation of the initial conditions is also quite simple. For example, $C[0] \sim N(0,5)$ and $D[0] \sim N(0,5)$. Although $Position_1[t=0]$ and $Position_2[t=0]$ are independent, $C[t=0]$ and $D[t=0]$ are not, because $cov(C[t=0], D[t=0]) = cov(2 \times Position_1[t=0] + Position_2[t=0], Position_1[t=0] + 2 \times Position_2[t=0]) = 2 \times var(Position_1[t=0]) + 2 \times var(Position_2[t=0]) + 5 \times cov(Position_1[t=0], Position_2[t=0]) = 4$. More generally, if the density function of $\mathbf{V}$ is $f(\mathbf{V})$, then the density function of $\mathbf{V}' = g(\mathbf{V})$ is $f(g^{-1}(\mathbf{V}')) \times |J|$, where $|J|$ is the absolute value of the determinant of the Jacobean of the transformation.

2.3 Translating Structural Equations

The structural equations for $C[u,t]$ and $D[u,t]$ can be found by a 3-step process:

1. Translate C and D into $Position_1[u,t]$ and $Position_2[u,t]$.
2. Write the structural equations for $Position_1[u,t]$ and $Position_2[u,t]$ in terms of $Position_1[u,t-1]$ and $Position_2[u,t-1]$.
3. Translate $Position_1[u,t-1]$ and $Position_2[u,t-1]$ back into $C[u,t-1]$ and $D[u,t-1]$.

For example,

$$
\begin{aligned}
C[u,t] &= 2 \times Position_1[u,t] + Position_2[u,t] \\
&:= 2 \times Position_1[u,t-1] + Position_2[u,t-1] \\
&= C[u,t-1] \\
D[u,t] &= Position_1[u,t] + 2 \times Position_2[u,t] \\
&:= Position_1[u,t-1] + 2 \times Position_2[u,t-1] \\
&= D[u,t-1]
\end{aligned}
$$

It is important to understand how the translated structural equations should be interpreted. The coefficient 1 of $C[u,t-1]$ in the equation for $C[u,t]$ is understood as giving the results when $C[u,t-1]$ is manipulated, but $D[u,t-1]$ is left the same, i.e. $Set(C[u,t-1], D[u,t-1] := null)$. These are different actions than the actions assumed by the structural equations expressed in terms of $Position_1$ and $Position_2$.

2.4 Translating Actions and Manipulations

The actions that can be performed upon the system can be translated by translating the distribution imposed upon $Position_1$ and $Position_2$ in the case the distribution is set at the population level, or by translating the values imposed upon $Position_1$ and $Position_2$ in the case that values are set at the level of units. So $Set(f(Position_1[t], Position_2[t]) := h(Position_1[t], Position_2[t]))$ translates into $Set(f(C[t], D[t])) := h(g^{-1}(<C[t], D[t]>) \times |J|)$. In this case, g^{-1} is $Position_1[t] = (2 \times C[t] - D[t])/3$ and $Position_2[t] = (2 \times D[t] - C[t])/3$.

In the case where actions set the values of individual units, then again one can translate $Position_1$ and $Position_2$ into C and D, and then solve the equations for values of C and D. Suppose for example that $Position_1[u,t] = 2$, and $Position_2[u,t] = 3$ and the action $Set(Position_1[u,t] = 0, Position_2[u,t] := null)$ is performed. Because $Position_2[u,t] = 3$, this is equal to $Set(Position_1[u,t] := 0, Position_2[u,t] := 3)$. Now substitute $(2 \times C[t] - D[t])/3$ for $Position_1[t]$ and $(2 \times D[t] - C[t])/3$ for $Position_2[t]$. This leads to $Set((2 \times C[t] - D[t])/3 := 0, (2 \times D[t] - C[t])/3 := 3)$. These equations can be solved, leading to $Set(C[u,t] := 3, D[u,t] := 6)$. If $Position_2[u,t]$ had some value other than 3, then there would be a different translation of $Set(Position_1[u,t] := 0, Position_2[u,t] := null)$.

The translation of actions does not preserve whether or not an action is an unconditional manipulation of a single variable, or even whether it is a manipulation of that single variable at all. For example, the action $Set(Position_1[u,t] := 0, Position_2[u,t] := null)$ changes the values of both $C[u,t]$ and $D[u,t]$; it is a manipulation of the pair $\langle C[u,t], D[u,t]\rangle$, but not of $\langle C[u,t]\rangle$ nor of $\langle D[u,t]\rangle$ individually. This captures the sense in which something that is a "local" action for the $\langle Position_1, Position_2\rangle$ system is a non-local action for the $\langle C, D\rangle$ system.

There is however, symmetry to this translation that implies that an unconditional manipulation of C and a null manipulation of D is not a manipulation of either $Position_1$ or $Position_2$ alone. For example, if $C[u,t] = 1, D[u,t] = 3$, then $Position_1[u,t] = -1/3$ and $Position_2[u,t] = 5/3$. But $Set(C[u,t] := 0, D[u,t] := null)$ translates into $Set(Position_1[u,t] := -1, Position_2[u,t] := 2)$, which changes the values of both $Position_1[u,t]$ and $Position_2[u,t]$. Thus, a manipulation of $C[u,t]$ is a manipulation of $\langle Position_1[u,t], Position_2[u,t]\rangle$, but not of $\langle Position_1[u,t]\rangle$ or $\langle Position_2[u,t]\rangle$ alone. This captures the sense in which something that is a "local" action for the $\langle C, D\rangle$ system is a non-local action for the $\langle Position_1, Position_2\rangle$ system.

Although there is symmetry in the translation between $Position_1$ and $Position_2$, there can be important asymmetries as well. In this example, there is clearly an intuition that a manipulation of $Position_1$ alone (i.e. the setting of $Position_2$ is $null$) is "really" local, while a manipulation of C alone (i.e. the setting of $Position_2$ is $null$) is not "really" local (because it is a manipulation of $Position_1$ based on the value of $Position_2$.) This is because $Position_1$ and $Position_2$ are properties of spatio-temporally non-overlapping regions of space-time, while C and D are not. Furthermore, a manipulation of $C[u,t]$ that is supposed to have no effect on $D[u,t]$ implies that they occur simultaneously, and hence have a spacelike separation. Barring some bizarre quantum mechanical distant coupling of the $Position_1$ and $Position_2$ there is no *reliable* mechanism for producing a manipulation of $C[u,t]$ alone (because this would involve sending a faster than light signal between the regions of spacelike separated regions of space-time).

These arguments might establish that in this example, $Position_1$ and $Position_2$ are the "right" variables, and C and D are the "wrong" variables. Policy predictions about policies that it is impossible to reliably implement are not very interesting. Nevertheless, I will argue that this metaphysical sense of "local" manipulation is not the one relevant for the range of phenomena that causal inference is performed on.

First, nothing in the standard account of predicting the effects of a manipulation depends upon a manipulation being produced intentionally, or knowingly, or reliably. There is nothing magical about actions taken by people. If an unconditional manipulation occurs without our knowledge or by coincidence, the prediction about the effects of the manipulation still applies. Moreover, typically the actions that are actually taken in the real world are complex, and are not known to be manipulations. Passing a law, administering a drug, giving someone a therapy, etc., are often assumed to be manipulations of some variable. However, it is typically not known whether they are actually manipulations prior to their administration. So while in the case of $Position_1$ and $Position_2$ it is intuitively the case that the actions that we can take are "really" manipulations of $Position_1$ or $Position_2$, and not of C or D individually, there are many other cases where the actions that are quite naturally taken correspond to manipulations of different variables. It is perfectly possible that drug A administered to control cholesterol does indeed change LDL cholesterol levels; but it would not be surprising if some drug B manipulated total cholesterol, and even left the difference between the levels of cholesterol the same. Because these problems are related to the concept of manipulation, they raise problems for randomized experiments as well as causal inference from observational studies.

Second, in many cases the kind of spatio-temporal non-overlap of the relevant properties such as $Position_1$ and $Position_2$, simply will not occur. Especially in the case of psychological, biological, or social causation, the relevant properties will occur in the same spatio-temporal regions. *IQ* and *Authoritarian-Conservative Personality*, or *GDP* and *Inflation* are not properties of non-overlapping spatio-temporal regions.

2.5 Translating Causal Graphs

Given the translation of the structural equations, the causal graph can be constructed from the structural equations and the manipulations in the same way. For example, in the causal system depicted in Fig.1 the causal graph

$$C[t=0] \to C[t=1] \to \cdots \to C[t=n] \cdots$$
$$D[t=0] \to D[t=1] \to \cdots \to D[t=n] \cdots$$

Figure 5:

Does the translated causal system also satisfy the conditions for constructing a time-index free DAG? The first condition, that there is an ordering of the variables such that each variable is not affected by a variable later in the ordering, is satisfied.

The second condition is that there not be any hidden common causes of C and D. Is that the case here? The question is what would this hidden common cause be? There is no hidden common cause in the <$Position_1, Position_2$> description of the causal system. The obvious candidate for a latent common cause of C and D is $Position_1$ and $Position_2$ (because $C[u,t] = 2 \times Position_1[u,t] + Position_2[u,t]$ and $D[u,t] = Position_1[u,t] + 2 \times Position_2[u,t]$). However, on closer examination $Position_1$ and $Position_2$ cannot be considered hidden common causes of C and D, because they are not causes of C and D at all. In the causal models under discussion, all causal influences go from earlier times to later times, and there are no causal influences between variables at the same time. Since $C[u,t]$ and $D[u,t]$ are functions of $Position_1[u,t]$ and $Position_2[u,t]$, all at the same time t, the relationship is not a causal one. Rather, C and D are just re-descriptions of $Position_1$ and $Position_2$. But there is nothing else in the $Position_1$ - $Position_2$ system that could plausibly be re-described as playing the role of a common latent cause in the <C, D> description of the system. Hence, there is no hidden common cause of C and D.

The third condition is also satisfied. If $Position_1$ and $Position_2$ are not changing, neither are C and D, so they will remain stationary.

Hence the conditions for C and D to be represented by a time-index free DAG are satisfied. The time-index free DAG in this case is the DAG C D (with the variables C and D, but no edges.) Note, however, that this is now a DAG in which C and D are correlated, but not because C causes D, D causes C, or there is a hidden common cause of C and D. The fact that the variables can

Probability and Structure in Econometric Models

be correlated without any hidden common causes does not expand the possible set of <distribution, manipulation> pairs that are typically considered; it does however, expand the set of ways in which such manipulation distribution pairs can be generated, and expands the set of <distribution, manipulation> pairs when there are no hidden common causes.

However, in general, the translation of a causal system representable by a time-index free DAG will not in general be representable by a time-index free DAG. Consider the following example.

Initial Condition: $\varepsilon_A[0], \varepsilon_B[0], A[0], B[0]$, jointly independent, all $N(0,1)$

Structural Equations:

$$\begin{aligned}
\varepsilon_A[u,t] &:= \varepsilon_A[u, t-1] \\
\varepsilon_B[u,t] &:= \varepsilon_B[u, t-1] \\
A[u,t] &:= \varepsilon_A[t-1] \\
B[u,t] &:= 3 \times A[t-1] + \varepsilon_B[t-1]
\end{aligned}$$

Causal DAG:

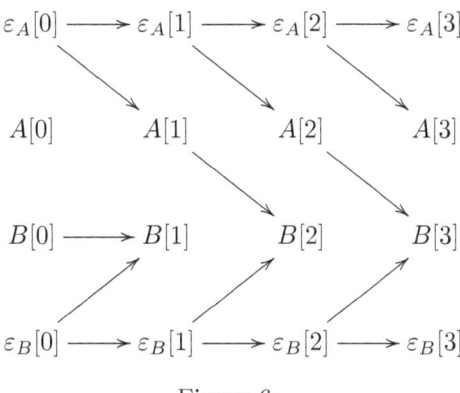

Figure 6:

Time-Index Free Causal DAG: $A \to B$

Transformation:

$$\begin{aligned}
C[u,t] &= 2 \times A[u,t] + B[u,t] \\
D[u,t] &= A[u,t] + 2 \times B[u,t]
\end{aligned}$$

Transformed Initial Condition:

$$\varepsilon_C[0] \sim N(0,5) \quad \varepsilon_D[0] \sim N(0,5) \quad C[0] \sim N(0,5) \quad D[0] \sim N(0,5)$$
$$cov(\varepsilon_C[0], \varepsilon_D[0]) = 4 \qquad cov(C[0], D[0]) = 4$$

Transformed Structural Equations:

$$\varepsilon_C[u,t] := \varepsilon_C[u,t-1]$$
$$\varepsilon_D[u,t] := \varepsilon_D[u,t-1]$$
$$C[u,t] = 2 \times A[u,t] + B[u,t]$$
$$:= 2 \times \varepsilon_A[t-1] + 3 \times A[t-1] + \varepsilon_B[t-1]$$
$$= 2 \times \varepsilon_A[t-1] + \varepsilon_B[t-1] + 2 \times C[u,t-1] - D[u,t-1]$$
$$= \varepsilon_C[t-1] + 2 \times C[u,t-1] - D[u,t-1]$$
$$D[u,t] = A[u,t] + 2 \times B[u,t]$$
$$:= \varepsilon_A[t-1] + 2 \times (3 \times A[t-1] + \varepsilon_B[t-1])$$
$$= \varepsilon_A[t-1] + 2 \times \varepsilon_B[t-1] + 2 \times (2 \times C[u,t-1] - D[u,t-1])$$
$$= 4 \times C[u,t-1] - 2 \times D[u,t-1]) + \varepsilon_D[t-1]$$

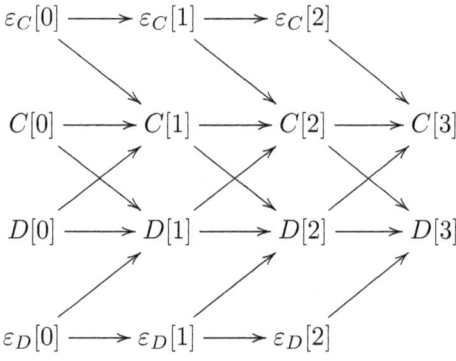

Figure 7:

Note that the <A, B> causal system reaches equilibrium after the first time step because it is a feed-forward network and the longest directed path is of length 1. It follows that the <C, D> system of variables, which are functions of the <A, B> system of variables reach equilibrium after 1 step as well. This means that after the first time step $C[u,t] = C[u,t-1]$, and $D[u,t] = D[u,t-1]$. Hence, in the equilibrium state, the transformed structural equations can be rewritten in such a way that $C[u,t]$ depends only on $\varepsilon_C[t-1]$ and $D[u,t-1]$, while $D[u,t]$ depends only on $\varepsilon_D[t-1]$ and $C[u,t-1]$.

Rewritten Transformed Structural Equations - in Equilibrium

$$\varepsilon_C[u] := c$$
$$\varepsilon_D[u] := d$$
$$C[u] := D[u] - \varepsilon_C[u]$$
$$D[u] := 4/3 \times C[u] + 1/3 \times \varepsilon_D[u]$$

This does not have a time-index free graphical representation as a DAG. However, it satisfies three of the four assumptions for a time-index free representation as a DAG, and if the ordering of variables assumption is dropped, then the equilibrium state has a time-index free representation as a cyclic graph.

Figure 8:

Again, equations of this form are commonly used in linear Gaussian structural equation modeling. Fisher (1970) gave the standard interpretation of them. He showed how a time series of the kind we are considering in this paper, with constant exogenous shocks, can generate such equations in the limit, if the system reaches equilibrium. What is different here is that the system reaches equilibrium after a finite number of steps. For a fixed unit, a manipulation of C to a particular value c should be interpreted as setting C to the value c at all times after some given time.

2.6 Translating the Causal Markov Assumption

However, this system of translation produces a problem for the Weak Causal Markov Assumption. In the $<Position_1, Position_2>$ description of the causal system, there was no problem. However in the $<C, D>$ description of the same causal system, there is an apparent violation of the Weak Causal Markov Assumption. C and D are associated, and yet C does not cause D, D does not cause C, and there is no hidden common cause of C and D, even one from the more remote past. If that is correct, the correct causal DAG for C and D is simply the empty DAG - but C and D are nonetheless associated.

One could try to amend the Causal Markov Assumption by limiting it to sets of variables that are not "logically related". This would not help in this example. It is clear that intuitively $\{C, D\}$ is "logically related" to $\{Position_1, Position_2\}$ in the sense that $\{C, D\}$ is just a re-description of $\{Position_1, Position_2\}$, and it is not possible to manipulate $\{Position_1, Position_2\}$ without simultaneously manipulating $\{C, D\}$. However, this does not suffice to make C and D "logically related" to each other. C and D are independently manipulable in the sense that C can be manipulated to any value while manipulating D to any value, or performing a null manipulation of D, and vice-versa. If the "logical relationship" between $\{C, D\}$ and $\{Position_1, Position_2\}$ is enough to establish a "logical relationship" between C and D, then $Position_1$ and $Position_2$ are also "logically related", because $\{Position_1, Position_2\}$ is logically related to $\{C, D\}$ as well. But $Position_1$ and $Position_2$ are intuitively not logically related at all.

On this view, the causal system has a "natural" evolution given by the structural equations. For the equations of evolution to pick out a unique state of the system, this requires either some initial boundary conditions, or the equations of evolution to be independent of the initial boundary conditions after a sufficiently long time. The equations of the natural evolution of the system correspond to the structural equations, and the setting (or re-setting) of the initial boundary conditions correspond to the manipulations. The distinction between the natural evolution of the system and the manipulations are not features of the causal system alone, but is an idealization that depends upon the description of the system. The manipulations are themselves causal processes, but are causal processes outside of the description of the causal system. If the causal system is embedded in a large causal system, then the manipulations become part of the "natural" evolution of the system. However, because manipulations can be the result of actions by human beings, embedding the manipulations in a large causal system and including them in the structural equations is impractical in many cases.

For the $<C,D>$ description of the causal system, unlike the $<Position_1, Position_2>$ description of the causal system, the population dependence structure does not reflect the structural equations. This is because the boundary conditions for $C[t=0]$ and $D[t=0]$ are based on a manipulation that makes $C[t=0]$ and $D[t=0]$ dependent, and the structural equations preserve that dependence. The population dependence structure does not reflect the structural equation structure - rather it is due to the initial conditional manipulation. If the initial conditions are such that the causally exogenous variables are independent, and the causal graph is acyclic, that is enough to guarantee that the Causal Markov Assumption is satisfied. The problem is that in the $<C,D>$ system the exogenous variables are not independent. Thus the $<C,D>$ system is a violation of the Weak Causal Markov Assumption in its usual formulation.

However, a modification of the Causal Markov Assumption can be applied to an extension of the directed acyclic graphical framework. In structural equation modeling, a double-headed arrow is drawn between exogenous variables whenever they are associated.[5] One reason that two exogenous variables are associated is that there is a hidden common cause of the two variables. What I have argued here is that another reason that two exogenous variables might be correlated is that they are a transformation of two unassociated variables. When double-headed arrows are used in this way, it is easy to extend the definition of d-separation to extended DAGs, and even to cyclic graphs. (See for example, [3, 7]).[6] Then, given the usual structural equation interpretation of the extended graph, the assumptions that the variables are Gaussian and the equations are linear, together with the assumption that exogenous variables are independent if there is no double-headed arrow between them, the extended graph entails $I(\mathbf{A}, \mathbf{B}|\mathbf{C})$ if and only if $\mathbf{A}$ is d-separated from $\mathbf{B}$ conditional on $\mathbf{C}$.

Theorem 1 *If there exists a system of Gaussian variables* $\mathbf{V}$ *with linear structural equations satisfying the Weak Causal Markov Assumption, then under a linear transformation of the variables* $g(\mathbf{V})$, *the Global Causal Markov Assumption is satisfied.*

I conjecture that this theorem is also true for non-linear equations, when the concept of d-separation is modified as described in [7].

Under some parametric assumptions, the associations that are due to transformations of variables that satisfy the Weak Causal Markov Assumption are limited in number. For example, suppose there are two binary variables A and B with values 0 and 1, for which the Weak Causal Markov Assumption is true, and A and B are causally unconnected (A does not cause B, B does not cause A, and there are no common causes of A and B). Because there are only a finite number of invertible transformations of A and B into other binary variables C

---

[5]There is a technical complication here that I will not go into detail about. If the distribution over the variables does not satisfy the property of composition, i.e. that $I(A,B)$ and $I(A,C)$ does not entail $I(A,\{B,C\})$ then the rule for adding double-headed arrows must be slightly more complicated. However, for linear Gaussian systems, the property of composition is satisfied.

[6]As shown in [7], for cyclic graphs, in contrast to acyclic graphs, d-separation for linear systems and d-separation for non-linear systems have to be defined somewhat differently.

Probability and Structure in Econometric Models 533

and D that "mix" A and B together, there are only a finite number of different associations between C and D that can be produced by a transformation of independent A and B, as opposed to causal relations existing between C and D. Hence in this case, the Weak Causal Markov Assumption does limit the possible associations between variables that don't cause each other and have no hidden common cause.

However, because there are an infinite number of invertible transformations of Gaussian variables A and B into other Gaussian variables C and D that "mix" A and B together, there are an infinite number of different associations between C and D that can be produced by a transformation of independent A and B, as opposed to causal relations existing between C and D. In this case, the Weak Causal Markov Assumption does not limit the possible associations between variables that don't cause each other and have no hidden common cause.

Suppose that the interpretation of a double-headed arrow between variables X and Y is changed, to include the possibility that X and Y are causally unconnected but associated because they are transformations of other causally unrelated variables that are independent. This does not have many implications for the calculation of the effects of manipulations from a causal graph. This is because if X and Y are associated solely due to a hidden confounder Z, then manipulating X has no effect on Y and manipulating Y has no effect on X. Similarly, if X and Y are associated solely because they are transformations of a causally unrelated pair of variables R and S that are also independent, then it is still the case that manipulating X has no effect on Y and manipulating Y has no effect on X.

2.7 Translating the Causal Faithfulness Assumption

The Weak Causal Faithfulness Assumption still plays a very important role in causal inference. In the previous section it was noted that one couldnt draw any causal conclusion from an association between A and B (which was true even before the possibility of variable transformation was taken into account.) On the other hand, the Weak Causal Faithfulness Assumption does allow one to draw causal conclusions from the absence of an association between A and B. Suppose that A and B are unassociated. One possible explanation is that A does not cause B, B does not cause A, and there is no correlated error for A and B. Note that this is not the only explanation of the lack of association between A and B. It could be that A causes B, B causes A, and A and B have correlated errors, and that all of these effects cancel. The Weak Causal Markov Assumption says to pick the simpler explanation, that A and B are causally unconnected and do not have correlated errors. (In general, however, it is the combination of Theorem 1 and the Weak Causal Faithfulness Assumption, which allows one to draw causal conclusions.)

It is easy to show that there are natural violations of the Causal Faithfulness Assumption introduced when variables transformations are performed. For example, consider the following causal DAG shown in Fig.9(a): Assume the distribution is faithful to Fig.9(a). A and B are d-separated conditional on C and D, so A and B are independent conditional on C and D. Now suppose that E and F are translations of C and D. The translated DAG is shown in Fig.9(b). A and B are d-connected conditional on E and F in Fig.9(b). However, because

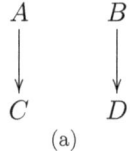
(a)

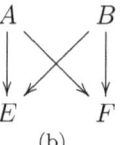
(b)

Figure 9:

C and D are functions of E and F, it follows from the fact that A and B are independent conditional on C and D, that A and B are independent conditional on E and F. However A and B are d-connected conditional on E and F. This is a violation of the Causal Faithfulness Assumption, but not of the Weak Causal Faithfulness Assumption. I have pointed out that satisfying the Weak Causal Faithfulness Assumption is sufficient for the existence of algorithms that reliably infer (classes of) causal DAGs in the large sample limit. However, the translation of variables naturally introduces cyclic causal graphs, and it is not known whether the Weak Causal Faithfulness Assumption is sufficient for the existence of algorithms that reliably infer cyclic causal graphs in the large sample limit. So at this time, it is not known how important examples of violations of the Causal Faithfulness Assumption (but not of the Weak Causal Faithfulness Assumption) are for causal inference.

I also do not know whether there are examples in which the translation of a system of variables in which there is no violation of the Weak Causal Faithfulness naturally introduces a violation of the Weak Causal Faithfulness Assumption in the transformed system. However, I suspect that such cases do exist.

2.8 Consistency for the Translated Causal System

Given the system of translation just proposed, even very simple causal systems can be translated into much more complex causal systems. The net effect of this is that even if the PC algorithm is reliable given the "right" choice of variables, it is not reliable given the "wrong" choice of variables. The introduction of correlated errors by themselves can be handled by the FCI algorithm (even though they are introduced for a reason other than hidden common causes), at the cost of increased computational complexity and decreased informativeness. Also, the introduction of cycles by themselves can be handled by the Cyclic Causal Discovery algorithm, at the cost of increased computational complexity and decreased informativeness. Unfortunately, there is no known provably reliable (in the large sample limit) causal discovery algorithm for graphs with both correlated errors and cycles. And as I have shown, such graphs are the natural result of translations of causal systems that are quite simple given the "right" choice of variables.

2.9 Ambiguity

There is one other major problem that describing a causal system in the "wrong" set of variables can lead to. See also [9]. (Interestingly enough, the problem I will describe here does not seem to arise for linear Gaussian models.) Consider two independent, causally unconnected binary variables A and B. Now let C be 0 when A and B are the same, and 1 when A and B are different, and let $D = A$.

It is clear that there is a translation back and forth between A and B on the one hand and C and D on the other hand. First use C to find out whether A and B are the same, and then use D to determine the value of A. Then if $C = 0$, A and B are the same, so the value of B is determined from knowing the value of A. And if $C = 1$, A and B have different values, and hence the value of B is again determined from knowing the value of A. A manipulation of C alone in the <C, D> system keeps D constant, and this can be translated into an action on A and B.

Now consider a third system of variables, C and E, where E is equal to B. There is also a translation of A and B into C and E and vice versa, using the same reasoning as above. A manipulation of C alone in the <C, E> system keeps E constant, and this can be translated into an action on A and B, which is different than the manipulation of C alone in the <C, D> system.

So far, no new problem has been introduced. But now suppose that the only variable in our causal system is C. There are two ways of augmenting C into a description of the <A, B> causal system, one of which consists of adding D to the system, and the other of which consists of adding E to the system. It is clear from the symmetry of the example that there is no sense in which either of these augmentations is more natural than the other. So given just the C variable, there is no natural way of choosing what action amounts to a manipulation of C is. This is analogous to the cholesterol example. In that case, a manipulation of total cholesterol could be done in many different ways (in terms of low density cholesterol and high density cholesterol), which would have different effects on heart disease. There is no natural way to choose which of these manipulations is meant. Note that this is not just a problem for causal inference from passive observation. There is nothing in a randomized experiment that indicates that a combination of translation and marginalization has led to a system of causal variables in which the concept of manipulation is ambiguous.

3 Conclusion

What I have argued in this paper is that the basic effect of using the "wrong" variables instead of the "right" variables to describe a causal system is that causal inference becomes much more difficult in a number of different ways. The Causal Markov Assumption for the "wrong" variables becomes a much weaker assumption, and for some parametric families is vacuous. Certain kinds of violations of the Weak Causal Markov Assumption become natural.

In addition, causal graphs that are very simple can be quite naturally turned into much more complicated graphs. These more complicated graphs will typically have more edges, more double-headed arrows, and introduce cycles. These graphs can only be reliably inferred by causal inference algorithms that are slower and less informative than the algorithms that are reliable for the more simple graphs among the "right" variables. (In addition, not all of these graphs can even be reliably estimated, and those that can introduce extra estimation problems.) For the most general kind of graph that can naturally be generated there is no known reliable causal inference algorithm.

Finally, under marginalization, the "wrong" variables lead to cases where the meaning of a manipulation becomes ambiguous.

It is not clear whether the problem of "wrong" variables is an actual practical problem for statistical (and experimental) practice, or whether like the problem of "grue" predicates it is a theoretical problem without much impact on actual practice. I know of no examples where a set of the "wrong" variables was actually used, and it was later shown that there was a "right" set of variables that could have been used in their place. This does not show that it has not happened.

Appendix

In order to define the d-separation relation, the following graph terminology is needed. A graph consists of two parts, a set of vertices **V** and a set of edges **E**. Each edge in **E** is between two distinct vertices in **V**. There are two kinds of edges in **E**, directed edges $A \to B$ or $A \leftarrow B$, and double-headed edges $A \leftrightarrow B$; in either case A and B are **endpoints** of the edge; further, A and B are said to be **adjacent**. There may be multiple edges between vertices. For a directed edge $A \to B$, A is the **tail** of the edge and B is the **head** of the edge, A is a **parent** of B, and B is a **child** of A.

An **undirected path** U between X_a and X_b is a sequence of edges <$E_1, \cdots, E_m$> such that one endpoint of E_1 is X_a, one endpoint of E_m is X_b, and for each pair of consecutive edges E_i, E_{i+1} in the sequence, $E_i \neq E_{i+1}$, and one endpoint of E_i equals one endpoint of E_{i+1}. A **directed path** P between X_a and X_b is a sequence of directed edges <$E_1, \cdots, E_m$> such that the tail of E_a is X_1, the head of E_m is X_b, and for each pair of edges E_i, E_{i+1} adjacent in the sequence, $E_i \neq E_{i+1}$, and the head of E_i is the tail of E_{i+1}. For example, $B \to C \to D$ is a directed path. A **vertex occurs on a path** if it is an endpoint of one of the edges in the path. The set of vertices on $A \leftrightarrow B \to C \leftarrow D$ is $\{A, B, C, D\}$. A path is **acyclic** if no vertex occurs more than once on the path. $C \to D \to C$ is a cyclic directed path.

A vertex A is an **ancestor** of B (and B is a **descendant** of A) if and only if either there is a directed path from A to B or $A = B$. Thus the ancestor relation is the transitive, reflexive closure of the parent relation. A vertex X is a **collider** on undirected path U if and only if U contains a subpath $Y \leftrightarrow X \leftrightarrow Z$, or $Y \to X \leftrightarrow Z$, or $Y \to X \leftarrow Z$, or $Y \to X \leftarrow Z$; otherwise if X is on U it is a **non-collider** on U. For example, C is a collider on $B \to C \leftarrow D$ but a non-collider on $B \to C \to D$. X is an **ancestor of a set** of vertices **Z** if X is an ancestor of some member of **Z**.

For disjoint sets of vertices, **X**, **Y**, and **Z**, **X** is **d-connected** to **Y** given **Z** if and only if there is an acyclic undirected path U between some member X of **X**, and some member Y of **Y**, such that every collider on U is an ancestor of **Z**, and every non-collider on U is not in **Z**. For disjoint sets of vertices, **X**, **Y**, and **Z**, **X** is **d-separated** from **Y** given **Z** if and only if **X** is not d-connected to **Y** given **Z**.

Bibliography

[1] Chickering, D. Optimal structure identification with greedy search. *Journal of Machine Learning Research*, 3:507–554, 2002.

[2] Demiralp, S., Hoover, K. Searching for the causal structure of a vector autoregression. *Oxford Bulletin of Economics and Statistics*, 65(issue 1):745–767, 2003.

[3] Koster, J. On the validity of the markov interpretation of path diagrams of gaussian structural equation systems with correlated errors. *Scandinavian Journal of Statistics*, 26:413–431, 1999.

[4] Pearl, J. *Causality: Models, Reasoning, and Inference*. Cambridge, UK: Cambridge University Press, 2000.

[5] Richardson, T. A discovery algorithm for directed cyclic graphs. In Horvitz, E., Jensen, F. (Eds.), *Proceedings of the 12th Conference on Uncertainty in Artificial Intelligence*. Portland, Oregon, 1996, San Francisco, CA.: Morgan Kaufmann, 1996.

[6] Shimizu, S., Hoyer, P., Hyvarinen, A., Kerminen, A. A linear non-gaussian acyclic model for causal discovery. *Journal of Machine Learning Research*, 7:2003–2030, 2006.

[7] Spirtes, P. Directed cyclic graphical representation of feedback models. In Besnard, P., Hanks, S. (Eds.), *Proceedings of the Eleventh Conference on Uncertainty in Artificial Intelligence,*. San Mateo: Morgan Kaufmann Publishers, Inc., 1995.

[8] Spirtes, P., Glymour, C., Scheines, R. *Causation, Prediction, and Search*. New York: MIT Press, 2nd edn., 2000.

[9] Spirtes, P., Scheines, R. Causal inference of ambiguous manipulations. *Philosophy of Science*, 71:833–845, 2004.

[10] Woodward, J. *Making Things Happen*. USA: Oxford University Press, 2005.

D
SCIENCE AND SOCIETY

National Ethics Committees: Diversity in Judgements and the Advice Function[1]

Kirsten B. Endres

Center for Philosophy and Ethics of Science

endres@ww.uni-hannover.de

ABSTRACT. Nearly each democratic country has at least one national ethics committee. The main objective of the committees should be to help politicians to introduce new bills. In the majority of cases national ethics committees do not reach unanimous judgements about the ethical questions they discuss. Therefore, many people argue that the committees fail to reach their aim of reducing moral uncertainty. In this paper I examine different causes of disagreement and give reasons why disagreement, which is due to moral pluralism, is important. I furthermore point out why striving for complete consensus will endanger other important characteristics of national committees, for example their composition and their aspiration towards transparency. I will set forth that the committees can yet fulfil their advice function vis-à-vis politicians despite their disagreement. Therefore, I have developed a specific interpretation of the advice function. According to the proposed perception, advice should provide the necessary tools to form moral judgements. I finally come to the conclusion that national ethics committees should be regarded as think tanks of a special kind.

1 Introduction

Nearly each democratic country established during the last twenty years has at least one national ethics committee or comparable body. National ethics committees have two main functions. On one side they should stimulate and structure the public debate about ethical questions that arise in the course of new developments in life sciences. On the other side their ethical opinions should support politicians to introduce new bills. In the following I will focus on the second aim that is the advice function of ethics committees vis-à-vis politicians.[2] To be more precise, I want to discuss whether this service function can be fulfilled and if so, how it can be achieved. Different from conventional groups of experts, national ethics committees display a special kind of expertise that is marked by an explicit reference to values. Some of the evaluations given by ethics committees had considerable influence on public discussions and the political decision making process. The institutionalisation of moral deliberation coincides with a pluralism of values. On one side ethical committees try to mirror pluralism whereas on the other side they want to give precise moral advice. Moral pluralism represents a

---

[1] Center for Philosophy and Ethics of Science, Hanover, Germany.
[2] See for the aim to facilitating the societal debate [15].

general problem of legitimation within the committees. Politicians expect clear solutions for morally problematic issues. On the contrary, pluralism claims that the questions ethics committees elaborate have no clear and distinct answers.[3]

My paper is structured as follows: I begin with a short explanation of the problem. National ethics committees normally fail to reach one clear judgement on the ethical problems they discuss. The leading question will be: Should they nevertheless try to agree on one opinion? Consequently, I will point out the pros and cons of the committee's ambition to reach unanimity. I begin by mentioning quality failures of opinions and turn to disagreement and how it is connected to other aims of ethics committees. Thereafter I will explain why some divergences in judgment speak in favour of and not against the committee's working method. The rest of the paper deals with the possibilities of ethical expertise and questions concerning ethic experts. Further on, I will elucidate the related problem about how national ethics committees can fulfil their advice function despite and/or because of their diverging opinions. I will end with a short résumé.

2 Explanation of the problem

When I say that national ethics committees diverge in their judgements, I mean that the members of *one* committee cannot agree on one opinion about the ethical topic in question. This is by the way not about a pure national issue; there is disagreement even within national committees of different countries. How to interpret this diversity is an interesting question, too. Some ethical evaluations entail legal regulations. Especially those regulations which affect more than one country – as it is the case within the European Union, for instance - call for ethical evaluations shared by several countries. I will leave this important question for another occasion. Hereafter, I will only address disagreements within one committee.

Very often national ethics committees fail to reach a unanimous judgement about an ethical question. Some members believe that certain areas of research are unmoral or false and should therefore be prohibited by law. In contrast, other members believe that this very area of research should be permitted for moral reasons. One consequence of this disagreement is to question the aim of national ethics committees in general. Which function do they have if they fail to reach clear opinions? In general, they represent think tanks that analyse complex and difficult questions and problems in order to develop clear opinions and definite positions. National ethics committees aim at reducing moral uncertainty which very often arises in connection with new areas of research in life sciences. It is widely believed that they cannot reach this aim by showing uncertainty in their opinions.[4] In other words, committees cannot fulfil their advice function if they fail to reach agreements.[5] Hence, critics report that despite their disagreement committees should strive for unanimous opinions.[6] There are different possi-

---

[3]See for the tension between the desire of politicians to hear definite answers and the impossibility to give them by means of applied ethics, [14, 144ff].

[4]For the problems to reach ethical agreement in pluralistic societies, see for example [4, 162ff].

[5]See for similar questions concerning disagreement of research ethics committees and whether they should look for consensus [7]. They argue that disagreement in itself isn't a quality failure.

[6]Inside the committees the question about consensus and what to do if they fail to reach

bilities at hand to reach agreement.[7] In an ideal situation they should find a consensus.[8] In case of need they should come to a compromise. If unanimity is impossible to reach, they should decide by means of an election which opinion can be given on behalf of the committee. Each of the methods has its own strengths and weaknesses. In the following I will focus on general quality failures of opinions before addressing some of above mentioned strengths and weaknesses.

3 Quality failures of opinions

Irregardless of a judgement being reached unanimously or not, false recommendations given by the members of a committee do not reflect the committee's good mode of practice. Likewise the results of the committee's work must not be based on prejudices. Members of national ethics committees should be open for new positions, even if they differ from their own point of view. Furthermore, the quality of an opinion is strongly affected by judgements based on conflicts of interest and infighting. Some members of national ethics committees apparently do argue political, try to build fractions and follow personal aims by stressing certain claims.[9] In addition, members and their opinions should always be neutral towards establishments, parties or persons. The committees are expected to give an independent knowledgeable judgement. This includes the claim that ethical committees should not argue in favour of the political attractiveness of their opinions. The aim of national ethics committees is to give independent moral evaluations of problematic matters. I therefore question: Is disagreement in general – irrespective of its causes – a deficiency? And should a committee always pursue unanimous judgements?

4 Disagreement and important aims of national ethics committees

Of course it is helpful for the work of a committee if its members find consensus on their positions. Likewise it is beneficial for their work to acknowledge alikeness and similarities of positions. By doing so they stress sameness and not differences. Looking for unanimous decisions in this sense is one of the advantages of the operating principle of a committee. However, according to my view it is not correct to accuse the committee for not reaching a concordant opinion. For example, in the United States it was regarded for a long time as a quality failure of an opinion when its evaluations weren't unanimous.(See [7]) Irrespective of its sources, disagreement itself was considered as a deficiency. This view is problematic because important aims or claims of ethics committees might be endangered when reaching agreement at their cost. In the following I will mention some of these costs.

---

one is discussed as well. See for example [18, pp.131–135].
  [7]See for an overview about different modes to give opinions [9, pp.1–5].
  [8]See for the hopes related to consensus formation and the impossibility to reach consensus about controversial issues [10]. For advocates of consensuses see [20] and [16].
  [9]See for those and other unwanted behaviour of committee members [21, 100ff].

4.1 Constitution of ethics committees

The members of national ethics committees dispose of different knowledge domains. Generally, they represent those scientific disciplines necessary to investigate the ethical problems in question. The areas of medicine, biology, law and philosophy, only to name some of the necessary qualifications, should be covered by a committee. Without going into detail, I will explain the role of some experts. Scientists of medicine and biology are responsible for informing the rest of the committee about the scientific aspects and developments of the discussed topic. The experts of law should link the problem to the constitutional system of the country. The role of the philosophers in the committees consists in clarifying terms, assessing logical coherence of ethical arguments, drawing distinctions and identifying ethical issues and drawing conclusions from a particular set of premises. Furthermore they should describe the ethical or moral attitudes of particular individuals, perspectives or theories. And above all, they should all give statements as to whether a particular conduct or an area of research is ethical approvable or not.

This heterogeneous composition enables the committees to give a new kind of advice. The distinctiveness of their expertise consists in interrelating scientific factual information with ethical and normative discussions. The committees begin their work by clarifying the scientific, legal, sociological and other relevant basics. In addition, they take into account a pluralism of values while the necessary expertise isnt related to a certain scientific discipline. In fact it can be developed by an interaction of experts of different scientific fields and areas of life. Ethical committees give opinions, for example, on reproductive and therapeutic cloning, prenatal screening, in vitro fertilization, import of human embryonic stem cells, patents of human genes, abortion and euthanasia. With regard to such a diversity of topics it is a given that only a sophisticated view can generate a well-founded ethical consultation.

According to this view, national committees should be designed as experts committees compared with a concurrent view that demands ethics committees to be laymen committees or interest committees.[10] I won't discuss these concurrent views at length. I only want to name some reasons why I think that national ethics committees should be composed of experts. The questions national committees have to evaluate are of a very complex kind. Very often they have to deal with new medical developments or perspectives of new areas of research. Even experts in these fields cannot thoroughly survey the potential of successful practices and unintended side effects or consequences for the society and important moral values. For laymen it is even more difficult to give well-founded judgements. Concerning interest groups, the aim of national ethics committees is to give an independent moral judgment. They should not favour the rights of certain social groups. Therefore, I believe that national ethics committees should be designed as expert commissions. In reality they are hybrids. Some members of a national ethics committee are experts in the sense mentioned above, others are laymen and some are representatives of certain interest groups.

The multidisciplinary composition of the committees is a condition to give well-founded and brilliant moral judgements about highly complex (scientific)

---

[10] See for an overview about different compositions of ethics committees and arguments for their design as experts groups [25].

facts and circumstances. Besides their qualifications to be generated by the members' expert knowledge, a broad composition of the committee ensures that many different value judgements enter into the work of national ethics committees. This assumption is based on the conviction that moral value judgements of one special individual are also characterized by many personal experiences such as job, social affiliation and life style. Following this line of argument it is important that the committee members' personal and professional backgrounds are as diverging as possible.(see [8]) On the one hand this enables a committee to represent an adequate part of the controversial and varying moral accounts preferred by the public. On the other hand this very fact makes it nearly impossible for the committees to give concordant judgements of the highly controversial topics they discuss.

4.2 The working methods of national ethics committees

Sometimes it is unclear how national ethics committees reach their judgements. There are opinions which contain statements like "The committee *feels* that x or y should be done." or "The committee *thinks* that a special course of action is morally wrong." To be fair, most statements of national ethics committees are much more sophisticated. In particular the committees normally give reasons why they believe or think that a certain moral evaluation is adequate.[11] Generally this critique can be faced by more aspiration towards transparency. There are at least two ways to reach more transparency: a) hold the meetings of the committees in public and b) give reasons for the positions and evaluations stated in the opinions.[12]

Some national ethics committees have public meetings or put the minutes of their meetings in the net, for example the German National Ethics Council and the American Presidents Council on Bioethics. However, not all meetings are public. Above all meetings are closed to public when their members have to decide about new matters. In addition public meetings are accompanied and prepared by working groups that aren't public. The members of national ethics committees usually stress the importance of non-public meetings whenever new topics are discussed initially. Members can discuss frankly their different views. Furthermore a closed meeting makes it much easier for them to be open for new and different moral attitudes. Hence, there is a tension between the desire for more transparency on one side and the desire for a free atmosphere of discussion and openness for new positions and arguments on the other side.

This line of reasoning is particularly relevant with regard to the committee's attempt to reach a consensus or compromise. Closed meetings that are not communicated to the public make it much easier for the members to express their own positions and to find consensuses or compromises. To put it the other way round, open meetings make it more difficult for members of a committee to leave their "official" positions and therefore consensuses are more difficult to reach. The claim for more transparency in the work of a committee by opening the meetings to public counteracts the aspiration towards consensuses and compromises. At the same time the opening of meetings has another effect. The urgency

---

[11] See for example the opinions of the French National Consultative Bioethics Committee for Health and Life Sciences.
[12] See for the importance of transparency in the bioethical debate [27].

to approach each other and to reach a compromise is higher because it is known which members block consensuses and hamper an agreement. Therefore, it is not clear whether open meetings deliver or hinder agreements.

If agreements are obtained by vote, how much information about the election should be given to the public?[13] Should the public know for example how many members voted for the "shared" opinion? Should the members who favoured the opinion be mentioned by name? More transparency will be reached by more information. Again the other side of the coin is that it will be easier for the members of a committee to vote for an option that doesn't wholly represent their own position if it is not publicized that they supported the judgement. However, as mentioned above, the transparency of votes can urge members to leave their convictions and favour the opinion which is supported by the majority. Therefore, it is unclear whether more public information encourages or rather prevents members from voting according to their convictions.

There is at least one more possibility at hand to bring more transparency in the decision making procedure of committees. In addition to the course of action to be favoured the opinions of a committee should also mention the reasons for the positions in order to make it more understandable what led the members to their judgements. Giving reasons has another advantage. The judgements of national ethics committees are supposed to have no special authority.(see [19]) Committees aren't democratically elected. They don't represent the general public. And last but not least, of course they don't have special access to moral truth or insight. Therefore, their judgments are at best as good and important as those of other qualified persons or groups. A special moral authority cannot be deduced. I believe that this is right. The opinions of national ethics committees and with them their judgements havent got any special authority qua being a national ethics committee. Good arguments, the rationality of reasoning and the consistency of substantiations vest authority to opinions. I acknowledge that the judgements of other ethics committees, commissions or persons can be of equal authority. However, they do have authority only by good and sound reasons and arguments. The moral authority of an opinion originates in the power of given arguments and reasons and not by the fact that the opinion was generated by a national ethics committee.

This is the reason why reasoning and arguments are important for both the transparency and authority of an opinion. Regarding the aim to give unanimous opinions, reasons and arguments are problematic. As it is already difficult to reach a shared evaluation it is even more difficult to reach a shared reasoning or substantiation because there are different and sometimes contradicting reasons and substantiations for one evaluation. Let me illustrate this by an example. Even if members of a committee agree on the import of human embryonic stem cells that doesn't automatically mean that they also agree on the substantiation of this position. Someone who argues for the import considering the given legal position wouldn't argue for an amendment of the German act of embryo protection in order to legalize the extraction of stem cells in Germany. On the contrary, someone who believes that early stages of human development dont require absolute protection of life would favour a change in the act of embryo protection as well. The arguments mentioned in the reasoning can have consequences which are

---

[13] Furthermore one could question the reasonableness of votes about moral problems. I have discussed this problem in another paper. See [9]

beyond the scope of the topic in question. Members of a committee can have different attitudes concerning the arguments given in the reasoning. Furthermore, it happens that the reasoning or substantiations of consensuses or compromises become inconsistent if every member of a committee agrees upon the reasoning. An inconsistency of substantiation will be at the costs of the authority of an opinion and in the long run of the committee itself. Besides these negative effects of trying to reach shared substantiations there are positive consequences as well. The attempt to agree on the substantiation of a position can be of use for the quality of opinions because thereby conflicts of reasons and problems of arguments can come to light which otherwise would have possibly been overlooked by the members.

Recapitulating, I can say that no matter whether a committee succeeds in establishing unity, shared judgements are favourable. The methods of finding agreement and the search for similarities are particularly suitable in order to disclose gaps of arguments, problems of reasoning and inconsistencies in substantiations. Furthermore and according to my deliberation, main objectives such as a balanced composition of the committee, transparency of the working method and independent judgements shouldn't be sacrificed by the aspiration towards unity of opinions.

In the following, I will elaborate how disagreement can be beneficial for a committee's work. Thereby, I will clarify the question whether disagreement is a deficiency of opinions in general.

5 Useful disagreement

As mentioned before, the most important source of disagreement between committee members is moral pluralism. Questions that are under the focus of national ethics committees are both highly controversial inside a committee as well as between experts of ethics and within the population. Members of ethics committees advocate different moral views. Each view can be rational and well-founded and yet conflicting and contradicting. Of course in reality we find irrational, dogmatic or narrow-minded positions, too. Very often members of a committee agree about values that are affected by an act. What they disagree about is how one should weigh the values. This phenomenon is often illustrated by the metaphor that they enter unknown moral territory. By doing this we as well as the committee members loose the solid moral intuition which could indicate to the morally preferable actions.

For instance, members of national ethics committees repeatedly argue about the question whether it is morally right to use early stages of human development such as fertilised ovum for scientific research in order to develop medical treatment for serious illnesses. Coming to this question we can see a whole range of possible positions. Some argue that research should be done for moral reasons. Those people negate protection of life in early stages of human development. Others attribute absolute life protection to these stages and therefore judge the proposals of research as morally wrong. Besides these two extreme positions, there are several others which ascribe protection of life in various gradations to these early stages of human development. Apart from the problem of balancing the values it is unclear how successful in fact the research will be. Furthermore, there is the notorious problem of misusing the results of that research with morally questionable or unintended effects. Keeping this problem in mind, members of ethics committees give different judgements about what should be done from a moral point of view and which course of action is morally favourable. There are good and sound reasons for all the different positions of the committee members. Of course those positions are also reflected in the population. And there are certainly bad, dogmatic and irrational arguments as well. Important to know is that there isn't one true or right answer at hand. It always depends on each individual moral standpoint and some views are better founded than others. Nonetheless, we will end up with more than one position based on good and sound reasons. And that is the way it goes in Bioethics. The vast majority of questions national ethics committees focus on hasn't got one correct or adequate answer but many different, reasonable and consistent assessments and substantiations.

In my opinion several voices of a committee's opinion do not stand for a quality failure as long as different and reasonable positions are adequately represented. The disagreement between the committee members is one effect of moral pluralism. National ethics committees give proof of their seriousness regarding many different moral attitudes especially through diverging and sophisticated opinions. Their judgements don't contradict moral intuitions and don't fail to meet the moral sentiments of the public. Very often disagreement is inherent to the subject in question.

The following describes positive side effects of disagreement. Many people say that the quasi socialisation of ethical discourses and the development of ethical management in the end lead to a trivialisation of moral philosophy. (See [26, p.86]) However, if national ethics committees generally illustrate more than one morally approved strategy, this objection looses its foundations. The reproach of trivializing moral problems is void if the opinions show the complexity and diversity of moral judgements.

6 Ethical expertise and ethic experts

Some hundred years ago it wasn't difficult to identify ethic experts in the population, for example, the clergy. As the forgoing paragraphs already illustrated, in connection with national ethics committees questions about the possibility of ethical expertise arise in pluralistic societies. Who could be an expert? What qualifications are needed? In respect to the proclaimed expertise of ethics committees the objection is valid that not only certain referees but every individual is competent enough to give moral judgements. Assuming that every individual has at least some moral competence, the question concerning the possibility of ethic experts and ethical expertise is a controversial issue which I cannot discuss in detail here.[14]

Apart from the fundamental question about the possibility of ethical expertise, discussions are held about what ethical expertise could be.[15] Is ethical

---

[14]Questions about the possibility of moral experts and moral expertise are discussed in the literature. See for example [3, 5, 30].

[15]Here I follow the differentiation done in literature. Ethical experts are those who are experienced in ethical theories. If there are moral experts, these are people who have an expertise about moral matters. Whether those expertises are related or can come apart is a

expertise the advice to undertake a certain kind of action? Is it the evaluation that some action is morally wrong or right? Or is ethical expertise the neutral mentioning of arguments and the sorting of actions according to different appraisal systems?[16] Those who argue that ethical expertise has advising force have to face a practical and normative problem. How should commissions overcome their disagreement? Why should national ethics committees and their opinions have a special moral authority?

If there isn't one right or favourable solution and therefore disagreement of moral judgements of committees reflect moral uncertainty, how could moral advice be possible? Can moral uncertainty nevertheless decrease? Can national ethics committees fulfil their advice function towards policy if they cannot recommend evaluations unanimously? In order to reach fulfilment of their advice function, groups of experts shouldn't only outline how one could act but say which course of action would be the right one according to the experts. It is expected that they succeed in giving unanimous advice even if the topic is difficult and complex. If the topic is easily to overlook, there is no need for think tanks in the first place. Why should national ethics committees be an exception?

According to my view, ethics committees shouldn't give unanimous advices and thereby count as an exception. On the one side the claim that expert groups should give clear statements is based upon the idea that there are only right or wrong answers for the questions the groups discuss. At best there is one right answer which politicians wish to receive from such a bundled know-how. Whether or not this idea is naïve concerning other expert groups cannot be answered in this paper. Expert groups that are expected to give advice about how to reduce unemployment, to reform a school system or to design a transparent and fair tax system face similar problems. They discuss different strategies and usually fail to get to one answer. Again, I come to the conclusion that there isn't one right answer for national ethics committees to produce. Rather there are many different views which can be consistent and rational and at the same time contradicting each other. Furthermore, even future development in life sciences might not show which evaluation was the right one.[17] Hence, their working methods differ a lot from those of scientific expert groups that have the opportunity to seize a certain scientific set of tools to work out definite solutions. The German national ethics council for instance evaluates its own work as the concentration of the interdisciplinary discourse of natural sciences, medicine, theology, social and legal sciences and philosophy.(see [12])

On the other side according to many people moral advices are dictations. Everyone knows how to morally judge a situation and therefore doesn't need help from an expert group. There is no group of people that is more knowledgeable than others.(see [2, p.244]) National ethics committees are supposed to have too much influence on the political discussion.[18] Many people say that they anticipate the decisions and dominate responsible people and politicians with their opinions. There is indeed the danger of exploiting ethics committees in view of desired outputs or through involvement of certain groups of interest in order to

---

matter I won't discuss here. See for example [3]

[16]For advocates of the view that ethical expertise doesn't include giving an advice, see for example [1]. For the opposite view see for example [28].

[17]For the specific of ethical expertise see [23, 107f].

[18]See for example [11].

compensate the negligence of other political decisions. It is inevitable that some people think the commissions act as a fig leaf for politicians. Politicians either decorate their decisions with appropriate opinions of national ethics committees or acquit themselves of the responsibility for their decisions by hiding behind the judgements of ethics committees.[19] There are indeed concerns about democratic decision making processes. In particular national ethics committees are supposed to have a questionable influence on political decisions about important questions that concern the whole public because they structure or even block preliminary decisions by giving opinions to politicians. (see [6, 105ff])

If national ethics committees either merely list or actually weight the most important positions of a moral debate than above mentioned reproaches are unfounded. The committees don't favour one course of action. They suggest different moral positions which can be reasonably held. Nobody will be dominated and they don't anticipate political decisions. A further positive aspect of giving more than one opinion is that national ethics committees can circumvent the impression to be a general binding wise council as some philosophical utopian literature propagates.[20] Furthermore, national ethics committees can antagonize concerns about political prejudices.

7 The advice function

Still the question remains. Can they fulfil their advice function by giving more than one possible moral position? I believe they can. If they give weight to the substantiations of the different moral positions they can give an advice. The advice doesn't consist in saying which course of action should be furthered according to the committee. Many different positions and their substantiations will be presented to the reader of the opinions. The problems and the strengths of each position will be explained. If the substantiations are well understandable in turn they enable the reader to make his or her own moral judgements. The output of national ethics committees is to arrange, to allocate and to illustrate different moral arguments and evaluations. Thereby, they reduce the confusion about moral arguments and positions. In turn, although moral uncertainty won't vanish it will be reduced. The advice comes down to explaining the possible moral attitudes and outlining how they belong to each other. The reader will learn about different evaluations and substantiations. Thereby, the reader can learn to form own opinions about moral problems. According to this view it isn't important for the fulfilment of the advice function how a committee judges the situation. More important is that the committee describes several positions with contextual arguments. The reader learns which moral attitudes can be held on which reasons. He or she learns which attitudes and positions are compatible with each other and what contradicts them. By this means the reader can build his or her own opinion. If committees deliver the answer this learning effect usually disappears.[21]

---

[19] See for these and other objections [1, 279f], [29], [13, 167f].

[20] For examples see [22, 30ff].

[21] I have to admit that this understanding of moral advice only partly coincides with the meaning of advice in other contexts. Guidance or direction might be alternatives which come near to the indented meaning. However they have wrong connotations as well. In lack of a better alternative I stick to advice.

National ethics committees can fulfil their advice function because not only the conclusion of their deliberation but also the mentioning and illustration of different moral positions form part of the advice. Admittedly, for this type of advice it is important that the opinions are read. Up to what extent this is realistic for the political everyday life is a different question, which I cannot answer here. Furthermore, the aim to give a full map of all possible and reasonable moral positions and their substantiations will result in a stack of paper. Obviously short opinions are more likely to be read by policy-makers than long ones. A way out of these conflicting aims could be to compile a long version and a short summary. The short summary should be easily understandable and is meant for those who just need a quick overview of the topic. Nevertheless the long version is of importance. Even if it doesn't play an important role in political dayly life it is significant for the standing of the committee. As mentioned before the authority of ethical committees is rooted in the power of arguments given for moral positions and the completeness of reasonable moral positions.

8 Conclusion

I argued that disagreement in judgements of national ethics committees isn't necessarily a quality failure. Insofar as disagreement has its roots in moral pluralism, it represents neither a problem for the competence of a committee nor for the fulfilment of their advice function vis-à-vis policy-makers. National ethics committees don't have to assume the job of the politicians. They shouldn't anticipate political decisions. Therefore national committees shouldn't formulate their opinions with an eye on their political attractiveness. Whether or not the judgements of a committee are capable of winning a majority in parliament shouldn't be of interest for the committee. The purpose of an ethics committee is to give an independent judgement. They shouldn't give moral justifications on request. At the same time the opinions of the committees should have influence on political decisions. Otherwise one could question their advice function and somehow even their existence in the first place. If the opinions would be of no difference to political decisions why do we need ethical committees? According to my view ethical committees can fulfil their advice function without coming to an agreement about the moral problem they discuss. By giving arguments and showing different sensible options about how to act they can educate politicians and other addressees. In the end the readers can built their own informed moral judgements.

National ethics committees should be seen as think tanks of a special kind because they handle special topics that are moral questions. In most cases there is more than one adequate answer to moral questions. Therefore, disagreement in judgement isn't a quality failure of a national ethics committee. Because there isn't one adequate answer it is sensible to fulfil the advice function by suggesting more than one way of action or evaluation. Furthermore, the advice function can be fulfilled more effectively by giving the necessary tools to form moral judgements, a breakdown of different reasonable positions including their substantiations. Politicians and other readers of the opinions thereby will be enabled to give well-founded judgements themselves. A positive side effect is that moral advices understood in this sense lose their taste of paternalism.

Bibliography

[1] Birnbacher, D. Für was ist der 'Ethik-Experte' Experte? In [24], 267–283. 1999.

[2] Broad, C. D. *Ethics and the History of Philosophy*. London: Routledge & K. Paul, 1952.

[3] Caplan, A. L. Moral experts and moral expertise: Does either exist? In *If I were a rich man, could I buy a pancreas?*, 18–39. Bloomington: Indiana University Press, 1992.

[4] Charlesworth, M. *Bioethics in a Liberal Society*. Cambridge: Cambridge University Press, 1993.

[5] Cowley, C. A New Rejection of Moral Expertise. *Medicine, Health Care and Philosophy*, 8(3):273–279, 2005.

[6] Düwell, M. Die Bedeutung ethischer Diskurse in einer wertepluralen Welt. In [17], 76–114. 2000.

[7] Edwards, S. J. L., Ashcroft, R., Kirchin, S. Research Ethics Committees: Differences and Moral Judgement. *Bioethics*, 18(5):408–427, 2004.

[8] Endres, K., Kellermann, G. Nationale Ethikkommissionen – Funktionen und Wirkungsweisen. In Weingart, P. (Ed.), *Nachrichten aus der Wissensgesellschaft. Analysen zur Veränderung der Wissenschaft*. Velbrück: Weilerswist, 2007.

[9] Endres, K. B. Funktion und Form der Stellungnahmen nationaler Ethikkommissionen. In [12]. 2002. URL http://www.ethikrat.org/texte/pdf/expertengespraech_aufgaben/Referat_Endres.pdf.

[10] Engelhardt Jr., H. T. Consensus Formation: The Creation of an Ideology. *Cambridge Quarterly of Healthcare Ethics*, 11(1):7–16, 2002.

[11] Engelhardt Jr., H. T. The Ordination of Bioethicists as Secular Moral Experts. *Social Philosophy and Policy*, 19(2):59–82, 2002.

[12] German National Ethics Council. *Terms of Reference*. 2007. URL http://www.ethikrat.org/_english/about_us/function.html.

[13] Gmeiner, R., Körtner, U. H. J. Die Bioethikkommission beim Bundeskanzleramt – Aufgaben, Arbeitsweise, Bedeutung. In *Recht der Medizin*, 164–173. 2002.

[14] Kaminsky, C. 'Angewandte Ethik' zwischen Moralphilosophie und Politik. In [24], 143–159. 1999.

[15] Kelly, S. E. Public Bioethics and Publics: Consensus, Boundaries, and Participation in Biomedical Science Policy. *Science, Technology & Human Values*, 28(3):339–364, 2003.

[16] Kettner, M. Zur moralischen Qualität klinischer Ethik-Komitees. Eine diskursethische Perspektive. In [24], 335–357. 1999.

[17] Kettner, M. (Ed.). *Angewandte Ethik als Politikum*. Frankfurt a. M.: Suhrkamp, 2000.

[18] Koch, L., Zahle, H. Ethik für das Volk. Dänemarks Ethischer Rat und sein Ort in der Bürgergesellschaft. In [17], 117–139. 2000.

[19] Kuhlmann, A. Kommissionsethik – Zur neuen Institutionalisierung der Moral. In *Merkur – Deutsche Zeitschrift für europäisches Denken*, 26–37. 2002.

[20] Moreno, J. D. *Deciding Together: Bioethics and Moral Consensus*. Oxford: Oxford University Press, 1995.

[21] Pettit, P. Institutionalising a research ethic: chilling and cautionary tales. *Bioethics*, 6(2):89–112, 1992.

[22] Pieper, A. Das ausgelagerte Gewissen – Der Boom der Ethikkommissionen. In *Krankenhaus und Recht*, 30–35. 1998.

[23] Rasmussen, L. M. Bioethics Consultation for Pharmaceutical Corporations. *AMA Medicine & Society*, 8:105–108, 2006.

[24] Rippe, K.-P. (Ed.). *Angewandte Ethik in der pluralistischen Gesellschaft*. Freiburg, Schweiz: Universitätsverlag Freiburg Schweiz, 1999.

[25] Rippe, K.-P. Ethikkommissionen in der deliberativen Demokratie. In [17], 140–164. 2000.

[26] Sommermann, K.-P. Ethisierung des öffentlichen Diskurses und Verstaatlichung der Ethik. In *Archiv für Rechts- und Sozialphilosophie*, 75–86. 2003.

[27] Teuwsen, R. Repräsentanz- und Transparenzdefizite in der Bioethik-Diskussion. *Biopolitik*, 1(1):34–42, 2002.

[28] van Willigenburg, T. Soll ethische Fachberatung, moralisch neutral' sein. In [24], 285–305. 1999.

[29] Weber-Hassemer, K. Politische Entscheidung und Politikberatung in der "konsensualen Demokratie" am Beispiel des Nationalen Ethikrates. In [12]. 2002. URL http://www.ethikrat.org/texte/pdf/elmau_referate/Referat_Weber-Hassemer.pdf.

[30] Yoder, S. D. The Nature of Ethical Expertise. *The Hastings Center Report*, 28(6):11–19, 1998.

Global Aging and Intergenerational Equity[1]

Norman Daniels

Harvard School of Public Health

ndaniels@hsph.harvard.edu

ABSTRACT. Societal aging is the result of declines in both the mortality rate and fertility rate for a society – both the result of the success of 20th century social policy. Societal aging intensifies issues of intergenerational equity. Two concepts, age groups and birth cohorts, are distinguished; distinct distributive issues arise for each. A proper solution to these problems must be integrated, so that institutions addressing them solve both and adjusted to do so over time. What this means in the context of societal aging is addressed with special attention to China.

1 The Challenge of Global Aging

The phenomenon of global aging creates what may be the most important public health problem of the 21st century. It thus adds urgency to the question of age bias and the just distribution of factors affecting health. Many Americans might assume that aging in their society is simply the result of the large post-war Baby Boom cohort they so often hear about. But the demographic roots of societal aging go much deeper. A society ages when there is a significant decline in its birth (or fertility) rate and a simultaneous drop in its mortality rate at all ages. Fewer children then are produced while people live longer. Sometimes, when there is a sudden, temporary increase in the birth rate followed by a significant drop, as in the case of the Baby Boom, the maturation and aging of that group temporarily accelerates the aging of society. But societal aging can take place without the effect of a super-large cohort. In its general form, we can depict such aging as the shift from a typical "population pyramid" from earlier periods, wide with many children at the bottom and narrow with few elderly at the top, into a shape that is more like a column or even a column with some bulges at higher ages (representing an enlarged cohort from some earlier period.)

Societal, indeed global, aging is greeted as a crisis in many recent book titles (which refer to an "age quake" [17], "age wave" [14], or "generational storm" [8] to note a few), even though it is a result of *the success*, not *the failure*, of longstanding and widely pursued health and family planning policies aimed at reducing mortality and fertility rates. In the United States, the fastest growing

---

[1]This paper is a much shortened version of Chapter 6 of *Just Health: Meeting Health Needs Fairly* (Cambridge University Press, 2008). In the longer version I develop more fully how my account of fair process is to be integrated with the solutions to the age group and birth cohort problems distinguished here.

age groups in the country are people over 75 and 85. These older age groups will grow even faster when the Baby Boom cohort reaches those ages beginning in 2020 and "walkers replace strollers" [8]. According to 1996 U.S. census figures, by 2040 there will be more people in the US over age 80 than there are preschoolers [14]. Furthermore, the aging of society is much more advanced in other developed countries. In Italy, for example, the fertility rate (1.2) is well below the level at which a population can reproduce itself (2.1), and the working age population is already shrinking (as it also is in Japan). Indeed, all the European G-7 countries are below the replacement level in fertility rates. By 2050, half of Continental Europe will be 49 or older, and well before that, by 2030, one of every two adults in developed countries will have reached retirement age [2].

What may be more surprising to people in developed countries, used to seeing mass media pictures of developing countries teeming with children, is that many developing countries are aging now at much faster rates than developed countries did. While the proportion of the elderly in developed countries is due to double over the next 50 years it is due to triple in East Asia. By 2050, there will be 332 million Chinese 65 years or over, equivalent to the world's elderly population in 1990 [7]. The two billion people over age 60 who will live in our aging world by 2050 will mostly be living in developing countries. Unfortunately, this rapid societal aging takes place without the wealth and the sophisticated economic institutions that exist in developed countries, so that China will grow old before it grows rich [7].

The primary effects of societal aging are domestic for each nation, for the changed profile of societal needs it produces requires a national response. But global aging will also have some international effects. When only developed countries faced societal aging, a shrinking labor force could be partly relieved by increased migration of plentiful young workers from other countries. (Only in the U.S. has the scale of such immigration had a significant impact on the workforce and on fertility rates.) But as aging impacts countries worldwide, more countries will find they have a shrinking work force. Immigration from younger to older countries may greatly intensify recent changes in immigration compared to those a century ago. In addition, societal aging may affect economic growth rates, leading to shifts in the flow of investment capital internationally. Since all of these factors act as social determinants of health, there can be significant but not easily predictable effects on health.

1.1 A changing profile of needs

The graying of societies changes dramatically the profile of health and social support needs. Consider the problem of increased disability and dependency with very old age. Over thirty years ago Bernice Neugarten [12] introduced the distinction between the "young old," those age 65-74, who are generally quite healthy and free of significant disability, and the "old old," those over age 75 who are more likely to have significant disabilities and dependencies for care. The rapid growth of those over 75, unless it is offset by dramatic reductions in the prevalence of disease and disability among that age group, will bring with it increased burdens for the management of chronic disease and long term care.

Since long-term care is generally provided by family members, much of the burden will fall on adult working children. Yet, nearly a quarter of all the elderly in the US in 1989 had no children, and another 20% had only one child. With more women in the workforce, the problem of providing family care is intensified, since women have traditionally been the primary care givers. In addition, if benefits for elderly retirees are reduced, which is one strategy advocated by many to cope with societal aging in the United States, the elderly will become more economically dependent on their adult children.

The care-giver problem for frail and dependent elderly arises in developing countries as well. In them, the problem is not the sustainability of the kinds of publicly supported social and medical services provided for the elderly in developed countries, but the sustainability of informal, social structures of support, such as the traditional patterns of care that involve aged parents living with adult children, as in Japan and China. China, for example, must face the specific consequence of the success of its very strict population policy: one child for urban couples, two for rural ones. Like the United States, China will have many elderly with no children. It will find that it has even more elderly who have only one child to support them than is the case in the United States. The Chinese refer to this as the "1-2-4" problem, one child must care for two parents and four grandparents.

In 1996 the Chinese government made it a legal requirement that adult children support their elderly parents, obviously anticipating that traditional filial obligations would be strained to the breaking point by the new demographic realities. The law is not going to solve the problem, given its scale. Instead, China will have to construct public supports that fill the gap left by eroding family supports and to do so without further undermining those family supports. China's problem is the problem of all developing countries where rapid aging, extensive urbanization and industrialization, and a lack of existing health care and income support systems of the elderly all collide with traditional family values.

The point is simple: no aging society, with or without public systems of support, escapes the problems created by societal aging for sustaining and improving institutions that provide care for elderly dependents.

The increase in medical needs with societal aging is much broader than the problem of long term care for frail elderly people. With aging there are increases in the prevalence of cardiovascular disease, chronic pulmonary disease, diabetes, arthritis, and cancer, as well as Alzheimer's and other dementias. The increased cost of treating the greater prevalence of these illnesses imposes great strains on resources and intensified competition for them in developed countries. The problem will be even more severe in developing country health care systems, many of which have barely begun to gear up to meet the needs posed by chronic diseases. In poorly funded systems, beefing up medical services for the chronic illnesses of middle and older age means stealing resources from primary care and preventive care for the whole population. The prospect of vast unmet health needs for a growing elderly population is why I earlier suggested that societal aging may be the greatest public health problem of this century.

The aging of a society also changes the profile of needs that arise outside the health sector and yet affect population health. For example, as people typically live longer lives, they must plan for income support over a longer period of retirement years, and they must do so in a global economy that has altered long-standing career patterns and attendant retirement plans. With typically longer lives there may be a need for education beyond childhood and adolescence, including novel forms of continuing education for adults. Such education has unknown effects on population health, but if it affects people in the ways education earlier in life does, it may have profound effects on population health.

2 Sustainability of institutions and competition for resources

Societal aging not only shifts the basic pattern of needs in a society, but it also creates problems in sustaining traditional institutions that meet those needs. Those institutions emerged in response to a particular pattern of needs and a particular population age structure. Societal aging, especially rapid aging, strains their adequacy and sustainability. One way to look at the source of the strain is to measure the change in the ratio of workers to dependents in a society. Actually, we need to look at two distinct ratios: workers to retirees and workers to all dependents, retirees and children. Reductions in the numbers of children, for example, can offset, at least temporarily and partially, increases in the numbers of retirees. If retirees are more expensive to support than children, especially if they retire early, live long lives, and develop significant dependencies, then fewer children to support may provide little relief. Moreover, the problems interact: if younger families of child-bearing age are heavily burdened with the cost of supporting a large elderly population, it may further suppress their own fertility rates, intensifying the problem later on.

Why think of support for the elderly as a burden on current workers, one that increases as the dependency ratios worsen? The simple answer is that all health care systems that include the elderly are pay-as-you-go. They are funded by some combination of general tax revenues, payroll taxes, sales or value-added taxes, and premiums. Most pension and income support systems in developed countries are as well. The explanation for this fact varies with the good supplied.

With health care insurance, there is probably no real alternative to pay-as-you-go, since the far greater medical needs of the elderly would make stratified insurance pools by age prohibitively expensive.[2] The same is true for long-term care needs. Private insurance markets for them fail to provide good protection (except for those economically best off) because people would have to buy into them at very early ages in order to spread their risk over their lifespan. Buying such plans late in life, when expected needs are greater, would be impossible for most people because premiums would be too high, given their levels of income and wealth. So health care insurance coverage for the elderly – universal coverage for all ages in other developed countries, universal coverage for those over 65 in the U.S. – spreads the risk between younger, healthier workers and older, less healthy retirees, with the promise that the system will remain stable and useful to younger cohorts when they age. In effect, the systems are "actuarially unfair" to the young and an "actuarial boon" to the elderly, but the goal is to make them actuarially fair over the lifespan.[3] The challenge posed by societal aging is

---

[2] Since individual health savings accounts of the sort that have emerged in several countries are usually backed by a shared catastrophic insurance plan, my points above would apply to them as well.

[3] In the United States, we can construe insured workers health care "premiums" to be the total of their contribution (and their employers') toward private insurance plus their payroll tax

whether the system will be stable over the lifespan of younger, working cohorts so that they will be supported at benefit levels approximating those they now support among the elderly.

The story is more complicated for pensions and income support, largely because of history and politics. In most developed countries, pensions were introduced through payroll taxes as "pay as you go" systems in which current workers paid for the benefits of current retirees, expecting in turn to have their benefits paid by later retirees. Such systems offer the political advantage of allowing them to provide benefits for retirees as soon as the system starts up. This was crucial to the political appeal of Social Security in the US, where many older workers who had faced the Great Depression needed support immediately on retirement and could not wait to accrue vested savings over a lifetime of work. In addition, as long as working populations were stable or increasing, as they were through most of the twentieth century in developed countries, then the ratio of workers to retirees was favorable. Ongoing population growth promised ongoing stability to financing.

Enter population aging. In Italy and Japan, the working-age population is already shrinking. There will be 28% fewer working-age Germans and 36% fewer working-age Japanese by 2050 than there are today [2]. The UN projects that the ratio of working-age adults to elderly in the developed world will drop from 4.5 to 1 today to 2.2 to 1 in 2050. The IMF estimates the ratios in 2050 for Japan, France, Germany, and Italy will be 1.5, 1.4, 1.2 and less than 1 to 1. Public retirement spending in typical developed countries will grow from 11 to 23% of GDP by 2050 [2].

Other factors also enter, such as the trend toward earlier retirement. In the U.S, France, Germany, and Italy, the age of retirement has decreased, rather than increased, as one might have expected with people living longer, healthier lives. The combination of longer lifespan and earlier retirement, along with shrinking working age populations in countries like Italy and Germany greatly intensifies the strain on resources in existing transfer schemes. Developing countries, including China, have a longer window of opportunity in which to rely on large working age populations and rapid economic growth rates to prepare the way for their own health care and income support later in retirement. But the absence of existing institutions to build on, aside from informal family support, will make this a very difficult task.

My point is not to side with critics of pay-as-you-go systems who insist that we turn to systems that rest on vested savings. There are some advantages, outside the context of rapid aging of society, to systems that share risks across cohorts. The wellbeing of one cohort is less dependent on its own economic history and can be protected against great bad luck by sharing risks with other cohorts. The problem is how to stabilize the system in a context of rapid aging. I return to this point later. Here I only address the source of the strain for traditional solutions to the problem of meeting the needs of the old and the young.

---

contribution (supporting Medicare) plus their contribution to general tax revenues (supporting Medicaid and public hospitals). They will pay out more than they will receive in medical benefits, on average, which is what I meant by saying the arrangement was "actuarially unfair" to them. Retirees, however, pay out less than they receive as benefits, even when we include their high out of pocket expenses (thus the "actuarial boon"). Over the lifespan of each, however, the system meets needs and approximate actuarial fairness (except that progressive or regressive taxes complicate the point).

3 Intergenerational Equity: Two Problems, Not One

Aside from some confused calls for intergenerational equity and some passing attention to the threat to traditional familial values,[4] the global aging literature is strikingly silent on basic questions of value, even as it discusses policy steps that are needed. One thing that makes the intensified conflict or competition between the old and the young especially difficult is that it is the basic needs of different groups that are at issue. The old and the young both need health care, often of different types. We are not talking about privileges for one group and necessities for another. Were that the case, justice would quite easily rule in favor of giving priority to the more important needs. The problem is much harder. What do we do when the competition between the old and the young is for life-extending health care or for income support necessary to meet basic needs? Can we choose between health care for the elderly and education for their grandchildren? What is striking about the policy literature is its silence on these issues.

The rest of this paper is aimed at distinguishing and answering two questions about justice between the old and the young that are often conflated – *equity among age groups and equity among birth cohorts*. As we shall see, separate concepts and distinct problems of distributive justice are involved in the appeal to "intergenerational equity." The policy debates about modifying medical and income support systems for retirees require that we be careful in distinguishing these ethical issues. Properly integrated solutions to both problems must also consider the role of the broad determinants of health across the lifespan and the need to rely on fair process in making the many resource allocation decisions such solutions involve.

4 Distinguishing Between Age Groups and Birth Cohorts

The question posed by many about intergenerational equity readily leads us to confuse two distinct problems, What is a fair distribution among age groups? And, What is a fair distribution among birth cohorts? Part of the confusion comes from the fact that the term "generation," and thus "intergenerational equity" is multiply ambiguous. When we worry if we will destroy our atmosphere or oceans or forests, we are concerned about equity between relatively distant generations. Will we leave enough and as good for our great-great grandchildren? I shall not address this problem, important as it is. I focus instead on *adjacent* generations, the young and the old in any society, and I concentrate on equity in distributing health and health care (though this may require equity with regard

---

[4]Traditional rural, agricultural settings that involved large families supporting their own elderly parents have long ago disappeared in developed countries and are rapidly disappearing in developing ones. Migration of adult children to cities for work often involves leaving the elderly behind, sometimes to care for grandchildren. But even where the elderly also are brought to the city, they can no longer engage in productive work in the ways they did in rural settings, and so dependency is increased. In addition, there are many fewer children to care for aged parents.

to the distribution of other goods as well). Still, even with the restriction to adjacent generations, there remains an ambiguity: "generation" can refer to age groups, as in the perennial conflict between the young and the old, or it can refer to birth cohorts, as when we contrast experiences of the World War II generation and the Vietnam generation. Another part of the confusion comes from the fact that the problems interact: a solution to the age group problem must be compatible with a solution to the birth cohort problem, and vice versa.

Despite the ambiguity in "generation," the concepts are clearly different. We individuate them differently. Age groups do not age, but birth cohorts do. The age group comprised of people 65 to 75 is always comprised of people 65 to 75 years of age. The birth cohort of people that happens to be 65 to 75 in 2010 will, of course, age ten years by 2020. At any given time, an age group consists of one or more birth cohorts, depending how we identify the cohorts. Over time, it consists of a succession of birth cohorts. The birth cohort of people born between 1935 and 1945 that comprises the 65 to 75 age group in 2010 will be replaced by 2020 by the cohort born between 1945 and 1955. As we age, we pass through different age groups but not different birth cohorts.

A birth cohort is a distinct group of people with a distinctive history and composition. The question, "What is a just distribution of social goods between birth cohorts?" thus carries with it the assumption that we are focused on differences between distinct groups of people. For example, special questions of fairness may arise because of particular facts about the socioeconomic history and composition of particular birth cohorts. In contrast, the notion of an age group abstracts from the distinctiveness of birth cohorts and considers people solely by reference to their place in the lifespan. Consequently our question about justice between age groups also abstracts from the particular differences between the current elderly and the current young that arise because of the distinctive features of the birth cohorts which happen to make up those age groups. We are concerned with a common problem about justice between the old and the young that persists through the succession of aging birth cohorts.

5   Equity between age groups and birth cohorts

Not only are these different concepts, but we must think about fair distribution differently for them. We should think of fairness between different age groups as a problem of prudential allocation over the lifespan. Institutions that allocate resources in a way that meets needs at each stage of life in a way that makes our lives as a whole go better will treat age groups fairly. Fairness to different birth cohorts, however, requires that cohorts enjoy a rough equity in the benefits they receive given the contributions they make as they pass through institutions that appropriately solve the age group problem. Though the concepts and problems of distribution are different, they interact and need to be solved together.

Our tendency to confuse these questions is not simply the result of collective stupidity, including the failure to notice a semantic ambiguity in the term "generation." Rather, the confusion results primarily from our perception of competition between groups. We see the age-group question in competitive terms: the old and the young compete at any given moment for scarce public welfare resources. Whose side should we take? At the personal level, where adult children must care for elderly parents while also raising their own children, the sense of competition is muted by love and obligation. Limits to resources still mean hard choices for many who struggle without adequate societal support.

Still, the perceptions of the age-group problem at both the personal and social levels share a common framework. At both levels, the issue is one of determining which transfers of goods should take place between distinct groups of individuals, the young and the old, adult children and elderly parents. Which transfers are just or fair? What does one group owe the other? The sharp competition felt at the social level may be tempered by love and a desire to care at the personal level. But the problem is still seen as one between "us" and "them".

This way of construing the age-group problem is reinforced by our tendency to see public policy as something that operates in the here and now. We scrutinize its effects at a particular time. We rarely think of public policies as instruments that operate over long periods of time, indeed, over our whole lifespan. When we do think about the long run, for example, about what life will be like later in this century, we shift questions. We stop thinking about the age-group question and substitute the question about equity between birth cohorts.

In what follows, I reject the perception of the age-group problem as a problem of competition between groups viewed at a given moment and adopt instead the lifespan approach promised earlier.

6 The Prudential Lifespan Account

What is a just distribution of resources between the young and the old? The key to answering this question lies in the humbling fact that we all age. In contrast, we do not change sex or race. The relevance of these banal observations needs some explanation.

If we treat blacks and whites or men and women differently, then we produce an inequality between persons, and such inequalities raise questions about justice. For example, if we hire and fire on the basis of race or sex rather than talents and skills, then we create inequalities that are objectionable on grounds of justice. If we treat the old and the young differently, however, we may or may not produce an inequality between persons. If we treat them differently just occasionally and arbitrarily, then we will be treating different persons unequally. But if we treat the young one way as a matter of policy and the old another, and we do so over their whole lives, then we treat all persons the same way. No inequality between persons is produced since each person is treated both ways in the course of a complete life. Thus the banal fact that we age means age is different from race or sex for purposes of distributive justice.

My account of justice between age groups builds on this basic point: Unequal treatment at different stages of life may be exactly what we want from institutions that operate over a lifetime. Since our needs vary at different stages of our lives, we want institutions to be responsive to these changes. In many industrialized countries, we defer income from our working lives to our post-work retirement period through some combination of individual savings and employee or government pension or social security plan. In many such schemes there are no vested savings, but a direct transfer from the working young to the retired old. Viewed at a moment, it appears that "we" young workers are taxed to benefit "them," the old. Similarly, working people pay for far more than the cost of their own health care, for they contribute to the costs of children and the elderly alike whether in the American mixed insurance system or in universal coverage systems elsewhere. If these systems are stable over the lifespan, then we have designed systems that treat us appropriately – differently – at different ages.

Age groups are treated differently in such schemes. The old pay less and get more while the young pay more and get less. If this system continues as we age, others will pay "inflated" premiums or taxes that will cover our higher costs when we are elderly. Such a system allows us to defer the use of resources from stages in our lives when we need them less into ones in which we need them more. In general, budgeting these transfers prudently enables us to take from some parts of our lives in order to make our lives as a whole better.

Two important lessons emerge about the unequal treatment of different age groups. First, treating the young and old differently does not mean that persons are treated unequally over their lifespan.[5] Second, unequal treatment of the young and old may have effects that benefit everyone.[6] These two points provide the central intuition behind the Prudential Lifespan Account of justice between age groups: Prudent allocation among stages of our lives is our guide to what is just between the young and the old.

The lifespan account involves a fundamental shift of perspective. Rather than seeing distinct groups in competition with each other, for example, working adults who pay high premiums and the frail elderly who consume so many services, we must view each group as *representing* a stage of our lives. The prudent allocation of resources through the stages of life is our guide to justice between groups. From the perspective of stable institutions operating over time, unequal treatment of people by age appears to be budgeting within a life. If we are concerned with net benefits within a life, we can appeal to a standard principle of individual rational choice: It is rational and prudent that a person take from one stage of his life to give to another in order to make his life as a whole go better. If the transfers made by an income support or health care system are prudent, they improve individual well-being. Different individuals in such schemes are each made better off, even when the transfers involve unequal treatment of the young and the old. This means that neither old nor young have grounds for complaint

---

[5] Mckerlie [9–11] suggests abandoning the standard concern for equality between complete lives in favor of a view that emphasizes equality between simultaneous segments of lives. Consider the inequality between simultaneous segments that shows up for Betty and Connie:

Decade	1	2	3	4	5	6	7	8	9
Betty	10	5	10	5	10	5	10	5	
Connie		10	5	10	5	10	5	10	5

Connie has the same pattern of wellbeing as Betty, but, because she is born ten years later, on McKerlie's view, there are thirty-five objectionable units of inequality in decades 2 through 8. McKerlie's view thus runs counter to a basic intuition: simply changing Connie's birth date, with no other effects on either life, should not make the situation more or less objectionable. I respond more fully to McKerlie's view in Daniels 1996.

[6] Raising the age of eligibility for income support benefits under Social Security, which arguably is a prudent and fair way to address both the age group and birth cohort problems, might leave African-Americans, who have lower life expectancy, worse off than whites or Asians. Where such effects take place, they may constitute good reasons for not adopting such a rationing policy, or they might give us reasons to link the rationing to facts about group life expectancy. The general point is that the Prudential Lifespan Account presupposes that solutions to the age group problem will not disturb more general requirements of justice (see [3, 4]).

that the system is unfair. Prudent allocation establishes fair terms of cooperation among age groups.

The contrast of age with race or sex should now be clear.

Differential treatment of people by age, when part of a prudent lifetime scheme, involves treating people equally over their whole lives. There are no losers. Differential treatment by sex or race always creates inequalities, benefiting some at the expense of others. Losers will have legitimate complaints about unfairness or injustice.[7]

To illustrate the basic idea, consider how I might design a health care insurance policy that operated over my whole life span. Suppose I am willing to spend only a certain amount of my lifetime resources insuring myself against health care risks – health care, however important, is not the only good in my life. In any case, I accept the fact that the benefits I can buy with that lifetime premium will not meet every conceivable medical need I will have. Therefore, I must be willing to trade coverage for some needs at certain stages of my life for coverage at others. I also believe that I should give equal consideration to my interests at all points in my life.

Unfortunately, if I know how old I am and think about things only from the perspective of what I consider important at that point in my life, then I risk biasing the design of my insurance package, for example by underestimating the importance of things I will need much later in life. To compensate for this bias, I should pretend that I do not know how old I am and will have to live through all the trade-offs I impose at each stage of my life. For example, I know that if I give myself too much acute health care when I am dying, I do so at the expense of other services, e.g. long-term care services, or preventive services earlier in life, that might improve my quality of life over a considerable period late in life.

Just as individuals set reasonable limits on their lifetime insurance premiums, prudent planners acting on behalf of society in general – in the kind of fair deliberative process developed in [6]– are limited by what counts as a "fair share" of health care. This share is not simply a dollar allotment per person. It consists of entitlements to services that are contingent on our having certain medical needs. Their problem is to find the reasons relevant to allocating this fair share over the whole lifespan. Accountability for reasonableness applies, then, to this lifespan allocation and not simply to setting limits at a given moment. The goal is a distribution that people in each age group would think is fair because they would all agree it makes their lives as a whole better than alternatives. To ensure that our planners avoid biasing the design in favor of their own age group, we shall require them to pretend that they do not know how old they are, and we require that they accept a distribution only if they are willing to live with what it does to them at each stage of their lives. Each stage of their own lives thus

---

[7] Anti-discrimination legislation concerning gender or race in the United States is premised on the idea that unequal treatment, measured by unequal outcomes (as opposed to racist or sexist intent), is prohibited behavior. Anti-discrimination regarding age in the United States is modeled on race and gender legislation, but it must allow for reasonable, age-based distinctions in treatment. It accepts as non-discriminatory age regulation of voting, military service, and alcohol use, as well as of eligibility for Medicare and Social Security and for "senior discounts." Its prohibition on age-based discrimination in hiring and firing is compatible with the view I propose: these are not prudent features of allocation of jobs over the lifespan but arbitrary, often stereo-typically driven or economically driven treatment of individuals because they belong to an identifiable group. Whether a prudent lifespan version of compulsory retirement could be defended is a complex issue I leave for another occasion.

stands in as proxy for an age group, and they will age from conception to death in the system of trade-offs to which they agree.[8]

Before saying more about application of the Prudential Lifespan Account to health, I want to emphasize that the Prudential Lifespan Account is quite general. It gives us a way of thinking about the distribution of many important goods, not just health care. As I remarked earlier, as life expectancy increases and career trajectories are modified, the importance of continuing education or income support for adults may increase. A prudent allocation of entitlements to such goods over a lifespan should accommodate such effects of societal aging, especially since both of these goods are themselves determinants of population health. The prudential lifespan account asks us to think about how planners who do not know their age would allocate a lifetime fair share of entitlements to income support or education to each stage of life. These entitlements are specified relative to what justice in general permits in the way of inequalities between persons.

7 Lifespan Allocation of Health Care

To illustrate the approach, albeit without the extra complexity raised by the lifespan distribution of such social determinants, consider how the Prudential Lifespan Account might be applied to the case of health care, given the account of just health developed earlier. How should prudent planners think about the entitlements to care that people can reasonably claim at each stage of life, given that meeting these health needs is intended to protect opportunity by protecting normal functioning? Remember, such planners are constrained by pretending that they do not know their own age and must plan for each stage of life on the assumption that they will live through it. Protecting opportunity at each stage of life is especially important since all have a fundamental interest in being able to revise their views about what is valuable in life as they age. Because impairments of normal functioning clearly restrict the portion of the normal opportunity range open to individuals at any stage of their lives, health care services should be rationed throughout a life in a way that respects the importance of the age-relative normal opportunity range. With this refinement, the fair equality of opportunity account of just health can avoid the charge of age bias.

Reasonable people will disagree about just what this principle means in the context of specific trade-offs that must be made. For example, if we allocate resources to prevention early in life in order to reduce the risk of loss of functioning at a much later stage of life, we may not be able to meet other health needs early or late in life. Versions of the unsolved rationing problems we earlier saw

---

[8] In [3, chap.3] I give a more detailed statement of these and some other qualifications on the concept of "prudent deliberation" appropriate for solving the age group problem. Appropriate considerations of prudence require even further restrictions on the knowledge of the deliberators, making them even less like the standard, fully-informed consumer of economic theory. For example, they should judge their well-being by reference to all-purpose goods, like income and opportunity, rather than through the very specific lens of the "plan of life" they happen to have at a given stage of life; otherwise the design of the lifetime allocation may be biased by a conception of what is good which just happens to be held at a given point in life; see also [1, 15, 16]. In [3, Appendix A](reprinted in Daniels 1996), I also defend a key feature of classical prudential reasoning, that we value each part of our lives equally ("the requirement of equal concern"), against criticism of this principle by [13].

arise in this context as well, even though they take place within stages of a life. Just as reasonable people may disagree about how much priority to give those who are worse off than others, a distribution problem across lives, so too they may disagree about how much priority to give being worse off in health at some stage of life compared to being better off at others.[9] Since we are designing a prudent social scheme, fair, deliberative process is needed to resolve such disagreements so that outcomes are accepted as fair and legitimate. Just as, in *Just Health Care*, I mistakenly thought the degree of impact of pathology on normal functioning would suffice as a measure of the importance of a health need, so too I underestimated the sources of disagreement about how to set priorities in meeting needs at different stages of a life by thinking prudence alone could inform our choices. In both cases, we must supplement the opportunity principle – even relativized to stage of life – with fair process.

Depending on the organization of health care in specific systems, this model of prudent allocation over the lifespan becomes either more or less hypothetical. In the United States, where a publicly funded universal coverage system exists for the elderly but most children and adults are covered by private insurance plans, there is little incentive to adopt a lifespan approach to the design of health care entitlements. Private plans have little economic incentive to invest in preventive measures that will protect the elderly, and decision-makers in Medicare never have to reconcile their resource allocation decisions with the needs of younger age groups. In universal coverage systems that fund benefits across the lifespan, the model has a natural applicability. [10]

8 Equity Between Birth Cohorts

Societal aging on a global scale, we have seen, changes the profile of needs in a society, forcing anew consideration about to how to allocate resources to different age groups. The Prudential Lifespan Account, coupled with fair procedures for resource allocation, provides a framework for thinking about how to address the age group problem. Societal aging, more alarmingly, increases burdens on the working-age population to support a much larger population of people who are retired and elderly, though some of this burden is offset by the decreasing burden of raising children as fertility rates decline. I noted earlier the dramatic decreases in the ratio of workers to elderly in all developed countries and in many developing countries over the next several decades. This demographic change strains resources and intensifies competition for them. It also gives rise to complaints about intergenerational inequity.

A special form of the problem of equity between birth cohorts is the question, Which inequalities in benefit ratios are fair or equitable? The benefit ratio is the ratio of benefits to contributions made over the lifespan to these schemes. More generally, what inequalities in the treatment of different cohorts are just or fair as these cohorts pass through institutions intended to meet the requirements of

---

[9] Lives containing equal numbers of quality adjusted life years (QALYs) may have them distributed differently; reasonable people may disagree about which distribution makes the life as a whole go best. Some may be willing to trade away total QALYs against some distributive considerations.

[10] In the longer version of this paper, I discuss implications of the lifespan view for children's health, long-term care, and the controversial issue of rationing by age. See [5, chap.6]

justice between age groups? Remember, I said both problems must be solved together.

Since each birth cohort ages, it has a fundamental interest in securing institutions that solve the age group problem effectively. Unfortunately, institutions or transfer schemes that solve the age-group problem operate under considerable uncertainty. There is uncertainty about population and economic growth rates, as well as about technological change, which further affects productivity. Errors are likely to abound, and inequalities in benefit ratios between cohorts will arise as a result. Despite these sources of error, institutions that solve the age-group problem must remain stable over time.

The problem goes beyond uncertainty and errors, for there is considerable certainty about the strains that rapid societal aging will place on existing transfer schemes in developed countries and the obstacle that aging imposes on establishing new schemes in many developing countries. Transfer schemes, such as Medicare and Social Security in the United States and their analogues elsewhere, must be able to weather the political struggle that results if errors are allowed to produce unjustifiable or unacceptable inequalities in benefit ratios. They must also be able to weather the rapid aging described earlier, including the emergence of worker to retiree ratios in some developed countries, like Italy, that may require one worker to support one elderly person. In China, which lacks formal transfer schemes and relies on children to support the parents, the problem takes the stark form of the "1-2-4 problem" – one child supporting two parents and four grandparents.

Formal, societal transfer schemes will be able to survive the struggle among coexisting birth cohorts only if each cohort feels it has a stake in preserving them. Each will feel it has such a commitment only if it believes these institutions work to its benefit within the limits of fairness. Such commitment will be sustained, then, only if the practical target of policy is to aim for *approximate equality* in benefit ratios.

One objection to this suggestion is that it ignores the fact that some cohorts may be wiser or more prudent than others and may therefore contribute more to productivity. Since many believe that people should be rewarded for their contributions, they insist that benefit ratios should reflect *desert*. Specifically, they urge that each cohort should depend on its own savings. But this appeal to desert (contribution) would require disentangling the many sources of change that contribute to rising or falling economic fortune. It would not justify simply relying on individual or cohort savings, for they result from many factors other than moral desert.

Since it is hard to see how a stable system could incorporate such factors in its scheme of benefits, it seems reasonable for cohorts to aim for approximate equality in benefit ratios and to seek other ways of persuading each other to act prudently over time. Each cohort, after all, has an interest in securing stable institutions that solve the age group problem. Cohorts must therefore cooperate to achieve such stability. But cooperation will require some *sharing of risks* across cohorts. In general, the burdens of economic declines and of living through unfavorable retiree/employee ratios must be shared, as must the benefits of economic growth and favorable retiree/employee ratios. This suggests again that approximate equality in benefit ratios should be the practical target of public policy, if not a hard and fast rule. My point, however, does not mean that where there are purely pay-as-you-go schemes, no effort should be made to find ways of supplementing them with vested savings. For health care, however, there seems to be little recourse but to find ways to make pay-as-you-go schemes stable.

Objections to unequal benefit ratios should not lead us to eliminate inter-cohort transfer schemes, at least not on the grounds that "equality" will then result. Making each cohort solely responsible for its own well-being over the lifespan will by no means assure that different cohorts will fare equally well. Inequalities will come about because of uneven economic growth rates. It is not at all obvious that inequalities of benefit ratios in inter-cohort schemes will generate more intolerable forms and degrees of inequality than the inequalities that result when each cohort must depend on its own resources and good luck. Cooperation may be a better strategy than 'go-it-alone', and the problem becomes one of institutional design and of securing a long-term commitment to schemes that are fair.

Several strategies are available for adjusting benefits so that we can achieve approximate equality in benefit ratios despite demographic shifts and other sources of uncertainty and error. One strategy is to build a cushion of unexpended benefits while the ratio of workers to retirees is still relatively high. This strategy has been adopted in financing reforms of the U.S. social security system, though there is always a risk that these benefits will be a target of convenience for politicians seeking to relieve budget deficits. In the 2000 U.S. election campaign, both candidates said they would not break into the "lock-box" of enlarged social security funds that were needed to make the US system stable through the Baby Boom retirement. But the budget deficits created by the Bush Administration tax cuts and war expenditures clearly threaten that lock-box.

A second strategy is more basic, for it involves rethinking some of the policies toward retirement that have dominated developed welfare systems in recent decades. Many current policies provide considerable incentives for older workers to withdraw from the workforce well before any disability actually makes such withdrawal necessary. It is also quite difficult for older workers to find flexible, part-time employment that can reduce the need for drawing on income support benefits. Underlying these incentives and policies are both economic and moral considerations. Pushing older workers out of the workforce in periods of unemployment, when there are large numbers of young workers seeking employment opportunities, may have seemed an acceptable way to ration jobs by age, or it may have seemed an appropriate way to make room for better educated and potentially more productive workers in technologically advancing economies. These economic considerations may have been reinforced by the view that the elderly want to enjoy more leisure time. These underlying considerations should be reassessed.

Health status for the elderly remains quite good well into the mid 70s. Millions of elderly who would be happier with some form of meaningful work, at least on a part-time basis, find themselves facing forced withdrawal from the workforce. At the same time, as I noted, both in the U.S. and in other developed countries, there is a surprising trend toward earlier retirement despite better health and longer life expectancy. With shrinking workforces and earlier retirement, many European economies face a shortage of workers in the next few decades. Under these conditions, it may well be wise to consider revising the existing benefits and incentives that lead workers to withdraw from the workforce early. The new shape of a life, with many vigorous and healthy years extending well beyond standard retirement age, means that we must revise our antiquated conception of the typical course of life.

In the United States, compulsory retirement ages have been raised or eliminated, at least for large categories of employment; similar legislation is being considered within the European Union. This development may encourage reassessment of the employability of older workers. It may not be enough, however, simply to eliminate legal or quasi-legal barriers to the employment of willing, elderly workers. Rather, we may have to encourage the emergence of flexible employment practices that accommodate the needs of older workers. Such practices may become an increasingly important way of assuring the welfare rights of an aging population.

9 Intergenerational Equity and Hard Cases

My main thesis in this paper has been that there are two problems of intergenerational equity, the age group and birth cohort problems, that interact and must be solved together. The solutions must also be integrated with a broader view of health needs and with the use of a fair, deliberative process to address pervasive disagreements about distributive fairness. To solve the age group problem, we should think about prudent allocation over the lifespan of the good in question. For health care that means protecting the age-relative opportunity range for individuals by promoting normal functioning at each stage of life. Yet a scheme that aims to do that must be sustainable and must be fair to different birth cohorts passing through it, assuring them roughly comparable benefit ratios. In a period of rapid societal aging, both are especially difficult to accomplish.

What do these general solutions to the distributive problems tell us about the challenging problems faced by a rapidly aging developing country like China?

China must develop a health system and insurance scheme that can meet health needs as they arise across the lifespan. As the lifespan is extended, and as typical lives will include periods of higher risk of chronic illness and disability later in life, health systems in developing countries must be prepared to meet those needs. In China there is a poorly funded urban health care insurance system and an even less adequately funded rural insurance system. Neither insurance system is prepared to address the chronic care needs that arise in later stages of life. Without appropriate insurance that shares risk widely over the population, there will be great demands on people to cover very high medical costs, and families with ill elderly relatives will face terrible choices about health care versus falling into poverty.

China will have to expand the capacity of its health system to meet chronic health needs of an aging population. It cannot simply allow the hospitals to expand services in ways that maximize revenues without attention to unmet needs or major inequities will emerge between people with different conditions. Confidence in the private sector to plan prudent lifespan allocation is misplaced, a lesson China should have learned from the collapse of its ability to meet rural health needs through purely private financing.

Especially problematic will be the intense burdens of caring for the frail elderly that were traditionally met by adult children, generally living extended households, largely in rural settings. As family structures in China have shifted to single households, with fewer elderly living with adult children, the strain of providing care is much greater. Earlier I noted that nearly a quarter of elderly Americans have no adult children to depend on. With the strict population policies China has used over several decades, comparable numbers of elderly Chinese will also have no family supports to draw on.

Fair treatment across age groups means that there is a societal responsibility to provide this care. Responsibility cannot simply be dumped on families, especially when they do not exist, or where they are too separated geographically to provide the care, or where changed family structures mean that cross-household transfers cannot substitute for the internal transfers that sustained the elderly in earlier times, when they lived in multiple family structures. China, like other countries that have strong cultural roots for filial obligation, cannot simply demand that families care for their elderly. Instead, it must develop a universal care system that assures all elderly that a lack of dependents will not mean they are abandoned by society in old age. The delicate policy task is to integrate such a solution with informal, familial supports, so that family support is made less burdensome – through incentives rather than sanctions– at the same time all are protected. This policy approach will be prudent from the perspective of the young as well as the old, since it will make it easier for the young to carry out family obligations they still believe they have, despite complex internal migration patterns. Just what form these policies should take depends on existing social structures. In some developed countries, forms of adult day care or respite care, that relieve adult children part time of the burden, have been found to lead to better and more sustained care for the elderly.

China, despite rapid aging, still has adequate time to put in place institutions that can meet these demands. It can even do so in a way that remains fair to different cohorts. Indeed, any system that is developed now in anticipation of the retirement of the post-revolution Chinese baby boom should be seen as meeting the interests of all cohorts. The huge Chinese workforce that would now be saddled with the cost of financing that insurance scheme will itself be drawing benefits from it only it if is put in place now in a way that can be sustained for the decades after 2015, when that cohort starts to retreat from the workforce, and after 2029, when Chinas population will peak and the numbers of workers will clearly start to shrink. Once the very large working age population begins to shrink, relative to the elderly, it will be much harder, if not impossible, to put in place a system in which that working age population helps prepare for its own old age. Nevertheless, it is not easy politically to do so now: for current workers to prepare for their own retirements in a period when fewer workers will follow them, they must contribute to savings (or taxes) at very high rates, rates that will be resisted.

Some policy proposals call for China to develop a mixed system, with a pay-as-you-go universal coverage for basic health and income support protections, combined with fully funded schemes to provide greater sustainability for income support for an elderly, retired population. As I noted earlier, my solutions to the two distributive problems do not rule out or in specific combinations of financing for institutions that solve them. The solutions do, however, impose ethical restrictions on what is done.

The hard cases posed by global aging do not, after all, show it is impossible simultaneously to address the age group and birth cohort problems fairly. Still, the challenge they pose is difficult to meet, in part because political discussion of it is not adequately informed by the kind of lifespan view of population health articulated here. Properly integrating approaches to it with accountability for reasonableness can create the social learning curve that improves political discussion and the resulting institutions.

Bibliography

[1] Buchanan, A. Revisability and rational choice. *Canadian Journal of Philosophy*, 5:395–408, 1975.

[2] CSIS. Center for strategic and international studies. In *Meeting the Challenge of Global Aging*. 2002. URL http://www.csis.org/gai.

[3] Daniels, N. *Am I My Parents' Keeper? An Essay on Justice between the Young and the Old*. New York: Oxford University Press, 1988.

[4] Daniels, N. *Justice and Justification: Reflective Equilibrium in Theory and Practice*. New York: Cambridge University Press, 1996.

[5] Daniels, N. *Just Health: Meeting Health Needs Fairly*. Cambridge University Press, 2008.

[6] Daniels, N., Sabin, J. *Setting Limits Fairly: Can We Learn to Share Medical Resources?* New York: Oxford University Press, 2002.

[7] Jackson, R., Howe, N. *Global Aging: The Challenge Of The New Millennium*. CSIS and Watson Wyatt Worldwide, 1999. URL http://www.csis.org/component/option,com_csis_pubs/task,view/id,892/type,1/.

[8] Kotlikoff, L. J., Burns, S. *The Coming Generational Storm*. Cambridge, MA: MIT Press, 2004.

[9] McKerlie, D. Equality and time. *Ethics*, 99:475–491, 1989.

[10] McKerlie, D. Justice between age groups: A comment on norman daniels. *Journal of Applied Philosophy*, 6:227–234, 1989.

[11] McKerlie, D. Justice between neighboring generations. In Cohen, L. M. (Ed.), *Justice Across Generations: What does it mean?*, 215–226. Washington DC: Public Policy Institute, AARP, 1993.

[12] Neugarten, B. L. Age groups in american society and the rise of the young-old. *Annals of the American Academy of Political and Social Science*, 415:189–198, 1974.

[13] Parfit, D. *Reasons and Persons*. Oxford: Oxford University Press, 1984.

[14] Peterson, P. *Gray Dawn: How The Coming Age Wave Will Transform AmericaAnd The World*. New York: Three Rivers Press, 1999.

[15] Rawls, J. Social unity and the primary goods. In Sen, A. K., Williams, B. (Eds.), *Utilitarianism and Beyond*, 159–185. Cambridge University Press, 1982.

[16] Rawls, J. *Justice as Fairness: A Restatement*. Cambridge MA: Harvard University Press, 2001.

[17] Wallace, P. *Agequake: Riding the Demographic Rollercoaster Shaking Business, Finance and Our World*. London: Nicolas Brealey Publishing Ltd, 2001.

Living Science and Public Scientific Literacy[1]

Guoping Zeng
Tsinghua University
Sts001@tsinghua.edu.cn

Honglin Li
Tsinghua University
Sts001@tsinghua.edu.cn

ABSTRACT. Based on the review of current situation of public scientific literacy in China and international comparative studies, the paper puts forward the concept of "living science" and analyses the main characteristics of it: connecting tightly with basic living demands; giving importance to accessibility and perception; integrating with social knowledge; giving priority to instrumental and practical results; connecting inherently with cultural tradition. If we could say that science is refinement of everyday experience and thinking, then approaching scholarly "academic science" from "living science," which particularly emphasizes on common sense becomes an essential way for public understanding of science. Also, it has an important implication: "academic science"(R. K. Merton, **UCDOS**) should be linked with "post-academic science" (J. Ziman, **PLACE**) and "living science" (**BASIC**) to conduct the activities of science popularization in order to guide the public understanding and use of science in daily life.

1 Preface

A number of historic, geographic, economic and other factors, such as the lag of science development, the zigzag of historical development and "government driving" in the construction of public scientific literacy and so on, result in the particularity of Chinese public scientific literacy – from unilaterally "science going into public" to bilaterally "science communicating with public", from directly "improving production and life" to "improving human's scientific and cultural literacy", and so on [13]. In 2006, the enaction of *National Action Program for Scientific Literacy of All Citizens* in China reveals peoples' new recognition of public scientific literacy under the new conditions inside and outside of China. Within this context, we now review the status quo and particularity of Chinese (hereinafter the mainland of China) public scientific literacy nowadays.

---

[1]The authors are with Center of Science, Technology and Society, Tsinghua University, Beijing, 100084.

2 The status review of the public scientific literacy in China and international comparison

The idea and concept of scientific literacy was introduced to China from western countries until later 1980s. [6] In 1990, adopting Miller's system [10, 11] and referring to the indicators and questionnaires of the measurement of scientific literacy widely used in the west, a modified version of a questionnaire was designed which was thought to be fitted for the Chinese social context and condition in order to survey Chinese adults'(aged 18~69) scientific literacy, and the survey was finished in 1992.[2] After that, five countrywide surveys had been done, which were respectively executed in 1994, 1996, 2001, 2003 and 2005.

The 2003 survey showed that Chinese public scientific literacy was 1.98%,[3] which rapidly rose from 0.2% in 1996 and 1.4% in 2001 but still kept a large gap with the level of developed countries in 1990s, comparing to 12% of USA in 1995, 5% of EU in 1992, 4% of Canada in 1989 and 3% of Japan in 1991. [2]

Considering the essential three dimensions of scientific literacy – understanding of scientific concepts, scientific methods and the relationship between science and society, we make international comparisons with the results of the 2003 survey in China and the 2001 survey in EU, the USA and Japan.

2.1 Understanding of scientific concepts

Generally, the degree of the Chinese public's understanding to most of the concepts was lower than that of the USA, EU and Japan. Meanwhile all of these countries showed the same trend of difficulty for questions. For example, they all seemed to have difficulty in answering test items No. 5, 6, 7 and 8 correctly, found No.2, 3, 11 and 14 more easier.

There were some particularities. The Chinese public had the lowest level of understanding to most of the test items (No.1, 3, 5, 6, 7, 8, 9, 12, 13, 15, and 16). But the level of understanding was the highest in the four countries for No.2, and, second, for No.10 and 11, as shown in Fig.1[4].

Note 1: 1-The earth's core temperature is very high;2-The earth turns around the sun;3-The oxygen we breathe comes from plants;4-Father's genes determines children's sex;5-Laser is not produced by accumulation of sound waves;6-Electrons are smaller than atoms; 7-Antibiotics cannot kill a virus; 8-The universe comes from a Big-Bang;9-The continent we live on has been drifting slowly for millions of years and will continue the process;10- According to our current knowledge, human beings evolved from early animals;11-Smoking will lead to lung cancer;12-The earliest human beings and dinosaurs did not live in the same era;13-Boiled milk that contains radioactive material remains harmful to human;14-The speed of light is faster than the speed of sound;15-Radioactive phenomena are not always caused artificially;16-The earth turns around the sun in an one-year cycle.

---

[2]More details about the results of the investigation in 1992 can refer to [17].

[3]More details about the results of the investigation in 2003 can refer to [3] or [7].

[4] Source: [3, p.7](Originally: Science and Engineering Indicators 2002, Volume l, National Science Board, 2002, NSB02-1, US Government Printing Office, Washington, DC 20402; The 2001 Survey for Public Attitudes Towards and Understanding of Science & Technology in Japan, December 2001, NISTEP RRPORT No.72, Shinji OKAMTO, Fujio NIWA, Kenya SHIMIZU, Toshio SUGIMAN, National Institute of Science and Technology Policy; EUROBAROMETER55.2, Europeans, Science and Technology, December 2001, the European Opinion Research Group EEIG, European Coordination Office.)

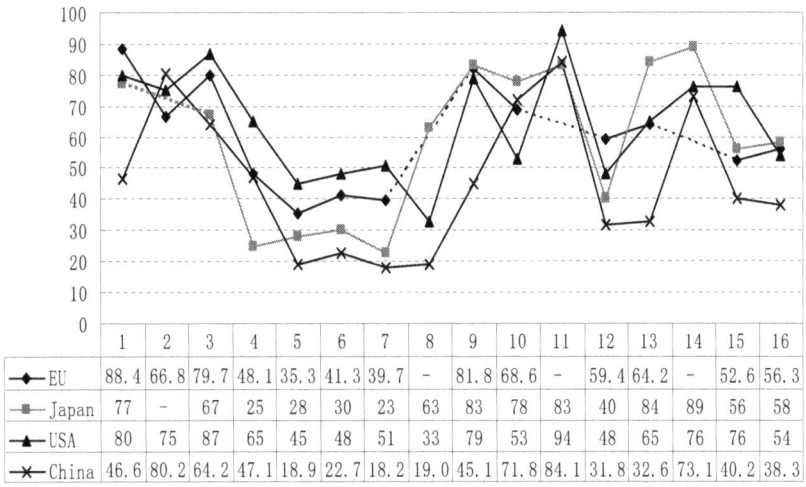

Figure 1: the international comparison of understanding of scientific concepts (%)

Note 2: *2003 Survey Report of China Public scientific literacy* points out: The three countries (referring to the European Union, Japan and the United States) and international organization member countries are still using Miller's test subjects. The test subjects used by China, United States and Japan are similar; however, the EU's are of a little different. There is no test data about "The universe comes from a Big-Bang", "Smoking will lead to lung cancer" and "The speed of light is faster than the speed of sound" for EU's test subjects. There is no data about "The earth turns around the sun" for subjects in Japan. The dotted lines reflect the missing data.

2.2 Understanding of scientific method

On understanding of scientific method, USA, EU and China all did better on "contrast experiment" than on "probability." Japanese subjects did not.

Turning to China separately, the proportion of public who could understand "contrast experiment" well was much lower than other countries, while the proportion correctly responding to questions on "probability" was close to Japan and lower than USA and the gap was not so distinct as "contrast experiment" (As shown in Fig.2[5]).

2.3 Understanding of the relationship between science and society

On this dimension, in fact, three countries measured the recognition the distinction between science and pseudo-science in similar ways. US, EU and Japan all took "astrology" as the test item, while according to the characteristics of the Chinese traditional culture, China took five prevailing fortune-telling methods as test items, such as QiuQian, physiognomy, fortune-telling by constellation, plates moving and pen drawing for an oracle and dream explanation by Zhou Gong. Owing to the different items and methods of each country, this dimension was incomparable.

---

[5] Source: [3, p.7]

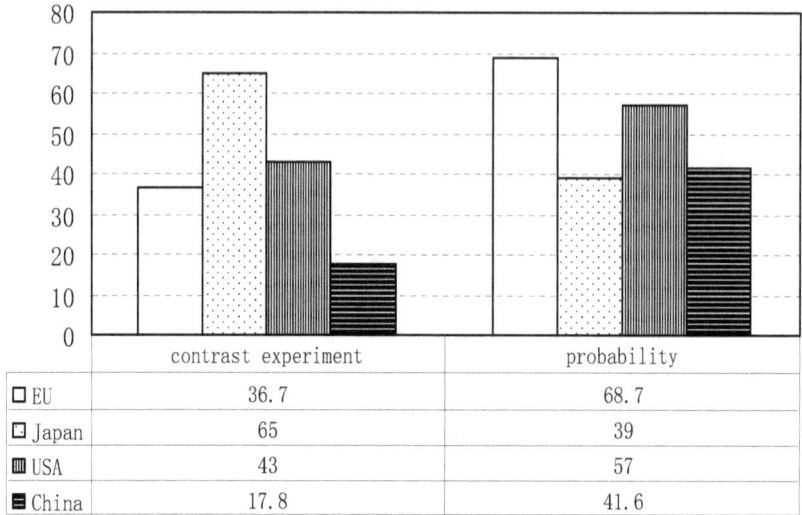

Figure 2: the international comparison of understanding of scientific method (%)

According to the result of the 2003 survey in China, the proportion of the public who believed in superstition was still not low. For instance, one out of five of the citizens strongly or slightly believed in "QiuQian", 1/4 believed in "physiognomy" and about 14% of them thought "fortune-telling by constellation" was reliable, 5% of respondents said "plates moving and pen drawing for an oracle" could be trusted, and 1/5 believed that "dreams could be explained by Zhou Gong" [3, p.17].

It could be seen that Chinese adults' scientific literacy far lagged behind the developed countries. But at the same time, as shown in Fig.3[6] below, Chinese adults' interest in science and technology information was not low, they even showed more interest than people in other countries: the total degree of interest was higher than EU's; the percentage of people who were interested in scientific discovery and new technology was higher than that of Japan and the USA, while the percentage of people who were interested in medical development was higher than Japan's and a little lower than America's.

There appears a strong contrast between the "high" interest in science and technology and "low" scientific literacy level of adults, which reflects the particularity of the public scientific literacy in China. We can't just simply use the word "low" to summarize the status of Chinese public scientific literacy level but have to analyze more deeply the particularity from the perspective of locality, which involves several factors such as, cultural characteristics, value system, the development level of economy and society, and so on.

---

[6]Source: [3, p.57]

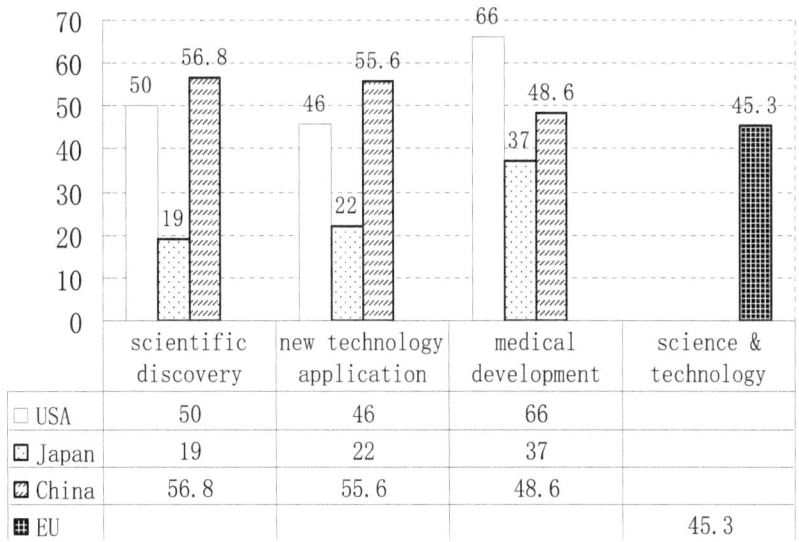

Figure 3: the international comparison of interest level of S&T information (%)

3 Living science: analysis on the particularity of Chinese public scientific literacy

Considering the locality of the public scientific literacy in China, and reviewing deeply the status of the scientific literacy, we find that the science which Chinese adults were interested in and able to understand has some typical traits. We use "living science" to summarize these traits as following.

3.1 Connecting closely with basic living demands

Living science first connects with people' basic necessities of life, which is the basic and principal demand of people' survival. According to Maslow Motivation Theory [9], it's also the basic condition of the existence and development of human society.

Combined with the comparison shown in Fig.3, we can see that the USA and Japanese citizens were much more interested in medical development than in scientific discovery and new technology, while Chinese adults' interest in medicine was very low. But in fact, the proportion of interest in health care information was 75%. Showing that the information that interested the Chinese public was about health in the sense of everyday life, but not the complicated progress or the frontier of contemporary medical science.

Also combined with the results of the survey in Fig.1, the Chinese public had better understanding of the concepts related to everyday life closely such as No.2, 3, 11 and 14, but had some difficulty in understanding those removed from daily life, such as No.5, 6, 7, 8. Especially on No.1, 13 and 15, the degree of understanding was much lower than in developed countries.

So it is obvious that what the Chinese public better understands connects with the basic demands of life, even though the basic demands will alter with the change of living conditions and social development.

3.2 Giving importance to accessibility and perception

Living science gives importance to perceptual cognition, which is usually based on accessibility and perception, and even intuition.

In the 2003 survey, about 12.5% of the respondents understood the scientific constructs, 30% understood scientific concepts, 8% understood scientific method, and 46.7% understood the relationship between science and technology. [3, p.18] The four data points reflect citizens' cognitive hierarchy for science: they could well understand the things just in terms of perceptional judgment, and had a little difficulty in understanding scientific constructs and concepts. Even on the understanding of scientific method, they owned much lower ability to understand "contrast experiment" than "probability" (as shown in Fig.2), which is meaningful and worthy of deliberation. Note:"0" means

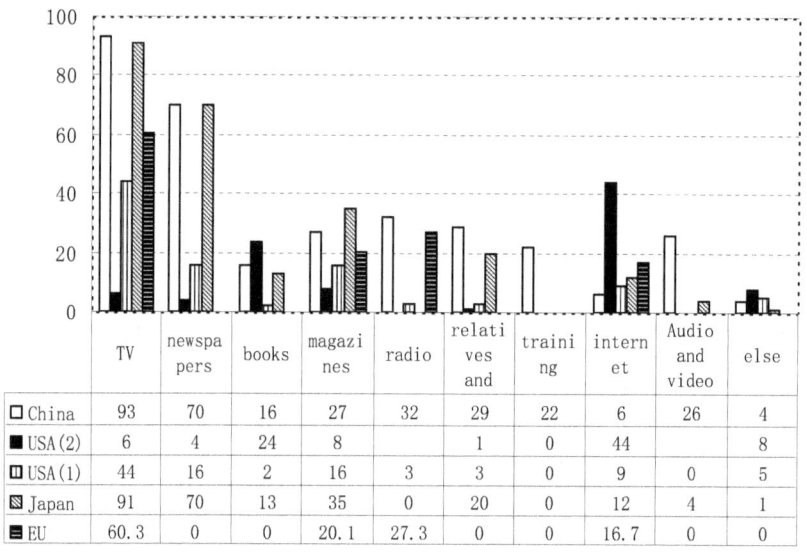

	TV	newspapers	books	magazines	radio	relatives and	training	internet	Audio and video	else
☐ China	93	70	16	27	32	29	22	6	26	4
■ USA (2)	6	4	24	8		1	0	44		8
◫ USA (1)	44	16	2	16	3	3	0	9	0	5
▨ Japan	91	70	13	35	0	20	0	12	4	1
▤ EU	60.3	0	0	20.1	27.3	0	0	16.7	0	0

Figure 4: the international comparison of sources of S&T information (%)

"none". There are multi-choice items in the surveys of China, Japan and the EU, so the sum is more than 100%. The items in USA are single-choice. The item "magazines" in the EU referred to science journals. USA (1) means the proportion of the sources the public that understood general S&T information; USA (2) means the proportion of the sources the public that understood special S&T information. Percentages may not sum up to 100% because "Don't know" responses are not shown (Originally: Science and Engineering Indicators 2002, Volume l, Chapter7–34).The items of sources of information in EU are: TV (60.3%), press (37%), radio (27.3%), school (22.3%), science journals (20.1%), internet (16.7%) (Originally: Europeans, science and technology, EUROBAROMETER 55.2, December 2001, p13). The items for Japan are unknown.

Of the sources for scientific information in China, TV programs took the first place and the proportion was much higher than that of EU and the USA. The sources that followed were newspapers, radio and relatives and friend; internet played an inessential role (as shown in Fig.4[7]). An important reason might be the low operational difficulty and easy acceptance of TV, newspapers, radio, relatives and friends as the sources of scientific information besides considering the development level of the country and other factors.

---
[7]Source: [3, p.34]

3.3 Integrating with social knowledge

Social knowledge means the knowledge involved in people's social living. The society as an assembly of individuals usually refers to the common critera and morality formed in the intercourse from day to day, so social knowledge may focus on the knowledge derived from people's daily experience and community life, such as economic, legal, mental, anthropic, political knowledge and etc. It may refer to social sciences or humanity but not to the natural sciences.

For instance, a person who doesn't know the natural mechanism of some medicine still can make a sensible choice to buy the right one just according to the "effect of tending to the same" in psychology. Or we can say, the public shouldn't get to the bottom of the knowledge or principle of something with respect to physical science, they can get to the purpose just through appealing to the knowledge or method of social sciences. In fact, this way is much more direct and effective in everyday life.

Take the "scientific popularization of mentality" for example; some communities in China supply necessary mental tutorship to the vulnerable groups such as the elders, single mothers or the unemployed to help them get back of confidence and overcome the problems [4]. This is a typical example of popularizing social knowledge to make the public feel better in the society.

Nowadays we usually hear "scientifically satisfy the higher education entrance examination" in China. The "scientifically" here doesn't only mean keeping to the procedures and criteria of selection, but also involves making choices after fully consulting several facts, such as the history of applying and matriculating, others' experiences, social reviews and so on. This is also the sign that the public pays attention to the social knowledge in the process of dealing with regular social matters.

3.4 Putting instrumental and practical result in the priority

Living science reveals that people usually pay more attention to the applicability, validity and the direct effects in resolving some matters.

Take the objective of the construction of Chinese public scientific literacy for example, which nowadays extends to "improving citizens' life quality and realizing the overall development" in nowadays in contrast to "improving the production and citizens' life" in earlier days. Though there is an upgrade, we can still trace the instrumental focus.

Meanwhile, the result of survey 2003 showed that Chinese citizens' were most interested in information about how to get rich, abiyt health care and about education, in turn. Obviously these kinds of information are useful and can bring visible, substantial results. The information not only connects with basic demands of life, but also links to instrumental and practical results.

Moreover, we can see "scientifically exercise", "scientifically preserve one's health", "scientifically bite and sup" everywhere in everyday life. The "scientifically" here refers to a useful method to get an actual effect in practice, which is equal to "effectively" and "reasonably". This is also the representation of the instrumental notion of living science in Chinese citizens' consciousness.

3.5 Connecting inherently with cultural tradition deposition

Generally speaking, culture is understood in three levels, material, institutional, and notional. Besides resulting in the material level of real-life, the Chinese traditional culture deeply influenced the forming of Chinese citizens' conception of "living science" from the institutional and notional dimensions, especially the latter.

The institutional and notional dimensions mainly point to the value system, manner of thinking and habits formed in people's long-term social practice and consciousness. In Chinese traditional cultural, humanity spirit was regarded as the soul, [16] which canonizes individual's culture, ethics and morality, social hierarchy and so on, but lacks scientific culture in the modern sense, such as rational criticism, strict logics, mathematical method, experimental means and so on. All of this affected the Chinese public' understanding of science, and we can see from the following aspects.

According to the understanding of scientific concepts in Fig.1, a high proportion of Chinese citizens had correctly answered test items No.2 and 10, which can be well linked to the traditional culture. The story that Confucius couldn't answer the problem of "Two children identified the sun" was well known in the traditional mainstream culture. Also, in recently modern culture, "sun" was regarded as liberation, brightness and source everything, and endowed with special political meanings which influenced each aspect of people's lives for several decades. So, it is quite reasonable for us to suppose the Chinese citizens gave more attention to questions such as "The earth turns around the sun". And linked to the second issue, as is well known, the mainstream of Chinese traditional culture is atheism, which obviously contrasts with the cultural tradition in the western world. In consequence, it is reasonable to think that Chinese find it to accept the theory of biological Darwinism rather than the doctrine of Creation. [14]

On understanding of scientific method, we can be aware of effects of traditional culture. In the traditional mode of thinking in China, the substantive thinking had been existence for a long time, especially in traditional Chinese medicine and health care from antiquity. Substantive thinking is different from visual thinking and abstract thinking. It is the mind's activity mediated by the embodied perception towards the image of things [8]. This manner of thinking plays a critical role in the forming of methodology, and which might result in Chinese citizens' better understanding of "contrast experiment" than "probability".

On understanding of the relationship between science and society, the effect is even more obvious. Witchcraft in the traditional culture and its prevalence in political history caused the diverse development of superstition in China, which is the direct reason why China set such items to measure citizens' third dimension of scientific literacy, as well as the reason why a high proportion of Chinese citizens showed belief in superstition.

4 Between living science, academic science and post-academic science

From the analysis above, we can see that living science is very different from academic science and post-academic science, and which invites the reflection on the relationship of academic science, post-academic science and living science and how to promote the public scientific literacy especially for later developer countries.

Britain scholar J.D. Bernal once said in his magnum opus "science in history": "science is (1) an institution (2)a method; (3)an accumulative knowledge tradition;(4)a main element to maintain and develop production; and (5)one of the strongest forces that composes our beliefs and the configuration of the universe and the human." [1] This generalization actually includes the two scientific institutions so-called nowadays – "academic science" and "post-academic science".

Academic science is "the stereotype of science in its purest form". In the academic science community, scientists "work together for the benefit of humanity in the attainment of truth" [18] and produce knowledge just based on their curiosity. Academic science is located in the core of the architectonic sciences. Those who engage this work are mostly the scientists or researchers in universities or research institutions; they keep far away from their personal interests, have enough self-determination, and obey an unwritten norm. The norm is summarized by R.K. Merton as the scientific ethos – universalism, communalism, disinterestedness, originality and organized skepticism, shorted for UCDOS.

John Ziman noticed the "undramatic revolution" in science since 1980s, and he regarded this new scientific and social institution as post-academic science. During the period of post-academic science, the relationship between science, society and economy became more and more complex, and the production of knowledge came to be tightly connected with the interests of the nation and industry, so post-academic science was also called industrial science. In this new institution, the behavioral norm of scientists has been changed, which is generalized by Ziman as – proprietary, local, authoritarian, commissioned, expert, shorted for PLACE.

Living science is regarded as forming from the process of requiring, understanding, acquiring and using knowledge based on the demand of people's common life. This knowledge is different from academic science or post-academic science, and is the perceptual, intuitionistic and useful but unsystematic common sense formed in people's ordinary life – empirical cognition. As mentioned forward, the characteristics of living science – connecting tightly with **basic living demands**; giving importance to **accessibility and perception**; integrating with **social knowledge**; putting **instrumental and practical result** in the priority; connecting inherently with **cultural tradition deposition**; which can be shortly as **BASIC**.

If we compare objectives of living science with those of academic science and post-academic science, we can consider that academic science corresponds to the cognitive world, independent from interest or benefit and just for learning (indicated as papers etc.). while post-academic science corresponds to the productive world, together with industry and economics, and points to for wealthy(indicated as GDP). It is the living science that is corresponding to the living world and from the consideration of utility and effect for quality of life and well-being(indicated as index of well-being).

In fact, it has become true that the special and formal "living science" has gradually integrated with formal science in everyday life. Now, we can see many press columns like "life and science" and various research centers for living science. Also, probably owing to the influence of Chinese traditional culture and the social status, the Taiwan National Open University founded the department of living science in 1993, which may indicate that living science is on the way to the educational system.[8]

From the doctrinal viewpoint, the UCDOS of academic science and the PLACE of post-academic science are both the generalizations of the images of science in different phases, but they don't point out the origin of science. The saying of Bernal following the sentence above gives us some inspiration; he pointed out "in all of the images listed above, the two images science as an institution and element of production belong to a contemporary stage, scientific method and its effect on beliefs at least appeared in Greece age. As for the knowledge tradition, it was handed down from parents to offspring and masters to prentices, which is the real root of science. The knowledge tradition existed in the earliest period of human history, long before science became an institution, or even before it got rid of common sense or legend to become a method." [1, p.6]

We can also say, far before the science became institutional, it was in the process of gestation, and one of its important sources is common sense, which is contained in living science. Here we have to notice that living science is not only a description of real status but also a basic source of science. In another word, living science relates to the actual and direct feeling and also involves the recognition of what science is.

Einstein once said that science is "nothing more than a refinement of everyday thinking", which means science comes from the criticism and upgrade of common sense.

---

[8] http://www.nou.edu.tw/~div05/department.htm, 2007-07-31

Also in this sense, Susan Hacck puts forward the concept of "critical common-sensism". She points out that essentially scientific evidence is similar to the evidence related to everyday empirical judgment; scientific inquiry is recognizably continuous with everyday empirical inquiry of the most familiar kind; and, science makes the procedure of ordinary inquiry intensified and refined. For example, auto mechanics, plumbers, cooks and scientists all make controlled experiments, but in many areas of science techniques of experimental control have been developed to a fine art. [5]

Relative to common sense, scientific knowledge is more systematic. Common sense is scattered and may be inconsistent from beginning to end, while scientific knowledge is inherently required to be systematic and logically consistent [15]. It's considered that the activities that remove the inconsistency and integrate the disorder of common sense drive the development of science, so science finishes the upgrade from common sense to sense. Or in another word, it's an effective approach to upgrade common sense to sense through scientific method such as reasoning, testifying and explaining etc., to distinguish, refine, remove falsehood and prereserve truths of common sense. At the same time, common sense is mutable and fallible.

In this respect, since a process of refinement from common sense to sense, as well as from living science to academic science and post-academic science. Combined with the construction of public scientific literacy, it's also a process of refinement from people's cognition towards science to understanding science and using science. For the later developed countries, it's very important to pay attention to this process, and it's a tentative and effective way to improve the public scientific literacy based upon the reality and follows the frontier, as well as combines with common sense.

Viewed from the current developed countries, the "2061 project" of American public scientific literacy construction stands on the scientific education of elementary and secondary school. "Benchmarks for scientific literacy" transfers the objectives of scientific literacy in "science for all Americans" to concrete study objectives, and directly influences the establishment of "national science education standards"; In Canada, the "Common Framework of Science Learning Outcomes K-12" also makes science education the main way to improve the public scientific literacy. Formal education as the main channel to the construction of public scientific literacy has been recognized all over the world, especially in developed countries which have put into practice widely.

But for later developed countries, while popularizing the systematic knowledge of academic science is the main channel through formal education, it's equally important to popularize "knowledge" of post-academic science as well as living science through informal education. Because for most citizens, what they most concerned is the practical application and consequences of knowledge related to daily demands, rather than the top technology or knowledge itself which is obscure and relatively far away from the ordinary lives of people. The "minimum science" in India contains the following essential components: (1) acquiring knowledge of certain scientific principles and facts; (2) internalized application of the method of science, and (3) acquiring the ability to continue to learn. [12] They still realize that resolving five kinds of problems is related to the public's lives and work, such as health issues, environment issues, measurement issues, agricultural science and technology and the technology used for the city and city population. Indeed, the Indian criteria, which insists each citizen has to acquire minimum and basic knowledge of science and technology as well as be familiar with or understand the operation and practice of scientific method. All of these are included in "living science".

As a late developing country with a centuries-old culture, for the construction of Chinese public scientific literacy we should recognize the importance of "academic science", but also focus on "post-academic science" as well as "living science" linked closely with the productive world and living world to conduct the activities of science popularization in order to direct the public understanding and to use science fully.

Bibliography

[1] Bernal, J. D. *Science in history, Trans. by Wu, Kuangpu.* Beijing: Science Press, 1983.

[2] Chinese Science and Technology Association, Chinese Public Scientific Literacy Investigation Panel. *2001 Survey Report of China Public scientific literacy.* Beijing: Science Popularization Press, 2002.

[3] Chinese Science and Technology Association, Chinese Public Scientific Literacy Investigation Panel. *2003 Survey Report of China Public scientific literacy.* Beijing: Science Popularization Press, 2004.

[4] Gao, J., Zeng, g. Science popularization close to living: investigation to the practice of social science popularization in Beijing City. In *S&T communication and the Construction of Public Scientific Literacy: Theory and Practice*, 165–190. Beijing, 2007.

[5] Hacck, S. *Defending Science: Within Reason – between Scientism and Cynicism.* New York: Prometheus Books, 2003.

[6] Li, D. What's the purpose of improving citizens' scientific literacy? *Democracy & Science*, 3:22–26, 2006.

[7] Li, D., He, W. Public understanding of science and attitudes towards science & technology in china. In *Achieving Public Scientific Literacy: Theory and Practice*, 64–76. Changsha: Hunan Science and Technology Press, 2006.

[8] Liu, T. Substantial thinking is the basic thinking pattern of chinese medicine. *Chinese Medicine and Basic Medical Journal*, 1:33, 1995.

[9] Maslow, A. H. A theory of human motivation. *Psychological Review*, 50:370–396, 1943.

[10] Miller, J. D. Scientific literacy: A conceptual and empirical review. *Daedalus*, 112:29–48, 1983.

[11] Miller, J. D. The measurement of civic scientific literacy. *Pubic Understanding of Science*, 7:203–223, 1998.

[12] Sehgal, N. K. Scientific literacy: Minimum science for everyone. In *Achieving Public Scientific Literacy: Theory and Practice.* The Office of the leading Group of National Scheme for Scientific Literacy Making, Changsha: Hunan Science and Technology Press, 2004.

[13] Wu, T. e. *The historical experiences and lessons of citizens' science literacy building and development in China.* Beijing: Science Popularization Press, 2005. The Office of the leading Group of National Scheme for Scientific Literacy Making.

[14] Zeng, G., Tan, X. Two puzzles of late developer countries entering globalization of science – the case of china. *Democracy & Science*, 4:6, 2006.

[15] Zeng, G. e. (Ed.). *The Tutorial of the Dialectics of Nature in Contemporary Era[.* Beijing: Tsinghua University Press, 2004.

[16] Zhang, K. talks about chinese traditional culture. 2007. URL http://www.xawb.com/gb/news/2007-07/30/content_1265842.htm. 2007-07-31.

[17] Zhang, Z., Zhang, J. A survey of public scientific literacy in china. *Public Understanding of Science*, 2:21–38, 1993.

[18] Ziman, J. *Real Science: What it is, and what it means.* London: Cambridge University Press, 2000.

Special Symposium on
COSMOLOGY

Observational Selection Effects in Quantum Cosmology[1]

Don N. Page

University of Alberta

don@phys.ualberta.ca

ABSTRACT. Scientific theories need to be testable by observations, say using Bayes' theorem. A complete theory needs at least the three parts of dynamical laws for specified physical variables, the correct solution of the dynamical laws (boundary conditions), and the connection with observations or experience or conscious perceptions (laws of psycho-physical parallelism). Principles are proposed for Bayesian meta-theories. One framework that obeys these principles is Sensible Quantum Mechanics (SQM), which is discussed. In principle, it allows one to test between single-history and many-worlds theories, and to discuss threats to certain theories from fake universes and Boltzmann brains. The threat of fake universes may be dismissed if one doubts the substrate-independence of consciousness, which seems very implausible in the SQM framework. Boltzmann brains seem more problematic, though there are many conceivable solutions. SQM also suggests the possibility that past steps along our evolutionary ancestry may be so rare that they have occurred nowhere else within the part of the universe that we can observe.

1 Goals and criteria for scientific theories

I see science as having the following goals:

1. Theories to explain observations
2. Observation-weighted probabilities for these theories
3. Predictions for these theories
4. Understanding of these theories

A complete theory or model of the universe needs at least three parts:

1. Complete set of physical variables (e.g., the arguments of a wavefunction or quantum state) and dynamical laws (e.g., the Schrödinger equation, or the action for a path integral)
2. The correct solution of the dynamical laws (e.g., the wavefunction or quantum state of the universe)
3. The connection with observation or experience (e.g., the laws of psycho-physical parallelism)

Item 1 alone is called a TOE or 'theory of everything,' but it is not complete. Even 1 and 2 alone are not complete, since they by themselves do not logically determine what, if any, conscious experiences occur in a universe.

One can do a Bayesian analysis for the epistemic probabilities of theories, which include the following elements:

---

[1] The author is with Institute for Theoretical Physics, Department of Physics, University of Alberta, Room 238 CEB, 11322 – 89 Avenue, Edmonton, Alberta, Canada T6G 2G7.

1. Prior probabilities $p_i \equiv p(T_i)$ for theories T_i
    - Necessarily subjective (in my view)
    - Perhaps favoring simplicity, e.g., $p_i = 2^{-n(T_i)}$, where $n(T_i)$ is an ordering of the theories in increasing order of complexity
    - Simplicity itself seems subjective
2. Conditional probabilities $L_{ij} \equiv P(O_j|T_i)$ of observations O_j given theories T_i (the likelihoods of the theories given the observations)
3. Posterior probabilities

$$P_{ij} \equiv P(T_i|O_j) = \frac{P(T_i \& O_j)}{P(O_j)}$$
$$= \frac{p(T_i)P(O_j|T_i)}{\sum_k p(T_k)P(O_j|T_k)}$$
$$= \frac{p_i L_{ij}}{\sum_k p_k L_{kj}} \quad (1)$$

In theories giving many observations, it is controversial what the conditional probabilities $L_{ij} \equiv P(O_j|T_i)$ should be taken to be [18, 25]. Is L_{ij} the probability that O_j occurs somewhere in theory T_i [18]? In classical mechanics, this would be $L_{ij} = 1$ if O_j occurs and $L_{ij} = 0$ if not. In conventional quantum mechanics, this would be $L_{ij} = \langle \psi_i | \mathbf{P}_{O_j} | \psi_i \rangle$, where $|\psi_i\rangle$ is the quantum state given by the theory T_1 and $\mathbf{P}_{O_j}$ is the projection operator onto the observation O_j.

This proposal for L_{ij} would not give observational distinctions between different theories all giving $L_{ij} = 1$. If T_i gives a universe large enough, many O_j's would give $L_{ij} \approx 1$. Then $P_{ij} \equiv P(T_i|O_j)$ would be highest for theories with the highest $p_i \equiv p(T_i)$, e.g., for the simple theory that all O_j's certainly exist, with essentially no influence from observations.

But if some observations occur more in some theory, surely they should be assigned higher conditional probabilities in that theory. E.g., suppose T_1 predicts 1 observation of O_1 and 10^6 observations of O_2, whereas T_1 predicts 10^6 observations of O_1 and 1 observation of O_2. Then surely $L_{11} < L_{21}$. To accomplish this [25], one might restrict to theories T_i giving

$$\sum_j L_{ij} \equiv \sum_j P(O_j|T_i) = 1, \quad (2)$$

so the total probability of all observations is 1 for each theory T_1.

One might expand this rule to a set of principles [25] for Bayesian meta-theories:

1. *Prior Alternatives Principle*: The set of alternatives to be assigned conditional probabilities by theories should be chosen logically prior to the observation to be used to test the theories.
2. *Normalization Principle*: The sum of the conditional probabilities each theory assigns to all of the alternatives in the chosen set should be unity.
3. *Conscious Anthropic Principle*: The alternatives ideally should be chosen to be conscious perceptions or observer-moments.

2 Sensible Quantum Mechanics

Conventional quantum mechanics does not seem to obey these principles [18], but *Sensible Quantum Mechanics* or *Mindless Sensationalism* [24, 26, 28] does. In it, there is a *quantum world* (what one might call the physical world, excluding consciousness) with quantum amplitudes and expectation values, but no probabilities, and there is a *conscious world* (what one might call the mental world) with frequency-type measures or statistical probabilities. Interpreting the unconscious quantum world itself probabilistically is taken to be *probabilism*, an *aesthemamorphic myth*.

The axioms of Sensible Quantum Mechanics are the following:

1. *Quantum World Axiom*: The unconscious "quantum world" Q is completely described by an appropriate algebra of operators and by a suitable state σ (a positive linear functional of the operators) giving the expectation value $\langle A \rangle \equiv \sigma[A]$ of each operator A.
2. *Conscious World Axiom*: The "conscious world" M, the set of all conscious perceptions, has a fundamental measure $\mu(S)$ for each subset S of M.
3. *Quantum-Consciousness Connection*: The measure $\mu(S)$ for each set S of conscious perceptions is given by the expectation value of a corresponding "awareness operator" $B(S)$, a positive-operator-valued (POV) measure, in the state σ of the quantum world:

$$\mu(S) = \langle B(S) \rangle \equiv \sigma[B(S)]. \qquad (3)$$

If for simplicity we suppose that there is a discrete set of quantum operators and a discrete set of conscious perceptions or observations O_j, then each Sensible Quantum Mechanics (SQM) theory T_i has

1. Set of operators $A_j^{(i)}$ and algebra $A_j^{(i)} A_k^{(i)} = \sum_l c_{jk}^{(i)\ l} A_l^{(i)}$
2. Quantum state σ_1 giving $\langle A_j^{(i)} \rangle = \sigma_i[A_j^{(i)}]$.
3. Normalized positive operators $B_j^{(i)}$ for observations O_j,

$$\sum_j \sigma_i[B_j^{(i)}] = 1.$$

4. $L_{ij} \equiv P(O_j|T_i) = \sigma_i[B_j^{(i)}]$ = normalized measure of O_j in T_i.

Then if one assigns epistemic prior probabilities p_i for T_i and has the observation O_j, the epistemic posterior probability for T_i is, by Bayes' theorem,

$$P_{ij} \equiv P(T_i|O_j) = \frac{p_i \sigma_i[B_j^{(i)}]}{\sum_k p_k \sigma_k[B_j^{(k)}]}. \qquad (4)$$

The principles of Bayesian meta-theories proposed here and exemplified by Sensible Quantum Mechanics may be related to various other principles that have been advocated:

1. *Copernican Principle*: We are not specially privileged.
2. *Weak Anthropic Principle*: What we observe is conditioned upon our existence as observers.
3. *Principle of Mediocrity*: We are a "typical" civilization [34].
4. *Strong Self-Sampling Assumption (SSSA)*: "One should reason as if one's present observer-moment were a random sample from the set of all observer-moments in its reference class" [6].
5. *Conditional Aesthemic Principle (CAP)*: "Unless one has compelling contrary evidence, one should reason as if one's conscious perception were a random sample from the set of all conscious perceptions" [24–26, 28].

3 Application to Single-History versus Many-Worlds Quantum Theory

One can apply these Bayesian meta-theory principles in general, and the framework of Sensible Quantum Mechanics in particular, to various issues. For example, in principle (if the quantum state of the universe is suitable) one can make statistical tests to distinguish between single-universe and many-worlds versions of quantum theory [2, 22, 27].

Consider a set of quantum theories giving 99.9% probability for a small universe with $N_1 = 1$ civilization seeing O_1, say with negative cosmological constant $\Lambda < 0$, and 0.1% probability for a large universe, $N_2 = 10^6$ civilizations seeing O_2, say with positive cosmological constant $\Lambda > 0$. Let T_1 be a single-history theory with just a small universe, T_2 be a single-history theory with just a large universe, and T_3 a many-worlds theory with both a small and a large universe. Say the respective prior probabilities for these three theories are p_1, p_2, and p_3, with $p_1 + p_2 = 0.9$ and $p_3 = 0.1$, so that initially one gives only 10% probability for many-worlds. Since the small universe is supposed to have 999 times the probability of a large universe, for the single-universe theories say $p_1 = 0.999(p_1 + p_2) = 0.8991$ and $p_2 = 0.001(p_1 + p_2) = 0.009$.

In T_1, O_1 is certain and O_2 is impossible, so

$$L_{11} \equiv P(O_1|T_1) = 1, \ L_{12} \equiv P(O_2|T_1) = 0. \tag{5}$$

Conversely, in T_2, O_1 is impossible and O_2 is certain, so

$$L_{21} \equiv P(O_1|T_2) = 0, \ L_{22} \equiv P(O_2|T_2) = 1. \tag{6}$$

On the other hand, in T_3, observations are made of both a small universe and a large universe, so if we take the probabilities of the observations to be weighted by the number of civilizations making them, we get

$$\begin{aligned}
L_{31} &\equiv P(O_1|T_3) \\
&= \frac{N_1(0.999)}{N_1(0.999) + N_2(0.001)} \\
&= \frac{0.999}{1000.999} \approx 0.001, \\
L_{32} &\equiv P(O_2|T_3) \\
&= \frac{N_1(0.001)}{N_1(0.999) + N_2(0.001)} \\
&= \frac{1000}{1000.999} \approx 0.999.
\end{aligned} \tag{7}$$

Suppose then we see O_2 (large universe, $\Lambda > 0$). Then applying Bayes' theorem gives the posterior probabilities of the three theories as

$$\begin{aligned}
P(T_1|O_2) &= \frac{p_1 L_{12}}{p_1 L_{12} + p_2 L_{22} + p_3 L_{32}} = 0, \\
P(T_2|O_2) &= \frac{p_2 L_{22}}{p_1 L_{12} + p_2 L_{22} + p_3 L_{32}} \approx 0.009, \\
P(T_3|O_2) &= \frac{p_3 L_{32}}{p_1 L_{12} + p_2 L_{22} + p_3 L_{32}} \approx 0.991.
\end{aligned} \tag{8}$$

Thus, in this hypothetical case, even though initially one gave the many-worlds theory only a 10% prior probability of being true, after the observation one would then give it more than 99% posterior probability of being true. Therefore, in principle one can gain observational evidence of whether or not the many-worlds version of quantum theory is correct; it is not just an equivalent interpretation of a single quantum theory.

4 Application to Fake Universes

Another application is to fake universes. Theories with posthuman civilizations seem in danger of producing too many fake universes [7]. Taking substrate-independence of consciousness as given, Nick Bostrom argues, "... *at least one* of the following propositions is true:

1. the humans species is very likely to go extinct before reaching a 'posthuman' stage;
2. any posthuman civilization is extremely unlikely to run a significant number of simulations of their evolutionary history (or variations thereof);
3. we are almost certainly living in a computer simulation."

Paul Davies [12] draws the following conclusion from this argument: "For every original world, there will be a stupendous number of available virtual worlds—some of which would even include machines simulating virtual worlds of their own, and so on ad infinitum." John Barrow [4] makes the further point, "So we suggest that, if we live in a simulated reality, we should expect occasional sudden glitches, small drifts in the supposed constants and laws of nature over time, and a dawning realization that the flaws of nature are as important as the laws of nature for our understanding of true reality." If much of posthuman computer simulation is done by the analogue of today's teenagers, I myself would expect even more chaos in simulations.

Since we have not observed these postulated sudden glitches, drifts, and chaos, we might seek explanations of why they are not seen, incantations to ward off fake universes. Bostrom himself [7] noted that humans may go extinct before becoming posthuman, or that posthumans may be unlikely to run simulations. But Davies [12] is arguing that these explanations would not ward off fake universes if there is a multiverse, which he takes as a *reductio ad absurdum* suggesting no multiverse.

In my mind, the weakest assumption leading to fake universes is the assumption [7] of the substrate-independence of consciousness. If that is not correct, then it could well be that most simulated beings simply are not conscious (and so, as conscious beings, we are not likely to be simulated). One might analyze this question within the framework of Sensible Quantum Mechanics, which postulates that there is a precise positive operator B_j for each observation or conscious perception O_j. Substrate-independence seems vague to me but might be taken to mean that B_j is invariant under some set of transformations (perhaps unitary), such as the assumption that $B_j = \hat{B}_j = U B_j \tilde{U}$ for some set of $(U, \tilde{U})$ pairs. But why should this be true? It seems highly implausible to me.

I am reminded of the South Pacific cargo cults, which arose out of observations during World War II that after airfields and conning towers were built on their islands, aircraft landed with cargo. Thinking that the airfields and conning towers were the sufficient conditions for such cargo to arrive, they themselves built other airfields and conning towers, but the expected cargo did not arrive. It seems to me that we are at a similar primitive stage concerning what conditions are sufficient for consciousness (e.g., what the operators B_j are in Sensible Quantum Mechanics), so guesses that they are substrate independent are very likely to be quite wrong.

5 Application to Boltzmann Brains

Another application is to Boltzmann brains. Theories in which spacetime can last too long seem in danger of producing too many Boltzmann brains (BBs) with mostly disordered observations, which would make our ordered observations highly atypical [20, 21, 23].

The germ of the idea goes back to Boltzmann [5], who said that his "old assistant" Dr. Schuetz had suggested that our observed universe might be a giant thermal fluctuation. Although this would be extremely improbable for any particular large region of the universe, it would certainly occur somewhere if the universe were large enough.

However, more recently it has been noticed [1, 30] that it would be much less improbable for a particular region of the universe to have just a small fluctuation give our observations, rather than the entire observable region of the universe. For example, one might postulate that a human brain arose from a thermal fluctuation in a state in which it had the same conscious awareness *as if* it had been making observations of a large region of the universe, even if the large region were *actually* not at all what the conscious awareness thought it was. (E.g., the conscious perception of the brain might be that observations had been made of distant stars and galaxies, whereas actually the surrounding universe might be empty, without any stars and galaxies, but just having the brain in the same state it would have been if there had been stars and galaxies it had observed.)

For a brain of the rough mass of a human brain of one kilogram to arise from a thermal fluctuation in empty deSitter space at its Gibbons-Hawking [16] temperature $T_{\rm dS}$, the expected number per 4-volume (3-volume of space multiplied by the time interval in spacetime) for such long-lasting brains or 'long brains' (lbs) would be

$$\mathcal{N}_{\rm lb} \sim e^{-(1\ {\rm kg})c^2/T_{\rm dS}} \sim e^{-10^{69}}. \tag{9}$$

This is very tiny but would apparently be important if the universe lasted long enough to make the total 4-volume large in comparison with $e^{10^{69}}$ times the 4-volume of the region of ordinary observers (OOs).

I realized [20, 21, 23] that the rate would be relatively much larger if the brains were not required to come into long-lasting existence by a thermal fluctuation but were just required to exist momentarily as vacuum fluctuations ('brief brains' or bbs). Then if the brain were in the right state or configuration, it could have a brief conscious perception and then disappear into the vacuum again. The minimum requirement presumably would be roughly that the matter of the brain be separated from the corresponding amount of antimatter that would appear during the virtual loops of the Feynman diagrams of the vacuum fluctuation.

For a one-kilogram human brain to become separated by its size of say 30 centimeters from the antimatter would require an action of $(1\ {\rm kg})c(0.3\ {\rm m})/\hbar \sim 10^{42}$, so the expected number of brief brains per 4-volume would be of the rough order of

$$\mathcal{N}_{\rm bb} \sim e^{-(1\ {\rm kg})c(0.3\ {\rm m})/\hbar} \sim e^{-10^{42}}. \tag{10}$$

Although this again is extremely tiny, so that the expected number of these vacuum fluctuation Boltzmann brains or brief brains (bb BBs) would be much less than unity over the whole past history of the part of the universe we can now observe (the 'observable universe'), it would dominate over ordinary observers if the universe lasted long enough to make the total 4-volume much larger than $e^{10^{42}}$ times the 4-volume of the region of ordinary observers, which would apparently be the case if the universe expanded forever.

If some theory predicted that Boltzmann brains greatly dominated over ordinary observers, we would expect that in that theory we would most likely be BBs rather than OOs. Only a very tiny fraction of BBs would give consciousness (though say still much larger than the number of OOs), but since we must be conscious to observe anything, this condition is a necessary selection effect, so that if such conscious BBs and/or OOs exist, it would not be at all improbable that we are a conscious BB or OO (and most likely a BB if there are far more conscious experiences produced by them, say in a universe that lasts much longer and grows much bigger than the region where OOs can exist).

However, once we include the necessary selection effect that we are conscious, the conditional probabilities for the various possible natures of the conscious perceptions should be given by the theory. If BBs enormously dominate over OOs, we should expect that we are BBs rather than OOs (or rather you should, since then most probably I am just a figment of your imagination). But only a very tiny fraction of BB conscious perceptions would be expected to be so ordered as yours generally are, with coherent visual images and memories, so the fact that yours are highly ordered counts as strong statistical evidence against any theory that implies that almost all conscious perceptions are produced by BBs (since they would produce mostly disordered observations).

If one includes more details of the order that you observe, then the likelihood that your observation comes from a theory in which BBs dominate is even lower. For example, even among the tiny fraction of BB conscious perceptions that are ordered, only a tiny fraction would give ordered observations of, say, galaxy-galaxy correlations. And of these, only a tiny fraction would give a galaxy-galaxy correlation similar to what you observe or may be aware of from what you think are the observations of others. So if you put in all the detail of the order that you observe or are aware of, it would be very unlikely that it is a conscious perception produced in a theory in which Boltzmann brains greatly dominate.

Although the next step would be going beyond what could be observationally checked, one might note that even among the extremely tiny fraction of BB conscious perceptions that included all the details of the order that you perceive, only a very tiny fraction of these would be non-illusory, say of actual galaxies, since there would be far more BBs in the entire huge spacetime to have brains in just the right states to have illusory perceptions that they have observed galaxies with all the correlations that you observe or are aware of, than for all the 'observed' galaxies actually to exist.

In conclusion, theories in which Boltzmann brains greatly dominate over ordinary observers would make your or my observations highly atypical. Some might argue [18] that this atypicality does not imply a correspondingly small likelihood for these theories, but I would respond [25] that it should.

Let me give some numerical estimates for the conditions that Boltzmann brains might dominate ordinary observers in our universe. Basically, a sufficient condition would seem to be that the expectation value of the total 4-volume of our comoving region of the universe is infinite when one includes the future.

After the period of radiation domination ended when the linear size of a comoving region of our universe was thousands of times smaller than today, our observed comoving region (that given by fixed markers such as galaxies that move apart with the expansion of the universe) is apparently dominated by a cosmological constant Λ and by low-velocity massive particles with negligible pressure (called 'dust' in cosmology), a cold-dark-matter-Λ model or CDMΛ model. Its spacetime geometry is apparently well described by the $k = 0$ (spatially flat) Friedmann-Robertson-Walker metric with unit-jerk ($J \equiv a^2(d^3a/dt^3)/(da/dt)^3 = 1$, where $a(t)$ is the linear size of a fixed comoving region as a function of the time t),

$$ds^2 = T^2[-d\tau^2 + (\sinh^{4/3}\tau)(dr^4 + r^2 d\theta^2 + r^2 \sin^2\theta d\varphi^2)]. \qquad (11)$$

The Hubble Space Telescope key project [15] gives that the present Hubble constant of our universe, today's value of $H \equiv d\ln a/dt$, is $H_0 = 72 \pm 8$ km/s/Mpc. From the third-year results of the Wilkinson Microwave Anisotropy Probe (WMAP), the fraction of the critical energy density given by the cosmological constant is $\Omega_\Lambda = 0.72 \pm 0.04$. Putting these numbers together [20, 21] gives that the asymptotic value of the Hubble expansion rate in the distant future is

$$H_\Lambda = \sqrt{\Lambda/3} = H_0\sqrt{\Omega_\Lambda} \approx (16 \pm 2 \text{ Gyr})^{-1}, \qquad (12)$$

the value of the constant parameter T in the metric above is

$$T = (2/3)H_\Lambda^{-1} = 11 \pm 1.5 \text{ Gyr}, \qquad (13)$$

the present value of the dimensionless time coordinate τ in the metric above is

$$\tau_0 = \tanh^{-1}\sqrt{\Omega_\Lambda} \approx 1.25 \pm 0.09, \qquad (14)$$

and the present value of the Hubble constant, H_0, multiplied by the present value of the age, t_0, or the present value of $d\ln a/d\ln t$, is

$$H_0 t_0 = \frac{2}{3}\frac{\tanh^{-1}\sqrt{\Omega_\Lambda}}{\sqrt{\Omega_\Lambda}} \approx 0.99 \pm 0.04. \qquad (15)$$

It is interesting that this last value is consistent with unity, up to a fraction of the current observational uncertainties. This means that when one plots the linear size a versus the time t since the beginning in this model (applicable long after very early inflation, and also after radiation domination ended), the $a(t)$ curve bends downward during the first part of the history (deceleration, when dark matter dominated and made gravity attractive) and then bends upward during the latter part of the past history (acceleration, when the cosmological constant or dark energy dominated and made gravity repulsive on cosmological scales) in just such a way that if one draws a straight line tangent to the $a(t)$ curve at the present time t_0, it will have slope da/dt that is very nearly the same as a/t and hence will very nearly go through the origin. To put it another way, not counting the very early inflationary period, the value of linear size over age, a/t, is, within the current observational uncertainties, minimized at the present time. This coincidence, along with the observed spatial flatness, plus the fact that the current best estimate [31] of the age 13.7 Gyr (Gigayears or billions of years) has the same first three digits as the reciprocal of the fine structure constant, $1/\alpha \approx 137.036$, allows one to write the metric and parameters above in an easy-to-remember form that I might call the *mnemonic universe*: CDMΛ $k = 0$ FRW $t_0 = H_0^{-1} = 10^8$ yr/α.

If this expanding universe lasts forever, then for any fixed comoving volume (e.g., the physically expanding volume occupied by some fixed set of galaxies and whatever massive particles they may eventually decay into, say electrons and positrons) that presumably contains only a finite number of ordinary observers (say if they cannot exist after the stars burn out in a finite time), one will have an infinite amount of 4-volume in the infinite future, and hence apparently an infinite number of Boltzmann brains, infinitely dominating over the presumably finite number of ordinary observers. One might however postulate that the universe will not expand forever but perhaps decay to some entirely different configuration obeying quite different effective laws of physics (say not giving any conscious brains at all) by the quantum formation of a bubble that then expands at the speed of light and destroys all that it engulfs. If the bubble nucleation rate per 4-volume is A (for 'annihilation,' which is what the bubble would do to the present form and laws of physics of the universe), then the expectation value of the total 4-volume per fixed comoving 3-volume would be finite only if the bubble formation rate is greater than some calculable minimum [20, 21]:

$$A > A_{\min} = \frac{9}{4\pi}H_\Lambda^4 = \frac{\Lambda^4}{4\pi} \approx (18 \pm 2 \text{ Gyr})^{-4} \sim e^{-563}, \qquad (16)$$

where the last number is in Planck units, setting $\hbar = c = G = 1$.

For a given value of A, the survival probability of the CDMΛ $k = 0$ FRW model above is

$$P(t) = \exp\left[-\frac{16}{27}\frac{A}{A_{\min}}\int_0^{\frac{3}{2}H_\Lambda t} dx \sinh^2 x \left(\int_x^{\frac{3}{2}H_\Lambda t}\frac{dy}{\sinh^{2/3} y}\right)^3\right]. \qquad (17)$$

Given that we have lasted until today, the probability per year for being annihilated by a bubble that has formed just outside our past light cone (and hence which will engulf us within the coming year, though we cannot see any signal of its coming since it would be coming at essentially the speed of light) would be greater than one part in one hundred billion. With the present earth population of nearly 7 billion, this would give an minimal expected death rate of about 7 persons per century. (Of course, it could not be 7 persons in one century, but all 7 billion with a probability of about one in a billion per century.) It also gives an upper limit on the present half-life of our universe of 19 Gyr [21].

Since we do not see that our observations are so disordered as would be statistically predicted if Boltzmann brains greatly dominated over ordinary observers, we might seek an explanation of why they apparently do not dominate, incantations to ward off Boltzmann brains:

1. The universe may be decaying classically if the quintessence potential slowly slides negative, so that the universe rolls into oblivion [29].
2. The universe may be decaying quantum mechanically with $e^{-563} \lesssim A \lesssim e^{-556}$, though this seems to require unnatural fine tuning [20, 21, 23].
3. Observers might conceivably fill the entire universe and be too large to form by vacuum fluctuations [20, 21, 23].
4. The local view restricts the 3-volume to the causal patch [8].
5. Use a regularization effectively cutting off before the Boltzmann brains [19]
6. Boltzmann brains should be lumped with bubble formation [36].
7. Holographic cosmology gives a time-dependent Hamiltonian [3].
8. 'Constants' of physics change so that the rate per 4-volume of Boltzmann brains asymptotically decays to zero [9].
9. Restrict to only certain group-averaged quantum states [17].
10. Quantum field theory may not apply to brains [20, 21].
11. We may not be typical [18].

Since an astronomical decay rate is suggested by one way of regularizing Boltzmann brains, and since it is not completely ruled out, perhaps we should take it seriously as one possibility (though not the only one). So let us see what are implications of the possible astronomical decay rate.

If ordinary observers could last until engulfed by a bubble:

1. They could not see it coming and so would not dread it.
2. They would be destroyed so fast they would feel no pain.
3. There would be no grieving survivors left behind.

So it would be the most humanely possible execution.

Furthermore, the universe will always persist in some decreasing fraction or measure of Everett worlds. Thus one could never absolutely rule out a decaying universe by observations at any finite time, though at sufficiently late times observations would be strong statistical evidence against the astronomical decay rate.

The main point of my discussing this scenario is not to advocate a particular solution to the Boltzmann brain problem, but rather to stimulate more research on the huge scientific mystery of the measure for the string/M landscape [32, 35] or other multiverse theory [10].

6 Application to Biological Evolution

Sensible Quantum Mechanics suggests we are effectively selected randomly by our measure of consciousness, like winners in a cosmic jackpot [13]. No further selection would be needed *within* a particular SQM theory.

This does not explain *which* SQM theory is correct. One might suppose it was chosen by a highly lawful and yet benevolent God (perhaps ultimately simple [33], say the greatest possible being, *contra* Dawkins' idea that God is complex [14]), who wanted to create understanding conscious beings in His image within a highly ordered and elegant multiverse.

Within Darwinian evolution on earth, our ancestors would not just be random life but conditionalized by being our ancestors. One or more correlated steps might be highly improbable on a per-planet basis [11] (say $P \sim 10^{-n}$ with $n \gg 24$ so that the probability would be very low for any others within the observable universe of probably not more than 10^{24} planets). Only steps off our ancestral line can be argued to be probable to occur on a random planet in which the steps up to that point have occurred. It would be interesting to see what features of complexity and intelligence developed off our ancestral line (e.g., to octopi from our last common ancestor), and what features have developed only within our ancestral line. Some of the latter features could be so rare that they have not occurred anywhere else within the observable universe (the part we can see), even though the entire universe may be so vast that all of these features have occurred many times very much further away than we can possibly see.

7 Conclusions

1. Scientific theories need to be testable by observations.
2. Multiverse theories can in principle be testable if they give probabilities for observations that depend upon the theory.
3. Sensible Quantum Mechanics (SQM) gives probabilities that are expectation values of positive operators associated with observations or conscious perceptions, and in principle they can be tested observationally.
4. One can test between single-history and many-worlds theories.
5. Fake universe and Boltzmann brains are threats to certain multiverse (and single-universe) theories.
6. Sensible Quantum Mechanics suggests that fake universes may not be a threat, since there is no compelling reason to believe in the substrate independence of consciousness.
7. There are many suggested solutions for Boltzmann brains, but it just is not clear what the correct one is.
8. No further life principle is needed *within* a Sensible Quantum Mechanics theory, but *which* Sensible Quantum Mechanics theory is correct needs explaining.
9. From the view of foresight, our biological ancestors may be highly improbable, post-selected by our existence as beings with a high measure of conscious perceptions.

Acknowledgments

I am indebted to discussions with Andreas Albrecht, Denis Alexander, John Barrow, Nick Bostrom, Raphael Bousso, Andrew Briggs, Peter Bussey, Bernard Carr, Sean Carroll, Brandon Carter, Kelly James Clark, Gerald Cleaver, Francis Collins, Robin Collins, Gary Colwell, William Lane Craig, Paul Davies, Richard Dawkins, William Dembski, David Deutsch, Michael Douglas, George Ellis, Debra Fisher, Charles Don Geilker, Gary Gibbons, J. Richard Gott, Thomas Greenlee, Alan Guth, James Hartle, Stephen Hawking, Rodney Holder, Richard Hudson, Chris Isham, Renata Kallosh, Denis Lamoureux, John Leslie, Andrei Linde, Robert Mann, Don Marolf, Greg Martin, Alister McGrath, Gerard Nienhuis, Gary Patterson, Alvin Plantinga, Chris Polachic, John Polkinghorne, Martin Rees, Hugh Ross, Peter Sarnak, Henry F. Schaefer III, James Sinclair, Lee Smolin, Mark Srednicki, Mel Stewart, Jonathan Strand, Leonard Susskind, Richard Swinburne, Max Tegmark, Donald Turner, Neil Turok, Bill Unruh, Alex Vilenkin, Steven Weinberg, and others whom I don't recall right now, on various aspects of this general issue, though the opinions expressed are my own. My scientific research on the quantum cosmology and the multiverse is supported in part by the Natural Sciences and Research Council of Canada.

Bibliography

[1] Albrecht, A., Sorbo, L. *Physical Review*, D70(063528), 2004. URL http://arxiv.org/abs/hep-th/0405270.

[2] Antia, M. *The Economist*, 145, 1999. May 22.

[3] Banks, T. Entropy and initial conditions in cosmology. URL http://arxiv.org/abs/hep-th/0701146. Eprint arXiv:hep-th/0701146.

[4] Barrow, J. D. Living in a simulated universe. In Carr, B. (Ed.), *Universe or Multiverse?*, 481–486. Cambridge: Cambridge University Press, 2007.

[5] Boltzmann, L. *Nature*, 51:413–415, 1895.

[6] Bostrom, N. *Anthropic Bias: Observation Selection Effects in Science and Philosophy*. New York: Routledge, 2002.

[7] Bostrom, N. Are you living in a computer simulation? *Philosophical Quarterly*, 53:243–255, 2003. URL http://www.simulation-argument.com.

[8] Bousso, R., Freivogel, B. A paradox in the global description of the multiverse. *Journal of High Energy Physics*, 0706(018), 2007. URL http://arxiv.org/abs/hep-th/0610132. Eprint arXiv:hep-th/0610132.

[9] Carlip, S. Transient observers and variable constants, or repelling the invasion of the boltzmann's brains. *Journal of Cosmological and Astroparticle Physics*, 0706(001), 2007. URL http://arxiv.org/abs/hep-th/0703115. Eprint arxiv:hep-th/0703115.

[10] Carr, B. (Ed.). *Universe or Multiverse?* Cambridge: Cambridge University Press, 2007.

[11] Carter, B. The anthropic principle and its implications for biological evolution. *Philosophical Transactions of the Royal Society of London*, A310:347–363, 1983.

[12] Davies, P. A brief history of the multiverse. *New York Times*, 2003. April 12.

[13] Davies, P. *Cosmic Jackpot: Why Our Universe Is Just Right for Life*. Boston: Houghton Mifflin, 2007.

[14] Dawkins, R. *The God Delusion*. Boston: Houghton Mifflin, 2006.

[15] Freedman, W. L., Madore, B. F., Gibson, B. K., Ferrarese, L., Kelson, D. D., Sakai, S., Mould, J. R., Kennicutt, R. C. Ford, H. C., Graham, J. A., Huchra, J. P., Hughes, S. M. G., Illingworth, G. D., Macri, L. M., Stetson, P. B. Final results from the hubble space telescope key project to measure the hubble constant. *Astrophysical Journal*, 553:47–72, 2001. URL http://arxiv.org/abs/astro-ph/0012376. Eprint arXiv:astro-ph/0012376.

[16] Gibbons, G. W., Hawking, S. W. *Physical Review*, D15:2738–2751, 1977.

[17] Giddings, S. B., Marolf, D. A global picture of quantum de sitter space. *Physical Review*, D76(064023), 2007. URL http://arxiv.org/abs/0705.1178. Eprint: arXiv:hep-th/0705.1178.

[18] Hartle, J. B., Srednicki, M. Are we typical? *Physical Review*, D75(123523), 2007. URL http://arxiv.org/abs/0704.2630. Eprint arxiv:0704.2630.

[19] Linde, A. Sinks in the landscape, boltzmann brains, and the cosmological constant problem. *Journal of Cosmological and Astroparticle Physics*, 0701(022), 2007. URL http://arxiv.org/abs/hep-th/0611043. Eprint arXiv:hep-th/0611043.

[20] Page, D. N. Is our universe decaying at an astronomical rate? URL http://arxiv.org/abs/hep-th/0612137. Eprint arXiv:hep-th/0612137.

[21] Page, D. N. Is our universe likely to decay within 20 billion years? URL http://arxiv.org/abs/hep-th/0610079. Eprint arXiv:hep-th/0610079.

[22] Page, D. N. Observational consequences of many-world quantum theory. URL http://arxiv.org/abs/quant-ph/9904004. Eprint arXiv:quant-ph/9904004.

[23] Page, D. N. Return of the boltzmann brains. URL http://arxiv.org/abs/hep-th/0611158. Eprint arXiv:hep-th/0611158.

[24] Page, D. N. Sensible quantum mechanics: Are only perceptions probabilistic? URL http://arxiv.org/abs/quant-ph/9506010. Eprint arXiv:quant-ph/9506010.

[25] Page, D. N. Typicality defended. URL http://arxiv.org/abs/0707.4169. Eprint arXiv:0707.4169.

[26] Page, D. N. Sensible quantum mechanics: Are probabilities only in the mind? *International Journal of Modern Physics*, D5:583–596, 1996. URL http://arxiv.org/abs/gr-qc/9507024. Eprint arXiv:gr-qc/9507024.

[27] Page, D. N. Can quantum cosmology give observational consequences of many-worlds quantum theory? In Burgess, C. P., Myers, R. C. (Eds.), *General Relativity and Relativistic Astrophysics*, 225–232. Eighth Canadian Conference, Montreal, Quebec, 1999, New York: American Institute of Physics, Melville, 1999. URL http://arxiv.org/abs/gr-qc/0001001. Eprint arXiv:gr-qc/0001001.

[28] Page, D. N. Mindless sensationalism: A quantum framework for consciousness. In Smith, Q., Jokic, A. (Eds.), *Consciousness: New Philosophical Perspectives*, 468–506. Oxford: Oxford University Press, 2003. URL http://arxiv.org/abs/quant-ph/0108039. Eprint arXiv:quant-ph/0108039.

[29] Page, D. N. The lifetime of the universe. *Journal of the Korean Physical Society*, 49:711–714, 2006. URL http://arxiv.org/abs/hep-th/0510003. Eprint arXiv:hep-th/0510003.

[30] Rees, M. J. *Before the Beginning: Our Universe and Others*. New York: Simon and Schuster, 1997.

[31] Spergel, D. N., Verde, L., Peiris, H. V., Komatsu, E.and Nolta, M. R., Bennett, C. L., Halpern, M., Hinshaw, G., Jarosik, N., Kogut, A., Limon, M., Meyer, S. S., Page, L., Tucker, G. S., Weiland, J. L., Wollack, E., Wright, E. L. First year wilkinson microwave anisotropy probe (wmap) observations: Determination of cosmological parameters. *Astrophysical Journal Supplement Series*, 148:175–194, 2003. URL http://arxiv.org/abs/astro-ph/0302209. Eprint arXiv:astro-ph/0302209.

[32] Susskind, L. *The Cosmic Landscape: String Theory and the Illusion of Intelligent Design*. New York: Little, Brown and Company, 2006.

[33] Swinburne, R. *The Existence of God*. Oxford: Clarendon Press, 2nd edn., 2004.

[34] Vilenkin, A. Predictions from quantum cosmology. *Physical Review Letters*, 74:846–849, 1995. URL http://arxiv.org/abs/gr-qc/9406010. Eprint arXiv:gr-qc/9406010.

[35] Vilenkin, A. *Many Worlds in One: The Search for Other Universes*. New York: Hill and Wang, 2006.

[36] Vilenkin, A. Freak observers and the measure of the multiverse. *Journal of High Energy Physics*, 0701(092), 2007. URL http://arxiv.org/abs/hep-th/0611271. Eprint arXiv:hep-th/0611271.

Special Symposium on
FREUD AND PSYCHOANALYSIS

Is Freud Back? [1]

Edward Erwin
University of Miami
eerwin@miami.edu

ABSTRACT. Some have recently argued that "Freud is back": meaning that reevaluated old evidence or new evidence, principally from neuroscience, provides support for some of Freud's central views. In Section 2, I argue that attempts to rehabilitate the pre-1996 Freudian evidence fail to vindicate Freud. In Section 3, I examine the new evidence and conclude that it too fails to support Freud. Skepticism about his theories and therapy are still warranted.

1 Preface

Some argue that Freud is back [6, 37], meaning that new evidence or a re-evaluation of the old evidence vindicates Freudian theory or therapy. Using "evidence" in a neutral way to refer to data allegedly giving us reason to believe a certain proposition, leaving it open whether the data do or do not have any probative value, the main Freudian evidence falls into one of three categories: 1) evidence from clinical case studies; 2) experimental evidence; and 3) evidence from commonsense psychology.

In my (1996) book, *A Final Accounting*, I examined all three types of Freudian evidence and concluded that virtually none of Freud's theories had strong empirical support; that his therapy had not been shown to be generally effective in treating any clinical disorder; and that these twin verdicts were likely to be final. This last point, however, was not meant to suggest that my arguments were unassailable, but rather that if they were cogent, new evidence would not likely change either verdict.

In fact, some of the arguments I relied on, including some of Adolf Grünbaum's seminal arguments [19, 21, 22], (1984, 1993; see as well his 2002 paper), have been attacked and new evidence has emerged, some of the most intriguing being recent evidence from neuroscience. In Section 2, I examine arguments about the pre-1996 Freudian evidence; in Section 3, I look at some of the new evidence.

2 The Old Freudian Evidence

Some of my arguments [11] presuppose certain evidential principles, one of which I have referred to as "Salmon's principle", alluding to a principle discussed by the American philosopher Wesley Salmon. A preliminary version says that for X to be causally relevant to Y, X must divide some reference class C into two subclasses such that the frequency of Y's occurrence is greater in the subclass where X occurs compared to the subclass where it does not. For example, if thimerosal ("thermasol") in children's vaccines causes autism, the incidence of autism, in the reference class consisting of children

---
[1] The author is with the Department of Philosophy, University of Miami.

who have been vaccinated, will be greater in the subclass of those receiving vaccines with thimerosal compared to children receiving thimerosal-free vaccines.

Call this version of Salmon's principle "the ontological version". It tells us what is necessary for one factor to be causally relevant to another. Nancy Cartwright (2009) raises problems for the principle in her discussion of Grünbaum's views about Freud's Ratman case. Although most of Grünbuam's arguments on other Freudian matters do not presuppose this principle, some do, as do some of mine.

One point Cartwright makes bears on the very idea of causal relevance as that which makes a difference. She argues that the view that a factor must make a difference to be causally relevant lies at the core of counterfactual accounts of causality, but these accounts give wrong verdicts in cases of causal preemption. To take her example, a trainee assassin shoots and kills the president, but if the trainee had not fired, the experienced assassin standing beside him would have fired a deadly shot. The firing of the shot by the trainee contributed to the death of the president and, therefore, was a causally relevant factor, but it made no difference, Cartwright argues, since the president in these circumstances would have died anyway.

I am not convinced by Cartwright's argument. Requiring that an event make a difference in order to be causally relevant does not presuppose any counter-factual account of causation; it is based merely on an analysis of what a cause is: namely, something that makes a difference. Second, Cartwright's interpretation of preemption cases turns, in my view, on a misinterpretation of "making a difference". To say that an event made a difference to the occurrence of a certain outcome does not entail that, in those very circumstances, the result would have been different without the factor. Even where there is no preemption but there are alternative causal routes to the same outcome, an event may make a difference, indeed it may be the main cause of the effect, and yet the same effect might have occurred without the cause.

My psychiatrist gives me a choice between cognitive therapy and a drug treatment for my depression. I choose the first and it cures my depression; the cognitive therapy made a very significant difference to my depression even though the drug, let us stipulate, would have produced the same result in exactly the same time had I not received the cognitive therapy. I have a headache and will take aspirin or Advil to cure it. Taking the aspirin may make a difference to the continuance of my headache even though the result would have been the same had I not taken it. Without taking the aspirin, I would have taken the Advil with the same result. An event's making a difference to an outcome does not require that the result would have been different without the event.

However, the issue of whether causation requires making a difference is relatively minor here. The primary question of interest to Cartwright is: How do we tell if one factor is causally relevant to another? Speaking of "making a difference" as being necessary for causal relevance but leaving the first notion unanalyzed is of little help in answering Cartwright's primary question.

In discussing the main issue, Cartwright distinguishes between singular causation and what she calls "generic level causation". We might illustrate the distinction by someone's claiming that Freud's therapy helped in one case, say, the Ratman case, as opposed to claiming that it generally contributes to a certain outcome (the generic case). In the generic case, Cartwright argues, Salmon's principle sometimes gives the wrong result. It does so in cases exemplifying Simpson's paradox, cases where causes may either reduce the probability of their effects or at least not increase them due to the presence of another variable. It also gives the wrong result in cases where a cause has opposing tendencies: both to enhance an effect and to retard it. Cartwright's example is a non-drowsiness inducing decongestant that contains a powerful soporific that fails to cause drowsiness only because it is always packaged with a powerful stimulant.

Salmon's principle (the ontological version) also runs into trouble in cases of singular causation. Suppose that a diet generally does not work but is effective for one individual, say Nancy Cartwright. There may be some description of her that fixes in a law-like fashion the efficacy of the diet for people who satisfy the description, but then again, there might not be such a description.

I agree with Cartwright's criticisms, but I believe Salmon's principle can be reformulated to make it more relevant to Cartwright's main question. Following Salmon, she states the principle, as noted earlier, as an onotological principle stating a necessary condition for a factor being causally relevant to the occurrence of Y. My suggestion is to reformulate it as an epistemological principle. As reformulated, the principle is more relevant to Cartwright's original question, which was not what is a causally relevant factor, but rather how do we tell if a factor fits that description, than is the ontological version. This epistemological question is also the one of primary interest to both Grünbaum and me in evaluating the Freudian evidence.

For the generic case, the epistemic version of Salmon's principle would talk about the sort of statistical evidence needed to show that X is causally relevant to Y, but account must be taken of "countervailing factors", factors that prevent the effects of x showing up in a statistical comparison even though x is generally causally relevant to the production of y. To handle such cases, I suggest the following modification of Salmon's principle: "Generic Causation: Except where there is reason to believe that there are countervailing factors, to establish that X is generally causally relevant in a positive way to the occurrence of Y, we need to establish for some reference class, c, that the frequency of Y's is greater in the subset of c in which X occurs than in the subset where X does not occur".

What about cases of singular causation, such as Cartwright's diet example? Her case is not a real one, but if there were an actual case, how would we get evidence of the diet's efficacy for one person given that there is no statistical support for its general effectiveness? The way to proceed would be to use a single subject design, repeatedly alternating diet-no diet, diet-no diet conditions and finding weight loss only under the diet condition, or by using some other variant of a single subject experimental design. No statistical comparison with others would be needed, but this sort of case can be handled by choosing a reference class with care; the comparison must be not with groups of individuals but with different experiential phases for the same individual. A separate principle, then, needs to be stated for singular causation to handle cases where group statistical evidence is not available and is not needed.

Before formulating it, an additional problem needs to be considered. There are some cases of possible relevance to psychoanalysis where evidence of causation can be obtained without a group statistical comparison or the employment of anything resembling a single subject experimental design. Where certain conditions are present, such as the absence of episodic symptoms and the occurrence of very dramatic changes (so-called "slam bang" effects), plus favorable background evidence which I will not try to specify, evidence that a treatment was causally relevant can sometimes be obtained without any sort of comparison [10, 26]. Such cases are rare, at least in the field of psychotherapy. Absent a better solution, I would simply add the following for the rare cases: "Singular Causation: Except where our background evidence or the presence of dramatic effects obviates the need for any sort of comparison, to obtain evidence in a single case that X was causally relevant to the occurrence of Y, we need to establish for some reference class, c, that the frequency of Y's is greater in the subset of c in which X occurs than in the subset where X does not occur." In cases where a statistical comparison is needed, the more usual cases, the reference class, once again, need not refer to multiple subjects. It can be formulated so that it describes different phases of the experience of only one person.

Making the above addition about exceptional cases runs the risk of trivializing the principle. It invites the defender of Freud to point out that appeal to Freudian theory is enough to satisfy the exception clause; if this is so, then no statistical evidence of any kind would be needed to show, for example, that the treatment of Little Hans by his Freud-instructed-father improved the child's psychological condition. This reply, however, is cogent only if the bit of Freudian theory that is appealed to is itself well confirmed and, I shall argue, this requirement is not met. For general doubts about cases of Freudian treatment meeting the exception clause, see [10, pp.205–215].

Neither Grünbaum nor I contend that satisfying Salmon's principle is sufficient for confirmation of causal hypotheses. Even with statistical evidence, we might be in no position to infer causal relevance. A second principle needs to be satisfied, which Harvey Siegel and I [13] have dubbed the "Differential principle of confirmation": For any causal hypothesis H, evidence E confirms H only if E provides reason for believing that H is true and does not provide equal (or better) reason for believing some rival, incompatible hypothesis that is at least as plausible as H.

This principle will be regarded as fairly trivial by many, but it does have counter-intuitive implications leading some philosophers and psychoanalysts to complain that it sets "an unrealistically high standard of confirmation" [27, p.269]. But none of those who reject the principle ([14], appear to be committed to its rejection) answer the argument for it. If we opt for a lower standard, if we say that evidence E can confirm H even though E is neutral between H and a logically incompatible and equally plausible H2, then we allow for confirmation of H by E even though E gives us just as much reason to believe that H is false as reason to believe that H is true. After all, if E confirms H non-differentially, it also confirms, under the stipulated conditions, the competitor H2 to the same degree as H, and H2 logically entails that H is false. Whatever reason we would have, then, for believing H to be true would be an equally good reason to think that H is false.

Another issue about evidential standards concerns what are sometimes called "meaning connections", also known as "thematic affinities". There are still many supporters of Freud who argue that connection in psychological content is generally a mark of causal connection [24, 25]. To take a standard example, there is an overlap in content between Anna O's watching a dog drink water from a glass and her later development of hydrophobia in that both concerned drinking water, disgust, anger, and refusal. This overlap in content, or thematic affinity, is said to be evidence of a causal connection between Anna O's experience of watching the dog drink and the development of her hydrophobia.

Adolf Grünbaum [20] and I [11, pp.26–40] have argued that a thematic affinity is not by itself evidence of a causal connection, although I allow for the possibility of developing an empirical argument that generally where there is a thematic affinity between two events, there is also a causal connection. Such an empirical argument, however, has so far not been made.

Donald Levy [30] sees the reliance on thematic affinities to prove causation in both commonsense psychology and Freudian psychology as the foundation for the claim of a continuity between the two. Although he rejects my claim about thematic affinities, he misconstrues it to say that even when combined with other evidence, thematic affinities cannot support causal conclusions, despite my making it clear that I do not hold or anywhere presuppose such an implausible view.

Virtually anything can be evidence of a causal connection if combined with the right sort of additional evidence, but pointing out this triviality is not likely to strengthen the Freudian case. To illustrate: Take the conditional proposition: "If there is a thematic affinity between Anna O's experience of watching the dog drink water from a glass and her subsequent water phobia, then there is a causal connection between the two events". If, contrary to fact, we had evidence for this conditional, then we could combine it with evidence that in fact there was such a thematic affinity, and, lo and behold, we would have evidence of a causal connection. But of what use would this argument be if, as is the case, we had no evidence for the conditional? What Grünbaum[20] and I are skeptical of is the idea, advocated by Levy himself, that thematic affinities are evidence by themselves of causal connections.

Not everyone who tries to defend Freud by appealing to a continuity between commonsense psychology and Freudian psychology rests the case on thematic affinities. Thomas Nagel[34] substitutes for thematic affinities an appeal to what he terms "the fundamental principle of commonsense psychology", which directs one to discover causally relevant conditions for a human action by fitting it into a rationally coherent interpretation of the whole person and seeing how from the agent's point of view the action is justified. I have argued [11] in reply that Nagel's principle is not a fundamental principle of commonsense psychology and that it is incorrect if it implies that this "understanding from within", as he calls it, is in and of itself evidence for any particular causal explanation of human action. We may figure out that a certain action or thought was justified from the agent's point of view by a certain reason, but we would still need empirical evidence that this reason was in fact a causally relevant factor, a factor that made a difference as opposed to being one that the agent could have justifiably acted on but did not.

I have so far referred to four evidential principles, two positive and two negative, that have played a key role in discussions of the Freudian evidence. The first, Salmon's principle, requires that except for special cases that can be specified in advance, to confirm a causal claim, we need a statistical comparison or at least, for the single case, a comparison of different states of the same individual. Getting such evidence, however, is not sufficient for confirmation. The second principle, the Differential Principle, requires that any logically incompatible competitor to our causal hypothesis must be ruled out if, given all of our evidence, it is at least as likely to be true as the hypothesis we are testing. Does this principle require that we also rule out rivals that no has thought of? No, because such unknown hypotheses, on our current evidence, have no likelihood of being true.

The third principle says that thematic affinities and meaning similarities are never evidence in and of themselves of a causal connection. The fourth principle says that the sort of understanding from within that Nagel talks about is also not in itself evidence of a causal connection. The arguments for all four of these principles are given in Erwin[11] and in Grünbaum [20]. None of these principles has been successfully challenged, although Cartwright has refuted the ontological version of Salmon's principle, which is not needed in assessing the Freudian evidence.

Disagreements about evidential principles are only part of the overall disagreement about how to interpret the Freudian evidence, but if these principles are accepted, then it can be shown that the pre-1996 Freudian evidence is of little evidential worth. Take, first, the clinical evidence. Many opponents of Freud have contributed to the rational downgrading of this sort of Freudian evidence (see Macmillan [31]), but apart from my own arguments on this topic, I have relied most on the arguments of Adolf Grünbaum; his arguments, in my view, are powerful and decisive. One need not conclude that there is absolutely nothing of value in the clinical evidence, but, to take Grünbaum's principal conclusion in his 1984 book[19], in so far as the evidence for Freudian theory is derived from the production of patients in analysis, this warrant is remarkably weak.

Some Freudians agree with Grünbaum's conclusion, but argue that the second kind of evidence, the experimental evidence, is sufficient to support central parts of Freudian theory. The late Paul Kline, in his 1981 book[28], and Seymour Fisher and Roger Greenburg, in their 2002 [15]contribution to the Freud Encyclopedia, and in their earlier work, criticize the designs of many of the 1300 plus Freudian experiments, but also argue that certain parts of Freudian theory have been confirmed by the experimental evidence. I have examined their arguments and the very same experimental evidence they rely on, plus evidence from newer experiments, and have shown, I believe, that none of this evidence is confirmatory to any appreciable degree for any distinctively Freudian theoretical hypothesis [11, pp.145–236].

If I am right about the poverty of both the clinical and experimental evidence, this leaves primarily whatever evidence can be taken from commonsense psychology. But concerning this last type of evidence, I have argued, as Grünbaum has, that it is of little help to Freud unless it is backed up by either Nagel's methodological principle or by thematic affinities, neither of which survives scrutiny (see [11, chap.3] for additional discussion of the continuity argument).

I have so far not mentioned the pre-1996 evidence for the effectiveness of Freud's therapy. For many years, the only such evidence came from case studies of individual patients; none of this evidence meets the requirements set by Salmon's principle or the differential principle. There have been, however, some formal studies of the outcomes of psychoanalysis favorably evaluated by a Subcommittee on Efficacy Research of the American Psychoanalytic Association [1] and some earlier studies (discussed in [9]).

Although some of these studies provide statistical data on outcomes, not one is a randomized clinical trial with a placebo control. For this reason and others, they run afoul of the differential principle; they fail to rule out a standard credible alternative hypothesis: that the favorable outcomes were due not to psychoanalysis but to spontaneous remission or placebo factors.

To conclude: I have seen no new arguments about the old Freudian evidence that would require revision of either of my 1996 verdicts: that the effectiveness of Freudian therapy for any clinical disorder had not been established; and with respect to Freudian theory, virtually none of it is confirmed by the old evidence. There is, however, new evidence to consider.

3 The new Freudian evidence

Since 1996, there have been some developments worth noting with respect to psychoanalytic treatment. In 1999, the Research Committee of the International Psychoanalytic Association prepared a comprehensive review of North American and European outcome studies of psychoanalytic treatment and concluded that existing studies failed to demonstrate that psychoanalysis is efficacious relative to either an alternative treatment or an active placebo [16]. This negative verdict agrees with my 1996 conclusion[11].

There have, however, been two large scale studies of psychoanalytic treatment in recent years, one published in [17], the second, published in [35], the Stockholm study of 756 patients. Both studies tried to obtain information by sending questionnaires to patients in the same way that the psychologist, Martin Seligman, did in 1995 [36] when he conducted his well known "Consumer Reports" survey of various types of psychotherapeutic treatments. Neither Seligman's study, nor either of the two psychoanalytic studies, were experimental studies. None involved a random assignment of patients to treatment and no treatment control groups; none, furthermore, controlled for placebo factors. For these reasons and others, none of these studies, as I show in detail in a forthcoming paper [12], provide any credible evidence of effectiveness. The "Consumer Reports" type of study should not serve as a model for future outcome studies of psychoanalysis; such studies are primarily useful for gaining information about consumer satisfaction, but they are incapable of establishing causal connections between treatments and therapeutic outcomes.

As to Freudian theory, there have been reports published after 1996 allegedly supporting Freud, but there remains the question of whether these newer studies overcome the epistemological problems of the earlier ones. One that has been cited as impressive is a study by Baumeister, Dale, and Sommer [2], which claims to provide support for Freudian defense mechanisms. The authors, however, define "defense" so broadly that many types of defensive behaviors qualify as instances of defense regardless of the causal mechanisms that account for the behavior. For example, they count the Dutton and Lake paper[8] as providing a "convincing demonstration" of reaction formation. Subjects who were given false feedback indicating they had racist tendencies gave more money to a black panhandler than did subjects not given the feedback. Yet no evidence was provided that any subject was reacting to a repressed wish [29, p.376].

[2, p.1009] anticipate this objection when they note "Although the mechanism underlying reaction formation may not conform precisely to Freud's model, the human phenomena that he characterized with that term do appear to be real". I agree. The phenomena are real, but the reactions themselves to which Freud gave various names, such as "reaction formation," "projection," and "displacement," were known before Freud ever developed his theories. Freud was not the first to notice, for example, that someone who cannot openly criticize his boss might "displace" his anger and mistreat his wife without being aware of his motivation, or that someone uncomfortable with her lack of generosity might "project" that characteristic to someone who is generous. Freud's distinctive contribution was to postulate particular causal mechanisms, particularly the operation of repressed wishes, to explain such behavior. Providing evidence for the existence of a certain type of defensive behavior but not Freud's explanation of it is not to confirm any part of Freudian theory.

Another much discussed paper is due to Drew Westen[40], who tries to defend the scientific legacy of Freud by marshalling evidence for five Freudian propositions. The first, and most important, is the claim that much of our mental life is unconscious, but is this a distinctively Freudian hypothesis? As is well known, the existence and prevalence of unconscious mental events and states were widely debated in the 19th Century [41]. By the time we get to the publication of Freud's Theory of Dreams, asserting the existence of the unconscious was hardly new. What was distinctive about Freud's view was his theory of the dynamic unconscious.

On this dynamic view, we can distinguish the unconscious from the preconscious as follows. Unconscious ideas have a dynamic quality in that they have effects; they are also typically prevented from entering consciousness due to the operation of repression; and they can enter consciousness only when the repression is "lifted", which requires overcoming the person's resistances through the use of psychoanalysis. Preconscious ideas may also have effects but they lack the other features of unconscious ideas.

To cite studies providing evidence for unconscious mental events, as Westin does, without supporting Freud's specific ideas about the dynamic unconscious is not to support any distinctive part of Freudian theory. The same problem arises for the other four hypotheses cited by Westin: Not one is distinctively Freudian.

The Westin studies require more discussion than I can provide here, but rather than examine any more studies which are more of the same in that they fail to overcome the old epistemic problems, I will mention a fourth type of evidence that some say now provides solid scientific support for certain parts of Freudian theory. This new evidence is experimental; so, it falls into my second category, but it deserves a category of its own: I am referring to evidence from neuroscience.

In recent years, there has been a trend for psychoanalysts to look to such findings to finally put Freudian theory and therapy on a firm scientific foundation (see, for example, [6, 37, chap.1]; and various papers published in the *Journal of Neuro-psychoanalysis*). In a carefully worked out, powerful challenge to this trend, Blass and Carmelli [4] argue that neuroscientific findings neither refute nor confirm psychoanalytic hypotheses and in fact are powerless to do either.

Two issues need to be separated. The first is: Are findings from neuroscience even relevant to psychoanalysis? The second is: If they sometimes are, do the current findings support Freud's theory or therapy?

As to the first issue, Blass and Carmelli offer both general and specific reasons for concluding that neuroscience is irrelevant to psychoanalysis. One of their general reasons is worth quoting in full:

"Neuroscience can describe the neural networks underlying psychological phenomena, patterns, and tendencies, but these phenomena, patterns, and tendencies are recognized and their laws specified without any information regarding the neurons that function concomitantly. Only once these are recognized on the psychological level can neuroscience proceed with its description, but it does so without adding anything to the psychological knowledge already obtained." [4, p.10].

Here is a clear challenge not only to the new field of neuropsychoanalysis but to neurocognitive models as well, indeed to almost any attempt to enrich psychology by appeal to neuroscientific findings. For this reason, it raises issues of general philosophical interest about the relations between psychology and neuroscience. Does the challenge succeed? An obvious problem is that it is question begging to assume without argument that the psychological patterns and laws (if there be any) are to be specified without any information regarding the neurons that function concomitantly. Some psychoanalysts and others once tried to specify the causes of schizophrenia in terms of malfunctioning family relationships or in other psychological terms, but evidence of brain malfunctioning in schizophrenics was relevant to deciding whether their attempts would succeed. The evidence made it likely that such attempts would fail.

To take another example, Freud's dream theory does not logically say or imply anything about the brain. This might suggest that neuroscience is irrelevant for deciding, for example, whether or not repressed infantile wishes, together with day residues in combination with somatic factors, are the causes of manifest dream content. Yet it is relevant. Suppose we add to Freud's hypothesis about the instigation of dreams the following premises: Dreams are instigated by unconscious wishes only if higher level motivational brain processes are responsible for the appearance of the dream; and these higher level processes cannot be responsible if all dreaming occurs only during REM sleep, assuming that REM processes are derived from automatic activity coming from the brain stem. These combined assumptions appear to be falsified by the finding that all dreaming does occur during REM sleep and that REM is automatically generated by brain stem mechanisms, or so Hobson and McCarley argued in a well known paper [33].

Here is a clear example where neurological findings are relevant to Freudian theory; indeed they refute part of the theory if all dreaming occurs only doing REM sleep (which is contentious; see below). Bass and Carmeli dismiss this finding, even if Hobson and McCarley are right about when all dreaming occurs, partly because they hold that what is central to Freudian dream theory is that dreams are meaningful, but this claim, they say [33, p.7], does not presuppose that the meaning is derived from a motivational source.

This point, however, raises a different issue, one about the centrality of Freud's wish fulfillment thesis. If one downgrades this thesis as being unimportant, then the Hobson-McCarley criticism is neutralized or at least is shown to be of marginal importance, but this does not answer the point that there is a version of Freud's dream theory, indeed the standard version, which is a purely psychological theory and yet is partly vulnerable to neuroscientific disconfirmation.

In brief, whether certain types of psychological events have psychological rather than neurological or genetic causes, or whether specifying laws of psychological phenomena can safely proceed without paying attention to neuroscientific findings, are empirical questions. The issues cannot be settled in advance by a priori theorizing. For this reason, one has to consider the neuropsychoanalytic arguments on a case by case basis. Bass and Carmeli do exactly that; for this reason, their overall case does not depend on whether or not their general argument begs the question.

In giving their specific arguments, Blass and Carmeli challenge the relevance of recent neuroscientific work to the recoverability of lost memories, Freud's theories of the mind, and motivation and effect. The authors demonstrate that the neuroscientific evidence they discuss fails to support Freud and for the most part is of only limited relevance to his theories or therapy, but not that it is altogether irrelevant. For example, some have argued that recent neuroscientific evidence shows that many traumatic memories are not coded explicitly and thus are not recoverable as memories per se. To this, the authors reply that objective recovery of lost memories is not necessarily the aim of psychoanalytic treatment. Whether it is or not has been a debated issue. Those who contend that recoverability of memories has always been crucial to the analytic process are simply taking one side in the dispute; their giving dominance to the one side is not itself supported by neuroscientific evidence. I agree with both points, but still the evidence does tell against the view of those psychoanalysts who hold that memories of traumatic experiences are always recoverable. Yet the evidence is only of marginal interest in that psychoanalytic treatment can proceed pretty much as before even if certain memories of traumatic events are not recoverable.

The key upshot of Blass and Carmeli's arguments is that, contrary to what some have claimed, the neuroscienctific evidence does not support Freudian theory or therapy, and in most cases, do not tell against either. I turn now to the claim made by many supporters of Freud that neuroscience not only can in principle support his theory or therapy, but in important respects has already done so.

Mark Solms is one of the leading proponents of the idea that neuroscience supports Freud. He points out [37] that interdisciplinary groups devoted to psychoanalysis and neuroscience have been formed in almost every major city of the world, that a new society has been formed, the International Neuro-Psychoanalysis Society, and a journal, Neuro-Psychoanalysis, is now publishing papers on topics of common interest to neuroscientists and psychoanalysts.

Solms also claims that neuroscientists are uncovering proof for some of Freud's key theories, including his theories about unconscious motivation, repression, the pleasure principle, the idea that dreams have meaning, and even the Id-Ego-Superego hypothesis. One point he makes is that a diverse group of neuroscientists are coming to the same conclusion as the eminent neuroscientist Eric Kandel, who says that psychoanalysis is "still the most coherent and intellectually satisfying view of the mind". Solms is presumably not intending to be making an argument here for Freud, but if he were, it would be worth mentioning that the neuroscientists he mentions are experts in their field but not necessarily experts on whether the neuroscientific evidence supports Freud. In any event, it is the evidence and not merely expert opinion that matters in deciding whether Freud was right. Citing their opinions without evidence adds nothing of substance to the debate.

One of the areas of most concern to Solms has been the bearing of neuroscience on Freudian dream theory. In a section headed "Dreams Have Meaning", Solms points out that the dream theory was seemingly discredited with the discovery of a strong correlation between dreaming and REM sleep, plus the finding that REM sleep is driven by brain chemicals in the brain stem that have nothing to do with emotion or motivation. More recent work, he points out, shows that dreaming and REM sleep are dissociable states, controlled by distinct mechanisms. In short, dreams also occur outside of REM sleep.

Although these findings answer a serious objection to Freud's dream theory, they do not show that dreams have meaning. Discrediting an argument against dreams having meaning is insufficient to show that they do have it, although there may be good reason of a non-neurological kind to hold that dreams have meaning. Furthermore, the claim that dreams have meaning of some kind or other was accepted long before Freud wrote, as he himself points out in the Interpretation of Dreams. The thesis is also widely accepted by Freud's critics [7, 23]; what many of the critics deny is not that dreams have meaning but that dreams have the sort of hidden meaning postulated by Freud. It is important to note, then, that the assigning of meaning to dreams does not support anything specifically Freudian. In addition, the thesis that dreams have meaning is not confirmed by evidence that dreaming occurs outside of REM sleep – removing one counter-argument is not confirmation – and is not confirmed by any other evidence from neuroscience.

What parts of Freudian dream theory, then, are confirmed by recent neuroscientific work? Does it confirm the hypothesis that unconscious wishes are the main instigators of dreams, or that dream censorship occurs, or that free association can routinely determine the latent meaning of dreams, or that dreams even have a latent meaning? Solms [38] claims that contemporary knowledge of the dreaming brain is "broadly consistent" with Freudian theory, but he does not claim that any of the neuroscientific evidence empirically confirms that dreams are meaningful or that they require interpretation or that dreams are motivated or that dream censorship occurs or that free association is useful in uncovering the hidden meaning of dreams. If he were to claim neurological confirmation in addition to mere consistency, it would be reasonable to ask for a citation of the evidence. He does not cite any such evidence in either [37] or [38]. All of this should be disappointing to those who have looked to neuroscience to bring back Freud, given that his dream theory looked like the best bet for this endeavor.

It is possible that future neuroscientific research will support Freudian dream theory, but in considering even this mere possibility, one needs to look at research that has already been done in dream laboratories that bears on this possibility. The extensive evidence discussed by Domhoff [7] makes it likely that central parts of the theory, including most of the theses mentioned above, are not just unfounded but false.

Concerning Freud's structural theory of the mind, Solms [38, p.6] claims that the core brain stem and limbic system correspond roughly to the id; the ventral frontal system, the dorsal frontal region, and the posterior cortex amount to the ego and superego. But how does he know that any of these postulated identities hold? There is an instinctual part of the brain, but without evidence that there is an unconscious mental agency or structure that has the basic properties which Freud attributes to the Id, such as being the source of much mental conflict and seeking its own gratification, why believe that the id even exists? Before one obtains evidence that certain mental structures, such as the id, ego, and superego are identical to brain structures, evidence is needed that there are such mental structures. Without evidence for their existence, evidence for their being identical with specific parts of the brain is unlikely to be found.

In a section "Repression Vindicated", Solms cites a (1994) study by Ramachandran of anosognosic patients who have damage to the right parietiel region of the brain making them unaware of certain gross physical defects such as paralysis of a limb. One such patient with a paralyzed left arm consistently denied that she had a problem. After Ramachandran artificially activated her right hemisphere, the woman suddenly became aware that her arm was paralyzed and that it had been paralyzed continuously since she suffered a stroke eight days earlier. Solms takes these facts as showing that the woman was capable of recognizing her deficits – at least, it should be noted, when her brain was being artificially stimulated – and that she had unconsciously registered her physical deficit during the previous eight days. His interpretation, however, is open to challenge. Based on Ramachandran's description of what occurred, it is not clear that the woman unconsciously registered the fact of her disability at any time prior to having her right hemisphere artificially stimulated.

When the stimulation wore off, the woman once again believed her arm was normal and forgot that part of the earlier interview in which she acknowledged that the arm was paralyzed. Based on his observations, Ramachandran concluded that memories can be selectively repressed and that observing this patient convinced him of the reality of repression phenomena. It is not clear why this interpretation is mandated by his findings. The observed events could be taken as evidence of the woman selectively repressing her memories of what happened immediately after her stroke, but, unless there is additional evidence, the events could equally be interpreted by skipping the psychology: The brain impairment causes both an incapacity to recognize the paralysis of the arm except when the parietiel region is being artificially stimulated and an inability to remember being aware of the paralysis during the stimulation period; there is no storing of memories in the unconscious and hence no Freudian repression.

A recent study by Anna Berti and her colleagues [3] appears to bear on this issue. Patients suffering damage to one brain hemisphere often exhibit both neglect (a deficit in attention to a limb or other bodily part on the opposite side) and ansognosia involving a denial of hemiplegia (total or partial paralysis), but neglect and ansognosia can appear separately. Berti and her colleagues compared the distribution of brain lesions in patients showing left spatial neglect, left hemiplegia, and anosognosia with patients showing neglect, left hemiplegia but not ansognosia. The authors found differences in brain lesions between the two patients and conclude (p. 489) that ansognosia for hemiplegia is best explained by the involvement of motor and premotor areas.

Even if the denial involved in ansognosia were motivated as opposed to being directly caused by specific brain lesions, there are findings that would raise questions about the postulation of repression. On Freud's theory of repression, repressed material is stored in the unconscious and can be returned to consciousness, but only if the anxiety associated with the repressed memory is removed. On this basis, one would predict that repressed memories would stay repressed without a removal of the anxiety and that the repression would not be lifted spontaneously. Yet studies of ansognosia find that it often does remit spontaneously. For example, in the study by Berti et al. [32, p.696], anosognosia disappeared within 3 months in all cases.

There are a number of theories being debated concerning the mechanisms involved in ansognosia and it is possible that the theory that repression is implicated might yet be proven. If this were to happen, then this would be important and it would show that to some extent the neurological evidence vindicates Freud. But to what extent? During the 1960's, there were debates about whether so-called "true conditioning" ever occurs in normal adult humans. Proving that it did occur would have aided the cause of behaviorists such as Ullmann and Krasner [39], who argue that all psychological disorders, including autism and schizophrenia, can be explained in terms of conditioning. One barrier to making their argument would have been knocked down. Yet discovering cases of true conditioning would not by itself have been evidence for the behaviorist grand etiological scheme. Likewise for Freud's theory of repression, which is not merely the theory that repression occurs; it also includes important theses about the etiology of psychoneuroses, the causes of many slips of the tongue and forgetting; and the causes of the manifest content of dreams. Proving that repression occurs is not evidence for the other theses that make up Freud's repression theory.

Neuroscience has also shown, Solms points out, that major brain structures essential for forming conscious (explicit) memories are not functional during the first two years of life. Yet our infantile memories, despite not being consciously encoded, can affect adult feelings and behavior. This claim, however, is so general and vague that it says nothing which critics of Freud typically deny, nor does it warrant acceptance of any major thesis that is specifically Freudian.

Solms also argues that neuroscience confirms Freud's views about the pleasure principle. Identifying exactly what Freud's theory was is difficult because he changed his mind a number of times. In an early formulation, Freud tries to explain the tendency to seek pleasure in terms of the reduction of unpleasurable drive tension. After the formulation of the structural theory, Freud links the pleasure principle to the operation of the ego. The principle, operating at an unconscious level, guides the ego as it defends against unpleasures from the superego and the external world [18].

Solms argues that if Freud was right about the operation of the pleasure principle, then damage to the inhibitory structures of the brain will release wishful, irrational modes of mental functioning; this prediction, he claims, has been confirmed. Patients suffering from Korsakoff's psychosis are unaware that they are amnesiac and fill in memory gaps with confabulations. Such patients, Solms continues, maintain false beliefs that are generated by the pleasure principle, but he cites no evidence that the ego or superego are involved. All that is claimed is that the patients, once their cognitive mechanisms are damaged as the result of brain lesions, construct confabulations to recast reality as they want it to be. Some investigators interpret such confabulating activity in Freudian terms, but it is not clear what evidence justifies this interpretation. People often see the world the way they wish it to be; there is nothing new in saying this. The finding that victims of a certain type of brain disorder are not just randomly making up false beliefs about the world, but are constructing visions of the world as they wish it to be is interesting, but not specifically Freudian. What we are left with is: If Freud is right about the pleasure principle, then damage to a certain area of the brain will cause wishful irrational modes of thinking. The prediction is correct, but no evidence is provided that only Freudian theory makes this prediction.

Conclusion: **Freud's therapy**: There is still no study that overcomes the epistemological problems of the earlier outcome studies of psychoanalysis discussed in Erwin [11]. My 1996 verdict about long term psychoanalytic treatment still stands: There is no credible evidence of psychoanalytic effectiveness for any clinical problem, nor are satisfactory experimental studies likely to be conducted in the future, partly because of ethical and practical problems in running randomized clinical trials with subjects spending years in therapy. It is hard to see, then, how this verdict is likely to be upended any time in the near future.

This is not the same as saying that the effectiveness of various psychoanalytically oriented short term therapies cannot be demonstrated experimentally. These therapies are often characterized using psychoanalytic concepts, but all of these short term therapies differ in significant respects from psychoanalysis; proof of the efficacy of any one of them is not proof of the efficacy of psychoanalysis.

Freudian theory: In conjecturing in [11] that a verdict of "not proven" for virtually all parts of Freudian theory would be the final (justified) verdict, I was relying on several suppositions argued for earlier in that book: that Grübaum had shown that in so far as Freud's theories rested on clinical case studies, the evidence was, and was likely to remain, extremely weak; that the continuities between Freudian theory and commonsense psychology, continuities that are mainly conceptual, would not be an important source of justification for Freudian theory; that the sorts of Freudian experiments done in the 20th century were not likely to yield anything that would substantially alter the current verdicts; and that neuroscience, although a possible source of new evidence, was not likely to change matters. About the last item, however, I was much too quick. I agree with Solms and others that the relevance of neuroscience to Freud's work cannot be dismissed without a careful examination of the relevant arguments. But based on the study by Blass and Carmelli [4] and my all too brisk examination of the neuroscientific evidence cited by Solms, neuroscientific vindication of Freudian theory has not yet materialized. If Freud's being back means not just acceptance of his theories or therapy but acceptance based on solid evidence, then Freud is not back. For some of his theories, such as his theory of dreams, we should not expect future confirmation from neurological evidence; too much evidence accumulated by dream researchers makes it likely that major parts of the dream theory are wrong.

Bibliography

[1] Bachrach, H., Galatzer-Levy, R., Skolnikoff, A., Waldron, S. On the efficacy of psychoanalysis. *Journal of the American Psychoanalytic Association*, 39:871–916, 1991.

[2] Baumeister, R., Dale, K., Sommer, K. Freudian defense mechanisms and empirical findings in modern social psychology: Reaction formation, projection, displacement, undoing, isolation, sublimation, and denial. *Journal of Personality*, 66:1081–1124, 1998.

[3] Berti, A., Bottini, G., Gandola, M., Pia, L., Smania, N., Stracciari, A., Castiglioni, I., Vallar, G., Paulesu, E. Shared cortical anatomy for motor awareness and motor control. *Science*, 309:488–491, 2005.

[4] Blass, R., Carmelli, Z. The case against neuropsychoanalysis: On fallacies underlying psychoanalysis' latest scientific trend and its negative impact on psychoanalytic discourse. *International Journal of Psychoanalysis*, 88:1–15, 2007.

[5] Cartwright, N. D. How can we know what made the Ratman sick? philosophy of religion, physics, and psychology. In Amherst, J. (Ed.), *Essays in Honor of Adolf Grünbaum*. Amherst NY: Prometheus Books, 2009. In press. Also as Causality: Metaphysics and Methods Technical Report CTR 08-03, CPNSS, LSE.

[6] Cavell, M. *Becoming a Subject: Reflections in Philosophy and Psychoanalysis*. New York: Oxford University Press, 2006.

[7] Domhoff, G. W. Why did empirical dream researchers reject freud? a critique of historical claims made by Mark Solms. *Dreaming*, 14:3–17, 2004.

[8] Dutton, D., Lake, R. Threat of own prejudice and reverse discrimination in interracial situations. *Journal of Personality and Social Psychology*, 28:94–100, 1973.

[9] Erwin, E. Psychoanalytic therapy: The Eysenck argument. *American Psychologist*, 35:115–128, 1980.

[10] Erwin, E. Psychoanalysis: Clinical versus experimental evidence. In Clark, P., Wright, C. (Eds.), *Mind, Psychoanalysis and Science*. New York: Basil Blackwell, 1988.

[11] Erwin, E. *A Final Accounting: Philosophical and Empirical Issues in Freudian Psychology*. MIT Press: Cambridge, Massachusetts, 1996.

[12] Erwin, E. Research methods in psychoanalysis. In Jokić, A. (Ed.), *Philosophy of Religion, Physics, and Psychology: Essays in Honor of Adolf Grünbaum*. Amherst, New York: Prometheus Books, 2009. In press.

[13] Erwin, E., Siegel, H. Is confirmation differential? *British Journal for the Philosophy of Science*, 40:105–119, 1989.

[14] Fine, A., Forbes, M. Grünbaum on Freud: Three grounds for dissent. *The Behavioral and Brain Sciences*, 9:237–238, 1986.

[15] Fisher, S., Greenberg, R. Scientific tests of Freud's theory and therapy. In Erwin, E. (Ed.), *The Freud Encyclopedia: Theory, Therapy, and Culture*. New York: Routledge, 2002.

[16] Fonagy, P. Psychoanalysis in clinical psychology. In Smelser, N., Miamisburg, P. B. (Eds.), *International Encyclopedia of the Social and Behavioral Sciences*. OH: Elsevier, 2001.

[17] Freedman, N., Hoffenberg, J., Vorus, N., Frosch, A. The effectiveness of psychoanalytic psychotherapy: The role of treatment duration, frequency of sessions, and the therapeutic relationship. *Journal of the American Psychoanalytic Association*, 47:741–772, 1999.

[18] Glick, R. Pleasure principle. In Erwin, E. (Ed.), *The Freud Encyclopedia: Theory, Therapy, and Culture*. New York: Routledge, 2002.

[19] Grünbaum, A. *The Foundations of Psychoanalysis: A Philosophical Critque*. Berkeley: University of California Press, 1984.

[20] Grünbaum, A. "meaning" connections and causal connections in the human sci-

ences; the poverty of hermeneutic philosophy. *Journal of the American Psychoanalytic Association*, 38:559, 1990.

[21] Grünbaum, A. *Validation in the Clinical Theory of Psychoanalysis: A Study in the Philosophy of Psychoanalysis*. CT.: International Universities Press, 1993.

[22] Grünbaum, A. Critique of psychoanalysis. In Erwin, E. (Ed.), *The Freud Encyclopedia: Theory, Therapy, and Culture*. New York: Routledge, 2002.

[23] Hobson, A. Freud returns? like a bad dream. *Scientific American*, 290:89, 2004.

[24] Hopkins, J. Epistemology and depth psychology: Critical notes on the foundations of psychoanalysis. In Clark, P., Wright, C. (Eds.), *Mind, Psychoanalysis and Science*. Basil Blackwell, 1988.

[25] Hopkins, J. The interpretation of dreams. In Neu, J. (Ed.), *The Cambridge Companion to Freud*. New York: Cambridge University Press, 1991.

[26] Kazdin, Z. Drawing valid inferences form case studies. *Journal of Consulting and Clinical Psychology*, 49:183–92, 1981.

[27] Kitcher, P. Review of Edward Erwin. *A Final Accounting: Philosophical and Empirical Issues in Freudian Psychology. Philosophy and Phenomenological Research*, LIX:268–271, 1999.

[28] Kline, P. *Fact and Fantasy in Freudian Theory*. New York: Methuen, 1981.

[29] Laplanche, J., Pontalis, J. *The Language of Psychoanalysis*. New York: Norton, 1973.

[30] Levy, D. Review of E. Erwin. *A Final Accounting: Philosophical and Empirical Issues in Freudian Psychology. Mind*, 110:740–746, 2001.

[31] Macmillan, M. *Freud Evaluated: The Completed Arc*. Cambridge MA: MIT Press, 1997.

[32] Maeshima, S., Dohi, N., Funahashi, K., Nakai, K., Itakura, T., Komai, N. Rehabilitation of patients with anosognosia for hemiplegia due to intracebral haemorraghage. *Brain Injury*, 11:691–697, 1997.

[33] McCarley, R., Hobson, J. A. The neurobiological orgins of psychoanalytic dream theory. *American Journal of Psychiatry*, 134:1211–1221, 1977.

[34] Nagel, T. *Other Minds: Critical Essays 1969-1994*. New York: Oxford University Press, 1995.

[35] Sandell, R., Blomberg, J., Lazar, A., Carlsson, J., Broberg, J., Schubert, J. Varieties of long-term outcome among patients in psychoanalysis and long-term psychotherapy: A review of findings in the stockholm outcome of psychoanalysis and psychotherapy project (stopp). *Int. J. Psychoanal*, 81:921–942, 2000.

[36] Seligman, M. The effectiveness of psychotherapy: The consumer reports study. *American Psychologist*, 50:965–974, 1995.

[37] Solms, M. Freud returns. *Scientific American*, 290:82–88, 2004.

[38] Solms, M. Reply to Domhoff (2004): Dream research in the court of public opinion. *Dreaming*, 14:18–20, 2004.

[39] Ullmann, L., Krasner, L. *Psychological Approach to Abnormal Behavior*. Englewood Cliffs, N. J.: Prentice-Hall, 1969.

[40] Westen, D. The scientific legacy of Sigmund Freud: Toward a psychodynamically informed psychological science. *Psychological Bulletin*, 124:333–371, 1998.

[41] Zentner, M. Nineteenth century precursors of freud. In Erwin, E. (Ed.), *The Freud Encyclopedia: Theory, Therapy, and Culture*. New York: Routledge, 2002.

Special Symposium on CHINESE TRADITIONAL MEDICINE

The Concept of Disease in Traditional Chinese Medicine: In Comparison with Modern Western Medicine[1]

Zhai Xiaomei, Liu Junxiang, Qiu Renzong

Peking Union Medical College

xmzhai@hotmail.com

ABSTRACT. In this paper the authors discuss the issues including conceptual framework of traditional Chinese medicine, its metaphysical basis, concepts of disease, modern western medicine and traditional Chiense medicine as two paradigms of medicine and the debate on whether traditional Chinese medicine is pseudo-scientific.

Recently while western scientists and pharmaceutical companies are more and more interested in Traditional Chinese Medicine (TCM)[2], a debate developed in the Chinese mainland as to whether TCM is pseudo-scientific and should be abolished.[3] Exploring the difference in the concepts of disease between TCM and Modern Western Medicine (MWM) from the perspective of philosophy of science may help us to treat the debate properly.

---

[1] The authors are with Centre for Bioethics, Peking Union Medical College.
[2] Please see, for example, [4]
[3] Hundreds of articles published in journals and websites in 2006–2007 to debate on whether TCM is pseudo-scientific and should be abolished or not, such as
http://pinglun.eastday.com/p/20070413/u1a2762883.html
http://health.163.com/special/00181NNJ/pseudoscience.html
http://www.haoo.cn/Html/Article/2006/03/28/1514364.html
http://yiqianwx.blog.sohu.com/25940241.html
http://www.gotoread.com/article/?NewID=DF480D40-6B51-49AC-9AB7-3DAC13395A6C
http://news.qq.com/a/20061013/001812.htm
http://ks.cn.yahoo.com/question/1407091404074.html
http://blog.bioon.com/user1/4033/archives/2007/123274.shtml
http://zhidao.baidu.com/question/17523131.html?fr=qrl3
http://blog.voc.com.cn/sp1/jinxiaoxia/21262478193.shtml
http://www.elyoo.com/ask/q/q3.htm
http://bbs.tongji.net/viewthread.php?tid=355232
http://www.zhuaxia.com/item/465755596
http://ido.3mt.com.cn/Article/200610/show522475c30p1.html
http://cul.beelink.com.cn/20050324/1811091.shtml
http://zyx.7139.com/Article_Show.asp?ArticleID=427
among the others. One of these websites had a poll on it: about 25% say yes, TCM is pseudo-scientific; about 75% say no, TCM is not pseudo-scientific. See
http://bbs.tongji.net/viewthread.php?tid=355232

1 Conceptual Framework of TCM

The term TCM[4] refers to the paradigm of medicine which is rooted in Chinese culture and has a set of characteristic features distinct from any other paradigms of medicine in the world. The aimedicine is to solve patients' problems, and it presupposes a helping healer to predict the outcome of illness and disease and to take action to restore the patient to normal health.

The conceptual framework of TCM lies embedded in a characteristic theory - Yin-Yang and Five Agents theory. Chinese doctors use the principles of these theories to explain the structure of human body, physiological functions and pathological changes and guide their action in the diagnosis, treatment and prevention. So in TCM and its Western counterpart play the game by quite different rules: concepts of disease, methods of collecting data, diagnosis and prognosis, methods of treatment, etc., are all substantially different.

The character 'medicine' in ancient Chinese language is composed of two parts: The upper part of the character is the sound uttered from a patient with pain or suffering; and the lower part of it is magic. But later the character was replaced by another character the lower part of which is wine. It indicated that the inventor and users of this character believed in the curative power of wine rather than that of magic.

In 780 B.C. a man called Bo Yang Fu began to explain earthquakes by imbalance of Yin and Yang (*Words of States*[1, p.28]) . The first doctor who explained disease with Yin, Yang and other factors was Yi He (*Zuo's Biography*[7, p.5]) . This development led to a series of conflicts between medicine-men and ancient doctors who even refused to treat the patients who believed in magic. For instance, the great ancient Chinese doctor Flat Magpie (Bian Que) said: "I refuse to treat the patients who believe in magic, not in medicine." (*Records of History*[3, p.44]) . The author of *Interior Classics of Yellow Emperor* wrote: "The truth cannot be told to those who believe in spirits-gods." (*Plain Inquiries*, On Organs)[2, p.67] . We can say that at that time medicine began to separate from and become independent of magic.

Now what are the things called Yin and Yang (two forms of Qi) is still a disputable issue among Chinese philosophers and historians of science and medicine. According to the Yin-Yang theory, all things in the universe consist of Yin and Yang. All changes in the universe are explained by the growth and decline, or ebb and flow of Yin and Yang, and the interrelation and interaction between them. Qi may be interpreted as air or rather all pervading essence of air, creative and animating spirits, something like pneuma in ancient Greek philosophy and medicine. So it is plausible to take Qi and its two forms Yin-Yang as vital force which maintains all forms of life and organ functions as well as all kinds of change and transformations in the universe and in human body.

The Yin-Yang theory is invoked to explain the structure of the human body, physiological functions and pathological changes, to make diagnose, and to manage treatment. From the view point of Yin-Yang, health is the dynamic equilibrium between Yin and Yang, and disease is the imbalance between them, one of which is in excess or deficiency beyond a certain limit. The first step in diagnosis is to recognize whether the disease belongs to Yin or to Yang. Diagnosis is also required to distinguish between 'Exterior' and 'Interior', 'Heat' and 'Cold', 'Excess and Deficiency' as supplements to Yin and Yang. The purpose of treatment is to restore the patient to balance of Yin and Yang. So the principle of treatment is "Treating Yin for the Yang disease (Yang Excess), and treating Yang for the Yin disease (Yin Excess)." The doctor has to select the medicine which belongs to Yin or to Yang to prescribe for the patient to nourish Yin in the former and support Yang in the latter.

The introduction of Wu Xing (five agents) theory into medicine was later than that

---

[4]As for theories of TCM, concept of disease and the rationale of clinical practice below, please see [2, 8]

	wood	fire	earth	metal	water
5 tastes	sour	bitter	sweat	hot	salty
5 colours	green	red	yellow	white	black
5 changes	generate	grow	transform	shrink	hide
5 airs	windy	hot	humid	dry	cold
5 directions	east	south	central	west	north
5 seasons	spring	summer	long summer	fall	winter

Table I: Wu Xing in the Nature

	wood	fire	earth	metal	water
5 parenchymatous viscera	liver	heart	spleen	lung	kidney
5 hollow viscera	gallbladder	small intestine	stomach	large intestine	bladder
5 senses	eye	tongue	mouth	nose	ear
5 shapes	tendon	pulse	flesh	fair& skin	bone
5 emotions	anger	joy	misgivings	sadness	fear

Table II: Wu Xing in Human Body

of Yin-Yang. According to the classics *The Book of Records*, as early as in Shang Dynasty (1,000 B.C.) the Chinese already realized that water, fire, metal, wood and earth are indispensable for human life. Here Xing means change. In Wu Xing theory all things and all changes in the universe are produced by the interaction of these five agents.

It is natural for us, as modern persons to feel it arbitrary to fill all these things into Wu Xing framework. And it also makes one think the theory of TCM is the kind of natural philosophy in which all phenomena in the universe can be explained by this theory, and there is no gap left for future scientists to fill. So the theory of TCM is perfect and complete, not open-ended.

The Wu Xing theory in medicine is mainly applied to account for physiological and pathological changes of internal organs, diagnose and treat diseases in these organs. In diagnosis if doctor finds a patient's face is green (wood), and likes eating sour food (wood), he will infer from these that the patient suffers from a disease of liver (wood). The treatment is no more than to help one agent that is deficient or inhibit the other that is over flourishing.

2 Metaphysical Basis in TCM

Contemporary philosophy of science no longer rejects the important role of metaphysics in scientific theory: It helps to constitute the hard core of a scientific theory. The hard core of TCM formed with the help of Yin-Yang and Wu Xing theories is substantially different from that of MWM. In sharp contrast with MWM, the cosmos of TCM is a continuous whole, all beings including human beings in the universe consist of Qi with its two forms Yi and Yang. For MWM the cosmos is discontinuous and consists of compartmental entities such as atoms and molecules and their complexes. From TCM's cosmos is derived the concept of unity between nature and human (Tian Ren He Yi): close interrelationship, interdependent, and interaction between nature and humans. Any physiological/pathological change in human body depends on meteorological, geographical and human (physical, mental, behavioural and social) factors. The success/failure of any human action (including treatment and prevention of disease) depends on the perfect combination of three kinds of factors: heavenly opportunities (Tian

	Yin	Yang
Inspection (colour)	dark	bright
Auscultation (sound)	light	loud
Feeling pulse	deep, slow, small, irregular, deficient	superficial, rapid, large, slippery, substantive

Table III: Ying-Yang

Shi); advantages of the geographical location (Di Li); and harmony of the people (Ren He).

For TCM the universe is a holistic macro-organism, and a human is a holistic microorganism. A human consists of Qi and has body, psyche, rationality and emotion as its different forms. As a whole, human body should not de reduced to the sum of its organ-tissue systems, and could not be known by dissection, but only by its outputs. In contrast, for MWM the human body is deemed a machine; it can be disassembled and re-assembled, and reduced to its parts or systems.

For TCM the human body is generative when a human is born, the Qi in it is imperfect (physically, mentally and morally), and so needs to be nourished and cultivated. The Confucian term 'person-making' refers to making efforts to transform an imperfect being into a perfect being physically, mentally and morally. Just as 'person making' is important for Confucianism, so is 'preserving life' for TCM: "A sage-doctor treats those who have not fallen ill, not those who have already have."[2, p.9] In contrast for MWM human body is constitutive: it was constituted by organ systems, tissues, cells and molecules.

3 The Concept of Disease in TCM

For TCM health is the balance between Yin and Yang and disease is the imbalance of them, one of which excessive or deficient. The imbalance may be caused by internal and external elements: Zheng Qi (disease resistant or immune) weakened and Xie Qi (pathogenic) enhanced. Meteorological, geographical, and human (physical, mental, behavioural and social) factors all have an impact on these disease resistant or pathogenic elements; the combination of these factors makes a patient's disease unique.

TCM requires doctors to collect clinical data with four methods of diagnosis: inspection (Wang), auscultation and olfaction (Wen), interrogation (Wen), feeling pulse and palpation (Qie). The symptoms and signs are also categorized into Yin or Yang:

However, it is too vague to only invoke the imbalance of Yin and Yang to explain the clinical manifestations of a patient's disease. TCM introduces Zheng as the pathological basis with the help of additional assumptions. There are eight Zheng (Ba Gang): Exterior (Yang) and Interior (Yin), Cold (Yin) and Heat (Yang), Deficiency (Yin) and Excess (Yang), Yin and Yang. The clinical data collected from the four diagnostic methods are to be classified into these eight Zheng. And each Zheng corresponds with certain signs or symptoms, such as:

Exterior high temperature, fear of cold, headache, stuffy nose, cough, thin and white coating of the tongue, superficial pulse

Interior high temperature without fear of cold, abnormal changes of coating of the tongue, deep pulse

Cold like warm and fear cold, pale, cold extremities, clear urine, thin stool, white, moist and gloomy coating of the tongue, slow and tense pulse

Heat high temperature and like cold, thirst, like drinking cold, red face and eyes, agitated, short and red urine, dry stool, red tongue and yellow coating of it, rapid pulse

And Excess, Deficiency, Yin, Yang have their own corresponding signs and symptoms too.

After a TCM doctor differentiates a certain Zheng (such as Yin Deficiency Zheng) in a given patient, then he has to identify the organ in which this Yin Deficiency Zheng exists. If the doctor collected data such as rapid heartbeat, bad memory, dream, night sweat, mouth and throat dry, tongue red, dry mouth, weak pulse, the data suggest Yin Deficiency Zheng, and he may identify the site of the Zheng at the heart.

It is evident for a TCM doctor that his observations on a patient are not theory-free. He is required to reach Zheng. And Zheng is heavily laden with Yin-Yang and Wu Xing theories. Clinical data or signs and symptoms of patient's disease and patient's complaints are also theory-laden, because the conceptual framework tells him that each outward manifestation which belongs to Yin or Yang, or one of the five agents is an indicator of inward pathological changes. For example, if it is confirmed that the patient is afflicted with Kidney Yin Deficiency, the doctor should use the medicine which belongs to 'water' and Yin to support 'kidney' and Yin or reduce Yang to treat the patient.

Now what is the thing called 'Zheng'? Is it an entity, thing or only a symbol? So long as Zheng was taken an independence entity or thing, it would lead to a series of misinterpretations and misunderstandings. Zheng is nothing but the pattern of relationships between observable clinical manifestations (signs and symptoms) and unobservable underlying pathological changes. The same disease (e.g. pulmonary tuberculosis) can have different Zheng in different individuals in the sense that the patterns of relationship between clinical manifestations and underlying pathological changes may be different for individuals who are afflicted with the same disease. Even for the same patient the patterns may be different if he is afflicted with it in different places, at different times and under different circumstances. So TCM is both holistic and individualized. Does 'Zheng' serve as an explanation of clinical manifestations?

In Qiu's article "Causal Explanation in Traditional Chinese Medicine"[5] he argued that Zheng does serve as an explanation of clinical manifestations, but its explanations do not fit in with the Deductive-Nomological (D-N) nor with the Inductive-Statistical (I-S) models. However, he further argued that Zheng also does not fit in with the pragmatic model of explanation. K. Sadegh-zadeh[6] put forward a pragmatic model of explanation which may be briefly outlined as follows:

EXP is a causal explanation of the event e relative to SL, x, and t if and only if there are an EX and a state description E such that

1. $EXP =< EX, E >$;
2. EX is a causal explanans of the event e relative to SL, x, and t;
3. E describes e;

$EX =< G_1, \cdots, G_m, A_1, \cdots A_n >$; $G_1, \cdots G_m$ are causal laws in SL; $A1, \cdots A_n$ are state descriptions (initial conditions) in SL. $G_1, \cdots G_m, A_1, \cdots A_n$ and E are accepted by x at t. S is a semantical system formulated in a first-order language. L is any system of predicate logic added to S. E is SL-deducible from EX.

Qiu argued that in TCM E is not SL-deducible for EX. Because it has not reached causal laws coherent enough to deduce E from the system of these laws and initial conditions. Now we think Qiu's claim is too strong in denying that explanation in TCM fits the pragmatic or contextual model of explanation. TCM can be looked as a kind of pragmatic or contextual model of explanation, because it does provide an individualized explanation of why the clinical manifestation of an individual patient is

---

[5] See [5, pp.149–170].The author rejected the view that the explanation in TCM does not fit in with pragmatic or contextual explanation.

so unique, and it is not very proper to treat all patients, even those who suffer same disease, as totally equal.

However, we agree that the explanation Zheng provides in Chinese medicine is metaphysical rather scientific. Its explanatory function is rather a heuristic one – that is, helping doctor to retrieve his experience from his long-term memory and provide an outlined orientation for treating patient – rather than a normative one.

Does the 'Zheng' really help doctor to predict the outcome of disease and treat the patient properly? Within the given TCM conceptual framework it is impossible to verify or falsify the explanation it provides. It is unfalsifiable because it can explain sway any of its anomalies with Yin-Yang and Wu Xing theory. It is unverifiable because there is no method within TCM to let people know which factor really acts to make patient feel better or recover from the abnormal.

4 Two Paradigm of Medicine

It is plausible to take TCM and MWM as different paradigms of medicine: They have different metaphysical base. TCM is holistic, while MWM is reductionist. In methodology, TCM prefers a black box method, while MWM prefers to open and change the box to white or grey. The concept of disease Zheng in TCM, the explanans of clinical data (explanadum), is constructive, and the explanation cannot be verified or falsified, while the concept of disease in MWM is built on the basis of physical, chemical and biological achievement, the explanation it provides can be verified or falsified.

MWM researchers have sought single compounds from TCM that might have a role in treating specific diseases, e.g. Artemisinin which is currently the most effective treatment for malaria. In clinical research TCM and MWM doctors can enter randomized controlled trials to compare the effect of respective therapies on patient groups and provide their explanation on the basis of respective theory. When we use TCM's compound formulae, e.g. a complex mixture of herbs[6] to compare with MWM's therapy, we can ignore its components for the time being focusing on the effect of the complex recipe. And systems biology may bring TCM and MWM together. Systems biology, the study of the interactions between proteins, genes, metabolites and components of cells or organisms may provide a way to understand TCM's holistic treatment, and even a measure of the entire body's response to compound formulae.

5 On the Debate "Whether TCM Is Pseudo-Scientific"

The concept of science is confused. For the public or decision makers all good things are called "scientific". "Scientific decision-making" means the decision made is correct, good, right, effective or practical. However, the term "scientific decision-making" covers other more important values, such as just, fair, respecting human rights and human dignity etc. Even some famous philosophers confused "philosophy" and "science", they said "philosophy is science, because it provides knowledge" etc. If so why we need different terms "philosophy" and "science"? They have no knowledge of the demarcation issue between science and non-science.

It is important to distinguish between science and non-science. The unique feature of science and why science develops so rapidly is its testability or empirical evidence base. It is normative not only to scientists but also to editors of scientific journals and sponsors and organizers of science. However, the term "scientific" or "pseudo-scientific"

---

[6]TCM medicines may combine extracts of several plants, sometimes containing as many as 50 species and many thousands of compounds for dealing with the complex of Zheng.

is not a term which can be properly applied to TCM as a knowledge system. Can we simply label the system of Aristotle's an knowledge as "scientific" or "pseudo-scientific"? Of course we cannot. We can identify certain statements in this system scientific or non-scientific, but not the system as a whole. Both in the system of TCM and that of Aristotle knowledge, we can identify certain statements as scientific, and the others non-scientific. And even these concepts (e.g. Yin-Yang and Wu Xing in TCM and Entelechy in the Aristotelian system) that are identified as non-scientific are not naturally useless or meaningless. For instance, the visceral manifestation theory in TCM on the basis of the Yin-Yang and Wu Xing framework provides a non-monofactorical and non-linear web model of causation for pathogenesis and treatment. TCM was founded 2000 years ago, and its evolution does not need the scientific laurel, its major proponents have never sought this laural, so how can we label TCM "pseudo-scientific"?

On the other hand, it is unnecessary to argue for TCM being scientific to persuade people of the merits of TCM and its contributions to the health of the Chinese population. About one third of the Chinese population and an increasing number of people in other countries prefer to be treated with TCM, because they have been sick of MWMs reductionist, disease-centered, machine-dependent and psycho-somatic dualistic approach and prefer TCM's holistic, patient-centered and more personalised and human approach.

Bibliography

[1] Department of Philosophy, Peking University and Institute of Philosophy. *Brief Edition of Materials of Chinese History of Philosophy, Chinese Academy of Social Sciences, Pre-Qin,* vol. 1. Beijing: Chinese Bookstore, 1962.

[2] He, W. B., et al (Eds.). *Plain Inquiries.* Beijing: Chinese Medical Science and Technology Press.

[3] Qian, C. C., et al. *Fundamentals of Ancient Medical Texts.* Beijing: Peoples Health Press, 1980.

[4] Qiu, J. A culture in balance. *Nature,* 448:126–128, 2007. July 12, 2007.

[5] Qiu, R. Z. Causal explanation in traditional chinese medicine. *Ludus Vitalis,* 2:149–170.

[6] Sadegh-zadeh, K. A pragmatic concept of causal explanation. In Nordenfelt, L., et al (Eds.), *Health, Disease and Causal Explanations in Medicine,* 201–210. Dordrecht: Reidel, 1984.

[7] Shanghai College of Traditional Medicine. *Selections of Ancient Medical Texts,.* Shanghai: Shanghai Press of Science and Technology, 1980.

[8] Sun, G. R. (Ed.). *Fundamental Theories of TCM.* Beijing: Science Press, 1996.

Western Concept of Pulse Diagnosis: Théophile de Bordeu and Organicism[1]

Hee-Jin Han

Institute of Humanities

heejinhan@snu.ac.kr

ABSTRACT. In their quest for diagnostic certainty, both western and eastern medicines explored pulse diagnosis for centuries. However, why does western medicine no longer pay attention to the 'qualitative' aspects of pulse? French doctors in the eighteenth century seriously doubted the reliability of pulse diagnosis. By analyzing Théophile de Bordeu's organicist doctrine of pulse, I point out the epistemological obstacles of pulse diagnosis commonly found in western and eastern medicines. I then conclude that recent studies attempting to objectify Traditional Chinese Medicine (TCM) deal with the same old problems Bordeu faced, which led western medicine to limit the validity and the use of pulse diagnosis.

1 Introduction: debates over the pulse diagnosis in Europe

In their quest for diagnostic certainty, both western and eastern medicines explored pulse diagnosis for centuries. In the French medical tradition followed since Galen, diverse theories of pulse diagnosis had been developed at least until the end of the eighteenth century. In Traditional Chinese Medicine (TCM), pulse diagnosis is still used as one of the most reliable methods of clinical examination. Although French and Chinese medicines have different theoretical backgrounds, they are aimed at the same goal of achieving precise diagnostic data.

However, among French doctors – physicians as well as surgeons – in the eighteenth century serious doubts were raised about the objectivity of pulse diagnosis and this led them to distance themselves from Chinese medicine. Pulse diagnosis increasingly lost its clinical importance in the French medical society and was replaced by other diagnostic methods that seemed to provide more constant and significant clinical data about the state of the patient and her pathology.

Nowadays in western medicine, pulse is nothing more than the throbbing of arteries. The term 'pulse' is also used, although incorrectly, to denote the frequency of heartbeat, usually measured in beats per minute. As a result, the concept of pulse has lost all its 'qualitative' meanings. This theoretical and practical orientation of western medicine is reinforced and definitively acknowledged as authentic scientific medicine in the nineteenth century by Claude Bernard and his experimental medicine.

Why does western medicine no longer pay attention to the 'qualitative' aspects of pulse? Since when and in which scientific context did western physicians begin to limit

---

[1] This work was supported by the Korea Research Foundation Grant funded by the Korean Government(MEST). (KRF-2007-361-AL0016)

the use of pulse diagnosis? What are the epistemological implications of this constraint that caused an important divergence between western and eastern medicines? By analyzing Bordeu's organicist doctrine of pulse, this paper points out the epistemological obstacles confronted by both western and eastern medicines in pulse diagnosis.

2 Impact of Bordeu's *Researches on the pulse in relation to the crisis* (1756)

Théophile de Bordeu (1722–1776) was at the center of European debates on pulse diagnosis in his time. He was the founder of a biomedical organicism in France and one of the most important representatives of the Montpellier Medical School along with the well-known vitalist Paul-Joseph Barthez (1734–1806). Bordeu noted that physicians of western medical culture had formerly neglected the tactile elements of diagnosis. Consequently, he established an organicist doctrine of pulse based upon the pioneering observations of a Spanish physician, Francisco Solano de Luque, whose work, *Lapis Lydos Apollinis* (1731), was translated into English by James Nihell (1741) and later into French by Louis Anne Lavirotte (1748). Bordeu examined the nature of various types of pulse in relation to the apparition of crisis that he previously treated in his article titled "Crisis" (1754), published in Diderot's *Encyclopédie*. The topic was of interest and Bordeu's book on pulse, *Researches on the pulse in relation to the crisis* (1756), [2] enjoyed great success. Within a decade his work had inspired three major works by a coterie that came to be called the "poulsistes": *Essai sur le pouls, par rapport aux affections des principaux organes* (1767) by Henri Fouquet, *Nouveau traité du pouls* (1768) by Ménuret de Chambaud, and *Recherches sur la cause de la pulsation des artères* (1769) by François de Lamure – all of them earned their degrees from the Faculty of Medicine in Montpellier.

Bordeu emphasized that his doctrine of pulse, since it was purely the result of observation, had to be confirmed or disproved by many more observations. In terms of the contemporary philosophy of science Bordeu had pursued some sort of verificationism or falsificationism. This was what his self-proclaimed disciples set out to do, verify his teaching on the pulse with observations from their own practice and publish the results in learned journals from the late 1750s onward. The influence of Bordeu was not limited to his self-proclaimed disciples. Many of his colleagues in other parts of Europe were also convinced of the utility of his pulse diagnosis doctrine.

3 Bordeu's organicism

Bordeu's pulse diagnosis doctrine is firmly rooted in his organicist perspective, which consists mainly of a two-fold assertion: an attack on the mechanist medicine and the interventionist therapy, on the one hand, and a defense of the scientific and autonomous nature of medicine, on the other. Against the mechanistic understanding of the human body and of disease, on which is founded the interventionist therapy, Bordeu acknowledged the limits of the so-called 'hard medicine.' As a partisan of Hippocratic naturalist therapy, he promoted 'soft medicine' as a complementary, yet indispensable, part of official medicine.

Due to the complexity and spontaneity of interrelated organs, the interventionist therapy that introduces poisonous substances or brutal operations with little prognostic certainty of their consequences appeared to be too dangerous and even irresponsible for Bordeu. He decried the heedless resorting to bleeding, purgatives, and strong remedies such as *ipecacuanha* in instances where the pulse would indicate the natural progress of disease and thus obviate the need for useless remedies. In this way Bordeu refuted many claims of science committed to physical and chemical determinism. It is true that all diseases follow a natural course that must not be interrupted, but there are no such things as regular critical days, as he had already demonstrated in his article "Crisis" in *Encyclopédie*.

His clinical approaches relied on an organicist concept of life which emphasizes the interrelationship of different organs composing the human body. He described it in his *Researches on the glands and their action* in terms of 'department,' general and particular circulation. [1] According to his organicism, a certain number of organs constitute a department that carries out a particular circulation. Like a swarm of bees, an organism is nothing more than a general circulation composed of several particular circulations. Each department is some kind of machine contributing toward the general life of a living body. With the help of this explanatory model, Bordeu accounted for the controversial problem of his time – apparently spontaneous actions of the gland – and tried to overcome both the mechanism of Herman Boerhaave (1668-1738) and the animism of Georg-Ernst Stahl (1660-1734).

Bordeu framed his work on pulse as an attack on mechanistic and deterministic medicines. In this instance he focused especially on the damage done by physicians who isolated the circulatory system as the origin of illness and unthinkingly relied on protracted rounds of bleeding and purging. Bordeu did not seek to undermine circulation theory as such, for he acknowledged Harvey's discovery as a great step in medicine. But he did argue that medicine had suffered greatly from ignoring the causes of illness other than circulatory obstruction and therapeutic responses beyond routine bleeding and purging.

As a result, Bordeu could not accept mere quantification of pulse, in particular the idea of measuring only the frequency of pulse for clinical purposes. He believed that the cause of pulse is more than the simple movement of cardiac muscles and argued that pulse is the reflection of all organic activities. For this reason each type of pulse designates a particular state of an organ or a sum of organs in a 'department.'

4 Difficulties in the objectification of pulse descriptions

In the opening of his *Researches on the pulse in relation to the crisis*, Bordeu refused to attempt an elementary definition of pulse. He asserted that such questions were purely speculative and did not belong in any work founded solely on practice. The pulse could not be defined but only described, in action, in its varied manifestations. Adopting the language of Boissier de Sauvages (1701-1767) and his nosology, Bordeu proposed to examine the many 'species' of pulse, as they were encountered in the observation of illness.

According to Bordeu, Chinese physicians of old had already developed an intricate doctrine of the pulse that might have been known to Galen. They also developed remedies without risky surgeries – for example, acupuncture, cauteries, vesicatories, sudden applications of hot and cold – that were appropriate to illnesses revealed by the pulse. However, "the Chinese ... have talked about 'rolling' pulse and other pulse which moves like a frog or which resembles the wriggling of a fish, a boiling pot, and a bill of a hen." [2, p. 257] These descriptions were precise but subjective; so it was quite difficult to achieve even a basic list of principal species of pulse acceptable to all physicians.

To surmount the subjective nature of pulse descriptions and to enhance the diagnostic certainty of pulse detection Bordeu tried to simplify the denominations of pulse, employing expressions like 'simple,' 'composed,' and 'complicated' pulse. He then associated them with the newly defined terms of modern anatomy and physiology. "The denominations or the terms of pectoral, capital, and intestinal are derived from the anatomy; they are expressions acknowledged and employed everyday in medicine." [2, p. 258] In this way, Bordeu examined the problems of semiotics that he himself called "a science of signs" (*une science des signes*), a relatively new discipline in the eighteenth century, so as to provide his fellow physicians with a diagnostic and prognostic tool that would save them from fatal errors of intervention.

Undoubtedly, Bordeu did not confine himself to the anatomical descriptions of pulse, when he examined species of pulse associated not with local but with general evacuations such as suppuration and sweat as well as pulses reflecting the work of sensibility, the passions, and such influences as age, temperament, and time of day. Especially difficult to recognize were pulses that were composites of these factors, or that indicated changing phases or directions of maladies.

At this point Bordeu met an important obstacle in his practice of pulse diagnosis: the absence of a 'fixed healthy pulse' that could be used as a datum point and that would allow him to identify pathological pulses. In fact, no diagnosis method is free from this fundamental problem as far as it is dependent on the method of comparison. In the case of pulse diagnosis, the physician must consider at least two measurements of pulse at the same time, the current one and the past one, and the past one is inevitably restored from personal memory, which is often faulty.

Bordeu was aware that far from appearing simple and certain his pulse doctrine might seem to require a lifetime to learn. Accordingly, he did not try to mask the subtlety of the art of the pulse; opening the section on abdominal pulses in his treatise, he observed that an infinitude of observations was required to recognize even a few simple species of pulse. Furthermore, his own descriptive practice could not help but suggest complexity with its proliferating dichotomies between pulses like hard and soft, fast and slow, full and thin, strong and weak, firm and feeble, supple and tense, equal and unequal, free and constrained, and lively and sluggish. Such a practice suggested that language could scarcely capture or reflect the spontaneous work of nature.

Now Bordeu, seeking a 'certain sign,' found instead variability, multiplicity, and experience beyond the usual frames of science. Bordeu wrote to fortify his fellow physicians faced with the vastnesses of nature, but this very effort forced physicians to confront their limitations: some diseases followed no regular course; signs and symptoms would always leave some uncertainty; and 'perfect health' did not exist.

Nevertheless, Bordeu reasserted that the pulse is the most constant symptom and could be relied on even in dangerous and strange cases, if physicians would cultivate a finesse of touch. This did not mean that physicians could expect to employ a standard measure; knowledge of pulse could be gained only by repeated experience with young and old, healthy and sick, male and female. And, finally, Bordeu ended on a hopeful note: if disease as the work of nature was sometimes obscure, nature herself gave the signs that could guide any healer willing to listen, watch, and wait. So, touch and listen, he urged his fellow physicians, and certainties will emerge from seemingly obscure, fleeting phenomena. Under this ambiance of hope it was possible later on to have the invention by René Laennec (1781-1826) of the stethoscope, a medical instrument to help us listen with more finesse and objectivity. Indeed no body sign or symptom could ever be relied on exclusively; the physician must never fail to consider all the symptoms and all the circumstances.

5 Conclusion: contemporary attempts in the objectification of pulse diagnosis

No doubt any physician, whether trained in a western or eastern medical school, can 'feel' and distinguish certain types of pulse. But can he 'know' and properly interpret on physiological grounds what thess pulse signs really mean? Moreover, can this knowledge about the pulse be objectified and integrated into a medical system that educates future physicians? Although Bordeu underlined these problems at the end of his treatise, he was not able to work them out.

We can easily find in today's literature, by searching PubMed or Google, articles concerning the same old problems that Bordeu met in his time. Recent studies attempting to quantify and hence objectify TCM's pulse diagnosis, for example in Korea ("Pulse type classification by varying contact pressure," 2000), in China ("A quantitative system for pulse diagnosis in Traditional Chinese Medicine," 2005), in Taiwan ("The quantization of Chinese deep pulse," 2005), and in Japan ("An objective method of pulse diagnosis," 1987), face the same old difficulties that Bordeu encountered, which led western medicine to restrict the validity and the use of pulse diagnosis. It seems obvious for the moment that we are still far from solving these fundamental and thus recurrent problems that are common to western and eastern medicines.

Bibliography

[1] Bordeu, T. d. Recherches anatomiques sur la position des glandes et sur leur action, 1751. In *Œuvres complètes de Bordeu*. Paris: chez Caille et Ravier, 1818. 45–208.

[2] Bordeu, T. d. Recherches sur le pouls par rapport aux crises, 1756. In *Œuvres complètes de Bordeu*. Paris: chez Caille et Ravier, 1818. 253–421.

Is Traditional Chinese Medicine a Pseudo-Science? – Debates on TCM in China[1]

Ruipeng Lei

Huazhong University of Science & Technology

Lxp73615@163.com

ABSTRACT. It has been a highly controversial question whether Traditional Chinese Medicine (TCM) is a pseudo-science or not. This paper is a preliminary attempt to figure out what the question is about. Firstly, the historical and current debates on TCM in China will be briefly reviewed. Secondly, the existing criteria of demarcating science from pseudo-science will be examined. And then, basic theories of TCM will be explored in comparison with Modern Western Medicine. Finally, a conclusion will be drawn that Traditional Chinese Medicine is the only system of medical theory and practice in the world which is parallel with or independent from Modern Western Medicine, and some comments will be made on the recent attempt at the modernization of TCM.

1 Historical debates on TCM

The first advocate of abolishing Traditional Chinese Medicine in Modern China is Yu Yue, living in Qing Dynasty. He published *On Abolishing Medicine* in 1879, and then *On Medicine and Remedy*, proposed that TCM should be abolished and its remedies should be retained to some extent. This is the outset of the idea of abolishing TCM and retaining its remedies. Till 1900, the ethos of denying the *Wu Xing* (five elements) theory emerged and began to spread. It had been the highly controversial debate in the intellectual and political circles during the course of time whether TCM should be abolished or not, which was regarded as one part of the conflict between Chinese and Western Culture. As a part of Chinese culture, TCM should be denied and rejected as old tradition and culture, which was a widely-held belief within some radical intellectuals during the period of "new culture" movement.

In the early twentieth century the expansion of western medicine in China went hand in hand with the building of the first modern Chinese state. From the outbreak of the Manchurian Plague in 1910 to the establishment of the first Ministry of Health in 1928, western-style doctors gradually developed the strategy of recruiting the Chinese state by translating public health measures and medical administration into tools for state building.[4, pp.1–25] The emergence of the modern Chinese state fundamentally transformed the logic of competition between western-style doctors and traditional Chinese doctors in China, causing them to struggle against each other in the field of the state. In 1912~1913, during this period, TCM was not included as a discipline in the New Education Act issued by the Minkuo state. In 1929, in the name of Chinese Medical Revolution, western-style doctors proposed to the Kuomintang state a wholesale abolition of TCM.

---

[1]Department of Philosophy, Huazhong University of Science & Technology.

The 1929 confrontation constitued an "epistemological event" that led many traditional Chinese medicine practitioners both to embrace the discourse of modernity and begin to "scientize" TCM on the basis of the discourse. Because TCM practitioners adopted the strategy of assimilating Chinese medicine into the emerging national medical system, their efforts radically transformed the theories, practice and social network of TCM. What is now called "traditional Chinese medicine" began to be re-constituted at the moment.[4, vii–viii] In 1950, Yu Yun-xiu, who actively participated in the 1929 debate as the advocate of the abolition of TCM, submitted to the Communist Party state a draft guideline on modernizing and scientizing TCM. He, however, changed the word of abolition into transformation. The so-called conjunction of TCM and Western Medicine (WM) began to creep into medical vocabularies and caused the competing groups to struggle against each other in a new form. Tracing the history of medical developments in China after 1950, we can find that the existing theories of Chinese medicine have been established by reference to western medicine, which take western medicine as criterion for evaluation and assessment. The current systems of Chinese medicine deviate from traditional conceptual framework and method, and cut its tie with traditional Chinese culture.

2 Current debates on TCM in China

China now is fiercely embracing modernity. Chinese medicine hospitals have been advocating the conjunction of TCM and WM. On the other hand, some clinics of TCM, which have been operating the same practice for thousands of years, seem vulnerable and out of place. Indeed, attitudes on TCM still have divided the country. In 2006, Zhang Gong-yao, from the Central South University in Changsha, Hunan, published an article in a Chinese Journal calling TCM a pseudo-science that should be banished from public healthcare and research.[13, pp.14–17] The article caused uproar in the country, and earlier this year the government announced an ambitious plan to modernize the millennia-old practice.

Fang Shi-min, a US-trained biochemist who now runs a society called New Threads that is known for fighting pseudoscience and research misconduct in China said that it is exactly those TCM theories that should be abolished, and some effective remedies in TCM which are compound formulae that should be retained. He is in favor of scientific research into Chinese herbal remedies, but thinks the emphasis on testing the theories of TCM is misplaced. Concepts such as *yin-yang*, *wu xing* and *qi*, "are inaccurate descriptions of the human body that verge on imagination", he says.[10, pp.126–128]

Zhu Qing-shi, an academician and the president of Chinese University of Science and Technology made the following points in a TCM forum, firstly, we should analyze the respective advantages of TCM and WM theories and practices in their own medical paradigm; Secondly, we should make judgment on the scientific status of TCM from the perspective of contemporary complex science, rather than modern mechnistic science. Thirdly, the main task of TCM is not to compete with WM, but to develop its own theories and construct its own medical paradigm on its own terms.[14, p.27]

It is significant to note that the current debates shift the focus from the modernization of TCM, or the conjunction of TCM and WM during 1950s back to the wholesale abolition of TCM. Zhang and Fang characterize TCM as a pseudo-science. The criteria for being a pseudo-science, however, are far from precise. In unhealthy academic circumstances, especially where an enforced orthodoxy exists, all other schools can be condemned without justification as pseudo-sciences.

3 Science and Pseudo-science

"Pseudo" from Greek means false or unreal. Literally it refers to a doctrine or set of views that falsely claims the status of science or knowledge.[1, p.832] Determining whether a doctrine is a pseudo-science is a controversial issue in the philosophy of science and requires that one first determines what counts as scientific knowledge. For logical positivism, if a thesis satisfies the verifiability criterion, it is scientific; otherwise it is pseudo-scientific. Accordingly, traditional metaphysics is pseudo-scientific and meaningless because none of its claims can be verified. Popper claimed that the criterion is that of falsifiability. To be scientific, a theory must be falsifiable, that is, in principle there must be some observation statements that would contradict the theory. According to Popper the ideal strategy for scientific investigations is roughly as follows: Start with an explanatory problem; Propose a bold conjecture as a solution to the problem; Severely test the conjecture; If the theory fails a test, propose a new theory which will give a non-ad hoc account of not only the original phenomena to be explained plus any successful prediction of the first theory, but also the result of the test which the first theory failed; Repeat the testing procedure.

There are many excellent criticisms which have been made of the above criteria proposed by the two schools of philosophy of science. They can be summarized as follows: take modern empirical science as analogy, neglect the role of metaphysics, overemphasize the role of direct empirical testing, lead to the premature abandonment of interesting theories.

Someone may doubt about whether modern empirical science is the only legitimate reference or analogy for knowledge about natural world, further, whether modern philosophy of science can provide a coherent and widely-accepted criterion for demarcating science from pseudo-science. Modern western science is highly esteemed. Apparently it is a widely held belief worldwide that there is something special about modern western science and its methods. As Alan Chalmers rightly pointed out, the naming of some claim or line of reasoning or piece of research "scientific" is done in a way that is intended to imply some kind of merit or special kind of reliability.[?, pp.8–10] But what, if anything, is so special about modern western science? Philosophers make their attempts to answer or doubt about questions of that kind, and still disagree with each other.

More recent scholars rely on the idea that a theory, to be scientific, must be testable, i.e. refutable in principle, and the holder of the theory must take conscious steps to refute it. Now, it is in fact not the case that all theories in science are meant to be refutable. Whether a theory is refutable or not depends on the job it is intended to do. Theories in crystallography are immune to discrepant observations since the crystallographers categorize solids such that any deviation from the theory is taken as an imperfection of the solid, not the theory. Any classificatory theory is a closed system of categories aimed a priori at logical completeness. Psychological theories as psychoanalytic and behavioristic accounts may be incontrovertible simply by reference to the facts, amendment of the hypothesis is not simply a matter of empirical testing. But it doesn't mean that the existence of Freudian psychology rules out the invention of a rival body of ideas which could compete with psychoanalysis on its own terms. It is the same case that TCM can compete with WM as a different medical paradigm.

Certainly all scientific theories must in some way "fit the facts". An alternative theory may make new "facts" available or emphasize different phenomena. Feyerabend shows that many empirical theories do not yield refutable consequences and it may only be when an alternative is imagined that an earlier theory is questioned. TCM and WM theories are each a body of conventions which the practitioner may or may not apply according to his experience in much the same way as the physicist may have to decide whether phenomena are, for example, gravitational or magnetic. It is no harmful matter if each particular theory taken in isolation yields only confirming observations.

Observations are necessarily theory-linked so that it is not a sign of self-deception that TCM theories produce TCM data and WM, WM data. Observation is never passive registration; it is always an active skill involving an interaction between observer and phenomena which may be changed by being observed. TCM practice is no difference in this respect from any other empirical enterprise.

Any methodological principle is like any other human institutions—they are not fixed and immutable laws but guidelines subject to interpretation and reinterpretation. Popper, of course, has argued otherwise. On his classical notion of disconfirmation theories the deductive predictions of which are falsified must be rejected. The notion of disconfirmation which the recent scholars have implied in talking about the role of alternative theories is rather less austere. A theory may still do useful work even though predictions from it have been falsified (in any case theories are usually amended not rejected if they cannot be confirmed). In some cases as Feyerabend notes a theory may not conceivably yield falsifiable predictions so that an alternative can open up new possibilities. Popper does not deny that the statements of a pseudo-science in his opinion are meaningful. Some critics of Popper claim that pseudo-sciences are typically both falsifiable and falsified, but are still retained by their supporters. From this perspective, the integrity of individual investigators and of the institutions of the scientific community is a more important consideration in distinguishing science from pseudo-science. Open-mindedness is certainly one of the scientific virtues but most scientists, important or not, are not always models of rectitude. Dogmatic rationalization of failure is in fact a normal part of a scientists behavior.

Maybe someone will respond that it's not necessary to exhort us to suspend judgment on the scientific status of TCM until we have a precise sense of the expression pseudo-science. It is a familiar comment in philosophy that however insufficiently defined some concept may be for decisions to be made about borderline cases, we have no difficulty in picking out the clear-cut instances of it. However, it is evident that TCM is not a clear-cut case of pseudo-science.

Medicine is meant to solve a patient's problem, and presuppose an explanation to help the healer to predict the outcome of that problem and to take some action to restore the patient to normal. The term TCM refers to the paradigm of medicine which is rooted in Chinese culture and has a set of characteristic features distinct from any other paradigms of medicine in the world.[2] TCM is a system that consists of the philosophical, ethical and psychological, and diagnostic and clinical levels of representation, which are related as a nested hierarchy, showing an understanding of human life from a variety of perspectives. Traditional Chinese treatments are mixtures of ingredients, concocted on the spot on the basis of a patient's symptoms and characteristics and using theories passed down through generations. The conceptual framework of TCM lies embedded in a characteristic theory – *Yin-Yang* and *Wu Xing* theory.[3] Chinese doctors use the principles of the theory to explain the structure of human body, physiological functions and pathological changes and guide their action in the diagnosis and treatment. TCM is millennia-practice which has achieved clinical effectiveness in many diseases and maintained, improved Chinese people's health. It shows that whether from the west or the east medical practice share the same goal. TCM as Chinese local knowledge should be one component of medical (scientific) knowledge. Open-mindedness, tolerance and diverse strategies are justifiable when we try to achieve the same goal, that is, maintaining and improving human health.

---

[2] Paradigm may refer to practical mastery of concrete scientific achievements, not to commonly accepted theoretical position. To accept a paradigm, it is not so much to understand and believe a statement as to grasp a practical skill.

[3] Ted J. Kaptchuk translated Wu Xing into "five phases". In his opinion, the misconception has long been embodied in the common translation "five elements" and exemplifies the problems that arise from looking at things Chinese with a western frame of reference. The five phases (Wu Xing) are not in any way ultimately constituents of matter.[3, p.465]

Contemporary philosophy of science is undergoing a process of evolution, different schools have been developing with criticisms. They provide the contradictory criteria for demarcating science and competitive accounts of science. The recent development of philosophy of science is inclined to regard science as a process, a kind of social practice, a way of interaction between man and natural world. The emerging practical philosophy of science characterizes science a filed of practical skill and action, not merely the field of belief and rationality. More importantly, the practical philosophy of science proposed a new account of science which is totally different from traditional accounts of science. It can be stated as follows: Science is not the network of meaningful statement and proposition, but the field of practical activities, and practice-dominated. Science is not simply the way of representing and observing the world, more importantly, the way of working on the world. Scientific practice is local and social. Scientific knowledge which is based on local practice is not universal, but rather essentially with a feature of local knowledge.[12, p.27,38,129] According to traditional accounts of science, science is a rational activity operating according to some special methods. The distinctiveness of scientific knowledge lies in the abstraction from practice to universal and generalizable knowledge. Scientific knowledge is derived from practice, and finally, more importantly, constitutes the network of abstract, general propositions and statements. In the view of practical philosophy of science, this is not the result of de-locality, but rather one of standardization. Locality not only refers to specific area, but also involves the context in which knowledge yields and is justified. However, logically speaking, science is a practical activity and scientific knowledge is a kind of local knowledge. It does not follow from these statements that all practical activities are science and local knowledge is scientific knowledge.

4 Basic Theories of TCM

As the global expansion of western biomedicine, there have been collisions and conflicts between the two medical paradigms, which are seemingly not commensurable. In the past few years, the pharmaceutical industry has seen a shift from the search for "magic bullets" that specifically target a single disease-causing molecule to the pursuit of combination therapies that comprise more than one active ingredient. More and more researchers worldwide have come to realize that TCM is a unique alternative to modern western biomedicine. TCM and WM, however, face almost irreconcilable differences. So a couple of questions come up with: Is it possible to bring them together? Is it possible to gauge the true potential of TCM? and How? To cope with theses questions, we'd have to explore and examine the conceptual framework and basic theories of TCM at first.

TCM is not just a medical system, but a branch of philosophy and healing arts that is an important part of Chinese Culture. The fundamental concepts in Chinese philosophy as *Qi, Yin and Yang, Wu Xing* which have been applied in TCM, produced the unique TCM theories, such as *Zang Xiang, Jing Luo, Bian Zheng Lun Zhi*, etc. Chinese philosophy underlying the theories of TCM, as Needham profoundly pointed out, is organic philosophy, which presupposes organic relationship between the whole and its parts, organism and its environment, as well as strong analogy between nature and living organism. Meanwhile, it places correlative thinking in high esteem. Accordingly, TCM have adopted a holistic approach to solve a patient's problem, which is nonreductionistic, nonanalytic and nonmechanistic, as an opposite to modern Western medicine. What's more, TCM also provides an explanation of a clinical manifestation, and helps the doctor to predict the outcome of a patient's problem and treat him/her properly. The explanatory model of TCM, Prof. Qiu Ren-zong gave a detailed account and tenable argument in the article, *Causal Explanation in Traditional Chinese Medicine*, published in the journal of *LUDUS VITALIS* in English in 1994, is one we may call a "web (nonlinear) model" of causation, contrasted with a linear model of causation which modern western medicine has employed.[11, pp.149–170]

5 Yin-Yang Theory

What are the things called *"yin"* and *"yang"* is still a disputable issue among Chinese philosophers and historians of science and medicine. According to the Yin-Yang theory, all things in the universe consist of *yin* and *yang*. All changes in the universe are explained by the growth and decline, or ebb and flow, of *yin* and *yang*, and the interrelation and interaction between them. Not few Chinese philosophers interpreted them as two kinds or forms of *qi* or *"air"*, but some think it is rather something like "pneuma", as their use in medicine indicated, because ancient Chinese medical writers used *"qi"* to refer to a vital force or energy that maintains all forms of life and organic functions, as well as any kind of change and transformation in the universe.

"The intercourse between male and female causes all things generating." [The Book of Changes]. "The intercourse between male and female is the secret of change." [Interior Classics: Questions and Answers about Living Matter]. "Qi is interacted between Heaven and Earth, all things are automatically generated, just like qi is interacted between husband and wife, the child is automatically born." [On Measurement] The concept of *yin* and *yang* may be derived from the worship of sex. We can say "yin" and *"yang"* are symbols of vital forces. After deriving from female and male sex, they were extended to all things in the universe that generate, grow, change and transform within these two vital forces. In a further step of abstraction, *"yin"* and *"yang"* become something like the Hegelian dialectical concept-opposites, positive and negative, and not simply refer to vital forces. However, in TCM this step has never been completed. All things in the universe, no matter if they are living or non-living, even nature or the universe itself, are like an organism, they consist of *yin* and *yang*, and are generated from the interaction between *yin* and *yang*.[11, pp.149–170] Life is the condensation of *qi*, and death is the dispersion of *qi*. [Zhuang Zi] *Yin-Yang* is the Tao of Heaven and Earth, the law of all things the cause of changes and the basis of life and death. [*Interior Classics: Questions and Answers about Living Matter*] *Yin-yang* causes the beginning and end of all things, and is the root of life and death. If you run counter against them, you will suffer from catastrophe, and if you follow them, you will prevent disease from you. [*Interior Classics: Questions and Answers about Living Matter*]

The three principles of *yin-yang* theory are the following:[11, pp.149-170]

1. The principle of mutual conflict: The succession of four reasons is the outcome of struggle between *"yin"* and *"yang"*. *"Yin"* wins in the winter and *"yang"* in the summer. Health and disease are the same: "Excess of *yin* cause the disorder of *yang* and excess of *yang* causes the disorder of *yin*." [*Interior Classics: Plain Inquiries*]
2. The principle of interdependence: *"Yang"* refers to upper, left, heat, excess; *"Yin"* to lower, right, cold, deficiency. But there would not be the former without the latter, and vice versa. "The internal *yin* is the source of *yang*, the external *yang* is the representation of *yin*." There would be nothing in the world if *yin*, or *yang*, exists alone.
3. The principle of mutual transformation: The change of all things is caused by ebb and flow of *yin* and *yang*. The excess of *yin*, or *yang*, reaches a certain limit beyond where the change will be reversed. The *yang* that increases to its climax will be transformed into *yin*, as we can see in a patient with an acute infectious disease, who, after several days of continuous fever exhibits a sudden decrease of temperature, pale, cold extremities and weak pulse, among other manifestations of *yin* character.

The *Yin-Yang* theory is invoked to explain the structure of the human body, physiological function and pathological changes, and to make diagnosis and manage treatment of disease.[11, pp.49-170] *Yin* refers to lower parts, inside of body, anterior side, *interior* side, hollow viscera, stomach, intestines, bladder, gall bladder. *Yang* otherwise. Health is the dynamic equilibrium between *yin* and *yang*, and disease is the imbalance between them, when one is in excess or deficiency beyond a certain limit. What is this certain limit? There are no precise defined criteria, everything depends on the individual patient, afflicted with a particular disease in a particular place, at a particular time and under particular circumstance. Moreover, the genesis of a disease involves Zheng (resistance, immunity) and Xie (evil, pathogens, pathogenetic factors). The wide variety of diseases is classified in two general categories: *Yin Zheng* and *Yang Zheng*. The first part of diagnosis is to recognize whether the disease belong to *yin* or *yang*. Diagnosis is also required to distinguish between Exterior and Interior, Heat and Cold, Excess and Deficiency. They are called eight guiding principle (Ba Gang). Exterior, Heat and Excess belong to *Yang*; Interior, Cold and Deficiency to *Yin*. The principle of treatment is "Treating *Yin* for the *Yang* disease (*Yang* excess), and treating *Yang* for the *Yin* disease (*Yin* excess)". The doctor will select the medicine that belongs to *yin* or *yang*, to prescribe the patient in order to nourish *yin* in the former and support *yang* in the later.

6 Wu Xing Theory

According to the classic *Book of Records*, as early as the Shang Dynasty (1000BC), Chinese people already realized that *water*, *fire*, *metal*, *wood* and *earth* are essential for human life. They are *Wu Xing*, "*Wu*" is the number of five, and "*Xing*" means walk or move here, and perhaps most pertinently, it implies a process. *Wu Xing* (five phases), therefore, are five kinds of process, not five elements. The theory of phases is a system of correspondences and patterns that subsume events and things, especially in relation to their dynamics.[3, p.465] The principles of *Wu Xing* theory are as follows:[11, pp.149–170]

1. Mutual generation means that one phase generates another and promotes the growth of even another phase. The order of mutual generation is: *wood* → *fire* → *earth* → *metal* → *water* → *wood* ⋯
2. Mutual inhibition (restriction) means that one phase inhibits, restrains or expels another. The order of mutual inhibition is: *wood* → *earth* → *water* → *fire* → *metal* → *wood* ⋯ The relationship of each phase with one another is reciprocal.
3. Mutual invasion refers to that if one phase inhibits another one beyond a normal extent, it will become evident as a disturbance of normal harmony.
4. Mutual insult (reverse restriction) refers to a reversal of mutual inhibition and is another manifestation of a disturbance of normal harmony.

The application of the *Wu Xing* theory in medicine is *Zang Xiang* theory, which is aimed mainly to account for the physiological and pathological changes of internal organs and to diagnose and treat diseases in these organs. TCM emphasizes *Zang Xiang*, not *Zang Fu*, the latter means solid organs and hollow organs. Liver (*wood*), heart (*fire*), kidney (*water*), lung (*metal*), spleen (*earth*) are not the entity of viscera and bowels, or the unit of anatomy, or (quasi-) empirical hypotheses, but rather a kind of theoretical model and construct, sign of function, introduced by TCM as a concept by reference to which various empirical hypotheses can be collected into a unified theory. A German scholar, Manfred Porkert, proposed that they are energetic sphere (functional configuration) in TCM theory.[5] *Zang Xiang* Theory is aimed mainly to account for the correlation of liver, heart, kidney, lung and spleen; of their *exterior* and *interior*; of their physiological and pathological changes; and their relationship to the structure of human body, to the *Jing Luo*, to the different emotions, etc. For instance, the heart controls blood and vessels, as well as mental activities. Meanwhile, it coordinates the activities of all other internal organs to maintain normal functions of human body through blood and mind. Its functions manifest on pulse, face and the tongue. Sweat is the juice of the heart. It is also correlated with the emotion of joy. Kidney keeps $jing^4$ *qi* and control growth, development and reproduction. Kidney yin and kidney yang are the root of yin-yang in human body. Kidney is responsible for receiving *qi*. *Jing qi* in kidney can be transformed into bone marrow, spine and brain tissue, nurture skeleton, and promote the growth of teeth and hair. Saliva is the fluid of kidney. Kidney is related with the emotion of fear. The spirit of the kidney (that belongs to *water*) provides the liver (*wood*) with nourishment and the liver in turn keeps blood to nourish the heart (*fire*). The heart in turn conveys heat to the spleen (*earth*) which nourishes the lung (*metal*) with the spirit of *water*-grain and so on. A disease of the liver (*wood*) can invade the spleen (*earth*) and the latter disease in turn can infect the liver (*wood*), etc.[11, pp.149–170] In diagnosing, a doctor relates the heart to the tongue, for by observing the tongue he learns about physiological functions and pathological changes of the heart; he also relates liver disease (*wood*) to a green face (*wood*) and preference for sour food (*wood*), and to spring (*wood*) as well. However, if a patient with a spleen disease (*earth*) has a green face (*wood*), it can be explained by the *earth* being invaded by *wood*. The treatment consists simply to help the deficient phase, or to inhibit the excessive one.

7 TCM vs. Modern Western Medicine

Now we can see TCM and Modern Western Medicine face almost irreconcilable differences, which can be roughly summed up as follows:

Differences in metaphysical basis: WM holds that human life system consists of concrete entities (different elements), which leads to a positivistic medicine aiming at microscopic research. TCM is based on ideas such as *qi*, in which illness is caused by blocked energy channels; *yin-yang*, which emphasizes the balance of energy; and *Wu Xing*, in which people's organs and health status are categorized according to their elemental characteristics: *fire*, *wood*, *water*, *earth* and *metal*. These characteristics appear in the process of change. Here, as Needham profoundly pointed out, Chinese people avoid replying the ontological questions. Is there any real entity called *yin* or *yang*, *exterior* or *interior*, *cold* or *heat*, *deficiency* or *excess*? No. Are heart, lung, kidney, liver etc. in the terminology of Chinese medicine, the same as those terms in modern medicine that refer to certain organs in human body? Again, no. In all theses cases, they are theoretical construct or models, which can not be empirically tested. We cannot expect statements of how the inward pathological condition is to be explained in terms of a particular theoretical construct to be empirical hypotheses. In sum, modern WM emphasizes material entity, TCM correlation and coordination.

Differences in methodology: By the methods of reductionism, WM dissects human body into different parts and elemental particles, successively finds morphological and functional changes that leads to man's physiological and pathological courses at different level, and then reduces life phenomena to general physical and chemical process. But according to the theory of *Yin-Yang* and *Wu Xing*, TCM focuses on the whole functional states of human life activities, the rules of man's physiological and pathological changes, and the correlation of intervenient factors and functional states. TCM views the various parts of the human body as an organic whole, emphasizes the harmony and coordination of the internal organs with other parts or structures, and also stresses the unity of the human body with the external environment. The methods of diagnosis and treatment are based on overall analysis of syndromes and signs, the cause, nature and location of

---

[4] "Jing is the root of human body." [*Interior Classics: Vital Axis*]

the illness and the patients physical and emotional condition. It is a kind of holistic approach.

Differences in objects of study: We can see a shift in the development of WM from the holistic mastery of human life to the imitation of human physiological and pathological process by reference to anatomic human body and nonhuman animal model. For modern WM human body is deemed as a machine, it can be disassembled and re-assembled, and reduced to its parts or systems. For TCM, the study and observation are never separated from the living human body. The universe is a holistic macro-organism, and human is a holistic micro-organism. As a whole human body should not be reduced to the sum of its organ-tissue systems, couldn't be known by dissection. Its internal changes could be known only by its outputs. This reflects the holistic, organic view of human body. In sum, we can say that for modern WM the major concern is the disease of man, for TCM it is human in disease.

Differences in differentiation of disease and *Zheng*: Modern western medicine generally prescribes treatments for specific diseases, often on the basis of their physiological cause. TCM, however, focuses on *Zheng* by analyzing relevant information, signs and symptoms through the four methods of diagnosis (observation, listening and smelling, inquiring, pulse feeling and palpation) in the light of the theory of TCM, having a good idea of the cause, nature and location of a disease, and the relationship between pathogenic factors and vital energy, and summarizing them into syndrome of a certain nature. *Zheng* is nothing but a pattern of the relationship between observable clinical manifestations and unobservable underlying pathological changes – imbalance between yin and yang. The same disease, e.g. pulmonary tuberculosis, can have different *Zheng* in different individuals who suffer from it, in the sense that the patterns of relationship between clinical manifestations and the underlying pathological changes may vary among individuals afflicted with the same disease. Even for the same patient, the patterns can be also different if he is afflicted with them in different places, at different times and under different circumstances. So TCM is very individualized.[11, pp.139–170] A diagnostic process as involved in TCM is cross-level cognitive movement from the natural level to the biological, ethical, psychological and philosophical ones. More specifically, diagnosis embodies a movement from empirical experience as explained by the philosophical notions of *Yin-Yang* and *Wu Xing*, to speculation at the diagnostic level; it is a movement from descriptions of outward symptoms and signs to speculation on the inward pathological condition, namely, on the emotions and mentality of the patient, on the functions and changes of the body, on the relationship of the five viscera to their corresponding body parts, natural environmental factors, etc., and on the relationship among the five viscera. The basic diagnostic method is determining the internal disturbance by observing external signs. It is not far from the principle of black box, which deals with the manifestations of the disease rather than the disease per se, and which aims at discovering what kind of input yields what kind of output.

Deng Yong-zhao explored the distinction between truth-oriented and goodness-oriented knowledge and science in the doctoral dissertation in 2003.[5] Dr. Deng made the claim that TCM is a typically goodness-oriented medical science, as distinguished from modern Western medicine, which is largely truth-oriented. Truth-oriented knowledge and science are based on empirically observable evidence and presented in a so-called literal language, goodness-oriented knowledge and science rely on speculation as portrayed in a language that is essentially metaphorical, and the validity of this speculation is confirmed indirectly, by inference or through the effect of this speculation. Both of

---

[5][2, pp.13–39] The dissertation aims at an exploration of Chinese correlative language, as exemplified in TCM. The central claim that the author makes is that TCM is able to attain functionally and qualitatively definite meanings, which provide the basis for diagnosis and treatment. It also examines how this claim works out with the epistemology of TCM and explores its possible implications to future medicine.

knowledge and science have to do with the nature of representation. Each type of knowledge is related to its specific representation and involves its specific confirmability and use of language. Truth-oriented knowledge may be called empirical knowledge, which is supported by a representation that is based on observable evidence, directly verifiable and immediately given in literal language. Goodness-oriented knowledge is the type of knowledge which is based on a representation that is essentially imaginative and speculative, and portrayed, indirectly and remotely, in a metaphorical language which, reflective of a value system, is referential to quality, function, or relationship rather than any object in reality. This distinction is due to the nature of human cognition, which involves perception and rationality and thus underlies the use of two types of language. Where and when we avail ourselves of literal language to represent paradoxes of reality directly, which are accessible to our senses, the result is our truth-oriented knowledge. And where and when we'd have to represent reality remotely, for example, contemporary theories of physics speculate on such "things" as fields and forces, or the nonperceptible events occurring at the microscopic level of the world, and explain elementary particles in terms of probability; a traditional Chinese physician describes physiological relationships and pathological influences among the five visceral systems with his correlative language of *yin-yang* and *wu xing*, the result is our goodness-oriented knowledge. Both types of knowledge are indispensable to our existential and intellectual survival.

To cure a patient, a traditional Chinese doctor needs first of all collect the information of the patient, by observing, listening and smelling, asking, pulse feeling and palpating. With the language of *yin-yang*, he differentiates between syndromes. Differing from a symptom, a syndrome summarizes a group of symptoms that characterize the pathological changes at a certain stage. Correlating them with *yin-yang*, he differentiates between syndromes, that is, between *yin* and *yang* syndromes, as well as *exterior* and *interior*, cold and heat, or deficiency and excess, syndromes, all of which are ascribed to *yin* and *yang* syndromes. Here "*yin*", "*yang*", "*cold*", "*heat*", and so on are metaphorical modifiers. On the other hand, with the language of *Wu Xing* the Chinese doctor relates the pathological changes of the five viscera to other body parts, colors, tastes, emotions, etc., and to environmental factors, insofar as he ascribes them to the five phases. He also describes the physiological relationships and pathological influences among the five viscera in terms of the five phases and their relationships, that is, generation, restriction, invasion, reverse restriction. In all cases, the doctor is expected to apply medications to removing specific syndrome so as to restore relative balance of *yin* and *yang*, and to return the relationships among the five phases back to normal. With the method of black box, a Chinese doctor tries to find out what kind of pathogenic factor yields what kind of signs and symptoms, or what kind of therapy yields what kind of effect. He verifies his diagnosis and treatment directly by looking at the therapeutic effect, which aims at the removal of signs and symptoms. Abstracted and generalized from the perceptual experience of reality in general, the philosophy of *yin-yang* and *wu xing* serves as the criterion of diagnostic procedure, which is goodness-based speculation cloaked in a correlative language that symbolizes and explains the psychological, physiological and pathological condition of the human body. Cognition eventually moves to the clinical level, where treatment is given and, along with diagnosis, tested indirectly but empirically (or "quasi-empirically", as Needham puts it), by examining the actual curative effect – a test which is truth-based.[2, pp.39–45] So confirmed indirectly, the diagnosis and therapy attain the status of a generally applicable medical theory or paradigm, which is parallel with and independent from modern Western Medicine. The clinical results are demonstrable in a controlled manner. With its "precision, verifiability, and communicability," which "rest on a qualitative foundation", TCM, as Manfred Porkert claims, is to a great extent a "exact science" rather than a "proto-science".[7, p.157], [6, pp.553–572] In fact, confirmation at the diagnostic and clinical level has historically corrected and helped shape the meanings of the philosophy of *yin-yang* and *wu xing*.

8 Back to the future vs. back to the tradition

Debates on TCM vs. Modern WM, accompanying with the collision and conflict between them, do not stop, and will not, not just in China, but around the world, in the so-called new century of biomedical science. It is more important for us to take the relationship between them into account at present, not just to stimulate struggling against each other, when researchers, practitioners and drug companies around the world are engaged in a complex, tentative dance over the best way to tap into the unknown potential of TCM. Pharmaceutical companies have become more interested in TCM over the past decade. The world market for Chinese herbal medicine has doubled during this period. But their approach has been characteristically western and reductionistic: isolate the active ingredients and test them one at a time.[10, pp.126–128] China also has announced an ambitious attempt to bring the ancient practice of TCM into line with modern standards. The government says it will expand basic and clinical research, and improve the testing and developing of TCM remedies for export. More controversial is the government's shift in approach. Previous attempts have focused on developing scientifically tested drugs from TCM, which has led to a handful of new treatments, such as artemisinin for treating malaria and the decongestant ephedrine. The new plan aims to develop methodologies to test TCM's more traditional features and principles.[9, pp.590–591] Although with the success of new treatments, there are still some concerns about the attempt at the modernization of TCM, from both advocates and skeptics of the practice. The advocates are uncomfortable with separating the study and development of Chinese herbal medicines from the theories that underlie its normal practice. Devoid of that cultural context, it would become a tree without roots. The skeptics are also unimpressed by the plan, but for opposite reasons. They point out that the government has already spent a lot of money trying to prove the mechanistic basis of TCM, which hasn't gone anywhere.

Should the formidable gap between TCM and WM be bridged? And how? Can Systems biology or metabonomics bridge them together? Systems biology is the study of the interactions between proteins, genes, metabolites and components of cells or organisms. Some researchers in China and elsewhere are advocating systems biology as a way to assess the usefulness of traditional medicine. The new emerging analytical tools can now dissect the complex physiological effects of Chinese herbal medicines with a new level of sophistication. They believe it is a powerful tool to tease out the effects of those multiple components in Chinese herbal medicine and offer a modern re-interpretation of this ancient tradition. This is, in a sense, back to the future.[8, pp.506–507] Constructive approaches to divining the potential usefulness of traditional therapies are to be welcomed. On the other hand, some researchers are concerned that modernization will simply distort the essence of traditional medicine. TCM is already close to perfect after thousands of years of practice and development. When we try to gauge the true potential of TCM, "back to the tradition" which means the development on its own terms and theories may be more a plausible strategy to adopt.

Bibliography

[1] *Dictionary of Western Philosophy: English-Chinese*. Beijing: The People Press, 2001.

[2] Deng, Y. Z. *Language, Culture and Traditional Chinese Medicine: A Fraserean Perspective*. Ph.D. thesis, University of Texas at Dallas, 2003.

[3] Kaptchuck, T. J. *Chinese Medicine: the web that has no weaver.* London: Rider, revised edn., 2000.

[4] Lei, H. L. *When Chinese Medicine Encountered the State:1910-1949.* Ph.D. thesis, University of Chicago, 1999.

[5] Porkert, M. *The Theoretical Foundations of Chinese Medicine: Systems of Correspondenc.* Cambridge, Massachusetts: The MIT Press, 1974.

[6] Porkert, M. The difficult task of blending chinese and western science: The case of modern interpretations of traditional chinese medicine. In Li, G., Zhang, M., Cao, T. (Eds.), *Explorations in the History of Science and Technology in China: Compiled in Honor of the 80th Birthday of Dr. Joseph Needham.* Shanghai: Shanghai Chinese Classic Publishing House, 1982.

[7] Porkert, M., Ullmann, C. *Chinese Medicine. Trans. Mark Hawson.* New York: Henry Holt and Company, 1988.

[8] Qiu, J. "back to the future" for chinese herbal medicines. *Nature Reviews*, 6, 2007.

[9] Qiu, J. China plans to modernize traditional medicin. *Nature*, 446, 2007.

[10] Qiu, J. A culture in the balance. *Nature*, 448, 2007.

[11] Qiu, R. Z. Causal explanation in traditional chinese medicine. *LUDUS VITALIS*, 2(2), 1994.

[12] Rouse, J. *Knowledge and Power: Toward a Political Philosophy of Science.* Ithaca: Cornell University Press, 1987.

[13] Yao, Z. G. Farewell to traditional chinese medicine and remedies. *Journal of Medicine and Philosophy*, 4, 2006.

[14] Zhu, Q. S. Traditional chinese medicine from the perspective of complex science. *Jiangsu Journal of Traditional Chinese Medicine*, 8, 2005.

Experimenting on innovative scientific vs. traditional treatments: the case of AIDS medical research in China[1]

Evelyne Micollier

Institut de Recherche pour le Développement

Evelyne.Micollier@ird.fr

ABSTRACT. Since the 1950s, rehabilitation and legitimacy of an ever-evolving and multi-faceted scholarly medical tradition tend to be gained through the use of the modern science explanatory model. Interfaces of knowledge and practice are approached through experiment in biomedicine and in traditional (empirical) medicine revealing a process of 'biomedicalisation' of the latter over time through an ongoing tentative process of modernisation, standardisation and means of legitimacy along the lines of biomedical sciences criteria even though their rationales are radically diverging.

Traditional Chinese Medicine (TCM, *zhongyi*), the traditional medicine integrated in the public health system promoted by the post-1949 Chinese government, can be defined as a 'neo-traditional' medicine insofar as references of modern biology and standardised biomedicine are inserted in its transmission, body of knowledge and practice. The question of clinical trials is relevant in biomedicine and traditional medicine in a growing number of developing countries: even though they raise controversial issues, such trials are often encouraged at local, national, and international levels. Trials in traditional medicine are promoted with a political will locally and globally, a trend initiated by the WHO recognition of the efficacy of some traditional medicines in treating a number of pathologies at the Alma Ata Conference in 1978. In Western countries, 'integrative/integrated' medicines are gradually inserted in treatments schemes of chronic diseases such as cancer and AIDS and in palliative care: this issue is addressed by health and research actors in collaboration with civic actors. A global trend in biomedical sciences is the expanding relevance attributed to 'evidence-based' medicine (EBM): in the early 1990s, focusing on the evidence of clinical research, the EBM founding group of scholars claimed that 'a new paradigm for medical practice' was emerging and its significance would reach the scale of a 'scientific revolution' in medicine ([16, 17]). Such trend is reflected in 'conventional' traditional medicine (TCM in China) as China secures a significant place in the 'global village' and in world affairs through an accelerated process of cultural and economic globalisation.

Therefore, from the 1990s and at a faster pace in the 2000s, the 'biomedicalisation' of TCM research reached another scale through the tentative adoption and increasing valuation of EBM practices and ideas. In my work, such process is approached through the case of AIDS research: a number of scientific practices and experiments in relation to HIV and AIDS treatments are ongoing while the new national scheme on AIDS treatment and care is gradually and tentatively been implemented since 2004 [13–15]. Acknowledging a global context of circulation of knowledge, the concepts of CAM

---

[1] The author is a permanent research fellow in Anthropology at IRD (French Research Institute for Development), UMR 145, Montpellier, France. Scientific Coordinator, IRD-PUMC/CAMS (Peking Union Medical College/Chinese Academy of Medical Sciences) Social Sciences Programme, Beijing.

(Complementary and Alternative Medicine) and evidence are both discussed as their elaboration and meanings offer a relevant glimpse of interacting/interlocking paradigms.

While societies in emerging countries are growingly based on knowledge, research and development following the route of developed countries, our approach to medical research draws on the broader field of the anthropology of sciences, building upon what are labelled 'new subjects' in anthropology in contrast to classical ones such as kinship or religious systems.

Using tools of social anthropology, the methodology is based on the collection of qualitative data, documentary and archival research, and qualitative research analysis. Still in an exploratory phase, this research is part of a social sciences programme on AIDS jointly run by IRD and PUMC/CAMS.

1 AIDS medical research in TCM: an overview

Along the lines of the national AIDS treatment and care scheme, the free TCM treatment project formally launched in June 2004, was designed to help in China's control of the epidemics. TCM has been accounted for a possible milestone in China's AIDS treatment and care strategies. According to the SATCM (State Administration of TCM) created in 2003, China will offer free treatment to approximately 30000 patients in 15 of its 21 provinces and autonomous regions. Initially, TCM-based treatment was provided for 2300 patients from rural areas under the TCM free treatment scheme in Hebei, Anhui, Henan, Hubei and Guangdong provinces.

Among TCM bodies involved in treating HIV and AIDS related symptoms, are included AIDS treatment centres such as that of Guang'anmen hospital (GAMH AIDS Clinical Centre) and research bodies such as the AIDS department and research centre of CATCM (Chinese Academy of TCM) in Beijing or the AIDS research centre of the TCM Institute in Zhengzhou, capital of Henan province, one of the most affected province by HIV infection.

From the content analysis of a number of Chinese publications on AIDS clinical trials, some results are reported in three areas: (1) in the process of identifying the herbs that are most effective, (2) in extracting the most active agents from the herbs, (3) in testing medicinal compounds based on TCM knowledge and practice. In those reports, herbs may be conceptualised consistently as regular drugs the same way they are conceptualised in the biomedical model, without explicit reference to TCM basics (theory and practice).

TCM trials have been conducted in the USA, in Tanzania and in China since the late 1980s. AIDS clinical research in TCM was initiated in 1985 in the US by an American team (Cohen, Abrams and Burack, Quan Yin Healing Arts Center in San-Francisco) and in 1989 in Africa by a Chinese team ([12], AIDS Department of the National TCM Research Institute in Beijing) whose results were controversial. In 1996, for the first time, a pilot randomized controlled trial of Chinese herbal treatment for HIV-associated symptoms was reported in an international journal ([2]). However, only one TCM drug (*Tangcao*) has been approved by the China SFDA (State Food and Drug Administration) as a treatment for AIDS lately in 2006.

Recent results were published in the *Chinese Journal of Integrative Medicine* [19]: '*Zhongyan*-4 has an immunity-protective and /or rebuilding function in HIV and AIDS patients in the early and middle stage, and also shows effects in lowering viral load, increasing body weight and improving symptoms and signs to a certain degree'. Over 20 years, four new compositions of TCM (innovative treatments in TCM for a new 'medical situation', namely HIV infection) labelled *Zhongyan*-1 to 4 were tested. These long term trials were conducted by the research team of the Chinese Academy of TCM at the Centre for AIDS research and at Guang'an men hospital in Beijing. However, as Professor Cao, the leading investigator in the China CIPRA (Comprehensive International Program for Research on AIDS) programme related to clinical research in TCM and immunologist underlines, no specific category of drugs has yet been found that could inhibit HIV replication and rebuild the body immunity[3]. Within the framework of China CIPRA project, a TCM drug, Chuankezhi injection, combined with ART (Anti-Retroviral Treatment) regimens was tested in order to evaluate its safety and efficacy on alleviating HIV-related symptoms, and on reducing side effects of ARVs (Anti-Retrovirals). The efficacy still needs to be evaluated. In addition to a dose of uncertainty, some results remain difficult to explain with current scientific tools and theories and are contradictory: a stabilization or an increase in CD4 cell count can be observed concomitantly with an increase in viral load.

A number of tested and used treatments are very common and polyvalent, namely mostly used in daily life for the preservation of health and for stimulating the overall immunity of the body either in the pharmacopoeia (*zhongyao*), in acupuncture-moxibustion (*zhenjiu*) or in corporal and health practice of *qigong*; all these methods are generic TCM treatment methods. Most frequently tested treatments are compositions and prescriptions of pharmacopoeia, eventually combined with another TCM method of treatment.

To conclude, according to the results of some clinical trials whose reliability is not clearly assessed, TCM herbal preparations might have an immunity protective and/or rebuilding function. However, until now results show no major breakthrough, and experimenting on TCM drugs raises controversy in the international academia.

2 Concept of CAM

In China, traditional medical treatments of AIDS outside the public health system are not yet documented. What kind of traditional medicine is used? In Chinese, the term *minjian zhongyi* (popular traditional medicine) designates traditional medicine practiced in non-official settings (family, locality and/or temple-related community/group) distinguishing it from the official (conventional) TCM. In other contexts out of China, what kind of traditional medicine of Chinese origin is used? Can Chinese medicine understood in a broadened meaning including TCM but not excluding unofficial traditional medicine, be labelled a CAM?

CAM is a methodological, operational category rather than a theoretical one needing to be conceptualised in the context of culture: for instance, TCM can be conceptualised as a 'conventional' rather than 'alternative' medicine in China. Nevertheless, the latter applies to unofficial traditional medicine.

The terms 'alternative', 'parallel', 'conventional', 'neo-traditional' may or may not be useful or not depending in contexts. 'Complementary' is not epistemologically a generic term but rather a contextual term: complementary to what? In most contexts, it is understood as complementary to the biomedical model which may vary greatly being always adjusted locally, and which remains dominant even in pluralistic health care systems such as those of India and China.

The term 'alternative' health care was devised when prevailing beliefs held that consumers sought other treatment modalities instead of conventional Western medical treatment [20]. The term 'complementary' was substituted to it when a number of studies demonstrated that most people use unconventional therapies in addition to conventional medicine [5, 20]. According to WHO [21] estimates, approximately eighty per cent of people who do not live in Western societies currently use forms of medicine which are labelled CAM in the Western context, as their primary health care. Use of traditional/popular medicines remains unchallenged in much of the world. In the White House Commission report [20], CAM is defined as follows: 'a group of diverse medical and health care systems, practices, and products that are not presently considered to be part of conventional medicine. The list of practices that are considered CAM changes continually, as those therapies that are proven to be safe and effective become adopted into conventional health care and as new approaches to health care emerge'.

Therefore, CAM labelling applies to any set of knowledge and practice that has yet to meet the lines of the scientific medical paradigm of cause and effect. The mandate for research to prove the usefulness of CAM is clear in the report which acknowledges the current research on these medicines but raises issues dealing with the size and rigor of clinical trials and underlines that safety and efficacy of CAM therapies still need to be evaluated [4]. As both in the US and in Europe, government funding institutions set aside amounts of money for CAM research, [6, p.198] heuristically stresses on research bias highlighting the difference between using the scientific method for proving (the misguided approach) and for testing (the correct one). Investigators tend to judge a trial report more positively if it confirms their previous findings [7]. Vickers *et al* [18] have demonstrated that virtually no negative trials of acupuncture have emerged from China. CAM researchers are usually self-taught. The combination of lack of scientific training and strong bias in favour of their intervention could be a recipe for bad research [10]. A researcher in CAM and any other area should primarily be enthusiastic about good science and not about the therapy under investigation [6].

3 'Biomedicalisation' of traditional medicine and evidence

Related to a cultural movement borrowing from the audit culture and applied to biomedicine since the early 1990s, the influence of EBM reveals an increasing valuation of the experimental method and tools of evaluation, the adoption of an agnostic and objectivistic approach when the proof is not yet reached within the framework of a research protocol: 'EBM de-emphasizes intuition, unsystematic clinical experience, and pathophysiological rationale as sufficient grounds for clinical decision-making and stresses the examination of evidence from clinical research' ([17], quoted by [8]). Even more controversially than in biomedicine, such set of practices and ideas is tentatively applied to other medical bodies of knowledge and practice such as TCM/Chinese medicine and other CAM.

An anthropological reading supposes a consideration of all types of evidence on the basis that all bodies of knowledge either popular/traditional/scientific are socially and culturally constructed. Subsequently, such critical stance means contrasting diverging types of evidence instead of valuating one single type, namely biomedical scientific evidence which excludes other rationales and types of rationalisation, as well as confronting the scientific paradigm to other epistemological models. For instance, extending 'evidence' to 'What works?' [1] in medicine, a practical but crucial question for anybody who has ever been in a patient role referring to whatever (eventually diverging) bodies of knowledge, is a mean to account for – emic/etic (insider, local) versus -etic (expert, global) views, therefore a way towards 'indigenizing' the type of evidence.

4 Concluding remarks

In her presentation based on a historical and philosophical approach of Western medicine, Fagot-Largeault [8] demonstrated that any body of medical knowledge and practice is the result of a merging of both traditional and modern lore and that the line between 'scientific versus traditional' may be difficult to draw: for instance, in the context of clinical research, the relevance of clinical case reports is partly based on their particularities, a fact that cannot be completely assessed with scientific tools.

Within the AIDS treatment/research framework and context, and while TCM treatment and research is not yet standardised, the choice in China's policy between the research and development of combined versus substitutive treatment is not always fully clarified nor acknowledged [13, 15]. Simultaneously, globalising/hybridised Chinese Medicine becomes a valued CAM in treatment schemes of chronic diseases such as cancer and AIDS outside of China: in a context of cultural globalisation, 'a certain degree of hybridisation (transformative and/or contextual) is the normal course of events, which does not prevent other aspects from being homogenized' [9, p.84]. In times of economic globalisation, Chinese remedies which appeal to the newly emerged middle-class in China and currently take a significant stake in the global market of CAM, have become obviously lucrative [11, p.237–238]. Therefore, one may notice that TCM/Chinese medicine may rank first among the most used traditional medicines in the world, and second, behind biomedicine, among the most used medicines within the healthcare systems of a significant number of countries.

Acknowledgements

The current research is conducted in partnership with the PUMC/CAMS in Beijing with a financial support provided by the CEFC (French Centre for Research on Contemporary China) in Hong Kong, China.

Bibliography

[1] Barry, C. A. The role of evidence in alternative medicine: Contrasting biomedical and anthropological approaches. *Social Science and Medicine*, 62:2646–2657, 2005.

[2] Burack, J. H., Cohen, M. R., Hahn, J. A., Abrams, D. I. Pilot randomized controlled trial of Chinese herbal treatment for HIV-associated symptoms. *Journal of AIDS*, 12(4):386–393, 1996.

[3] Cao, Y. Z. HIV treatment by Chinese medicine: Exploration and expectation. Oral Presentation, Fifth China CIPRA Annual Meeting. Beijing, 2007. April 16–18.

[4] Cushman, M. J., Hoffman, M. J. Complementary and alternative health care and the home care population. *Home Health Care Management & Practice*, 16(5):360–373, 2004.

[5] Eisenberg, D. M., Kessler, R. C., Foster, C., Norlock, F. E. Unconventional medicine in the US – prevalence, costs, and patterns of use. *New England Journal of Medicine*, 328(4):246–252, 1993.

[6] Ernst, E. A passion for complementary and alternative medicine research. *Complementary Health Practice Review*, 8(3):198–200, 2003.

[7] Ernst, E., Resch, K. L., Uher, E. M. Reviewer bias. *Annals of Internal Medicine*, 116:958, 1992.

[8] Fagot-Largeault, A. Scientific vs. traditional (empirical) medicine – a universal debate? Beijing: 13th LMPS Congress, 2007. August 9–15.

[9] Frank, R., Stollberg, G. Conceptualizing hybridization. on the diffusion of Asian medical knowledge to Germany. *International Sociology*, 19(1):71–88, 2004.

[10] Hrobjartsson, A., Brorson, S. Interpreting results from randomised clinical trials of complementary/alternative interventions: The role of trial quality and pre-trial beliefs. In Callaghan, D. (Ed.), *The role of complementary and alternative medicine*, 107–121. Washington DC: Georgetown Univ. Press, 2002.

[11] Hsu, E. La médecine chinoise traditionnelle en république populaire de chine, dans a. cheng dir. In *La pensée en Chine aujourd'hui*, 214–238. Paris: Gallimard, 2007.

[12] Lu, W. B. *Traitement du sida par la médecine et la pharmacopée traditionnelles chinoises. Huit cas de conversion séronègative*. Paris: Quimétao, 1998. Trad. Mo Xuqiang.

[13] Micollier, E. AIDS medical research in China, an exploratory enquiry Poster 535, Session 26.4 in AIDS impact International Conference. Marseilles, 2007. July 1–4.

[14] Micollier, E. Facettes de la recherche médicale et de la gestion du VIH-sida dans le système de santé chinois : un autre exemple d'adaptation locale de la biomédecine. *Sciences Sociales et Santé*, 25(3):31–39, 2007.

[15] Micollier, E. Neo-traditional treatments for AIDS in China: national AIDS treatment policy and local use of TCM (Traditional Chinese Medicine) Oral Presentation 345, session 8.5. Marseilles: AIDS impact International Conference, 2007. July 1–4.

[16] Sackett, D. L., Rosenberg, W. M., et al. Evidence based medicine: What it is and what it is n't? *British Medical Journal*, 312:71–72, 1996.

[17] The Evidence-Based Medicine Working Group. Evidence-based medicine. a new approach to teaching the practice of medicine. *JAMA*, 268(17):2420–2425, 1992.

[18] Vickers, A., Goyal, N., Harland, R., Rees, R. Do certain countries produce only positive results – a systematic review of controlled trials. *Controlled Clinical Trials*, 19:159–166, 1998.

[19] Wang, J., Yang, F. Z., Zhang, Y. X., Liu, Y., et al. Randomized double-blinded and controlled clinical trial on treatment of HIV/AIDS by zhongyan-4. *Zhongguo jiehe yixue zazhi (Chinese Journal of Integrative Medicine)*, 12(1):6–11, 2006.

[20] White House Commission. White House Commission on CAM policy. Tech. rep., Washington D. C., 2002.

[21] WHO. Traditional and alternative medicine. Tech. rep., 2002. Fact Sheet 271.

A philosophical analysis of the place of acupuncture in the French health care system[1]

Fabrice Gzil

Université Paris-7 Diderot & IHPST

fabricegzil@free.fr

ABSTRACT. This paper offers a philosophical analysis of the current place of acupuncture in the French health care system. It uses as a starting point a sociological study recently published by Triadou and his colleagues. In this study of the practice of acupuncture in France, the authors interviewed 850 acupuncture patients, in order to understand the reasons why the patients went for acupuncture. Triadou *et al.* propose two interpretations. The first is that acupuncture appears to French patients as an alternative to classical medicine, because classical medicine has become too reductionist, and because acupuncture offers a more global or "person-centred" approach. The second interpretation is quite different. Triadou et al. suggest that patients go for acupuncture when classical medicine fails to solve their medical problems, that is to say they use acupuncture as a complementary medicine, in a pragmatic quest for efficacy. After discussing these interpretations, this paper provides another interpretation, according to which acupuncture is neither to be regarded as an *alternative* nor as a *complementary* medicine, but is to be compared with an emerging field in modern medicine, the so-called rehabilitation medicine.

In France, Chinese traditional medicine often confines itself to the practice of acupuncture. Like homeopathy and herbal medicine, acupuncture is classified in the heterogeneous group of the so-called alternative medicines, as opposed to conventional or classical medicine. It is used for the treatment of a large number of conditions, for example dorsal and articular pain, headache, allergies, weariness, and sleeping disorders. It is also used for the treatment of three "French specialities": stress, anxiety and depression. One can currently observe in France (like in other European and North-American countries), an increasing recourse to acupuncture[1, 2]. My aim is to offer a philosophical analysis of the current place of acupuncture in the French health care system. I will use as a starting point a sociological study recently published by Triadou and his colleagues[5-9]. In this study of the practice of acupuncture in France, Triadou *et al.* interviewed 850 acupuncture patients, in order to understand the reasons why French patients went for acupuncture. Triadou et al. assumed that acupunctures growing success could be explained by the differences between acupuncture and classical medicine and come up with two interpretations. The first is that acupuncture appears to French patients as an *alternative* to classical medicine, because classical medicine has become too reductionist, and because acupuncture offers a more global or "person-centred" approach. The second interpretation is quite different. Triadou et al. suggest that patients go for acupuncture when classical medicine fails to solve their medical problems, i.e.

---

[1]The author is with Université Paris-7 Diderot (Départment d'Histoire et de Philosophie des Sciences) & Institut d'Histoire et de Philosophie des Sciences et des Techniques (CNRS/ENS/Paris1).

they use acupuncture as a *complementary* medicine, in a pragmatic quest for efficacy. After discussing these interpretations, I will propose another interpretation, in which acupuncture is neither regarded as an alternative nor as a complementary medicine. I will suggest that in order to understand the place currently occupied by acupuncture in the French health care system, one could compare acupuncture with an emerging field in modern medicine, the so-called rehabilitation medicine.

1 Acupuncture as an alternative to the reductionism of modern medicine

When French patients describe their experience of acupuncture, they emphasize the quality of the practitioner/patient relationship (most of the patients find that acupuncturists have a better contact than practitioners of classical medicine, they appreciate that a long time is devoted to the initial visit, they feel they are listened to and understood during the discussion). Patients also insist that acupuncture is a global approach (the acupuncturist considers the person as a whole, he pays attention not only to the physical symptomatology, but also to the patients' psychic life and personal history). What is more, acupuncture often appears as an alternative to medications (some patients use acupuncture in order to avoid the use of drugs, especially antidepressants, tranquillizers, sleeping pills, and antibiotics), and acupuncture patients are curious about a technique which is based on a different conception of diseases.

These results suggest that the choice of acupuncture could be understood as the result of a distrust of the recent evolutions of medicine. Patients may recourse to acupuncture because of the reductionism of contemporary medicine. Their interest for a global approach might be interpreted as a reaction against the focusing of current medicine on physical symptoms, physiological mechanisms, and diseases' biological dimension. It could also be interpreted as a reaction against the increase in the number of medical specialists, and the partition (in classical medicine) between somatic and psychic features. The importance ascribed to the quality of the acupuncturist's reception and listening might be explained by the fact that in modern medicine the patient/practitioner relationship has become less humane, because physicians tend to focus their attention on technical actions. Thus, patients might turn away from a therapeutic relation focused on drug prescription, and seek a richer therapeutic relation.

In this interpretation, the choice of acupuncture is the result of a distrust of recent evolutions in classical medicine. It could be interpreted as a form of nostalgia of a former state of medicine, when medicine offered a more global approach. Patients would not so much valuate acupuncture because it pertains to Chinese culture (as opposed to western culture), but because it pertains to a traditional medicine (as opposed to modern medicine). The choice of acupuncture should therefore not be interpreted as a distrust of medicine as such, but as a distrust of the "scientificization" of current medicine, i.e. of the reductionism and the technicism generally associated to it, and as a craving for a more "person-centred" approach.

This interpretation is interesting. However, it raises some questions, for it is not sure that modern medicine is blind to subjective and holistic features (let's think, for example, of the growing attention paid in modern medicine to the phenomenon of pain, and to the ways to alleviate it). What's more, acupuncture is not actually "person-centred", but "energy-centred" (the focus on the person, that is to say on the individual subject, is a western concern, which does not seem to be shared by Chinese traditional medicine). Thus, the previous interpretation does not so much explain the choice of acupuncture by invoking some of its positive features, than by emphasizing negative features of modern medicine. It underlines some problems raised by the scientificization of modern medicine, but it does not really tell us what the characteristic features of acupuncture are.

2 Acupuncture as a complementary medicine when conventional medicine is not efficient

The study conducted by Triadou *et al.* shows that the use of acupuncture is seldom an immediate process, but is generally the result of a long quest for solving a medical problem. Acupuncture appears as a possible remedy when all the possibilities of conventional medicine have been tried. Patients come to it when other treatments have dubious or adverse effects. What's more, the fact that conventional medicine proves to be inefficient for solving a particular health problem does not call into question the patients' trust in conventional medicine (most patients continue to consult their general practitioner and consider their acupuncturist as a specialist for specific health problems; only a few patients take the acupuncturist as their leading practitioner). Therefore, acupuncture is often considered as an additional treatment and is often added to a main treatment, which pertains to conventional medicine (acupuncture is used in severe conditions – like AIDS, or multiple sclerosis – for reducing the pain associated with the medical treatment, or for improving the general state of health, in order to maximize the effects of the medical treatment). Patients are all the more confident about acupuncture because their acupuncturist is competent in both acupuncture and conventional medicine.

These results suggest that the choice of acupuncture could be explained, on the one hand, by a kind of pragmatism (the quest for therapeutic efficiency), and on the other hand by a kind of eclecticism (patients use equally different forms of medicine). According to this second interpretation, acupuncture and classical medicine are not opposed at all, but regarded as complementary tools for treating a given condition. The differences between them are pushed into the background. Acupuncture is one medicine among others. This interpretation is corroborated by the growing interest, in the so-called classical medicine, to understand the working mechanism of acupuncture. It is also corroborated by the increasing number of studies which, because of the increasing cost of pharmacological treatments, compare the efficiency of drugs and acupuncture.

However, to argue that acupuncture can constitute both an alternative to modern medicine, and an auxiliary for conventional medicine is a good way for explaining why one *should* go for acupuncture, but it is not sure that it is the best way to explain why people actually go for acupuncture. Like the first interpretation, the second remains silent about the characteristics of acupuncture. It underlines acupuncture's efficiency but it does not really tell us which specificities of acupuncture make the French patients using it.

3 Acupuncture as an analogous of rehabilitation medicine

The study conducted by Triadou and colleagues suggest an interpretation which has not been explicitly developed by the authors, but which could both explain the choice of acupuncture and the specific place of acupuncture in the French health care system. For acupuncture patients are not only curious about acupuncture and other alternative medicines. Most of them have a medical education and are interested in health questions in general. Moreover, they do not expect that acupuncture will perform miracles when considering serious diseases. They are confident that acupuncture can help them to overcome stress, anxiety, pain, without respect for the transitory or chronic character of these disorders. However, when considering more serious disorders, like digestive or neurological diseases, patients generally regard acupuncture as an auxiliary therapy. Finally, patients generally consider that acupuncture can help them feeling better, regardless of the fact they suffer or not from an identified medical problem. Most of them consider that acupuncture not only alleviates the pain related to their disease, but also gives them a feeling of well-being (50%), a kind of relaxation (38%), and a psychological support (39%).

These results are interesting because they suggest a comparison between acupuncture and an emerging field in classical medicine, the field of rehabilitation. Rehabilitation is concerned with physical and mental disabilities. It constituted itself through a twofold process[3]. The first step was accomplished after World War Two, with the emergence of re-adaptation, as an alternative to the classical medical approach, which was based on a curative conception of medicine, and on an etiological conception of health problems. The curative/etiological model lost some of its relevance in the fifties, with the progressive disappearance of infectious diseases, the decreasing importance of acute pathologies with high mortality rates, the appearance in the forefront of chronic diseases associated with incapacities, and the ageing of population[4]. The curative/etiological model got progressively challenged by a readaptative/functional model, which made a distinction between pathology and functional limitation, where the emphasis is no longer made on the cause of the disease, but on the limitations it generates, and where the goal is a normalisation of the individual, through corrective, readaptative, or re-educative interventions (for example orthopaedic interventions). This readaptative model got subjected to criticism in the seventies. Disabled persons' movements underlined that this model continued to conceive disability according to a causal model (disease → impairment → disability). They also argued that disability is a natural condition, and that the right goal for interventions is not to "normalize" disabled people, but to maximize people's quality of life and participation in society. Thus, the re-educative model got progressively challenged by a rehabilitative model, which concern is neither diseases nor disabilities, but health and abilities. The rehabilitative model puts the emphasis on mobilizing the person's remaining capacities and on maximizing the person's abilities.

From my point of view, there are some analogies between acupuncture and rehabilitation. Like acupuncture patients, disabled persons are not laymen with respect to medical questions, they are often quasi-experts in the questions about their health state. Like acupuncture, rehabilitation is not conceived as a treatment but as an intervention, which can help people coping with their impairments. Like in acupuncture, the purpose is generally more global than in classical medicine: rehabilitation is intended to help the patient in his everyday life, and to improve his quality of life in general. What is more, acupuncture and rehabilitation face problems that are very similar. In both cases, one is confronted with difficult questions concerning the evaluation of interventions. Moreover, like for acupuncture, it is not easy to define the relationship between rehabilitation and classical medicine. Rehabilitation practitioners can either be physicians or other professionals. And rehabilitation interventions can either be associated with a medical treatment, or used because of the absence of a medical treatment, or used in the absence of a disease. Thus, one could argue that acupuncture occupies, in the French health care system, a place which is quite similar to the place of rehabilitation. Like rehabilitation, which constituted itself by differentiating on the one hand from an acute-disease-centred medicine, and on the other hand from an impairment-centred-medicine, acupuncture could be defined as a health-centred medicine, where health is not merely the absence of disease, but the maximal use of one person's capacities, and the maximal actualization of one person's abilities.

Bibliography

[1] Eisenberg, D. M., et al. Trends in alternative medicine use in the United States, 1990-1997. *JAMA*, 280:1569–1575, 1998.

[2] Fischer, P., Ward. A. complementary medicine in Europe. *BMJ*, 309:107–110, 1994.

[3] Gzil, F., Lefeve, C., Cammelli, M., Pachoud, B., Ravaud, J. F., Leplege, A. Why is rehabilitation not yet fully person-centred and should it be more person-centred? *Disability & Rehabilitation*, 29:20–21, 2007.

[4] Ravaud, J. F. Modèle individuel, modèle médical, modèle social: la question du sujet. *Handicap: Revue de Sciences Humaines et Sociales*, 81:64–75, 1999.

[5] Triadou, P., Desjeux, D., et al. L'acupuncture en France aujourd'hui (I) La consultation d'acupuncture. *Acupuncture & moxibustion*, 4(1):11–18, 2005.

[6] Triadou, P., Desjeux, D., et al. L'acupuncture en France aujourd'hui (II) Perceptions des effets et évaluation de l'acupuncture. *Acupuncture & moxibustion*, 4(2):107–112, 2005.

[7] Triadou, P., Desjeux, D., et al. L'acupuncture en France aujourd'hui (III) Médicaments et acupuncture. *Acupuncture & moxibustion*, 4(3):171–181, 2005.

[8] Triadou, P., Desjeux, D., et al. L'acupuncture en France aujourd'hui (IV) Représentations et rapports à la science. *Acupuncture & moxibustion*, 5(1):12–17, 2006.

[9] Triadou, P., Desjeux, D., et al. L'acupuncture en France aujourd'hui (V) Typologie des patients. *Acupuncture & moxibustion*, 5(2):15–26, 2006.

Scientific vs. traditional (empirical) medicine - a recurrent debate?[1]

Anne Fagot-Largeault

Collège de France

anne.fagot-largeault@college-de-france.fr

ABSTRACT. Western medicine has often oscillated between tradition (blind faith in dogmatic knowledge and empirical recipes) and scientific innovation (call for unbiased collection of data, comparative tests, criticism of old dogmas). This presentation focuses on three episodes in the history of 'scientific' medicine : numerical medicine, experimental medicine, evidence-based medicine. The hypothesis is that all medicine (either western or eastern) is a mixture of routinely accepted tradition (in case x, do y) and scientific rationality. The objective is to characterize the opposition between *traditional* and *scientific*. For that purpose, we choose to analyse historical examples of medicine "getting more scientific", and observe that there is in each case a difficulty in getting *completely* scientific.

1 Pierre Charles Alexandre Louis (1787-1872) and the " numerical method "

Around 1825 in France, the standard treatment of pneumonia, recommended by Dr. F.J.V. Broussais, was bloodletting (leeches were applied on the chest). Doctor P.C.A. Louis set about challenging the orthodoxy. On the basis of 78 cases of pneumonia, observed between 1821 and 1827 at the Charité hospital in Paris, he showed that no difference in mortality, or in the duration or severity of symptoms, could be evidenced between the patients who had been treated and those who had been left untreated[2]. As a consequence of his work, the leeches market collapsed. There is something disturbing in Louis' attitude. In spite of there being a reference treatment, he dared leave some of the patients untreated. And (even though his statistical analysis was crude : small samples, no formal randomization, no estimation of the error rate), he perseveringly based his conclusions on numerical data. His enthusiastic young students (William Farr, William Guy, Marc d'Espine, Henry Bowditch, George Shattuck, etc.) praised him for having promoted the numerical method. Elisha Bartlett wrote in the *American Journal of the Medical Sciences* (1836) : " With Louis' adoption of what is called the numerical system has commenced a new era in our science"[1].

However, in 1837, a raging battle between conservatives (led by Dr. Double) and progressives (led by Dr. Louis) took place at the French Academy of Sciences. Dr. Double argued that drawing general conclusions from having counted cases is absurd, because individual cases are always different, and it makes no sense to amalgamate a variety of cases. The good physician, he said, possesses both the talent of sensing

---

[1]The author is with Collège de France & Academy of Sciences, Paris.
[2][6] For more references, see jameslindlibrary.org

exactly the peculiarities of individual cases, and the skill to act accordingly for the benefit of his patient. Dr Louis counter-argued that medical teaching is meaningless if we cannot learn from experience, that is, draw conclusions from what has been observed in analogous cases. Knowledge requires cautious and sure generalization : " without the aid of statistics nothing like real medical science is possible "[3]. It remained that statistical knowledge does not allow to make predictions for the singular case. William A. Guy granted the point in 1839 : " Does the numerical method admit of application to individual cases ? It must be conceded that such application is limited "[4].

In this first example, the opposition between 'traditional' and 'scientific' may be characterized as follows. There exists a dominant and influential medical system or theory (here, Broussais'), teaching that all diseases with a fever are inflammatory, inflammation calls for bloodletting, leeches are to be applied on the surface of the body corresponding to the inflamed organ. Medical practitioners have learned the doctrine in school. The traditional physician does not question the theory, he conforms to doctrine and applies the rule. The scientifically minded physician challenges dogma. His attitude requires more than having doubts. He wants to perform an experiment. He will get data (observing many patients with pneumonia, treating them the traditional way or letting them untreated, counting the number of deaths in each group, and the number of days from onset to recovery) and draw conclusions from a statistical analysis of the data. Note that such a methodological strategy may imply considerable risk taking for the physician himself and for the patients.

2 Claude Bernard (1813-1878) and "experimental medicine"

Around 1860, the physiologist and physician Claude Bernard wanted to take medicine from a state of opportunistic empiricism (including hasty statistical reasoning) to the state of "an exact science based on experimental determinism " . He reckoned that "it is easier to statistically count cases pro and con than to conduct proper experimental reasoning ". He only published the Introduction to his planned monumental book on the Principles of Experimental Medicine, but he left notes from which the book was later reconstructed[4]. His methodology of medical research, and his project for the development of a science of general physiology, aimed at discovering universal laws of nature, and at making learned medical prediction scientifically certain.

Bernard distinguished three major steps in the history of medicine[2, b, chap.III, IV, VIII]. (1) *Pre-scientific* medicine is a medicine practiced by priests or monks, in which knowledge is kept secret, transmitted as a treasure inherited from wise (godlike) ancestors; divulgation is punished; any alteration in the ritual is sinful, a typical medical intervention is exorcism. (2) *Scientific-empirical* medical knowledge is acquired from careful observation. On the basis of observation (and inductive generalization) the physician learns how to identify the disease (diagnosis), and how to anticipate the issue (prognosis); personal experience cannot be easily transmitted; remedies are scarce, empirical recipes are used conservatively, the typical attitude is : *wait and see*. (3) *Scientific-experimental* medicine moves forward, from observation to experimentation ; physicians develop rational strategies for the acquisition of knowledge (formulating and testing hypotheses); such knowledge is (published and) available to everyone. Exact

---

[3]cited by [1], ibid., from an English translation of the debate published in the *London Medical Gazette*, 1837

[4][2], Digital version online
http://www.uqac.uquebec.ca/zone30/Classiques_des_sciences_sociales.
On statistical reasoning, see: *Principes* ..., chaps VI and VII.

experimental knowledge gives physicians the power to actively and efficiently interfere with the course of nature.

Bernard is crystal clear about the 'logic' of testing hypotheses : " In experimental reasoning there are two possible cases : either the experimenter's hypothesis will be invalidated, or it will be confirmed by the experiment. When the experiment invalidates his idea, the experimenter must reject the idea, or modify it. But whenever the experiment entirely confirms his idea, the experimenter must keep doubting "[2, a, Part 1, chap.2] – a popperian statement, written before Sir Karl was born. Bernard is also very specific about the aim of scientific knowledge : " The objective of experimental medicine is to determine the very mechanism of phenomena, in order to be able to manipulate them "[2, b, chap.VI] .

Yet, at the end of his life, Bernard acknowledged the fact of biological variability, which makes physiological knowledge and prediction irreducibly uncertain. Science indeed aims at generalizing and outlining typical things or situations, but real situations always differ from the type – " and that is important for the physician to realize : he does not deal with human types, he deals with individuals, placed in particular situations ". We scientists certainly hope that there may be general laws of the variation of types, but until we discover such laws, individual idiosyncrasies may not be overruled : "while generalizing we must retain a sense of the special[2, b, chap.XIV]".

In this second example, 'traditional' points out to an attitude of empirical habituation to regularities, passive expectation of recurrent phenomena, holistic approach of individuals ; 'scientific' evokes a rational methodology for the acquisition of new knowledge, active intervention to modify the course of nature, analytic approach of individuals. The scientific attitude finds its limit when acknowledging that the complexity of individual cases is not entirely reducible to a combination of generalities. " Don't we have to take into account everything there is, Bernard writes, and is there anything else over and above individual beings ? "[2, b, chap.XIV]

3 The Evidence-Based Medicine Working Group

In 1992, a Canadian initiative emerged (precursors had been, in the 1950s, the British statistician Austin Bradford Hill, who pioneered the Streptomycine in Tuberculosis clinical trials, and the physician and epidemiologist Archie Cochrane). The North-Americans David L. Sackett, Gordon H. Guyatt, and others, boldly meant to launch a " scientific revolution " in medicine. The Evidence-Based Working Group (that is how they called themselves) published a series of papers in the *Journal of the American Medical Association* (JAMA). Here is what their manifesto claimed : " A new paradigm for medical practice is emerging. Evidence-based medicine de-emphasizes intuition, unsystematic clinical experience, and pathophysiologic rationale as sufficient grounds for clinical decision making and stresses the examination of evidence from clinical research. Evidence-based medicine requires new skills of the physician, including efficient literature searching and the application of formal rules of evidence evaluating the medical literature "[7, 9] .

The " paradigm shift " was expected to introduce " a new philosophy of medical practice and teaching ". Physicians were to : (1) replace " unsystematic observations " with " systematic attempts to record observations in a reproducible and unbiased fashion ", (2) realize that knowledge of physiological mechanisms of disease " are necessary but insufficient guides for clinical practice ", (3) recognize that personal clinical experience and common sense is not enough : in order to provide pertinent diagnosis and optimal patient care, clinicians should search the literature, be able to appreciate the relevance and methodological rigor of new publications, and " make independent assessments of evidence " – that is, rather than always trust authority, " be ready to accept and live with uncertainty "[9, p.2421]. The supporters of the EBM movement deemed it *neither scientific nor ethical* to ignore the methodological advances brought into medicine by randomized trials. They wanted the research to effectively guide medical practice, for the benefit of patients : " Evidence-based medicine is based on a strong ethical and clinical ideal - that it allows the best evaluated methods of health care to be identified and enables patients and doctors to make better informed decisions "[5].

Some years later, computerized literature searching and clinical appraisal of experimental approaches have indeed become a substantial part of medical training. Evidence-based methods have proved very useful to clean up a stock of old therapeutic recipes and eliminate routinely applied procedures. However, there are paradoxical 'evidence-based' results ; and justification is not the whole of science. For example, case reports are rated very low, by EBM reference documents, among the methods of knowledge acquisition, on account of their reflecting subjective, uncontrolled experience. But, as was convincingly shown by O. Steichen[8], case reports have steadily constituted 10% of the medical literature for the last 40 years ; they have significant heuristic value : discovery of new pieces of knowledge may be based on careful analysis of single cases ; and they are essential to skilled teaching.

In this third example, 'traditional' means : going by the so-called " medical intuition ", trusting the doctor's unsystematic clinical experience, accepting the theoretical plausibility of hypotheses without test, having insufficient ethics. The 'scientific' or 'evidence-based' behavior, that presumably goes with good ethics, implies : doing clinical research (conducting randomized and double-blind medical trials) ; and/or searching the medical literature to learn about the results of clinical trials, learning how to evaluate the quality of the trial reports, and keeping up with scientific advances for the good of patients.

4 Conclusion

Tradition, in medicine, evokes empiricism and slow evolution ; some knowledge gets cumulated haphazardly, without planning ; the irreducible variability of cases is presupposed ; and ultimately nature knows better than man – as Cl. Bernard wrote: " When you don't know how to cure, let them cure themselves "[2, b, chap.XIV]. Science, in medicine, evokes rational behavior and fast changes ; systematic trials are conducted for the testing of hypotheses ; general knowledge is buildt, based on the presupposition of (statistical) uniformity of types ; and where nature errs and makes mistakes, human medicine corrects nature's aberrations. What is the right way ? Do we prefer the traditional or the scientific doctor ? As François Jacob liked to say : " What I prefer ? – a little bit of both ... " [5]

Bibliography

[1] Armitage, P. Trials and errors. the emergence of clinical statistics. *Journal of the Royal Statistical Society A*, 146(4):321–324, 1836.

[2] Bernard, C. (a) Introduction à l'étude de la médecine expérimentale. Paris, 1865. (b) *Principes de médecine expérimentale*, Paris:posthumous, 1947.

[3] Fagot-Largeault, Rahman, Torres (Eds.). *The Influence of Genetics on Contemporary Thinking*, vol. xxix-lvi. Dordrecht: Springer, 2007.

[4] Guy, W. A. On the value of the numerical method as applied to science, but especially to physiology and medicine. *Journal of the Royal Statistical Society*, 2:25–47, 1839.

---

[5] see the interview of F. Jacob in: [3]

[5] Kerridge, I., Lowe, M., Henry, D. Ethics and evidence-based medicin. *British Medical Journal*, 316:1151–1153, 1998.

[6] Louis, P. C. A. Recherche sur les effets de la saignée dans plusieurs maladies inflammatoires. *Archives générales de médecine*, 18:321–226, 1828. Engl. tr. *Researches on the effects of bloodletting in some inflammatory diseases*, Boston, 1836.

[7] Sackett, D. L., Rosenberg, W. M. C., Muir, G. J., Brian, H. R., Scott, Richardson, W. Evidence based medicine: What it is and what it isn't. *British Medical Journal*, 312:71–72, 1996. Editorial, 13 Jan.

[8] Steichen, O. *L'expérience clinique et les rapports de cas dans l'Evidence-Based Medicine*. Paris: Université Pierre & Marie Curie, 2006.

[9] The Evidence-Based Medicine Working Group. Evidence-based medicine. a new approach to teaching the practice of medicine. *JAMA*, 268:2420–2425, 1992. Nov 4.

Index

$(E_0, \prec)$, 88
$(\mathcal{L}_2\text{-}I_\in)$, 88
$Ord(L)$, 39
$Pre(L)$, 39
$Suc(L)$, 39
X_L, 37
Δ_2^0 stability, 39
Π_3 reflection, 75
$\mathcal{L}^\circ$, 77
$\mathcal{L}_1$, 77
$\mathcal{L}_2$, 87
COST, 75
$\mathsf{E}^r(\mathsf{ZFW})$, 90
$\mathsf{KP}(\mathcal{P})$, 81
$\mathsf{KP}\omega$, 81
MK, 75
NBG, 86
$\mathsf{NBG}_{<E_0}$, 86
OST, 74
$\mathsf{OST}(\mathbf{E}, \mathbb{P})$, 80
$\mathsf{OST}(\mathbb{P})$, 80
ZFC, 75
ZF, 75

a priori, 499
AA act, 294–296, 298–300
abstracta, 381
absurdity, 122
active, 348
Adams, 349
adjacency, 37
Adjacency Rule, 448
admissibility rule, 410
    ad hoc, 418
    empty, 418
    universal, 418
admissible evidence, 197–198, 202–203
aesthetic canon, 211, 217, 225–227, 232–234, 236
aesthetic progress, 211, 217, 226, 227, 231
Agent, 142, 145, 154
AIDS research, 640
alculus, 356
Algebra, 354, 372
algebra
    Boolean, 326, 327
    Heyting, 339, 340
Algebraic Geometry, 56

Alston, William P., 202
Analysis, 359, 360
Analytic geometry, 356, 358, 359, 361
ancestor, 414
Andersen, K., 355
Anscombe, F. J., 289
antirealism, 213, 226, 237
Apollonius, 124, 356
appetite, 345
application of areas, 364
applicative axioms, 78
approximation, 318, 357
Aquinas, 8
Archimedes, 359
Aristotle, 9, 356, 621
arithmetic, 356, 357, 381
artificial life, 310, 319
Atheism, 12
Aumann, R. J., 289
autism, 455, 461, 462, 466

Ba Gang, 618
Backward Induction, 150
Balaguer, Mark, 385
Barcan formula (BF), 170, 175
baryon number, 271, 272, 278
basic set-theoretic axioms, 78
Becker, O., 123, 132, 133
Belief, 145
belief, *see* degree of belief
belief revision, 145
Bell, John, 393
Bergson, H., 9
BHK-interpretation, 122, 123, 133
Bian Que, 616
Big Bang Universe (Friedmann), 14
Biomedicalisation of TCM, 642
Bo Yang Fu, 616
Bockstaele, P., 361
Bohm Theory, 396
Boolean homomorphism, 400
Borovik, Alexandre, 57
Bos, H. J. M., 354
Brouwer, L.E.J., 122–132, 134

C^* algebra, 394
C. elegans, 447, 448
calibration, 319

CAM (concept): Complementary and Alternative Medicine, 640
cancellation postulate, 290, 292–298, 301, 303
category, 326, 331, 336, 337
causal model, 410
    ad hoc, 419, 423
    complete, 419
        theorem of, 419
    deterministic , 419
    domain of, 410
    genuinely indeterministic, 419
    universal, 419
causal reasoning, 461, 465
causality postulate, 420
causation
    non-linear-, 621
    non-monofactorical-, 621
cause, 422
    direct, 422
cellular automata, 346
cerebral cortex, 447, 451
certainty and mental vision, 366
certainty and mental vision, 372
certainty of algebraic operations, 355, 362
certainty of mathematical reasoning, 354
Cesa-Bianchi, N, 246
chance, 198, 203–204, 206–208
    counterfeit, 203–204, 207–208
chance setups, 501
Cherlin, Gregory, 57
Cherniak, 443
Christian axiom of ontological dependence, 8
Church's Thesis, 156
Clavius, Ch., 357
Clifton, Rob, 394, 395, 406
Clinical Trials, 639, 642
closed world reasoning, 461, 462, 468
coding, 38, 347, 348
cold, 618
collapse interpretation, 396
collapsing function, 31
collective optimality, 262
combined logics, 156
common knowledge, 139
communication, 138
comparative realism, 211, 213, 217, 218, 226
comparative success hypothesis, 215, 228
complex analysis, 381
complex numbers, 381, 383
complexity, 148

component placement optimization, 447
computability, 381
computable
    linear ordering, 37
    presentation, 36
    structure, 36
computation
    limits of, 307, 322
computational intractability, 444
concept of disease, 620
conditional belief, 146
conditionalization
    conservative, 417
    theorem of, 417
confirmation, 316
conservation law, 267, 268, 271
    family, 277–279
Conservation of matter-density, 14
construction, 122–133
constructive empiricism, 237
constructive realism, 239
constructive set theory, 75
Content, 11
Contingent entity (logically), 6
contradiction, 122, 125, 129, 132, 133
Converse of
    Barcan formula (CBF), 170, 173
    Ghilardi formula (CGF), 171, 173
    RG (CRG), 171, 174, 175
    RG^v (CRG^v), 171
Correspondence, 175
Cosmological argument for the existence of God, 15, 16
Cournot, A. A., 199
Crapulli, G., 360
Creation (Divine) ex nihilo, 8, 12
Creationist theistic answer, 5, 15
Creative cause ex nihilo, 9
Creator ex nihilo, 12
credence, 205–206
crucial points sequence
    c.p.s., 41

Dalen, D.v., 122, 123, 125–127, 129, 133, 134
Dawkins, R., 16
de Bordeu (1722–1776), Théophile, 622
de dicto, 167
de re, 167
de Volder, 349
decision theory, 288
decoding, 348, 352
deduction, 366
default explanation, 211, 212, 215

INDEX

default prediction, 215, 226
deficiency, 616
definition
    implicit, 379
degree of belief, 201–202, 206
Der Spiegel (German magazine), 15
derivative force, 348–350
Descartes, 8
Descartes, R., 354
Designator
    rigid, 165, 168
    stable, 168
Destouches-Février, P., 123
determined, 149
Determinism, 13, 14
determinism, 194–195, 198, 199, 203, 204, 207
Di Li (geographical locations), 618
diagnostic certainty, 622, 624
diamond(lattice), 328
differential principle, 604
Direct Inference Principle (DI), 197, 201–203, 207, 208
discretization, 318
disjunctive syllogism, 123, 124
distinguished set, 33
Diversity, 143
division, 354, 357, 367, 369
dominance, 261
domination, 47
Douven, I., 232
Downey, R., 37, 39
Dummett, M., 123, 134
Dynamic ancestral graph, 490
dynamic logics, 143
dynamics of informational actions, 138
Dynamization, 156

earth, 617
economic theory, 498
Edwards, P., 7
eighteenth century, 622
Einstein's general theory of relativity and gravitation, 12
electron number, 271, 272, 278
empirical progress, 211, 215, 217, 219
entity realism, 230, 238
epimorphism, 332, 337
epistemic conservatism, 275
Epistemic logic, 140
epistemic-temporal logic, 156
Epistemization, 156
Epistemology, 157
ethic experts, 542, 548

ethical expertise, 542, 548, 549
Euclid, 356
Eudoxus, 371
Event models, 144
Evidence, 642
evidence, 143
Evidence-Based Medicine: EBM, 639
Ex Falso, 123, 124, 126, 130–134
exactness, 357
excess, 616
executive function, 459, 460, 466
experiment
    as simulation, 317
explanation
    inductive-statistical, 619
    nomological-deductive-, 619
    pragmatic or contextual model of -, 619
explicit mathematics, 74
exterior, 618

falsum ($\bot$), 122, 124, 133, 134
Fermat, R. de, 356
fictionalism, 377, 387
    realist, 383
fictionalists
    instrumentalist, 387
Field, Hartry
    nominalization programme, 384
fire, 617
fluid dynamics
    continuum hypothesis, 380
formal learning theory, 246
Foundations of mathematics, 157
France, 622
Franklin, A., 315
Freud's dream theory, 606, 607
Freud's therapy, 604
Freudenthal, H., 123, 128, 129, 132
fulfilment, 131–133
Full commutativity of substitution (FCS), 171, 174, 175, 177

Game, 149, 150
gamification, 156
genomic bottleneck, 452
geometry, 354–357
    Euclidean, 380
Ghilardi formula (GF), 170, 176
Gigerenzer, G., 246
gnomon, 364, 365
God, 6, 8
Goncharov, S., 37
Goodman, 275, 282

Granger causality, 482, 485
Grimm, V., 313
Griss, G., 123, 129
group, 326, 329, 331, 332
    symmetric, 331
Group Theory, 65, 67
groups, 155

Hacking, Ian, 199
Halvorson, Hans, 394, 395, 404, 405
Hamilton, W., 11
Hamiltonian mechanics, 382
hard information, 140
Harizanov, V., 37
Harris, S., 16
Hartmann, S., 309
heat, 618
Heath, Th. L, 364
heuristics, 444, 451
Heyting, A., 122, 123, 128, 129, 131–134
high, 38
Hilbert's Program, 130, 153
Hirschfeldt, D., 37, 38
history-theory partition, 203, 204, 207, 208
Hrushovski, Ehud, 58, 62, 63
Hume's problem, 241
Hume, D., 241
Hybridisation (knowledge and practice), 643
Hyperbolic groups, 59, 60
hypothetical construction, 126, 127, 129
hypothetical judgement, 123–130, 133

identification, 497
Ill-conceived non-starter, 5, 15
imagination, 354, 355, 362, 366, 367, 369
implication, 125, 127, 130–132
Incompleteness Theorems, 153
Independence Principle (IN), 196, 200–201
Indexed modalities, 163
induction, 274, 282, 284
    Problem of, 268
    Riddle of, 275
inductive extension, 90
inductive inference, 241
inference, 137
inference to the best explanation, 212, 217, 222, 223
informatical precedence, 409
information, 140
informational interpretation, 346
instrumental variables, 499
instrumentalism, 238
instrumentalist, 387
intellect, 354, 362, 366, 367

intelligent interaction, 152
intention, 131–133
Interaction, 148
interactive logic, 152
interior, 618
interpretation of a physical theory , 391
Interrogative ontological challenge (PEQ), 5
intervention, 423
introspective, 143
intuitionistic logic, 122, 123, 130, 131
invariant structure, 410
    atomic, 411
        dependent set of, 413
    composite, 413
irrational, 356, 357
isomorphism, 332

Jeffrey, Richard C., 207
Johansson, I., 123, 132
Jones, R., 235
Joseph Melia, 385

Kariera system, 379
Kelly, K., 246
Khoussainov, B., 37
King, D., 10
Klein Group, 330, 331
Klein group, 379
Kolmogorov, A., 122, 123, 130–134
Kripke-Platek set theory with infinity, 81
Kuhn, 275
Kuiper, J., 126
Kuipers, T., 211

Lütkehaus, L., 7
Lévi-Strauss, 379
Ladyman, J., 212, 218
Latent variable, 487
lattice, 326, 331, 399–400
    atom in, 400
    Boolean, 400
    distributive, 399
    orthocomplemented, 399
Lawvere, F. W., 329, 338
Learning, 492
learning, 147
learning theory, 275
Leibniz, 344
Leibniz's law (LBZ), 171, 172
Leibniz, Gottfried W, 5–7, 9, 10, 12–16
Lempp, S., 37
Lenghtening (LNGT), 171–174
Levi, Isaac, 198–203

INDEX 659

Lewis, C.I., 123, 124, 133
Lewis, D., 327, 328
Lewis, David, 203–208, 282
lexicographic upgrade, 147
linear space, 273
Liu Junxiang, 615
Loewer, Barry, 203
logic in cognitive science, 456, 457, 460
logic of partial terms, 78
logical operations axioms, 79
low, 38
low$_2$, 38
Lugosi, G., 246

magnitude, 356, 358, 360, 371
Mahlo, 75
Malament, David, 385
Marion, J.-L., 360
Markov equivalence, 490
Markov property, 486
Martin, T., 47
Martin-Löf, P., 123, 133
mathematical, 379
    objects, 378
    structures, 377
mathematical objects, 378
mathematics, 360, 361, 372
    applications, 381
    kinds applications of, 377
mathesis universalis, 360–362, 372
maximal beable approach, 391–395, 400–401, 405
maximum rule, 245
McAllister, J., 223, 232, 233
McLarty, Colin, 381
meaning, 157
Mellor, D. H., 203
memory, 143
mental vision, 366, 367, 372
mereology
    classical, 326
    general, 326, 328
    spatial, 329
merging algebra and geometry, 356, 357, 359, 360
meta-induction, 243
meta-inductivist, 243
metal, 617
Mill, J. S., 11
Miller, J., 38
mind change bounds, 275
minimal logic, 123, 132
modal μ-calculus, 151
modal interpretation, 397

Model theory, 56, 58, 67
models
    mathematical, 382
Modern Western Medicine (MWM), 615
monadology, 344
monomorphism, 332
Monte Carlo method, 308
moral advice, 541, 549
moral authority, 546, 549
moral judgements, 541, 544, 548–551
moral pluralism, 541, 547, 548, 551
moral uncertainty, 541, 542, 549, 550
Moravec, H., 309
Morgenbesser, Sidney, 199
Morley, Michael, 56, 57
morphism, 331
Morse-Kelley set theory, 75
Moses, M., 37
multiplication, 357–359
muon number, 271, 272, 278

Nagel's principle, 603
national ethics committee, 541, 542, 548, 551
natural kind, 269, 282
natural property, *see* natural kind
Necessity of
    distinction (ND), 171, 177
    identity (NI), 171, 174–176
network optimization, 447, 450
neuroanatomy, 446, 447, 450
neuroscientific evidence, 607, 610
Newton's theory of gravitation, 12
Newtonian gravitational theory, 385
Niiniluoto, I., 212, 216
no-miracles argument, 217, 219, 220, 223
nomic possibilities, 213, 232
nomic postulate, 230, 235
nomic theorem, 232, 234
nomic truth, 211
nominalist, 385
nomological machines, 500
non-constructive μ-operator, 74
non-genomic nativism, 448, 452
non-monotonicity, 459, 460, 467
NP-complete, 445, 447
Null possibility, 6
Null World, 9, 10
number, 356, 357, 362

object-induction, 242
observation, 138
observational equivalence, 499
observational truth, 215, 216, 227

Occam's Razor (Principle of Parsimony), 10, 11
Occam, W., 11
one-favorite meta-induction, 252
ontology, 277, 282
    particle, 268, 277, 279
operational existential quantification, 80
operational power set axiom, 80
operational set theory, 74
operational set-theoretic axioms, 79
optimality, 242, 246
ordinal analyses, 28
ordinary quantum mechanics, 391–392
organic body, 345
Organicism, 622, 623
outcome space, 288

Pappus, 359
parental investment, 319
    Concorde hypothesis, 320
    desertion hypothesis, 321
    paternal uncertainty hypothesis, 321
Parfit, D., 6
parsimony, 281
part type, 333, 334
partialization, 420
    theorem of, 421
parts
    structural, 330, 334
passive, 348
passive observation, 497
Path diagrams, 485
Peirce, 275
perceptions, 345
Permutation (PRM), 170, 172
person-making, 618
pessimistic meta-induction, 212, 217, 219
phenomena, 345, 348
physiology, 624
pigeonhole principle, 185
planning, 460, 467
Platonist, 378, 379
plausibility, 145
plenitude, 270, 284
Podzorov, S., 38
Popper, K. R., 12
possibilia, 382
possible world, 203–205
pre-encode, 142
prediction game, 244
prediction method, 243
predictive function, 412
Preference logic, 150
Preferences, 150

preservation of life, 618
prima facie, 508
primitive force, 348–350
primitive terms, 379
Primordial Existential Question (PEQ), 5–9, 12–16
    A priori justification of, 9, 10
Principal Principle, 203, 206, 207
Principle of Sufficient Reason (PSR), 12–15
principle of the common cause, 497
Principle of the Excluded Middle, 126, 129
probabilistic dependence, 502, 504
probabilistic model, 415
probability
    inductive, 193, 197, 201, 202, 206, 207
    physical, 193–208
probability distribution, 498
problem of induction, 241
Proclus, 360
Product update, 144
program, 347
program interpretation, 290
Program structures, 148
proof complexity, 185
proof-theoretic ordinal, 28–30
propositional dynamic logic, 148
propositional formulas, 185
Pseudo-issue (pseudo-problem), 5, 7, 15
pseudo-science, 621
Psillos, S., 220, 223
psychiatric disorders, 455
public announcement logic, 141
public announcements, 140
pulse, 622, 623
pulse diagnosis, 623
Putnam, 275, 282
Pythagoras, 370

Qiu Renzong, 615
$Q.K_{im}$, 172, 183
QM_∞, 391
quantity, 356, 360
quantum field theory, 401
quantum mechanics, 381
Quantum theory, 13
quantum theory, 381
question/answer, 139
Quine, 175, 282

R-proposition, 197, 202
Railsback, S., 313
ratio, 357
rational agent, 142

INDEX

Rational dynamics, 151
rationality, 444
real numbers, 381, 386
realism
    scientific, 385, 387
realist, 383
    scientific, 379
reality
    physical, 378, 381
    spatiotemporal, 384
recursive function theory, 381
reduction axioms, 142
referential realism, 230, 238
referential truth, 211, 217, 221, 230
reflection, 25, 29
Regulae, 354, 355, 358, 361, 362, 364, 366, 367, 371, 372
Reichenbach, H., 242, 275
relative state interpretation, 397
relevance logic, 123
reliability, 246
Ren He (harmony of the people), 618
Renaming (RNM), 170, 173
representation theorem, 287–291, 293, 295, 296, 298–302
repression, 607, 609
Rescher, N., 245
Rigidity of
    terms (RG), 170, 175, 179
    variables (RG^v), 170, 172, 183
$R.K_{im}$, 172, 177
Robert M, 349
Roomen A. van, 360
root extraction, 354, 362, 369, 370
Rosebrugh, R., 329

safe beliefs, 147
Salmon's principle, 599–603
Salmon, W., 245
Savage act, 289, 293–298
Savage, L. J., 287–289
Scaltsas, T., 342
Schaffer, Jonathan, 203
Schanuel, P., 329
Schurz, G., 241
Schuster, J., 372
selection rule, 270–272
self-correction, 154
set theory, 381
    with urelements, 387
Shapiro, Stewart, 377
Shelah, Saharon, 56, 58
Shore, R., 37
Shortening (SHRT), 171, 173, 175

similarity, 326, 339
similarity structure, 326, 340
Simplicity, 9–12
    Overall simplicity, 12
simplicity, 269, 282
    ontological, 282
simulation, 307
    as experiment, 314
    computer, 306
    definition, 308, 310
    epistemology of, 316
    homomorphic, 311
Size Law, 448
skepticism, 237
Sklar, 275
Skyrms, B., 243
Slinko, A., 37
Smart, J. J. C., 11
Soare, R., 39
social choice, 150
soft information, 146
SoN (the "ontological spontaneity of nothingness"), 8–10, 12–15
    A priori justification of, 9
    Empirical appraisal of SoN, 14
    As a pre-supposition of PEQ, 7
Space, Euclidean, 340
space, topological, 327, 341
Specification Principle (SP), 195, 199–200
spectrum, 36
Spurious causality, 484
Stability theory, 56
Standard Model of Particle Physics, 268, 272, 276–278
state space, 288
statistical population, 409, 410
Steady-state universe (Bondi & Gold, 1948), 14
Stewart Shapiro, 378
straight rule, 245
strategies, 149
Strong Rationality, 151
structuralist
    ante rem, 377, 378, 387, 388
    view of applications, 378, 381
Substitution, 168
Substitution that identifies variables (SIV), 171, 174
success theorem, 232, 234
suppression of inferences, 469
    closed world reasoning, 471
    in autists, 474
    neural considerations, 475

Suslin operator, 74
Swinburne, R., 5, 6, 9–11, 13, 15
Switching, 155
system
    mereological, 326

Take the Best, 249
Tangcao: approved AIDS TCM treatment, 640
tau number, 271, 272, 278
tautology, 185
TCM (Traditional Chinese Medicine), 615
TCM:Traditional Chinese Medicine, 639
temporal logic, 149
Théophile de Bordeu (1722–1776), 623
Theism, 8, 12, 15
thematic affinities, 602
theoretical truth, 215, 219, 227
theories
    mathematical, 380–384
    nonmathematical, 384
    nonmathematical content, 388
    phase space, 382
    physical, 379
theory
    coherent, 387
Theory of mind reasoning, 463, 464
thermodynamic limit, 401
Thompson, John, 63, 65
Tian Ren He Yi (unity of nature and human), 617
Tian Shi (heavenly opportunities), 618
time, 309
Traditional Chinese Medicine (TCM), 622, 626
Transition semantics, 164
Trivers, R., 320
Troelstra, A., 122, 123, 130, 133, 134
truth approximation, 211–213, 216, 217
truthlikeness, 211
Turing Machine, 153

ultrafilter, 403–405
uncertain factor, 412
Undecidability, 444
underdetermination, 267
    global, 267
    local, 268
unit segment, 358, 362, 363, 369
universal prediction, 246
unknown quantities, 360, 361
update, 138

validation, 312, 316

van Fraassen, B., 217, 224, 225, 237
Venn, John, 199
verification, 316
Verisimilitude, 11, 12
Viète, F., 359
visualization, 318
von Neumann, 346
von Neumann algebra
    non-atomic, 402–403
    type I, 391, 397, 399, 401, 405
        defined, 397–398
    non-type I, 399, 401–403
von Neumann-Bernays-Gödel set theory, 86

Wason selection task, 456
water, 617
wave phase, 381
Weak Rationality, 151
weighted-average meta-induction, 256
wellfounded, 22–26
winning strategy, 149
Wolff, Christian, 348, 350
wood, 617
Wu Xing, 620
Wu Xing (five agents), 616
Wu, Guohua, 37

Yang, 620
Yi He, 616
Yin, 620

Zajonc, R., 233
Zermelo-Fraenkel set theory, 75
Zhai Xiaomei, 615
Zheng, 618
Zhongyan 1-4: AIDS innovative treatment in TCM, 640

www.ingramcontent.com/pod-product-compliance
Lightning Source LLC
Chambersburg PA
CBHW071148230426
43668CB00009B/873